Physical Chemistry
of Surfaces

Physical Chemistry of Surfaces

Sixth Edition

ARTHUR W. ADAMSON

Department of Chemistry, University of Southern California
Los Angeles, California

and

ALICE P. GAST

Department of Chemical Engineering, Stanford University
Stanford, California

A WILEY-INTERSCIENCE PUBLICATION

John Wiley & Sons, Inc.

NEW YORK / CHICHESTER / WEINHEIM / BRISBANE / SINGAPORE / TORONTO

About the Front Cover Design

The fern-like shapes are two-dimensional dendritic crystals of the protein streptavidin bound to a lipid monolayer. These dendrites were produced by Andrew Ku with the help of Seth Darst, Roger Kornberg, and Channing Robertson and are described in Chapter XV; note Fig. XV-5.

The island pattern is from a marbled paper, for which we thank Mrs. Phoebe Easton. Marbling is an ancient process whereby colored oils with surfactant are spread on the surface of water stiffened by agar–agar and the pattern lifted onto the paper. The pattern could be that of domains of O_2 and of CO chemisorbed on Pt(100); note Fig. XVIII-28.

Library of Congress Cataloging in Publication Data:

Adamson, Arthur W.
    Physical chemistry of surfaces / Arthur W. Adamson and Alice P. Gast.—6th ed.
        p.   cm.
    "A Wiley-Interscience Publication."
    Includes bibliographical references and index.
    ISBN 0-471-14873-3 (cloth : alk. paper)
    1. Surface chemistry.   2. Chemistry, Physical and theoretical.
    I. Gast, Alice P. (Alice Petry), 1958–   . II. Title.
    QD506.A3 1997
    541.3′3—dc21

Printed in the United States of America

10 9 8 7 6 5 4

To Virginia,
without whose fifty-five years of steady support
this book and its previous editions would not have been written,
and
to Bradley,
whose devotion and encouragement
made this edition possible

Contents

Preface

The first edition of *Physical Chemistry of Surfaces* appeared in 1960—a long time ago and a long time for a book to be continuously in print through successive editions. Much has changed; much remains about the same. An important change—a most happy one—is that the senior author is joined by a co-author and one who may well carry the book into yet further editions. Unchanged is the purpose of the book. We hope that this 6th edition will continue to serve as a textbook for senior and graduate-level courses, of both academic and industrial venue, and that it will continue to be of value to practitioners in surface chemistry, especially those whose interests have only recently moved them toward that field. Some comments for special groups of users follow.

Students (and instructors). Each chapter presents first the basic surface chemistry of the topic, with optional material in small print. Derivations are generally given in full and this core material is reinforced by means of problems at the end of the chapter. A solutions manual is available to instructors. It is assumed that students have completed the usual undergraduate year course in physical chemistry. As a text for an advanced course, the basic material is referenced to fundamental, historical sources, and to contemporary ones where new advances have been incorporated. There are numerous examples and data drawn from both the older and from current literature.

Each chapter section will generally conclude with a heavily referenced review of more recent advances in its area. The typical course in surface chemistry might follow the chapter sequence of the book. The first two-thirds of the course, through Chapter XI, would typically stress fundamentals with frequent homework assignments. The student is taken through the basics of the physical chemistry of liquid–gas and liquid–solid interfaces, including electrochemistry, long-range forces, and the various methods of spectroscopic and structural study of surfaces. Chapters XII through XV are more descriptive and problem assignments might taper off, to be replaced by a term paper. The citations on recent advances in these chapters serve to give the student a good start on a literature survey for such a project and a basis for thoughtful discussion. Chapter XV is a new one, designed to give appropriate attention to the topic of macromolecular films.

The concluding chapters, Chapters XVI through XVIII, take up the important subjects of physical and chemical adsorption of vapors and gases, and heterogeneous catalysis. As with the earlier chapters, the approach is relatively quantitative and problem assignments regain importance.

While the Système International d'Unités (SI) system of units is not particularly relevant to physical chemistry and requires additional and sometimes awkward constants, its broad use deserves attention. The majority of the derivations are made in the cgs/esu (centimeter–gram–second/electrostatic unit) system of units; however, both the SI and cgs systems are explained and tables for their interconversion are given in Chapters V and VI.

Professional chemists. Surface chemistry is a broad subject, and it is hoped that even those established in some particular aspect will find the many references to contemporary work helpful in areas not in their immediate expertise. The subject is also a massively developing one, and many scientists whose basic experience has been in spectroscopy, photochemistry, biomimicking systems, engineering, and so on, have found themselves drawn into surface chemical extensions of their work. The book should serve the dual purpose of providing a fairly detailed survey of basic surface chemistry and an entrée into contemporary, important work on each aspect. Many of the references were chosen because of the extensive bibliography that they contain, as a means of helping people who need to get acquainted rapidly with a subject area. Also, the Index is unusually complete; it is intended to be helpful in chasing down related aspects of a given subject, often present in other than the principal chapter.

Those acquainted with the 5th edition. Some statistics on this new edition are the following. It is slightly longer (in spite of stringent efforts and the elimination of much material of diminished interest). About 30% of the text is new and about a third of the problems; there are now some 3400 references, of which about 30% are new.

There has been a general updating of the material in all the chapters; the treatment of films at the liquid–air and liquid–solid interfaces has been expanded, particularly in the area of contemporary techniques and that of macromolecular films. The scanning microscopies (tunneling and atomic force) now contribute more prominently. The topic of heterogeneous catalysis has been expanded to include the well-studied case of oxidation of carbon monoxide on metals, and there is now more emphasis on the "flexible" surface, that is, the restructuring of surfaces when adsorption occurs. New calculational methods are discussed.

In the Prefaces of both the 4th and the 5th editions the senior author commented on the tendency of "wet" and "dry" surface chemistry for differentiation into separate schools. This remains the case today; also, academic research in wet surface chemistry continues to move from chemistry departments to engineering ones. On the other hand, new connections between the two areas have been forming apace with the current prominence of scanning microscopies.

Also, many of the structural and spectroscopic techniques are now being applied to new types of systems, such as those involving the physical adsorption of vapors. Such bridging of methodologies will surely help to keep surface chemistry a single, broad field with good intercommunication between the various subareas.

We are both greatly indebted to the hundreds of authors who supplied us with thousands of reprints, to constitute libraries of important contemporary literature. One of us (AWA) wishes to acknowledge help in the preparation of the manuscript through a grant to the University of Southern California Emeriti College from the Rita H. Small Charitable Trust; also, the many hours spent by Virginia Adamson in reading proofs has made the book (if not the subject) at least partly hers. APG wishes to thank the numerous students who contributed to this book, in particular, Matthew Losey for his help on Chapter VIII and on the physical chemistry of art restoration. Lindi Bauman of the Stanford University Department of Chemical Engineering and Grace Baysinger, the Chemistry Librarian, deserve much appreciation for their help with the manuscript and bibliography. Finally, Bradley Askins, Rebecca and David Askins-Gast, and Dru Gast deserve tremendous thanks for their sacrifices during the many days and nights of work that this book required.

ARTHUR W. ADAMSON
ALICE P. GAST

January 1997

A solutions manual for the problems is available from either author; the request should be on institutional letterhead and from an authorized person.

ARTHUR W. ADAMSON

Department of Chemistry
University of Southern California
Los Angeles, California

ALICE P. GAST

Department of Chemical Engineering
Stanford University
Stanford, California

CHAPTER I

General Introduction

In this book we discuss the physical chemistry of surfaces in a broad sense. Although an obvious enough point, it is perhaps worth noting that in reality we will always be dealing with the *interface* between two phases and that, in general, the properties of an interface will be affected by physical or chemical changes in either of the two phases involved. We will address, to various degrees of detail, all the possible interfaces between the three states of matter—solid, liquid, and gas. At these interfaces, we will see some intriguing phenomena imparted by the constraints imposed by the surface.

A general prerequisite for the existence of a stable interface between two phases is that the free energy of formation of the interface be positive; were it negative or zero, fluctuations would lead to complete dispersion of one phase in another. As implied, thermodynamics constitutes an important discipline within the general subject. It is one in which surface area joins the usual extensive quantities of mass and volume and in which surface tension and surface composition join the usual intensive quantities of pressure, temperature, and bulk composition. The thermodynamic functions of free energy, enthalpy and entropy can be defined for an interface as well as for a bulk portion of matter. Chapters II and III are based on a rich history of thermodynamic studies of the liquid interface. The phase behavior of liquid films enters in Chapter IV, and the electrical potential and charge are added as thermodynamic variables in Chapter V.

The key physical elements in a molecular thermodynamic analysis are the interaction potentials between the molecules. The intermolecular forces have a profound influence on interfacial phenomena where properties change dramatically over molecular length scales. This is addressed in Chapters V and VI, where electrostatic and long-range forces are discussed; these intermolecular attractions and repulsions play a recurring role throughout the book. An important characteristic of an interface is that it is directional. Properties vary differently both along and perpendicular to an interface. This aspect is responsible for many of the fascinating phenomena occurring at interfaces and also provides leverage in the study of long-range forces. As described in Chapter VI, it is possible, for example, to measure *directly* the van der Waals force between two surfaces. This area is one in which surface physical chemists have made fundamental contributions to physical chemistry as a whole.

In Chapter VII, the solid surface is introduced. Structure is as important in surface physical chemistry as it is in chemistry generally. The structure of a crystalline solid can be determined by x-ray diffraction studies; the surface structure of a solid can, somewhat analogously, be determined by low-energy electron diffraction (LEED). Chapter VIII is devoted to the myriad ways of probing surface structure and chemistry at a molecular level. High-vacuum surface spectroscopy has become quite well developed, often involving well-defined single-crystal surfaces, such that the chemical state of adsorbed and reacting molecules can be investigated.

A remarkable development, discussed in Chapter VIII, has been the ability to "see" individual atoms and molecules through scanning probe microscopies. With this ability has come the documentation of the structures of solid surfaces, even refractory ones, that differ from the bulk phases. Surface structures may change drastically if there is chemical bonding to an adsorbate. This is particularly true in the case of chemisorption and catalysis, the subject of Chapter XVIII. One now speaks of the "flexible" solid surface and chemisorption has become a structure-intensive subject. Scanning probe microscopy has brought similar insight on the structure of films adsorbed from solution, including polymers, proteins, and the so-called self-assembled monolayers (SAMs) discussed in Chapter XI and Langmuir–Blodgett films described in Chapter XV.

Systems involving an interface are often *metastable*, that is, essentially in equilibrium in some aspects although in principle evolving slowly to a final state of global equilibrium. The solid–vapor interface is a good example of this. We can have adsorption equilibrium and calculate various thermodynamic quantities for the adsorption process; yet the particles of a solid are unstable toward a drift to the final equilibrium condition of a single, perfect crystal. Much of Chapters IX and XVII are thus thermodynamic in content.

The physical chemist is very interested in kinetics—in the mechanisms of chemical reactions, the rates of adsorption, dissolution or evaporation, and generally, in time as a variable. As may be imagined, there is a wide spectrum of rate phenomena and in the sophistication achieved in dealing with them. In some cases changes in area or in amounts of phases are involved, as in rates of evaporation, condensation, dissolution, precipitation, flocculation, and adsorption and desorption. In other cases surface composition is changing as with reaction in monolayers. The field of catalysis is focused largely on the study of surface reaction mechanisms. Thus, throughout this book, the kinetic aspects of interfacial phenomena are discussed in concert with the associated thermodynamic properties.

We attempt to delineate between surface physical chemistry and surface chemical physics and solid-state physics of surfaces. We exclude these last two subjects, which are largely wave mechanical in nature and can be highly mathematical; they properly form a discipline of their own.

We also attempt to distinguish between surface physical chemistry and colloid and polymer physical chemistry. This distinction is not always possible, and clearly many of the features of physical chemistry of surfaces, such as the electrostatic interactions and adsorption of macromolecules, have a significant impact on colloidal suspensions. The emphasis here is on the thermodynamics, structure, and rate processes involving an interface. In colloid and polymer physical chemistry the emphasis is more on the collective properties of a dispersed system. Light scattering by a suspension is not, for example, of central interest in this book; however, light scattering from liquid interfaces arises as an important tool in Chapter IV for the study of spread monolayers. Likewise, random coil configurations of a long-chain polymer in solution and polymer elasticity only enter the discussion if the polymer is adsorbed at an interface. The field of biophysics is beginning to merge with the world of physical chemistry of surfaces since so many interfacial phenomena are involved as seen, for example, in Chapter XV. As physical measurement techniques improve and theoretical analyses become more sophisticated, the boundaries between these disciplines blur.

There is a host of problems of practical importance that require at least a phenomenological, that is, macroscopic, view of surface physical chemistry. The contact angle (Chapter X), itself a manifestation of the thermodynamics of interfaces discussed in Chapters II and III, is of enormous importance to the flotation industry. Wetting, adhesion, detergency, emulsions, and foams all depend on the control of interfacial tensions, often through judicious use of surface active agents. These topics are covered in Chapters XII–XIV. Chapter XV takes up the now enormous subject of macromolecular surface films, including transferred Langmuir–Blodgett films, biological films and membranes. The emphasis in these chapters is on those aspects that have received sufficient attention to be somewhat established. Surface probe techniques are bringing important new molecular insight into these more applied areas of surface physical chemistry.

The solid–gas interface and the important topics of physical adsorption, chemisorption, and catalysis are addressed in Chapters XVI–XVIII. These subjects marry fundamental molecular studies with problems of great practical importance. Again the emphasis is on the basic aspects of the problems and those areas where modeling complements experiment.

Clearly, the "physical chemistry of surfaces" covers a wide range of topics. Most of these subjects are sampled in this book, with emphasis on fundamentals and important theoretical models. With each topic there is annotation of current literature with citations often chosen because they contain bibliographies that will provide detailed source material. We aim to whet the reader's appetite for surface physical chemistry and to provide the tools for basic understanding of these challenging and interesting problems.

CHAPTER II

Capillarity

The topic of capillarity concerns interfaces that are sufficiently mobile to assume an equilibrium shape. The most common examples are meniscuses, thin films, and drops formed by liquids in air or in another liquid. Since it deals with equilibrium configurations, capillarity occupies a place in the general framework of thermodynamics in the context of the macroscopic and statistical behavior of interfaces rather than the details of their molecular structure. In this chapter we describe the measurement of surface tension and present some fundamental results. In Chapter III we discuss the thermodynamics of liquid surfaces.

1. Surface Tension and Surface Free Energy

Although referred to as a *free energy per unit area*, surface tension may equally well be thought of as a force per unit length. Two examples serve to illustrate these viewpoints. Consider, first, a soap film stretched over a wire frame, one end of which is movable (Fig. II-1). Experimentally one observes that a force is acting on the movable member in the direction opposite to that of the arrow in the diagram. If the value of the force per unit length is denoted by γ, then the work done in extending the movable member a distance dx is

$$\text{Work} = \gamma l \ dx = \gamma dA \qquad \text{(II-1)}$$

where $dA = l \ dx$ is the change in area. In the second formulation, γ appears to be an energy per unit area. Customary units, then, may either be ergs per square centimeter (ergs/cm^2) or dynes per centimeter (dyn/cm); these are identical dimensionally. The corresponding SI units are joules per square meter (J/m^2) or Newtons per meter (N/m); surface tensions reported in dyn/cm and mN/m have the same numerical value.

A second illustration involves the soap bubble. We will choose to think of γ in terms of energy per unit area. In the absence of gravitational or other fields, a soap bubble is spherical, as this is the shape of minimum surface area for an enclosed volume. A soap bubble of radius r has a total surface free energy of $4\pi r^2\gamma$ and, if the radius were to decrease by dr, then the change in surface free energy would be $8\pi r\gamma \ dr$. Since shrinking decreases the surface energy, the tendency to do so must be balanced by a pressure difference across the film

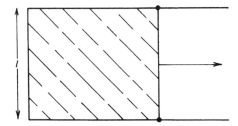

Fig. II-1. A soap film stretched across a wire frame with one movable side.

ΔP such that the work against this pressure difference $\Delta P\, 4\pi r^2\, dr$ is just equal to the decrease in surface free energy. Thus

$$\Delta P\, 4\pi r^2\, dr = 8\pi r\gamma\, dr \qquad\text{(II-2)}$$

or

$$\Delta P = \frac{2\gamma}{r} \qquad\text{. (II-3)}$$

One thus arrives at the important conclusion that the smaller the bubble, the greater the pressure of the air inside relative to that outside.

The preceding conclusion is easily verified experimentally by arranging two bubbles with a common air connection, as illustrated in Fig. II-2. The arrangement is unstable, and the smaller of the two bubbles will shrink while the other enlarges. Note, however, that the smaller bubble does not shrink indefinitely; once its radius equals that of the tube, its radius of curvature will increase as it continues to shrink until the final stage, where mechanical equilibrium is satisfied, and the two radii of curvature are equal as shown by the dotted lines.

The foregoing examples illustrate the point that equilibrium surfaces may be treated using either the mechanical concept of surface tension or the mathematically equivalent concept of surface free energy. (The derivation of Eq. II-3 from the surface tension point of view is given as an exercise at the end of the chapter). This mathematical equivalence holds everywhere in capillarity phenomena. As discussed in Section III-2, a similar duality of viewpoint can be argued on a molecular scale so that the decision as to whether *surface tension* or *surface free energy* is the more fundamental concept becomes somewhat a matter of individual taste. The term *surface tension* is the older of the two; it goes back to early ideas that the surface of a liquid had some kind of contractile "skin." *Surface free energy* implies only that work is required to bring molecules from the interior of the phase to the surface. Because of its connection to thermodynamic language, these authors consider the latter preferable if

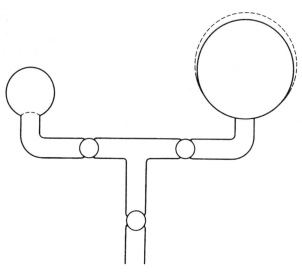

Fig. II-2. Illustration of the Young–Laplace equation.

a choice must be made; however, the two terms are used interchangeably in this book.

2. The Young–Laplace Equation

Equation II-3 is a special case of a more general relationship that is the basic equation of capillarity and was given in 1805 by Young [1] and by Laplace [2]. In general, it is necessary to invoke two radii of curvature to describe a curved surface; these are equal for a sphere, but not necessarily otherwise. A small section of an arbitrarily curved surface is shown in Fig. II-3. The two radii of curvature, R_1 and R_2,† are indicated in the figure, and the section of surface taken

---

†It is perhaps worthwhile to digress briefly on the subject of radii of curvature. The two radii of curvature for some arbitrarily curved surface are obtained as follows. One defines a normal to the surface at the point in question and then passes a plane through the surface containing the normal. The line of intersection in general will be curved, and the radius of curvature is that for a circle tangent to the line at the point involved. The second radius of curvature is obtained by passing a second plane through the surface, also containing the normal, but perpendicular to the first plane. This gives a second line of intersection and a second radius of curvature.

If the first plane is rotated through a full circle, the first radius of curvature will go through a minimum, and its value at this minimum is called the principal radius of curvature. The second principal radius of curvature is then that in the second plane, kept at right angles to the first. Because Fig. II-3 and Eq. II-7 are obtained by quite arbitrary orientation of the first plane, the radii R_1 and R_2 are not necessarily the principal radii of curvature. The pressure difference ΔP, cannot depend upon the manner in which R_1 and R_2 are chosen, however, and it follows that the sum $(1/R_1 + 1/R_2)$ is independent of how the first plane is oriented (although, of course, the second plane is always at right angles to it).

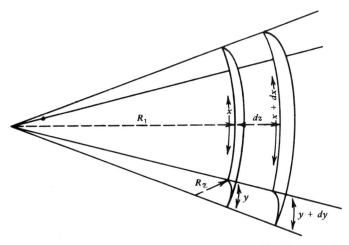

Fig. II-3. Condition for mechanical equilibrium for an arbitrarily curved surface.

is small enough so that R_1 and R_2 are essentially constant. Now if the surface is displaced a small distance outward, the change in area will be

$$\Delta\mathcal{A} = (x + dx)(y + dy) - xy = x\,dy + y\,dx$$

The work done in forming this additional amount of surface is then

$$\text{Work} = \gamma(x\,dy + y\,dx)$$

There will be a pressure difference ΔP across the surface; it acts on the area xy and through a distance dz. The corresponding work is thus

$$\text{Work} = \Delta P\,xy\,dz \qquad\qquad (\text{II-6})$$

Most of the situations encountered in capillarity involve figures of revolution, and for these it is possible to write down explicit expressions for R_1 and R_2 by choosing plane 1 so that it passes through the axis of revolution. As shown in Fig. II-7a, R_1 then swings in the plane of the paper, i.e., it is the curvature of the profile at the point in question. R_1 is therefore given simply by the expression from analytical geometry for the curvature of a line

$$1/R_1 = y''/(1 + y'^2)^{3/2} \qquad\qquad (\text{II-4})$$

where y' and y'' denote the first and second derivatives with respect to x. The radius R_2 must then be in the plane perpendicular to that of the paper and, for figures of revolution, must be given by extending the normal to the profile until it hits the axis of revolution, again as shown in Fig. II-7a. Turning to Fig. II-7b, the value of R_2 for the coordinates (x, y) on the profile is given by $1/R_2 = \sin\phi/x$, and since $\tan\phi$ is equal to y', one obtains the following expression for R_2:

$$1/R_2 = y'/x(1 + y'^2)^{1/2} \qquad\qquad (\text{II-5})$$

From a comparison of similar triangles, it follows that

$$\frac{x + dx}{R_1 + dz} = \frac{x}{R_1} \quad \text{or} \quad dx = \frac{x\, dz}{R_1}$$

and

$$\frac{y + dy}{R_2 + dz} = \frac{y}{R_2} \quad \text{or} \quad dy = \frac{y\, dz}{R_2}$$

If the surface is to be in mechanical equilibrium, the two work terms as given must be equal, and on equating them and substituting in the expressions for dx and dy, the final result obtained is

$$\Delta \mathbf{P} = \gamma \left(\frac{1}{R_1} + \frac{1}{R_2} \right) \tag{II-7}$$

Equation II-7 is the fundamental equation of capillarity and will recur many times in this chapter.

It is apparent that Eq. II-7 reduces to Eq. II-3 for the case of both radii being equal, as is true for a sphere. For a plane surface, the two radii are each infinite and $\Delta \mathbf{P}$ is therefore zero; thus there is no pressure difference across a plane surface.

3. Some Experiments with Soap Films

There are a number of relatively simple experiments with soap films that illustrate beautifully some of the implications of the Young–Laplace equation. Two of these have already been mentioned. Neglecting gravitational effects, a film stretched across a frame as in Fig. II-1 will be planar because the pressure is the same as both sides of the film. The experiment depicted in Fig. II-2 illustrates the relation between the pressure inside a spherical soap bubble and its radius of curvature; by attaching a manometer, $\Delta \mathbf{P}$ could be measured directly.

An interesting set of shapes results if one forms a soap bubble or liquid bridge between two cylindrical supports, as shown in Fig. II-4. In Fig. II-4a, the upper support is open to the atmosphere so that the pressure is everywhere the same, and ΔP must be zero. Although the surface appears to be curved, Eq. II-7 is not contradicted. The two radii of curvature indicated in Fig. II-4a, where R_1 swings in the plane of the paper and R_2 swings in the plane perpendicular to it, are equal in magnitude and opposite in sign because they originate on opposite sides of the film; hence they cancel each other in Eq. II-7. This is an example of a surface with zero mean curvature. Such surfaces are found in other situations such as static "dewetting holes" (see Chapter XIII).

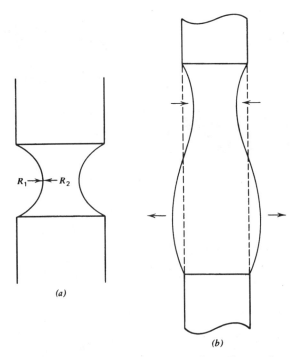

Fig. II-4. (*a*) A cylindrical soap film; (*b*) manner of a collapse of a cylindrical soap film of excessive length.

Instability of Cylindrical Columns. C. V. Boys, in his elegant little monograph of 1890 [3], discusses an important property of quasistatic cylindrical films that was first studied in cylindrical columns of fluids by Lord Rayleigh in 1879. If the soap film in Fig. II-4*a* were made to be cylindrical by adjusting the gas pressure inside, it, like a cylindrical thread of fluid, would be unstable to surface waves whose length exceeds the circumference of the cylinder. The column would contract at one end and bulge at the other, as illustrated in Fig. II-4*b*, before breaking up into a smaller and larger bubble (or drop) as shown in the photographs of a liquid stream in Fig. II-5 [4]. The mechanism is associated with the nonzero curvature of the static state and the fact that fluctuations establish capillary pressure gradients that drive the fluid away from the equilibrium. It is now recognized that capillary breakup is a particularly simple example of the geometric instability of states of static equilibrium in the presence of surface tension. For a general description dealing with pendant and sessile drops, finite cylinders (capillary bridges) and other capillary surfaces, see Michael [5]. A detailed discussion of the capillary break up of jets, including several interesting practical applications, is given by Bogy [6]. The case of one liquid in a second, immiscible one is discussed in Refs. 6a and 7. A similar instability occurring in a thin annular coating inside a capillary can have important consequences for capillary columns in chromatography [8].

Returning to equilibrium shapes, these have been determined both experimentally and by solution of the Young–Laplace equation for a variety of situations. Examples

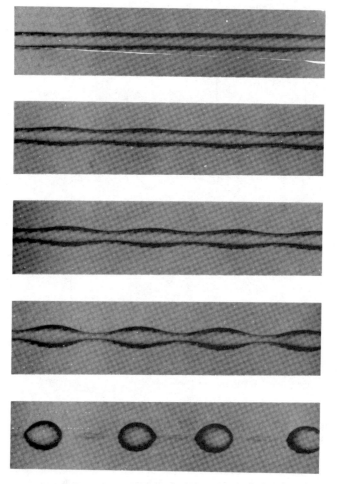

Fig. II-5. Necking in a liquid stream. [Courtesy S. G. Mason (4).]

include the shape of a liquid plug in capillary tubes of various shapes of cross sections (9) and of liquid bridges between spheres in a gravitational field [10]; see Refs. 11 to 12 for reviews.

4. The Treatment of Capillary Rise

A. Introductory Discussion

An approximate treatment of the phenomenon of capillary rise is easily made in terms of the Young–Laplace equation. If the liquid completely wets the wall of the capillary, the liquids surface is thereby constrained to lie parallel to the wall at the region of contact and the surface must be concave in shape. The

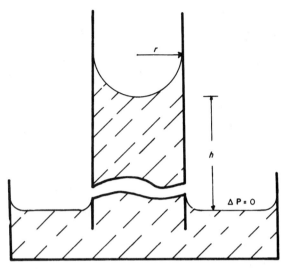

Fig. II-6. Capillary rise (capillary much magnified in relation to dish).

radii of curvature are defined in terms of an outward normal from the liquid; thus, it follows from Eq. II-7 that the pressure is lower in the liquid than in the gas phase. Small circular capillaries will have an approximately hemispherical meniscus as illustrated in Fig. II-6. Here the two radii of curvature are equal to each other and to the radius of the capillary. Eq. II-7 then reduces to

$$\Delta P = \frac{2\gamma}{r} \tag{II-8}$$

where r is the radius of the capillary. If h denotes the height of the meniscus above a *flat* liquid surface (for which ΔP must be zero), then ΔP in Eq. II-8 must also equal the hydrostatic pressure drop in the column of liquid in the capillary. Thus $\Delta P = \Delta \rho \, gh$, where $\Delta \rho$ denotes the difference in density between the liquid and gas phases and g is the acceleration due to gravity. Equation II-8 becomes

$$\Delta \rho \, gh = \frac{2\gamma}{r} \tag{II-9}$$

or

$$a^2 = \frac{2\gamma}{\Delta \rho \, g} = rh \tag{II-10}$$

The quantity a, defined by Eq. II-10 is known as the *capillary constant* or *cap-*

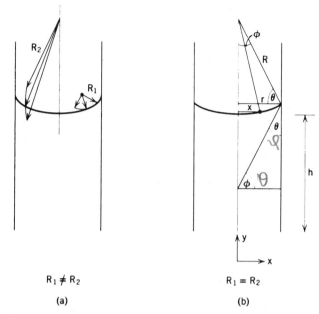

Fig. II-7. The meniscus in a capillary as a figure of revolution.

illary length. The factor of 2 in the definition of a arises from this particular boundary value problem; for many other situations the capillary length is defined by $a^2 = \gamma/\Delta\rho g$ (see Section X-6A).

Similarly, the identical expression holds for a liquid that completely fails to wet the capillary walls, where there will be an angle of contact between the liquid and the wall of 180°, a convex meniscus and a capillary depression of depth h.

A slightly more general case is that in which the liquid meets the circularly cylindrical capillary wall at some angle θ, as illustrated in Fig. II-7. If the meniscus is still taken to be spherical in shape, it follows from simple geometric consideration that $R_2 = r/\cos\theta$ and, since $R_1 = R_2$, Eq. II-9 then becomes

$$\Delta\rho\, gh = \frac{2\gamma\cos\theta}{r} \tag{II-11}$$

B. Exact Solutions to the Capillary Rise Problem

The exact treatment of capillary rise must take into account the deviation of the meniscus from sphericity, that is, the curvature must correspond to the $\Delta P = \Delta\rho\, gy$ at each point on the meniscus, where y is the elevation of that point above the flat liquid surface. The formal statement of the condition is obtained by writing the Young–Laplace equation for a general point (x, y) on the meniscus, with R_1 and R_2 replaced by the expressions from analytical geometry given in the footnote to Section II-2. We still assume that the capillary is circular in cross section so that the meniscus shape is that of a figure of revolution; as indicated in Fig. II-7, R_1 swings in the plane of paper, and R_2 in the plane perpendicular to the paper. One thus obtains

$$\Delta\rho\, gy = \gamma\left[\frac{y''}{(1+y'^2)^{3/2}} + \frac{y'}{x(1+y'^2)^{1/2}}\right] \tag{II-12}$$

where $y' = dy/dx$ and $y'' = d^2y/dx^2$, as in Eqs. II-4 and II-5. A compact alternative form is

$$\bar{y} = \frac{1}{\bar{x}}\frac{d}{d\bar{x}}(\bar{x}\sin\phi) \tag{II-13}$$

where a bar denotes that the quantity has been made dimensionless by multiplication by $\sqrt{2/a}$ (see Refs. 10 and 13).

Equations II-12 and II-13 illustrate that the shape of a liquid surface obeying the Young–Laplace equation with a body force is governed by *differential equations* requiring boundary conditions. It is through these boundary conditions describing the interaction between the liquid and solid wall that the contact angle enters.

The total weight of the column of liquid in the capillary follows from Eq. II-12:

$$W = 2\pi r\gamma\cos\theta \tag{II-14}$$

This is *exact*—see Problem II-8. Notice that Eq. II-14 is exactly what one would write, assuming the meniscus to be "hanging" from the wall of the capillary and its weight to be supported by the vertical component of the surface tension, $\gamma\cos\theta$, multiplied by the circumference of the capillary cross section, $2\pi r$. Thus, once again, the mathematical identity of the concepts of *surface tension* and *surface free energy* is observed.

While Eq. II-14 is exact, its use to determine surface tension from capillary rise experiments is not convenient. More commonly, one measures the height, h, to the *bottom* of the meniscus.

Approximate solutions to Eq. II-12 have been obtained in two forms. The first, given by Lord Rayleigh [13], is that of a series approximation. The derivation is not repeated here, but for the case of a nearly spherical meniscus, that is, $r \ll h$, expansion around a deviation function led to the equation

$$a^2 = r\left(h + \frac{r}{3} - \frac{0.1288r^2}{h} + \frac{0.1312r^3}{h^2}\cdots\right) \tag{II-15}$$

The first term gives the elementary equation (Eq. II-10). The second term takes into account the weight of the meniscus, assuming it to be spherical (see Problem II-3). The succeeding terms provide corrections for deviation from sphericity.

The general case has been solved by Bashforth and Adams [14], using an iterative method, and extended by Sugden [15], Lane [16], and Paddy [17]. See also Refs. 11 and 12. In the case of a figure of revolution, the two radii of curvature must be equal at the apex (i.e., at the bottom of the meniscus in the case of capillary rise). If this radius of curvature is denoted by b, and the elevation of a general point on the surface is denoted by z, where $z = y - h$, then Eq. II-7 can be written

$$\gamma\left(\frac{1}{R_1} + \frac{1}{R_2}\right) = \Delta\rho\, gz + \frac{2\gamma}{b} \tag{II-16}$$

Thus, at $z = 0$, $\Delta P = 2\gamma/b$, and at any other value of z, the change in ΔP is given by $\Delta\rho\, gz$. Equation II-16 may be rearranged so as to involve only dimensionless parameters

$$\frac{1}{R_1/b} + \frac{\sin\phi}{x/b} = \beta\,\frac{z}{b} + 2 \tag{II-17}$$

where R_2 has been replaced by its equivalent, $x/\sin\phi$, and the dimensionless quantity β is referred to as the Bond number, given by

$$\beta = \frac{\Delta\rho gb^2}{\gamma} = \frac{2b^2}{a^2} \tag{II-18}$$

where small Bond numbers indicate weak body forces or strong surface tensions. This parameter is positive for oblate figures of revolution, that is, for a sessile drop, a bubble under a plate, and a meniscus in a capillary. It is negative for prolate figures, that is, for a pendant drop or a clinging bubble.

Bashforth and Adams obtained solutions to Eq. II-17 (with R_1 replaced by the expression in analytical geometry), using a numerical integration procedure (this was before the day of high-speed digital computers, and their work required tremendous labor). Their results are reported as tables of values of x/b and z/b for closely spaced values of β and of ϕ. For a given β value, a plot of z/b versus x/b gives the profile of a particular figure of revolution satisfying Eq. II-17. By way of illustration, their results for $\beta = 80$ are reproduced (in abbreviated form) in Table II-1. Observe that x/b reaches a maximum at $\phi = 90°$, so that in the case of zero contact angle the surface is now tangent to the capillary wall and hence $(x/b)_{\max} = r/b$. The corresponding value of r/a is given by $(r/b)\sqrt{\beta/2}$. In this manner, Sugden compiled tables of r/b versus r/a.

TABLE II-1
Solution to Eq. II-17 for $\beta = 80$

ϕ (deg)	x/b	z/b	ϕ (deg)	x/b	z/b
5	0.08159	0.00345	100	0.33889	0.17458
10	0.14253	0.01133	110	0.33559	0.18696
20	0.21826	0.03097	120	0.33058	0.19773
⌄ 30	0.26318	0.05162	130	0.32421	0.20684
40	0.29260	0.07204	140	0.31682	0.21424
50	0.31251	0.09183	150	0.30868	0.21995
60	0.32584	0.11076	160	0.30009	0.22396
70	0.33422	0.12863	170	0.29130	0.22632
80	0.33872	0.14531			
90	0.34009	0.16067			

Lane improved on these tables with accurate polynomial fits to numerical solutions of Eq. II-17 [16]. Two equations result; the first is applicable when $r/a \leq 2$

$$b/r = 1 + [3327.9(r/a)^2 + 65.263(r/a)^3 - 473.926(r/a)^4 + 663.569(r/a)^5$$

$$- 300.032(r/a)^6 + 75.1929(r/a)^7 - 7.3163(r/a)^8]/10^4 \qquad \text{(II-19)}$$

and another is to be employed when $r/a \geq 2$

$$r/b = (r/a)^{3/2} \exp[-1.41222(r/a) + 0.66161 + 0.14681(a/r) + 0.37136(a/r)^2]$$

$$\text{(II-20)}$$

The use of these equations is perhaps best illustrated by means of a numerical example. In a measurement of the surface tension of benzene, the following data are obtained:

Capillary radius—0.0550 cm
Density of benzene—0.8785; density of air—0.0014 (both at 20°C);
hence $\Delta\rho = 0.8771$ g/ml
Height of capillary rise—1.201 cm

We compute a first approximation to the value of the capillary constant a_1 by means of Eq. II-10 ($a^2 = rh$). The ratio r/a_1 is then obtained and the corresponding value of r/b is determined from Eq. II-19 or II-20; in the present case, $a_1^2 = 1.201 \times 0.0550 = 0.660$; hence, $r/a_1 = 0.0550/0.2570 = 0.2140$. From Eq. II-19, r/b_1 is then 0.9850. Since b is the value of R_1 and of R_2 at the bottom of the meniscus, the equation $a^2 = bh$ is exact. From the value of r/b_1, we obtain a first approximation to b, that is, $b_1 = 0.0550/0.9850 = 0.05584$. This value of b gives a second approximation to a from $a_2^2 = b_1h = 0.05584 \times 1.201 = 0.06706$. A second round of approximations is not needed in this case but would be carried out by computing r/a_2; then from Eq. II-19, r/b_2, and so on. The value of 0.06706 for a_2^2 obtained here leads to 28.84 dyn/cm for the surface tension of benzene (at 20°).

The calculation may be repeated in SI units (see, however, Ref. 18). The radius is now 5.50×10^{-4} m, the densities become 878.5 and 1.4 kg/m^3, and h is 1.20×10^{-2} m. We find $a_1^2 = 6.60 \times 10^{-6}$ m^2; the dimensionless ratio r/a_1 remains unchanged. The final approximation gives $a_1^2 = 6.706 \times 10^{-6}$ m^2, whence

$$\gamma = \frac{877.1 \times 9.807 \times 6.706 \times 10^{-6}}{2} = 2.884 \times 10^{-2} \text{N/m(or J/m}^2) \qquad \text{(II-21)}$$

This answer could have been stated as 28.84 mN/m (or dyn/cm).

C. Experimental Aspects of the Capillary Rise Method

The capillary rise method is generally considered to be the most accurate means to measure γ, partly because the theory has been worked out with considerable exactitude and partly because the experimental variables can be closely controlled. This is to some extent a historical accident, and other methods now rival or surpass the capillary rise one in value.

Perhaps the best discussions of the experimental aspects of the capillary rise method are still those given by Richards and Carver [20] and Harkins and Brown [21]. *For the most accurate work, it is necessary that the liquid wet the wall of the capillary so that there be no uncertainty as to the contact angle.* Because of its transparency and because it is wet by most liquids, a glass capillary is most commonly used. The glass must be very clean, and even so it is wise to use a receding meniscus. The capillary must be accurately vertical, of accurately known and uniform radius, and should not deviate from circularity in cross section by more than a few percent.

As is evident from the theory of the method, h must be the height of rise above a surface for which ΔP is zero, that is, a flat liquid surface. In practice, then, h is measured relative to the surface of the liquid in a wide outer tube or dish, as illustrated in Fig. II-6, and it is important to realize that there may not be an appreciable capillary rise in relatively wide tubes. Thus, for water, the rise is 0.04 mm in a tube 1.6 cm in radius, although it is only 0.0009 mm in one of 2.7-cm radius.

The general attributes of the capillary rise method may be summarized as follows. It is considered to be one of the best and most accurate absolute methods, good to a few hundredths of a percent in precision. On the other hand, for practical reasons, a zero contact angle is required, and fairly large volumes of solution are needed. With glass capillaries, there are limitations as to the alkalinity of the solution. For variations in the capillary rise method, see Refs. 11, 12, and 22–26.

5. The Maximum Bubble Pressure Method

The procedure, as indicated in Fig. II-8, is to slowly blow bubbles of an inert gas in the liquid in question by means of a tube projecting below the surface. As also illustrated in the figure, for *small* tubes, the sequence of shapes assumed by the bubble during its growth is such that, while it is always a section of a sphere, its radius goes through a minimum when it is just hemispherical. At this point the radius is equal to that of the tube and, since the radius is at a minimum, ΔP is at a maximum. The value of ΔP is then given by Eq. II-3, where r is the radius of the tube. If the liquid wets the material of the tube, the bubble will form from the inner wall, and r will then be the inner radius of the tube. Experimentally, then, one measures the maximum gas pressure in the tube such that bubbles are unable to grow and break away. Referring again to Fig. II-8, since the tube is some arbitrary distance t below the surface of the liquid, ΔP_{max} is given by $(P_{max} - P_t)$, where P_{max} is the measured maximum pressure and P_t is the pressure corresponding to the hydrostatic head t.

If ΔP_{max} is expressed in terms of the corresponding height of a column of the liquid, that is, $\Delta P_{max} = \Delta \rho\, gh$, then the relationship becomes identical to that for the simp'e capillary rise situation as given by Eq. II-10.

It is important to realize that the preceding treatment is the limiting one for sufficiently small tubes and that significant departures from the limiting Eq. II-10 occur for r/a values as small as 0.05. More realistically, the situation is as shown in Fig. II-9, and the maximum pressure may not be reached until ϕ is considerably greater than $90°$.

As in the case of capillary rise, Sugden [27] has made use of Bashforth's and Adams' tables to calculate correction factors for this method. Because the figure is again one of revolution, the equation $h = a^2/b + z$ is exact, where b is the value of $R_1 = R_2$ at the origin and z is the distance of OC. The equation simply states that ΔP, expressed as height of a column of liquid, equals the sum of the hydrostatic head and the pressure

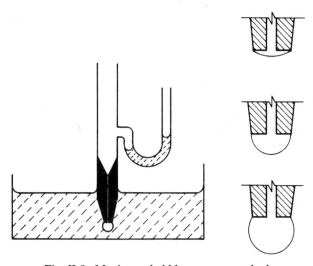

Fig. II-8. Maximum bubble pressure method.

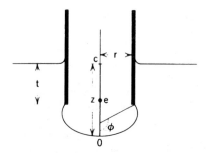

Fig. II-9

change across the interface; by simple manipulation, it may be put in the form

$$\frac{r}{X} = \frac{r}{b} + \frac{r}{a}\frac{z}{b}\left(\frac{\beta}{2}\right)^{1/2} \tag{II-22}$$

where β is given by Eq. II-18 and $X = a^2/h$. For any given value of r/a there will be a series of values of r/X corresponding to a series of values of β and of ϕ. For each assumed value of r/a, Sugden computed a series of values of r/b by inserting various values of β in the identity $r/b = (r/a)(2/\beta)^{1/2}$. By means of the Bashforth and Adams tables [14], for each β value used and corresponding r/b value, a value of z/b and hence of r/X (by Eq. II-22) was obtained. Since r/X is proportional to the pressure in the bubble, the series of values for a given r/a go through a maximum as β is varied. For each assumed value, Sugden then tabulated this maximum value of r/X. His values are given in Table II-2 as X/r versus r/a

The table is used in much the same manner as are Eqs. II-19 and II-20 in the case of capillary rise. As a first approximation, one assumes the simple Eq. II-10 to apply, that is, that $X = r$; this gives the first approximation a_1 to the capillary constant. From this, one obtains r/a_1 and reads the corresponding value of X/r from Table II-2. From the derivation of $X(X = a^2/h)$, a second approximation a_2 to the capillary constant is obtained, and so on. Some more recent calculations have been made by Johnson and Lane [28].

The maximum bubble pressure method is good to a few tenths percent accuracy, does not depend on contact angle (except insofar as to whether the inner or outer radius of the tube is to be used), and requires only an approximate knowledge of the density of the liquid (if twin tubes are used), and the measurements can be made rapidly. The method is also amenable to remote operation and can be used to measure surface tensions of not easily accessible liquids such as molten metals [29].

A pulsating bubble surfactometer, available commercially, allows one to measure the dynamic surface tension in solutions [30, 31]. The bubble is expanded and contracted to change its area by a factor of 1.5–2 at rates of 1–100 cycles per minute. Studying hexadecanol in water, an important component of a lung surfactant replacement drug, Franses and co-workers [30] illustrate the importance of the geometry of the measuring technique in the study of surfactant dispersions. In the pulsating bubble technique, hexadecanol particles rise to the surface enhancing flux and speeding the reduction in

TABLE II-2

Correction Factors for the Maximum Bubble Pressure Method (Minimum Values of X/r for Values of r/a from 0 to 1.50)

r/a	0.00	0.01	0.02	0.03	0.04	0.05	0.06	0.07	0.08	0.09
0.0	1.0000	9999	9997	9994	9990	9984	9977	9968	9958	9946
0.1	0.9934	9920	9905	9888	9870	9851	9831	9809	9786	9762
0.2	9737	9710	9682	9653	9623	9592	9560	9527	9492	9456
0.3	9419	9382	9344	9305	9265	9224	9182	9138	9093	9047
0.4	9000	8952	8903	8853	8802	8750	8698	8645	8592	8538
0.5	8484	8429	8374	8319	8263	8207	8151	8094	8037	7979
0.6	7920	7860	7800	7739	7678	7616	7554	7493	7432	7372
0.7	7312	7252	7192	7132	7072	7012	6953	6894	6835	6776
0.8	6718	6660	6603	6547	6492	6438	6385	6333	6281	6230
0.9	6179	6129	6079	6030	5981	5933	5885	5838	5792	5747
1.0	5703	5659	5616	5573	5531	5489	5448	5408	5368	5329
1.1	5290	5251	5213	5176	5139	5103	5067	5032	4997	4962
1.2	4928	4895	4862	4829	4797	4765	4733	4702	4671	4641
1.3	4611	4582	4553	4524	4496	4468	4440	4413	4386	4359
1.4	4333	4307	4281	4256	4231	4206	4181	4157	4133	4109
1.5	4085									

surface tension; in the pendant drop technique (see Section II-7A) the buoyant particles are depleted at the interface.

6. Detachment Methods

Several convenient ways to measure surface tension involve the detachment of a solid from the liquid surface. These include the measurement of the weight in a drop falling from a capillary and the force to detach a ring, wire, or thin plate from the surface of a liquid. In this section we briefly describe these methods and their use.

A. The Drop Weight Method

This is a fairly accurate and convenient method for measuring the surface tension of a liquid–vapor or liquid–liquid interface. The procedure, in its simplest form, is to form drops of the liquid at the end of a tube, allowing them to fall into a container until enough have been collected to accurately determine the weight per drop. Recently developed computer-controlled devices track individual drop volumes to $\pm = 0.1$ μl [32].

The method is a very old one, remarks on it having been made by Tate in 1864 (33), and a simple expression for the weight W of a drop is given by what

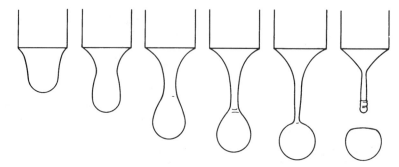

Fig. II-10. High-speed photographs of a falling drop.

is known as *Tate's law*† :

$$W = 2\pi r\gamma \tag{II-22}$$

Here again, the older concept of "surface tension" appears since Eq. II-22 is best understood in terms of the argument that the maximum force available to support the weight of the drop is given by the surface tension force per centimeter times the circumference of the tip.

In actual practice, a weight W' is obtained, which is less than the "ideal" value W. The reason for this becomes evident when the process of drop formation is observed closely. What actually happens is illustrated in Fig. II-10. The small drops arise from the mechanical instability of the thin cylindrical neck that develops (see Section II-3); in any event, it is clear that only a portion of the drop that has reached the point of instability actually falls—as much as 40% of the liquid may remain attached to the tip.

The usual procedure is to apply a correction factor f to Eq. II-22, so that W' is given by

$$W' = 2\pi r\gamma f$$

Harkins and Brown [21] concluded that f should be a function of the dimensionless ratio r/a or, alternatively, of $r/V^{1/3}$, where V is the drop volume. (See Refs. 34 and 35 for a more up-to-date discussion.) This they verified experimentally by determining drop weights for water and for benzene, using tips of various radii. Knowing the values of γ from capillary rise measurements, and thence the respective values of a, f could be determined in each case. The resulting variation of f with $r/V^{1/3}$ has been fitted to a smoothing function to allow tabulation at close intervals [36].

---

†The actual statement by Tate is "Other things being equal, the weight of a drop of liquid is proportional to the diameter of the tube in which it is formed." See Refs. 34 and 35 for some discussion.

It is desirable to use $r/V^{1/3}$ values in the region of 0.6 to 1.2, where f is varying most slowly. The correct value for the surface tension is then given by

$$\gamma = \frac{mg}{2\pi r f}$$ (II-25)

It is to be noted that not only is the correction quite large, but for a given tip radius it depends on the nature of the liquid. It is thus *incorrect* to assume that the drop weights for two liquids are in the ratio of the respective surface tensions when the same size tip is used. Finally, correction factors for $r/V^{1/3} < 0.3$ have been determined, using mercury drops [37].

In employing this method, an important precaution to take is to use a tip that has been ground smooth at the end and is free from any nicks. In the case of liquids that do not wet the tip, r is the inside radius. Volatile liquids are studied in a closed system as described by Harkins and Brown [21] to minimize evaporation losses.

Since the drop volume method involves creation of surface, it is frequently used as a dynamic technique to study adsorption processes occurring over intervals of seconds to minutes. A commercial instrument delivers computer-controlled drops over intervals from 0.5 sec to several hours [38, 39]. Accurate determination of the surface tension is limited to drop times of a second or greater due to hydrodynamic instabilities on the liquid bridge between the detaching and residing drops [40].

An empirically determined relationship between drop weight and drop time does allow surface tensions to be determined for small surface ages [41].

B. The Ring Method

A method that has been rather widely used involves the determination of the force to detach a ring or loop of wire from the surface of a liquid. It is generally attributed to du Noüy [42]. As with all detachment methods, one supposes that a first approximation to the detachment force is given by the surface tension multiplied by the periphery of the surface detached. Thus, for a ring, as illustrated in Fig. II-11,

$$W_{tot} = W_{ring} + 4\pi R\gamma$$ (II-26)

Harkins and Jordan [43] found, however, that Eq. II-26 was generally in serious error and worked out an empirical correction factor in much the same way as was done for the drop weight method. Here, however, there is one additional variable so that the correction factor f now depends on two dimensionless ratios. Thus

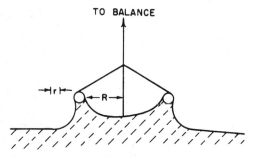

Fig. II-11. Ring method.

$$f = \frac{\gamma}{p} = f\left(\frac{R^3}{V}, \frac{R}{r} \right) \tag{II-27}$$

where p denotes the "ideal" surface tension computed from Eq. II-26, and V is the meniscus volume. The extensive tables of Harkins and Jordan, as recalculated by Huh and Mason [44] are summarized graphically in Fig. II-12, and it is seen that the simple equation may be in error by as much as 25%. Additional tables are given in Ref. 45.

Experimentally, the method is capable of good precision. Harkins and Jordan used

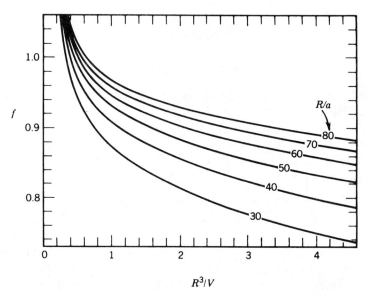

Fig. II-12. The factor f in the range of $R^3/V = 0.4$ to 4.5 and $R/a = 30$ to 80. (From Ref. 44.)

a chainomatic balance to determine the maximum pull, but a popular simplified version of the *tensiometer*, as it is sometimes called, makes use of a torsion wire and is quite compact. Among experimental details to mention are that the dry weight of the ring, which is usually constructed of platinum, is to be used, the ring should be kept horizontal (a departure of 1° was found to introduce an error of 0.5%, whereas one of 2.1° introduced an error of 1.6%), and care must be taken to avoid any disturbance of the surface as the critical point of detachment is approached. The ring is usually flamed before use to remove surface contaminants such as grease, and it is desirable to use a container for the liquid that can be overflowed so as to ensure the presence of a clean liquid surface. Additional details are given in Ref. 46.

A zero or near-zero contact angle is necessary; otherwise results will be low. This was found to be the case with surfactant solutions where adsorption on the ring changed its wetting characteristics, and where liquid-liquid interfacial tensions were measured. In such cases a Teflon or polyethylene ring may be used [47]. When used to study monolayers, it may be necessary to know the increase in area at detachment, and some calculations of this are available [48]. Finally, an alternative method obtains γ from the slope of the plot of W versus z, the elevation of the ring above the liquid surface [49].

C. Wilhelmy Slide Method

The methods so far discussed have required correction factors to the respective "ideal" equations. Yet there is one method, attributed to Wilhelmy [50] in 1863, that entails no such corrections and is very simple to use.

The basic observation is that a thin plate, such as a microscope cover glass or piece of platinum foil, will support a meniscus whose weight both as measured statically or by detachment is given very accurately by the "ideal" equation (assuming zero contact angle):

$$W_{tot} = W_{plate} + \gamma p \tag{II-28}$$

where p is the perimeter. The experimental arrangement is shown schematically in Fig. II-13. When used as a detachment method, the procedure is essentially the same as with the ring method, but Eq. II-28 holds to within 0.1% so that no corrections are needed [51, 52]. A minor, omitted term in Eq. II-28 allows for the weight of liquid directly under the plate (see Ref. 46).

It should be noted that here, as with capillary rise, there is an adsorbed film of vapor (see Section X-7D) with which the meniscus merges smoothly. The meniscus is not "hanging" from the plate but rather from a liquidlike film [53]. The correction for the weight of such film should be negligible, however.

An alternative and probably now more widely used procedure is to raise the liquid level gradually until it just touches the hanging plate suspended from a balance. The increase in weight is then noted. A general equation is

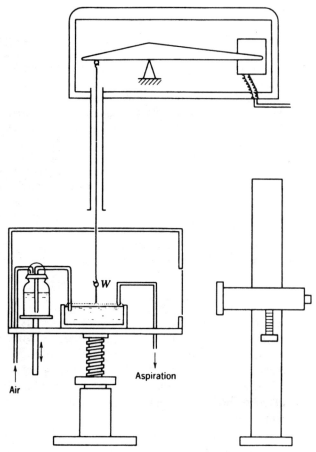

Fig. II-13. Apparatus for measuring the time dependence of interfacial tension (from Ref. 54). The air and aspirator connections allow for establishing the desired level of fresh surface. W denotes the Wilhelmy slide, suspended from a Cahn electrobalance with a recorder output.

$$\gamma \cos \theta = \frac{\Delta W}{p} \tag{II-29}$$

where ΔW is the change in weight of (i.e., force exerted by) the plate when it is brought into contact with the liquid, and p is the perimeter of the plate. The contact angle, if finite, may be measured in the same experiment [54]. Integration of Eq. II-12 gives

$$\left(\frac{h}{a}\right)^2 = 1 - \sin \theta \tag{II-30}$$

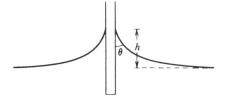

Fig. II-14. Meniscus profile for a nonwetting liquid.

where, as illustrated in Fig. II-14, h is the height of the top of the meniscus above the level liquid surface. Zero contact angle is preferred, however, if only the liquid surface tension is of interest; it may help to slightly roughen the plate, see Refs. 46 and 55.

As an example of the application of the method, Neumann and Tanner [54] followed the variation with time of the surface tension of aqueous sodium dodecyl sulfate solutions. Their results are shown in Fig. II-15, and it is seen that a slow but considerable change occurred.

A modification of the foregoing procedure is to suspend the plate so that it is partly immersed and to determine from the dry and immersed weights the meniscus weight. The procedure is especially useful in the study of surface adsorption or of monolayers, where a change in surface tension is to be measured. This application is discussed in some detail by Gaines [57]. Equation II-28 also applies to a wire or fiber [58].

The Wilhelmy slide has been operated in dynamic immersion studies to measure advancing and receding contact angles [59] (see Chapter X). It can also

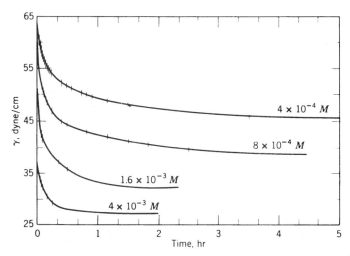

Fig. II-15. Variation with time of aqueous sodium dodecyl sulfate solutions of various concentrations (from Ref. 54). See Ref. 56 for later data with highly purified materials.

be used with a trapezoidal pulse applied to the barrier at a fluid–fluid interface to measure the transient response of the surface to a change in area [60].

7. Methods Based on the Shape of Static Drops or Bubbles

Small drops or bubbles will tend to be spherical because surface forces depend on the area, which decreases as the square of the linear dimension, whereas distortions due to gravitational effects depend on the volume, which decreases as the cube of the linear dimension. Likewise, too, a drop of liquid in a second liquid of equal density will be spherical. However, when gravitational and surface tensional effects are comparable, then one can determine in principle the surface tension from measurements of the shape of the drop or bubble. The variations situations to which Eq. II-16 applies are shown in Fig. II-16.

The general procedure is to form the drop or bubble under conditions such that it is not subject to disturbances and then to measure its dimensions or profile from a photograph or with digital image processing of video images (see Refs. 61, 62). The image analysis has recently been automated [62] to improve accuracy over manual analysis. In axisymmetric drop shape analysis of surface tension, the pendant drop geometry is preferable due to the ease with which large drops can be made axisymmetric. Sessile drops, however, are useful for studies of contact angles [63, 64] (see Chapter X). The greatest accuracy is achieved with fewer very accurate points on the drop surface rather than a large number of less reliable points [65].

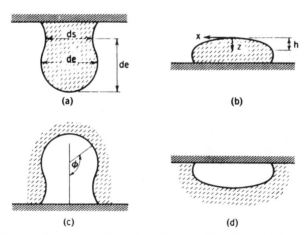

Fig. II-16. Shapes of sessile and hanging drops and bubbles: (*a*) hanging drop; (*b*) sessile drop; (*c*) hanging bubble; (*d*) sessile bubble.

A. Pendant Drop Method

A drop hanging from a tip (or a clinging bubble) elongates as it grows larger because the variation in hydrostatic pressure ΔP eventually becomes appreciable in comparison with that given by the curvature at the apex. As in the case of a meniscus, it is convenient to write Eq. II-12 in the form of Eq. II-17, where in the present case, the dimensionless parameter β is negative. A profile calculated from Eq. II-17 for the case of $\beta = -0.45$ is given in Table II-3 [66]. The best value of β for a given drop can be determined by profile matching (see further below), but some absolute quantity such as b must also be measured in order to obtain an actual γ value.

An alternative to obtaining β directly involves defining some more convenient shape-dependent function, and an early but still very practical method is the following. We define a shape-dependent quantity as $S = d_s/d_e$; as indicated in Fig. II-16, d_e is the equatorial diameter and d_s is the diameter measured at a distance d_e up from the bottom of the drop. The hard-to-measure size parameter b in Eq. II-17 is combined with β by defining the quantity $H = -\beta(d_e/b)^2$. Thus

$$\gamma = \frac{-\Delta\rho \, gb^2}{\beta} = \frac{-\Delta\rho \, gd_e^2}{\beta(d_e/b)^2} = \frac{\Delta\rho \, gd_e^2}{H} \qquad \text{(II-31)}$$

The relationship between the shape-dependent quantity H and the experimentally measurable quantity S originally was determined empirically [66], but a set of quite accurate $1/H$ versus S values were later obtained by Niederhauser and Bartell [67] (see also Refs. 34 and 68) and by Stauffer [69].

A set of pendant drop profiles is shown in Fig. II-17 as an illustration of the range of shapes that may be observed. It has been pointed out that for practical reasons, the size of the tip from which the drop is suspended should be such that r/a is about 0.5 or less [66].

A modern alternative procedure involves computer matching of the entire drop profile to a best fitting theoretical curve; in this way the entire profile is used, rather than just d_s and d_e, so that precision is increased. Also, drops whose d_s is not measurable (how does this happen?) can be used. References 61 and 71–74 provide examples of this type of approach.

The automated pendant drop technique has been used as a film balance to study the surface tension of insoluble monolayers [75] (see Chapter IV). A motor-driven syringe allows changes in drop volume to study surface tension as a function of surface areas as in conventional film balance measurements. This approach is useful for materials available in limited quantities and it can be extended to study monolayers at liquid–liquid interfaces [76].

B. Sessile Drop or Bubble Method

The cases of the sessile drop and bubble are symmetrical, as illustrated in Fig. II-16. The profile is also that of a meniscus; β is now positive and, as an

TABLE II-3
Solutions to Eq. II-17 for $\beta = -0.45$

ϕ^a	x/b	z/b
0.099944	0.099834	0.004994
0.199551	0.198673	0.019911
0.298488	0.295547	0.044553
0.396430	0.389530	0.078600
0.493058	0.479762	0.121617
0.588070	0.565464	0.173072
0.681175	0.645954	0.232352
0.772100	0.720657	0.298779
0.860590	0.789108	0.371635
0.946403	0.850958	0.460175
1.029319	0.905969	0.533649
1.109130	0.954013	0.621322
1.185644	0.995064	0.712480
1.258681	1.029190	0.806454
1.328069	1.056542	0.902619
1.393643	1.077347	1.000413
1.455242	1.091895	1.099333
1.512702	1.100530	1.198946
1.565856	1.103644	1.298886
1.614526	1.101667	1.398856
1.658523	1.095060	1.498630
1.697641	1.084311	1.598044
1.731653	1.069933	1.697000
1.760310	1.052460	1.795458
1.783338	1.032445	1.893432
1.800443	1.010466	1.990986
1.811310	0.987123	2.088223
1.815618	0.963039	2.185279
1.813050	0.938868	2.282314
1.803321	0.915293	2.379495
1.786207	0.893023	2.476982
1.761593	0.872791	2.574912
1.729517	0.855344	2.673373
1.690226	0.841424	2.772393

[a]The angle ϕ is in units of $360/2\pi$ or $57.295°$.

example, the solution to Eq. II-17 for $\beta = 0.5$ is given in Ref. 77 (note also Table II-1).

The usual experimental situation is that of a sessile drop and, as with the pendant drop, it is necessary to determine a shape parameter and some absolute length. Thus β may be determined by profile fitting, and z_e measured, where z_e is the distance from the plane at $\phi = 90$ to the apex. If the drop rests with

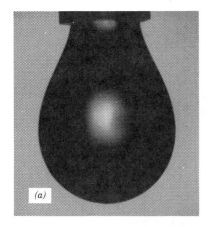
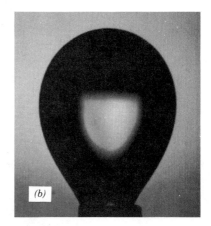

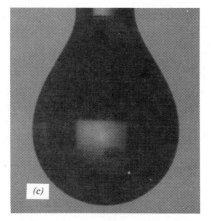

Fig. II-17. Pendant drops: (*a*) water; (*b*) benzene/water; (*c*) formamide. The measurements were at 21°. [Courtesy A. W. Neumann (see also Ref. 70).]

a contact angle of less than 90°, there is no z_e and, instead, the control angle and the total height of the drop can be measured. Some of the specific procedures that have been used are found in Refs. 71–74 and 77–79. The sessile drop method has been used to follow surface tension as a function of time, as with sodium laurate solutions [80], the surface tension of molten metals [81–83] and liquid-liquid interfacial tensions [70, 73].

The case of very large drops or bubbles is easy because only one radius of curvature (that in the plane of the drawings) is considered. Equation II-12 then becomes

$$\Delta \rho g y = \gamma \, \frac{y''}{(1 + y'^2)^{3/2}}$$

or

$$\frac{2y}{a^2} = \frac{p \; dp/dy}{(1+p^2)^{3/2}} \tag{II-32}$$

where $p = dy/dx$. Integration gives

$$\frac{y^2}{a^2} = \frac{-1}{(1+p^2)^{1/2}} + \text{const} \tag{II-33}$$

Since h denotes the distance from the apex to the equatorial plane, then at $y = h$, $p = \infty$, and Eq. II-33 becomes

$$\frac{y^2}{a^2} - \frac{h^2}{a^2} = \frac{-1}{(1+p^2)^{1/2}}$$

Furthermore, at $y = 0$, $p = 0$, from which it follows that $h^2/a^2 = 1$, or $h = a$,

$$\gamma = \frac{\Delta\rho \; gh^2}{2} \tag{II-34}$$

This very simple result is independent of the value of the contact angle because the configuration involved is only that between the equatorial plane and the apex.

Very small sessile drops have a shape that depends on the line tension along the circular contact line; if large enough it induces a dewetting transition detaching the drop from the surface [84].

C. Sources of Other Deformed Shapes

The discussion so far has been of interfaces in a uniform gravitational field. There are several variants from this situation, some of which are useful in the measurement of liquid-liquid interfacial tensions where these are very small. Consider the case of a drop of liquid A suspended in liquid B. If the density of A is less than that of B, on rotating the whole mass, as illustrated in Fig. II-18, liquid A will go to the center, forming a drop astride the axis of revolution. With increasing speed of revolution, the drop of A elongates, since centrifugal force increasingly opposes the surface tensional drive toward minimum interfacial area. In brief, the drop of A deforms from a sphere to a prolate ellipsoid. At a sufficiently high speed of revolution, the drop approximates to an elongated cylinder.

The general analysis, while not difficult, is complicated; however, the limiting case of the very elongated, essentially cylindrical drop is not hard to treat. Consider a section of the elongated cylinder of volume V (Fig. II-18b). The centrifugal force on a volume element is $\omega^2 r \Delta\rho$, where ω is the speed of revolution and $\Delta\rho$ the difference in density. The potential energy at distance r from the axis of revolution is then $\omega^2 r^2 \Delta\rho/2$, and the total potential energy for the

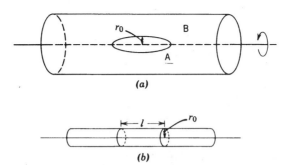

Fig. II-18. Illustration of the rotating drop method.

cylinder of length l is $l \int_0^{r_0} (\omega^2 r^2 \, \Delta\rho/2) 2\pi r \, dr = \pi \omega^2 \, \Delta\rho \, r_0^4 l/4$. The interfacial free energy is $2\pi r_0 l \gamma$. The total energy is thus

$$E = \frac{\pi \omega^2 \, \Delta\rho \, r_0^4 l}{4} + 2\pi r_0 l \gamma = \frac{\omega^2 \, \Delta\rho \, r_0^2 V}{4} + \frac{2V\gamma}{r_0}$$

since $V = \pi r_0^2 l$. Setting $dE/dr_0 = 0$, we obtain

$$\gamma = \frac{\omega^2 \, \Delta\rho \, r_0^3}{4} \tag{II-35}$$

Equation II-35 has been called Vonnegut's equation [85].

Princen and co-workers have treated the more general case where ω is too small or γ too large to give a cylindrical profile [86] (see also Refs. 87 and 88). In such cases, however, a correction may be needed for buoyancy and Coriolis effects [89]; it is best to work under conditions such that Eq. II-35 applies. The method has been used successfully for the measurement of interfacial tensions of 0.001 dyn/cm or lower [90, 91].

Small interfacial tensions may also be measured from the deformation of a drop suspended in a liquid having a similar density [92]. The distortion of drops and bubbles placed in shearing flows of liquids was first investigated theoretically by G. I. Taylor in 1934, who also conducted a series of careful experiments. He established that the parameter measuring the distorting forces due to the flow relative to the interfacial tension opposing the distortion is a capillary number, $Ca = U\mu/\gamma$, where U is the fluid velocity, μ the viscosity, and γ the interfacial tension. For unbounded simple shear flow, U is replaced by Ga, where G is the shear rate and a the radius of the undistorted drop. Taylor carried out a perturbation expansion for small Ca, showing the first effects of shape distortion. Many theoretical and experimental studies have extended his work to finite internal viscosity, more general flows, and large deformation or breakup. This work is reviewed by Rallison [93] and Stone [93a]. An initially spherical drop will deform to a spheroid of major axis ℓ and minor axis b; the degree of deformation is defined by $D = (\ell - b)/(\ell + b)$. The deformation of a drop of radius r in an electric field E is

$$D = \frac{9\epsilon_0 \epsilon r E^2}{16\gamma} \qquad \text{(II-36)}$$

where ϵ_0 is the permittivity of vacuum (8.854×10^{-12} $C^2N^{-1}m^{-2}$, and ϵ is the dielectric constant of the outer fluid (that for the drop is assumed to be high) [94–96]. This effect was noted by Lord Kelvin [97]. Finally, the profiles of nonaxisymmetric drops including inclined pendant [98] and sessile [99] drops have been calculated.

In the converse situation free of gravity, a drop assumes a perfectly spherical shape. At one point, the U.S. Space program tested this idea with the solidification of ball bearings from molten metal drops in microgravity conditions.

An interesting application of capillarity and drops in fields occurs in inkjet printing technology. In this process, illustrated in Fig. II-19, ink resides in a small square chamber with a meniscus balanced at the exit orifice by the pressure in the reservoir and capillary forces. In the wall opposite the orifice is a thin film resistor that, upon heating at $10^{8\circ}C/sec$, causes rapid growth of a vapor bubble that ejects a drop of ink through the orifice (Fig. II-19b). The chamber refills and the process is repeated. The newest printers achieve a repetition frequency of 8000 Hz by carefully controlling the refilling process [100].

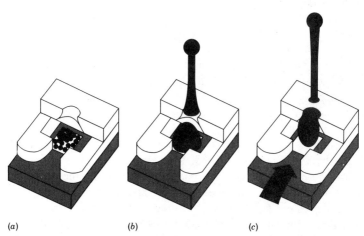

(a)   (b)   (c)

Fig. II-19. The drop ejection process in an inkjet printer: (a) bubble nucleation; (b) bubble growth and drop ejection; (c) refill. [From J. H. Bohórquez, B. P. Canfield, K. J. Courian, F. Drogo, C. A. E. Hall, C. L. Holstun, A. R. Scandalis, and M. E. Shepard, *Hewlett-Packard J.* **45**(1), 9–17 (Feb. 1994). Copyright 1994, Hewlett-Packard Company. Reproduced with permission.]

8. Dynamic Methods of Measuring Surface Tension

The profound effect of surface active agents on the surface tension of a liquid motivate the study of their adsorption at liquid surfaces through the dynamic measurement of surface tension. Recent computer-controlled devices have enabled such studies via the pulsating bubble method described in Section II-5 and the drop weight technique (Section II-6A). These techniques are generally limited to the study of surface tensions varying over time periods of seconds to minutes. It is of interest to study surface aging and relaxation effects on a very short time scale, and for this more rapid dynamic methods are needed. Two good reviews of dynamic surface tension techniques by Miller and co-workers and by Chang and Franses appear in Refs. 101 and 102. We briefly describe three of these techniques below.

A. Flow Methods

A jet emerging from a noncircular orifice is mechanically unstable, not only with respect to the eventual breakup into droplets discussed in Section II-3, but, more immediately, also with respect to the initial cross section not being circular. Oscillations develop in the jet since the momentum of the liquid carries it past the desired circular cross section. This is illustrated in Fig. II-20.

The mathematical treatment was first developed by Lord Rayleigh in 1879, and a more exact one by Bohr has been reviewed by Sutherland [103], who gives the formula

$$\gamma_{app} = \frac{4\rho v^2 (1 + 37b^2/24r^2)}{6r\lambda^2 (1 + 5\pi^2 r^2/3\lambda^2)} \tag{II-37}$$

where ρ is the density of the liquid, v is the volume velocity, λ is the wavelength, r is the sum of the minimum and maximum half-diameters, and b is their difference. The required jet dimensions were determined optically, and a typical experiment would make use of jets of about 0.03 cm in size and velocities of about 1 cm^3/sec, giving λ values of around 0.5 cm. To a first approximation, the surface age at a given node is

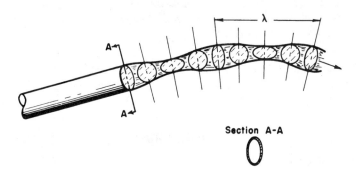

Section A-A

Fig. II-20. Oscillations in an elliptical jet.

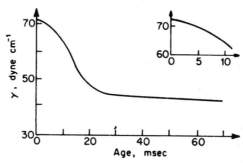

Fig. II-21. Surface tension as a function of age for 0.05 g/100 cm³ of sodium di(2-ethylhexyl)sulfosuccinate solution determined with various types of jet orifices [109].

just the distance from the orifice divided by the linear jet velocity and, in the preceding example, would be about 1 msec per wavelength.

It was determined, for example, that the surface tension of water relaxes to its equilibrium value with a relaxation time of 0.6 msec [104]. The oscillating jet method has been useful in studying the surface tension of surfactant solutions. Figure II-21 illustrates the usual observation that at small times the jet appears to have the surface tension of pure water. The slowness in attaining the equilibrium value may partly be due to the times required for surfactant to diffuse to the surface and partly due to chemical rate processes at the interface. See Ref. 105 for similar studies with heptanoic acid and Ref. 106 for some anomalous effects.

For times below about 5 msec a correction must be made to allow for the fact that the surface velocity of the liquid *in* the nozzle is zero and takes several wavelengths to increase to the jet velocity after emerging from the nozzle. Correction factors have been tabulated [107, 108]; see also Ref. 109.

The oscillating jet method is not suitable for the study of liquid-air interfaces whose ages are in the range of tenths of a second, and an alternative method is based on the dependence of the shape of a falling column of liquid on its surface tension. Since the hydrostatic head, and hence the linear velocity, increases with *h*, the distance away from the nozzle, the cross-sectional area of the column must correspondingly decrease as a material balance requirement. The effect of surface tension is to oppose this shrinkage in cross section. The method is discussed in Refs. 110 and 111. A related method makes use of a falling sheet of liquid [112].

Another oscillatory method makes use of a drop acoustically levitated in a liquid. The drop is made to oscillate in shape, and the interfacial tension can be calculated from the resonance frequency [113].

B. Capillary Waves

The wavelength of ripples on the surface of a deep body of liquid depends on the surface tension. According to a formula given by Lord Kelvin [97],

$$v^2 = \frac{g\lambda}{2\pi} + \frac{2\pi\gamma}{\rho\lambda}$$

$$\gamma = \frac{\lambda^3 \rho}{2\pi\tau^2} - \frac{g\lambda^2 \rho}{4\pi^2} \tag{II-38}$$

where v is the velocity of propagation, λ is the wavelength, and τ is the period of the ripples. For water there is a minimum velocity of about 0.5 mph (mi/hr) for $\lambda = 1.7$ cm; for $\lambda = 0.1$ cm, it is 1.5 mph, whereas for $\lambda = 10^5$ cm, it is 89 mph!

Experimentally, the waves are measured as standing waves, and the situation might be thought to be a static one. However, individual elements of liquid in the surface region undergo a roughly circular motion, and the surface is alternately expanded and compressed. As a consequence, damping occurs even with a pure liquid, and much more so with solutions or film-covered surfaces for which transient surface expansions and contractions may be accompanied by considerable local surface tension changes and by material transport between surface layers. Hansen has reviewed the subject [114]. A more detailed discussion is deferred to Chapter IV, but it should be mentioned here that capillary waves are spontaneously present because of small temperature and hence density fluctuations. These minute waves (about 5 Å amplitude and 0.1 mm wavelength) can be detected by laser light-scattering techniques. The details are beyond the scope of this text; they are discussed in Refs. 115–118. Both liquid-air and liquid-liquid surface tensions can be measured, as well as the rate of damping of waves.

C. Maximum Bubble Pressure Method

A recent design of the maximum bubble pressure instrument for measurement of dynamic surface tension allows resolution in the millisecond time frame [119, 120]. This was accomplished by increasing the system volume relative to that of the bubble and by using electric and acoustic sensors to track the bubble formation frequency. Miller and co-workers also assessed the hydrodynamic effects arising at short bubble formation times with experiments on very viscous liquids [121]. They proposed a correction procedure to improve reliability at short times. This technique is applicable to the study of surfactant and polymer adsorption from solution [101, 120].

9. Surface Tension Values as Obtained by Different Methods

The surface tension of a pure liquid should and does come out to be the same irrespective of the method used, although difficulties in the mathematical treatment of complex phenomena can lead to apparent discrepancies. In the case of solutions, however, dynamic methods, including detachment ones, often tend

TABLE II-4
Surface Tension Values[a]

Liquid	Temperature	γ (dyn/cm mN/m)
Liquid–Vapor Interfaces		
Water[b]	20°C	72.94
	21.5°C	72.75
	25°C	72.13
Organic compounds		
Methylene iodide[c]	20°C	67.00
	21.5°C	63.11
Glycerine[d]	24°C	62.6
	20°C	48.09
Ethylene glycol[e]	25°C	47.3
	40°C	46.3
Dimethyl sulfoxide[f]	20°C	43.54
Propylene carbonate[g]	20°C	41.1
1-Methyl naphthalene[h]	20°C	38.7
Dimethyl aniline[i]	20°C	36.56
Benzene[b]	20°C	28.88
	30°C	27.56
Toluene[b]	20°C	28.52
Chloroform[b]	25°C	26.67
Propionic acid[b]	20°C	26.69
Butyric acid[b]	20°C	26.51
Carbon tetrachloride[b]	25°C	26.43
Butyl acetate[j]	20°C	25.09
Diethylene glycol[k]	20°C	30.9
Nonane[b]	20°C	22.85
Methanol[b]	20°C	22.50
Ethanol[b]	20°C	22.39
	30°C	21.55
Octane[b]	20°C	21.62
Heptane[b]	20°C	20.14
Ether[b]	25°C	20.14
Perfluoromethylcyclohexane[b]	20°C	15.70
Perfluoroheptane[b]	20°C	13.19
Hydrogen sulfide[l]	20°C	12.3
Perfluoropentane[b]	20°C	9.89
Dodecane[gg]	22°C	25.44
Polydimethyl siloxane, MW 3900[ff]	20°C	20.47
	20°C	21.01

TABLE II-4   (*Continued*)

Liquid	Temperature	γ (dyn/cm mN/m)
Liquid–Vapor Interfaces		
Low-boiling substances		
$^4He^m$	1K	0.365
$H_2{}^n$	20 K	2.01
$D_2{}^n$	20 K	3.54
$N_2{}^o$	75 K	9.41
Ar^o	90 K	11.86
$CH_4{}^b$	110 K	13.71
$F_2{}^b$	85 K	14.84
$O_2{}^b$	77 K	16.48
Metals		
Hg^b	20°C	486.5
	25°C	485.5
	30°C	484.5
Na^s	130°C	198
Ba^t	720°C	226
Sn^u	332°C	543.8
Salts		
$NaCl^y$	1073°C	115
$KClO_3{}^z$	368°C	81
$KNCS^z$	175°C	101.5
$C_2H_6{}^p$	180.6 K	16.63
Xe^q	163 K	18.6
N_2O^p	182.5 K	24.26
$Cl_2{}^b$	−30°C	25.56
$NOCl^b$	−10°C	13.71
$Br_2{}^r$	20°C	31.9
Ag^v	1100°C	878.5
Cu^w	mp	1300
Ti^x	1680°C	1588
Pt^w	mp	1800
Fe^w	mp	1880
$NaNO_3{}^{aa}$	308°C	116.6
$K_2Cr_2O_7{}^z$	397°C	129
$Ba(NO_3)_2{}^{aa}$	595°C	134.8

to give high values. Padday and Russell discuss this point in some detail [122]. The same may be true of interfacial tensions between partially miscible liquids.

The data given in Table II-4 were selected with the purpose of providing a working stock of data for use in problems as well as a convenient reference to

TABLE II-4 *(Continued)*

Liquid	Temperature	γ (dyn/cm mN/m)
Liquid–Liquid Interface		
Liquid 1: water		
n-Butyl alcohol[bb]	20°C	1.8
Ethyl acetate[bb]	20°C	6.8
Heptanoic acid[cc]	20°C	7.0
Benzaldehyde[aa]	20°C	15.5
Liquid 1: mercury		
Water[dd]	20°C	415
	25°C	416
Ethanol[cc]	20°C	389
n-Hexane[cc]	20°C	378
Liquid 1: fluorocarbon polymer		
Benzene[ee]	25°C	7.8
Liquid 1: diethylene glycol		
n-Heptane[j]	20°C	10.6
Nitrobenzene[bb]	20°C	25.2
Benzene[cc]	20°C	35.0
Carbon tetrachloride[cc]	20°C	45.0
n-Heptane[cc]	20°C	50.2
n-Heptane[cc]	20°C	378
Benzene[dd]	20°C	357
Water[ee]	25°C	57
n-Decane[k]	20°C	11.6

[a]Extensive compilations are given by J. J. Jasper, *J. Phys. Chem. Ref. Data*, **1,** 841 (1972) and G. Korosi and E. sz. Kováts, *J. Chem. Eng. Data*, **26,** 323 (1981).

[b]A. G. Gaonkar and R. D. Neuman, *Colloids & Surfaces*, **27,** 1 (1987) (contains an extensive review of the literature); V. Kayser, *J. Colloid Interface Sci.*, **56,** 622 (1972).

[c]R. Grzeskowiak, G. H. Jeffery, and A. I. Vogel, *J. Chem. Soc.*, **1960,** 4728.

[d]Ref. 61.

[e]Ref. 41.

[f]H. L. Clever and C. C. Snead, *J. Phys. Chem.*, **67,** 918 (1963).

[g]M. K. Bernett, N. L. Jarvis, and W. A. Zisman, *J. Phys. Chem.*, **66,** 328 (1962).

[h]A. N. Gent and J. Schultz, *J. Adhes.*, **3,** 281 (1972).

[i]Ref. 20.

[j]J. B. Griffin and H. L. Clever, *J. Chem. Eng. Data*, **5,** 390 (1960).

[k]G. L. Gaines, Jr., and G. L. Gaines III, *J. Colloid Interface Sci.*, **63,** 394 (1978).

TABLE II-4   *(Continued)*

[l]C. S. Herrick and G. L. Gaines, Jr., *J. Phys. Chem.*, **77**, 2703 (1973).

[m]K. R. Atkins and Y. Narahara, *Phys. Rev.*, **138**, A437 (1965).

[n]V. N. Grigor'ev and N. S. Rudenko, *Zh. Eksperim. Teor. Fiz.*, **47**, 92 (1964) (through *Chem. Abstr.*, **61**, 12669[g] (1964)).

[o]D. Stansfield, *Proc. Phys. Soc.*, **72**, 854 (1958).

[p]A. J. Leadbetter, D. J. Taylor, and B. Vincent, *Can. J. Chem.*, **42**, 2930 (1964).

[q]A. J. Leadbetter and H. E. Thomas, *Trans. Faraday Soc.*, **61**, 10 (1965).

[r]M. S. Chao and V. A. Stenger, *Talanta*, **11**, 271 (1964) (through *Chem. Abstr.*, **60**, 4829[g] (1964)).

[s]C. C. Addison, W. E. Addison, D. H. Kerridge, and J. Lewis, *J. Chem. Soc.*, **1955**, 2262.

[t]C. C. Addison, J. M. Coldrey, and W. D. Halstead, *J. Chem. Soc.*, **1962**, 3868.

[u]J. A. Cahill and A. D. Kirshenbaum, *J. Inorg. Nucl. Chem.*, **26**, 206 (1964).

[v]I. Lauerman, G. Metzger, and F. Sauerwald, *Z. Phys. Chem.*, **216**, 42 (1961).

[w]B. C. Allen, *Trans. Met. Soc. AIME*, **227**, 1175 (1963).

[x]J. Tille and J. C. Kelley, *Brit. J. Appl. Phys.*, **14**(10), 717 (1963).

[y]J. D. Patdey, H. R. Chaturvedi, and R. P. Pandey, *J. Phys. Chem.*, **85**, 1750 (1981).

[z]J. P. Frame, E. Rhodes, and A. R. Ubbelohde, *Trans. Faraday Soc.*, **55**, 2039 (1959).

[aa]C. C. Addison and J. M. Coldrey, *J. Chem. Soc.*, **1961**, 468.

[bb]D. J. Donahue and F. E. Bartell, *J. Phys. Chem.*, **56**, 480 (1952).

[cc]L. A. Girifalco and R. J. Good, *J. Phys. Chem.*, **61**, 904 (1957).

[dd]E. B. Butler, *J. Phys. Chem.*, **67**, 1419 (1963).

[ee]F. M. Fowkes and W. M. Sawyer, *J. Chem. Phys.*, **20**, 1650 (1952).

[ff]Q. S. Bhatia, J. K. Chen, J. T. Koberstein, J. E. Sohn, and J. A. Emerson, *J. Colloid Interface Sci.*, **106**, 353 (1985).

[gg]P. Cheng, D. Li, L. Boruvka, Y. Rotenburg, and A. W. Neumann, *Colloids & Surf.*, **43**, 151 (1990).

surface tension values for commonly studied interfaces. In addition, a number of values are included for uncommon substances or states of matter (e.g., molten metals) to provide a general picture of how this property ranges and of the extent of the literature on it. While the values have been chosen with some judgment, they are not presented as critically selected best values. Finally, many of the references cited in the table contain a good deal of additional data on surface tensions at other temperatures and for other liquids of the same type as the one selected for entry in the table. A useful empirical relationship for a homologous series of alkane derivatives is [123]

$$\gamma = \gamma_\infty - \frac{k}{M^{2/3}} \qquad\qquad (\text{II-39})$$

Series of the type $C_nH_{2n+1}X$ were studied. For X = CH_2Cl, k and γ_∞ were 304 and 37.44 dyn/cm, respectively, and for X = $COOCH_3$, k and γ_∞ were 254 and 35.47 dyn/cm, again respectively.

1, 2, 10, 13, 16, 26

10. Problems

• 1. Derive Eq. II-3 using the "surface tension" point of view. *Suggestion:* Consider the sphere to be in two halves, with the surface tension along the join balancing the force due to ΔP, which would tend to separate the two halves.

• 2. The diagrams in Fig. II-22 represent capillaries of varying construction and arrangement. The diameter of the capillary portion is the same in each case, and all of the capillaries are constructed of glass, unless otherwise indicated. The equilibrium rise for water is shown at the left. Draw meniscuses in each figure to correspond to (*a*) the level reached by water rising up the clean, dry tube and (*b*) the level to which the water would recede after having been sucked up to the end of the capillary. The meniscuses in the capillary may be assumed to be spherical in shape.

3. Show that the second term in Eq. II-15 does indeed correct for the weight of the meniscus. (Assume the meniscus to be hemispherical.)

4. Calculate to 1% accuracy the capillary rise for water at 20°C in a 1.2-cm-diameter capillary.

5. Referring to the numerical example following Eq. II-18, what would be the surface tension of a liquid of density 1.423 g/cm³ (2-bromotoluene), the rest of the data being the same?

6. Derive Eq. II-5.

7. Derive Eq. II-14 from an exact analysis of the meniscus profile. *Hint:* Start with Eq. II-12 and let $p = y'$, where $y'' = p\,dp/dy$. The total weight W is then given by $W = 2\Delta\rho g\pi \int_0^r xy\,dx$.

8. Derive Eq. II-13. *Hint:* Use Eqs. II-4, II-5, and II-7 and note an alternative statement for R_2.

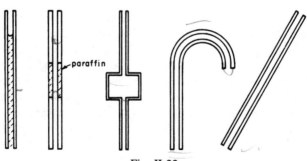

paraffin

Fig. II-22

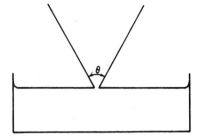

Fig. II-23.

9. Obtain Eq. II-14 from Eq. II-11. It is interesting that the former equation is exact although it has been obtained in this case from Eq. II-11, which is approximate.

10. The surface tension of a liquid that wets glass is measured by determining the height Δh between the levels of the two meniscuses in a U-tube having a small radius r_1 on one side and a larger radius r_2 on the other. The following data are known: $\Delta h = 1.90 \times 10^{-2}$ m, $r_1 = 1.00 \times 10^{-3}$ m, $r_2 = 1.00 \times 10^{-2}$ m, $\rho = 950$ kg/m^3 at 20°C. Calculate the surface tension of the liquid using (a) the simple capillary rise treatment and (b) making the appropriate corrections using Eqs. II-19 and II-20.

11. The surface tension of a liquid is determined by the drop weight method. Using a tip whose outside diameter is 5×10^{-3} m and whose inside diameter is 2.5×10^{-5} m, it is found that the weight of 20 drops is 7×10^{-4} kg. The density of the liquid is 982.4 kg/m^3, and it wets the tip. Using $r/V^{1/3}$, determine the appropriate correction factor and calculate the surface tension of this liquid.

12. Derive the equation for the capillary rise between parallel plates, including the correction term for meniscus weight. Assume zero contact angle, a cylindrical meniscus, and neglect end effects.

13. Derive, from simple considerations, the capillary rise between two parallel plates of infinite length inclined at an angle of θ to each other, and meeting at the liquid surface, as illustrated in Fig. II-23. Assume zero contact angle and a circular cross section for the meniscus. Remember that the area of the liquid surface changes with its position.

14. The following values for the surface tension of a $10^{-4}M$ solution of sodium oleate at 25°C are reported by various authors: (a) by the capillary rise method, $\gamma = 43$ mN/m; (b) by the drop weight method, $\gamma = 50$ mN/m; and (c) by the sessile drop method, $\gamma = 40$ mN/m. Explain how these discrepancies might arise. Which value should be the most reliable and why?

15. Derive Eq. II-30.

16. Molten naphthalene at its melting point of 82°C has the same density as does water at this temperature. Suggest two methods that might be used to determine the naphthalene-water interfacial tension. Discuss your suggestions sufficiently to show that the methods will be reasonably easy to carry out and should give results good to 1% or better.

17. Using Table II-3, calculate S and $1/H$ for $\beta = -0.45$ for a pendant drop. *Hint:* x/b in the table is at a maximum when x is the equatorial radius.

18. This problem may be worked as part (a) only or as part (b) only; it is instructive, however, to work all three parts.

(a) A drop of liquid A, of density 1.01 g/cm^3, rests on a flat surface that it does not wet but contacts with an angle θ (measured in the liquid phase). The height of the drop above the surface is 0.22 cm, and its largest diameter is 0.67 cm. Its shape corresponds to $\beta = 80$ (see Table II-1). Calculate the surface tension of liquid A and its value of θ.

(b) Plot the profile for the drop of liquid A as z (in cm) versus x (in cm).

19. Use the table in Ref. 36 to calculate f of Eq. II-25 for $r/V^{1/3}$ values of 0.40, 0.80, and 1.20.

20. Show how Eq. II-28 should be written if one includes the weight of liquid directly under the plate.

21. In a rotating drop measurement, what is the interfacial tension if the two liquids differ in density by 0.2 g/cm^3, the speed of rotation is 60 rpm, the volume of the drop is 0.4 cm^3, and the length of the extended drop is 6.5 cm?

22. For a particular drop of a certain liquid of density 0.83, β is -0.45 and d_e is 0.72 cm. (a) Calculate the surface tension of the drop and (b) calculate the drop profile from apex to the tip, assuming r_t/a to be 0.55, where r_t is the radius of the tip.

23. The surface tension of mercury is 471 dyn/cm at 24.5°C. In a series of measurements [37] the following drop weight data were obtained, (diameter of tip in centimeters, weight of drop in grams): (0.05167, 0.06407), (0.10068, 0.11535), (0.13027, 0.14245). Calculate the corresponding f and $r/V^{1/3}$ values.

24. Johnson and Lane [28] give the equation for the maximum bubble pressure:

$$h = \frac{a^2}{r} + \frac{2}{3}\, r + \frac{1}{6}\, \frac{r^3}{a^2}$$

For a certain liquid, $a^2 = 0.0780$ cm^2 and $r = 0.160$ cm. Calculate, using the equation, the values of X/r and r/a and compare with the X/r value given by Table II-2.

25. According to the simple formula, the maximum bubble pressure is given by $P_{max} = 2\gamma/r$ where r is the radius of the circular cross-section tube, and P has been corrected for the hydrostatic head due to the depth of immersion of the tube. Using the appropriate table, show what maximum radius tube may be used if γ computed by the simple formula is not to be more than 5% in error. Assume a liquid of $\gamma = 25$ dyn/cm and density 0.98 g/cm^3.

26. A liquid of density 2.0 g/cm^3 forms a meniscus of shape corresponding to $\beta = 80$ in a metal capillary tube with which the contact angle is 30°. The capillary rise is 0.063 cm. Calculate the surface tension of the liquid and the radius of the capillary, using Table II-1.

27. Equation II-30 may be integrated to obtain the profile of a meniscus against a vertical plate; the integrated form is given in Ref. 53. Calculate the meniscus profile for water at 20°C for (a) the case where water wets the plate and (b) the case where the contact angle is 40°. For (b) obtain from your plot the value of h, and compare with that calculated from Eq. II-28. [Hint: Obtain a^2 from II-15.]

28. An empirical observation is if one forms drops at a constant flow rate such that the drop time is t, then the observed drop mass, $M(t)$, varies with t according to the equation $M(t) = M_\infty + st^{-3/4}$ where M_∞ is the "equilibrium" value and s is a 9.44s + 5.37. Using these two relationships, it is possible to determine surface tension as a

function of surface age by means of the drop weight method. Combine these equations to obtain one relating $M_\infty(t)$ and t, where $M_\infty(t)$ is the equilibrium drop mass (and the one to which Eq. II-25 applies) and surface age t. For a particular surfactant solution, the observed drop mass is 80.4 mg for drops formed very slowly and 125.7 mg for drops formed every 2 sec. The radius of the tip used is 5 mm and the density of the solution was 1.03 g/cm^3. Calculate the equilibrium surface tension and that for a surface age of 2 sec given (a) $f = 0.6456$; and (b) $f = 0.6043$. (Note Refs. 41 and 124.)

29. Estimate the surface tension of n-decane at 20°C using Eq. 11-39 and data in Table II-4.

General References

N. K. Adam, *The Physics and Chemistry of Surfaces*, 3rd ed., Oxford University Press, London, 1941.

R. Aveyard and D. A. Haydon, *An Introduction to the Principles of Surface Chemistry*, Cambridge University Press, Cambridge, UK, 1973.

J. T. Davies and E. K. Rideal, *Interfacial Phenomena*, 2nd ed., Academic, New York, 1963.

W. D. Harkins, *The Physical Chemistry of Surface Films*, Reinhold, New York, 1952.

P. C. Hiemenz, *Principles of Colloid and Surface Chemistry*, 2nd ed., Marcel Dekker, New York, 1986.

S. R. Morrison, *The Chemical Physics of Surfaces*, Plenum, London, 1977.

H. van Olphen and K. J. Mysels, *Physical Chemistry: Enriching Topics from Colloid and Surface Science*, Theorex (8327 La Jolla Scenic Drive), La Jolla, CA, 1975.

L. I. Osipow, *Surface Chemistry, Theory and Industrial Applications*, Krieger, New York, 1977.

J. S. Rowlinson and B. Widom, *Molecular Theory of Capillarity*, Clarendon Press, Oxford, 1984.

D. J. Shaw, *Introduction to Colloid and Surface Chemistry*, Butterworths, London, 1966.

Textual References

1. T. Young, in *Miscellaneous Works*, G. Peacock, ed., J. Murray, London, 1855, Vol. I, p. 418.

2. P. S. de Laplace, *Mechanique Celeste*, Supplement to Book 10, 1806.

3. C. V. Boys, *Soap Bubbles and the Forces that Mould Them*, Society for Promoting Christian Knowledge, London, 1890; reprint ed., Doubleday Anchor Books, Science Study Series S3, Doubleday, Garden City, NY, 1959.

4. F. D. Rumscheit and S. G. Mason, *J. Colloid Sci.*, **17**, 266 (1962).

5. D. H. Michael, *Ann. Rev. Fluid Mech.*, **13**, 189 (1981).

6. D. B. Bogy, *Ann. Rev. Fluid Mech.*, **11**, 207 (1979).

6a. A. Hajiloo, T. R. Ramamohan, and J. C. Slattery, *J. Colloid Interface Sci.*, **117**, 384 (1987).

7. See R. W. Coyle and J. C. Berg, *Chem. Eng. Sci.*, **39,** 168 (1984); H. C. Burkholder and J. C. Berg, *AIChE J.*, **20,** 863 (1974).

8. K. D. Bartle, C. L. Woolley, K. E. Markides, M. L. Lee, and R. S. Hansen, *J. High Resol. Chromatog. Chromatogr. Comm.*, **10,** 128 (1987).

9. R. Sen and W. R. Wilcox, *J. Crystal Growth*, **74,** 591 (1986).

10. E. Bayramli, A. Abou-Obeid, and T. G. M. Van de Ven, *J. Colloid Interface Sci.*, **116,** 490, 503 (1987).

11. E. A. Boucher, *Rep. Prog. Phys.*, **43,** 497 (1980).

11a. F. Morgan, *Am. Scient.*, **74,** 232 (1986).

12. H. M. Princen, in *Surface and Colloid Science*, E. Matijevic, ed., Vol. 2, Wiley-Interscience, New York, 1969.

13. Lord Rayleigh (J. W. Strutt), *Proc. Roy. Soc.* (London), **A92,** 184 (1915).

14. F. Bashforth and J. C. Adams, *An Attempt to Test the Theories of Capillary Action*, University Press, Cambridge, England, 1883.

15. S. Sugden, *J. Chem. Soc.*, **1921,** 1483.

16. J. E. Lane, *J. Colloid Interface Sci.*, **42,** 145 (1973).

17. J. F. Padday and A. Pitt, *J. Colloid Interface Sci.*, **38,** 323 (1972).

18. A. W. Adamson, *J. Chem. Ed.*, **55,** 634 (1978).

19. T. A. Erikson, *J. Phys. Chem.*, **69,** 1809 (1965).

20. T. W. Richards and E. K. Carver, *J. Am. Chem. Soc.*, **43,** 827 (1921).

21. W. D. Harkins and F. E. Brown, *J. Am. Chem. Soc.*, **41,** 499 (1919).

22. S. S. Urazovskii and P. M. Chetaev, *Kolloidn. Zh.*, **11,** 359 (1949); through *Chem. Abstr.*, **44,** 889 (1950).

23. P. R. Edwards, *J. Chem. Soc.*, **1925,** 744.

24. G. Jones and W. A. Ray, *J. Am. Chem. Soc.*, **59,** 187 (1937).

25. W. Heller, M. Cheng, and B. W. Greene, *J. Colloid Interface Sci.*, **22,** 179 (1966).

26. S. Ramakrishnan and S. Hartland, *J. Colloid Interface Sci.*, **80,** 497 (1981).

27. S. Sugden, *J. Chem. Soc.*, **1922,** 858; **1924,** 27.

28. C. H. J. Johnson and J. E. Lane, *J. Colloid Interface Sci.*, **47,** 117 (1974).

29. Y. Saito, H. Yoshida, T. Yokoyama, and Y. Ogina, *J. Colloid Interface Sci.*, **66,** 440 (1978).

30. S. Y. Park, C. H. Chang, D. J. Ahn, and E. I. Franses, *Langmuir*, **9,** 3640 (1993).

31. C. H. Chang and E. I. Franses, *Chem. Eng. Sci.*, **49,** 313 (1994).

32. R. Miller, A. Hofmann, R. Hartmann, K. H. Schano, and A. Halbig, *Adv. Materials*, **4,** 370 (1992).

33. T. Tate, *Phil. Mag.*, **27,** 176 (1864).

34. E. A. Boucher and M. J. B. Evans, *Proc. Roy. Soc.* (London), **A346,** 349 (1975).

35. E. A. Boucher and H. J. Kent, *J. Colloid Interface Sci.*, **67,** 10 (1978).

36. J. L. Lando and H. T. Oakley, *J. Colloid Interface Sci.*, **25,** 526 (1967).

37. M. C. Wilkinson and M. P. Aronson, *J. Chem. Soc., Faraday Trans. I*, **69,** 474 (1973).

38. R. Miller and K. H. Schano, *Tenside Surf. Det.*, **27,** 238 (1990).

39. R. Miller, K. H. Schano, and A. Hofmann, *Colloids & Surfaces*, **A92,** 189 (1994).

40. V. B. Fainerman and R. Miller, *Colloids & Surfaces*, **A97**, 255 (1995).

41. C. Jho and M. Carreras, *J. Colloid Interface Sci.*, **99**, 543 (1984).

42. P. Lecomte du Noüy, *J. Gen. Physiol.*, **1**, 521 (1919).

43. W. D. Harkins and H. F. Jordan, *J. Am. Chem. Soc.*, **52**, 1751 (1930).

44. C. Huh and S. G. Mason, *Colloid Polym. Sci.*, **253**, 566 (1975).

45. H. W. Fox and C. H. Chrisman, Jr., *J. Phys. Chem.*, **56**, 284 (1952).

46. A. G. Gaonkar and R. D. Neuman, *J. Colloid Interface Sci.*, **98**, 112 (1984).

47. J. A. Krynitsky and W. D. Garrett, *J. Colloid Sci.*, **18**, 893 (1963).

48. F. van Zeggeren, C. de Courval, and E. D. Goddard, *Can. J. Chem.*, **37**, 1937 (1959).

49. B. Maijgren and L. Ödberg, *J. Colloid Interface Sci.*, **88**, 197 (1982).

50. L. Wilhelmy, *Ann. Phys.*, **119**, 177 (1863).

51. D. O. Jordan and J. E. Lane, *Austral. J. Chem.*, **17**, 7 (1964).

52. J. T. Davies and E. K. Rideal, *Interfacial Phenomena*, Academic Press, New York, 1961.

53. A. W. Adamson and A. Zebib, *J. Phys. Chem.*, **84**, 2619 (1980).

54. A. W. Neumann and W. Tanner, *Tenside*, **4**, 220 (1967).

55. H. M. Princen, *Austral. J. Chem.*, **23**, 1789 (1970).

56. J. Kloubek and A. W. Neumann, *Tenside*, **6**, 4 (1969).

57. G. L. Gaines, Jr., *Insoluble Monolayers at Liquid-Gas Interfaces*, Interscience, New York, 1966; *J. Colloid Interface Sci.*, **62**, 191 (1977).

58. S. K. Li, R. P. Smith, and A. W. Neumann, *J. Adhesion*, **17**, 105 (1984).

59. R. A. Hayes, A. C. Robinson, and J. Ralston, *Langmuir*, **10**, 2850 (1994).

60. G. Loglio, U. Tesei, N. Delgi Innocenti, R. Miller, and R. Cini, *Colloids & Surfaces*, **57**, 335 (1991).

61. S. H. Anastasiadis, J. K. Chen, J. T. Koberstein, A. F. Siegel, J. E. Sohn, and J. A. Emerson, *J. Colloid Interface Sci.*, **119**, 55 (1987).

62. P. Cheng, D. Li, L. Boruvka, Y. Rotenberg, and A. W. Neumann, *Colloids & Surf.*, **43**, 151 (1990).

63. D. Li, P. Cheng, and A. W. Neumann, *Adv. Colloid & Interface Sci.*, **39**, 347 (1992).

64. S. Rooks, L. M. Racz, J. Szekely, B. Benhabib, and A. W. Neumann, *Langmuir*, **7**, 3222 (1991).

65. P. Cheng and A. W. Neumann, *Colloids & Surfaces*, **62**, 297 (1992).

66. J. M. Andreas, E. A. Hauser, and W. B. Tucker, *J. Phys. Chem.*, **42**, 1001 (1938).

67. D. O. Niederhauser and F. E. Bartell, *Report of Progress—Fundamental Research on the Occurrence and Recovery of Petroleum*, Publication of the American Petroleum Institute, The Lord Baltimore Press, Baltimore, 1950, p. 114.

68. S. Fordham, *Proc. Roy. Soc.* (London), **A194**, 1 (1948).

69. C. E. Stauffer, *J. Phys. Chem.*, **69**, 1933 (1965).

70. J. F. Boyce, S. Schürch, Y. Rotenburg, and A. W. Neumann, *Colloids & Surfaces*, **9**, 307 (1984).

71. Y. Rotenberg, L. Boruvka, and A. W. Neumann, *J. Colloid Interface Sci.*, **93**, 169 (1983).

72. S. H. Anastastadis, I. Gancarz, and J. T. Koberstein, *Macromolecules*, **21,** 2980 (1988).

73. C. Huh and R. L. Reed, *J. Colloid Interface Sci.*, **91,** 472 (1983).

74. H. H. J. Girault, D. J. Schiffrin, and B. D. V. Smith, *J. Colloid Interface Sci.*, **101,** 257 (1984).

75. D. Y. Kwok, D. Vollhardt, R. Miller, D. Li, and A. W. Neumann, *Colloids & Surfaces*, **A88,** 51 (1994).

76. J. Li, R. Miller, W. Wustneck, H. Mohwald, and A. W. Neumann, *Colloids & Surfaces*, **A96,** 295 (1995).

77. J. F. Padday, *Phil. Trans. Roy. Soc.* (London), **A269,** 265 (1971).

78. J. F. Padday, *Proc. Roy. Soc.* (London), **A330,** 561 (1972).

79. L. M. Coucoulas and R. A. Dawe, *J. Colloid Interface Sci.*, **103,** 230 (1985).

80. G. C. Nutting and F. A. Long, *J. Am. Chem. Soc.*, **63,** 84 (1941).

81. P. Kosakévitch, S. Chatel, and M. Sage, *CR*, **236,** 2064 (1953).

82. C. Kemball, *Trans. Faraday Soc.*, **42,** 526 (1946).

83. N. K. Roberts, *J. Chem. Soc.*, **1964,** 1907.

84. B. Widom, *J. Phys. Chem.*, **99,** 2803 (1995).

85. B. Vonnegut, *Rev. Sci. Inst.*, **13,** 6 (1942).

86. H. M. Princen, I. Y. Z. Zia, and S. G. Mason, *J. Colloid Interface Sci.*, *23,* 99 (1967).

87. J. C. Slattery and J. Chen, *J. Colloid Interface Sci.*, **64,** 371 (1978).

88. G. L. Gaines, Jr., *Polym. Eng.*, **12,** 1 (1972).

89. P. K. Currie and J. Van Nieuwkoop, *J. Colloid Interface Sci.*, **87,** 301 (1982).

90. J. L. Cayias, R. S. Schechter, and W. H. Wade, in *Adsorption at Interfaces*, K. L. Mittal, ed., ACS Symposium Series, **8,** 234 (1975); L. Cash, J. L. Cayias, G. Fournier, D. MacAllister, T. Schares, R. S. Schechter, and W. H. Wade, *J. Colloid Interface Sci.*, **59,** 39 (1977).

91. K. Shinoda and Y. Shibata, *Colloids & Surfaces*, **19,** 185 (1986).

92. J. Lucassen, *J. Colloid Interface Sci.*, **70,** 335 (1979).

93. J. M. Rallison, *Ann. Rev. Fluid Mech.*, **16,** 45 (1984).

93a. H. A. Stone, *Ann. Rev. Fluid Mech.*, **26,** 65 (1994).

94. C. T. O'Konski and P. L. Gunter, *J. Colloid Sci.*, **10,** 563 (1964).

95. S. Koriya, K. Adachi, and T. Kotaka, *Langmuir*, **2,** 155 (1986).

96. S. Torza, R. G. Cox, and S. G. Mason, *Phil. Trans. Roy. Soc.* (London), **269,** 295 (1971).

97. Lord Kelvin (W. Thomson), *Phil. Mag.*, **42,** 368 (1871).

98. A. Lawal and R. A. Brown, *J. Colloid Interface Sci.*, **89,** 332 (1982).

99. H. V. Nguyen, S. Padmanabhan, W. J. Desisto, and A. Bose, *J. Colloid Interface Sci.*, **115,** 410 (1987).

100. J. H. Bohorquez, B. P. Canfield, K. J. Courian, F. Drogo, C. A. E. Hall, C. L. Holstun, A. R. Scandalis, and M. E. Shepard, *Hewlett-Packard J.*, **45,** 9 (1994).

101. R. Miller, P. Joos, and V. B. Fainerman, *Adv. Colloid Interface Sci.*, **49,** 249 (1994).

102. C. H. Chang and E. I. Franses, *Colloids & Surfaces*, **A100,** 1 (1995).

103. K. L. Sutherland, *Austral. J. Chem.*, **7,** 319 (1954).
104. N. N. Kochurova and A. I. Rusanov, *J. Colloid Interface Sci.*, **81,** 297 (1981).
105. R. S. Hansen and T. C. Wallace, *J. Phys. Chem.*, **63,** 1085 (1959).
106. W. D. E. Thomas and D. J. Hall, *J. Colloid Interface Sci.*, **51,** 328 (1975).
107. D. A. Netzel, G. Hoch, and T. I. Marx, *J. Colloid Sci.*, **19,** 774 (1964).
108. R. S. Hansen, *J. Phys. Chem.*, **68,** 2012 (1964).
109. W. D. E. Thomas and L. Potter, *J. Colloid Interface Sci.*, **50,** 397 (1975).
110. C. C. Addison and T. A. Elliott, *J. Chem. Soc.*, **1949,** 2789.
111. F. H. Garner and P. Mina, *Trans. Faraday Soc.*, **55,** 1607 (1959).
112. J. Van Havenbergh and P. Joos, *J. Colloid Interface Sci.*, **95,** 172 (1983).
113. C. Hsu and R. E. Apfel, *J. Colloid Interface Sci.*, **107,** 467 (1985).
114. R. S. Hansen and J. Ahmad, *Progress in Surface and Membrane Science*, Vol. 4, Academic Press, New York, 1971.
115. H. Löfgren, R. D. Newman, L. E. Scriven, and H. T. Davis, *J. Colloid Interface Sci.*, **98,** 175 (1984).
116. S. Hård and R. D. Newman, *J. Colloid Interface Sci.*, **115,** 73 (1987).
117. M. Sano, M. Kawaguchi, Y-L. Chen, R. J. Skarlupka, T. Chang, G. Zografi, and H. Yu, *Rev. Sci. Instr.*, **57,** 1158 (1986).
118. J. C. Earnshaw and R. C. McGivern, *J. Phys. D: Appl. Phys.*, **20,** 82 (1987).
119. V. B. Fainerman, R. Miller, and P. Joos, *Colloid Polym. Sci.*, **272,** 731 (1994).
120. R. Miller, P. Joos, and V. B. Fainerman, *Prog. Colloid Polym. Sci.*, **97,** 188 (1994).
121. V. B. Fainerman, A. V. Makievski, and R. Miller, *Colloids & Surfaces*, **A75** (1993).
122. J. F. Padday and D. R. Russell, *J. Colloid Sci.*, **15,** 503 (1960).
123. D. G. LeGrand and G. L. Gaines, Jr., *J. Colloid Interface Sci.*, **42,** 181 (1973).
124. J. Kloubek, *Colloid Polymer Sci.*, **253,** 929 (1975).

CHAPTER III

The Nature and Thermodynamics of Liquid Interfaces

It was made clear in Chapter II that the surface tension is a definite and accurately measurable property of the interface between two liquid phases. Moreover, its value is very rapidly established in pure substances of ordinary viscosity; dynamic methods indicate that a normal surface tension is established within a millisecond and probably sooner [1]. In this chapter it is thus appropriate to discuss the thermodynamic basis for surface tension and to develop equations for the surface tension of single- and multiple-component systems. We begin with thermodynamics and structure of single-component interfaces and expand our discussion to solutions in Sections III-4 and III-5.

1. One-Component Systems

A. Surface Thermodynamic Quantities for a Pure Substance

Figure III-1 depicts a hypothetical system consisting of some liquid that fills a box having a sliding cover; the material of the cover is such that the interfacial tension between it and the liquid is zero. If the cover is slid back so as to uncover an amount of surface $d\mathcal{A}$, the work required to do so will be $\gamma\, d\mathcal{A}$. This is reversible work at constant pressure and temperature and thus gives the increase in free energy of the system (see Section XVII-12 for a more detailed discussion of the thermodynamics of surfaces).

$$dG = \gamma\, d\mathcal{A} \tag{III-1}$$

The total free energy of the system is then made up of the molar free energy times the total number of moles of the liquid plus G^s, the surface free energy per unit area, times the total surface area. Thus

$$G^s = \gamma = \left(\frac{\partial G}{\partial \mathcal{A}} \right)_{T,P} \tag{III-2}$$

Because this process is a reversible one, the heat associated with it gives the *surface entropy*

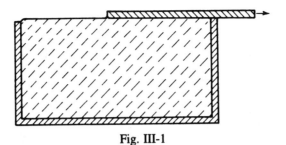

Fig. III-1

$$dq = T \, dS = TS^s \, d\mathcal{A} \tag{III-3}$$

where S^s is the surface entropy per square centimeter of surface.
Because $(\partial G/\partial T)_P = -S$, it follows that

$$\left(\frac{\partial G^s}{\partial T} \right)_P = -S^s \tag{III-4}$$

or, in conjunction with Eq. III-1,

$$\frac{d\gamma}{dT} = -S^s \tag{III-5}$$

Finally, the total surface enthalpy per square centimeter H^s is

$$H^s = G^s + TS^s \tag{III-6}$$

Often, and as a good approximation, H^s and the surface energy E^s are not distinguished, so Eq. III-6 can be seen in the form

$$E^s = G^s + TS^s \tag{III-7}$$

or

$$E^s = \gamma - T \frac{d\gamma}{dT} \tag{III-8}$$

The total surface energy E^s generally is larger than the surface free energy. It is frequently the more informative of the two quantities, or at least it is more easily related to molecular models.

Other thermodynamic relationships are developed during the course of this chapter. The surface specific heat C^s (the distinction between C_p^s and C_v^s is rarely made), is an additional quantity to be mentioned at this point, however. It is given by

$$C^s = \frac{dE^s}{dT} \qquad \text{(III-9)}$$

The surface tension of most liquids decreases with increasing temperature in a nearly linear fashion, as illustrated in Fig. III-2. The near-linearity has stimulated many suggestions as to algebraic forms that give exact linearity. An old and well-known relationship, attributed to Eötvös [3], is

$$\gamma V^{2/3} = k(T_c - T) \qquad \text{(III-10)}$$

where V is the molar volume. One does expect the surface tension to go to zero at the critical temperature, but the interface seems to become diffuse at a slightly lower temperature, and Ramsay and Shields [4] replaced T_c in Eq. III-10 by $(T_c - 6)$. In either form, the constant k is about the same for most liquids and has a value of about 2.1 ergs/K. Another form originated by van der Waals in 1894 but developed further by Guggenheim [5] is

$$\gamma = \gamma^\circ \left(1 - \frac{T}{T_c} \right)^n \qquad \text{(III-11)}$$

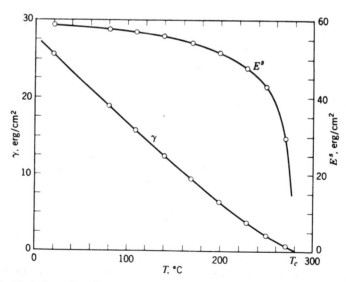

Fig. III-2. Variation of surface tension and total surface energy of CCl₄ with temperature. (Data from Ref. 2.)

where n is 11/9 for many organic liquids but may be closer to unity for metals [6]. A useful empirical relationship for metals is

$$\gamma_m = \frac{3.6T_m}{V_m^{2/3}} \qquad \text{(III-12)}$$

where subscript m denotes melting point and γ is in dynes per centimeter [7].

There is a point of occasional misunderstanding about dimensions that can be illustrated here. The quantity $\gamma V^{2/3}$ is of the nature of a surface free energy per mole, yet it would appear that its dimensions are energy per mole$^{2/3}$, and that k in Eq. III-10 would be in ergs per degree per mole$^{2/3}$. The term "mole$^{2/3}$" is meaningless, however, because "mole" is not a dimension but rather an indication that an Avogadro number of molecules is involved. Because Avogadro's number is itself arbitrary (depending, for example, on whether a gram or a pound molecular weight system is used), to get at a rational meaning of k one should compute it on the per molecule basis; this gives $k' = 2.9 \times 10^{-16}$ ergs/degree-molecule. Lennard-Jones and Corner [8] have pointed out that Eq. III-10 arises rather naturally out of a simple statistical mechanical treatment of a liquid, where k' is the Boltzmann constant (1.37×10^{-16}) times a factor of the order of unity.

Equations III-10 and III-11 are, of course, approximations, and the situation has been examined in some detail by Cahn and Hilliard [9], who find that Eq. III-11 is also approximated by regular solutions not too near their critical temperature.

B. The Total Surface Energy, E^S

If the variation of density and hence molar volume with temperature is small, it follows from Eqs. III-10 and III-8 that E^S will be nearly temperature-independent. In fact, Eq. III-11 with $n = 1$ may be written in the form

$$\gamma = E^S \left(1 - \frac{T}{T_C} \right) \qquad \text{(III-13)}$$

which illustrates the point that surface tension and energy become equal at 0 K. The temperature independence of E^S generally holds for liquids not too close to their critical temperature, although, as illustrated in Fig. III-2, E^S eventually drops to zero at the critical temperature.

Inspection of Table III-1 shows that there is a wide range of surface tension and E^S values. It is more instructive, however, to compare E^S values calculated on an energy per mole basis. The area per mole of spherical molecules of molecular weight M and radius r is

TABLE III-1
Temperature Dependence of Surface Tension[a]

Liquid	γ (ergs/ cm^2)	Temperature	$d\gamma/dT$	E^S (ergs/ cm^2)	$E^{S'}$ (cal/ mol)	Reference
He	0.308	2.5 K	−0.07	0.47	8.7	12
N$_2$	9.71	75 K	−0.23	26.7	585	13
Ethanol	22.75	20°C	−0.086	46.3	1,340	ICT
Water	72.88	20°C	−0.138	113	1,590	14
NaNO$_3$[b]	116.6	308°C	−0.050	146	2,150	15
C$_7$F$_{14}$[c]	15.70	20°C	−0.10	45.0	2,610	16
Benzene	28.88	20°C	−0.13	67.0	2,680	ICT
n-Octane	21.80	20°C	−0.10	51.1	2,920	ICT
Sodium	191	98°C	−0.10	228	3,850	17
Copper	1550	1083°C	−0.176	1,355	10,200	17
Silver	910	961°C	−0.164	1,234	11,050	17
Iron	1880	1535°C	−0.43	2,657	20,100	18

[a] An extensive compilation of γ and $d\gamma/dT$ data is given by G. Korosi and E. sz. Kováts, *J. Chem. Eng. Data*, **26**, 323 (1981).
[b] $E^{S'}$ computed on a per gram ion basis.
[c] Perfluoromethyl cyclohexane.

$$A = 4\pi N_A \left(\frac{3M}{4\pi\rho N_A} \right)^{2/3} \qquad (III\text{-}14)$$

where N_A is Avogadro's number. Only about one-fourth of the area of the surface spheres will be exposed to the interface; thus Eq. III-14 becomes $A = f N_A^{1/3} V^{2/3}$, where V denotes molar volume and f is a geometric factor near unity. Then, $E^{S'} = A E^S$ on an energy per mole basis. In Table III-1 one sees that the variation in $E^{S'}$ is much smaller than that of γ or E^S. Semiempirical equations useful for molten metals and salts may be found in Refs. 7 and 10, respectively.

Example. We reproduce the entry for iron in Table III-1 as follows. First $E^S = 1880-(1808) \; (-0.43) = 2657$ ergs/cm^2. Next, estimating V to be about 7.1 cm^3/mol and taking f to be unity, $A = (6.02 \times 10^{23})^{1/3}(7.1)^{2/3} = 3.1 \times 10^8$ cm^2/mol whence $E^{S'} = (2657)(3.1 \times 10^8)/(4.13 \times 10^7) = 20,100$ cal/mol.

One may consider a molecule in the surface region as being in a state intermediate between that in the vapor phase and that in the liquid. Skapski [11] has made the following simplified analysis. Considering only nearest-neighbor interactions, if n_i and n_s denote the number of nearest neighbors in the interior of the liquid and the surface region, respectively, then, per molecule

$$E^{S'} = \frac{N_A U}{2} (n_i - n_s) \qquad \text{(III-15)}$$

where U is the interaction energy. On this basis, the energy of vaporization should be $U n_i/2$. For close-packed spheres, $n_i = 12$ and $n_s = 9$ so that $E^{S'}$ should be about one-fourth of the energy of vaporization (not exactly because of f). On this basis the surface energy of metals is somewhat *smaller* than expected; however, over the range in Table III-1 from helium to iron, one sees that the variation in the surface energy *per area* depends almost equally on the var in intermolecular forces and on that of the density of packing or molecular

C. The Effect of Curvature on Vapor Pressure and Surface Tension

A very important thermodynamic relationship is that giving the effe surface curvature on the molar free energy of a substance. This is perhaps understood in terms of the pressure drop ΔP across an interface, as give Young and Laplace in Eq. II-7. From thermodynamics, the effect of a ch in mechanical pressure at constant temperature on the molar free energy substance is

$$\Delta G = \int V \, d\mathbf{P} \qquad \text{(III-16)}$$

or if the molar volume V is assumed constant and Eq. II-7 is used for $\Delta \mathbf{P}$

$$\Delta G = \gamma V \left(\frac{1}{R_1} + \frac{1}{R_2} \right) \qquad \text{(III-17)}$$

It is convenient to relate the free energy of a substance to its vapor pressure and, assuming the vapor to be ideal, $G = G^0 + RT \ln P$. Equation III-17 then becomes

$$RT \ln \frac{P}{P^0} = \gamma V \left(\frac{1}{R_1} + \frac{1}{R_2} \right) = \frac{\gamma V}{R_m} \qquad \text{(III-18)}$$

where P^0 is the normal vapor pressure of the liquid, P is that observed over the curved surface, and R_m is the mean radius of curvature (in a more exact version, the quantity γ/R_m becomes $\gamma/R_m - (P - P^0)$) [19]. Equation III-18 is frequently called the *Kelvin* equation and, with the Young–Laplace equation (II-7), makes the second fundamental relationship of surface chemistry.

For the case of a spherical surface of radius r, Eq. III-18 becomes

$$RT \ln \frac{P}{P^0} = \frac{2\gamma V}{r} \tag{III-19}$$

Here, r is positive and there is thus an increased vapor pressure. In the case of water, P/P^0 is about 1.001 if r is 10^{-4} cm, 1.011 if r is 10^{-5} cm, and 1.114 if r is 10^{-6} cm or 100 Å. The effect has been verified experimentally for several liquids [20], down to radii of the order of 0.1 μm, and indirect measurements have verified the Kelvin equation for R_m values down to about 30 Å [19]. The phenomenon provides a ready explanation for the ability of vapors to supersaturate. The formation of a new liquid phase begins with small clusters that may grow or aggregate into droplets. In the absence of dust or other foreign surfaces, there will be an activation energy for the formation of these small clusters corresponding to the increased free energy due to the curvature of the surface (see Section IX-2).

While Eq. III-18 has been verified for small droplets, attempts to do so for liquids in capillaries (where R_m is negative and there should be a pressure reduction) have led to startling discrepancies. Potential problems include the presence of impurities leached from the capillary walls and allowance for the film of adsorbed vapor that should be present (see Chapter X). There is room for another real effect arising from structural perturbations in the liquid induced by the vicinity of the solid capillary wall (see Chapter VI). Fisher and Israelachvili [19] review much of the literature on the verification of the Kelvin equation and report confirmatory measurements for liquid bridges between crossed mica cylinders. The situation is similar to that of the meniscus in a capillary since R_m is negative; some of their results are shown in Fig. III-3. Studies in capillaries have been reviewed by Melrose [20] who concludes that the Kelvin equation is obeyed for radii at least down to 1 μm.

Tolman [21] concluded from thermodynamic considerations that with sufficiently curved surfaces, the value of the *surface tension itself* should be affected. In reviewing the subject, Melrose [22] gives the equation

$$\gamma = \gamma^0 \left(1 - \frac{\delta}{R_m} \right) \tag{III-20}$$

where δ is a measure of the thickness of the interfacial region (about 5 Å for cyclohexane [23]) and R_m may be positive or negative. (See also Section III-2A.)

This effect assumes importance only at very small radii, but it has some applications in the treatment of nucleation theory where the excess surface energy of small clusters is involved (see Section IX-2). An intrinsic difficulty with equations such as III-20 is that the treatment, if not modelistic and hence partly empirical, assumes a continuous medium, yet the effect does not become important until curvature comparable to molecular dimensions is reached. Fisher and Israelachvili [24] measured the force due to the Laplace pressure for a pendular ring of liquid between crossed mica cylinders and concluded that for several organic liquids the effective surface tension remained unchanged

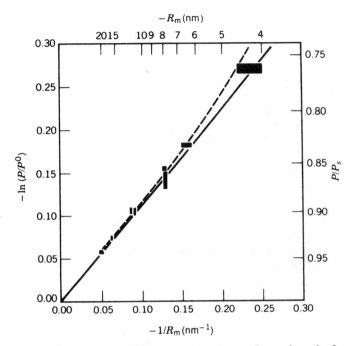

Fig. III-3. Comparison of Eq. III-18 (solid line) with experimental results for cyclohexane bridges formed between crossed mica cylinders; the dashed line is the calculation including Eq. III-20 (from Ref. 19).

down to radii of curvature as low as 0.5 nm. Christenson [25], in a similar experiment, found the Laplace equation to hold for water down to radii of 2 nm.

D. The Effect of Pressure on Surface Tension

The following relationship holds on thermodynamic grounds [26, 27]:

$$\left(\frac{\partial \gamma}{\partial P} \right)_{A,T} = \left(\frac{\partial V}{\partial A} \right)_{P,T} = \Delta V^S \qquad \text{(III-21)}$$

where A denotes area. In other words the pressure effect is related to the change in molar volume when a molecule goes from the bulk to the surface region. This change would be positive, and the effect of pressure should therefore be to increase the surface tension.

Unfortunately, however, one cannot subject a liquid surface to an increased pressure without introducing a second component into the system, such as some inert gas. One thus increases the density of matter in the gas phase and, moreover, there will be some gas adsorbed on the liquid surface with a corresponding volume change.

Studies by Eriksson [28] and King and co-workers [29] have shown that adsorption dominates. One may also study the effect of pressure on the surface tension of solutions [30, 31] and on interfacial tensions. As an example of the latter case, ΔV^S was found to be about 2.8×10^{-9} cm^3/cm^2 for the *n*-octane–water interface [32].

2. Structural and Theoretical Treatments of Liquid Interfaces

The surface free energy can be regarded as the work of bringing a molecule from the interior of a liquid to the surface, and that this work arises from the fact that, although a molecule experiences no net forces while in the interior of the bulk phase, these forces become unbalanced as it moves toward the surface. As discussed in connection with Eq. III-15 and also in the next sections, a knowledge of the potential function for the interaction between molecules allows a calculation of the total surface energy; if this can be written as a function of temperature, the surface free energy is also calculable.

The unbalanced force on a molecule is directed inward, and it might be asked how this could appear as a surface "tension." A mechanical analogy is shown in Fig. III-4, which illustrates how the work to raise a weight can appear as a horizontal pull; in the case of a liquid, an extension of the surface results in molecules being brought from the interior into the surface region.

The next point of interest has to do with the question of how deep the surface region or region of appreciably unbalanced forces is. This depends primarily on the range of intermolecular forces and, except where ions are involved, the principal force between molecules is of the so-called van der Waals type (see Section VI-1). This type of force decreases with about the seventh power of the intermolecular distance and, consequently, it is only the first shell or two of nearest neighbors whose interaction with a given molecule is of importance. In other words, a molecule experiences essentially symmetrical forces once it is a few molecular diameters away from the surface, and the thickness of the surface region is of this order of magnitude (see Ref. 23, for example). (Certain aspects of this conclusion need modification and are discussed in Sections X-6C and XVII-5.)

It must also be realized that this thin surface region is in a very turbulent state. Since the liquid is in equilibrium with its vapor, then, clearly, there is a two-way and balanced traffic of molecules hitting and condensing on the surface from the vapor phase and of molecules evaporating from the surface into the vapor phase. From the gas kinetic theory, the number of moles striking 1 cm^2 of surface per second is

$$Z = P \left(\frac{1}{2\pi MRT} \right)^{1/2} \tag{III-22}$$

For vapor saturated with respect to liquid water at room temperature, Z is about 0.02 mol/cm^2 · sec or about 1.2×10^{22} molecules/cm^2 · sec. At equilibrium, then, the evaporation rate must equal the condensation rate, which differs from

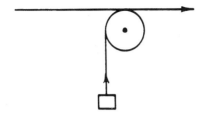

Fig. III-4. Mechanical analogy to surface tension.

the preceding figure by a factor less than but close to unity (called the condensation coefficient). Thus each square centimeter of liquid water entertains 1.2×10^{22} arrivals and departures per second. The traffic on an area of 10 Å^2, corresponding to the area of a single water molecule, is 1.2×10^7 sec^{-1}, so that the lifetime of a molecule on the surface is on the order of a tenth microsecond.

There is also a traffic between the surface region and the adjacent layers of liquid. For most liquids, diffusion coefficients at room temperature are on the order of 10^{-5} cm^2/sec, and the diffusion coefficient $\mathcal{D}$ is related to the time t for a net displacement x by an equation due to Einstein:

$$\mathcal{D} = \frac{x^2}{2t} \tag{III-23}$$

If x is put equal to a distance of, say, 100 Å, then t is about 10^{-6} sec, so that, due to Brownian motion, there is a very rapid interchange of molecules between the surface and the adjacent bulk region.

The picture that emerges is that a "quiescent" liquid surface is actually in a state of violent agitation on the molecular scale with individual molecules passing rapidly back and forth between it and the bulk regions on either side. Under a microscope of suitable magnification, the surface region should appear as a fuzzy blur, with the average density varying in some continuous manner from that of the liquid phase to that of the vapor phase.

In the case of solids, there is no doubt that a lateral tension (which may be anisotropic) can exist between molecules on the surface and can be related to actual stretching or compression of the surface region. This is possible because of the immobility of solid surfaces. Similarly, with thin soap films, whose thickness can be as little as 100 Å, stretching or extension of the film may involve a corresponding variation in intermolecular distances and an actual tension between molecules.

A case can be made for the usefulness of surface "tension" as a concept even in the case of a normal liquid–vapor interface. A discussion of this appears in papers by Brown [33] and Gurney [34]. The informal practice of using surface tension and surface free energy interchangeably will be followed in this text.

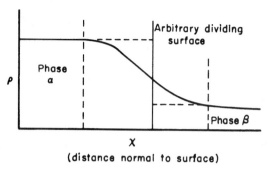

Fig. III-5

A. Further Development of the Thermodynamic
Treatment of the Surface Region

Consider a liquid in equilibrium with its vapor. The two bulk phases α and β do not change sharply from one to the other at the interface, but rather, as shown in Fig. III-5, there is a region over which the density and local pressure vary. Because the actual interfacial region has no sharply defined boundaries, it is convenient to invent a mathematical dividing surface [35]. One then handles the extensive properties (G, E, S, n, etc.) by assigning to the bulk phases the values of these properties that would pertain if the bulk phases continued uniformly up to the dividing surface. The actual values for the system as a whole will then differ from the sum of the values for the two bulk phases by an excess or deficiency assigned to the surface region.

The following relations will then hold:

$$\text{Volume}: V = V^\alpha + V^\beta \tag{III-24}$$

$$\text{Internal energy}: E = E^\alpha + E^\beta + E^\sigma \tag{III-25}$$

$$\text{Entropy}: S = S^\alpha + S^\beta + S^\sigma \tag{III-26}$$

$$\text{Moles}: n_i = n_i^\alpha + n_i^\beta + n_i^\sigma \tag{III-27}$$

We will use the superscript σ to denote surface quantities calculated on the preceding assumption that the bulk phases continue unchanged to an assumed mathematical dividing surface. For an arbitrary set of variations from equilibrium,

$$dE = T\,dS + \sum_i \mu_i\,dn_i - \mathbf{P}^\alpha\,dV^\alpha - \mathbf{P}^\beta\,dV^\beta + \gamma\,d\mathcal{A} + C_1\,dc_1 + C_2\,dc_2$$

$$\tag{III-28}$$

where c_1 and c_2 denote the two curvatures (reciprocals of the radii of curvature) and C_1 and C_2 are constants. The last two terms may be written as $\frac{1}{2}(C_1 + C_2)d(c_1 + c_2) + \frac{1}{2}(C_1 - C_2)d(c_1 - c_2)$, and these plus the term $\gamma \, d\mathcal{A}$ give the effect of variations in area and curvature. Because the actual effect must be independent of the location chosen for the dividing surface, a condition may be put on C_1 and C_2, and this may be taken to be that $C_1 + C_2 = 0$. This particular condition gives a particular location of the dividing surface such that it is now called the *surface of tension*.

For the case where the curvature is small compared to the thickness of the surface region, $d(c_1 - c_2) = 0$ (this will be exactly true for a plane or for a spherical surface), and Eq. III-28 reduces to

$$dE = T \, dS + \sum_i \mu_i \, dn_i - \mathbf{P}^\alpha \, dV^\alpha - \mathbf{P}^\beta \, dV^\beta + \gamma \, d\mathcal{A} \qquad \text{(III-29)}$$

Because

$$G = E - TS + \mathbf{P}^\alpha V^\alpha + \mathbf{P}^\beta V^\beta \qquad \text{(III-30)}$$

where G is the Gibbs free energy, it follows that

$$dG = -S \, dT + \sum_i \mu_i \, dn_i + V^\alpha \, d\mathbf{P}^\alpha + V^\beta \, d\mathbf{P}^\beta + \gamma \, d\mathcal{A} \qquad \text{(III-31)}$$

(Equation III-31 is obtained by differentiating Eq. III-30 and comparing with Eq. III-29.) At equilibrium, the energy must be a minimum for a given set of values of S and of n_i, and

$$-\mathbf{P}^\alpha \, dV^\alpha - \mathbf{P}^\beta \, dV^\beta + \gamma \, d\mathcal{A} = 0 \qquad \text{(III-32)}$$

but

$$dV = 0 = dV^\alpha + dV^\beta \qquad \text{(III-33)}$$

so

$$(\mathbf{P}^\alpha - \mathbf{P}^\beta) \, dV^\alpha = \gamma \, d\mathcal{A} \qquad \text{(III-34)}$$

Equation III-34 is the same as would apply to the case of two bulk phases separated by a membrane under tension γ.

If the surface region is displaced by a distance dt

$$d\mathcal{A} = (c_1 + c_2)\mathcal{A} \; dt \tag{III-35}$$

and, because

$$dV^\alpha = \mathcal{A} \; dt = -dV^\beta \tag{III-36}$$

then

$$(\mathbf{P}^\alpha - \mathbf{P}^\beta)\mathcal{A} \; dt = \gamma(c_1 + c_2)\mathcal{A} \; dt \tag{III-37}$$

or

$$\Delta\mathbf{P} = \gamma(c_1 + c_2) \tag{III-38}$$

Equation III-38 is the Young–Laplace equation (Eq. II-7).

The foregoing serves as an introduction to the detailed thermodynamics of the surface region; the method is essentially that of Gibbs [35], as reviewed by Tolman [36, 37]. An additional relationship is

$$\gamma = \int_{-a}^{0} (\mathbf{P}^\alpha - \mathbf{p}) \; dx + \int_{0}^{a} (\mathbf{P}^\beta - \mathbf{p}) \; dx \tag{III-39}$$

where x denotes distance normal to the surface and the points a and $-a$ lie in the two bulk phases, respectively.

For a plane surface, $\mathbf{P}^\alpha = \mathbf{P}^\beta$, and

$$\gamma = \int_{-a}^{a} (\mathbf{P} - \mathbf{p}) \; dx \tag{III-40}$$

Here, $\mathbf{P}$ is the bulk pressure, which is the same in both phases, and $\mathbf{p}$ is the local pressure, which varies across the interface.

Returning to the matter of the location of the dividing surface, the position defined by $(C_1 + C_2) = 0$ is in general such that $n^\alpha + n^\beta$, calculated by assuming the bulk phases to continue up to the dividing surface, will differ from the actual n. That is, even for a single pure substance, there will be a nonzero surface excess Γ that can be positive or negative. This convention, while mathematically convenient, is not pleasing intuitively, and other conventions are possible for locating the dividing surface, including one such that the surface excess is zero. In fact, the quantity δ in Eq. III-20 is just the distance between this dividing surface and the surface of tension and is positive for a drop. This general subject has been discussed by Kirkwood and Buff [38], Buff [39], Melrose [22,40], Mandell and Reiss [41], and also by Neumann and co-workers [42–44]. Neumann and co-workers have developed a generalized hydrostatic treatment of capillarity that is valid for highly curved surfaces [45] and have shown that in the context of this theory the free energy remains invariant to a shift in the dividing surface [46].

B. Calculation of the Surface Energy and Structure of Interfaces

The function of thermodynamics is to provide phenomenological relationships whose validity has the authority of the laws of thermodynamics themselves. One may proceed further, however, if specific models or additional assumptions are made. For example, the use of the van der Waals equation of state allows an analysis of how $\mathbf{P} - \mathbf{p}$ in Eq. III-40 should vary across the interface; Tolman [36,37] made an early calculation of this type. There has been a high degree of development of statistical thermodynamics in this field (see Ref. 47 and the General References and also Sections XV-4 and XVI-3). A great advantage of this approach is that one may derive thermodynamic properties from knowledge of the intermolecular forces in the fluid. Many physical systems can be approximated with model interaction potential energies; a widely used system comprises "attractive hard spheres" where rigid spheres of diameter b interact with an attractive potential energy, $u_{att}(r)$.

The hard-sphere treatment also suggested a relationship between surface tension and the compressibility of the liquid. In a more classic approach [48], the equation

$$\gamma = \frac{b}{8} \left(\frac{\partial E}{\partial V} \right)_T \qquad \text{(III-41)}$$

relates the internal pressure of the liquid, $(\partial E / \partial V)_T$, to the surface tension where b is the side of a cube of molecular volume. The internal pressure can be replaced by the expression $(\alpha T / \beta - P)$, where α and β are the coefficients of thermal expansion and of compressibility and P is the ambient pressure [49].

The classic theory due to van der Waals provides an important phenomenological link between the structure of an interface and its interfacial tension [50–52]. The expression

$$\gamma = m \int_{-\infty}^{\infty} \left(\frac{\partial \rho(z)}{\partial z} \right)^2 dz \qquad \text{(III-42)}$$

relates the surface tension to the square of the gradient in the equilibrium density profile across the interface. The equilibrium density profile is that minimizing the free energy of the interface. The parameter m is found from the intermolecular forces in the fluid, or in more precise theories, from the direct correlation function from statistical mechanics [53]. If one follows the attractive hard-sphere model of van der Waals, a simple result for m is

$$m = -\frac{1}{6} \int_{r>b} r^2 u_{\text{att}}(r) d\mathbf{r} \qquad \text{(III-43)}$$

One problem with this treatment is that it neglects higher-order terms depending on higher moments of u_{att} that become undefined for slowly decaying interaction potentials (see Problem III-9).

The gradient model has been combined with two equations of state to successfully model the temperature dependence of the surface tension of polar and nonpolar fluids [54]. Widom and Tavan have modeled the surface tension of liquid ^{4}He near the λ transition with a modified van der Waals theory [55].

Another statistical mechanical approach makes use of the radial distribution function $g(r)$, which gives the probability of finding a molecule at a distance r from a given one. This function may be obtained experimentally from x-ray or neutron scattering on a liquid or from computer simulation or statistical mechanical theories for model potential energies [56]. Kirkwood and Buff [38] showed that for a given potential function, $U(r)$

$$\gamma = \frac{\pi}{8} \rho^2 \int_0^\infty g(r) U'(r) r^4 dr \qquad \text{(III-44)}$$

$$E^S = -\frac{\pi}{2} \rho^2 \int_0^\infty g(r) U(r) r^3 dr \qquad \text{(III-45)}$$

where ρ is the average density and U' denotes dU/dr. A widely used and successful form for $U(r)$ is that due to Lennard-Jones:

$$U(r) = -4\epsilon_0 \left[\left(\frac{\sigma}{r} \right)^6 - \left(\frac{\sigma}{r} \right)^{12} \right] \qquad \text{(III-46)}$$

where, as shown in Fig. III-6, ϵ_0 is the potential energy at the minimum and σ is an effective molecular diameter; the two parameters can be obtained from the internal pressure of a liquid or from the nonideality of the vapor. One calculation along these lines gave γ for argon at 84.3 K as 15.1 ergs/cm^2 [57] as compared to the experimental value of 13.2 ergs/cm^2. The Kirkwood–Buff approach has been applied to ethanol–water mixtures [58] and molten salts [59].

The statistical mechanical approach, density functional theory, allows description of the solid–liquid interface based on knowledge of the liquid properties [60, 61]. This approach has been applied to the solid–liquid interface for hard spheres where experimental data on colloidal suspensions and theory [62] both indicate $\gamma b^2/kT \approx 0.6$; this verifies that no attraction is necessary for a solid–liquid interface to exist. The adhesive sphere and Lennard-Jones (Eq. III-46) solids have also been studied [61, 63]. In

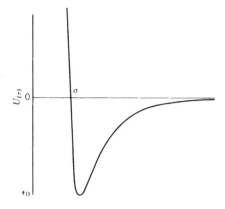

Fig. III-6. The Lennard-Jones potential function.

the Lennard-Jones system the influence of the disordered phase density is evident with the solid–vapor interfacial tension greatly exceeding that above the triple point where a solid coexists with a liquid.

Two simulation methods—Monte Carlo and molecular dynamics—allow calculation of the density profile and pressure difference of Eq. III-44 across the vapor–liquid interface [64, 65]. In the former method, the initial system consists of N molecules in assumed positions. An intermolecule potential function is chosen, such as the Lennard-Jones potential, and the positions are randomly varied until the energy of the system is at a minimum. The resulting configuration is taken to be the equilibrium one. In the molecular dynamics approach, the N molecules are given initial positions *and* velocities and the equations of motion are solved to follow the ensuing collisions until the set shows constant time-average thermodynamic properties. Both methods are computer intensive yet widely used.

In Fig. III-7 we show a molecular dynamics computation for the density profile and pressure difference **P** – **p** across the interface of an argonlike system [66] (see also Refs. 67, 68 and citations therein). Similar calculations have been made of δ in Eq. III-20 [69, 70]. Monte Carlo calculations of the density profile of the vapor–liquid interface of magnesium show stratification penetrating about three atomic diameters into the liquid [71]. Experimental measurement of the *transverse* structure of the vapor–liquid interface of mercury and gallium showed structures that were indistinguishable from that of the bulk fluids [72, 73].

3. Orientation at Interfaces

There is one remaining and very significant aspect of liquid–air and liquid–liquid interfaces to be considered before proceeding to a discussion of

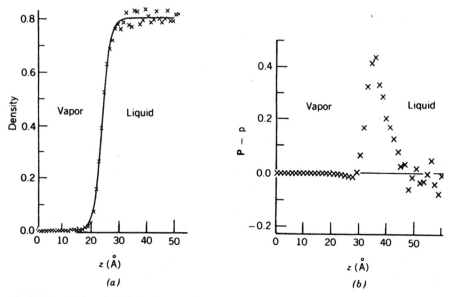

Fig. III-7. (*a*) Interfacial density profile for an argonlike liquid–vapor interface (density in reduced units); *z* is the distance normal to the surface. (*b*) Variations of **P**–**p** of Eq. III-40 (in reduced units) across the interface. [From the thesis of J. P. R. B. Walton (see Ref. 66).]

the behavior and thermodynamics of solution interfaces. This is the matter of molecular orientation at interfaces.

The idea that unsymmetrical molecules will orient at an interface is now so well accepted that it hardly needs to be argued, but it is of interest to outline some of the history of the concept. Hardy [74] and Harkins [75] devoted a good deal of attention to the idea of "force fields" around molecules, more or less intense depending on the polarity and specific details of the structure. Orientation was treated in terms of a principle of "least abrupt change in force fields," that is, that molecules should be oriented at an interface so as to provide the most gradual transition from one phase to the other. If we read "interaction energy" instead of "force field," the principle could be reworded on the very reasonable basis that molecules will be oriented so that their mutual interaction energy will be a maximum.

A somewhat more quantitative development along these lines was given by Langmuir [76] in what he termed the *principle of independent surface action*. He proposed that, qualitatively, one could suppose each part of a molecule to possess a local surface free energy. Taking ethanol as an example, one can employ this principle to decide whether surface molecules should be oriented according to Fig. III-8*a* or *b*. In the first case, the surface presented would comprise hydroxyl groups whose surface energy should be about 190 ergs/cm^2, extrapolating from water. In the second case, a surface energy like that of a

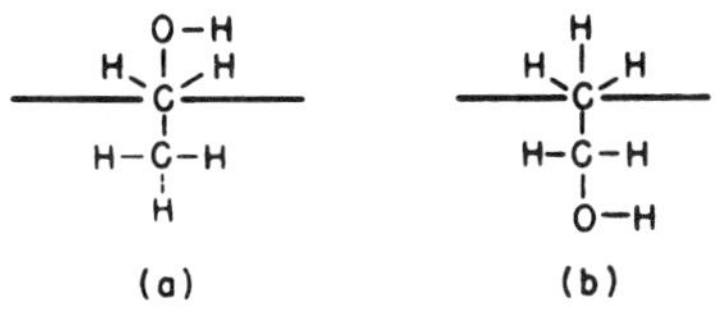

Fig. III-8

hydrocarbon should prevail, that is, about 50 ergs/cm^2 (see Table III-1). This is a difference of 140 ergs/cm^2 or about 30×10^{-14} ergs per molecule. Since kT is on the order of 4×10^{-14} ergs per molecule, the Boltzmann factor, $\exp(-U/kT)$, favoring the orientation in Fig. III-8b should be about 10^5. This conclusion is supported by the observation that the actual surface tension of ethanol is 22 ergs/cm^2 or not very different from that of a hydrocarbon. Langmuir's principle may sound rather primitive but, in fact, it is widely used and useful today in one or another (often disguised) form.

There is, of course, a mass of rather direct evidence on orientation at the liquid–vapor interface, much of which is at least implicit in this chapter and in Chapter IV. The methods of statistical mechanics are applicable to the calculation of surface orientation of assymmetric molecules, usually by introducing an angular dependence to the inter-molecular potential function (see Refs. 67, 68, 77 as examples). Widom has applied a mean-field approximation to a lattice model to predict the tendency of AB molecules to adsorb and orient perpendicular to the interface between phases of AA and BB [78]. In the case of water, a molecular dynamics calculation concluded that the surface dipole density corresponded to a tendency for surface–OH groups to point toward the vapor phase [79].

4. The Surface Tension of Solutions

A. Binary Solutions

The principal point of interest to be discussed in this section is the manner in which the surface tension of a binary system varies with composition. The effects of other variables such as pressure and temperature are similar to those for pure substances, and the more elaborate treatment for two-component sys-tems is not considered here. Also, the case of immiscible liquids is taken up in Section IV-2.

A fairly simple treatment, due to Guggenheim [80], is useful for the case of ideal or nearly ideal solutions. An abbreviated derivation begins with the free energy of a species

$$G_i = kT \ln a_i \qquad \qquad (\text{III-47})$$

where a_i is the absolute activity, $a_i = N_i g_i$, where N_i is the mole fraction of species i (unity for pure liquids), and g_i derives from the partition function Q_i. For a pure liquid 1, the surface tension may be written as

$$\gamma_1 \sigma_1 = -kT \ln \frac{a_1}{a_1^S} \tag{III-48}$$

or

$$\exp\left(\frac{-\gamma_1 \sigma_1}{kT}\right) = \frac{g_1}{g_1^S} \tag{III-49}$$

where the surface is viewed as a two-dimensional phase of molecular state corresponding to g_1^S and σ_1 is the molecular area. Thus, the work of bringing a molecule into the surface is expressed as a ΔG using Eq. III-47.

The same relations are then applied to each component of a solution

$$\exp\left(\frac{-\gamma \sigma_1}{kT}\right) = \frac{N_1 g_1}{N_1^S g_1^S} \tag{III-50}$$

$$\exp\left(\frac{-\gamma \sigma_2}{kT}\right) = \frac{N_2 g_2}{N_2^S g_2^S} \tag{III-51}$$

where N^S denotes the mole fraction in the surface phase. Equations III-50 and III-51 may be solved for N_1^S and N_2^S, respectively, and substituted into the requirement that $N_1^S + N_2^S = 1$. If it is assumed that $\sigma = \sigma_1 = \sigma_2$, one then obtains

$$\exp\left(\frac{-\gamma \sigma}{kT}\right) = \frac{N_1 g_1}{g_1^S} + \frac{N_2 g_2}{g_2^S} \tag{III-52}$$

and, in combination with Eq. III-49

$$\exp \frac{-\gamma \sigma}{kT} = N_1 \exp \frac{-\gamma_1 \sigma}{kT} + N_2 \exp \frac{-\gamma_2 \sigma}{kT} \tag{III-53}$$

Hildebrand and Scott [81] give an expansion of Eq. III-53 for $\sigma_1 \neq \sigma_2$.

Guggenheim [5] extended his treatment to the case of regular solutions, that is, solutions for which

$$RT \ln f_1 = -\alpha N_2^2 \qquad RT \ln f_2 = -\alpha N_1^2 \tag{III-54}$$

where f denotes the activity coefficient. A very simple relationship for such regular solutions comes from Prigogine and Defay [82]:

$$\gamma = \gamma_1 N_1 + \gamma_2 N_2 - \beta N_1 N_2 \tag{III-55}$$

where β is a semiempirical constant.

Figure III-9a shows some data for fairly ideal solutions [81] where the solid lines 2, 3, and 6 show the attempt to fit the data with Eq. III-53; line 4 by taking σ as a purely empirical constant; and line 5, by the use of the Hildebrand–Scott equation [81]. As a further example of solution behavior, Fig. III-9b shows some data on fused-salt mixtures [83]; the dotted lines show the fit to Eq. III-55.

An extensive formalism for various types of nonideal solutions has been developed by Prigogine, Defay and co-workers using a lattice model allowing for interacting molecules of different sizes [82]. Nissen [84] has applied the approach to molten-salt mixtures, as has Gaines [85]. Reiss and Mayer [10] developed an expression for the surface tension of a fused salt, using their hard-sphere treatment of liquids (Section III-2B), and have extended the approach to solutions. The same is true for a statistical mechanical model developed by Eyring and co-workers [86]. Goodisman [87] has commented on various treatments for molten-salt mixtures.

The theoretical treatments of Section III-2B have been used to calculate interfacial tensions of solutions using suitable interaction potential functions. Thus Gubbins and co-workers [88] report a molecular dynamics calculation of the surface tension of a solution of A and B molecules obeying Eq. III-46 with $\epsilon_{0,BB}/\epsilon_{0,AA} = 0.4$ and

$$\epsilon_{0,AB} = (\epsilon_{0,BB}\epsilon_{0,AA})^{1/2}. \tag{III-56}$$

We have considered the surface tension behavior of several types of systems, and now it is desirable to discuss in slightly more detail the very important case of aqueous mixtures. If the surface tensions of the separate pure liquids differ appreciably, as in the case of alcohol–water mixtures, then the addition of small amounts of the second component generally results in a marked decrease in surface tension from that of the pure water. The case of ethanol and water is shown in Fig. III-9c. As seen in Section III-5, this effect may be accounted for in terms of selective adsorption of the alcohol at the interface. Dilute aqueous solutions of organic substances can be treated with a semiempirical equation attributed to von Szyszkowski [89,90]

$$\frac{\gamma}{\gamma_0} = 1 - B \ln\left(1 + \frac{C}{A}\right) \tag{III-57}$$

where γ_0 is the surface tension of water, B is a constant characteristic of the homologous series of organic compounds involved, A is a constant characteristic of each compound, and C is its concentration. This equation may be derived on the basis of the Langmuir adsorption equation (see Problem III-8 and Sections XI-1A and XVII-3).

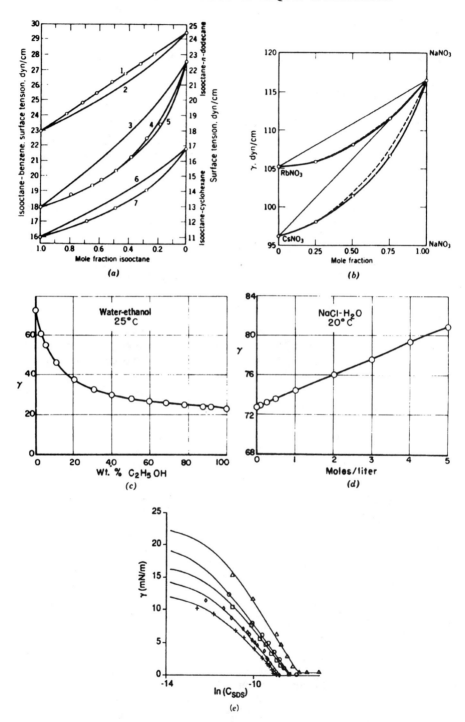

(a)

(b)

(c)

(d)

(e)

The type of behavior shown by the ethanol–water system reaches an extreme in the case of higher-molecular-weight solutes of the polar–nonpolar type, such as soaps and detergents [91]. As illustrated in Fig. III-9e, the decrease in surface tension now takes place at very low concentrations sometimes showing a point of abrupt change in slope in a γ/C plot [92]. The surface tension becomes essentially constant beyond a certain concentration identified with micelle formation (see Section XIII-5). The lines in Fig. III-9e are fits to Eq. III-57. The authors combined this analysis with the Gibbs equation (Section III-5B) to obtain the surface excess of surfactant and an alcohol cosurfactant.

B. The Surface Tension of Polymeric Systems

In polymer solutions and blends, it becomes of interest to understand how the surface tension depends on the molecular weight (or number of repeat units, N) of the macromolecule and on the polymer–solvent interactions through the interaction parameter, χ. In terms of a Flory lattice model, χ is given by the polymer and solvent interactions through

$$\chi = 6[\epsilon_{ps} - \tfrac{1}{2}(\epsilon_{pp} + \epsilon_{ss})]/kT \qquad \text{(III-58)}$$

where ϵ_{ps}, ϵ_{pp}, and ϵ_{ss} are the polymer–solvent, polymer–polymer, and solvent–solvent interaction energies, respectively. Much of the pioneering work in this area was done by Gaines and co-workers [94]. Szleifer and Widom [95] developed a lattice model of dilute phase-separated polymer solutions. Their derivation permits a simple expression for the surface tension

$$\frac{\gamma}{kT} = \int_{\phi}^{\phi'} \left[\left(\frac{2}{3}\chi + \frac{5}{12\phi} \right) h(\phi) \right]^{1/2} d\phi \qquad \text{(III-59)}$$

that requires knowledge of the coexisting phase compositions, ϕ and ϕ', but not the detailed composition profile through the interface. The function $h(\phi)$

Fig. III-9. Representative plots of surface tension versus composition. (a) Isooctane-n-dodecane at 30°C: 1 linear, 2 ideal, with $\sigma = 48.6$. Isooctane–benzene at 30°C: 3 ideal, with $\sigma = 35.4$, 4 ideal-like with empirical σ of 112, 5 unsymmetrical, with $\sigma_1 = 136$ and $\sigma_2 = 45$. Isooctane–cyclohexane at 30°C: 6 ideal, with $\sigma = 38.4$, 7 ideal-like with empirical σ of 109.3, (σ values in Å^2/molecule) (from Ref. 93). (b) Surface tension isotherms at 350°C for the systems (Na–Rb) NO$_3$ and (Na–Cs) NO$_3$. Dotted lines show the fit to Eq. III-55 (from Ref. 83). (c) Water–ethanol at 25°C. (d) Aqueous sodium chloride at 20°C. (e) Interfacial tensions between oil and water in the presence of sodium dodecylchloride (SDS) in the presence of hexanol and 0.20 M sodium chloride. Increasing both the surfactant and the alcohol concentration decreases the interfacial tension (from Ref. 92).

$$h(\phi) = \frac{1}{N} \phi \ln \frac{\phi}{\phi'} + (1 - \phi) \ln \frac{1 - \phi}{1 - \phi'}$$

$$+ \left(1 - \frac{1}{N}\right)(\phi - \phi') - \overline{\chi}(\phi - \phi')^2 \tag{III-60}$$

represents the distance of the free energy from its double tangent (at the coexisting phase compositions, ϕ and ϕ'). This requires the *local* interaction parameter

$$\overline{\chi} = 6\left[\frac{5}{6} \epsilon_{ps} - \frac{1}{2} \left(\frac{2}{3} \epsilon_{pp} + \epsilon_{ss}\right)\right] \Big/ kT \tag{III-61}$$

A scaling analysis [95–97] provides the temperature and molecular weight dependence

$$\gamma = N^{-1} f\left[\frac{1}{2} \sqrt{N}(1 - T/T_c)\right] \tag{III-62}$$

where $f(x)$ is a scaling function of the scaling variable $x = \frac{1}{2}\sqrt{N}(1 - T/T_c)$,

$$f(x) = \begin{cases} \text{const } x^\mu, x \to 0 \\ \text{const } x^2, x \to \infty \end{cases} \tag{III-63}$$

depending on the surface tension critical exponent, $\mu = 1.26$. Experimental studies have verified the $x \to 0$ limit with all of the available data falling within $x < 1$ [97]. One example of this scaling behavior is for polystyrene in methylcyclohexane [98, 99] where

$$\gamma = (54 \text{ dyn/cm})N^{-0.39}(1 - T/T_c)^{1.26} \tag{III-64}$$

The surface tension of polymer melts can be strongly influenced by the potential surface activity of the chain ends [100]. While the density of a polymer depends on its molecular weight, the primary effect on surface tension is not through density variation but rather due to preferential adsorption or depletion of the ends at the surface. Koberstein and co-workers [101] have demonstrated this effect with end-functionalized poly(dimethylsiloxane). Their pendant drop studies (see Section II-7) of low-molecular-weight polymers having *amine-*, *hydroxyl-* or *methyl-*terminal groups show surface tensions decreasing, independent of, or increasing with molecular weight due to the higher, intermediate, or lower surface energies, respectively, of the end groups. The end groups also alter the interfacial tension between immiscible polymer blends [102] in a similar way. The addition of block copolymers to immiscible polymer blends is analogous to adding a surface active agent to immiscible liquids. The interfacial tension is reduced by the adsorption of the block copolymer at the interface until it is saturated [103].

Finally, similar effects can be seen in miscible polymer blends where the surface tension correlates with the enrichment of the lower-energy component at the surface as monitored by x-ray photoelectron spectroscopy [104].

5. Thermodynamics of Binary Systems: The Gibbs Equation

We now come to a very important topic, namely, the thermodynamic treatment of the variation of surface tension with composition. The treatment is due to Gibbs [35] (see Ref. 49 for an historical sketch) but has been amplified in a more conveniently readable way by Guggenheim and Adam [105].

A. Definition of Surface Excess

As in Section III-2A, it is convenient to suppose the two bulk phases, α and β, to be uniform up to an arbitrary dividing plane S, as illustrated in Fig. III-10. We restrict ourselves to plane surfaces so that c_1 and c_2 are zero, and the condition of equilibrium does not impose any particular location for S. As before, one computes the various extensive quantities on this basis and compares them with the values for the system as a whole. Any excess or deficiency is then attributed to the surface region.

Taking the section shown in Fig. III-10 to be of unit area in cross section, then, if the phases were uniform up to S, the amount of the ith component present would be

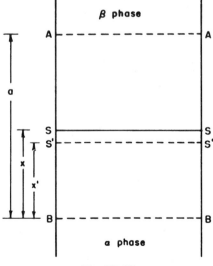

Fig. III-10

$$xC_i^\alpha + (a - x)C_i^\beta \tag{III-65}$$

Here, the distances x and a are relative to planes A and B located far enough from the surface region so that bulk phase properties prevail. The actual amount of component i present in the region between A and B will be

$$xC_i^\alpha + (a - x)C_i^\beta + \Gamma_i^\sigma \tag{III-66}$$

where Γ_i^σ denotes the surface excess per unit area.†
   For the case where the phase β is gaseous, C_i^β may be neglected, and quantities III-65 and III-66 become

$$xC_i^\alpha \quad \text{and} \quad xC_i^\alpha + \Gamma_i^\sigma \tag{III-67}$$

If one now makes a second arbitrary choice for the dividing plane, namely, S' and distance x', it must follow that

$$x'C_i + \Gamma_i^{\sigma\prime} = xC_i + \Gamma_i^\sigma \tag{III-68}$$

(dropping the superscript α as unnecessary), because the same total amount of the ith component must be present between A and B regardless of how the dividing surface is located. One then has

$$\frac{\Gamma_i^{\sigma\prime} - \Gamma_i^\sigma}{C_i} = x - x' \tag{III-69}$$

so that

$$\frac{\Gamma_1^{\sigma\prime} - \Gamma_1^\sigma}{C_1} = \frac{\Gamma_2^{\sigma\prime} - \Gamma_2^\sigma}{C_2} = \text{etc.} \tag{III-70}$$

or, in general,

$$\frac{\Gamma_i^{\sigma\prime} - \Gamma_i^\sigma}{N_i} = \frac{\Gamma_j^{\sigma\prime} - \Gamma_j^\sigma}{N_j} \tag{III-71}$$

†The term *surface excess* will be used as an algebraic quantity. If positive, an actual excess of the component is present and, if negative, there is a surface deficiency. An alternative name that has been used is *superficial density*.

where N denotes mole fraction, or

$$\Gamma_j^\sigma N_i - \Gamma_i^\sigma N_j = \Gamma_j^{\sigma\prime} N_i - \Gamma_i^{\sigma\prime} N_j \tag{III-72}$$

Since S and S' are purely arbitrary in location, Eq. III-72 can be true only if each side separately equals a constant

$$\Gamma_j^\sigma N_i - \Gamma_i^\sigma N_j = \text{constant} \tag{III-73}$$

B. The Gibbs Equation

With the preceding introduction to the handling of surface excess quantities, we now proceed to the derivation of the third fundamental equation of surface chemistry (the Laplace and Kelvin equations, Eqs. II-7 and III-18, are the other two), known as the *Gibbs* equation.

For a small, reversible change dE in the energy of a system, one has

$$dE = dE^\alpha + dE^\beta + dE^\sigma$$
$$= T\,dS^\alpha + \sum \mu_i\,dn_i^\alpha - P^\alpha\,dV^\alpha + T\,dS^\beta$$
$$+ \sum \mu_i\,dn_i^\beta - P^\beta\,dV^\beta + T\,dS^\sigma + \sum \mu_i\,dn_i^\sigma + \gamma\,d\mathcal{A} \tag{III-74}$$

Since

$$dE^\alpha = T\,dS^\alpha + \sum \mu_i\,dn_i^\alpha - P^\alpha\,dV \tag{III-75}$$

and similarly for phase β, it follows that

$$dE^\sigma = T\,dS^\sigma + \sum \mu_i\,dn_i^\sigma + \gamma\,d\mathcal{A} \tag{III-76}$$

If one now allows the energy, entropy, and amounts to increase from zero to some finite value, keeping T, $\mathcal{A}$ (area), and the n_i^σ constant, Eq. III-76 becomes

$$E^\sigma = TS^\sigma + \sum \mu_i n_i^\sigma + \gamma\mathcal{A} \tag{III-77}$$

Equation III-77 is generally valid and may now be differentiated in the usual manner to give

$$dE^\sigma = T\, dS^\sigma + S^\sigma\, dT + \sum \mu_i\, dn_i^\sigma$$

$$+ \sum n_i^\sigma\, d\mu_i + \gamma\, d\mathcal{A} + \mathcal{A}\, d\gamma \tag{III-78}$$

Comparison with Eq. III-76 gives

$$0 = S^\sigma\, dT + \sum n_i^\sigma\, d\mu_i + \mathcal{A}\, d\gamma \tag{III-79}$$

or, per unit area

$$d\gamma = -S^\sigma\, dT - \sum \Gamma^\sigma\, d\mu_i \tag{III-80}$$

For a two-component system at constant temperature, Eq. III-80 reduces to

$$d\gamma = -\Gamma_1^\sigma\, d\mu_1 - \Gamma_2^\sigma\, d\mu_2 \tag{III-81}$$

Moreover, since Γ_1^σ and Γ_2^σ are defined relative to an arbitrarily chosen dividing surface, it is possible in principle to place that surface so that $\Gamma_1^\sigma = 0$ (this is discussed in more detail below), so that

$$\Gamma_2^1 = -\left(\frac{\partial \gamma}{\partial \mu_2}\right)_T \tag{III-82a}$$

or

$$\Gamma_2^1 = -\frac{a}{RT}\frac{d\gamma}{da} \tag{III-82b}$$

where a is the activity of the solute and the superscript 1 on the Γ means that the dividing surface was chosen so that $\Gamma_1^\sigma = 0$. Thus if $d\gamma/da$ is negative, as in Fig. III-9c, Γ_2^1 is positive, and there is an actual surface excess of solute. If $d\gamma/da$ is positive, as in Fig. III-9d, there is a surface deficiency of solute.

C. The Dividing Surface

A schematic picture of how concentrations might vary across a liquid–vapor interface is given in Fig. III-11. The convention indicated by superscript 1, that is, the $\Gamma_1^\sigma = 0$ is illustrated. The dividing line is drawn so that the two areas shaded in full strokes are equal, and the surface excess of the solvent is thus zero. The area shaded with dashed strokes, which lies to the right of the dividing

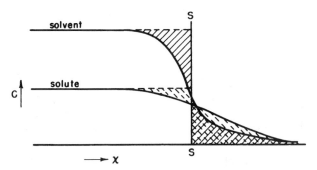

Fig. III-11. Schematic illustration of surface excess.

surface minus the smaller similarly shaded area to the left corresponds to the (in this case positive) surface excess of solute. The quantity Γ_2^1 may thus be defined as the (algebraic) excess of component 2 in a 1-cm^2 cross section of surface region over the moles that would be present in a bulk region containing the *same number of moles of solvent as does the section of surface region.*

Obviously, a symmetric definition Γ_1^2 also exists. Here $\Gamma_2^\sigma = 0$ and Γ_1^2 represents the excess of component 1 in a 1-cm^2 cross section of surface region over the moles that would be present in a bulk region containing the *same number of moles of solvent as does the section of surface region.* Another way of locating the dividing surface would be such that the algebraic *sum* of the areas in Fig. III-11 to the right of the dividing line is equal to the sum of the areas to the left. The surface excesses so defined are written Γ_i^N. Similarly, Γ_i^M and Γ_i^V are the excess of the ith component in the surface region over the moles that would be present in a bulk region of the *same total mass or volume as the surface region.*

We can construct a numerical illustration with a 0.5 mol fraction solution of ethanol in water. We take a slice of surface region deep enough to include some of the bulk solution. This is taken, let us say, from a surface region of area A cm^2, and contains 10 mol of water and 30 mol of ethanol. We obtain Γ_2^1 by comparing with a sample of bulk solution containing the same 10 mol of water. This sample would contain only 10 mol of ethanol, and Γ_2^1 is therefore $(30-10)/A = 20/A$. Γ_1^2 would be obtained by comparing with a bulk sample containing 30 mol of ethanol and hence 30 mol of water; therefore $\Gamma_1^2 = (10-30)/A = -20/A$. We obtain Γ_1^N and Γ_2^N by comparing with a bulk sample having the same total moles as the surface or 40 mol total (20 each of water and ethanol). Thus $\Gamma_1^N = (10-20)/A = -10/A$ and $\Gamma_2^N = (30-20)/A = 10/A$.

The surface excesses obey the relationship

$$P_1\Gamma_1 + P_2\Gamma_2 = 0 \qquad\qquad \text{(III-83)}$$

where P is determined by the specific property invoked in deciding how to choose the dividing surface. Thus for Γ_i^N, P is unity; for Γ_i^M, P_i is M_i, the molecular weight; and for Γ_i^V, P_i is V_i, the molar volume. We can summarize the entire picture as follows:

$$\frac{-N_1 d\gamma}{d\mu_2} = N_1 \Gamma_2^1 = \Gamma_2^N = \frac{\overline{M}}{M_1} \Gamma_1^M = \frac{\overline{V}}{V_1} \Gamma_1^V \qquad \text{(III-84)}$$

(handwritten annotation: *molar fraction* pointing to Γ_2^N)

where $\overline{M} = N_1 M_1 + N_2 M_2$ and $\overline{V} = N_1 V_1 + N_2 V_2$. (Note Problem III-17.)

An approach developed by Guggenheim [106] avoids the somewhat artificial concept of the Gibbs dividing surface by treating the surface region as a bulk phase whose upper and lower limits lie somewhere in the bulk phases not far from the interface.

D. Other Surface Thermodynamic Relationships

The preceding material of this section has focused on the most important phenomenological equation that thermodynamics gives us for multicomponent systems—the Gibbs equation. Many other, formal thermodynamic relationships have been developed, of course. Many of these are summarized in Ref. 107. The topic is treated further in Section XVII-13, but is worthwhile to give here a few additional relationships especially applicable to solutions.

Using the Gibbs convention for defining surface quantities, we define

$$G^\sigma = E^\sigma - T S^\sigma \qquad \text{(III-85)}$$

so that

$$dG^\sigma = dE^\sigma - T \, dS^\sigma - S^\sigma \, dT \qquad \text{(III-86)}$$

or, in combination with Eq. III-76,

$$dG^\sigma = -S^\sigma \, dT + \sum_i \mu_i \, dn_i^\sigma + \gamma \, d\mathcal{A} \qquad \text{(III-87)}$$

Alternatively, for the *whole* system (i.e., including the bulk phases),

$$dG = -S \, dT - P \, dV + \sum_i \mu_i \, dn_i + \gamma \, d\mathcal{A} \qquad \text{(III-88)}$$

Thus

$$\gamma = \left(\frac{\partial G^\sigma}{\partial \mathcal{A}} \right)_{T, n_i^\sigma} \qquad \text{(III-89)}$$

$$\gamma = \left(\frac{\partial G}{\partial \mathcal{A}} \right)_{T, V, n_i} \qquad \text{(III-90)}$$

Integration of Eq. III-87 holding constant the intensive quantities T, μ_i, and γ gives

$$G^\sigma = \sum_i \mu_i n_i^\sigma + \gamma \mathcal{A} \qquad \text{(III-91)}$$

or

$$G^\sigma = \gamma + \sum_i \mu_i \Gamma_i^\sigma \qquad \text{(III-92)}$$

where the extensive quantities are now on a per unit area basis. G^σ is the specific surface excess free energy, and, unlike the case for a pure liquid (Eq. III-2), it is not in general equal to γ. This last would be true only in the unlikely situation of no surface adsorption, so that the Γ^σ's were zero. (Some authors use the entirely permissible definition $G^\sigma = A^\sigma - \gamma \mathcal{A}$, in which case γ does not appear in the equation corresponding to Eq. III-92—see Ref. 107.)

6. Determination of Surface Excess Quantities

A. Experimental Methods

The most widely used experimental method for determining surface excess quantities at the liquid–vapor interface makes use of radioactive tracers. The solute to be studied is labeled with a radioisotope that emits weak beta radiation, such as ^{3}H, ^{14}C, or ^{35}S. One places a detector close to the surface of the solution and measures the intensity of beta radiation. Since the penetration range of such beta emitters is small (about 30 mg/cm^2 for ^{14}C, with most of the adsorption occurring in the first two-tenths of the range), the measured radioactivity corresponds to the surface region plus only a thin layer of solution (about 0.06 mm for ^{14}C and even less for ^{3}H).

As an example, Tajima and co-workers [108] used ^{3}H labeling to obtain the adsorption of sodium dodecyl sulfate at the solution–air interface. The results, illustrated in Fig. III-12, agreed very well with the Gibbs equation in the form

$$\Gamma_2^1 = -\frac{1}{2RT} \frac{d\gamma}{d \ln C} \qquad \text{(III-93)}$$

when corrected for activity coefficients. The factor of 2 in the denominator appears because of the activity of an electrolyte, in this case (Na^+, X^-), is given by $a_{Na^+} a_{X^-}$ or by C^2 if activity coefficients are neglected. The quantity $\partial\mu$ in Eq. III-82 thus becomes $RT d \ln C^2$ or $2RT d \ln C$. If, however, $0.1M$ sodium chloride was present at a swamping electrolyte, the experimental Γ_2^1 was twice

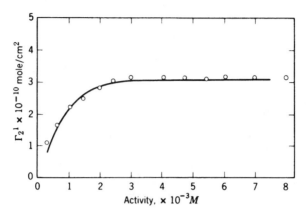

Fig. III-12. Verification of the Gibbs equation by the radioactive trace method. Observed (○) and calculated (line) values for Γ_2^1 for aqueous sodium dodecyl sulfate solutions. (From Ref. 108.)

that in Eq. III-93 [109] as expected since a_{Na+} is now constant and $\partial\mu$ is just $RT\,d\ln C$. A more elaborate treatment is given by Hall et al. [110].

Results can sometimes be unexpected. The first study of this type made use of ^{35}S labeled Aerosol OTN [111], an anionic surfactant, also known as di-n-octylsodium sulfosuccinate. The measured Γ_2^1 was twice that in Eq. III-93 and it was realized that hydrolysis had occurred, that is, $X^+ + H_2O = HX + OH^-$, and that it was the undissociated acid HX that was surface-active. Since pH was essentially constant, the activity of HX was just proportional to C. A similar behavior was found for aqueous sodium stearate [112].

A quite different means for the experimental determination of surface excess quantities is *ellipsometry*. The technique is discussed in Section IV-3D, and it is sufficient to note here that the method allows the calculation of the thickness of an adsorbed film from the ellipticity produced in light reflected from the film covered surface. If this thickness, τ, is known, Γ may be calculated from the relationship $\Gamma = \tau/V$, where V is the molecular volume. This last may be estimated either from molecular models or from the bulk liquid density.

Smith [113] studied the adsorption of n-pentane on mercury, determining both the surface tension change and the ellipsometric film thickness as a function of the equilibrium pentane pressure. Γ could then be calculated from the Gibbs equation in the form of Eq. III-106, and from τ. The agreement was excellent. Ellipsometry has also been used to determine the surface compositions of solutions [114,115], as well polymer adsorption at the solution-air interface [116].

The actual structure at a vapor–liquid interface can be probed with x-rays. Rice and co-workers [72,73,117] use x-ray reflection to determine the composition perpendicular to the surface and grazing incidence x-ray diffraction to study the transverse structure of an interface. In a study of bismuth gallium mixtures, they find a partial monolayer of bismuth at the surface with the structure of a supercooled liquid unmixed with gallium [117,118].

B. Historical Footnote and Commentary

Although Gibbs published his monumental treatise on heterogeneous equilibrium in 1875, his work was not generally appreciated until the turn of the century, and it was not until many years later that the field of surface chemistry developed to the point that experimental applications of the Gibbs equation became important.

It was of interest to many surface chemists to verify the Gibbs equation experimentally. One method, tried by several investigators, was to bubble a gas through the solution and collect the froth in a separate container. The solution resulting from the collapsed froth should differ from the original according to the value of the surface excess of the solute. Satisfactory results were not obtained, however, perhaps because of the difficulty in estimating the area of the bubbles. Probably the first successful experimental verification of the Gibbs equation is due to McBain and co-workers [119, 120]. They adopted the very direct approach of actually skimming off a thin layer of the surface of a solution, using a device called a *microtome*. A slice about 0.1 mm thick could be taken from about 1 m^2 of surface, so that a few grams of solution were collected, allowing surface excess determinations for aqueous solutions of *p*-toluidine, phenol, and *n*-hexanoic acid (see Problems III-24 and III-25).

At this point a brief comment on the justification of testing the Gibbs or any other thermodynamically derived relationship is in order. First, it might be said that such activity is foolish because it amounts to an exhibition of scepticism of the validity of the laws of thermodynamics themselves, and surely they are no longer in doubt! This is justifiable criticism in some specific instances but, in general, we feel it is not. The laws of thermodynamics are phenomenological laws about observable or operationally defined quantities, and where one of the more subtle deductions from these laws is involved it may not always be clear just what the operational definition of a given variable really is. This question comes up in connection with contact angles and the meaning of surface tensions of solid interfaces (see Section X-6). Second, thermodynamic derivations can involve the exercise of logic at a very rigorous level, and it is entirely possible for nonsequiturs to creep in, which escape attention until an experimental disagreement forces a reexamination. Finally, the testing of a thermodynamic relationship may reveal unsuspected complexities in a system. Thus, referring to the preceding subsection, it took experiment to determine that the surface active species of Aerosol OTN was HX rather than (Na$^+$, X$^-$) and that, Eq. III-93 was the appropriate form of the Gibbs equation to use. The difficulties in confirming the Kelvin equation for the case of liquids in capillaries have led people to consider various possible complexities (see Section III-1C).

C. Theoretical Calculation of Surface Excess Quantities

Both the Monte Carlo and the molecular dynamics methods (see Section III-2B) have been used to obtain theoretical density-versus-depth profiles for a hypothetical liquid-vapor interface. Rice and co-workers (see Refs. 72 and 121) have found that density along the normal to the surface tends to be a

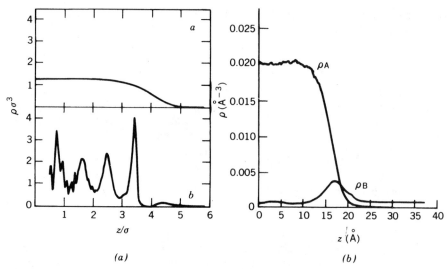

Fig. III-13. (a) Plots of molecular density versus distance normal to the interface; σ is molecular diameter. Upper plot: a dielectric liquid. Lower plot: as calculated for liquid mercury. (From Ref. 122.) (b) Equilibrium density profiles for atoms A and B in a rare-gas-like mixture for which $\epsilon_{0,\,BB}/\epsilon_{0,\,AA} = 0.4$ and $\epsilon_{0,\,AB}$ is given by Eq. III-56. Atoms A and B have the same σ (of Eq. III-46) and the same molecular weight of 50 g/mol; the solution mole fraction is $x_B = 0.047$. Note the strong adsorption of B at the interface. [Reprinted with permission from D. J. Lee, M. M. Telo de Gama, and K. E. Gubbins, *J. Phys. Chem.*, **89**, 1514 (1985) (Ref. 88). Copyright 1985, American Chemical Society.]

monotonic function in the case of a dielectric liquid, while for a liquid metal such as Na or Hg, the surface region is stratified as illustrated in Fig. III-13a. Such stratification carries a number of implications about the interpretation of surface properties of metals and alloys [123].

It was noted in connection with Eq. III-56 that molecular dynamics calculations can be made for a liquid mixture of rare gas-like atoms to obtain surface tension versus composition. The same calculation also gives the variation of density for each species across the interface [88], as illustrated in Fig. III-13b. The density profiles allow a calculation, of course, of the surface excess quantities.

7. Gibbs Monolayers

If the surface tension of a liquid is lowered by the addition of a solute, then, by the Gibbs equation, the solute must be adsorbed at the interface. This adsorption may amount to enough to correspond to a monomolecular layer of solute on the surface. For example, the limiting value of Γ_2^1 in Fig. III-12 gives an area per molecule of 52.0 Å^2, which is about that expected for a close-packed

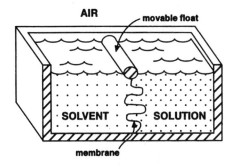

Fig. III-14. Cross section of the PLAWM (Pockels–Langmuir–Adam–Wilson–McBain) trough.

layer of dodecyl sulfate ions. It is thus a physically plausible concept to treat Γ_2^1 as giving the two-dimensional concentration of surfactant in a monomolecular film.

Such a monolayer may be considered to exert a film pressure π, such that

$$\pi = \gamma_{\text{solvent}} - \gamma_{\text{solution}} \qquad \text{(III-94)}$$

This film pressure (or "two-dimensional" pressure) has the units of dynes per centimeter and can be measured directly. As illustrated in Fig. III-14, if one has a trough divided by a thin rubber membrane into two compartments, one filled with solvent and the other with solution, then a force will be observed to act on a float attached to the upper end of the membrane. In the PLAWM† [124, 125], the rubber membrane was very thin, and the portion below the surface was so highly convoluted that it could easily buckle so as to give complete equalization of any hydrostatic differences between the two solutions. The force observed on the float was thus purely surface tensional in origin and resulted from the fact that a displacement in the direction of the surface of higher surface tension would result in a lower overall surface free energy for the system. This force could be measured directly by determining how much opposing force applied by a lever attached to a torsion wire was needed to prevent the float from moving.

In the preceding explanation π arises as a difference between two surface tensions, but it appears physically as a force per unit length on the barrier separating the two surfaces. It is a very fruitful concept to regard the situation as involving two surfaces that would be identical except that on one of them there are molecules of surface-adsorbed solute that can move freely in the plane of the surface but cannot pass the barrier. The molecules of the adsorbed film possess, then, two-dimensional translational energy, and the film pressure π can be regarded as due to the bombardment of the barrier by these molecules. This is analogous to viewing the pressure of a gas as due to the bombardment of molecules against the walls of the container. This interpretation of π allows

†Pockels–Langmuir–Adam–Wilson–McBain.

a number of very pleasing and constructive analogies to be made with three-dimensional systems, and the concept becomes especially plausible, physically, when one is dealing with the quite insoluble monolayers discussed in the next chapter.

It is not the only interpretation, however. Another picture, again particularly useful in the case of insoluble monolayers where the rubber diaphragm of the PLAWM trough is not needed, is to regard the barrier as a semipermeable membrane through which water can pass (i.e., go around actually) but not the surface film. The surface region can then be viewed as a relatively concentrated solution having an osmotic pressure π_{os}, which is exerted against the membrane.

It must be kept in mind that both pictures are modelistic and invoke extrathermodynamic concepts. Except mathematically, there is no such thing as a "two-dimensional" gas, and the "solution" whose osmotic pressure is calculated is not uniform in composition, and its average concentration depends on the depth assumed for the surface layer.

A. The Two-Dimensional Ideal-Gas Law

For dilute solutions, solute–solute interactions are unimportant (i.e., Henry's law will hold), and the variation of surface tension with concentration will be linear (at least for nonelectrolytes). Thus

$$\gamma = \gamma_0 - bC \qquad \text{(III-95)}$$

where γ_0 denotes the surface tension of pure solvent, or

$$\pi = bC \qquad \text{(III-96)}$$

Then, by the Gibbs equations,

$$-\frac{d\gamma}{dC} = \frac{\Gamma_2^1 RT}{C} \qquad \text{(III-97)}$$

By Eq. III-95, $-d\gamma/dC$ is equal to b, so that Eq. III-97 becomes

$$\pi = \Gamma_2^1 RT \qquad \text{(III-98)}$$

or

$$\pi\sigma = kT \qquad \pi A = RT \qquad \text{(III-99)}$$

where σ and A denote area per molecule and per mole, respectively. Equation III-99 is analogous to the ideal-gas law, and it is seen that in dilute solutions the film of adsorbed solute obeys the equation of state of a two-dimensional ideal gas. Figure III-15a shows that for a series of aqueous alcohol solutions π increases linearly with C at low concentrations and, correspondingly, Fig. III-15c shows that $\pi A/RT$ approaches unity as π approaches zero.

A sample calculation shows how Fig. III-15c is computed from the data of Fig. III-15a. Equation III-97 may be put in the form

$$A = \frac{RT}{d\pi/d \ln C} \qquad \text{(III-100)}$$

or, at 25°C and with σ in angstrom squared units, $\sigma = 411.6/[d\pi/d(\ln C)]$. For n-butyl alcohol π is 15.4 dyn/cm for $C = 0.1020$ and is 11.5 dyn/cm for $C = 0.0675$. Taking the slope of the line between these two points, we find

$$\sigma = \frac{411.6}{(11.5 - 15.4)/[-2.69 - (-2.28)]} = \frac{411.6}{3.9/0.41} = 43 \text{Å}^2 \text{ per molecule}$$

This locates a point at the approximate π value of $(11.5 + 15.4)/2$ or 13.5 dyn/cm. Thus $\pi A/RT = \pi\sigma/kT = 1.41$.

B. Nonideal Two-Dimensional Gases

The deviation of Gibbs monolayers from the ideal two-dimensional gas law may be treated by plotting $\pi A/RT$ versus π, as shown in Fig. III-15c. Here, for a series of straight-chain alcohols, one finds deviations from ideality increasing with increasing film pressure; at low π values, however, the limiting value of unity for $\pi A/RT$ is approached.

This behavior suggests the use of an equation employed by Amagat for gases at high pressure; the two-dimensional form is

$$\pi(A - A^0) = qRT \qquad \text{(III-101)}$$

where A^0 has the aspect of an excluded area per mole and q gives a measure of the cohesive forces. Rearrangement yields the linear form

$$\frac{\pi A}{RT} = \frac{A^0}{RT} \pi + q \qquad \text{(III-102)}$$

This form is obeyed fairly well above π values of 5–10 dyn/cm in Fig. III-15c. Limiting areas or σ^0 values of about 22 Å^2 per molecule result, nearly independent of chain length, as would be expected if the molecules assume a final orientation that is perpendicular to the surface. Larger A^0 values are found for longer-chain surfactants, such as sodium dodecyl sulfate, and this has been attributed to the hydrocarbon tails having a variety of conformations [127].

Various other non-ideal-gas-type two-dimensional equations of state have been proposed, generally by analogy with gases. Volmer and Mahnert [128,

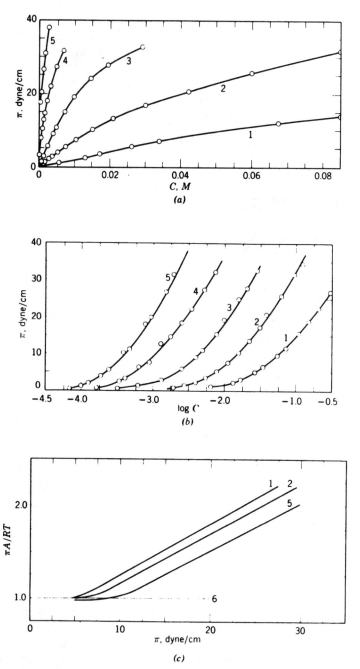

Fig. III-15. Surface tension data for aqueous alcohol: illustration of the use of the Gibbs equation. (1) n-butyl; (2) n-amyl; (3) n-hexyl; (4) n-heptyl; (5) n-octyl. (Data from Ref. 126).

129] added only the covolume correction to the ideal-gas law:

$$\pi(A - A^0) = RT \qquad \text{(III-103)}$$

One may, of course, use a two-dimensional modification of the van der Waals equation:

$$\left(\pi + \frac{a}{A^2} \right) (A - A^0) = RT \qquad \text{(III-104)}$$

The varying actual orientation of molecules adsorbed at an aqueous solution-CCl$_4$ interface with decreasing A has been followed by resonance Raman spectroscopy using polarized light [130]. The effect of *pressure* has been studied for fatty alcohols at the water-hexane [131] and water-paraffin oil [132] interfaces.

Adsorption may occur from the vapor phase rather than from the solution phase. Thus Fig. III-16 shows the surface tension lowering when water was exposed for various hydrocarbon vapors; P^0 is the saturation pressure, that is, the vapor pressure of the pure liquid hydrocarbon. The activity of the hydrocarbon is given by its vapor pressure, and the Gibbs equation takes the form

$$-d\gamma = d\pi = \Gamma RT \, d \ln P \qquad \text{(III-105)}$$

(for simplicity we have written just Γ instead of the exact designation Γ_2^1), and Γ may thus be calculated from the analogue of Eq. III-100

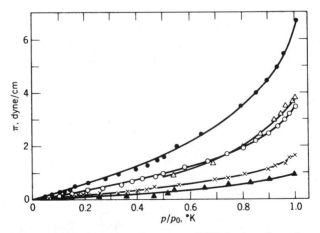

Fig. III-16. Surface tension lowering of water at 15°C due to adsorption of hydrocarbons. ●, *n*-pentane; △, 2,2,4-trimethylpentane; ○, *n*-hexane; ×, *n*-heptane; ▲, *n*-octane. (From Ref. 133.)

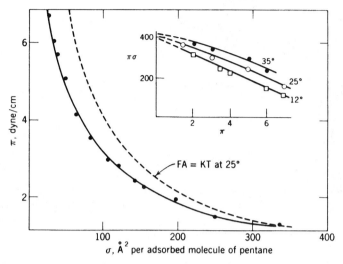

Fig. III-17. Adsorption of pentane on water. (From Refs. 134, 135.)

$$\Gamma = \frac{1}{RT}\frac{d\pi}{d\ln P} \tag{III-106}$$

The data may then be expressed in conventional π-versus-σ or $\pi\sigma$-versus-π plots, as shown in Fig. III-17. The behavior of adsorbed pentane films was that of a nonideal two-dimensional gas, as can be seen from the figure.

The data could be expressed equally well in terms of Γ versus P, or in the form of the conventional adsorption isotherm plot, as shown in Fig. III-18. The appearance of these isotherms is discussed in Section X-6A. The Gibbs equation thus provides a connection between adsorption isotherms and two-dimensional equations of state. For example, Eq. III-57 corresponds to the adsorption isotherm

$$\Gamma = \frac{aC}{1 + bC} \tag{III-107}$$

(where a and b are constants), which is a form of the Langmuir adsorption equation (see Section XI-1A). The reverse situation, namely, the adsorption of water vapor on various organic liquids, has also been studied [136].

C. The Osmotic Pressure Point of View

It was pointed out at the beginning of this section that π could be viewed as arising from an osmotic pressure difference between a surface region comprising an adsorbed film and that of the pure solvent. It is instructive to develop

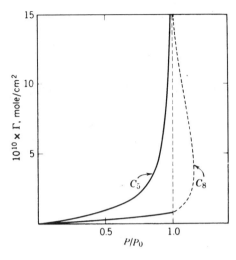

Fig. III-18. Adsorption isotherm for *n*-pentane and *n*-octane at 15°C. The dotted curve shows the hypothetical isotherm above $P/P_0 = 1$, and the arrows mark the Γ values corresponding to a monolayer. (From Refs. 128, 129.)

this point of view somewhat further. The treatment can be made along the line of Eq. III-52, but the following approach will be used instead.

To review briefly, the osmotic pressure in a three-dimensional situation is that pressure required to raise the vapor pressure of solvent in a solution to that of pure solvent. Thus, remembering Eq. III-16,

$$RT \ln \frac{a_1^0}{a_1} = \int V_1 \, dP = \pi_{os} V_1 \qquad \text{(III-108)}$$

where a_1 denotes the activity of the solvent; it is usually assumed that its compressibility can be neglected, so that the integral may be replaced by $\pi_{os} V_1$, where V_1 is the molar volume. For *ideal* solutions, the ratio a_1/a_1^0 is given by N_1, the solvent mole fraction, and for *dilute* solutions, $-\ln N_1$ is approximated by N_2, which in turn is approximately n_2/n_1, the mole ratio of solute to solvent. Insertion of these approximations into Eq. III-108 leads to the limiting form

$$\pi_{os} V_1 = \frac{n_2}{n_1} RT \quad \text{or} \quad \pi_{os} V = n_2 RT \qquad \text{(III-109)}$$

Let us now suppose that the surface region can be regarded as having a depth τ and an area $\mathcal{A}$ and hence volume V^s. A volume V^s of surface region, if made up of pure solvent, will be

$$V^s = n_1^{s0} V_1 \qquad \text{(III-110)}$$

and, if made up of a mixture of solvent and surface adsorbed solute, will be

$$V^s = n_1^s V_1 + n_2^s V_2 \tag{III-111}$$

assuming the molar volumes V_1 and V_2 to be constant. The solute mole fraction in this surface region is then

$$N_2^s = \frac{n_2^s}{n_1^s + n_2^s} = \frac{n_2^s V_1}{n_1^{s0} V_1 - n_2^s (V_2 - V_1)} \tag{III-112}$$

using Eqs. III-110 and III-111 to eliminate n_1^s. There will be an osmotic pressure given approximately by

$$\pi_{os} V_1 = RTN_2^s \tag{III-113}$$

and if this is viewed as acting against a semipermeable barrier, the film pressure will be the osmotic pressure times the depth of the surface region in which it is exerted, that is,

$$\pi = \pi_{os} \tau \tag{III-114}$$

Equations III-112 and III-113 may now be combined with III-114 to give

$$\pi = \frac{\tau RT n_2^s}{n_1^{s0} V_1 - n_2^s (V_2 - V_1)} \tag{III-115}$$

Now, $n_1^{s0} V_1 / \tau$ is just the surface area $\mathcal{A}$, and, moreover, V_1/τ and V_2/τ have the dimensions of molar area. *If* the surface region is considered to be just *one* molecule thick, V_1/τ and V_2/τ becomes A_1^0 and A_2^0, the actual molar areas, so that Eq. III-115 takes on the form

$$\pi = \frac{RT n_2^s}{\mathcal{A} - n_2^s (A_2^0 - A_1^0)} \tag{III-116}$$

or, on rearranging and remembering that $A = \mathcal{A}/n_2^s$,

$$\pi[A - (A_2^0 - A_1^0)] = RT \tag{III-117}$$

If further, A_1^0 is neglected in comparison with A_2^0, then Eq. III-117 becomes the same as the nonideal gas law, Eq. III-103.

This derivation has been made in a form calculated best to bring out the very considerable and sometimes inconsistent approximations made. However, by treating the surface region as a kind of solution, an avenue is opened for employing our considerable knowledge of solution physical chemistry in estimating association, interionic attraction, and other nonideality effects. Another advantage, from the writers' point of view, is the emphasis on the role of the solvent as part of the surface region, which helps to correct the tendency, latent in the two-dimensional equation of state treatment, to regard the substrate as merely providing an inert plane surface on which molecules of the adsorbed species may move freely. The approach is not really any more empirical than that using the two-dimensional nonideal gas, and considerable use has been made of it by Fowkes [137].

It has been pointed out [138] that algebraically equivalent expressions can be derived without invoking a surface solution model. Instead, surface excess as defined by the procedure of Gibbs is used, the dividing surface always being located so that the sum of the surface excess quantities equals a given constant value. This last is conveniently taken to be the maximum value of Γ_2^1. A somewhat related treatment was made by Handa and Mukerjee for the surface tension of mixtures of fluorocarbons and hydrocarbons [139].

D. Surface Elasticity

The *elasticity* (or the *surface dilatational modulus*) E is defined as

$$E = -\frac{d\gamma}{d \ln \mathcal{A}} \tag{III-118}$$

where $\mathcal{A}$ is the geometric area of the surface. E is zero if the surface tension is in rapid equilibrium with a large body of bulk solution, but if there is no such molecular traffic, $\mathcal{A}$ in Eq. III-118 may be replaced by A, the area per mole of the surface excess species, and an alternative form of the equation is therefore

$$E = \frac{d\pi}{d \ln \Gamma} \tag{III-119}$$

The reciprocal of E is called the *compressibility*.

It is not uncommon for this situation to apply, that is, for a Gibbs monolayer to be in only slow equilibrium with bulk liquid—see, for example, Figs. II-15 and II-21. This situation also holds, of course, for spread monolayers of insoluble substances, discussed in Chapter IV. The experimental procedure is illustrated in Fig. III-19, which shows that a portion of the surface is bounded by bars or floats, an opposing pair of which can be moved in and out in an oscillatory manner. The concomitant change in surface tension is followed by means of a Wilhelmy slide. Thus for dilute aqueous solutions of a methylcellu-

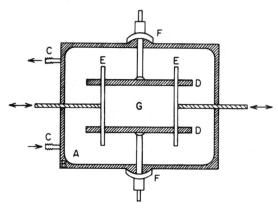

Fig. III-19. Trough for dynamic surface measurements: A, stainless-steel dish; B, aluminum mantle; C, inlet thermostatting water; D, lower PTFE bars; E, oscillating bars; F, attachment lower bars; G, Wilhelmy plate. (From Ref. 140.)

lose polymer, the equilibrium Gibbs monolayer was perfectly elastic with $E = 9.5$ dyn/cm for a 2.4×10^{-3} wt% solution [142]. If the period of oscillation in $\mathcal{A}$ is small compared to the bulk solution–surface equilibration time, then E may be determined as a function of surface *age*. If the period is comparable to the equilibration time, E becomes an E_{app}, which depends on the period. This last was the situation for aqueous solutions of a series of surfactants of the type $RO[CH_2CH_2O]_nH$ [140, 141].

Bianco and Marmur [143] have developed a means to measure the surface elasticity of soap bubbles. Their results are well modeled by the von Szyszkowski equation (Eq. III-57) and Eq. III-118. They find that the elasticity increases with the size of the bubble for small bubbles but that it may go through a maximum for larger bubbles. Li and Neumann [144] have shown the effects of surface elasticity on wetting and capillary rise phenomena, with important implications for measurement of surface tension.

The discussion of surface viscosity and other aspects of surface rheology is deferred to Section IV-3C.

E. Traube's Rule

The surface tensions for solutions of organic compounds belonging to a homologous series, for example, $R(CH_2)_nX$, show certain regularities. Roughly, Traube [145] found that for each additional CH_2 group, the concentration required to give a certain surface tension was reduced by a factor of 3. This rule is manifest in Fig. III-15b; the successive curves are displaced by nearly equal intervals of 0.5 on the log C scale.

Langmuir [146] gave an instructive interpretation to this rule. The work W to transfer one mole of solute from bulk solution to surface solution should be

$$W = RT \ln \frac{C^s}{C} = RT \ln \frac{\Gamma}{\tau C} \qquad \text{(III-120)}$$

where C^s is the surface concentration and is given by Γ/τ, where τ is the thickness of the surface region. For solutes of chain length n and $(n - 1)$, the difference in work is then

$$W_n - W_{n-1} = RT \ln \frac{\Gamma_n/\tau C_n}{\Gamma_{n-1}/\tau C_{n-1}} \qquad \text{(III-121)}$$

By Traube's rule, if $C_{n-1}/C_n = 3$, then $\gamma_n = \gamma_{n-1}$, and, as an approximation, it is assumed that the two surface concentrations are also the same. If so, then

$$W_n - W_{n-1} = RT \ln 3 = 2.67 \text{ kJ/mol} \qquad \text{(III-122)}$$

This value may be regarded as the work to bring one CH_2 group from the body of the solution into the surface region. Since the value per CH_2 group appears to be independent of chain length, it is reasonable to suppose that all the CH_2 groups are similarly situated in the surface, that is, that the chains are lying flat.

If the dependence on temperature as well as on composition is known for a solution, enthalpies and entropies of adsorption may be calculated from the appropriate thermodynamic relationships [82]. Nearn and Spaull [147] have, for example, calculated the enthalpies of surface adsorption for a series of straight-chain alcohols. They find an increment in enthalpy of about 1.96 kJ/mol per CH_2 group.

Van Oss and Good [148] have compared solubilities and interfacial tensions for a series of alcohols and their corresponding hydrocarbons to determine the free energy of hydration of the hydroxyl group; they find -14 kJ/mol per $-OH$ group.

F. Some Further Comments on Gibbs Monolayers

It is important to realize that there is, in principle, no necessary difference between the nature of the adsorbed films discussed so far and those formed by spreading monolayers of insoluble substances on a liquid substrate or by adsorption from either a gas or a liquid phase onto a solid (or a liquid) surface. The distinction that *does* exist has to do with the nature of the accessible experimental data. In the case of Gibbs monolayers, one is dealing with fairly soluble solutes, and the direct measurement of Γ is not easy to carry out. Instead, one measures the changes in surface tension and obtains Γ through the use of the Gibbs equation. With spread monolayers, the solubility of the material is generally so low that its concentration in solution is not easily measurable, but Γ is known directly, as the amount per unit area that was spread onto the surface, and the surface tension also can be measured directly. In the case of adsorption, Γ is known from the decrease in concentration (or pressure) of the adsorbate material, so that both Γ and the concentration or pressure (if it is gas adsorption) are known. It is not generally possible, however, to measure the surface tension of a solid surface. Thus, it is usually possible to measure only two out of the three quantities γ, Γ, and C or P.

The succeeding material is broadly organized according to the types of exper-imental quantities measured because much of the literature is so grouped. In the next chapter spread monolayers are discussed, and in later chapters the topics of adsorption from solution and of gas adsorption are considered. Irrespective of the experimental compartmentation, the conclusions as to the nature of mobile adsorbed films, that is, their structure and equations of state, will tend to be of a general validity. Thus, only a limited discussion of Gibbs monolayers has been given here, and none of such related aspects as the contact potentials of solutions or of adsorption at liquid–liquid interfaces, as it is more efficient to treat these topics later.

8. Problems

1. Given that $d\gamma/dT$ is -0.086 erg/cm$^2 \cdot$ K for ethanol at 20°C, calculate E^s and $E^{s'}$. Look up other data on physical properties, as needed.

2. Referring to Problem 1, calculate S^s and $S^{s'}$ for ethanol at 20°C. Do the same for n-octane and compare the results.

3. Calculate the vapor pressure of water when present in a capillary of 0.1 μm radius (assume zero contact angle). Express your result as percent change from the normal value at 25°C. Suppose now that the effective radius of the capillary is reduced because of the presence of an adsorbed film of water 100 Å thick. Show what the percent reduc-tion in vapor pressure should now be.
_{1000 Å}
_{6 1×10⁻⁷}

4. Calculate γ_m for sodium, using Eq. III-12.

5. Application of 150 MPa pressure increases the interfacial tension for n-hex-ane–water from 50.5 to 53.0 mN/m at 25°C. Calculate ΔV^s. What is ΔV^s for that area corresponding to a molecular size (take a representative molecular area to be 20 Å^2)? Convert this to cm^3/cm^2 mol.

6. Illustrate the use of Eq. III-44 as follows. Approximate $g(r)$ by a step function, $g(r) = 0$ for $r < \sigma$ and $g(r) = 1$ for $r \geq \sigma$; assume that $u(r)$ is given by Eq. III-46, and that the molecule is argonlike, with $\sigma = 3.4$ Å and $\epsilon_0 = 124k$, where k is the Boltzmann constant. Calculate γ from this information and the density of liquid argon.

7. Use Eq. III-15 and related equations to calculate E^s and the energy of vaporization of argon. Take u to be ϵ_0 of Problem 6, and assume argon to have a close-packed structure of spheres 3.4 Å in diameter.

8. Use Fig. III-7b to estimate the surface tension of the argonlike liquid. Pressure is given in units of ϵ_0/σ^3, where $\epsilon_0 = 119.8k$ and $\sigma = 3.4$ Å. Hint: remember Eq. III-40.

9. The gradient model for interfacial tension described in Eqs. III-42 and III-43 is limited to interaction potentials that decay more rapidly than r^{-5}. Thus it can be applied to the Lennard-Jones potential but not to a longer range interaction such as dipole–dipole interaction. Where does this limitation come from, and what does it imply for interfacial tensions of various liquids?

10. Plot the scaling behavior for the surface tension of polystyrene solutions using Eq. III-64, for $N = 1,000$ and T from zero to T_c. Now plot the behavior for $T = 0.8T_c$ for $N = 100$–1000. Comment on the influence of polymers on surface tension.

11. Derive Eq. III-21 from the first and second laws of thermodynamics and related definitions.

12. Estimate, by means of Eq. III-41, the surface tensions of CCl_4, $CHCl_3$ and of water at 20°C. Look up the necessary data on thermal expansion and compressibility.

13. The following two statements seem mutually contradictory and seem to describe a paradoxical situation. (a) The chemical potential of a species must be everywhere the same for an equilibrium system at constant temperature and pressure; therefore, if we have a liquid in equilibrium with its vapor (the interface is planar), the chemical potential of the species must be the same in the surface region as it is in the bulk liquid, and no work is required to move a molecule from the bulk region to the surface region. (b) It must require work to move a molecule from the bulk region to the surface region because to do so means increasing the surface area and hence the surface free energy of the system.

Discuss these statements and reconcile the apparent contradiction. (Note Ref. 138.)

14. Calculate, using the data of Fig. III-9a and Eq. III-53, the surface tension versus mole fraction plot for mixtures of cyclohexane and benzene.

15. Derive an equation for the heat of vaporization of a liquid as a function of drop radius r.

16. Using Langmuir's principle of independent surface action, make qualitative calculations and decide whether the polar or the nonpolar end of ethanol should be oriented toward the mercury phase at the ethanol–mercury interface.

17. Complete the numerical illustration preceding Eq. III-83 by calculating Γ_1^M and Γ_2^M.

18. As $N_2 \to 0$, $\Gamma_2^1 \to k_2 N_2$, where k_2 is a Henry law constant, and, similarly, $\Gamma_1^2 \to k_1 N_1$ as $N_1 \to 0$. Show that $\Gamma_2^1 \to -k_1$ as $N_2 \to 1$ and that $\Gamma_1^2 \to -k_2$ as $N_1 \to 1$. Note Eqs. III-83 and III-84.

19. A 2% by weight aqueous surfactant solution has a surface tension of 69.0 dyn/cm (or mN/m) at 20°C. (a) Calculate σ, the area of surface containing one molecule. State any assumptions that must be made to make the calculation from the preceding data. (b) The additional information is now supplied that a 2.2% solution has a surface tension of 68.8 dyn/cm. If the surface-adsorbed film obeys the equation of state $\pi(\sigma - \sigma_0) = kT$, calculate from the combined data a value of σ_0, the actual area of a molecule.

20. There are three forms of the Langmuir–Szyszkowski equation, Eq. III-57, Eq. III-107, and a third form that expresses π as a function of Γ. (a) Derive Eq. III-57 from Eq. III-107 and (b) derive the third form.

21. Tajima and co-workers [108] determined the surface excess of sodium dodecyl sulfate by means of the radioactivity method, using tritiated surfactant of specific activity 9.16 Ci/mol. The area of solution exposed to the detector was 37.50 cm². In a particular experiment, it was found that with $1.0 \times 10^{-2} M$ surfactant the surface count rate was 17.0×10^3 counts per minute. Separate calibration showed that of this count was 14.5×10^3 came from underlying solution, the rest being surface excess. It was also determined that the counting efficiency for surface material was 1.1%. Calculate Γ for this solution.

22. An adsorption isotherm known as the Temkin equation [149] has the form: $\pi = \alpha \Gamma^2 / \Gamma^\infty$ where α is a constant and Γ^∞ is the limiting surface excess for a close-packed

monolayer of surfactant. Using the Gibbs equation find π as a function of C and Γ as a function of C.

23. An adsorption equation known as the Frumkin isotherm has the form

$$\ln \Gamma - \ln(\Gamma_m - \Gamma) + a\Gamma/\Gamma_m = \ln bc \qquad \text{(III-123)}$$

when $\Gamma \ll \Gamma_m$, Γ_m is the limiting value of Γ. Show what the corresponding two-dimensional equation of state is, that is, show what corresponding relationship is between π and σ. (Suggested by W. R. Fawcett.)

24. McBain reports the following microtome data for a phenol solution. A solution of 5 g of phenol in 1000 g of water was skimmed; the area skimmed was 310 cm^2 and a 3.2-g sample was obtained. An interferometer measurement showed a difference of 1.2 divisions between the bulk and the scooped-up solution, where one division corresponded to 2.1×10^{-6} g phenol per gram of water concentration difference. Also, for 0.05, 0.127, and 0.268M solutions of phenol at 20°C, the respective surface tensions were 67.7, 60.1, and 51.6 dyn/cm. Calculate the surface excess Γ_2^1 from (a) the microtome data, (b) for the same concentration but using the surface tension data, and (c) for a horizontally oriented monolayer of phenol (making a reasonable assumption as to its cross-sectional area).

25. The thickness of the equivalent layer of pure water τ on the surface of a 3M sodium chloride solution is about 1 Å. Calculate the surface tension of this solution assuming that the surface tension of salt solutions varies linearly with concentration. Neglect activity coefficient effects.

26. The surface tension of an aqueous solution varies with the concentration of solute according to the equation $\gamma = 72 - 350C$ (provided that C is less than 0.05M). Calculate the value of the constant k for the variation of surface excess of solute with concentration, where k is defined by the equation $\Gamma_2^1 = kC$. The temperature is 25°C.

27. The data in Table III-2 have been determined for the surface tension of isooctane-benzene solutions at 30°C. Calculate Γ_1^2, Γ_2^1, Γ_1^N, and Γ_2^N for various concentrations and plot these quantities versus the mole fraction of the solution. Assume ideal solutions.

28. The surface tension of water at 25°C exposed to varying relative pressures of a

TABLE III-2
Data for the System Isooctane–Benzene

Mole Fraction Isooctane	Surface Tension (dyn/cm)	Mole Fraction Isooctane	Surface Tension (dyn/cm)
0.000	27.53	0.583	19.70
0.186	23.40	0.645	19.32
0.378	21.21	0.794	18.74
0.483	20.29	1.000	17.89

TABLE III-3
Adsorption of Water Vapor on Mercury

Water Vapor Pressure (atm)	Surface Tension (dyn/cm)	Water Vapor Pressure (atm)	Surface Tension (dyn/cm)
0	483.5	7.5	459
1×10^{-3}	483	10	450
3	482	15	442
4	481		438
5	477		436
6	467		

hydrocarbon vapor changes as follows:

P/P^0	0.10	0.20	0.30	0.40	0.50	0.60	0.70	0.80	0.90
π(dyn/cm)	0.22	0.55	0.91	1.35	1.85	2.45	3.15	4.05	5.35

Calculate and plot π versus σ (in Å^2 per molecule) and Γ versus P/P^0. Does it appear that this hydrocarbon wets water (note Ref. [133])?

29. Derive the equation of state, that is, the relationship between π and σ, of the adsorbed film for the case of a surface active electrolyte. Assume that the activity coefficient for the electrolyte is unity, that the solution is dilute enough so that surface tension is a linear function of the concentration of the electrolyte, and that the electrolyte itself (and not some hydrolyzed form) is the surface-adsorbed species. Do this for the case of a strong 1 : 1 electrolyte and a strong 1 : 3 electrolyte.

30. Some data obtained by Nicholas et al. [150] are given in Table III-3, for the surface tension of mercury at 25°C in contact with various pressures of water vapor. Calculate the adsorption isotherm for water on mercury, and plot it as Γ versus P.

31. Estimate A^0 in Eq. III-101 from the data of Fig. III-15 and thence the molecular area in $\text{Å}^2/\text{molecule}$.

32. The surface elasticity E is found to vary linearly with π and with a slope of 2. Obtain the corresponding equation of state for the surface film, that is, the function relating π and σ.

33. The following data have been reported for methanol-water mixtures at 20°C (*Handbook of Chemistry and Physics*, U.S. Rubber Co.):

Wt% methanol	7.5	10.0	25.0	50.0	60.0	80.0	90.0	100
γ (dyn/cm)	60.90	59.04	46.38	35.31	32.95	27.26	25.36	22.65

Make a theoretical plot of surface tension versus composition according to Eq. III-53, and compare with experiment. (Calculate the equivalent spherical diameter for water and methanol molecules and take σ as the average of these.)

General References

R. Defay, I. Prigogine, A. Bellemans, and D. H. Everett, *Surface Tension and Adsorption*, Longmans, Green, London. 1966.

R. H. Fowler and E. A. Guggenheim, *Statistical Thermodynamics*, Cambridge University Press, London, 1939.

W. D. Harkins, *The Physical Chemistry of Surfaces*, Reinhold, New York, 1952.

J. H. Hildebrand and R. L. Scott, *The Solubility of Nonelectrolytes*, 3rd ed., Van Nostrand-Reinhold, Princeton, New Jersey, 1950.

G. N. Lewis and M. Randall, *Thermodynamics*, 2nd ed. (revised by K. S. Pitzer and L. Brewer), McGraw-Hill, New York, 1961.

A. W. Neumann et al., *Applied Surface Thermodynamics. Interfacial Tension and Contact Angles*, Marcel Dekker, New York, 1996.

J. S. Rowlinson and B. Widom, *Molecular Theory of Capillarity*, Clarendon Press, Oxford, 1984.

H. van Olphen and K. J. Mysels, *Physical Chemistry: Enriching Topics from Colloid and Surface Science*, Theorex (8327 La Jolla Scenic Drive), CA, 1975.

Textual References

1. W. D. E. Thomas and L. Potter, *J. Colloid Interface Sci.*, **50,** 397 (1975).
2. K. L. Wolf, *Physik und Chemie der Grenzflächen*, Springer-Verlag, Berlin, 1957.
3. R. Eötvös, Weid Ann. **27,** 456 (1886).
4. W. Ramsay and J. Shields, *J. Chem. Soc.*, **LXIII,** 1089 (1893).
5. E. A. Guggenheim, *J. Chem. Phys.*, **13,** 253 (1945).
6. A. V. Grosse, *J. Inorg. Nucl. Chem.*, **24,** 147 (1962).
7. S. Blairs and U. Joasoo, *J. Colloid Interface Sci.*, **79,** 373 (1981).
8. J. E. Lennard-Jones and J. Corner, *Trans. Faraday Soc.*, **36,** 1156 (1940).
9. J. W. Cahn and J. E. Hilliard, *J. Chem. Phys.*, **28,** 258 (1958).
10. H. Reiss and S. W. Mayer, *J. Chem. Phys.*, **34,** 2001 (1961).
11. A. S. Skapski, *J. Chem. Phys.*, **16,** 386 (1948).
12. K. R. Atkins and Y. Narahara, *Phys. Rev.*, **138,** A437 (1965).
13. D. Stansfield, *Proc. Phys. Soc.*, **72,** 854 (1958).
14. W. V. Kayser, *J. Colloid Interface Sci.*, **56,** 622 (1976).
15. C. C. Addison and J. M. Coldrey, *J. Chem. Soc.*, **1961,** 468 (1961).
16. J. B. Griffin and H. L. Clever, *J. Chem. Eng. Data*, **5,** 390 (1960).
17. S. Blairs, *J. Colloid Interface Sci.*, **67,** 548 (1978).
18. B. C. Allen, *AIME Trans.*, **227,** 1175 (1963).
19. L. R. Fisher and J. N. Israelachvili, *J. Colloid Interface Sci.*, **80,** 528 (1981).
20. J. C. Melrose, *Langmuir*, **5,** 290 (1989).
21. R. C. Tolman, *J. Chem. Phys.*, **17,** 333 (1949).
22. J. C. Melrose, *Ind. Eng. Chem.*, **60,** 53 (1968).

23. D. S. Choi, M. J. Jhon, and H. Eyring, *J. Chem. Phys.*, **53**, 2608 (1970).

24. L. R. Fisher and J. N. Israelachvili, *Chem. Phys. Lett.*, **76**, 325 (1980).

25. H. K. Christenson, *J. Colloid Interface Sci.*, **104**, 234 (1985).

26. G. N. Lewis and M. Randall, *Thermodynamics and Free Energy of Chemical Substances*, McGraw-Hill, New York, 1923, p. 248.

27. M. Kahlweit, *Ber. Bunsen-Gesell. Phys. Chemie*, **74**, 636 (1970).

28. J. C. Eriksson, *Acta Chem. Scand.*, **16**, 2199 (1962).

29. C. Jho, D. Nealon, S. Shodbola, and A. D. King Jr., *J. Colloid Interface Sci.*, **65**, 141 (1978).

30. C. Jho and A. D. King, Jr., *J. Colloid Interface Sci.*, **69**, 529 (1979).

31. M. Lin, *J. Chem. Phys.*, **77**, 1063 (1980).

32. K. Motomura, H. Iyota, M. Arantono, M. Yamanaka, and R. Matuura, *J. Colloid Interface Sci.*, **93**, 264 (1983).

33. R. C. Brown, *Proc. Phys. Soc.*, **59**, 429 (1947).

34. C. Gurney, *Proc. Phys. Soc.*, **A62**, 639 (1947).

35. J. W. Gibbs, *The Collected Works of J. W. Gibbs*, vol. I, Longmans, Green, New York, 1931, p. 219.

36. R. C. Tolman, *J. Chem. Phys.*, **16**, 758 (1948).

37. R. C. Tolman, *J. Chem. Phys.*, **17**, 118 (1949).

38. J. G. Kirkwood and F. P. Buff, *J. Chem. Phys.*, **17**, 338 (1949).

39. F. Buff, "The Theory of Capillarity," in *Handbuch der Physik*, vol. 10, Springer-Verlag, Berlin, 1960.

40. J. C. Melrose, *Pure Appl. Chem.*, **22**, 273 (1970).

41. M. J. Mandell and H. Reiss, *J. Stat. Phys.*, **13**, 107 (1975).

42. L. Boruvka, Y. Rotenberg, and A. W. Neumann, *Langmuir*, **1**, 40 (1985).

43. L. Boruvka, Y. Rotenberg, and A. W. Neumann, *J. Phys. Chem.*, **89**, 2714 (1985).

44. J. Gaydos, L. Boruvka, and A. W. Neumann, *Langmuir*, **7**, 1035 (1991).

45. M. Pasandideh-Fard, P. Chen, M. Mostaghimi, and A. W. Neumann, *Adv. Colloid Interface Sci.*, **63**, 151 (1996).

46. A. Jyoti, P. Chen, and A. W. Neumann, *J. Math. Chem.*, **20**, 183 (1996).

47. T. L. Hill, *Statistical Mechanics*, McGraw-Hill, New York, 1956.

48. H. T. Davis and L. E. Scriven, *J. Phys. Chem.*, **80**, 2805 (1976).

49. A. W. Adamson, *Textbook of Physical Chemistry*, 3rd ed., Academic, Orlando, FL, 1986.

50. B. Widom, "Structure and Thermodynamics of Interfaces," in *Statistical Mechanics and Statistical Methods in Theory and Application*, Plenum, New York, 1977, pp. 33–71.

51. B. Widom, *Disc. Faraday Soc.*, **16**, 7 (1981).

52. B. Widom, "Interfacial Phenomena," in *Liquids, Freezing and Glass Transition, Les Houches Session LI, 1989*, Elsevier, 1991, pp. 507–546.

53. B. Widom, Physica **95A**, 1 (1979).

54. J. H. Perez-Lopez, L. J. Gonzalez-Oritz, M. A. Leiva, and J. E. Puig, *AIChE J.*, **38**, 753 (1992).

55. P. Tavan and B. Widom, *Phys. Rev. B*, **27**, 180 (1983).

56. D. A. McQuarrie, *Statistical Mechanics*, Harper Collins, 1976.

57. P. D. Shoemaker, G. W. Paul, and L. E. Marc de Chazal, *J. Chem. Phys.*, **52**, 491 (1970).

58. E. Tronel-Peyroz, J. M. Douillard, R. Bennes, and M. Privat, *Langmuir*, **5**, 54 (1988).

59. J. Goodisman and R. W. Pastor, *J. Phys. Chem.*, **82**, 2078 (1978).

60. W. A. Curtin, *Phys. Rev. B*, **39**, 6775 (1989).

61. D. W. Marr and A. P. Gast, *Phys. Rev. E*, **47**, 1212 (1993).

62. D. W. Marr and A. P. Gast, *Langmuir*, **10**, 1348 (1994).

63. D. W. Marr and A. P. Gast, *J. Chem. Phys.*, **99**, 2024 (1993).

64. F. F. Abraham, *Rep. Prog. Phys.*, **45**, 1113 (1982).

65. F. F. Abraham, *J. Vac. Sci. Technol. B*, **2**, 534 (1984).

66. J. P. R. B. Walton, D. J. Tidesley, and J. S. Rowlinson, *Mol. Phys.*, **50**, 1357 (1983).

67. K. E. Gubbins, *Fluid Interfacial Phenomena*, Wiley, New York, 1986.

68. J. Eggbrecht, S. M. Thompson, and K. E. Gubbins, *J. Chem. Phys.*, **86**, 2299 (1987).

69. S. M. Thompson, K. E. Gubbins, J. P. R. B. Walton, R. A. R. Chantry, and J. S. Rowlinson, *J. Chem. Phys.*, **81**, 530 (1984).

70. D. J. Lee, M. M. Telo de Gama, and K. E. Gubbins, *J. Chem. Phys.*, **85**, 490 (1986).

71. M. A. Gomez and S. A. Rice, *J. Chem. Phys.*, **101**, 8094 (1994).

72. B. N. Thomas, S. W. Barton, F. Novak, and S. A. Rice, *J. Chem. Phys.*, **86**, 1036 (1987).

73. E. R. Flom et al., *J. Chem. Phys.*, **96**, 4743 (1992).

74. W. B. Hardy, *Proc. Roy. Soc.* (London), **A88**, 303 (1913).

75. W. D. Harkins, *The Physical Chemistry of Surface Films*, Reinhold, New York, 1952.

76. I. Langmuir, *Colloid Symposium Monograph*, The Chemical Catalog Company, New York, 1925, p. 48.

77. J. H. Thurtell, M. M. Telo de Gama, and K. E. Gubbins, *Mol. Phys.*, **54**, 321 (1985).

78. B. Widom, *J. Phys. Chem.*, **88**, 6508 (1984).

79. M. A. Wilson, A. Pohorille, and L. R. Pratt, *J. Phys. Chem.*, **91**, 4873 (1987).

80. E. A. Guggenheim, *Trans. Faraday Soc.*, **41**, 150 (1945).

81. J. H. Hildebrand and R. L. Scott, *Solubility of Nonelectrolytes*, Reinhold, New York, 1950, Chapter 21.

82. R. Defay, I. Prigogine, A. Bellemans, and D. H. Everett, *Surface Tension and Adsorption*, Longmans, Green, London, 1966.

83. G. Bertozzi and G. Sternheim, *J. Phys. Chem.*, **68**, 2908 (1964).

84. D. A. Nissen, *J. Phys. Chem.*, **82**, 429 (1978).

85. G. L. Gaines, Jr., *Trans. Faraday Soc.*, **65**, 2320 (1969).

86. T. S. Ree, T. Ree, and H. Eyring, *J. Chem. Phys.*, **41**, 524 (1964).

87. J. Goodisman, *J. Colloid Interface Sci.*, **73,** 115 (1980).

88. D. J. Lee, M. M. Telo de Gama, and K. E. Gubbins, *J. Phys. Chem.*, **89,** 1514 (1985).

89. B. von Szyszkowski, *Z. Phys. Chem.*, **64,** 385 (1908).

90. H. P. Meissner and A. S. Michaels, *Ind. Eng. Chem.*, **41,** 2782 (1949).

91. M. Aratono, S. Uryu, Y. Hayami, K. Motomura, and R. Matuura, *J. Colloid Interface Sci.*, **98,** 33 (1984).

92. W. K. Kegel et al., *Langmuir*, **9,** 252 (1993).

93. H. B. Evans, Jr. and H. L. Clever, *J. Phys. Chem.*, **68,** 3433 (1964).

94. G. L. Gaines, Jr., *J. Polym. Sci.*, **10,** 1529 (1972).

95. I. Szleifer and B. Widom, *J. Chem. Phys.*, **90,** 7524 (1989).

96. P. G. de Gennes, *Scaling Concepts in Polymer Physics*, Cornell University Press, Ithaca, NY, 1979.

97. K. Q. Xia, C. Franck, and B. Widom, *J. Chem. Phys.*, **97,** 1446 (1992).

98. K. Shinozaki, T. V. Tan, Y. Saito, and T. Nose, *Polymer*, *23,* 728 (1982).

99. I. C. Sanchez, *J. Phys. Chem.*, **93,** 6983 (1989).

100. P. G. de Gennes, *CR (Comptes Rendus), Acad. Sci. Ser. 2*, **307,** 1841–1844 (1988).

101. C. Jalbert, J. T. Koberstein, I. Yilgor, P. Gallagher, and V. Krukonis, *Macromolecules*, **26,** 3069 (1993).

102. C. A. Fleischer, J. T. Koberstein, V. Krukonis, and P. A. Wetmore, *Macromolecules*, **26,** 4172 (1993).

103. W. Hu, J. T. Koberstein, J. P. Lingelser, and Y. Gallot, *Macromolecules*, **28,** 5209 (1995).

104. Q. S. Bhatia, D. H. Pan, and J. T. Koberstein, *Macromolecules*, **21,** 2166 (1988).

105. E. A. Guggenheim and N. K. Adam, *Proc. Roy. Soc.* (London), **A139,** 218 (1933).

106. E. A. Guggenheim, *Trans. Faraday Soc.*, **36,** 397 (1940).

107. D. H. Everett, *Pure Appl. Chem.*, **31,** 579 (1972).

108. K. Tajima, M. Muramatsu, and T. Sasaki, *Bull. Chem. Soc. Jpn.*, **43,** 1991 (1970).

109. K. Tajima, *Bull. Chem. Soc. Jpn.*, **43,** 3063 (1970).

110. D. G. Hall, B. A. Pethica, and K. Shinoda, *Bull. Chem. Soc. Jpn.*, **48,** 324 (1975).

111. D. J. Salley, A. J. Weith Jr., A. A. Argyle, and J. K. Dixon, *Proc. Roy. Soc.* (London), **A203,** 42 (1950).

112. K. Sekine, T. Seimiya, and T. Sasaki, *Bull. Chem. Soc. Japan*, **43,** 629 (1970).

113. T. Smith, *J. Colloid and Interface Sci.*, **28,** 531 (1968).

114. M. Privat, R. Bennes, E. Tronel-Peyroz, and J. M. Douillard, *J. Colloid and Interface Sci.*, **121,** 198 (1988).

115. E. Tronel-Peyroz, J. M. Douillard, L. Tenebre, R. Bennes, and M. Privat, *Langmuir*, **3,** 1027 (1987).

116. M. Kawaguchi, M. Oohira, M. Tajima, and A. Takahashi, *Polymer J.*, **12,** 849 (1980).

117. N. Lei, Z. Huang, and S. A. Rice, *J. Chem. Phys.*, **104,** 4802 (1996).

118. E. B. Flom, M. Li, A. Acero, N. Maskil, and S. A. Rice, *Science*, **260,** 332 (1993).

119. J. W. McBain and C. W. Humphreys, *J. Phys. Chem.*, **36,** 300 (1932).

120. J. W. McBain and R. C. Swain, *Proc. Roy. Soc.* (London), **A154,** 608 (1936).

121. M. P. D'Evelyn and S. A. Rice, *J. Chem. Phys.*, **78,** 5081 (1983).

122. S. W. Barton et al., *Nature*, **321,** 685 (1986).

123. S. A. Rice, *Proc. Natl. Acad. Sci.* (USA), **84,** 4709 (1987).

124. J. W. McBain, J. R. Vinograd, and D. A. Wilson, *J. Am. Chem. Soc.*, **62,** 244 (1940).

125. J. W. McBain, J. R. Vinograd, and D. A. Wilson, *Kolloid-A*, **78,** 1 (1937).

126. A. M. Posner, J. R. Anderson, and A. E. Alexander, *J. Colloid Interface Sci.*, **7,** 623 (1952).

127. M. J. Vold, *J. Colloid Interface Sci.*, **100,** 224 (1984).

128. M. Volmer and P. Mahnert, *Z. Phys. Chem.*, **115,** 239 (1925).

129. M. Volmer, *Z. Phys. Chem.*, 253 (1925).

130. T. Takenaka, N. Isono, J. Umemura, M. Shimomura, and T. Kunitake, *Chem. Phys. Lett.*, **128,** 551 (1986).

131. N. Matubayasi, K. Motomura, M. Aratono, and R. Matuura, *Bull. Chem. Soc. Jpn.*, **51,** 2800 (1978).

132. M. Lin, J. Firpo, P. Mansoura, and J. F. Baret, *J. Chem. Phys.*, **71,** 2202 (1979).

133. F. Hauxwell and R. H. Ottewill, *J. Colloid Interface Sci.*, **34,** 473 (1970).

134. C. L. Cutting and D. C. Jones, *J. Chem. Soc.*, **1955,** 4067 (1955).

135. M. Blank and R. H. Ottewill, *J. Phys. Chem.*, **68,** 2206 (1964).

136. K. Huang, C. P. Chai, and J. R. Maa, *J. Colloid Interface Sci.*, **79,** 1 (1981).

137. F. M. Fowkes, *J. Phys. Chem.*, **68,** 3515 (1964), and preceding papers.

138. E. H. Lucassen-Reynders, *J. Colloid Interface Sci.*, **41,** 156 (1972).

139. T. Handa and P. Mukerjee, *J. Phys. Chem.*, **85,** 3916 (1981).

140. J. Lucassen and D. Giles, *J. Chem. Soc., Faraday Trans. I*, **71,** 217 (1975).

141. E. H. Lucassen-Reynders, J. Lucassen, P. R. Garrett, D. Giles, and F. Hollway, *Adv. Chem.*, **144,** 272 (1975).

142. B. M. Abraham, J. B. Ketterson, and F. Behroozi, *Langmuir*, **2,** 602 (1986).

143. H. Bianco and A. Marmur, *J. Colloid Interface Sci.*, **158,** 295 (1993).

144. D. Li and A. W. Neumann, *Langmuir*, **9,** 50 (1993).

145. I. Traube, *Annalen*, **265,** 27 (1891).

146. I. Langmuir, *J. Am. Chem. Soc.*, **39,** 1848 (1917).

147. M. R. Nearn and A. J. B. Spaull, *Trans. Far. Soc.*, **65,** 1785 (1969).

148. C. J. van Oss and R. J. Good, *J. Disp. Sci. Technol.*, **17,** 433 (1996).

149. M. I. Temkin, *Zh. Fiz. Khim.*, **15,** 296 (1941).

150. M. E. Nicholas, P. A. Joyner, B. M. Tessem, and M. D. Olsen, *J. Phys. Chem.*, **65,** 1373 (1961).

CHAPTER IV

Surface Films on Liquid Substrates

1. Introduction

When a slightly soluble substance is placed at a liquid–air interface, it may spread out to a thin and in most cases monomolecular film. Although the thermodynamics for such a system is in principle the same as for Gibbs monolayers, the concentration of the substance in solution is no longer an experimentally convenient quantity to measure. The solution concentration, in fact, is not usually of much interest, so little use is made even of the ability to compute changes in it through the Gibbs equation. The emphasis shifts to more direct measurements of the interfacial properties themselves, and these are discussed in some detail, along with some of the observations and conclusions.

First, it is of interest to review briefly the historical development of the subject. Gaines, in his monograph [1], reminds us that the calming effect of oil on a rough sea was noted by Pliny the Elder and by Plutarch and that Benjamin Franklin in 1774 characteristically made the observation more quantitative by remarking that a teaspoon of oil sufficed to calm a half-acre surface of a pond. Later, in 1890, Lord Rayleigh (J. W. Strutt) [2] noted that the erratic movements of camphor on a water surface were stopped by spreading an amount of oleic acid sufficient to give a film only about 16 Å thick. This, incidentally, gave an upper limit to the molecular size and hence to the molecular weight of oleic acid so that a minimum value of Avogadro's number could be estimated. This comes out to be about the right order of magnitude.

About this time Miss Pockels† [3] showed how films could be confined by means of barriers; thus she found little change in the surface tension of fatty-acid films until they were confined to an area corresponding to about 20 Å^2 per molecule (the Pockels point). In 1899, Rayleigh [5] commented that a reasonable interpretation of the Pockels point was that at this area the molecules of the surface material were just touching each other. The picture of a surface film

---

†Agnes Pockels (1862–1935) had only a girl's high-school education, in Lower Saxony; the times were such that universities were closed to women and later, when this began to change, family matters prevented her from obtaining a higher formal education. She was thus largely self-taught. Her experiments were conducted in the kitchen, with simple equipment, and it is doubtful that they would have found recognition had she not written to Lord Rayleigh about them in 1881. It is to Rayleigh's credit that he sponsored the publication in *Nature* (Ref. 3) of a translation of her letter. For a delightful and more detailed account, see Ref. 4.

that was developing was one of molecules "floating" on the surface, with little interaction until they actually came into contact with each other. Squeezing a film at the Pockels point put compressive energy into the film that was available to reduce the total free energy to form more surface; that is, the surface tension was reduced.

Also, these early experiments made it clear that a monomolecular film could exert a physical force on a floating barrier. A loosely floating circle of thread would stretch taut to a circular shape when some surface-active material was spread inside its confines. Physically, this could be visualized as being due to the molecules of the film pushing against the confining barrier. Devaux [6] found that light talcum powder would be pushed aside by a film spreading on a liquid surface and that some films were easily distorted by air currents, whereas others appeared to be quite rigid.

Langmuir [7] in 1917 gave a great impetus to the study of monomolecular films by developing the technique used by Pockels. He confined the film with a rigid but adjustable barrier on one side and with a floating one on the other. The film was prevented from leaking past the ends of the floating barrier by means of small airjets. The actual force on the barrier was then measured directly to give π, the film pressure (see Section III-7). As had been observed by Miss Pockels, he found that one could sweep a film off the surface quite cleanly simply by moving the sliding barrier, always keeping it in contact with the surface. As it was moved along, a fresh surface of clean water would form behind it. The floating barrier was connected to a knife-edge suspension by means of which the force on the barrier could be determined. The barriers were constructed of paper coated with paraffin so as not to be wet by the water.

A sketch of Langmuir's film balance is shown in Ref. 7 and a modern version of a film balance, in Fig. IV-5.

Langmuir also gave needed emphasis to the importance of employing pure substances rather than the various natural oils previously used. He thus found that the limiting area (at the Pockels point) was the same for palmitic, stearic, and cerotic acids, namely, 21 Å^2 per molecule. (For convenience to the reader, the common names associated with the various hydrocarbon derivatives most frequently mentioned in this chapter are given in Table IV-1.)

This observation that the length of the hydrocarbon chain could be varied from 16 to 26 carbon atoms without affecting the limiting area could only mean that at this point the molecules were oriented vertically. From the molecular weight and density of palmitic acid, one computes a molecular volume of 495 Å^3; a molecule occupying only 21 Å^2 on the surface could then be about 4.5 Å on the side but must be about 23 Å long. In this way one begins to obtain information about the shape and orientation as well as the size of molecules.

The preceding evidence for orientation at the interface plus the considerations given in Section III-3 make it clear that the polar end is directed toward the water and the hydrocarbon tails toward the air. On the other hand, the evidence from the study of the Gibbs monolayers (Section III-7) was that the smaller molecules tended to lie flat on the surface. It will be seen that the orientation

TABLE IV-1
Common Names of Long-Chain Compounds

Formula	Name	Geneva Name
$C_{10}H_{21}COOH$	Undecoic acid	Undecanoic acid
$C_{11}H_{23}OH$	Undecanol	1-Hendecanol
$C_{11}H_{23}COOH$	Lauric acid	Dodecanoic acid
$C_{12}H_{25}OH$	Lauryl alcohol, dodecyl alcohol	1-Dodecanol
$C_{12}H_{25}COOH$	Tridecylic acid	Tridecanoic acid
$C_{13}H_{27}OH$	Tridecyl alcohol	1-Tridecanol
$C_{13}H_{27}COOH$	Myristic acid	Tetradecanoic acid
$C_{14}H_{29}OH$	Tetradecyl alcohol	1-Tetradecanol
$C_{15}H_{31}COOH$	Palmitic acid	Hexadecanoic acid
$C_{16}H_{33}OH$	Cetyl alcohol	1-Hexadecanol
$C_{16}H_{33}COOH$	Margaric acid	Heptadecanoic acid
$C_{17}H_{35}OH$	Heptadecyl alcohol	1-Heptadecanol
$C_{17}H_{35}COOH$	Stearic acid	Octadecanoic acid
$C_{18}H_{37}OH$	Octadecyl alcohol	1-Octadecanol
$C_8H_{17}CH{=}CH(CH_2)_7COOH$	Elaidic acid	*trans*-9-Octadecenoic acid
$C_8H_{17}CH{=}CH(CH_2)_7COOH$	Oleic acid	*cis*-9-Octadecenoic acid
$CH_3(CH_2)_7CH{=}CH(CH_2)_8OH$	Oleyl alcohol	*cis*-9-Octadecenyl alcohol
$CH_3(CH_2)_7CH{=}CH(CH_2)_8OH$	Elaidyl alcohol	*trans*-9-Octadecenyl alcohol
$C_{18}H_{37}COOH$	Nonadecylic acid	Nonadecanoic acid
$C_{19}H_{39}OH$	Nonadecyl alcohol	1-Nonadecanol
$C_{19}H_{39}COOH$	Arachidic acid	Eicosanoic acid
$C_{20}H_{41}OH$	Eicosyl alcohol, arachic alcohol	1-Eicosanol
$C_{21}H_{45}COOH$	Behenic acid	Docosanoic acid
$CH_3(CH_2)_7CH{=}CH(CH_2)_{11}COOH$	Erucic acid	*cis*-13-Docosenoic acid
$CH_3(CH_2)_7CH{=}CH(CH_2)_{11}COOH$	Brassidic acid	*trans*-13-Docosenoic acid
$C_{25}H_{51}COOH$	Cerotic acid	Hexacosanoic acid

Note: Notice that it is generally those acids containing an even number of carbon atoms that have special common names. This is because these are the naturally occurring ones in vegetable and animal fats.

depends not only on the chemical constitution but also on other variables, such as the film pressure π.

To resume the brief historical sketch, the subject of monolayers developed rapidly during the interwar years, with the names of Langmuir, Adam, Harkins, and Rideal perhaps the most prominent; the subject became one of precise and mature scientific study. The post-World War II period was one of even greater quantitative activity.

A belief that solid interfaces are easier to understand than liquid ones shifted emphasis to the former; but the subjects are not really separable, and the advances in the one are giving impetus to the other. There is increasing interest in films of biological and of liquid crystalline materials; because of the importance of thin films in microcircuitry (computer "chips"), there has been in recent years a surge of activity in the study of deposited mono- and multilayers. These "Langmuir-Blodgett" films are discussed in Section XV-7.

On the environmental side, it turns out that the surfaces of oceans and lakes are usually coated with natural films, mainly glycoproteins [8]. As they are biological in origin, the extent of such films seems to be seasonal. Pollutant slicks, especially from oil spills, are of increasing importance, and their cleanup can present interesting surface chemical problems.

A final comment on definitions of terms should be made. The terms *film* and *monomolecular film* have been employed somewhat interchangeably in the preceding discussion. Strictly speaking, a film is a layer of substance, spread over a surface, whose thickness is small enough that gravitational effects are negligible. A molecular film or, briefly, monolayer, is a film considered to be only one molecule thick. A *duplex film* is a film thick enough so that the two interfaces (e.g., liquid-film and film-air) are independent and possess their separate characteristic surface tensions. In addition, material placed at an interface may form a *lens*, that is, a thick layer of finite extent whose shape is constrained by the force of gravity. Combinations of these are possible; thus material placed on a water surface may spread to give a monolayer, the remaining excess collecting as a lens.

2. The Spreading of One Liquid on Another

Before proceeding to the main subject of this chapter—namely, the behavior and properties of spread films on liquid substrates—it is of interest to consider the somewhat wider topic of the spreading of a substance on a liquid surface. Certain general statements can be made as to whether spreading will occur, and the phenomenon itself is of some interest.

A. Criteria for Spreading

If a mass of some substance were placed on a liquid surface so that initially it is present in a layer of appreciable thickness, as illustrated in Fig. IV-1, then two possibilities exist as to what may happen. These are best treated in terms of what is called the spreading coefficient.

At constant temperature and pressure a small change in the surface free energy of the system shown in Fig. IV-1 is given by the total differential

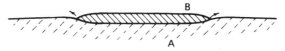

Fig. IV-1. The spreading of one liquid over another.

$$dG = \frac{\partial G}{\partial \mathcal{A}_A} d\mathcal{A}_A + \frac{\partial G}{\partial \mathcal{A}_{AB}} d\mathcal{A}_{AB} + \frac{\partial G}{\partial \mathcal{A}_B} d\mathcal{A}_B \qquad \text{(IV-1)}$$

but

$$d\mathcal{A}_B = -d\mathcal{A}_A = d\mathcal{A}_{AB}$$

where liquid A constitutes the substrate, and

$$\frac{\partial G}{\partial \mathcal{A}_A} = \gamma_A$$

and so on. The coefficient $-(dG/d\mathcal{A}_B)_{\text{area}}$ gives the free energy change for the spreading of a film of liquid B over liquid A and is called the *spreading coefficient* of B on A. Thus

$$S_{B/A} = \gamma_A - \gamma_B - \gamma_{AB} \qquad \text{(IV-2)}$$

$S_{B/A}$ is positive if spreading is accompanied by a decrease in free energy, that is, is spontaneous.

The process described by Eq. IV-2 is that depicted in Fig. IV-1, in which a thick or *duplex* film of liquid B spreads over liquid A. This typically happens when a liquid of low surface tension is placed on one of high surface tension. Some illustrative data are given in Table IV-2; it is seen, for example, that benzene and long-chain alcohols would be expected to spread on water, whereas CS_2 and CH_2I_2 should remain as a lens. As an extreme example, almost any liquid will spread to give a film on a mercury surface, as examination of the data in the table indicates. Conversely, a liquid of high surface tension would not be expected to spread on one of much lower surface tension; thus $S_{A/B}$ is negative in all of the cases given in Table IV-2d.

A complication now arises. The surface tensions of A and B in Eq. IV-2 are those for the pure liquids. However, when two substances are in contact, they will become mutually saturated, so that γ_A will change to $\gamma_{A(B)}$ and γ_B to $\gamma_{B(A)}$. That is, the convention will be used that a given phase is saturated with respect to that substance or phase whose symbol follows in parentheses. The corresponding spreading coefficient is then written $S_{B(A)/A(B)}$.

For the case of benzene on water,

TABLE IV-2a
Spreading Coefficients at 20°C of Liquids on Water (erg/cm^2)

Liquid B	$S_{B/A}$	Liquid B	$S_{B/A}$
Isoamyl alcohol	44.0	Nitrobenzene	3.8
n-Octyl alcohol	35.7	Hexane	3.4
Heptaldehyde	32.2	Heptane (30°C)	0.2
Oleic acid	24.6	Ethylene dibromide	−3.2
Ethyl nonanoate	20.9	o-Monobromotoluene	−3.3
p-Cymene	10.1	Carbon disulfide	−8.2
Benzene	8.8	Iodobenzene	−8.7
Toluene	6.8	Bromoform	−9.6
Isopentane	9.4	Methylene iodide	−26.5

TABLE IV-2b
Liquids on Mercury [9]

Liquid B	$S_{B/A}$	Liquid B	$S_{B/A}$
Ethyl iodide	135	Benzene	99
Oleic acid	122	Hexane	79
Carbon disulfide	108	Acetone	60
n-Octyl alcohol	102	Water	−3

TABLE IV-2c
Initial versus Final Spreading Coefficients on Water [9,10]

Liquid B	γ_B	$\gamma_{B(A)}$	$\gamma_{A(B)}$	γ_{AB}	$S_{B/A}$	$S_{B(A)/A(B)}$	$S_{A/B}$	$S_{A(B)/B(A)}$
Isoamyl alcohol	23.7	23.6	25.9	5	44	−2.7	−54	−1.3
Benzene	28.9	28.8	62.2	35	8.9	−1.6	−78.9	−68.4
CS$_2$	32.4	31.8		48.4	−7	−9.9	−89	
n-Heptyl alcohol	27.5			7.7	40	−5.9	−56	
CH$_2$I$_2$	50.7			41.5	−27	−24	−73	

$$S_{B/A} = 72.8 - (28.9 + 35.0) = 8.9 \qquad \text{(IV-3)}$$

$$S_{B(A)/A} = 72.8 - (28.8 + 35.0) = 9.0 \qquad \text{(IV-4)}$$

$$S_{B(A)/A(B)} = 62.2 - (28.8 + 35.0) = -1.6 \qquad \text{(IV-5)}$$

The final or equilibrium spreading coefficient is therefore negative; thus if ben-

TABLE IV-2d
Initial versus Final Spreading Coefficients on Mercury[a]

Liquid B	γ_B	$\gamma_{B(A)}$	$\gamma_{A(B)}$	γ_{AB}	$S_{B/A}$	$S_{B(A)/A(B)}$	$S_{A/B}$	$S_{A(B)/B(A)}$
Water	72.8	(72.8)	448	415	−3	−40	−817	−790
Benzene	28.8	(28.8)	393	357	99	7	−813	−721
n-Octane	21.8	(21.8)	400	378	85	0	−841	−756

[a]Data for equilibrium film pressures on mercury are from Ref. 11.

zene is added to a water surface, a rapid initial spreading occurs, and then, as mutual saturation takes place, the benzene retracts to a lens. The water surface left behind is not pure, however; its surface tension is 62.2, corresponding to the Gibbs monolayer for a saturated solution of benzene in water (or, also, corresponding to the film of benzene that is in equilibrium with saturated benzene vapor).

The situation illustrated by the case of benzene appears to be quite common for water substrates. Low-surface-tension liquids will have a positive initial spreading coefficient but a near-zero or negative final one; this comes about because the film pressure π of the Gibbs monolayer is large enough to reduce the surface tension of the water–air interface to a value below the sum of the other two. Thus the equilibrium situation in the case of organic liquids on water generally seems to be that of a monolayer with any excess liquid collected as a lens. The spreading coefficient $S_{B(A)/A}$ can be determined directly, and Zisman and co-workers report a number of such values [12].

B. Empirical and Theoretical Treatments

Various means have been developed for prediciting or calculating a γ_{AB} or a work of adhesion. Two empirical ones are the following. First, an early relationship is that known as *Antonow's* rule [13],

$$\gamma_{AB} = |\gamma_{A(B)} - \gamma_{B(A)}| \qquad (IV-6)$$

This rule is approximately obeyed by a large number of systems, although there are many exceptions; see Refs. 15–18. The rule can be understood in terms of a simple physical picture. There should be an adsorbed film of substance B on the surface of liquid A. If we regard this film to be thick enough to have the properties of bulk liquid B, then $\gamma_{A(B)}$ is effectively the interfacial tension of a *duplex* surface and should be equal to $\gamma_{AB} + \gamma_{B(A)}$. Equation IV-6 then follows. See also Refs. 14 and 18.

An empirical equation analogous to Eq. II-39 can be useful:

$$\gamma_{AB} = I + AM_A^{-2/3} + BM_B^{-2/3} \qquad (IV-7)$$

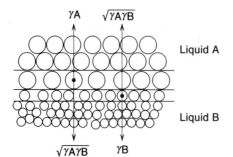

Fig. IV-2. The Good-Fowkes model for calculating interfacial tension. (From Ref. 20.)

where M denotes molecular weight. For liquid A an α, ω-diol and liquid B an n-alkane, the constants I, A, and B were -5.03, 408.2, and -82.38, respectively, with γ_{AB} in dynes per centimeter. The same equation applied for water–n-alkane interfacial tensions if M_A was taken to be that of water [19].

There has been considerable theoretical development in the treatment of interfacial tension and work of adhesion. The approach used is similar to that of Eq. III-44 but using average densities rather than the actual radial distribution functions—these are generally not available for systems such as those of Table IV-2. As illustrated in Fig. IV-2, γ_{AB} may be regarded as the sum of the work to bring molecules A and B to their respective liquid vapor interfaces less the free energy of interaction across the interface. This is determined through the work of adhesion, w_{AB}, between two phases given by

$$w_{AB} = \gamma_A + \gamma_B - \gamma_{AB} \qquad \text{(IV-8)}$$

$$S_{B/A} = \gamma_A - \gamma_B - \gamma_{AB} \qquad (IV\text{-}2)$$

which is the work necessary to separate one square centimeter of interface AB into two liquid–vapor interfaces A and B. Calculation of this work can be carried out if the potential energy function for A–B interactions is known. Girifalco and Good [10] assumed the geometric mean rule, $U_{AB}(r) = [U_A U_B]^{1/2}$, referring to Eq. III-44, to obtain

$$w_{AB} = 2\Phi(\gamma_A\gamma_B)^{1/2} \qquad \text{(IV-9)}$$

where Φ is a function of the molar volumes of the two liquids; empirically, its value ranges from 0.5 to 1.15 if calculated from experimental w_{AB} values in Eq. IV-9. Interestingly, Eq. IV-9 also follows if the Skapski type of approach is used, where all except nearest neighbor interactions are neglected (see Eq. III-15) [21].

The different kinds of intermolecular forces (dispersion, dipole–dipole, hydrogen bonding, etc.; see Section VI-1) may not equally contribute to A–A, B–B, and A–B

TABLE IV-3
Calculation of γ_W^d at 20°C [25] (ergs/cm^2)

Hydrocarbon	γ_H	γ_{WH} (ergs/cm^2)	γ_W^d	
n-Hexane	18.4	51.1	21.8	
n-Heptane	20.4	50.2	22.6	
n-Octane	21.8	50.8	22.0	
n-Decane	23.9	51.2	21.6	
nTetradecane	25.6	52.2	20.8	21.8 ± 0.7
Cyclohexane	25.5	50.2	22.7	
Decalin	29.9	51.4	22.0	
White oil	28.9	51.3	21.3	

interactions. Consider, for example, the water–hydrocarbon system. The potential function $U_W(r)$ for water contains important hydrogen bonding contributions that are absent in $U_{WH}(r)$, the function for water–hydrocarbon interactions. A more general form of Eq. IV-9 would use the surface tensions that liquids A and B would have if their intermolecular potentials contained only the same kinds of interactions as those involved between A and B (see Refs. 20, 22–24). For the hydrocarbon–water system, Fowkes [20] assumed that U_H arose solely from dispersion interactions leaving

$$w_{WH} = 2(\gamma_W^d \gamma_H)^{1/2} \qquad (IV-10)$$

where Φ has been approximated as unity and γ_W^d is the contribution to the surface tension of water from the dispersion effect only. A number of systems obey Eq. IV-10 very well, with $\gamma_W^d = 21.8$ ergs/cm^2 at 20° [25] as illustrated in Table IV-3. Variations from this value are attributed to anisotropic dispersion interactions in hydrocarbons and have been discussed by Fowkes [26].

When both liquids are polar, Fowkes [26] has written

$$w_{AB} = w_{AB}^d + w_{AB}^p + w_{AB}^h \qquad (IV-11)$$

where superscripts p and h denote dipolar and hydrogen bonding interaction, respectively. That is, different kinds of van der Waals forces (see Chapter VI) are assumed to act independently. This type of approach has also been taken by Tamai and co-workers [27] and Panzer [28]. Good and co-workers [29] have suggested dividing the polar interactions into electron-donor and electron-acceptor components and write

$$w_{AB}^p = 2(\sqrt{\gamma_A^+ \gamma_B^-} + \sqrt{\gamma_A^- \gamma_B^+}) \qquad (IV-12)$$

Finally, Newmann and co-workers [30] (see also Ref. 31) have argued that while free energy contributions may not be strictly additive as in Eq. IV-11, there should, in principle, be an equation of state relating the work of adhesion to the separate liquid surface tensions such as

$$w_{AB} = (2 - 0.015\gamma_{AB})\sqrt{\gamma_A\gamma_B} \tag{IV-13}$$

Equation IV-13 is also applicable to polar as well as nonpolar systems. (See also Section X-7B.)

These approaches (Eqs. IV-9–IV-13) have found much use in the estimation of the surface tension of a solid surface and are discussed further in Section X-7B, but some reservations might be noted here. The calculation of w_{AB} in Eq. IV-8 has been questioned [32–34]; a reversible separation process would recognize that surface density and composition changes occur as two materials are separated and the analogue of Eq. IV-9 would now have $\gamma_{A(B)}$ and $\gamma_{B(A)}$ in place of γ_A and γ_B. The Girifalco–Good equation (Eq. IV-9) and related expressions may work as well as they do as a consequence of some cancellation of errors in the derivations. The assumption that the hydrocarbon interaction in Eq. IV-10 is due to dispersion forces only has been questioned [21]. There is evidence that strong polar forces are involved in the interaction of the first hydrocarbon layer with water [32–37]. In the reversible process of separating the two phases, a thin film of hydrocarbon remains on the water surface (note Fig. III-17). Much of the work is done in separating the hydrocarbon from this thin film rather than from the water itself. The effective γ_W would be that of the film–air interface and should approximate that of a hydrocarbon–air interface, or about 20 erg/cm^2.

C. Kinetics of Spreading Processes

The spreading process itself has been the object of some study. It was noted very early [38] that the disturbance due to spreading was confined to the region immediately adjacent to the expanding perimeter of the spreading substance. Thus if talc or some other inert powder is sprinkled on a water surface and a drop of oil is added, spreading oil sweeps the talc back as an accumulating ridge at the periphery of the film, but the talc further away is entirely undisturbed. Thus the "driving force" for spreading is localized at the linear interface between the oil and the water and is probably best regarded as a steady bias in molecular agitations at the interface, giving rise to a fairly rapid net motion.

Spreading velocities v are on the order of 15–30 cm/sec on water [39], and v for a homologous series tends to vary linearly with the equilibrium film pressure, π^e, although in the case of alcohols a minimum π^e seemed to be required for v to be appreciable. Also, as illustrated in Fig. IV-3, substrate water is entrained to some depth (0.5 mm in the case of oleic acid), a compensating counterflow being present at greater depths [40]. Related to this is the observation that v tends to vary inversely with substrate viscosity [41–43]. An analysis of the stress–strain situation led to the equation

$$x = \left(\frac{4\pi^e}{3}\right)^{1/2}(\rho\eta)^{-1/4}t^{3/4} \tag{IV-14}$$

where x is the distance traveled by the spreading film in time t and ρ and η are the substrate density and viscosity, respectively [44]. For the spreading of a thin layer rather than a film, π^e is replaced by the spreading coefficient. Fractionation has been observed if the spreading liquid is a mixture [45].

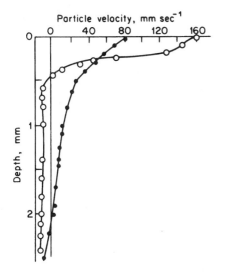

Fig. IV-3. Velocity profiles for particles suspended in water with elapsed time, due to spreading of oleic acid. Time after onset of spreading: $\bigcirc$, $\frac{1}{8}$ sec, $\bullet$, $\frac{1}{2}$ sec. (From Ref. 31.)

If the spreading is into a limited surface area, as in a laboratory experiment, the film front rather quickly reaches the boundaries of the trough. The film pressure at this stage is low, and the now essentially uniform film more slowly increases in π to the final equilibrium value. The rate of this second-stage process is mainly determined by the rate of release of material from the source, for example a crystal, and the surface concentration Γ [46]. Franses and co-workers [47] found that the rate of dissolution of hexadecanol particles sprinkled at the water surface controlled the increase in surface pressure; here the slight solubility of hexadecanol in the bulk plays a role.

The topic of spreading rates is of importance in the technology of the use of mono-layers for evaporation control (see Section IV-6); it is also important, in the opposite sense, in the lubrication of fine bearings, as in watches, where it is necessary that the small drop of oil remain in place and not be dissipated by spreading. Zisman and co-workers have found that spreading rates can be enhanced or reduced by the presence of small amounts of impurities; in particular, strongly adsorbed surfactants can form a film over which the oil will not spread [48].

D. The Marangoni Effect

The dependence of spreading rates on substrate viscosity, as in Eq. IV-14, indicates that films indeed interact strongly with the bulk liquid phase and cannot be regarded as merely consisting of molecules moving freely in a two-dimensional realm. This interaction complicates the interpretation of monolayer viscosities (Section IV-3C). It is also an aspect of what is known as the *Marangoni effect* [49], namely, the carrying of bulk material through motions energized by surface tension gradients.

A familiar (and biblical [50]) example is the formation of tears of wine in a glass. Here, the evaporation of the alcohol from the meniscus leads to a local raising of the surface tension, which, in turn, induces a surface and accompanying bulk flow upward, the accumulating liquid returning in the form of drops, or tears. A drop of oil on a surfactant solution may send out filamental streamers as it spreads [51].

Interesting pattern formations also occur in surfactants spreading on water due to a hydrodynamic instability [52]. The spreading velocity from a crystal may vary with direction, depending on the contour and crystal facet. There may be sufficient imbalance to cause the solid particle to move around rapidly, as does camphor when placed on a clean water surface. The many such effects have been reviewed by Sternling and Scriven [53].

The Marangoni effect has been observed on the rapid compression of a monolayer [54] and on application of an electric field, as in Ref. [55]; it occurs on evaporation [56].

The effect can be important in mass-transfer problems (see Ref. 57 and citations therein). The Marangoni instability is often associated with a temperature gradient characterized by the *Marangoni number* Ma:

$$\text{Ma} = -\frac{d\gamma}{dT}\frac{dT}{dz}h^2\frac{1}{\eta\kappa} \tag{IV-15}$$

This definition is in terms of a pool of liquid of depth h, where z is distance normal to the surface and η and κ are the liquid viscosity and thermal diffusivity, respectively [58]. (Thermal diffusivity is defined as the coefficient of thermal conductivity divided by density and by heat capacity per unit mass.) The critical Ma value for a system to show Marangoni instability is around 50–100.

E. Lenses–Line Tension

The equilibrium shape of a liquid lens floating on a liquid surface was considered by Langmuir [59], Miller [60], and Donahue and Bartell [61]. More general cases were treated by Princen and Mason [62] and the thermodynamics of a liquid lens has been treated by Rowlinson [63]. The profile of an oil lens floating on water is shown in Fig. IV-4. The three interfacial tensions may be represented by arrows forming a Newman triangle:

$$\gamma_{A(B)}\cos\gamma = \gamma_{B(A)}\cos\beta + \gamma_{AB}\cos\alpha \tag{IV-16}$$

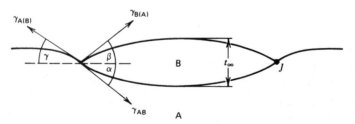

Fig. IV-4. Profile of a lens on water.

Donahue and Bartell verified Eq. IV-16 for several organic alcohol–water systems.

For very large lenses, a limiting thickness t_∞ is reached. Langmuir [59] gave the equation

$$t_\infty^2 = -\frac{2S\rho_A}{g\rho_B\Delta\rho} \tag{IV-17}$$

relating this thickness to the spreading coefficient and the liquid densities ρ_A and ρ_B.

The three phases of Fig. IV-4 meet at a line, point J in the figure; the line is circular in this case. There exists correspondingly a *line tension* λ, expressed as force or as energy per unit length. Line tension can be either positive or negative from a theoretical point of view; experimental estimates have ranged from -10^{-4} to $+10^{-4}$ dyn for various systems (see Ref. 64). A complication is that various authors have used different defining equations (see Ref. 65 and also Section X-5B). Neumann and co-workers have proposed a means to measure the line tension from the shape of the meniscus formed near a wall comprising vertical stripes of different wettability [66]. Kerins and Widom [67] applied three models including the van der Waals theory (Section III-2B) to the density profiles near the three-phase contact line and find both possitive and negative line tensions.

As a general comment, one can arrange systems according to a hierarchy of size or curvature. The capillary constant a (of Eq. II-10) is typically between 0.1 and 1 cm, and surface tension effects can be neglected for systems (meniscuses, lenses, etc.) much larger than this size but increasingly dominate as one goes down in size. The next scale is that given by the ratio λ/γ, whose value is not apt to be larger than about 2×10^{-6} cm. Thus line tension should become important only for systems approaching this size; this could be the case, for example, with soap films and microemulsions (Chapter XIV). Finally curvature effects on γ (or on λ) (e.g., Eq. III-20) become important only for curvatures of molecular dimension.

3. Experimental Techniques for the Study of Monomolecular Films

There has been tremendous activity in the study of spread films with many different experimental techniques. Here we limit our focus to the more basic methods giving classic information regarding film pressure, electrical potential, viscosity (through viscometry and dynamic light scattering), thickness, and molecular arrangement. We restrict our attention to methods used to study spread films; transferred or Langmuir–Blodgett film techniques are covered in Chapter XV.

A. Measurement of π

The *film pressure* is defined as the difference between the surface tension of the pure fluid and that of the film-covered surface. While any method of surface tension measurement can be used, most of the methods of capillarity are, for one reason or another, ill-suited for work with film-covered surfaces with the principal exceptions of the Wilhelmy slide method (Section II-6) and the pendant drop experiment (Section II-7). Both approaches work very well with fluid films and are capable of measuring low values of pressure with similar precision of 0.01 dyn/cm. In addition, the film balance, considerably updated since Langmuir's design (see Section III-7) is a popular approach to measurement of π.

In the Wilhelmy slide method, it seems best to partly immerse the slide and determine the change in height with constant upward pull or the change in pull at constant position of the slide. If the latter procedure is used, then

$$\pi = \Delta w/p$$

where Δw is the change in pull, for example, as measured by means of a balance (see Fig. II-13) and p is the perimeter of the slide. Thin glass, mica, platinum, and filter paper [68] have been used as slide materials.

This method suffers from two disadvantages. Since it measures γ or changes in γ rather than π directly, temperature drifts or adventitious impurities can alter γ and be mistakenly attributed to changes in film pressure. Second, while ensuring that zero contact angle is seldom a problem in the case of pure liquids, it may be with film-covered surfaces as film material may adsorb on the slide. This problem can be a serious one; roughening the plate may help, and some of the literature on techniques is summarized by Gaines [69]. On the other hand, the equipment for the Wilhelmy slide method is simple and inexpensive and can be just as accurate as the film balance described below.

Neumann has adapted the pendant drop experiment (see Section II-7) to measure the surface pressure of insoluble monolayers [70]. By varying the droplet volume with a motor-driven syringe, they measure the surface pressure as a function of area in both expansion and compression. In tests with octadecanol monolayers, they found excellent agreement between axisymmetric drop shape analysis and a conventional film balance. Unlike the Wilhelmy plate and film balance, the pendant drop experiment can be readily adapted to studies in a pressure cell [70]. In studies of the rate dependence of the molecular area at collapse, Neumann and co-workers found more consistent and reproducible results with the actual area at collapse rather than that determined by conventional extrapolation to zero surface pressure [71]. The collapse pressure and shape of the pressure–area isotherm change with the compression rate [72].

Film pressure is often measured directly by means of a *film balance*. The principle of the method involves the direct measurement of the horizontal force on a float separating the film from clean solvent surface. The film balance has been considerably refined since the crude model used by Langmuir and in many laboratories has been made into a precision instrument capable of measuring film pressures with an accuracy of hundredths of a dyne per centimeter. Various methods of measuring film pressure have been described by Gaines [1, 69], Costin and Barnes [73], Abraham et al. [74], and Mysels [75]; a perspective drawing of a film balance is shown in Fig. IV-5. Kim and Cannell [76] and Munger and Leblanc [77] describe film balances for measuring very small film pressures.

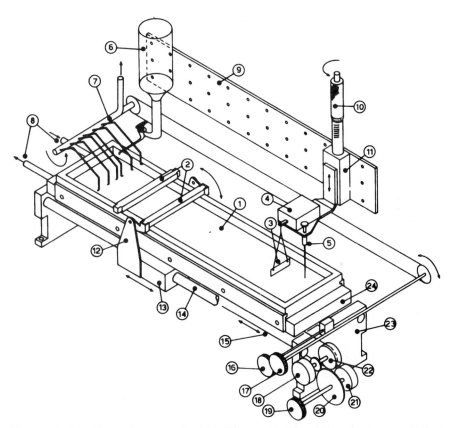

Fig. IV-5. A modern film balance: (1) PTFE trough; (2) barriers of reinforced PTFE; (3) Wilhelmy plate; (4) force transducer; (5) pointer for water level; (6) reservoir for aqueous subphase; (7) suction tubes for cleaning the surface; (8) inlet and outlet in trough base for thermostatted water; (9) accessory rack; (10) micrometer; (11) vertical slide for Wilhelmy plate assembly; (12) barrier drive and lifting assembly; (13) linear motion assembly for barriers; (14) shafts for 13; (15) cable for barrier movement; (16) control knob for barrier lift mechanism; (17) knob for raising and lowering suction tubes; (18) optical shaft encoder for barrier position; (19) knob for manual barrier drive; (20) clutch and gear assembly for motorized barrier drive; (21) barrier drive motor; (22) grooved drum for barrier drive cable; (23) support plate; (24) trough base. (Courtesy of G. T. Barnes.)

The material of interest is dissolved in a volatile solvent, spread on the surface and allowed to evaporate. As the sweep moves across, compressing the surface, the pressure is measured providing π versus the area per molecule, σ. Care must be taken to ensure complete evaporation [1] and the film structure may depend on the nature of the spreading solvent [78]. When the trough area is used to calculate σ, one must account for the area due to the meniscus [79]. Barnes and Sharp [80] have introduced a remotely operated barrier drive mechanism for cleaning the water surface while maintaining a closed environment.

The limiting compression (or maximum π value) is, theoretically, the one that places the film in equilibrium with the bulk material. Compression beyond this point should force film material into patches of bulk solid or liquid, but in practice one may sometimes compress past this point. Thus in the case of stearic acid, with slow compression collapse occurred at about 15 dyn/cm [81]; that is, film material began to go over to a three-dimensional state. With faster rates of compression, the π–σ isotherm could be followed up to 50 dyn/cm, or well into a metastable region. The mechanism of collapse may involve folding of the film into a bilayer (note Fig. IV-18).

Some recommendations on reporting film balance π–σ data have been made to the International Union of Pure and Applied Chemistry [82].

B. Surface Potentials

A second type of measurement that may be made on films, usually in conjunction with force–area measurements is that of the contact or surface potential. One essentially measures the Volta potential between the surface of the liquid and that of a metal probe.

There are two procedures for doing this. The first makes use of a metal probe coated with an α emitter such as polonium or [241]Am (around 1 mCi) and placed above the surface. The resulting air ionization makes the gap between the probe and the liquid sufficiently conducting that the potential difference can be measured by means of a high-impedance dc voltmeter that serves as a null indicator in a standard potentiometer circuit. A submerged reference electrode may be a silver–silver chloride electrode. One generally compares the potential of the film-covered surface with that of the film-free one [83, 84].

A popular alternative method employs a vibrating electrode illustrated in Fig. IV-6 [1,85]. Here an audiofrequency current drives a Rochelle salt or loudspeakder magnet and the vibrations are transmitted mechanically to a small electrode mounted parallel to and about 0.5 mm above the surface. The electrode vibrations cause a corresponding variation in the capacity across the airgap so that an alternating current is set up in the second circuit, whose magnitude depends on the voltage difference across the gap. The potentiometer is adjusted to minimize the current. This method is capable of measuring potentials to about 0.1 mV and is somewhat more precise than the α emitter, although it is more susceptible to malfunctionings.

One ordinarily attributes the difference in surface potentials between that of the substrate and that for the film-coated surface to the film. Two conducting

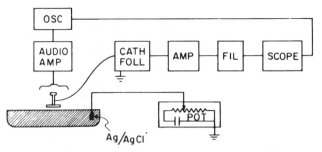

Fig. IV-6. Vibrating electrode method for measuring surface potentials. (From Ref. 1.)

plates separated by a distance d and enclosing a charge density σ will have a potential difference ΔV given by a formula attributed to Helmholtz:

$$\Delta V = \frac{4\pi\sigma d}{\epsilon} \qquad (IV\text{-}18)$$

where ϵ is the dielectric constant. Actually, one supposes that charge separation due to the presence of an effective dipole moment $\bar{\mu}$ simulates a parallel-plate condenser. If there are $\mathbf{n}/cm^2$ (perhaps corresponding to Γ polar molecules per square centimeter of film), then $\sigma d = \mathbf{n}ed = \mathbf{n}\bar{\mu}$ and Eq. IV-18† becomes

$$\Delta V = \frac{4\pi\bar{\mu}\mathbf{n}}{\epsilon} = 4\pi\mu \cos\theta \qquad (IV\text{-}19)$$

Customarily, it is assumed that ϵ is unity and that $\bar{\mu} = \mu\cos\theta$, where θ is the angle of inclination of the dipoles to the normal. Harkins and Fischer [86] point out the empirical nature of this interpretation and prefer to consider only that ΔV is proportional to the surface concentration Γ and that the proportionality constant is some quantity characteristic of the film. This was properly cautious as there are many indications that the surface of water is structured and that the structure is altered by the film (see Ref. 37). Accompanying any such structural rearrangement of the substrate at the surface should be a change in its contribution to the surface potential so that ΔV should not be assigned too literally to the film molecules.

While there is some question about the interpretation of absolute ΔV values, such measurements are very useful as an alternative means of determining the concentration of molecules in a film (as in following rates of reaction or

†Equations IV-18 and IV-19 are for the cgs/esu system of units; σ is in esu/cm^2 and ΔV is in volts esu (1 V_{esu} = 300 V). In the SI system, the equations become $\Delta V = \sigma d/\epsilon_0\epsilon = \mathbf{n}\mu\cos\theta/\epsilon_0\epsilon$, where $\epsilon_0 = 1 \times 10^7/4\pi c^2 = 8.85 \times 10^{-12}$. Charge density is now in Coulombs per square meter and ΔV in volts. (See Section V-3.)

dissolution) and in ascertaining whether a film is homogeneous. Fluctuations in ΔV with position across the film may occur if there are two phases present. McConnell and Benvegnu find that the variation in ΔV over a two phase film correlates well with the surface fraction of the two phases [87], but the dipole density difference determined from this measurement is below that found from Brownian motion of a trapped domain [88,89] and electrophoretic mobility of the domains [90].

There are some theoretical complications discussed in Refs. 91 and 92. Experimental complications include adsorption of solvent or of film on the electrode [93,94]; the effect may be used to detect atmospheric contaminants. The atmosphere around the electrode may be flushed with dry nitrogen to avoid condensation problems [87].

C. Measurement of Surface Viscosity

The subject of surface viscosity is a somewhat complicated one; it has been reviewed by several groups [95,96], and here we restrict our discussion to its measurement via surface shear and scattering from capillary waves.

The shear viscosity is an important property of a Newtonian fluid, defined in terms of the force required to shear or produce relative motion between parallel planes [97]. An analogous two-dimensional *surface shear viscosity* η^s is defined as follows. If two line elements in a surface (corresponding to two area elements in three dimensions) are to be moved relative to each other with a velocity gradient dv/dx, the required force is

$$f = \eta^s l \frac{dv}{dx} \qquad (IV\text{-}20)$$

where l is the length of the element.

The surface viscosity can be measured in a manner entirely analogous to the Poiseuille method for liquids, by determining the rate of flow of a film through a narrow canal under a two-dimensional pressure difference $\Delta\gamma$. The apparatus is illustrated schematically in Fig. IV-7, and the corresponding equation for calculating η^s is analogous to the Poiseuille equation [99,100]

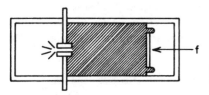

Fig. IV-7. Canal-type viscometer. (From Ref. 1; see also Ref. 98.)

$$\eta^s = \frac{\Delta\gamma a^3}{12l(dA/dt)} - \frac{a\eta}{\pi} \qquad \text{(IV-21)}$$

where a is the width of the canal, l is its length, dA/dt is the areal flow rate and η is the subphase viscosity. Several applications of this method [94–96] show film viscosities on the order of 10^{-2} to 10^{-4} g/sec or surface poises (sp). A recent study of the shear viscosity of lipid monolayers by Gaub and McConnell [104] shows that while solid phases have large viscosities, the fluid phase has a viscosity below that of the pure water subphase, implying a disruption to the surface structure of water. As the pressure is further lowered and the fluid phase becomes discontinuous the viscosity increases back to that of water.

While the canal viscometer provides absolute viscosities and the effect of the substrate drag can be analyzed theoretically, the shear rate is not constant and the measurement cannot be made at a single film pressure as a gradient is required. Another basic method, more advantageous in these respects, is one that goes back to Plateau [105]. This involves the determination of the damping of the oscillations of a torsion pendulum, disk, or ring such as illustrated in Fig. IV-8. Gaines [1] gives the equation

$$\eta^s = \left(\frac{\tau I}{r\pi^2}\right)^{1/2}\left(\frac{1}{a^2} - \frac{1}{b^2}\right)\left(\frac{\lambda}{4\pi^2 + \lambda^2} - \frac{\lambda_0}{4\pi^2 + \lambda_0^2}\right) \qquad \text{(IV-22)}$$

where a is the radius of the disk or ring; b is the radius of the (circular) film-covered area; λ and λ_0 are the natural logarithms of the ratio of successive amplitudes in the presence and absence of the film, respectively; and I is the moment of inertia of the pendulum. The torsion constant is given by

$$\tau = \frac{4\pi^2 I}{P_a^2} \qquad \text{(IV-23)}$$

where P_a is the period of the pendulum in air. Tschoegl [106] has made a detailed analysis of torsion pendulum methods, including treatment of substrate drag. Some novel variants have been described by Krieg et al. [107] and Abraham and Ketterson [108]. A new modification of the torsion pendulum employs a knife-edge ring hanging from

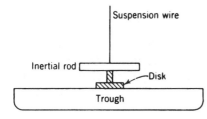

Fig. IV-8. Torsion pendulum surface viscometer. (From Ref. 1.)

a torsion wire [109]. The damped oscillations of the pendulum are detected optically to provide simultaneous measurement of the interfacial viscosity and elastic modulus.

In the *viscous traction viscometer* the film is spread in a circular annular canal formed by concentric cylinders; either the canal walls or floor may be made to rotate at a constant velocity or in an oscillatory manner, allowing determination of various viscoelastic coefficients [110]. In an automated variation of this method, Gaub and McConnell developed a rotating-disk viscometer that is driven by an external magnetic field allowing easy and sensitive measurement of the torque and hence the viscosity [104]. The problem of substrate drag is greatly reduced in the case of a rotating ring making a knife-edge contact with the interface [111,112].

Theoretical models of the film viscosity lead to values about 10^6 times smaller than those often observed [113, 114]. It may be that the experimental phenomenology is not that supposed in derivations such as those of Eqs. IV-20 and IV-22. Alternatively, it may be that virtually all of the measured surface viscosity is developed in the substrate through its interactions with the film (note Fig. IV-3). Recent hydrodynamic calculations of shape transitions in lipid domains by Stone and McConnell indicate that the transition rate depends only on the subphase viscosity [115]. Brownian motion of lipid monolayer domains also follow a fluid mechanical model wherein the mobility is independent of film viscosity but depends on the viscosity of the subphase [116]. This contrasts with the supposition that there is little coupling between the monolayer and the subphase [117]; complete explanation of the film viscosity remains unresolved.

Another important property is the *surface dilational viscosity*, κ

$$\Delta\gamma = \kappa \, \frac{1}{A} \, \frac{dA}{dt} \tag{IV-24}$$

where $1/\kappa$, relates the fractional change in area per unit time to the applied surface pressure. The equilibrium quantity corresponding to κ is the *modulus of surface elasticity E* defined by Eq. III-118; the film compressibility, K is just $1/E$.

As a very direct method for measuring κ, the surface is extended by means of two barriers that move apart at a velocity such that $d \ln \mathcal{A} /dt$ is constant. The dilation or depletion of the film results in a higher surface tension, measured by means of a Wilhelmy slide positioned at the center between the two barriers, where no liquid motion occurs. The procedure was applied to surfactant solutions [118]. An alternative approach is that of generating longitudinal waves by means of an oscillating barrier and observing the amplitude and phase lag of the motion of a small test particle [119,120] or of the film pressure [121]. On analysis, the data yield both E and the *sum* of η^s and κ. One may also measure the rate of change of surface dipole orientation, as obtained from the change in contact potential, following a change in surface area [122].

While bulk viscosity is a relatively obscure property of a liquid or a solid because its effect is not usually of importance, κ and E are probably the most important rheological properties of interfaces and films. They are discussed further in connection with foams (Section XIV-8), and their importance lies in the fact that interfaces are more often subjected to dilational than to shear strains. Also, κ is often numerically larger than η^s; thus for a stearic acid monolayer at 34 Å^2/molecule area, κ was found to be about 0.3 g/sec, and the ratio κ/η^s was about 300 [120].

Another approach to measurement of surface tension, density, and viscosity is the analysis of capillary waves or ripples whose properties are governed by surface tension rather than gravity. Space limitations prevent more than a summary presentation here; readers are referred to several articles [123,124].

The mathematical theory is rather complex because it involves subjecting the basic equations of motion to the special boundary conditions of a surface that may possess viscoelasticity. An element of fluid can generally be held to satisfy two kinds of conservation equations. First, by conservation of mass,

$$\frac{\partial u}{\partial x} + \frac{\partial v}{\partial y} = 0 \tag{IV-25}$$

where x and y denote the horizontal and vertical coordinates for an element of fluid, as illustrated in Fig. IV-9, and u and v are their time derivatives, that is, the velocities of the element. Equation IV-25 can be derived by considering a unit cube and requiring that the sum of the net flows in the x and y directions be zero. The second conservation relationship is the force balance in the Navier–Stokes equations:

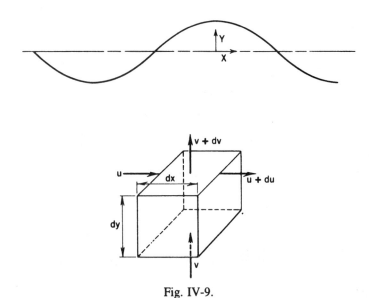

Fig. IV-9.

$$\rho \frac{\partial u}{\partial t} + \rho u \frac{\partial u}{\partial x} + \rho v \frac{\partial u}{\partial y} = -\frac{\partial P}{\partial x} + \eta \, \Delta u$$

and

$$\rho \frac{\partial v}{\partial t} + \rho u \frac{\partial v}{\partial x} + \rho v \frac{\partial v}{\partial y} = -\frac{\partial P}{\partial y} + \eta \, \Delta v - \rho g \qquad \text{(IV-26)}$$

$$(1) \qquad (2) \qquad (3) \qquad (4) \quad (5) \quad (6)$$

where ρ denotes the density. The three terms on the left are inertial terms, that is, force $= d(mv)/dt = m \, dv/dt + v \, dm/dt$, where mass is changing; term 1 corresponds to the $m \, dv/dt$ component and terms 2 and 3 to the $v \, dm/dt$ component. Term 4 gives the balancing force component due to any pressure gradient; term 5 takes account of viscous friction; and term 6 gives the force due to gravity. The boundary conditions are that at the surface: the vertical pressure component is that of the gas phase plus the Laplace pressure (Eq. II-7 or γ/r for a plane wave), while the horizontal component is given by the gradient of surface tension with area $\partial \gamma / \partial \mathcal{A}$, which is taken to be zero for a pure liquid but involves the surface elasticity of a film-covered surface and any related time-dependent aspects, whereby phase lags may enter. It is customary to assume that no slippage or viscosity anomaly occurs between the surface layer and the substrate.

Capillary waves may be generated mechanically by means of an oscillating bar, and for this case one writes the solutions to Eqs. IV-25 and IV-26 in the form

$$u = U_1 \exp(iwt) + U_2 \exp(2iwt) \cdots$$
$$v = V_1 \exp(iwt) + V_2 \exp(2iwt) \cdots \qquad \text{(IV-27)}$$

These expressions are inserted in the conservation equations, and the boundary conditions provide a set of relationships defining the U and V coefficients [125–129].

The detailed mathematical developments are difficult to penetrate, and a simple but useful approach is that outlined by Garrett and Zisman [130]. If gravity is not important, the first of Eqs. II-38 reduces to

$$v^2 = \frac{2\pi\gamma}{\rho\lambda} \qquad \text{(IV-28)}$$

This is an approximation to the complete dispersion equation [131]. The amplitude of a train of waves originating from an infinitely long linear source decays exponentially with the distance x from the source

$$A = A_0 e^{-\Gamma x} \tag{IV-29}$$

and a relationship due to Goodrich [126] gives

$$\Gamma = \frac{8\pi\eta\,\omega}{3\gamma} \tag{IV-30}$$

where ω is the wave frequency.

Figure IV-10 illustrates how Γ may vary with film pressure in a very complicated way although the π–σ plots are relatively unstructured. The results correlated more with variations in film elasticity than with its viscosity and were explained qualitatively in terms of successive film structures with varying degrees of hydrogen bonding to the water substrate and varying degrees of structural regularity. Note the sensitivity of k to frequency; a detailed study of the dispersion of k should give information about the characteristic relaxation times of various film structures.

The experimental procedure used by Hansen and co-workers involved the use of a loudspeaker magnet to drive a rod touching the surface (in an up-and-down motion) and

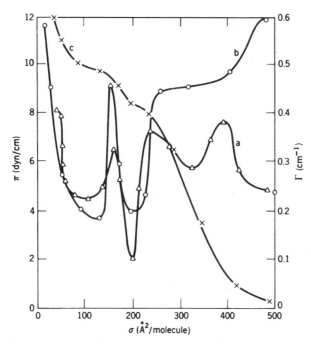

Fig. IV-10. Wave-damping behavior of polydimethylsiloxane heptadecamer on water at 25°C at (a) 60 cps and (b) 150 cps. Curve (c) gives the π–σ behavior. (From Ref. 130.)

a detector whose sensitive element was a phonograph crystal cartridge. Zisman and co-workers used an electromechanical transducer to drive a linear knife-edge so that linear waves were produced [130]. The standing-wave pattern was determined visually by means of stroboscopic illumination.

It is not necessary to generate capillary waves mechanically since small wavetrains are always present at a liquid interface due to thermal agitation. Such *thermal waves*, or ripplons, are typically on the order of 5–10 Å high and several hundred per centimeter in q where $q = 2\pi/\lambda$. A wavetrain will act as a grating scattering laser light, and while the effect is small, it can be isolated by beating against light diffracted by a reference grating. The experimental arrangement is illustrated in Fig. IV-11; the *wavevector k* is determined by the wavelength of light used and the offset angle $\delta\theta$. Thus for 5400-Å light, a $\delta\theta$ of 10′ will provide a wavevector of 360 cm^{-1}.

The scattering techniques, *dynamic light scattering* or *photon correlation spectroscopy* involve measurement of the fluctuations in light intensity due to density fluctuations in the sample, in this case from the capillary wave motion. The light scattered from thermal capillary waves contains two observables. The Doppler-shifted peak propagates at a rate such that its frequency follows Eq. IV-28 and

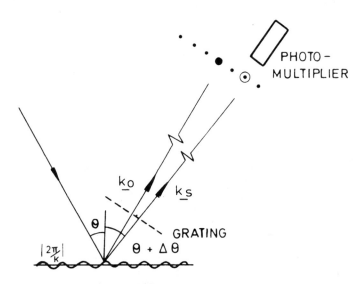

Fig. IV-11. A laser beam incident on the liquid surface at angle θ is scattered by angle $\Delta\theta$ by surface thermal waves of wave vector k. (From Ref. 132.)

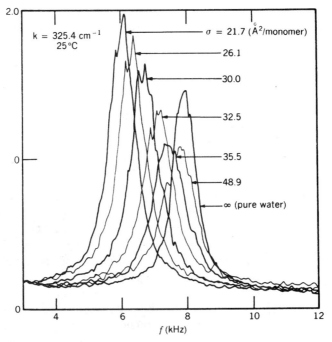

Fig. IV-12. Power spectra for monolayers of the polymer

$$(CH_2 - CH - CH - CH - CH),$$
$$\qquad\quad | \qquad | \qquad\quad |$$
$$\qquad\quad R \quad\ COOH \quad COOH$$

R = $(CH_2)_{15}CH_3$, for various surface coverages at pH 2 and 25°C; average molecular weight was 3960 g/mol. (See also Ref. 133. Courtesy of H. Yu.)

$$f_s = \left(\frac{\gamma}{\rho} \right)^{1/2} k^{3/2} \qquad\qquad\qquad \text{(IV-31)}$$

depends on the square root of the surface tension divided by the density, ρ. The peak broadens as a result of viscous damping. Thus its width

$$\Delta f = \left(\frac{2\eta}{\rho\pi} \right) k^2 \qquad\qquad\qquad \text{(IV-32)}$$

is proportional to the fluid viscosity, η. This shift and broadening in the power spectrum are illustrated in Fig. IV-12 for polymer monolayers at different densities. Equivalent information may be obtained from the autocorrelation function of the scattered light (see Ref. 134). Several papers cite theoretical work

[124,135,136] and show experimental applications [137]. Photon correlation spectroscopy has been used quite successfully to give the surface tensions, densities, and viscosities of various pure liquids (Refs. 138, 139, e.g.). These quantities are easily measured by alternative means, of course, and of much greater interest is the matter of obtaining surface viscoelastic properties of film-covered surfaces. Here, the situation is more difficult, and approximations are made in the treatment (e.g., Ref. 133). Mann and Edwards [140] make some cautionary remarks on this matter, as does Earnshaw [141].

D. Optical Properties of Monolayers

The detailed examination of the behavior of light passing through or reflected by an interface can, in principle, allow the determination of the monolayer thickness, its index of refraction and absorption coefficient as a function of wavelength. The subjects of ellipsometry, spectroscopy, and x-ray reflection deal with this goal; we sketch these techniques here.

In ellipsometry monochromatic light such as from a He–Ne laser, is passed through a polarizer, rotated by passing through a compensator before it impinges on the interface to be studied [142]. The reflected beam will be elliptically polarized and is measured by a polarization analyzer. In null ellipsometry, the polarizer, compensator, and analyzer are rotated to produce maximum extinction. The phase shift between the parallel and perpendicular components Δ and the ratio of the amplitudes of these components, $\tan \psi$, are related to the polarizer and analyzer angles p and a, respectively. The changes in Δ and ψ when a film is present can be related in an implicit form to the complex index of refraction and thickness of the film.

In the case of Langmuir monolayers, film thickness and index of refraction have not been given much attention. While several groups have measured Δ versus σ, [143–145], calculations by Knoll and co-workers [146] call into question the ability of ellipsometry to unambiguously determine thickness and refractive index of a Langmuir monolayer. A small error in the chosen index of refraction produces a large error in thickness. A new microscopic imaging technique described in section IV-3E uses ellipsometric contrast but does not require absolute determination of thickness and refractive index. Ellipsometry is routinely used to successfully characterize thin films on solid supports as described in Sections X-7, XI-2, and XV-7.

Interferometry is based on the fact that light reflected from the front and back interfaces of a film travels different distances, producing interference effects. The method has been applied to Langmuir–Blodgett films (Section XV-7) and to soap films (Section XIV-8) [147–149].

Absorption spectroscopy provides a means to study particular details about a monolayer. Transmission spectroscopy is difficult because the film, which is thin, absorbs little. Gaines [1] describes multiple-pass procedures for overcoming this problem. Reflection spectroscopy in the UV–visible range has been reported for lipid monolayers [150,151] and in the IR range for oleic acid [152].

The external reflection of infrared radiation can be used to characterize the thickness and orientation of adsorbates on metal surfaces. Buontempo and Rice [153–155] have recently extended this technique to molecules at dielectric surfaces, including Langmuir monolayers at the air–water interface. Analysis of the dichroic ratio, the ratio of reflectivity parallel to the plane of incidence (*p*-polarization) to that perpendicular to it (*s*-polarization) allows evaluation of the molecular orientation in terms of a tilt angle and rotation around the backbone [153]. An example of the *p*-polarized reflection spectrum for stearyl alcohol is shown in Fig. IV-13. Unfortunately, quantitative analysis of the experimental measurements of the antisymmetric CH_2 stretch for heneicosanol [153,155] stearly alcohol [154] and tetracosanoic [156] monolayers is made difficult by the scatter in the IR peak heights.

Resonance Raman reflection spectroscopy of monolayers is possible, as illustrated in Fig. IV-14 for cetyl orange [157]. The polarized spectra obtained with an Ar ion laser allowed estimates of orientational changes in the cetyl orange molecules with σ.

Photoexcited fluorescence from spread monolayers may be studied [158,159] if the substance has both a strong absorption band and a high emission yield as in the case for chlorophyll [159]. Gaines and co-workers [160] have reported on the emission from monolayers of $Ru(bipyridine)_3^{2+}$, one of the pyridine ligands having attached C_{18} aliphatic chains. Fluorescence depolarization provides information about the restriction of rotational diffusion of molecules in a monolayer [161]. Combining pressure–area

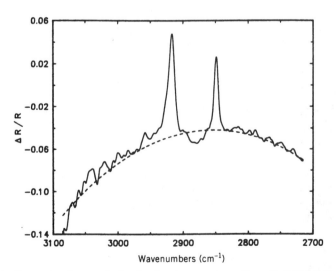

Fig. IV-13. Example of a *p*-polarized reflection spectrum from Ref. [154] for a stearyl alcohol monolayer on water. The dashed line is the baseline to be subtracted from the spectra. [Reprinted with permission from Joseph T. Buontempo and Stuart A. Rice, *J. Chem. Phys.* **98**(7), 5835-5846 (April 1, 1993). Copyright 1993, American Institute of Physics.]

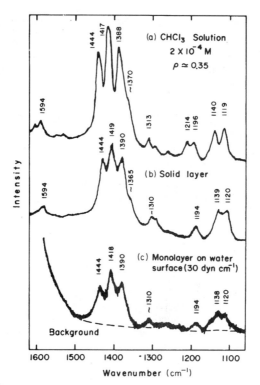

Fig. IV-14. Resonance Raman Spectra for cetyl orange using 457.9-nm excitation. [From T. Takenaka and H. Fukuzaki, "Resonance Raman Spectra of Insoluble Monolayers Spread on a Water Surface," *J. Raman Spectr.*, **8,** 151 (1979) (Ref. 157). Copyright Heyden and Son, Ltd., 1979; reprinted by permission of John Wiley and Sons, Ltd.]

measurements with fluorescence emission spectroscopy shows the onset of aggregation and multilayer formation as the subphase pH is reduced under acidic spectroscopic lipids [162].

The attachment of pyrene or another fluorescent marker to a phospholipid or its addition to an insoluble monolayer facilitates their study via fluorescence spectroscopy [163]. Pyrene is often chosen due to its high quantum yield and spectroscopic sensitivity to the polarity of the local environment. In addition, one of several amphiphilic quenching molecules allows measurement of the pyrene lateral diffusion in the monolayer via the change in the fluorescence decay due to the bimolecular quenching reaction [164,165].

E. Microscopic Evaluation of Monolayers

The ability to image lateral heterogeneity in Langmuir monolayers dates back to Zocher and Stiebel's 1930 study with divergent light illumination [166]. More recently the focus shifted toward the use of fluorescence microscopy of monolayers containing a small amount of fluorescent dye [167]. Even in single-com-

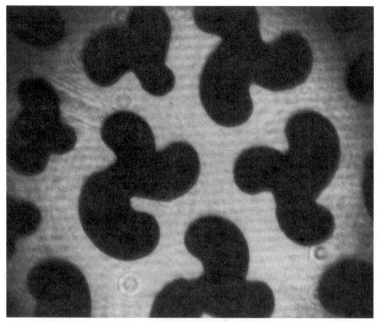

Fig. IV-15. A fluorescence micrograph showing the chiral solid domains formed in a mixture of the two enantiomers of dipalmitoylphosphatidylcholine (DPPC) at a pressure of 9 dyn/cm and average molecular area of 70 Å^2. (From Ref. 169.)

ponent monolayers, the dye generally has a different solubility or fluorescence quantum yield in coexisting phases, providing ample contrast to image phase domains as shown in Fig. IV-15. Studies of this kind have elucidated the nature of the phase transitions [168], domain shapes [169], and domain organization into periodic arrays [170]. A concern in these studies is the influence of the dye on the phase behavior. Confirming studies with electron microscopy on transferred films and ellipsometric and Brewster angle microscopy (described below) prove that the observations are not affected by the small amounts of dye used. The dye, however, does influence the area fraction of the dense phase near the melting line because of its enrichment in the coexisting fluid phase due to its virtual exclusion from the dense phase.

A newer and perhaps more useful application of ellipsometry to Langmuir films is their lateral characterization via ellipsometric microscopy [146]. A simple modification of the null ellipsometer allows one to image features down to 10-μm resolution. Working with a fixed polarizer and analyzer, some domains are at extinction while others are not and appear bright. This approach requires no fluorescent label and can be applied to systems on reflective supports.

Brewster angle microscopy takes advantage of the reflectivity behavior of light at an interface. This method relies on the fact that light passing from a material of lower refractive index, n_0 into a medium of higher index n_1 will have zero p-polarized reflected light at the Brewster angle defined by $\tan \theta_B = n_1/n_0$. The reflectivity around the Brewster angle

$$\frac{I_p(\theta)}{I_0(\theta)} = R_p = |r_p|^2 = \left[\frac{\tan(\theta_0 - \theta_1)}{\tan(\theta_0 + \theta_1)} \right]^2 \qquad \text{(IV-33)}$$

depends on the angle of incidence, θ_0 and the angle of refraction, θ_1, related through Snell's law ($n_0 \sin \theta_0 = n_1 \sin \theta_1$). Adding a thin film to the interface creates multiple reflections at the air–film (0–1) and film–water (1–2) interfaces, producing a reflectivity

$$R = \frac{r_{01} + r_{12}e^{-2i\beta}}{1 + r_{01}r_{02}e^{-2i\beta}} \qquad \text{(IV-34)}$$

that depends on the film index, n_1 and thickness, d through $\beta = (2\pi \, dn_1/\lambda) \cos \theta_1$. Thus, operating a Brewster angle microscope at the Brewster angle for a clean air–water interface allows one to image monolayers; the brightness of the reflected light will increase with n_1 and d [171]. In addition, this method yields information on the tilt and bond orientational order in monolayer domains through analysis of the optical anisotropy observed in the depolarization of the reflected light [172,173].

F. Diffraction Studies

Grazing incidence x-ray diffraction has been used to characterize a variety of monolayers [174–177]. In this experiment, synchrotron radiation (wavelength around 1.4–2 Å) is directed at a Langmuir trough at a small angle below the critical angle for total external reflection; typical angles are several milliradians [176]. The resulting diffraction peaks due to the molecular packing in the monolayer can be used to determine the degree of tilt and the intermolecular distances. Lever-rule analysis provides a measure of the coexistence between ordered (and tilted) domains and the disordered phase [178]. Recent studies of fluorinated fatty acids show a smaller degree of tilt than in the hydrogenated analogs and a first-order transition between the ordered condensed phase and disordered dilute phase [176].

Electron diffraction studies are usually limited to transferred films (see Chapter XV). One study on Langmuir films of fatty acids has used cryoelectron microscopy to fix the structures on vitrified water [179]. Electron diffraction from these layers showed highly twinned structures in the form of faceted crystals.

The ability to contrast match the air with a mixture of water and heavy water makes neutron reflectivity an attractive technique [180,181]. Under these contrast conditions the scattering arises from the monolayer alone and combining specular and off-specular measurements provides information about surface texture in collapsed monolayers [181].

4. States of Monomolecular Films

There has been much activity in the study of monolayer phases via the new optical, microscopic, and diffraction techniques described in the previous section. These experimental methods have elucidated the unit cell structure, bond orientational order and tilt in monolayer phases. Many of the condensed phases have been classified as mesophases having long-range correlational order and short-range translational order. A useful analogy between monolayer mesophases and the smectic mesophases in bulk liquid crystals aids in their characterization (see [182]).

The three general states of monolayers are illustrated in the pressure-area isotherm in Fig. IV-16. A low-pressure gas phase, G, condenses to a liquid phase termed the *liquid*-expanded (LE or L_1) phase by Adam [183] and Harkins [9]. One or more of several more dense, liquid–condensed phase (LC) exist at higher pressures and lower temperatures. A solid phase (S) exists at high pressures and densities. We briefly describe these phases and their characteristic features and transitions; several useful articles provide a more detailed description [184–187].

A. Gaseous Films

A film at low densities and pressures obeys the equations of state described in Section III-7. The available area per molecule is large compared to the cross-sectional area. The film pressure can be described as the difference in osmotic pressure acting over a depth, τ, between the interface containing the film and the pure solvent interface [188–190].

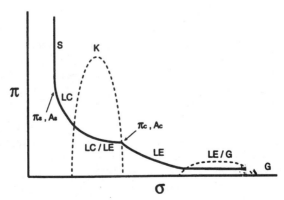

Fig. IV-16. A schematic pressure area isotherm illustrating the general states of monolayers.

In this model, a gaseous film is considered to be a dilute surface solution of surfactant in water and Eq. III-108 can be put in the form

$$\pi_{os}\tau = \pi = -\frac{RT}{A_1}\, \ln f_1^s N_1^s \qquad (IV\text{-}35)$$

where τ is the thickness of the surface phase region and A_1 is the molar area of the solvent species (strictly $\overline{A}_1 = (\partial A/\partial n_1^s)_T$); f_1^s is the activity coefficient of the solvent in the surface phase (near unity – Gaines gives values of 1.01 – 1.02 for pentadecanoic acid films). We assume the geometric area $A = A_1 n_1^s + A_2 n_2^s$ to obtain

$$\pi = \frac{RT}{A_1}\left[\ln\left(1 + \frac{A_1\Gamma_2}{1 - A_2\Gamma_2} \right) - \ln f_1^s \right] \qquad (IV\text{-}36)$$

where Γ_2 is moles of film material per unit area; A_1 is known from the estimated molecular area of water, 9.7 Å^2, and A_2 can either be estimated, thus allowing calculation of f_1^s, or left as an empirical parameter. A regular solution model using statistical mechanics has been described by Popielawski and Rice [191].

The alternative approach is to treat the film as a nonideal two-dimensional gas. One may use an appropriate equation of state, such as Eq. III-104. Alternatively, the formalism has been developed for calculating film activity coefficients as a function of film pressure [192].

B. Gas–Liquid Transitions

On compression, a gaseous phase may condense to a liquid-expanded, L_1 phase via a first-order transition. This transition is difficult to study experimentally because of the small film pressures involved and the need to avoid any impurities [76,193]. There is ample evidence that the transition is clearly first-order; there are discontinuities in π–σ plots, a latent heat of vaporization associated with the transition and two coexisting phases can be seen. Also, fluctuations in the surface potential [194] in the two phase region indicate two-phase coexistence. The general situation is reminiscent of three-dimensional vapor–liquid condensation and can be treated by the two-dimensional van der Waals equation (Eq. III-104) [195] or statistical mechanical models [191].

McConnell et al. [196] and Andelman and co-workers have predicted [197,198] an ordered array of liquid domains in the gas–liquid coexistence regime caused by the dipole moment difference between the phases. These superstructures were observed in monolayers of dipalmitoyl phosphatidylcholine monolayers [170].

One may apply the two-dimensional analogue of the Clapeyron equation

$$\frac{d\pi}{dT} = \frac{\Delta H}{T \Delta A} \tag{IV-37}$$

where ΔH is now the latent heat of vaporization of the L_1 state. Representative values are 2, 3.2, and 9.5 kcal/mol for tridecylic, myristic and pentadecylic acid, respectively [199]. Since the polar groups remain solvated in both the G and L_1 phases, the latent heat of vaporization can be viewed as arising mainly from the attraction between the hydrocarbon tails. It is understandable that these latent heats are lower than those of the bulk materials.

At lower temperatures a gaseous film may compress indefinitely to a liquid-condensed phase without a discernable discontinuity in the π–σ plot.

C. Condensed Phases

A rich family of condensed states of monomolecular films have been revealed by optical and x-ray analysis of phases appearing only as subtle slope changes in the π–σ plot. The generalized phase diagram presented by several groups [184–187] contains the qualitative features presented in Fig. IV-17. The phases in this diagram have been characterized according to liquid crystalline smectic phases having stratified planes with varying degrees of orientational order. We will describe these phases briefly; several reviews provide additional information [184,186,187].

1. L_1. The liquid-expanded, L_1 phase is a two-dimensionally isotropic arrangement of amphiphiles. This is in the *smectic* A class of liquidlike in-plane structure. There is a continuing debate on how best to formulate an equation of state of the liquid-expanded monolayer. Such monolayers are fluid and coherent, yet the average intermolecular distance is much greater than for bulk liquids. A typical bulk liquid is perhaps 10% less dense than its corresponding solid state,

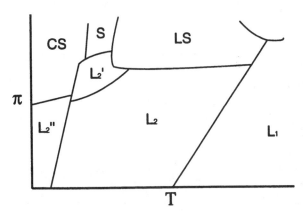

Fig. IV-17. A schematic phase diagram illustrating the condensed mesophases found in monolayers of fatty acids and lipids.

yet a liquid-expanded monolayer may exist at molecular areas twice that for the solid-state monolayer. As noted by Gershfeld [200], various modifications of the two-dimensional van der Waals equation can be successful in fitting data. A statistical regular solution model [191] and Monte Carlo simulations based on a lattice model [201] allow treatment of the first-order L_1–G transition. A mean-field model illustrates the influence of polymer molecules in the subphase on the liquid-expanded L_1 to liquid-condensed L_2 phase transition.

2. L_2. This is the primary liquid-condensed phase first considered by Adam [183] and Langmuir [59] as a semisolid film having hydrated polar heads. Now it is identified as the *smectic* I, or rotator phase, having short-range positional order yet enough cross-sectional area to allow free rotation. The molecules are tilted relative to the normal, and the tilt angle varies with pressure. L_2 films generally exhibit a nearly linear regime in π–σ plots.

3. L_2'. This region has been divided into two subphases, L_2^* and S^*. The L_2^* phase differs from the L_2 phase in the direction of tilt. Molecules tilt toward their nearest neighbors in L_2 and toward next nearest neighbors in L_2^* (a *smectic* F phase). The S^* phase comprises the higher-π and lower-T part of L_2'. This phase is characterized by *smectic* H or a tilted herringbone structure and there are two molecules (of different orientation) in the unit cell. Another phase having a different tilt direction, L_1', can appear between the L_2 and L_2' phases. A new phase has been identified in the L_2' domain. It is probably a *smectic* L structure of different azimuthal tilt than L_2' [185].

4. LS. In the LS phase the molecules are oriented normal to the surface in a hexagonal unit cell. It is identified with the hexatic *smectic* BH phase. Chains can rotate and have axial symmetry due to their lack of tilt. Cai and Rice developed a density functional model for the tilting transition between the L_2 and LS phases [202]. Calculations with this model show that amphiphile-surface interactions play an important role in determining the tilt; their conclusions support the lack of tilt found in fluorinated amphiphiles [203].

5. S. Chains in the S phase are also oriented normal to the surface, yet the unit cell is rectangular possibly because of restricted rotation. This structure is characterized as the *smectic* E or herringbone phase. Schofield and Rice [204] applied a lattice density functional theory to describe the second-order rotator (LS)–herringbone (S) phase transition.

6. CS. The true two-dimensional crystal with chains oriented vertically exists at low T and high π in the CS phase. This structure exhibits long-range translational order.

7. *Surface Micelles*. The possibility of forming clusters of molecules or micelles in monolayer films was first proposed by Langmuir [59]. The matter of surface micelles and the issue of equilibration has been the subject of considerable discussion [191,201,205–209]. Nevertheless, many π–σ isotherms exhibit "nonhorizontal lines" unexplained by equations of state or phase models. To address this, Israelachvili [210] developed a model for π–σ curves where the amphiphiles form surface micelles of N chains. The isotherm

$$\pi = \frac{kT}{N}\left[\frac{1}{(\sigma - \sigma_0)} + \frac{(N-1)}{(\sigma_1 - \sigma_0)}\right] \qquad \text{(IV-38)}$$

depends on σ_0, the molecular area of an amphiphile in a micelle, and σ_1, the area of an amphiphile as a monomer defined by

$$\sigma_1 - \sigma_0 = (\sigma - \sigma_0)\left[1 + \left(\frac{2(\sigma_c - \sigma_0)}{\sigma_1 - \sigma_0}\right)^{N-1}\right] \qquad \text{(IV-39)}$$

where σ_c is the critical micelle area where micelles first appear and can be estimated from a model of micellization [210].

D. The Solid State

While most of the phases described above were "mesophases" [186] having long range orientational order with short-range translational order, the CS phase represented a truly crystalline solid, that is, one exhibiting long-range translational order. Solid films generally appear as high-density rigid or plastic layers. Most fatty acids and alcohols exhibit this type of film at sufficiently low temperatures or with sufficiently long chain lengths. The $\pi - \sigma$ plots are linear, extrapolating to 20.5 Å^2 per molecule at $\pi = 0$, in the case of fatty acids. This exceeds the value of 18.5 Å^2 obtained from the structure of three-dimensional crystals but can be accounted for as the preferred surface packing [211,212]. Structure within the solvent subphase should also be important; Garfias [213,214] has suggested that in solid monolayers of long-chain alcohols, half of the water molecules in the surface layer are replaced by film molecules, the whole forming a highly ordered structure.

Since the development of grazing incidence x-ray diffraction, much of the convincing evidence for long-range positional order in layers has come from this technique. Structural relaxations from distorted hexagonal structure toward a relaxed array have been seen in heneicosanol [215]. Rice and co-workers combine grazing incidence x-ray diffraction with molecular dynamics simulations to understand several ordering transitions [178,215–219].

In particular, comparisons between fluorinated amphiphiles and their hydrogenated counterparts reveal the influence of chain stiffness in the former producing solids having molecules with constant tilt from close packing to coexistence [178]. Molecular flexibility is further probed in simulations of partially fluorinated alkane amphiphiles where disorder in the hydrocarbon portion of the chain produces more disordered monolayers than their purely hydrogenated or fluorinated counterparts [216]. Finally, a nonpolar long-chain molecule, perfluoro-n-eicosane, forms stable monolayers much like their amphiphilic counterparts [220]. Rice and co-workers suggest that van der Waals forces (see Chapter VI) are sufficient to stabilize such a monolayer. Gao and Rice [216,221] have studied the phase behavior of long-chain heterogeneous amphiphiles via grazing incidence x-ray scattering and molecular dynamics simulations. These monolayers, composed of half-fluorinated alcoxybenzoic acid chains, form a phase exhibiting translational order between that of a liquid and a solid.

Grazing incidence excitation of a fluorescent probe in a phospholipid monolayer can also be used to indicate order. The collective tilt of the molecules in a domain inferred from such measurements is indicative of long-range orientational order [222].

E. Monolayer Collapse

At sufficiently high surface pressure a monolayer will "collapse" into three-dimensional multilayer films. For example, 2-hydroxytetradecanoic acid collapses at 68 dyn/cm with no observed decrease in pressure. Electron micrographs such as those shown in Fig. IV-18a show ridges up to 2000 A in height [223]. Collapsed films of calcium stearate show crystalline platelets whose appearance depends somewhat on the shadowing method [224]. Possible collapse sequences are shown in Fig. IV-18b. The collapse may follow a homogeneous nucleation mechanism followed by continued growth of bulk fragments [225].

F. Domain Shapes and Interactions

The use of fluorescence and Brewster angle microscopy to study Langmuir monolayers has revealed a rich morphology of coexisting phases in both single-component and binary layers (see Section IV-3 and Refs. [167,168,184]. Circular domains sometimes form ordered arrays [196,197], while under different conditions the circular shapes are unstable to higher harmonic shapes such as those illustrated in Fig. IV-19 [226–230]. Another supercrystalline structure in coexisting domains is the "stripe phase" or alternating parallel stripes [168,198,231]. Finally, the presence of chiral amphiphiles produces curved, spiral domains as shown in Fig. IV-19 [168,170,232,233]. We briefly summarize the physical basis for these shape transitions and refer interested readers to the references cited above.

The free energy of a monolayer domain in the coexistence region of a phase transition can be described as a balance between the dipolar electrostatic energy and the line tension between the two phases. Following the development of McConnell [168], a monolayer having n circular noninteracting domains of radius R has a free energy

$$F = 2\pi Rn \left(\mu^2 \ln \frac{e^2 \delta}{4R} + \lambda \right) \tag{IV-40}$$

where the first term is proportional to μ the difference in dipole density between the domain and the surrounding phase, e is the base of the natural logarithm, and δ is a distance comparable to the separation between dipoles that represents an excluded region surrounding the domain. The second term is proportional

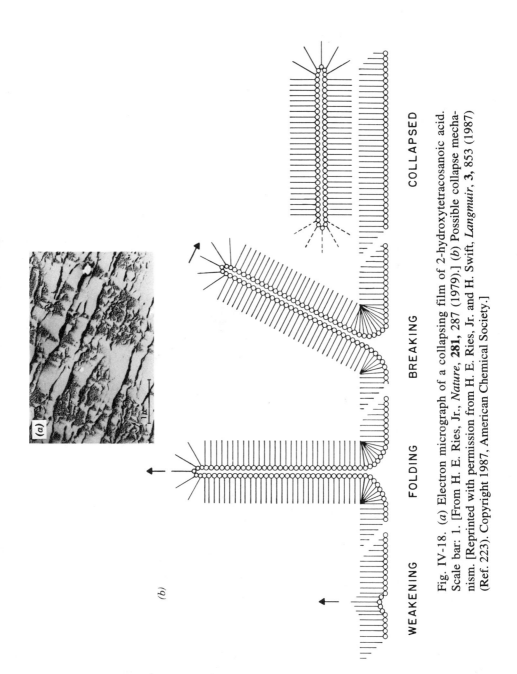

Fig. IV-18. (*a*) Electron micrograph of a collapsing film of 2-hydroxytetracosanoic acid. Scale bar: 1. [From H. E. Ries, Jr., *Nature*, **281**, 287 (1979).] (*b*) Possible collapse mechanism. [Reprinted with permission from H. E. Ries, Jr. and H. Swift, *Langmuir*, **3**, 853 (1987) (Ref. 223). Copyright 1987, American Chemical Society.]

WEAKENING   FOLDING   BREAKING   COLLAPSED

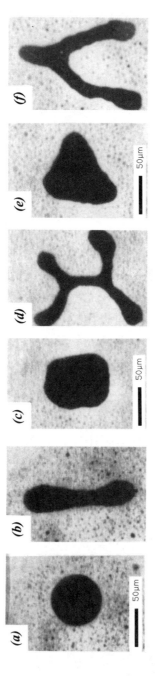

Fig. IV-19. Fluorescence micrographs showing the shape transitions in monolayers of dimyristoylphosphatidylcholine (DMPC) (84%) and dihydrocholesterol (15%) and a lipid containing the dye, Texas Red. (From Ref. 228.)

to the line tension, λ. The free energy is minimized at the equilibrium radius

$$R_{eq} = \frac{e^3 \delta}{4} \exp[\lambda/\mu^2] \qquad \text{(IV-41)}$$

showing explicitly the balance between dipolar interactions through μ^2 and the line tension. The dipole strength can be determined from the surface potential as outlined in section IV-3B. The line tension, λ, can be measured by studying the relaxation of flow distortions to domains [234]. The distance δ cannot be measured directly but can be inferred from observations of domain sizes and shapes [168].

In principle, compressing a monolayer should produce more domains of radius R_{eq}, however; McConnell shows that in practice, domains grow to exceed this radius. Once a domain grows to $R = e^{1/3}R_{eq}$, it becomes unstable toward spontaneous deformation into an ellipse [227,235]. Larger domains are unstable toward transitions to shapes of n-fold symmetry as illustrated in Fig. IV-19, where the instability occurs at a radius

$$R_n = e^{Z-3}R_{eq} \qquad \text{(IV-42)}$$

with $Z_3 = 11/3$, $Z_4 = 11.76/3$, $Z_5 = 12.37/3$.

As circular domains grow in size or number, the dipolar interactions between them increase until they form a hexagonal array of spacing

$$D = \left\{ \left[\frac{(\pi R^2 \mu)^2}{kT} \right]^{1/2} \frac{1}{kT} \right\}^{1/3} \qquad \text{(IV-43)}$$

thus if $\mu = 1$ debye/100 A^2, $R = 10$ μm, and $T = 300$ K, the spacing $D \approx 130$ μm [168].

Several groups have studied the structure of chiral phases illustrated in Fig. IV-15 [167,168]. These shapes can be understood in terms of an anisotropic line tension arising from the molecular symmetry. The addition of small amounts of cholesterol reduces λ and produces thinner domains. Several studies have sought an understanding of the influence of cholesterol on lipid domain shapes [168,196].

The effects of electric fields on monolayer domains graphically illustrates the repulsion between neighboring domains [236, 237]. A model by Stone and McConnell for the hydrodynamic coupling between the monolayer and the subphase produces predictions of the rate of shape transitions [115,238].

5. Mixed Films

The study of mixed films has become of considerable interest. From the theoretical side, there are pleasing extensions of the various models for single-component films; and from the more empirical side, one moves closer to modeling biological membranes. Following Gershfeld [200], we categorize systems as follows:

1. Both components form insoluble monolayers.
    a. Equilibrium between mixed condensed and mixed vapor phases can be observed.
    b. Only condensed phases are observed.
2. One component forms an insoluble monolayer, while the other is soluble. Historically, the phenomenon is known as *penetration*.
3. Both components are soluble.

Category 3 was covered in Chapter III, and category 2 is treated later in this section.

Condensed phases of systems of category 1 may exhibit essentially ideal solution behavior, very nonideal behavior, or nearly complete immiscibility. An illustration of some of the complexities of behavior is given in Fig. IV-20, as described in the legend.

The thermodynamics of relatively ideal mixed films can be approached as follows. It is convenient to define

$$A_{av} = N_1 A_1 + N_2 A_2 \tag{IV-44}$$

where A_1 and A_2 are the molar areas at a given π for the pure components. An *excess* area A_{ex} is then given by

$$A_{ex} = A - A_{av} \tag{IV-45}$$

If an ideal solution is formed, then the actual molecular A is just A_{av} (and $A_{ex} = 0$). The *same* result obtains if the components are completely immiscible as illustrated in Fig. IV-21 for a mixture of arachidic acid and a merocyanine dye [116]. These systems are usually distinguished through the mosaic structure seen in microscopic evaluation.

We may also define a free energy of mixing [240]. The alternative (and equally acceptable) definition of G^σ given in Eq. III-87 is

$$G^{*\sigma} = E^\sigma - TS^\sigma - \gamma \mathcal{A} \tag{IV-46}$$

Differentiation and combination with Eq. III-73 yields

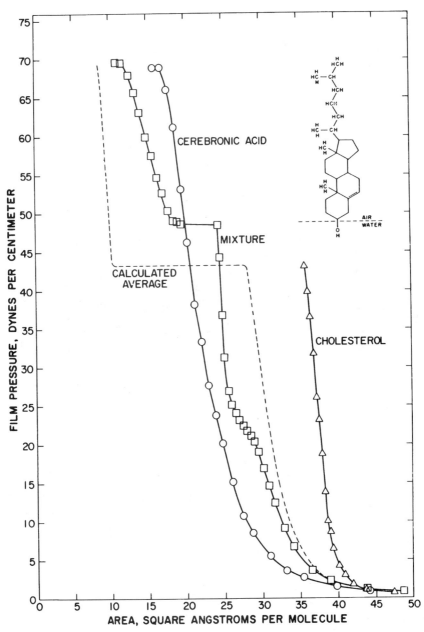

Fig. IV-20. Film pressure–area plots for cerebronic acid (a long-chain α-hydroxy carboxylic acid) and cholesterol (see insert) and for an equimolar mixture. At low pressures
the π–σ plot is close to that of the average (dashed line), an unanticipated kink then
appears, and finally, the horizontal portion probably represents ejection of the cholesterol. (From Ref. 239.)

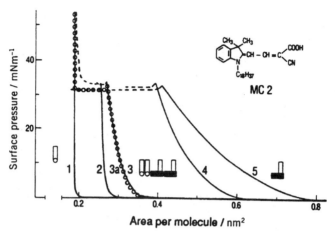

Fig. IV-21. Surface pressure versus area for monolayers of immiscible components: a monolayer of pure cadmium arachidate (curve 1) and monolayers of mixed merocyanine dye, MC2, and cadmium arachidate of molar ratio r = 1:10 (curve 2); 1:5 (curve 3), 1:2 (curve 4), and pure MC2 (curve 5). The subphase is $2.5 \times 10^{-4}M$ CdCl$_2$, pH = 5.5 at 20°C. Curve 3a (○) was calculated from curves 1 and 5 using Eq. IV-44. (From Ref. [116].)

$$dG^{*\sigma} = -S^{\sigma}\,dT + \sum \mu_i\,dn_i - \mathcal{A}\,d\gamma \qquad \text{(IV-47)}$$

At constant temperature and mole numbers

$$dG^{*\sigma} = -\mathcal{A}\,d\gamma \qquad \text{(IV-48)}$$

Consider the mixing process

[n_1 moles of film (1) at π] + [n_2 moles of film (2) at π]
= (mixed film at π)

First, the films separately are allowed to expand to some low pressure, π^*, and by Eq. IV-48 the free energy change is

$$\Delta G^{*\sigma}_{1,2} = -N_1 \int_{\pi^*}^{\pi} A_1\,d\pi - N_2 \int_{\pi^*}^{\pi} A_2\,d\pi$$

The pressure π^* is sufficiently low that the films behave ideally, so that on mixing

$$\Delta G^{*\sigma}_{\text{mix}} = RT(N_1 \ln N_1 + N_2 \ln N_2)$$

The mixed film is now compressed back to π:

$$\Delta G^{*\sigma}_{12} = \int_{\pi^*}^{\pi} A_{12} \, d\pi$$

$\Delta G^{*\sigma}$ for the overall process is then

$$\Delta G^{*\sigma} = \int_{\pi^*}^{\pi} (A_{12} - N_1 A_1 - N_2 A_2) d\pi$$
$$+ RT(N_1 \ln N_1 + N_2 \ln N_2) \tag{IV-49}$$

and the excess free energy of mixing is thus

$$\Delta G^{*\sigma}_{\text{ex}} = \int_{\pi^*}^{\pi} (A_{12} - N_1 A_1 - N_2 A_2) d\pi \tag{IV-50}$$

A plot of G^*_{ex} versus composition is shown in Fig. IV-22 for condensed films of octadecanol with docosyl sulfate. Gaines [241] and Cadenhead and Demchak [242] have extended the above approach, and the subject has been extended and reviewed by Barnes and co-workers (see Ref. 243).

Barnes cautions about using the appropriate units (molecular area with mole fraction, or area per unit mass with mass fraction) when analyzing area data [244].

In the alternative surface phase approach, Eq. IV-36 may be expanded for mixed films to give [245]

$$\pi = \frac{RT}{A_1} \left\{ \ln \left[1 + \frac{A_1(\Gamma_2 + \Gamma_3)}{1 - A_2\Gamma_2 - A_3\Gamma_3} \right] - \ln f_1^s \right\} \tag{IV-51}$$

Many groups, including Ries and Swift [247], Cadenhead and Müller-Landau [248], and Tajima and Gershfeld [249], have studied mixed films. A case of immiscibility at low π is discussed by Cadenhead and co-workers [250]. Mixed films of a phospholipid and a polysoap have been treated by Ter-Minassian-Saraga in terms of adsorption on a linear adsorbent [251]. Motomura and co-workers [252] have developed a related thermodynamic approach to mixed films. Hendrikx [253] reports on the three-component system of anionic soap–cationic soap–cetyl alcohol. A more complex ternary system containing two lipids and palmitic acid is designed to mimic phospholipid biomembranes [254]. Shah and Shiao [255] discuss chain length compatibility of long-chain alcohols. Möbius [256] observed optical and π–σ properties of monolayers incorporating dye molecules.

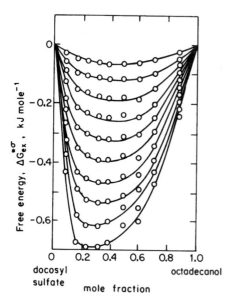

Fig. IV-22. Excess free energy of mixing of condensed films of octadecanol-docosyl sulfate at 25°C, at various film pressures. Top curve: $\pi = 5$ dyn/cm; bottom curve: $\pi = 50$ dyn/cm; intermediate curves at 5-dyn/cm intervals. The curves are uncorrected for the mixing term at low film pressure. (From Ref. 246.)

There has been extensive activity in the study of lipid monolayers as discussed above in Section IV-4E. Coexisting fluid phases have been observed via fluorescence microscopy of mixtures of phospholipid and cholesterol where a critical point occurs near 30 mol% cholesterol [257].

Category 2 mixed films, or those formed by penetration, have also been of some interest. Here, a more or less surface-active constituent of the substrate enters into a spread monolayer, in some cases to the point of diluting it extensively. Thus monolayers of long-chain amines and of sterols are considerably expanded if the substrate contains dissolved low-molecular-weight acids or alcohols. A great deal of work has been carried out on the penetration of sodium cetyl sulfate and similar detergent species into films of biological materials (e.g., Ref. 258). Pethica and co-workers [259] and Fowkes [260] studied the penetration of cetyl alcohol films by sodium dodecyl sulfate, and Fowkes [261] mixed monolayers of cetyl alcohol and sodium cetyl sulfate, using an aqueous sodium chloride substrate to reduce the solubility of the detergent. Lucassen-Reynders has applied equations of the type of Eq. IV-51 to systems such as sodium laurate-lauric acid [262,263] and ones involving egg lecithin films [264].

In actual practice the soluble component usually is injected into the substrate solution *after* the insoluble monolayer has been spread. The reason is that if one starts with the solution, the surface tension may be low enough that the monolayer will not spread easily. McGregor and Barnes have described a useful injection technique [265].

A difficulty in the physicochemical study of penetration is that the amount of soluble component present in the monolayer is not an easily accessible quantity. It may be measured directly, through the use of radioactive labeling (Section III-6) [263, 266], but the technique has so far been used only to a limited extent.

Two alternative means around the difficulty have been used. One, due to Pethica [267] (but see also Alexander and Barnes [268]), is as follows. The Gibbs equation, Eq. III-80, for a three-component system at constant temperature and locating the dividing surface so that Γ_1 is zero becomes

$$d\pi = RT\Gamma_f^1 \, d \ln \, a_f + RT\Gamma_s^1 \, d \ln \, a_s \qquad \text{(IV-52)}$$

where subscripts f and s denote insoluble monolayer and surfactant, respectively, and a is the rational activity. We can eliminate the experimentally inaccessible $\ln a_f$ quantity from Eq. IV-52 as follows. The definition of partial molal area is $(\partial \mathcal{A} /\partial n_i)_{T,n_j} = \overline{A}_i$, and we obtain from Eq. IV-48 (by differentiating with respect to n_f)

$$RT\left(\frac{\partial \ln \, a_f}{\partial \pi} \right)_{T,n_j} = \overline{A}_f \qquad \text{(IV-53)}$$

Combination with Eq. IV-52 gives

$$d\pi = \frac{A_f}{A_f - \overline{A}_f} \, RT\Gamma_s^1 \, d \ln \, m_s \qquad \text{(IV-54)}$$

where $A_f = 1/\Gamma_f^1$, and since the surfactant solution is usually dilute, a_s is approximated by the molality m_s. (A factor of 2 may appear in Eq. IV-54 if the surfactant is ionic and no excess electrolyte is present—see Section III-6B.) Pethica assumed that at a given film pressure $\overline{A}_f$ would be the same in the mixed film as in the pure film; A_f is just the experimental area per mole of film material in the mixed film. The coefficient $(\partial \pi/\partial \ln m_s)_{T,n_f}$ could be obtained from the experimental data, and Γ_s^1 follows from Eq. IV-54. Some plots of Γ_s^1 versus m_s are shown in Fig. IV-23, as calculated from these relationships [269]; notice that for a given $\overline{A}_f$, the amount of penetration reaches a maximum. Barnes and coworkers have pointed out certain approximations in the above treatment [268,270]. See also Hall [271].

Still another manifestation of mixed-film formation is the absorption of organic vapors by films. Stearic acid monolayers strongly absorb hexane up to a limiting ratio of 1:1 [272], and data reminiscent of adsorption isotherms for gases on solids are obtained, with the surface density of the monolayer constituting an added variable.

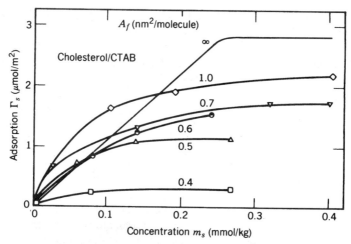

Fig. IV-23. Penetration of cholesterol monolayers by CTAB (hexadecyl-trimethylam-monium bromide. [From D. M. Alexander, G. T. Barnes, M. A. McGregor, and K. Walker, "Phenomena in Mixed Surfactant Systems," in J. F. Scamehorn, ed., ACS Symposium Series 311, p. 133, 1986 (Ref. 269). Copyright 1986, American Chemical Society.]

6. Evaporation Rates through Monomolecular Films

An interesting consequence of covering a surface with a film is that the rate of evaporation of the substrate is reduced. Most of these studies have been carried out with films spread on aqueous substrates; in such cases the activity of the water is practically unaffected because of the low solubility of the film material, and it is only the rate of evaporation and not the equilibrium vapor pressure that is affected. Barnes [273] has reviewed the general subject.

One procedure makes use of a box on whose silk screen bottom powdered desiccant has been placed, usually lithium chloride. The box is positioned 1–2 mm above the surface, and the rate of gain in weight is measured for the film-free and the film-covered surface. The rate of water uptake is reported as $v = m/t\mathcal{A}$, or in g/sec·cm^2. This is taken to be proportional to $(C_w - C_d)/R$, where C_w and C_d are the concentrations of water vapor in equilibrium with water and with the desiccant, respectively, and R is the diffusional resistance across the gap between the surface and the screen. Qualitatively, R can be regarded as actually being the sum of a series of resistances corresponding to the various diffusion gradients present:

$$R_{\text{total}} = R_{\text{surface}} + R_{\text{film}} + R_{\text{desiccant}} = R_0 + R_{\text{film}} \qquad (\text{IV-55})$$

Here R_0 represents the resistance found with no film present. We can write

$$\frac{R_{film}}{A} = r = (C_w - C_d)\left(\frac{1}{v_f} - \frac{1}{v_w}\right) \simeq C_w\left(\frac{1}{v_f} - \frac{1}{v_w}\right) \qquad \text{(IV-56)}$$

where r is the specific evaporation resistance (sec/cm), and the subscripts f and w refer to the surface with and without film, respectively. The general subject can be treated in the theory of mass transfer across interfaces [274].

Some fairly typical results, obtained by LaMer and co-workers [275] are shown in Fig. IV-24. At the higher film pressures, the reduction in evaporation rate may be 60–90%—a very substantial effect. Similar results have been reported for the various fatty acids and their esters [276,277]. Films of biological materials may offer little resistance, as is the case for cholesterol [278] and dimyristoylphosphatidylcholine (except if present as a bilayer) [279].

One application of evaporation retardation involves the huge water loss due to evaporation from reservoirs [280]. The first attempts to use monolayers to reduce reservoir evaporation were made by Mansfield [281] in Australia; since then moderately successful tests have been made in a number of locations. In most of these tests, cetyl alcohol was used as an available surfactant offering a good compromise between specific resistance and rate of spreading. High spreading rates are extremely important; a film must not only form easily but must do so against wind friction and must heal ruptures caused by waves or boat wakes. Also the material must not be rapidly biodegraded and it must not interfere with aquatic

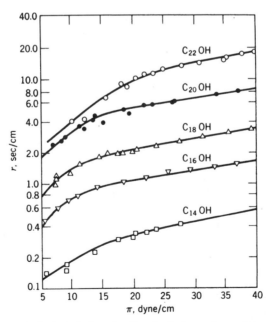

Fig. IV-24. Effect of alkyl chain length of n-alcohols on the resistance of water evaporation at 25°C. (From Ref. 275.)

life through prevention of areation of the water. An analysis of the effect of environmental conditions on evaporation control has been made by Barnes [282].

Barnes and co-workers have studied mixed-monolayer systems [278,281,283,284] and found some striking nonidealities. Mixed films of octadecanol and cholesterol, for example, show little evaporation resistance if only 10% cholesterol is present [278] apparently due to an uneven granular microstructure in films with cholesterol [284]. Another study of cellulose decanoate films showed no correlation between holes in the monolayer and permeation rate [285]. Polymerized surfactants make relatively poor water evaporation retarders when compared to octadecanol [286]. There are problems in obtaining reproducible values for r [287] due to impurities in the monolayer material or in the spreading solvent.

A potentially serious problem to quantitative analysis is that evaporation cools the water layers immediately below the surface, the cooling being less if there is retardation (see Ref. 288). There is also a problem in determining absolute v values, that is, evaporation rates into vacuum. Ideally, this is given by Eq. III-22, but this involves the assumption that molecules hitting the surface from the gas phase stick with unitary efficiency or, alternatively, that the *evaporation coefficient* α is unity (see Refs. 273, 289); α values for water have ranged from 10^{-3} to 1! It is easier to deal with net evaporation rates or r values. The temperature dependence of r gives an apparent activation energy, and the increment per CH_2 group is about 200 cal/mol in the case of long-chain alcohols, but the variation is pressure dependent and in a way that suggests that the energy requirement is one of forming a hole in a close-packed monolayer [273, 280]. The "accessible" area, $a = \mathcal{A} - n_f A_f^0$ is a [273, 280] good correlating parameter, where n_f is the moles of film material of actual molar area A_f^0, suggesting that $v_f = a/\mathcal{A}$, or

$$r = \frac{\mathcal{A} C_w}{v_w} \left(\frac{\mathcal{A}}{a} - 1 \right) \qquad (IV-57)$$

Barnes and Hunter [290] have measured the evaporation resistance across octadecanol monolayers as a function of temperature to test the appropriateness of several models. The experimental results agreed with three theories; the energy barrier theory, the density fluctuation theory, and the accessible area theory. A plot of the resistance times the square root of the temperature against the area per molecule should collapse the data for all temperatures and pressures as shown in Fig. IV-25. A similar temperature study on octadecylurea monolayers showed agreement with only the accessible area model [291].

7. Dissolution of Monolayers

The rate of dissolving of monolayers constitutes an interesting and often practically important topic. It affects, for example, the rate of loss of monolayer

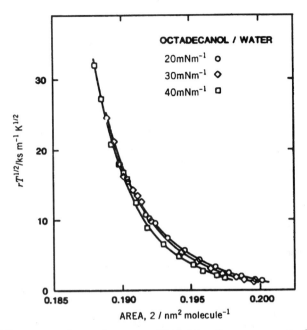

Fig. IV-25. The evaporation resistance multiplied by the square root of temperature versus area per molecule for monolayers of octadecanol on water illustrating agreement with the accessible area model. (From Ref. 290.)

material used in evaporation control. From the physicochemical viewpoint, the topic represents a probe into the question of whether insoluble monolayers are in equilibrium with the underlying bulk solution. Film dissolution also represents the reverse of the process that gives rise to the slow aging or establishment of equilibrium surface tension in the cases of surfactant solutions discussed in Chapter II (see Fig. II-15).

The usual situation appears to be that after a short initial period the system obeys the equation

$$\mathcal{A} = \mathcal{A}^0 e^{-kt} \tag{IV-58}$$

if the film is kept at a constant film pressure. The differential form of Eq. IV-58 may be written as

$$\left(\frac{dn}{dt} \right)_\pi = -kn \tag{IV-59}$$

where n is the number of moles of film present. As discussed by Ter Minassian-Saraga [292], this form can be accounted for in terms of a steady-state diffusion

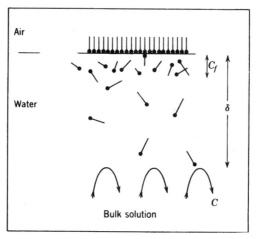

Fig. IV-26. Steady-state diffusion model for film dissolution. (From Ref. 293.)

process. As illustrated in Fig. IV-26, we suppose the film to be in equilibrium with the immediately underlying solution, to give a concentration C_f. The rate-limiting process is taken to be the rate of diffusion across a thin stagnant layer of solution of thickness δ. According to Fick's law and remembering that $n = \mathcal{A}\,\Gamma$,

$$\frac{dn}{dt} = \mathcal{A}\mathcal{D}\,\frac{dC}{dx} = -\frac{n}{\Gamma}\,\frac{\mathcal{D}}{\delta}\,(C_f - C) \tag{IV-60}$$

Here $\mathcal{D}$ is the diffusion coefficient and C is the concentration in the general bulk solution. For initial rates C can be neglected in comparison to C_f so that from Eqs. IV-59 and IV-60 we have

$$k = \frac{\mathcal{D}\,C_f}{\delta\Gamma} = \frac{\mathcal{D}}{\delta K} \tag{IV-61}$$

where

$$K = \frac{\Gamma}{C_f} \tag{IV-62}$$

Some further details are the following. Film nonideality may be allowed for [192]. There may be a chemical activation barrier to the transfer step from monolayer to sub-surface solution and hence also for monolayer formation by adsorption from solution [294–296]. Dissolving rates may be determined with the use of the radioactive labeling technique of Section III-6A, although precautions are necessary [297].

As a general comment, it is fortunate for the study of monolayers that dissolving pro-cesses are generally slow enough to permit the relatively unperturbed study of equilibrium film properties, because many films are inherently unstable in this respect. Gaines [1] notes that the equilibrium solubility of stearic acid is 3 mg/l and that films containing only a tenth of the amount that should dissolve in the substrate can be studied with no evidence of solution occurring.

8. Reactions in Monomolecular Films

The study of reactions in monomolecular films is rather interesting. Not only can many of the usual types of chemical reactions be studied but also there is the special feature of being able to control the orientation of molecules in space by varying the film pressure. Furthermore, a number of processes that occur in films are of special interest because of their resemblance to biological systems. An early review is that of Davies [298]; see also Gaines [1].

A. Kinetics of Reactions in Films

In general, it is not convenient and sometimes not possible to follow reactions in films by the same types of measurements as employed in bulk systems. It is awkward to try to make chemical analyses to determine the course of a reaction, and even if one of the reactants is in solution in the substrate, the amounts involved are rather small (a micromole at best). Such analyses are greatly facilitated, of course, if radioactive labeling is used. Also, film collapsed and collected off the surface has been analyzed by infrared [299] or UV-visible [300] spectroscopy and chromatographically [301]. In situ measurements can be made if the film material has a strong absorption band that is altered by the reaction, as in the case of chlorophyll [302], or if there is strong photoexcited emission [303]. Reactions have been followed by observing changes in surface viscosity [300] and with radioactive labeling if the labeled fragment leaves the interface as a consequence of the reaction.

The most common situation studied is that of a film reacting with some species in solution in the substrate, such as in the case of the hydrolysis of ester monolayers and of the oxidation of an unsaturated long-chain acid by aqueous permanganate. As a result of the reaction, the film species may be altered to the extent that its area per molecule is different or may be fragmented so that the products are soluble. One may thus follow the change in area at constant film pressure or the change in film pressure at constant area (much as with homogeneous gas reactions); in either case concomitant measurements may be made of the surface potential.

Case 1. A chemical reaction occurs at constant film pressure. To the extent that area is an additive property, one has

$$\frac{n_A}{n_A^0} = \frac{\mathcal{A} - \mathcal{A}^\infty}{\mathcal{A}^0 - \mathcal{A}^\infty} \tag{IV-63}$$

where n_A denotes moles of reactant. In the case of soluble product(s), Eq. IV-63 is exact, with $\mathcal{A}^\infty = 0$. If the product(s) form monolayers which are completely immiscible with the starting material, Eq. IV-63 should again be accurate. Reactant and products may form a mixed film, however, and the behavior may be quite nonideal (see Section IV-5) so that molar areas are not additive, and Eq. IV-63 becomes a poor approximation. Monolayer reactions are often first-order, particularly if the reaction is between a substrate species and the insoluble one, in which case we have

$$\frac{\mathcal{A} - \mathcal{A}^\infty}{\mathcal{A}^0 - \mathcal{A}^\infty} = e^{-kt} \qquad \text{(IV-64)}$$

Case 2. The surface potential is measured as a function of time. Here, since by Eq. IV-19 $\Delta V = 4\pi \mathbf{n}\bar{\mu}/\epsilon$, then

$$\mathcal{A} \, \Delta V = (4\pi) n_A \bar{\mu}_A = \alpha_A n_A \qquad \text{(IV-65)}$$

Because mole numbers are additive, it follows that the product $\mathcal{A} \, \Delta V$ will be an additive quantity provided that α for each species remains constant during the course of the reaction. This last condition implies, essentially, that the effective dipole moments and hence the orientation of each species remain constant, which is most likely to be the case at constant film pressure. Then

$$\frac{\mathcal{A} \, \Delta V - \mathcal{A}^\infty \, \Delta V^\infty}{\mathcal{A}^0 \, \Delta V^0 - \mathcal{A}^\infty \, \Delta V^\infty} = e^{-kt} \qquad \text{(IV-66)}$$

if, again, the reaction is first-order.

B. Kinetics of Formation and Hydrolysis of Esters

An example of an alkaline hydrolysis is that of the saponification of monolayers of α-monostearin [304]; the resulting glycerine dissolved while the stearic acid anion remained a mixed film with the reactant. Equation IV-64 was obeyed, with $k = k'(OH^-)$ and showing an apparent activation energy of 10.8 kcal/mol. O'Brien and Lando [305] found strong pH effects on the hydrolysis rate of vinyl stearate and poly(vinyl stearate) monolayers. Davies [306] studied the reverse type of process, the lactonization of γ-hydroxystearic acid (on acid substrates). Separate tests showed that ΔV for mixed films of lactone and acid was a linear function of composition at constant π, which allowed a modification of Eq. IV-66 to be used. The pseudo-first-order rate constant was proportional to the hydrogen ion concentration and varied with film pressure, as shown in Fig. IV-27. This variation of k with π could be accounted for by supposing that the γ-hydroxystearic acid could assume various configurations,

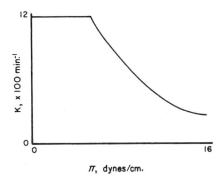

Fig. IV-27. Rate of lactonization of γ-hydroxystearic acid as a function of film pressure.

as illustrated in Fig. IV-28, each of which was weighted by a Boltzmann factor. The steric factor for the reaction was then computed as a function of π, assuming that the only configurations capable of reacting were those having the hydroxy group in the surface. Likewise, the change in activation energy with π was explainable in terms of the estimated temperature coefficient of the steric factors; thus

$$k = \phi z p e^{-E/RT} \qquad\qquad \text{(IV-67)}$$

where ϕ is the variable steric factor, so that

$$-\frac{d \ln k}{d(1/T)} = \frac{E}{R} - \frac{d \ln \phi}{d(1/T)}$$

and

$$E_{\text{apparent}} = E + E_{\text{steric}} \qquad\qquad \text{(IV-68)}$$

An interesting early paper is that on the saponification of 1-monostearin monolayers, found to be independent of surface pressure [307].

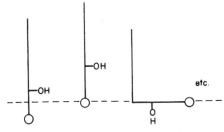

Fig. IV-28. Possible orientations for γ-hydroxystearic acid.

A more elaborate treatment of ester hydrolysis was attempted by Davies and Rideal [308] in the case of the alkaline hydrolysis of monolayers of monocetyl-succinate ions. The point in mind was that since the interface was charged, the local concentration of hydroxide ions would not be the same as in the bulk substrate. The surface region was treated as a bulk phase 10 Å thick and, using the Donnan equation, actual concentrations of ester and hydroxide ions were calculated, along with an estimate of their activity coefficients. Similarly, the Donnan effect of added sodium chloride on the hydrolysis rates was measured and compared with the theoretical estimate. The computed concentrations in the surface region were rather high (1–3 M), and since the region is definitely not isotropic because of orientation effects, this type of approach would seem to be semiempirical in nature. On the other hand, there was quite evidently an electrostatic exclusion of hydroxide ions from the charged monocetyl-succinate film, which could be predicted approximately by the Donnan relationship.

An example of a two-stage hydrolysis is that of the sequence shown in Eq. IV-69. The kinetics, illustrated in Fig. IV-29, is approximately that of successive first-order reactions but complicated by the fact that the intermediate II is ionic [301]

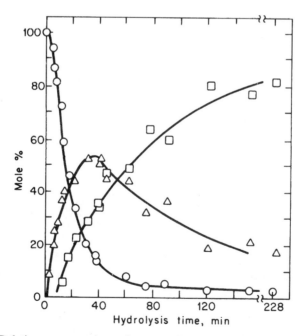

Fig. IV-29. Relative concentrations of reactant I ($\bigcirc$) and products II ($\triangle$) and III ($\square$) (Eq. IV-69) during the hydrolysis on 0.1 M NaHCO$_3$/NaCl substrate at 10 dyn/cm film pressure and 22°C. (Reproduced with permission from Ref. 301.)

$$\left[(bpy)_2Ru \underset{N}{\overset{N}{\diagdown}} \begin{matrix} \bigcirc\!-COOR \\ \bigcirc\!-COOR \end{matrix} \right]^{2+} \xrightarrow[-ROH]{OH^-} \left[(bpy)_2Ru \underset{N}{\overset{N}{\diagdown}} \begin{matrix} \bigcirc\!-COOR \\ \bigcirc\!-COO^- \end{matrix} \right]^{1+} \xrightarrow[-ROH]{OH^-} \left[(bpy)_2Ru \underset{N}{\overset{N}{\diagdown}} \begin{matrix} \bigcirc\!-COO \\ \bigcirc\!-COO \end{matrix} \right]^{0}$$

I. $R = R = C_{18}H_{37}$ II. $R = C_{18}H_{37}$ III.

(IV-69)

C. Other Chemical Reactions

Another type of reaction that has been studied is that of the oxidation of a double bond. In the case of triolein, Mittelmann and Palmer [309] found that, on a dilute permanganate substrate, the area at constant film pressure first increased and then decreased. The increase was attributed to the reaction

$$-CH_2-CH{=}CH-CH_2-\longrightarrow-CH_2-CHOH-CH_2-$$

with a consequent greater degree of anchoring of the middle of the molecule in the surface region. The subsequent decrease in area seemed to be due to fragmentation into two relatively soluble species. The initial reaction obeyed Eq. IV-63, and the pseudo-first-order rate constant was proportional to the permanganate concentration. The rate constant also fell off with increasing film pressure, and this could be accounted for semiquantitatively by calculating the varying probability of the double bond being on the surface.

Reactions in which a product remains in the film (as above) are complicated by the fact that the areas of reactant and product are not additive, that is, a nonideal mixed film is formed. Thus Gilby and Alexander [310], in some further studies of the oxidation of unsaturated acids on permanganate substrates, found that mixed films of unsaturated acid and dihydroxy acid (the immediate oxidation product) were indeed far from ideal. They were, however, able to fit their data for oleic and erucic acids fairly well by taking into account the separately determined departures from ideality in the mixed films.

Photopolymerization reactions of monolayers have become of interest (note Chapter XV). Lando and co-workers have studied the UV polymerization of 16-heptadecenoic acid [311] and vinyl stearate [312] monolayers. Particularly interesting is the UV polymerization of long-chain diacetylenes. As illustrated in Fig. IV-30, a zipperlike process can occur if the molecular orientation in the film is just right (e.g., polymerization does not occur readily in the neat liquid) (see Refs. 313–315).

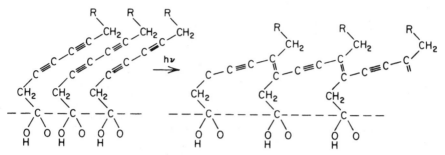

Fig. IV-30. Photopolymerization of an oriented diacetylene monolayer.

Photodegradation as well as fluorescence quenching has been observed in chlorophyll monolayers [302,316]. Whitten [317] observed a substantial decrease in the area of mixed films of tripalmitin and a *cis*-thioindigo dye as isomerization to the *trans*-thioindigo dye occurred on irradiation with UV light.

9. Problems

1. Benjamin Franklin's experiment is mentioned in the opening paragraphs of this chapter. Estimate, from his results, an approximate value for Avogadro's number; make your calculation clear. The answer is a little off; explain whether more accurate measurements on Franklin's part would have helped.

2. According to Eq. IV-9 (with $\Phi = 1$), the spreading coefficient for a liquid of lower surface tension to spread on one of higher surface tension is always negative. Demonstrate whether this statement is true.

3. In detergency, for separation of an oily soil O from a solid fabric S just to occur in an aqueous surfactant solution W, the desired condition is: $\gamma_{SO} = \gamma_{WO} + \gamma_{SW}$. Use simple empirical surface tension relationships to infer whether the above condition might be met if (a) $\gamma_S = \gamma_W$, (b) $\gamma_O = \gamma_W$, or (c) $\gamma_S = \gamma_O$.

4. If the initial spreading coefficient for liquid b spreading on liquid a is to be just zero, what relationship between γ_a and γ_b is implied by the simple Girifalco–Good equation?

5. $S_{B/A}$ is 40 dyn/cm for a particular alcohol on water, with $\gamma_B = 27.5$ dyn/cm. Calculate γ_{AB} (assume 20°C).

6. Calculate γ_{AB} of Problem 5 using the simple Girifalco–Good equation, assuming Φ to be 1.0. How does this compare with experimental data?

7. Derive the expression (in terms of the appropriate works of adhesion and cohesion) for the spreading coefficient for a substance C at the interface between two liquids A and B.

8. Calculate γ_{WH} at 20°C for toluene using the Good–Fowkes approach.

9. Calculate γ_{WH} for the cyclohexane–water interface using the Good–Fowkes approach. Repeat the calculation using Eq. IV-13. Compare both results with the experimental value and comment.

10. Show what $S_{B(A)/A(B)}$ should be according to Antonow's rule.

11. Calculate γ_{WH} for the heptane–water interface using Eq. IV-7. Compare the result with the experimental value and comment.

12. A surfactant for evaporation control has an equilibrium film pressure of 15 dyn/cm. Assume a water surface and 25°C and calculate the distance traveled by the spreading film in 8 sec.

13. Referring to Fig. IV-4, the angles α and β for a lens of isobutyl alcohol on water are 42.5° and 3°, respectively. The surface tension of water saturated with the alcohol is 24.5 dyn/cm; the interfacial tension between the two liquids is 2.0 dyn/cm, and the surface tension of isobutyl alcohol is 23.0 dyn/cm. Calculate the value of the angle γ in the figure. Which equation, IV-6 or IV-9, represents these data better? Calculate the thickness of an infinite lens of isobutyl alcohol on water.

14. Should there be Marangoni instability in a layer of water 1.5 mm deep, with the surface 0.3°C cooler than the bottom? What dimensions does the Marangoni number (Ma) have? (Assume $T \sim 25°C$.)

15. Calculate the limiting thickness of a lens of n-heptyl alcohol on water at 20°C.

16. An alternative defining equation for surface diffusion coefficient $\mathcal{D}_s$ is that the surface flux J_s is $J_s = -\mathcal{D}_s \, d\Gamma/dx$. Show what the dimensions of J_s must be.

17. Derive Eq. IV-36 from Eq. IV-35. Assume that areas are additive, that is, that $\mathcal{A} = A_1 n_1^s + A_2 n_2^s$, where A_1 and A_2 are the molar areas. A particular gaseous film has a film pressure of 0.25 dyn/cm, a value 5% smaller that the ideal gas value. Taking A_2 to be 25 Å^2 per molecule, calculate f_1^s. (Assume 20°C.)

18. Assume that an aqueous solute adsorbs at the mercury–water interface according to the Langmuir equation: $x/x_m = bc/(1 + bc)$, where x_m is the maximum possible amount and $x/x_m = 0.5$ at $C = 0.3M$. Neglecting activity coefficient effects, estimate the value of the mercury–solution interfacial tension when C is $0.2M$. The limiting molecular area of the solute is 20 Å^2 per molecule. The temperature is 25°C.

19. It was commented that surface viscosities seem to correspond to anomalously high bulk liquid viscosities. Discuss whether the same comment applies to surface diffusion coefficients.

20. The film pressure of a myristic acid film at 20°C is 10 dyn/cm at an area of 23 Å^2 per molecule; the limiting area at high pressures can be taken as 20 Å^2 per molecule. Calculate what the film pressure should be, using Eq. IV-36 with $f = 1$, and what the activity coefficient of water in the interfacial solution is in terms of that model.

21. Use Eqs. IV-38 and IV-39 to calculate σ_1 and plot $\pi - \sigma$ for systems having surface micelles of $N = 30, 100, 300$; for $\sigma_c = 50$ Å^2, $\sigma_0 = 20$Å^2, and T = 300 K (*Hint:* Equation IV-38 is an nth-order binomial and must be solved to obtain σ_1.)

22. A monolayer undergoes a first-order reaction to give products that also form monolayers. An equation that has been used under conditions of constant total area is $(\pi - \pi^\infty)/(\pi^0 - \pi^\infty) = \exp(-kt)$. Discuss what special circumstances are implied if this equation holds.

23. Under what physical constraints or assumptions would the rate law

$$\frac{\Delta V - \Delta V^\infty}{\Delta V^0 - \Delta V^\infty} = e^{-kt}$$

be expected to hold?

General References

N. K. Adam, *The Physics and Chemistry of Surfaces*, 3rd ed., Oxford University Press, London, 1941.

B. J. Berne and R. Pecora, *Dynamic Light Scattering*, Wiley, New York, 1976.

S. Chandrasekhar, *Liquid Crystals*, 2nd ed., Cambridge University Press, 1992.

J. T. Davies and E. K. Rideal, *Interfacial Phenomena*, Academic, New York, 1963.

D. H. Everett, ed., *Specialist Periodical Reports*, Vols. 2 and 3, The Chemical Soceity London, 1975, 1979.

G. L. Gaines, Jr., *Insoluble Monolayers at Liquid-Gas Interfaces*, Interscience, New York, 1966.

E. D. Goddard, ed., *Monolayers*, Advances in Chemistry Series 144, American Chemical Society, 1975.

R. J. Good, R. L. Patrick, and R. R. Stromberg, eds., *Techniques of Surface and Colloid Chemistry*, Marcel Dekker, New York, 1972.

V. K. LaMer, ed., *Retardation of Evaporation by Monolayers: Transport Processes*, Academic Press, New York, 1962.

V. G. Levich, *Physicochemical Hydrodynamics*, translated by Scripta Technica, Inc., Prentice-Hall, Englewood Cliffs, NJ, 1962.

A. W. Neumann and J. K. Spelt, eds., *Applied Surface Thermodynamics. Interfacial Tension and Contact Angles*, Marcel Dekker, New York, 1996.

L. I. Osipow, *Surface Chemistry, Theory and Industrial Applications*, Krieger, New York, 1977.

H. Van Olphen and K. J. Mysels, *Physical Chemistry: Enriching Topics from Colloid and Surface Chemistry*, Theorex (8327 La Jolla Scenic Drive), La Jolla, CA, 1975.

E. J. W. Verey and J. Th. G. Overbeek, *Theory of the Stability of Lyophobic Colloids*, Elsevier, New York, 1948.

Textual References

1. G. L. Gaines, Jr., *Insoluble Monolayers at Liquid-Gas Interfaces*, Interscience, New York, 1966.

2. Lord Rayleigh, *Proc. Roy. Soc.* (London), **47,** 364 (1890).

3. A. Pockels, *Nature*, **43,** 437 (1891).

4. M. C. Carey, in *Enterohepatic Circulation of Bile Acids and Sterol Metabolism*, G. Paumgartner, ed., MTP Press, Lancaster, Boston, 1984.

5. Lord Rayleigh, *Phil. Mag.*, **48,** 337 (1899).

6. H. Devaux, *J. Phys. Radium*, **699,** No. 2, 891 (1912).

7. I. Langmuir, *J. Am. Chem. Soc.*, **39,** 1848 (1917).

8. R. E. Baier, D. W. Goupil, S. Perlmutter, and R. King, *J. Rech. Atmos.*, **8,** 571 (1974).

9. W. D. Harkins, *The Physical Chemistry of Surface Films*, Reinhold, New York, 1952, Chapter 2.

10. L. A. Girifalco and R. J. Good, *J. Phys. Chem.*, **61,** 904 (1957).

11. F. E. Bartell, L. O. Case, and H. Brown, *J. Am. Chem. Soc.*, **55,** 2769 (1933).

12. P. Pomerantz, W. C. Clinton, and W. A. Zisman, *J. Colloid Interface Sci.*, **24,** 16 (1967); C. O. Timons and W. A. Zisman, ibid., **28,** 106 (1968).

13. G. Antonow, *J. Chim. Phys.*, **5,** 372 (1907).

14. G. L. Gaines, Jr., *J. Colloid Interface Sci.*, **66,** 593 (1978).

15. W. D. Harkins, *Colloid Symposium Monograph*, Vol. VI, The Chemical Catalog Company, New York, 1928, p. 24.

16. G. L. Gaines, Jr., and G. L. Gaines III, *J. Colloid Interface Sci.*, **63,** 394 (1978).

17. M. P. Khosla and B. Widom, *J. Colloid Interface Sci.*, **76,** 375 (1980).

18. J. C. Lang, Jr., P. K. Lim, and B. Widom, *J. Phys. Chem.*, **80,** 1719 (1976).

19. G. L. Gaines, Jr., *Colloids & Surfaces*, **8,** 383 (1984).

20. F. M. Fowkes, *Adv. Chem.*, **43,** 99 (1964).

21. A. W. Adamson, *Adv. Chem.*, **43,** 57 (1964).

22. F. M. Fowkes, *J. Colloid Interface Sci.*, **28,** 493 (1968).

23. R. J. Good, *Int. Eng. Chem.*, **62,** 54 (1970).

24. J. C. Melrose, *J. Colloid Interface Sci.*, **28,** 403 (1968).

25. F. M. Fowkes, *J. Phys. Chem.*, **67,** 2538 (1963).

26. F. M. Fowkes, *J. Phys. Chem.*, **84,** 510 (1980).

27. Y. Tamai, T. Matsunaga, and K. Horiuchi, *J. Colloid Interface Sci.*, **60,** 112 (1977).

28. J. Panzer, *J. Colloid Interface Sci.*, **44,** 142 (1973).

29. C. J. van Oss, R. J. Good, and M. K. Chaudhury, *Separ. Sci. Technol.*, **22,** 1 (1987); C. J. van Oss, R. J. Good, and M. K. Chaudhury, *Langmuir*, **4,** 884 (1988).

30. A. W. Neumann, R. J. Good, C. J. Hope, and M. Sejpal, *J. Colloid Interface Sci.*, **49,** 291 (1974).

31. J. Kloubek, *Collection Czechoslovak Chem. Commun.*, **52,** 271 (1987).

32. A. W. Adamson, *J. Phys. Chem.*, **72,** 2284 (1968).

33. J. F. Padday and N. D. Uffindell, *J. Phys. Chem.*, **72,** 1407 (1968); see also Ref. 34.

34. F. M. Fowkes, *J. Phys. Chem.*, **72,** 3700 (1968).

35. R. H. Ottewill, thesis, University of London, Queen Mary College, 1951; see also Ref. 36.

36. F. Hauxwell and R. H. Ottewill, *J. Colloid Interface Sci.*, **34,** 473 (1970).

37. M. W. Orem and A. W. Adamson, *J. Colloid Interface Sci.*, **31,** 278 (1969).

38. O. Reynolds, *Works*, **1,** 410; *Br. Assoc. Rept.*, 1881.

39. R. N. O'Brien, A. I. Feher, and J. Leja, *J. Colloid Interface Sci.*, **56,** 474 (1976).

40. R. N. O'Brien, A. I. Feher, and J. Leja, *J. Colloid Interface Sci.*, **51,** 366 (1975).

41. A. Cary and E. K. Rideal, *Proc. Roy. Soc.* (London), **A109,** 301 (1925).

42. J. Ahmad and R. S. Hansen, *J. Colloid Interface Sci.*, **38,** 601 (1972).

43. D. G. Sucio, O. Smigelschi, and E. Ruckenstein, *J. Colloid Interface Sci.*, **33,** 520 (1970).

44. P. Joos and J. van Hunsel, *J. Colloid Interface Sci.*, **106,** 161 (1985).

45. J. W. Peterson and J. C. Berg, *I & EC Fundamentals*, **25,** 668 (1986).

46. J. E. Saylor and G. T. Barnes, *J. Colloid Interface Sci.*, **35,** 143 (1971).

47. S. Y. Park, C. H. Chang, D. J. Ahn, and E. I. Franses, *Langmuir*, **9,** 3640 (1993).

48. M. K. Bernett and W. A. Zisman, *Adv. Chem., Ser.* **43,** 332 (1964); W. D. Bascom, R. L. Cottington, and C. R. Singleterry, ibid., p. 355.

49. C. G. M. Marangoni, *Annln PHys.* (Poggendorf), **143,** 337 (1871).

50. Proverbs 23:31.

51. F. Sebba, *J. Colloid Interface Sci.,* **73,** 278 (1980).

52. S. M. Troian, X. L. Wu, and S. A. Safran, *Phys. Rev. Lett.*, **62,** 1496 (1989).

53. C. V. Sternling and L. E. Scriven, *AIChE J.*, 514 (Dec. 1959).

54. A. G. Bois, M. G. Ivanova, and I. I. Panaiotov, *Langmuir*, **3,** 215 (1987).

55. L. S. Chang and J. C. Berg, *AIChE J.*, **31,** 551 (1985).

56. H. K. Cammenga, D. Schreiber, G. T. Barnes, and D. S. Hunter, *J. Colloid Interface Sci.*, **98,** 585 (1984).

57. M. V. Ostrovsky and R. M. Ostrovsky, *J. Colloid Interface Sci.*, **93,** 392 (1983).

58. J. Berg, *Canad. Metallurg. Quart.*, **21,** 121 (1982).

59. I. Langmuir, *J. Chem. Phys.*, **1,** 756 (1933).

60. N. F. Miller, *J. Phys. Chem.*, **45,** 1025 (1941).

61. D. J. Donahue and F. E. Bartell, *J. Phys. Chem.*, **56,** 480 (1952).

62. H. M. Princen and S. G. Mason, *J. Colloid Sci.* **20,** 246 (1965).

63. J. S. Rowlinson, *J. Chem. Soc., Faraday Trans. 2*, **79,** 77 (1983).

64. B. V. Toshev, D. Platikanov, and A. Scheludko, *Langmuir*, **4,** 489 (1988).

65. V. G. Babak, *Colloids & Surfaces*, **25,** 205 (1987).

66. L. Boruvka, J. Gaydos, and A. W. Neumann, *Colloids & Surfaces*, **43,** 307 (1990).

67. J. Kerins and B. Widom, *J. Chem. Phys.*, **77,** 2061 (1982).

68. G. L. Gaines, Jr., *J. Colloid Interface Sci.*, **62,** 191 (1977).

69. G. L. Gaines, Jr., in *Surface Chemistry and Colloids* (MTP International Review of Science), Vol. 7, M. Kerker, ed., University Park Press, Baltimore, 1972.

70. D. Y. Kwok, D. Vollhardt, R. Miller, D. Li, and A. W. Neumann, *Colloids & Surfaces*, **A88,** 51 (1994).

71. D. Y. Kwok, M. A. Cabrerizo-Vilchez, D. Vollhardt, R. Miller, and A. W. Neumann, *Langmuir* (in press).

72. D. Y. Kwok, B. Tadros, H. Deol, D. Vollhardt, R. Miller, M. A. Cabrerizo-Vilchez, and A. W. Neumann, *Langmuir* (in press).

73. I. S. Costin and G. T. Barnes, *J. Colloid Interface Sci.*, **51,** 94 (1975).

74. B. M. Abraham, K. Miyano, K. Buzard, and J. B. Ketterson, *Rev. Sci. Instrum.*, **51,** 1083 (1980).

75. K. J. Mysels, *Langmuir*, **6,** (in press).

76. M. W. Kim and D. S. Cannell, *Phys. Rev. A*, **13,** 411 (1976).

77. G. Munger and R. M. Leblanc, *Rev. Sci. Instrum.*, **51,** 710 (1980).

78. H. E. Ries, Jr., *Nat. Phys. Sci.*, **243,** 14 (1973).

79. G. L. Gaines, Jr., *J. Colloid Interface Sci.*, **98,** 272 (1984).

80. G. T. Barnes and J. C. W. Sharp, *J. Colloid Interface Sci.*, **144,** 595 (1991).

81. W. Rabinovitch, R. F. Robertson, and S. G. Mason, *Can. J. Chem.*, **38,** 1881 (1960).

82. L. Ter-Minassian-Saraga, *Pure Appl. Chem.*, **57,** 621 (1985).

83. J. A. Bergeron and G. L. Gaines, Jr., *J. Colloid Interface Sci.*, **23,** 292 (1967).

84. M. Plaisance and L. Ter-Minassian-Saraga, *CR*, **270,** 1269 (1970).

85. A. Noblet, H. Ridelaire, and G. Sylin, *J. Phys. E*, **17,** 234 (1984).

86. W. D. Harkins and E. K. Fischer, *J. Chem. Phys.*, **1,** 852 (1933).

87. D. J. Benvegnu and H. M. McConnell, *J. Phys. Chem.*, **97,** 6686 (1993).

88. H. M. McConnell, P. A. Rice, and D. J. Benvegnu, *J. Phys. Chem.*, **94,** 8965 (1990).

89. H. M. McConnell, *Biophys. J.*, **64,** 577 (1993).

90. J. F. Klingler and H. M. McConnell, *J. Phys. Chem.*, **97,** 2962 (1993).

91. J. R. MacDonald and C. A. Barlow, *J. Chem. Phys.*, **43,** 2575 (1965) and preceding papers.

92. B. A. Pethica, M. M. Standish, J. Mingins, C. Smart, D. H. Iles, M. E. Feinstein, S. A. Hossain, and J. B. Pethica, *Adv. Chem.*, **144,** 123 (1975).

93. M. Blank and R. H. Ottewill, *J. Phys. Chem.*, **68,** 2206 (1964).

94. K. W. Bewig and W. A. Zisman, *J. Phys. Chem.*, **68,** 1804 (1964).

95. M. van den Tempel, *J. Non-Newtonian Fluid Mech.*, **2,** 205 (1977).

96. F. C. Goodrich, *Proc. Roy. Soc.* (London) **A374,** 341 (1981).

97. R. B. Bird, W. E. Stewart, and E. N. Lightfoot, *Transport Phenomena*, Wiley, New York, 1960.

98. D. W. Criddle, in *Rheology*, Vol. 3, F. R. Eirich, Academic, New York, 1960.

99. R. J. Myers and W. D. Harkins, *J. Chem. Phys.*, **5,** 601 (1937).

100. W. D. Harkins and J. G. Kirkwood, *J. Chem. Phys.*, **6,** 53 (1938).

101. G. C. Nutting and W. D. Harkins, *J. Am. Chem. Soc.*, **62,** 3155 (1940).

102. W. E. Ewers and R. A. Sack, *Austral. J. Chem.*, **7,** 40 (1954).

103. R. S. Hansen, *J. Phys. Chem.*, **63,** 637 (1959).

104. H. E. Gaub and H. M. McConnell, *J. Phys. Chem.*, **90,** 6830 (1986).

105. J. Plateau, *Phil Mag.*, **38**(4), 445 (1869).

106. N. W. Tschoegl, *Kolloid Z.*, **181,** 19 (1962).

107. R. D. Krieg, J. E. Son, and R. W. Flumerfelt, *J. Colloid Interface Sci.*, **79,** 14 (1981).

108. B. M. Abraham and J. B. Ketterson, *Langmuir*, **1,** 461 (1985).

109. J. Krägel, S. Siegel, R. Miller, M. Born, and K. H. Schano, *Colloids & Surfaces*, **A91,** 169 (1994).

110. J. W. Gardner, J. V. Addison, and R. S. Schechter, *AIChE J.*, **24,** 400 (1978).

111. F. C. Goodrich, L. H. Allen, and A. Poskanzer, *J. Colloid Interface Sci.*, **52,** 201 (1975).

112. F. C. Goodrich and D. W. Goupil, *J. Colloid Interface Sci.*, **75,** 590 (1980).

113. M. Blank and J. S. Britten, *J. Colloid Sci.*, **20,** 789 (1965).

114. E. R. Cooper and J. A. Mann, *J. Phys. Chem.*, **77,** 3024 (1973).

115. H. A. Stone and H. M. McConnell, *Proc. Roy. Soc.* (London) **A448,** 97 (1995).

116. H. Kuhn and D. Möbius, "Monolayer Assemblies," in *Investigations of Surfaces and Interfaces—Part B*, 2nd ed., B. W. Rossiter and R. C. Baetzold, eds., Vol. IXB of *Physical Methods of Chemistry*, Wiley, 1993.

117. K. C. O'Brien, J. A. Mann, Jr., and J. B. Lando, *Langmuir*, **2**, 338 (1986).

118. F. van Voorst Vader, Th. F. Erkens, and M. van den Tempel, *Trans. Faraday Soc.*, **60**, 1170 (1964).

119. J. Lucassen and M. van den Tempel, *Chem. Eng. Sci.*, **27**, 1283 (1972).

120. H. C. Maru, V. Mohan, and D. T. Wasan, *Chem. Eng. Sci.*, **34**, 1283 (1979); H. C. Maru and D. T. Wasan, ibid., **34**, 1295 (1979).

121. K. C. O'Brien and J. B. Lando, *Langmuir*, **1**, 301 (1985).

122. L. Ter-Minassian-Saraga, I. Panaiotov, and J. S. Abitboul, *J. Colloid Interface Sci.*, **72**, 54 (1979).

123. H. Yu, in *Physics of Polymer Surfaces and Interfaces*, I. Sanchez ed., Butterworth-Heinemann, 1992, Chapter 12.

124. D. Langevin, *J. Colloid Interface Sci.*, **80**, 412 (1981).

125. M. van den Tempel and R. P. van de Riet, *J. Chem. Phys.*, **42**, 2769 (1965).

126. F. C. Goodrich, *Proc. Roy. Soc. (London)*, **A260**, 503 (1961).

127. R. S. Hansen and J. A. Mann, Jr., *J. Appl. Phys.*, **35**, 152 (1964).

128. J. A. Mann, Jr., and R. S. Hansen, *J. Colloid Sci.*, **18**, 805 (1963).

129. J. A. Stone and W. J. Rice, *J. Colloid Interface Sci.*, **61**, 160 (1977).

130. W. D. Garrett and W. A. Zisman, *J. Phys. Chem.*, **74**, 1796 (1970).

131. V. G. Levich, *Physicochemical Hydrodynamics*, translated by Scripta Technica, Inc., Prentice-Hall, Englewood Cliffs, NJ, 1962, p. 603.

132. D. Byrne and J. C. Earnshaw, *J. Phys. D: Appl. Phys.*, **12**, 1133 (1979).

133. Y. Chen, M. Sano, M. Kawaguchi, H. Yu, and G. Zographi, *Langmuir,* **2**, 349 (1986). (Note also p. 683.)

134. G. D. J. Phillies, in *Treatise in Analytical Chemistry*, P. J. Elving and E. J. Meehan, eds., Wiley, New York, 1986, Part 1, Section H, Chapter 9.

135. M. A. Bouchiat and J. Meunier, *J. Physique*, **32**, 561 (1971).

136. J. A. Stone and W. J. Rice, *J. Colloid Interface Sci.*, **61**, 160 (1977).

137. H. Löfgren, R. D. Neuman, L. E. Scriven, and H. T. Davis, *J. Colloid Interface Sci.*, **98**, 175 (1984).

138. J. C. Earnshaw and R. C. McGivern, *J. Phys. D: Appl. Phys.*, **20**, 82 (1987).

139. R. B. Dorshow, A. Hajiloo, and R. L. Swofford, *J. Appl. Phys.*, **63**, 1265 (1988).

140. J. A. Mann and R. V. Edwards, *Rev. Sci. Instrum.*, **55**, 727 (1984).

141. J. C. Earnshaw, in *The Application of Laser Light Scattering to the Study of Biological Motion*, J. C. Earnshaw and N. W. Steer, eds., Plenum Press, New York, 1983, p. 275.

142. R. M. A. Azzam and N. M. Bashara, *Ellipsometry and Polarized Light*, North-Holland, 1977.

143. D. Ducharme, C. Salesse, and R. M. Leblanc, *Thin Solid Films*, **132**, 83 (1985).

144. D. Ducharme, A. Tessier, and R. M. Leblanc, *Rev. Sci. Instrum.*, **58**, 571 (1987).

145. M. Kawaguchi, M. Tohyama, Y. Mutch, and A. Takahashi, *Langmuir*, **4**, 407 (1988).

146. R. Reiter, H. Motschmann, H. Orendi, A. Nemetz, and W. Knoll, *Langmuir*, **8**, 1784 (1992).

147. R. N. O'Brien, in *Physical Methods of Chemistry*, A. Weissberger and B. W. Rossiter, eds., Part 3A, Wiley, New York, 1972.

148. K. B. Blodgett and I. Langmuir, *Phys. Rev.*, **51**, 964 (1937).

149. R. E. Hartman, *J. Opt. Soc. Am.*, **44**, 192 (1954).

150. Z. Kozarac, A. Dhathathreyan, and D. Möbius, *Eur. Biophys. J.*, **15**, 193 (1987).

151. H. Grüniger, D. Möbius, and H. Meyer, *J. Chem. Phys.*, **79**, 3701 (1983); D. Möbius, M. Orrit, H. Grüniger, and H. Meyer, *Thin Solid Films*, **132**, 41 (1985).

152. R. A. Dluhy and D. G. Cornell, *J. Phys. Chem.*, **89**, 3195 (1985).

153. J. T. Buontempo and S. A. Rice, *J. Chem. Phys.*, **98**, 5825 (1993); **98**, 5835 (1993).

154. J. T. Buontempo and S. A. Rice, *J. Chem. Phys.*, **99**, 7030 (1993).

155. J. T. Buontempo, S. A. Rice, S. Karaborni, and J. I. Seipmann, *Langmuir*, **9**, 1604 (1993).

156. M. Li and S. A. Rice, *J. Chem. Phys.*, **104**, 6860 (1996).

157. T. Takenaka and H. Fukuzaki, *J. Raman Spectroscopy*, **8**, 151 (1979).

158. A. G. Tweet, G. L. Gaines, Jr., and W. D. Bellamy, *J. Chem. Phys.*, **40**, 2596 (1964).

159. T. Trosper, R. B. Park, and K. Sauer, *Photochem. Photobiol.*, **7**, 451 (1968).

160. S. J. Valenty, D. E. Behnken, and G. L. Gaines, Jr., *Inorg. Chem.*, **13**, 2160 (1979).

161. P. A. Anfinrud, D. E. Hart, and W. S. Struve, *J. Phys. Chem.*, **92**, 4067 (1988).

162. R. A. Hall, D. Hayes, P. J. Thistlewaite, and F. Grieser, *Colloids & Surfaces*, **56**, 339 (1991).

163. E. Wistus, E. Mukhtar, M. Almgren, and S. Lindquist, *Langmuir*, **8**, 1366 (1992).

164. F. Caruso, F. Grieser, A. Murphy, P. Thistlewaite, M. Almgren, and E. Wistus, *J. Am. Chem. Soc.*, **113**, 4838 (1991).

165. F. Caruso, F. Grieser, and P. J. Thistlewaite, *Langmuir*, **9**, 3142 (1993).

166. H. Zocher and F. Stiebel, *Z. Phys. Chem.*, **147**, 401 (1930).

167. H. Möhwald, *Annu. Rev. Phys. Chem.*, **41**, 441 (1990).

168. H. M. McConnell, *Annu. Rev. Phys. Chem.*, **42**, 171 (1991).

169. R.M. Weis and H. M. McConnell, *Nature*, **310**, 47 (1984).

170. H. M. McConnell, L. K. Tamm, and R. M. Weis, *Proc. Natl. Acad. Sci.* (USA), **81**, 3249 (1984).

171. R. C. Ahuja, P. L. Caruso, D. Hönig, J. Maack, D. Möbius, and G. A. Overbeck, "Morphology of Organized Monolayers by Brewster Angle Microscopy," in *Microchemistry*, H. Masuhara, ed., Elsevier, 1994.

172. G. A. Overbeck, D. Hönig, and D. Möbius, *Thin Solid Films*, **242**, 213 (1994).

173. G. A. Overbeck, D. Hönig, L. Wolthaus, M. Gnade, and D. Möbius, *Thin Solid Films*, **242**, 26 (1994).

174. P. Dutta, J. B. Peng, B. Lin, J. B. Ketterson, M. Prakash, P. Georgopolous, and S. Ehrlich, *Phys. Rev. Lett.*, **58**, 2228 (1987).

175. S. W. Barton, B. N. Thomas, E. B. Flom, S. A. Rice, B. Lin, and J. B. Peng, *J. Chem. Phys.*, **89**, 2257 (1988).

176. A. A. Acero, M. Li, B. Lin, S. A. Rice, M. Goldmann, I. B. Azouz, A. Goudot, and F. Rondelez, *J. Chem. Phys.*, **99**, 7214 (1993).

177. M. Goldmann, P. Nassoy, F. Rondelez, A. Renault, S. Shin, and S. A. Rice, *J. Phys. II*, **4**, 773 (1994).

178. S. W. Barton, A. Goudot, O. Bouloussa, F. Rondelez, B. Lin, F. Novak, A. A. Acero, and S. A. Rice, *J. Chem. Phys.*, **96**, 1343 (1992).

179. J. Majewski, L. Margulis, D. Jacquemain, F. Leveiller, C. Böhm, T. Arad, Y. Talmon, M. Lahav, and L. Leiserowitz, *Science*, **261**, 899 (1993).

180. I. R. Gentle, P. M. Saville, J. W. White, and J. Penfold, *Langmuir*, **9**, 646 (1993).

181. P. M. Saville, I. R. Gentle, J. W. White, J. Penfold, and J. R. P. Webster, *J. Phys. Chem.*, **98**, 5935 (1994).

182. S. Chandrasekhar, *Liquid Crystals*, 2nd ed., Cambridge University Press, 1992.

183. N. K. Adam, *The Physics and Chemistry of Surfaces*, 3rd ed., Oxford University Press, London, 1941.

184. C. M. Knobler and R. C. Desai, *Annu. Rev. Phys. Chem.*, **43**, 207 (1992).

185. G. A. Overbeck and D. Möbius, *J. Phys. Chem.*, **97**, 7999 (1993).

186. G. A. Lawrie and G. T. Barnes, *J. Colloid Interface Sci.*, **162**, 36 (1994).

187. D. Andelman, F. Brochard, C. Knobler, and F. Rondelez, *Micelles, Membranes, Microemulsions and Monolayers*, Springer-Verlag, 1994, Chapter 12.

188. L. Ter-Minassian-Saraga and I. Prigogine, *Mem. Serv. Chim. Etat*, **38**, 109 (1953).

189. F. M. Fowkes, *J. Phys. Chem.*, **66**, 385 (1962).

190. G. L. Gaines, Jr., *J. Chem. Phys.*, **69**(2), 924 (1978).

191. J. Popielawski and S. A. Rice, *J. Chem. Phys.*, **88**, 1279 (1988).

192. N. L. Gershfeld and C. S. Patlak, *J. Phys. Chem.*, **70**, 286 (1966).

193. N. R. Pallas and B. A. Pethica, *J. Chem. Soc., Faraday Trans.*, **83**, 585 (1987).

194. M. W. Kim and D. S. Cannell, *Phys. Rev. A.* **14**, 1299 (1976).

195. B. Stoeckly, *Phys. Rev. A*, **15**, 2558 (1977).

196. D. J. Keller, H. M. McConnell, and V. T. Moy, *J. Phys. Chem.*, **90**, 2311 (1986).

197. D. Andelman, F. Broçhard, P. G. de Gennes, and J. F. Joanny, *CR Acad. Sci.* (Paris), **30**, 675 (1985).

198. D. Andelman, F. Broçhard, and J. F. Joanny, *J. Chem. Phys.*, **86**, 3673 (1987).

199. J. F. Baret, H. Hasmonay, J. L. Dupin, and M. Dupeyrat, *Chem. Phys. Lipids*, **30**, 177 (1982).

200. N. L. Gershfeld, *Annu. Rev. Phys. Chem.*, **27**, 350 (1976).

201. J. Harris and S. A. Rice, *J. Chem. Phys.*, **88**, 1298 (1988); see also Z. Wang and S. A. Rice, ibid., **88**, 1290 (1988).

202. Z. Cai and S. A. Rice, *J. Chem. Phys.*, **96**, 6229 (1992).

203. S. Shin and S. A. Rice, *J. Chem. Phys.*, **101**, 2508 (1994).

204. J. Schofield and S. A. Rice, *J. Chem. Phys.*, **103**, 5792 (1995).

205. W. D. Harkins and E. Boyd, *J. Phys. Chem.*, **45**, 20 (1941); G. E. Boyd, *J. Phys. Chem.*, **62**, 536 (1958).

206. D. A. Cadenhead and R. J. Demchak, *J. Chem. Phys.*, **49,** 1372 (1968).

207. F. Müller-Landau and D. A. Cadenhead, *J. Colloid Interface Sci.*, **73,** 264 (1980).

208. A. G. Bois, I. I. Panaiotov, and J. F. Baret, *Chem. Phys. Lipids*, **34,** 265 (1984).

209. S. A. Safran, M. O. Robbins, and S. Garoff, *Phys. Rev. A*, **33,** 2186 (1986).

210. J. Israelachvili, *Langmuir*, **10,** 3774 (1994).

211. M. J. Vold, *J. Colloid Sci.*, **7,** 196 (1952).

212. J. J. Kipling and A. D. Norris, *J. Colloid Sci.*, **8,** 547 (1953).

213. F. J. Garfias, *J. Phys. Chem.*, **83,** 3126 (1979).

214. F. J. Garfias, *J. Phys. Chem.*, **84,** 2297 (1980).

215. B. Lin, J. B. Peng, J. B. Ketterson, P. Dutta, B. N. Thomas, J. Buontempo, and S. A. Rice, *J. Chem. Phys.*, **90,** 2393 (1989).

216. S. Shin and S. A. Rice, *Langmuir*, **10,** 262 (1994).

217. Z. Huang, M. L. Schlossman, A. A. Acero, Z. Zhang, N. Lei, and S. A. Rice, *Langmuir*, **11,** 2742 (1995).

218. M. E. Schmidt, S. Shin, and S. A. Rice, *J. Chem. Phys.*, **104,** 2104 (1996).

219. M. E. Schmidt, S. Shin, and S. A. Rice, *J. Chem. Phys.*, **104,** 2114 (1996).

220. M. Li, A. A. Acero, Z. Huang, and S. A. Rice, *Nature*, **367,** 151 (1994).

221. J. Gao and S. A. Rice, *J. Chem. Phys.*, **99,** 7020 (1993).

222. V. T. Moy, D. J. Keller, H. E. Gaub, and H. H. McConnell, *J. Phys. Chem.*, **90,** 3198 (1986).

223. H. E. Ries, Jr., and H. Swift, *Langmuir*, **3,** 853 (1987).

224. R. D. Neuman, *J. Microscopy*, **105,** 283 (1975); *J. Colloid Interface Sci.*, **56,** 505 (1976).

225. R. D. Smith and J. C. Berg, *J. Colloid Interface Sci.*, **74,** 273 (1980).

226. D. J. Keller, J. P. Korb, and H. M. McConnell, *J. Phys. Chem.*, **91,** 6417 (1987).

227. H. M. McConnell, *J. Phys. Chem.*, **94,** 4728 (1990).

228. K. Y. C. Lee and H. M. McConnell, *J. Phys. Chem.*, **97,** 9532 (1993).

229. R. de Koker and H. M. McConnell, *J. Phys. Chem.*, **97,** 13419 (1993).

230. R. de Koker and H. M. McConnell, *J. Phys. Chem.*, **98,** 5389 (1994).

231. R. de Koker, W. Jiang, and H. M. McConnell, *J. Phys. Chem.*, **99,** 6251 (1995).

232. H. E. Gaub, V. T. Moy, and H. M. McConnell, *J. Phys. Chem.*, **90,** 1721 (1986).

233. D. Andelman and H. Orland, *J. Am. Chem. Soc.*, **115,** 12322 (1993).

234. D. J. Benvegnu and H. M. McConnell, *J. Phys. Chem.*, **96,** 6820 (1992).

235. T. K. Vanderlick and H. Möhwald, *J. Phys. Chem.*, **94,** 886 (1990).

236. K. Y. C. Lee, J. Klingler, and H. M. McConnell, *Science*, **263,** 655 (1994).

237. K. Y. C. Lee and H. M. McConnell, *Biophys. J.*, **68,** 1740 (1995).

238. H. A. Stone and H. M. McConnell, *J. Phys. Chem.*, **99,** 13505 (1995).

239. H. E. Ries, Jr., *Colloids & Surfaces*, **10,** 283 (1984).

240. R. E. Pagano and N. L. Gershfeld, *J. Phys. Chem.*, **76,** 1238 (1972).

241. G. L. Gaines, Jr., *J. Colloid Interface Sci.*, **21,** 315 (1966).

242. D. A. Cadenhead and R. J. Demchak, ibid., **30,** 76 (1969).

243. K. J. Bacon and G. T. Barnes, *J. Colloid Interface Sci.*, **67,** 70 (1978), and preceding papers.

244. G. T. Barnes, *J. Colloid Interface Sci.*, **144,** 299 (1991).

245. E. H. Lucassen-Reynders and M. van den Tempel, *Proceedings of the IVth International Congress on Surface Active Substances, Brussels, 1964,* Vol. 2, J. Th. G. Overbeek, ed., Gordon and Breach, New York, 1967; E. H. Lucassen-Reynders, *J. Colloid Interface Sci.*, **42,** 554 (1973); **41,** 156 (1972).

246. I. S. Costin and G. T. Barnes, *J. Colloid Interface Sci.*, **51,** 106 (1975).

247. H. E. Ries, Jr. and H. Swift, *J. Colloid Interface Sci.*, **64,** 111 (1978).

248. D. A. Cadenhead and F. Müller-Landau, *Chem. Phys. Lipids*, **25,** 329 (1979).

249. K. Tajima and N. L. Gershfeld, *Adv. Chem.*, **144,** 165 (1975).

250. D. A. Cadenhead, B. M. J. Kellner, and M. C. Phillips. *J. Colloid Interface Sci.*, **57,** 1 (1976).

251. L. Ter-Minassian-Saraga, *J. Colloid Interface Sci.*, **70,** 245 (1979).

252. H. Matuo, N. Yosida, K. Motomura, and R. Matuura, *Bull. Chem. Soc. Jpn.*, **52,** 667 (1979), and preceding papers.

253. Y. Hendrikx, *J. Colloid Interface Sci.*, **69,** 493 (1979).

254. K. M. Maloney and D. W. Grainger, *Chem. Phys. Lipids*, **65,** 31 (1993).

255. D. O. Shah and S. Y. Shiao, *Adv. Chem.*, **144,** 153 (1975).

256. D. Möbius, *Accts. Chem. Res.*, **14,** 63 (1981); *Z. Physik. Chemie Neue Folge*, **154,** 121 (1987).

257. S. Subramaniam and H. M. McConnell, *J. Phys. Chem.*, **91,** 1715 (1987).

258. E. D. Goddard and J. H. Schulman, *J. Colloid Sci.*, **8,** 309 (1953).

259. P. J. Anderson and B. A. Pethica, *Trans. Faraday Soc.*, **52,** 1080 (1956).

260. F. M. Fowkes, *J. Phys. Chem.*, **66,** 385 (1962).

261. F. M. Fowkes, *J. Phys. Chem.*, **67,** 1982 (1963).

262. E. H. Lucassen-Reynders, *J. Colloid Interface Sci.*, **42,** 563 (1973).

263. Y. Hendrikx and L. Ter-Minassian-Saraga, *Adv. Chem.*, **144,** 177 (1975).

264. L. Ter-Minassian-Saraga, *Langmuir*, **2,** 24 (1986).

265. M. A. McGregor and G. T. Barnes, *J. Colloid Interface Sci.*, **60,** 408 (1977).

266. Y. Hendrikx and L. Ter-Minassian-Saraga, *CR*, **269,** 880 (1969).

267. B. Pethica, *Trans. Faraday Soc.*, **51,** 1402 (1955).

268. D. M. Alexander and G. T. Barnes, *J. Chem. Soc. Faraday I*, **76,** 118 (1980).

269. D. M. Alexander, G. T. Barnes, M. A. McGregor, and K. Walker, in *Phenomena in Mixed Surfactant Systems*, J. F. Scamehorn, ed., ACS Symposium Series 311, 1976, p. 133.

270. M. A. McGregor and G. T. Barnes, *J. Pharma. Sci.*, **67,** 1054 (1978); *J. Colloid Interface Sci.*, **65,** 291 (1978 and preceding papers.

271. D. G. Hall, *Langmuir*, **2,** 809 (1986).

272. R. B. Dean and K. E. Hayes, *J. Am. Chem. Soc.*, **73,** 5583 (1954).

273. G. T. Barnes, *Adv. Colloid Interface Sci.*, **25,** 89 (1986).

274. M. V. Ostrovsky, *Colloids & Surfaces*, **14,** 161 (1985).

275. V. K. LaMer, T. W. Healy, and L. A. G. Aylmore, *J. Colloid Sci.*, **19,** 676 (1964).

276. H. L. Rosano and V. K. LaMer, *J. Phys. Chem.*, **60,** 348 (1956).

277. R. J. Archer and V. K. LaMer, *J. Phys. Chem.*, **59,** 200 (1955).

278. G. T. Barnes, K. J. Bacon, and J. M. Ash, *J. Colloid Interface Sci.*, **76,** 263 (1980).

279. L. Ginsberg and N. L. Gershfeld, *Biophys. J.*, **47,** 211 (1985).

280. V. K. LaMer and T. W. Healy, *Science*, **148,** 36 (1965).

281. W. W. Mansfield, *Nature*, **175,** 247 (1955).

282. G. T. Barnes, *J. Hydrology*, **145,** 165 (1993).

283. I. S. Costin and G. T. Barnes, *J. Colloid Interface Sci.*, **51,** 122 (1975).

284. M. Matsumoto and G. T. Barnes, *J. Colloid Interface Sci.*, **148,** 280 (1992).

285. G. T. Barnes and D. S. Hunter, *J. Colloid Interface Sci.*, **129,** 585 (1989).

286. C. J. Drummond, P. Elliot, D. N. Furlong, and G. T. Barnes, *Thin Solid Films*, **210/211,** 69 (1992).

287. G. T. Barnes, K. J. Bacon, and J. M. Ash, *J. Colloid Interface Sci.*, **76,** 263 (1980).

288. R. J. Vanderveen and G. T. Barnes, *Thin Solid Films*, **134,** 227 (1985).

289. G. T. Barnes and H. K. Cammenga, *J. Colloid Interface Sci.*, **72,** 140 (1979).

290. G. T. Barnes and D. S. Hunter, *J. Colloid Interface Sci.*, **136,** 198 (1990).

291. D. S. Hunter, G. T. B. J. S. Godfrey, and F. Grieser, *J. Colloid Interface Sci.*, **138,** 307 (1990).

292. L. Ter-Minassian-Saraga, *J. Chim. Phys.*, **52,** 181 (1955).

293. N. L. Gershfeld, in *Techniques of Surface and Colloid Chemistry*, R. J. Good, R. L. Patrick, and R. R. Stromberg, eds., Marcel Dekker, New York, 1972.

294. R. Z. Guzman, R. G. Carbonell, and P. K. Kilpatrick, *J. Colloid Interface Sci.*, **114,** 536 (1986).

295. I. Panaiotov, A. Sanfeld, A. Bois, and J. F. Baret, *J. Colloid Interface Sci.*, **96,** 315 (1983).

296. Z. Biikadi, J. D. Parsons, J. A. Mann, Jr., and R. D. Neuman, *J. Chem. Phys.*, **72,** 960 (1980).

297. M. L. Agrawal and R. D. Neuman, *Colloids & Surfaces*, **32,** 177 (1988).

298. J. T. Davies, *Adv. Catal.*, **6,** 1 (1954).

299. J. Bagg, M. B. Abramson, M. Fishman, M. D. Haber, and H. P. Gregor, *J. Am. Chem. Soc.*, **86,** 2759 (1964).

300. S. J. Valenty, in *Interfacial Photoprocesses: Energy Conversion and Synthesis*, M. S. Wrighton, ed., Advances in Chemistry Series No. 134, American Chemical Society, Washington, DC, 1980.

301. S. J. Valenty, *J. Am. Chem. Soc.*, **101,** 1 (1979).

302. W. D. Bellamy, G. L. Gaines, Jr., and A. G. Tweet, *J. Chem. Phys.*, **39,** 2528 (1963).

303. F. Grieser, P. Thistlethwaite, and R. Urquhart, *J. Phys. Chem.*, **91,** 5286 (1987).

304. H. H. G. Jellinek and M. H. Roberts, *J. Sci. Food Agric.*, **2,** 391 (1951).

305. K. C. O'Brien and J. B. Lando, *Langmuir*, **1,** 533 (1985).

306. J. T. Davies, *Trans. Faraday Soc.*, **45,** 448 (1949).

307. H. H. G. Jellinek and M. R. Roberts (The Right-Honorable Margaret Thatcher), *J. Sci. Food Agric.*, **2,** 391 (1951).

308. J. T. Davies and E. K. Rideal, *Proc. Roy. Soc.* (London), **A194,** 417 (1948).

309. R. Mittelmann and R. C. Palmer, *Trans. Faraday Soc.*, **38,** 506 (1942).

310. A. R. Gilby and A. E. Alexander, *Austral. J. Chem.*, **9,** 347 (1956).

311. K. C. O'Brien, C. E. Rogers and J. B. Lando, *Thin Solid Films*, **102,** 131 (1983).

312. K. C. O'Brien, J. Long, and J. B. Lando, *Langmuir*, **1,** 514 (1985).

313. D. Day and J. B. Lando, *Macromolecules*, **13,** 1478 (1980).

314. R. Rolandi, R. Paradiso, S. Z. Xu, C. Palmer, and J. H. Fendler, *J. Am. Chem. Soc.*, **111,** 5233 (1989).

315. D. Day and H. Ringsdorf, *J. Polym. Sci., Polym. Lett. Ed.*, **16,** 205 (1978).

316. A. G. Tweet, G. L. Gaines, Jr., and W. D. Bellamy, *J. Chem. Phys.*, **41,** 1008 (1964).

317. D. G. Whitten, *J. Am. Chem. Soc.*, **96,** 594 (1974).

CHAPTER V

Electrical Aspects of Surface Chemistry

1. Introduction

The influence of electrical charges on surfaces is very important to their physical chemistry. The Coulombic interaction between charged colloids is responsible for a myriad of behaviors from the formation of opals to the stability of biological cells. Although this is a broad subject involving both practical application and fundamental physics and chemistry, we must limit our discussion to those areas having direct implications for surface science.

The discussion focuses on two broad aspects of electrical phenomena at interfaces; in the first we determine the consequences of the presence of electrical charges at an interface with an electrolyte solution, and in the second we explore the nature of the potential occurring at phase boundaries. Even within these areas, frequent reference will be made to various specialized treatises dealing with such subjects rather than attempting to cover the general literature. One important application, namely, to the treatment of long-range forces between surfaces, is developed in the next chapter.

One complication is the matter of units; while the Système International d'Unités (SI) requires additional and sometimes awkward constants, its broad use requires attention [1]. Hence, while we present the derivation in the cgs/esu system, we show alternative forms appropriate to the SI system in Tables V-1 and V-2.

2. The Electrical Double Layer

An important group of electrical phenomena concerns the nature of the ion distribution in a solution surrounding a charged surface. To begin with, consider a plane surface bearing a uniform positive charge density in contact with a solution phase containing positive and negative ions. The electrical potential begins at the surface as ψ_0 and decreases as one proceeds into the solution in a manner to be determined. At any point the potential ψ determines the potential energy $ze\psi$ of an ion in the local field where z is the valence of the ion and e is the charge on the electron. The probability of finding an ion at a particular point will depend on the local potential through a Boltzmann distribution, $e^{-ze\psi/kT}$ in analogy to the distribution of a gas in a gravitational field where the potential is mgh, and the variation of concentration with altitude is given by

TABLE V-1
Conversions between cgs/esu and SI Constants and Their Units

Function	cgs/esu	SI
Force	f in dynes	f in newtons (N) = 10^5 dynes
Potential energy	U in ergs	U in joules (J) = 10^7 ergs
Electrostatic energy	$e\psi$ in ergs	$e\psi$ in joules (J)
Electrostatic charge	$e = 4.803 \times 10^{-10}$ esu	$e = 1.602 \times 10^{-19}$ Coulombs (C)
Permittivity	1	$\epsilon_0 = 8.854 \times 10^{-12}$ C^2 J^{-1} m^{-1}
Boltzmann constant	$k = 1.38 \times 10^{-16}$ ergs/K	$k = 1.38 \times 10^{-23}$ J/K
Capacitance	C in statfarads/cm^2 $= esu/V_{esu}$	C in farads/m^2 = C V^{-1} m^{-2}
Voltage	V_{esu}	300 V

$n = n_0 e^{-mgh/kT}$, where n_0 is the concentration at zero altitude. A symmetrical electrolyte, comprising two kinds of ions of equal and opposite charge, $+z$ and $-z$, will have a distribution

$$n^- = n_0 e^{ze\psi/kT} \qquad n^+ = n_0 e^{-ze\psi/kT} \qquad \text{(V-1)}$$

that is more complicated than the gravitational case. First, positive charges are repelled from the surface whereas negative ones are attracted; second, the system as a whole should be electrically neutral so that far away from the surfaces $n^+ = n^-$. Close to the surface there will be an excess of negative ions so that a new charge exists; the total net charge in the solution is balanced by the equal and opposite net charge on the surface. Finally, a third complication over the gravitational problem is that the local potential is affected by the local charge density, and the interrelation between the two must be considered.

The net charge density at any point is given by

$$\rho = ze(n^+ - n^-) = -2n_0 ze \sinh \frac{ze\psi}{kT}. \qquad \text{(V-2)}$$

The integral of ρ over all space gives the total excess charge in the solution, per unit area, and is equal in magnitude but opposite in sign to the surface charge density σ:

$$\sigma = -\int \rho \, dx \qquad \text{(V-3)}$$

This produces a double layer of charge, one localized on the surface of the plane and the other developed in a diffuse region extending into solution.

The mathematics is completed by one additional theorem relating the divergence of the gradient of the electrical potential at a given point to the charge density at that point through Poisson's equation

TABLE V-2

Electrostatic Formulas in cgs/esu and SI

Function	cgs/esu	SI
Dimensionless potential	$e\psi/kT = \psi/25.69$ mV at 25°C	$e\psi/kT = \psi/25.69$ mV at 25°C
Coulomb force	$f = \dfrac{q_1 q_2}{x^2 \epsilon}$	$f = \dfrac{q_1 q_2}{4\pi\epsilon_0 x^2 \epsilon}$
Coulomb energy	$U = \dfrac{q_1 q_2}{x\epsilon}$	$U = \dfrac{q_1 q_2}{4\pi\epsilon_0 x\epsilon}$
Poisson equation	$\nabla^2\psi = -\dfrac{4\pi\rho}{\epsilon}$	$\nabla^2\psi = -\dfrac{\rho}{\epsilon_0\epsilon}$
Poisson–Boltzmann equation	$\nabla^2\psi = \dfrac{8\pi n_0 ze}{\epsilon}\sinh\dfrac{ze\psi}{kT}$	$\nabla^2\psi = \dfrac{2n_0 ze}{\epsilon_0\epsilon}\sinh\dfrac{ze\psi}{kT}$
Debye–Hückel equation	$\nabla^2\psi = \dfrac{8\pi n_0 z^2 e^2\psi}{\epsilon kT} = \kappa^2\psi$	$\nabla^2\psi = \dfrac{2n_0 z^2 e^2\psi}{\epsilon_0\epsilon kT} = \kappa^2\psi$
Debye–length	$\kappa^2 = \dfrac{4\pi e^2}{\epsilon kT}\sum_i n_i z_i^2$	$\kappa^2 = \dfrac{e^2}{\epsilon_0\epsilon kT}\sum_i n_i z_i^2$
Condenser capacity	$C = \dfrac{\epsilon}{4\pi d}$	$C = \dfrac{\epsilon\epsilon_0}{d}$

$$\nabla^2\psi = -\frac{4\pi\rho}{\epsilon} \tag{V-4}$$

where ∇^2 is the Laplace operator $(\partial^2/\partial x^2 + \partial^2/\partial y^2 + \partial^2/\partial z^2)$ and ϵ is the dielectric constant of the medium.

Substitution of Eq. V-2 into Eq. V-4 gives

$$\nabla^2\psi = \frac{8\pi n_0 ze}{\epsilon}\sinh\frac{ze\psi}{kT} \tag{V-5}$$

which is the Poisson–Boltzmann equation. Solutions to Eq. V-5 have been studied by Gouy [2], Chapman [3], and Debye and Hückel [4]; these have been summarized by Verwey and Overbeek [5], Kruyt [6], James and Parks [7], Blum [8], and Russel et al. [2]. Perhaps the best-known treatment is the linearization due to Debye and Hückel. If $ze\psi/kT \ll 1$, the exponentials in Eq. V-5 may be expanded and only the first terms retained to obtain the Debye–Hückel equation

$$\nabla^2\psi = \frac{8\pi n_0 z^2 e^2\psi}{\epsilon kT} = \kappa^2\psi \tag{V-6}$$

When ions of various charges are involved,

$$\kappa^2 = \frac{4\pi e^2}{\epsilon kT} \sum_i n_i z_i^2 \qquad \text{(V-7)}$$

and solution to Eq. V-6 for the jth type of ion

$$\psi_j(r) = \frac{z_j e}{\epsilon r} e^{-\kappa r} \qquad \text{(V-8)}$$

illustrates the screened Coulombic $(1/r)$ nature of the potential in an electrolyte solution. The quantity κ provides the length scale for the screening and $1/\kappa$ is associated with the thickness of the ionic atmosphere around each ion and is called the *Debye length*. The work of charging an ion in its atmosphere leads to an electrical contribution to the free energy of the ion, usually expressed as an activity coefficient correction to its concentration. The detailed treatment of interionic attraction theory and its various complications is outside the scope of interest here; more can be found in the monograph by Harned and Owen [10].

   The treatment in the case of a plane charged surface and the resulting diffuse double layer is due mainly to Gouy and Chapman. Here $\nabla^2 \psi$ may be replaced by $d^2\psi/dx^2$ since ψ is now only a function of distance normal to the surface. It is convenient to define the quantities y and y_0 as

$$y = \frac{ze\psi}{kT} \quad \text{and} \quad y_0 = \frac{ze\psi_0}{kT} \qquad \text{(V-9)}$$

Equations V-2 and V-4 combine to give the simple form

$$\frac{d^2 y}{dx^2} = \kappa^2 \sinh y \qquad \text{(V-10)}$$

Using the boundary conditions ($y = 0$ and $dy/dx = 0$ for $x = \infty$), the first integration gives

$$\frac{dy}{dx} = -2\kappa \sinh \frac{y}{2} \qquad \text{(V-11)}$$

and with the added boundary condition ($y = y_0$ at $x = 0$), the final result is

$$e^{y/2} = \frac{e^{y_0/2} + 1 + (e^{y_0/2} - 1)e^{-\kappa x}}{e^{y_0/2} + 1 - (e^{y_0/2} - 1)e^{-\kappa x}} \qquad \text{(V-12)}$$

(note Problem V-4).

For the case of $y_0 \ll 1$ (or, for singly charged ions and room temperature, $\psi_0 \ll 25$ mV), Eq. V-12 reduces to

$$\psi = \psi_0 e^{-\kappa x} \tag{V-13}$$

The quantity $1/\kappa$ is thus the distance at which the potential has reached the $1/e$ fraction of its value at the surface and coincides with the center of action of the space charge. The plane at $x = 1/\kappa$ is therefore taken as the effective thickness of the diffuse double layer. As an example, $1/\kappa = 30$ Å in the case of 0.01 M uni-univalent electrolyte at 25°C.

For $y_0 \gg 1$ and $x \gg 1/\kappa$, Eq. V-12 reduces to

$$\psi = \frac{4kT}{ze} e^{-\kappa x} \tag{V-14}$$

This means that the potential some distance away appears to follow Eq. V-13, but with an apparent ψ_0 value of $4kT/ze$, which is independent of the actual value. For monovalent ions at room temperature this apparent ψ_0 would be 100 mV.

Once the solution for ψ has been obtained, Eq. V-2 may be used to give n^+ and n^- as a function of distance, as illustrated in Fig. V-1. Illustrative curves for the variation of ψ with distance and concentration are shown in Fig. V-2. Furthermore, use of Eq. V-3 gives a relationship between σ and y_0:

$$\sigma = -\int_0^\infty \rho \, dx = \frac{\epsilon}{4\pi} \int_0^\infty \frac{d^2\psi}{dx^2} \, dx = -\frac{\epsilon}{4\pi} \left(\frac{d\psi}{dx} \right)_{x=0} \tag{V-15}$$

Insertion of $(d\psi/dx)_{x=0}$ from Eq. V-11 gives

$$\sigma = \left(\frac{2n_0\epsilon kT}{\pi} \right)^{1/2} \sinh \frac{y_0}{2} \tag{V-16}$$

Equation XV-17 is the same as Eq. V-16 but solved for ψ_0 and assuming 20°C and $\epsilon = 80$. For *small* values of y_0, Eq. V-16 reduces to

$$\sigma = \frac{\epsilon \kappa \psi_0}{4\pi} \tag{V-17}$$

By analogy with the Helmholtz condenser formula, for small potentials the diffuse double layer can be likened to an electrical condenser of plate distance $1/\kappa$. For larger y_0 values, however, σ increases more than linearly with ψ_0, and the capacity of the double layer also begins to increase.

Several features of the behavior of the Gouy-Chapman equations are illus-

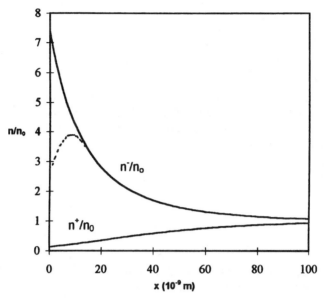

Fig. V-1. Variation of n^-/n_0 and n^+/n_0 with distance for $\psi_0 = 51.38$ mV and 0.01 M uni-univalent electrolyte solution at 25°C. The areas under the full lines give an *excess* of 0.90×10^{-11} mol of anions in a column of solution of 1-cm² cross section and a deficiency of 0.32×10^{-11} mol of cations. There is, correspondingly, a compensating positive surface charge of 1.22×10^{-11} mol of electronic charge per cm². The dashed line indicates the effect of recognizing a finite ion size.

trated in Fig. V-2. Thus the higher the electrolyte concentration, the more sharply does the potential fall off with distance, Fig. V-2b, as follows from the larger κ value. Figure V-2c shows that for a given equivalent concentration, the double-layer thickness decreases with increasing valence. Figure V-2d gives the relationship between surface charge density σ and surface potential ψ_0, assuming 0.001M 1:1 electrolyte, and illustrates the point that these two quantities are proportional to each other at small ψ_0 values so that the double layer acts like a condenser of constant capacity. Finally, Figs. V-2d and e show the effects of electrolyte concentration and valence on the charge-potential curve.

We present the constants and fundamental forces in cgs/esu and SI in Tables V-1 and V-2.

Marmur [12] has presented a guide to the appropriate choice of approximate solution to the Poisson–Boltzmann equation (Eq. V-5) for planar surfaces in an *asymmetrical* electrolyte. The solution to the Poisson–Boltzmann equation around a spherical charged particle is very important to colloid science. Explicit solutions cannot be obtained but there are extensive tabulations, known as the LOW tables [13]. For small values of ψ_0, an approximate equation is [9, 14]

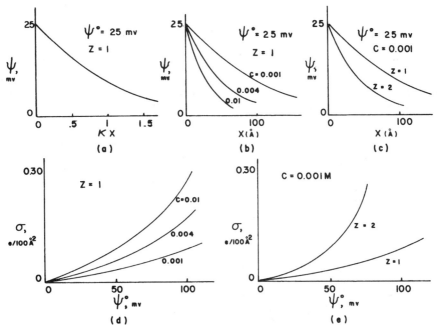

Fig. V-2. The diffuse double layer. (From Ref. 11.)

$$\psi = \psi_0 \frac{a}{r} e^{-\kappa(r-a)} \tag{V-18}$$

where r is the distance from the center of a sphere of radius a.

3. The Stern Layer

The Gouy–Chapman treatment of the double layer runs into difficulties at small κx values when ψ_0 is large. For example, if ψ_0 is 300 mV, y_0 is 12 and if C_0 is, say, 10^{-3} mol/l, then the local concentration of negative ions near the surface, given by Eq. V-1, would be $C^- = 10^{-3}e^{12} = 160$ mol/l! The trouble lies in the assumption of point charges and the consequent neglect of ionic diameters. The treatment of real ions is difficult. Several groups have presented statistical mechanical analyses of the double layer for finite-sized ions [15–19]; one effect is to exclude ions from the solid–liquid interface as illustrated by the dashed line in Fig. V-1. Experimental analysis of the effect of finite ion sizes, through fluorescence spectroscopy indicates it may be less important than expected [20]. Experimental confirmations of the Gouy–Chapman ion profile have been made by fluorescence quenching [21] and neutron scattering [22].

One approach to handling the double layer is to divide the region near the surface into two or more parts. This was first suggested by Stern [23], who designated a compact layer of adsorbed ions as distinct from the diffuse Gouy layer. The potential decays with distance as illustrated in Fig. V-3. One must estimate the extent of ion adsorption into the compact layer and the degree to which ψ is reduced. Stern divided the surface region and the bulk solution region into occupiable sites and assumed that the fractions of sites occupied by ions were determined by a Boltzmann distribution. If S_0 denotes the number of occupiable sites on the surface, then $\sigma_0 = zeS_0$ and $\sigma_S/(\sigma_0 - \sigma_S)$ is the ratio of occupied to unoccupied sites. For a dilute bulk solution phase, the corresponding ratio is just the mole fraction N_S of the solute. Stern considered these two ratios to be related as follows:

$$\frac{\sigma_S}{\sigma_0 - \sigma_S} = N_S e^{ze\Psi_\delta + \phi/kT} \tag{V-19}$$

where ψ_δ is the potential at the boundary between the compact and diffuse layers and ϕ allows for any additional chemical adsorption potential. The charge density for the compact layer is then

$$\frac{\sigma_S}{\sigma_0} = \frac{N_S e^{ze\Psi_\delta + \phi/kT}}{1 + N_S e^{ze\Psi_\delta + \phi/kT}} \tag{V-20}$$

an expression often simplified by neglecting the second term in the denominator. Within the compact layer, of thickness δ the potential gradient $-d\psi/dx$ is approximated by $(\psi_0 - \psi_\delta)/\delta$; hence

$$\sigma_S = \frac{\epsilon'}{4\pi\delta} (\psi_0 - \psi_\delta) \tag{V-21}$$

where ϵ' is a local dielectric constant that may differ from that of the bulk solvent. The capacity of the compact layer $\epsilon'/4\pi\delta$ can be estimated from electrocapillarity and related studies (see Section V-5). Outside the compact layer is the diffuse Gouy layer given by Eqs. V-12 and V-15, with ψ_0 replaced by ψ_δ.

The compact layer must be further subdivided, especially when considering the dissociation or dissolution of charge groups determining the surface potential on many sols and biological materials. First, consider a sol of silver iodide in equilibrium with a saturated solution. There are essentially equal concentrations of silver and iodide ions in solution, although the particles themselves are negatively charged due to a preferential adsorption of iodide ions. If, now, the Ag^+ concentration is increased 10-fold (e.g., by the addition of silver nitrate), the thermodynamic potential of the silver increases by

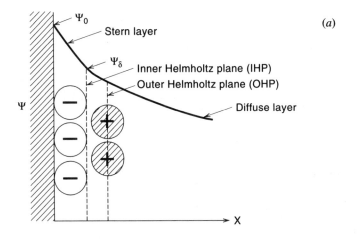

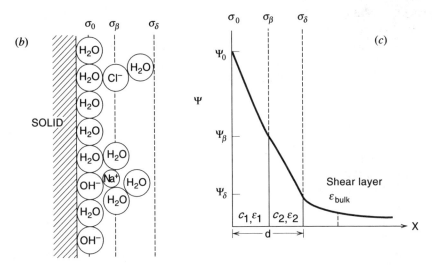

Fig. V-3. Schematic representation of (a) the Stern layer; (b) the potential-determining ions at an oxide interface; (c) the potential-determining and Stern layers together.

$$\mu = kT\ln\frac{C}{C_0} \tag{V-22}$$

which, expressed as a potential amounts to 57 mV at 25°C. As a result, some Ag^+ ions are adsorbed on the surface of the silver iodide, but these are few (on a mole scale) and are indistinguishable from other surface Ag^+ ions. Thus the *chemical* potential of Ag^+ in the silver iodide is virtually unchanged. On the other hand, the total, or *electrochemical*, potential of Ag^+ must be the same in both phases (see Section V-9), and this can be true only if the surface potential has increased by 57 mV. Thus, changes to ψ_0 can be computed from Eq. V-22. Since the Ag^+ concentration in solution can be changed by many powers of 10, ψ_0 can therefore be easily changed by hundreds of millivolts. In this case, Ag^+ is called the *potential-determining* ion. Furthermore, measurements of the electromotive force (emf) of a cell having a silver–silver iodide electrode allows the determination of the amount of Ag^+ adsorbed and hence the surface charge, as well as the change in surface potential [24].

In proteins, biological materials and metal oxides, H^+ is frequently a potential-determining ion because of the pH dependence of the degree of dissociation of acid or basic groups. Hydrogen ion is not part of the Stern layer in the sense discussed above. Also, the potential determining ion need not be present in the colloidal particle. Thus, Cl^- ion is potential-determining for gold sols, apparently through the formation of highly stable complexes with the surface. It seems reasonable to consider that such potential-determining ions completely leave the solution phases and become desolvated and in tight chemical association with the solid.

Clearly another division of the surface region is in order. As illustrated in Figs. V.3*b* and *c*, the surface of the solid has a potential ψ_0, which may be controlled by potential-determining ions indistinguishable from those already in the solid (as in the case of AgI). Next there may be a layer of chemically bound, desolvated ions such as H^+ or OH^- on an oxide surface or Cl^- on Au. If such a layer is present, ψ_0 will be associated with it as indicated in the figure, along with the surface charge density σ_0. Next will be the Stern layer of compactly bound but more or less normally solvated ions at potential ψ_β and net surface charge density σ_β. We thus have an inner, potential-determining compact layer of dielectric constant ϵ_1; an outer compact (Stern) layer of dielectric constant ϵ_2; and, finally, the diffuse layer. The Stern layer is rather immobile in the sense of resisting shear, thus there is no reason for the shear plane to coincide exactly with the Stern layer boundary and, as suggested in Fig. V-3*c*, it may be located somewhat farther out. The potential at this shear layer, known as the *zeta potential*, is involved in electrokinetic phenomena discussed in Section V-4.

One can write acid–base equilibrium constants for the species in the inner compact layer and ion pair association constants for the outer compact layer. In these constants, the concentration or activity of an ion is related to that in the bulk by a term $\exp(-e\psi/kT)$, where ψ is the potential appropriate to the layer [25]. The charge density in both layers is given by the algebraic sum of the ions present per unit area, which is related to the number of ions removed from solution by, for example, a pH titration. If the capacity of the layers can be estimated, one has a relationship between the charge density and potential and thence to the experimentally measurable zeta potential [26].

The compact layer can be structured into what is called an *inner Helmholtz plane*

(IHP) (the Helmholtz condenser formula is used in connection with it), located at the surface of the layer of Stern adsorbed ions, and an *outer Helmholtz plane* (OHP), located on the plane of centers of the next layer of ions marking the beginning of the diffuse layer. These planes, marked IHP and OHP in Fig. V-3 are merely planes of average electrical property; the actual local potentials, if they could be measured, must vary wildly between locations where there is an adsorbed ion and places where only water resides on the surface. For liquid surfaces, discussed in Section V-7C, the interface will not be smooth due to thermal waves (Section IV-3). Sweeney and co-workers applied gradient theory (see Chapter III) to model the electric double layer and interfacial tension of a hydrocarbon–aqueous electrolyte interface [27].

4. The Free Energy of a Diffuse Double Layer

The calculation involved here is conceptually a complex one, and for the necessarily detailed discussion needed to do it justice, the reader is referred to Verwey and Overbeek [5] and Kruyt [6] or to Harned and Owen [10]. Qualitatively, what must be done is to calculate the reversible electrostatic work for the process:

Charged surface plus diffuse double layer of ions $\longrightarrow$

uncharged surface plus normal solution of uncharged particles

One way of doing this makes use of the Gibbs equation (Eq. III-81) in the form (see also Section V-7A)

$$dG^s(= d\gamma) = -\sum \Gamma_i \, d\mu_i = -\sigma \, d\psi_0 \qquad \text{(V-23)}$$

Here, the only surface adsorption is taken to be that of the charge balancing the double-layer charge, and the electrochemical potential change is equated to a change in ψ_0. Integration then gives

$$G^s_{\text{elect}} = G_d = -\int_0^{\psi_0} \sigma \, d\psi_0 \qquad \text{(V-24)}$$

and with the use of Eq. V-16 one obtains for the electrostatic free energy per square centimeter of a diffuse double layer

$$G_d = -\frac{8n_0 k T}{\kappa} \left(\cosh \frac{y_0}{2} - 1 \right) \qquad \text{(V-25)}$$

The ordinary Debye–Hückel interionic attraction effects have been neglected and are of second-order importance.

We note that G_d differs from the mutual electrostatic energy of the double layer by an entropy term. This term represents the difference in entropy between the more random arrangement of ions in bulk solution and that in the double layer. A very similar mathematical process arises in the calculation of the electrostatic part of the free energy of an electrolyte due to interionic attraction effects. The fact that G_d is negative despite the adverse entropy term illustrates the overriding effect of the electrostatic energy causing adsorption or concentration of ions near a charged surface (note Ref. 28).

5. Repulsion between Two Planar Double Layers

The repulsion between two double layers is important in determining the stability of colloidal particles against coagulation and in setting the thickness of a soap film (see Section VI-5B). The situation for two planar surfaces, separated by a distance $2d$, is illustrated in Fig. V-4, where two ψ versus x curves are shown along with the actual potential.

This subject has a long history and important early papers include those by Derjaguin and Landau [29] (see Ref. 30) and Langmuir [31]. As noted by Langmuir in 1938, the total force acting on the planes can be regarded as the sum of a contribution from osmotic pressure, since the ion concentrations differ from those in the bulk, and a force due to the electric field. The total force must be constant across the gap and since the field, $d\psi/dx$ is zero at the midpoint, the total force is given the net osmotic pressure at this point. If the solution is dilute, then

$$\pi_{os}^{net} = P = n_{excess}kT = [n_0(e^{ze\psi_M/kT} + e^{-ze\psi_M/kT}) - 2n_0]kT \qquad (V-26)$$

where the last term corrects for the osmotic pressure of the bulk solution having an electrolyte concentration, n_0 and ψ_M is evaluated at the midpoint. For large values of ψ_M, Eq. V-26 reduces to [6]

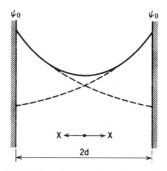

Fig. V-4. Two interacting double layers.

$$P = n_0 kT e^{e\psi_M/kT}. \tag{V-27}$$

Integration of Eq. V-11 with the new boundary conditions and combination with Eq. V-27 gives

$$P = \left(\frac{2\pi}{\kappa d}\right)^2 n_0 kT = \frac{\pi}{2} \epsilon \left(\frac{kT}{ed}\right)^2 \tag{V-28}$$

or for water at 20°C, $P = 8.90 \times 10^{-7}/d^2$ dyn/cm^2. The interaction potential energy between the two plates is given by

$$U_d = -2 \int_{\infty}^{d} P \, dx = \frac{1.78 \times 10^{-6}}{d} \text{ ergs/cm}^2 \tag{V-29}$$

This treatment is restricted to large ψ_M (> 50 mV) and κd larger than ~3. When the surfaces are far apart, where ψ_M is small and the interaction weak, Kruyt [6] gives the equation

$$U_d = \frac{64 n_0 kT}{\kappa} \left(\frac{e^{y_0/2} - 1}{e^{y_0/2} + 1}\right)^2 e^{-2\kappa d} \tag{V-30}$$

The equations are transcendental for the general case, and their solution has been discussed in several contexts [32–35]. One important issue is the treatment of the boundary condition at the surface as d is changed. Traditionally, the constant surface potential condition is used where ψ_0 is constant; however, it is equally plausible that σ_0 is constant due to the behavior of charged sites on the surface.

In the preceding derivation, the repulsion between overlapping double layers has been described by an increase in the osmotic pressure between the two planes. A closely related but more general concept of the *disjoining pressure* was introduced by Derjaguin [30]. This is defined as the difference between the thermodynamic equilibrium state pressure applied to surfaces separated by a film and the pressure in the bulk phase with which the film is equilibrated (see section VI-5).

A number of refinements and applications are in the literature. Corrections may be made for discreteness of charge [36] or the excluded volume of the hydrated ions [19, 37]. The effects of surface roughness on the electrical double layer have been treated by several groups [38–41] by means of perturbative expansions and numerical analysis. Several geometries have been treated, including two eccentric spheres such as found in encapsulated proteins or drugs [42], and biconcave disks with elastic membranes to model red blood cells [43]. The double-layer repulsion between two spheres has been a topic of much attention due to its importance in colloidal stability. A new numerical scheme [44] and simple but accurate analytical formulas [45] have been presented by Carnie, Chan and co-workers. These analyses allow analytic representation of the interactions between spheres that are valid for all sphere separations. The interactions between unlike spheres have also been addressed [46–48] as well as that between patch-wise heterogeneous surfaces [49].

Double-layer repulsions can be measured experimentally via the surface force apparatus discussed in Chapter VI. Here the forces between crossed, mica cylinders immersed in aqueous [50] or nonaqueous [51] electrolyte or surfactant [52] solutions are measured. Agreement between theory and experiment is excellent as seen in Fig. V-5. The interactions between smaller surfaces can be measured by atomic force microscopy (AFM) where a colloidal particle is glued to the tip (see Chapter VIII). The electrostatic interaction between two SiO_2 spheres and between an SiO_2 sphere and a TiO_2 surface in an electrolyte indicates that the constant charge boundary condition may be more appropriate than constant potential [53]. The interaction between a sphere and a flat plate has been treated theoretically [48] and experimentally [54]. Interactions between spheres can be investigated experimentally via small angle neutron scattering [55]. Finally, as an illustration of the complexity of the subject, like-charged surfaces bearing surface potentials of different magnitudes may actually attract one another as first predicted by Derjaguin (see Ref. 56).

6. The Zeta Potential

A number of electrokinetic phenomena have in common the feature that relative motion between a charged surface and the bulk solution is involved. Essen-

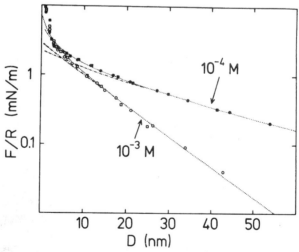

Fig. V-5. The repulsive force between crossed cylinders of radius R (1 cm) covered with mica and immersed in propylene carbonate solutions of tetraethylammonium bromide at the indicated concentrations. The dotted lines are from double-layer theory (From Ref. 51).

tially, a charged surface experiences a force in an electric field, and conversely, a field is induced by the relative motion of such a surface. In each case the plane of slip between the double layer and the medium is involved, and the results of the measurements may be interpreted in terms of its charge density. The ζ *potential* is not strictly a phase boundary potential because it is developed wholly within the fluid region. It can be regarded as the potential difference in an otherwise practically uniform medium between a point some distance from the surface and a point on the plane of shear. The importance of the slip velocity and associated fluid dynamics in electrophoresis and other phoretic processes is reviewed by Anderson [57].

The relationship between the various electrokinetic effects are summarized in Table V-3.

A. Electrophoresis

The most familiar type of electrokinetic experiment consists of setting up a potential gradient in a solution containing charged particles and determining their rate of motion. If the particles are small molecular ions, the phenomenon is called ionic *conductance*; if they are larger units, such as protein molecules, or colloidal particles, it is called *electrophoresis*.

In the case of small ions, Hittorf transference cell measurements may be combined with conductivity data to give the mobility of the ion, that is, the velocity per unit potential gradient in solution, or its *equivalent conductance*. Alternatively, these may be measured more directly by the moving boundary method.

For ions, the charge is not in doubt, and the velocity is given by

$$v = ze\omega \mathbf{F} \tag{V-31}$$

where $\mathbf{F}$ is the field in volts per centimeter (here, esu/cm) and ω is the intrinsic mobility (the Stoke's law value of ω for a sphere is $1/6\pi\eta r$). If $\mathbf{F}$ is given in ordinary volts per centimeter, then

TABLE V-3
Electrokinetic Effects

Potential	Nature of Solid Surface	
	Stationary[a]	Moving[b]
Applied	Electroosmosis	Electrophoresis
Induced	Streaming potential	Sedimentation potential

[a]For example, a wall or apparatus surface.
[b]For example, a colloidal particle.

$$v = \frac{ze\omega \mathbf{F}}{300} = zu\mathbf{F} \tag{V-32}$$

where u is now an electrochemical mobility. From the definition of equivalent conductance, it follows that

$$\lambda = \mathcal{F} uz \tag{V-33}$$

where $\mathcal{F}$ is Faraday's number.

Thus in the case of ions, measurements of this type are generally used to obtain values of the mobility and, through Stoke's law or related equations, an estimate of the effective ionic size.

In the case of a charged particle, the total charge is not known, but if the diffuse double layer up to the plane of shear may be regarded as the equivalent of a parallel-plate condenser, one may write

$$\sigma = \frac{\epsilon \zeta}{4\pi\tau} \tag{V-34}$$

where τ is the effective thickness of the double layer from the shear plane out, usually taken to be $1/\kappa$. The force exerted on the surface, per square centimeter, is $\sigma \mathbf{F}$, and this is balanced (when a steady-state velocity is reached) by the viscous drag $\eta v/\tau$, where η is the viscosity of the solution. Thus

$$v = \frac{\mathbf{F}\tau\sigma}{\eta} = \frac{\mathbf{F}\sigma\tau}{300\eta} \tag{V-35}$$

where the factor $\frac{1}{300}$ converts $\mathbf{F}$ to practical volts, or

$$v = \frac{\zeta\epsilon\mathbf{F}}{4\pi\eta} \tag{V-36}$$

and the velocity per unit field is proportional to the ζ potential, or, alternatively, to the product $\sigma\tau$, sometimes known as the *electric moment per square centimeter*.

There are a number of complications in the experimental measurement of the electrophoretic mobility of colloidal particles and its interpretation; see Section V-6F. The experiment itself may involve a moving boundary type of apparatus, direct microscopic observation of the velocity of a particle in an applied field (the *zeta-meter*), or measurement of the conductivity of a colloidal suspension.

A more detailed theory of electrophoresis is found in Refs. 9 and 58. The motion of the ions in the double layer due to the field F and due to the relative motion of the particle, cause a retardation of the electrophoretic motion that must be considered to determine the correct ζ potential [9]. In addition, the Stern and double-layer regions can be sources of conductance (see Section V-6E). In suspensions of polymer latex particles, electrophoresis measurements by de las Nieves and co-workers indicate that a layer of short polymer chains on the surface contributes to surface conductance [59, 60] and colloidal stability [61].

Rowell and co-workers [62–64] have developed an *electrophoretic fingerprint* to uniquely characterize the properties of charged colloidal particles. They present contour diagrams of the electrophoretic mobility as a function of the suspension pH and specific conductance, pλ. These "fingerprints" illustrate anomalies and specific characteristics of the charged colloidal surface. A more sophisticated electroacoustic measurement provides the particle size distribution and ζ potential in a polydisperse suspension. Not limited to dilute suspensions, in this experiment, one characterizes the sonic waves generated by the motion of particles in an alternating electric field. O'Brien and co-workers have an excellent review of this technique [65].

Electrophoresis is an important technique to use on nonspherical particles. Fair and Anderson solved the electrophoretic problem for ellipsoids exactly for an arbitrary ζ-potential distribution on the surface; interestingly, particles carrying zero net charge can have a significant electrophoretic mobility [66]. Velegol and co-workers tracked the electrophoretic rotation of an aggregated doublet to determine the rigidity of the doublet and the difference in ζ potentials between the two spheres [67]. A related phenomenon, *diffusiophoresis*, occurs when a colloidal particle resides in a solute gradient. In charged systems, the diffusiophoretic mobility comprises electrophoretic and chemiphoretic parts driving the particle either up or down the electrolyte gradient [68].

B. Electroosmosis

The effect known either as *electroosmosis* or *electroendosmosis* is a complement to that of electrophoresis. In the latter case, when a field **F** is applied, the surface or particle is mobile and moves relative to the solvent, which is fixed (in laboratory coordinates). If, however, the surface is fixed, it is the mobile diffuse layer that moves under an applied field, carrying solution with it. If one has a tube of radius r whose walls possess a certain ζ potential and charge density, then Eqs. V-35 and V-36 again apply, with v now being the velocity of the diffuse layer. For water at 25°C, a field of about 1500 V/cm is needed to produce a velocity of 1 cm/sec if ζ is 100 mV (see Problem V-14).

The simple treatment of this and of other electrokinetic effects was greatly clarified by Smoluchowski [69]; for electroosmosis it is as follows. The volume flow V (in cm^3/sec) for a tube of radius r is given by applying the linear velocity v to the body of liquid in the tube

$$V = \pi r^2 v \qquad \qquad (\text{V-37})$$

or

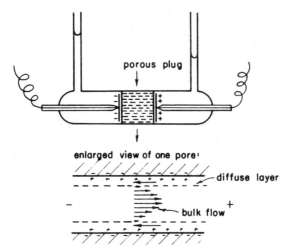

Fig. V-6. Apparatus for the measurement of electroosmotic pressure.

$$V = \frac{r^2 \zeta \epsilon \mathbf{F}}{4\eta} \qquad (\text{V-38})$$

As illustrated in Fig. V-6, one may have a porous diaphragm (the pores must have radii greater than the double-layer thickness) separating the two fluid reservoirs. Here, the return path lies through the center of each tube or pore of the diaphragm, and as shown in the figure, the flow diagram has solution moving one way near the walls and the opposite way near the center. When the field is first established, electroosmotic flow occurs, and for a while, there is net transport of liquid through the diaphragm so that a difference in level develops in the standing tubes. As the hydrostatic head develops, a counterflow sets in, and finally, a steady-state electroosmotic pressure is reached such that the counterflow just balances the flow at the walls. The pressure P is related to the counterflow by Poiseuille's equation

$$V_{\text{counterflow}} = \frac{\pi r^4 P}{8\eta l} \qquad (\text{V-39})$$

where l is the length of the tube. From Eqs. V-38 and V-39, the steady-state pressure is

$$P = \frac{2\zeta \epsilon \mathbf{E}}{\pi r^2} \qquad (\text{V-40})$$

where $\mathbf{E} = \mathbf{F}l$ and is the total applied potential. For water at 25°C,

$$v = 7.8 \times 10^{-3} \mathbf{F} \zeta \ (\text{cm/sec}) \tag{V-41}$$

$$P = \frac{4.2 \times 10^{-8} \mathbf{E} \zeta}{r^2} \ (\text{cm Hg}) \tag{V-42}$$

where potentials are in practical volts and lengths are in centimeters. For example, if ζ is 100 mV, then 1 V applied potential can give a pressure of 1 cm of mercury only if the capillary radius is about 1 μm.

An interesting biological application of electroosmosis is in the analysis of flow in renal tubules [70].

C. Streaming Potential

The situation in electroosmosis may be reversed when the solution is caused to flow down the tube, and an induced potential, the *streaming potential*, is measured. The derivation, again due to Smoluchowski [69], begins with the assumption of Poiseuille flow such that the velocity at a radius x from the center of the tube is

$$v = \frac{P(r^2 - x^2)}{4\eta l} \tag{V-43}$$

The double-layer is centered at $x = r - \tau$, and substitution into Eq. V-43 gives the double-layer velocity

$$v_d = \frac{\tau r P}{2\eta l} \tag{V-44}$$

if the term in τ^2 is neglected. The current due to the motion of the double layer is then

$$i = 2\pi r \sigma v_d = \frac{\pi r^2 \sigma \tau P}{\eta l} \tag{V-45}$$

If k is the specific conductance of the solution, then the actual conductance of the liquid in the capillary tube is $C = \pi r^2 k / l$, and by Ohm's law, the streaming potential, E, is given by $E = i/C$. Combining these equations (including Eq. V-44)

$$E = \frac{\tau \sigma P}{\eta k} = \frac{\zeta P \epsilon}{4\pi \eta k} \tag{V-46}$$

shows the relationship between the streaming potential E and the ζ potential. A more detailed derivation is found in Kruyt [6].

Streaming potentials, like other electrokinetic effects, are difficult to measure reproducibly. One means involves forcing a liquid under pressure through a porous plug or capillary and measuring E by means of electrodes in the solution on either side [6, 23, 71–73].

The measurement of the streaming potential developed when a solution flows through two parallel plates [74–76] allows the characterization of macroscopic surfaces such as mica.

The flow can be radial, that is, in or out through a hole in the center of one of the plates [75]; the relationship between E and ζ (Eq. V-46) is independent of geometry. As an example, a streaming potential of 8 mV was measured for 2-cm-radius mica disks (one with a 3-mm exit hole) under an applied pressure of 20 cm H_2 on $10^{-3}M$ KCl at $21°C$ [75]. The ζ potentials of mica measured from the streaming potential correspond well to those obtained from force balance measurements (see Section V-6 and Chapter VI) for some univalent electrolytes; however, important discrepancies arise for some monovalent and all multivalent ions. The streaming potential results generally support a single-site dissociation model for mica with σ_o, σ_β, and σ_δ defined by the surface site equilibrium [76].

Quite sizable streaming potentials can develop with liquids of very low conductivity. The effect, for example, poses a real problem in the case of jet aircraft where the rapid flow of jet fuel may produce sparks.

D. Sedimentation Potential

The final and less commonly dealt-with member of the family of electrokinetic phenomena is the sedimentation potential. If charged particles are caused to move relative to the medium as a result, say, of a gravitational or centrifugal field, there again will be an induced potential E. The formula relating E to ζ and other parameters is [72, 77]

$$E = \frac{C_m}{k} \frac{\epsilon\zeta}{6\pi\eta} \frac{\omega^2}{2} (R_2^2 - R_1^2) \qquad \text{(V-47)}$$

where C_m is the apparent density of the solution, ω is the angular velocity of the centrifuge, and R_2 and R_1 are the distances from the axis of rotation of the two points between which the sedimentation potential is measured.

Marlow and Rowell discuss the deviation from Eq. V-47 when electrostatic and hydrodynamic interactions between the particles must be considered [78]. In a suspension of glass spheres, beyond a volume fraction of 0.018, these interparticle forces cause nonlinearities in Eq. V-47, diminishing the induced potential E.

E. Interrelationships in Electrokinetic Phenomena

In electroosmosis, the volumetric flow and current are related through

$$\frac{V}{i} = \frac{\epsilon\zeta}{4\pi\eta k} \qquad \text{(V-48)}$$

and an old relation attributed to Saxén [79] provides that V/i at zero pressure is equal to the streaming potential ratio E/P at zero current. A further generalization of such cross effects and the general phenomenological development of electrokinetic and hydrodynamic effects is given by Mazur and Overbeek [80] and Lorenz [81]. The subject constitutes an important example of Onsager's reciprocity relationships; this reciprocity has been verified in electroosmosis and streaming potential measurements in binary alcohol mixtures [82].

The presence of surface conductance behind the slip plane alters the relationships between the various electrokinetic phenomena [83, 84]; further complications arise in solvent mixtures [85]. Surface conductance can have a profound effect on the streaming current and electrophoretic mobility of polymer latices [86, 87]. In order to obtain an accurate interpretation of the electrostatic properties of a suspension, one must perform more than one type of electrokinetic experiment. One novel approach is to measure electrophoretic mobility and dielectric spectroscopy in a single instrument [88].

F. Potential, Surface Charge, and Colloidal Stability

The Stern and double-layer systems play an important role in stabilizing colloidal suspensions, and conversely, the study of suspensions has provided much information about the electrical nature of the interface. In particular, lyophobic sols are often stabilized by the presence of a surface charge. Two particles approaching closely will flocculate because of the dispersion interaction discussed in Chapter VI. The electrostatic repulsion discussed in Section V-6 deters this approach; hence, variations in ionic strength and ψ_0 or ζ will have a marked effect on the stability of such sols. Some classic data [89] shown in Table V-4, illustrate the general relation between stability and ζ potential for a gold sol in a solution containing added Al^{3+}. A review of these stability issues in ceramic materials is given by Pugh [90].

The ζ potential of silver iodide can be varied over the range ± 75 mV, by varying the Ag^+ or I^- concentration again demonstrating that varying the concentration of potential-determining ions can reverse the sign of the ζ potential.

TABLE V-4
Flocculation of Gold Sol

Concentration of Al^{3+} (eq/l[a] $\times 10^6$)	Electrophoretic Velocity ($cm^2/Vs \times 10^6$)	Stability
0	3.30 (toward anode)	Indefinitely stable
21	1.71	Flocculated in 4 hr
—	0	Flocculates spontaneously
42	0.17 (toward cathode)	Flocculated in 4 hr
70	1.35	Incompletely flocculated in 4 days

[a]Equivalents per liter.

Thus, when titrating iodide with silver nitrate, coagulation occurs as soon as a slight excess of silver ion has been added (so that a point of zero charge has been surpassed).

Electrolytes have a flocculating effect on charge stabilized sols, and the flocculation "value" may be expressed as the concentration needed to coagulate the sol in some given time interval. There is a roughly 10- to 100-fold increase in flocculation effectiveness in going from mono- to di- to trivalent ions; this is due in part to the decreasing double-layer thickness and partly to increasing adsorption of ions into the Stern layer [39, 91]. The effect of valence can be stated by the *Schulze–Hardy* rule, where the critical flocculation concentration decreases as z^{-6} [9] (see Section VI-4). There is an order of effectiveness of ions within a given valence known as the *Hofmeister* series. See Section VI-4 for more on such observations.

These effects can be illustrated more quantitatively. The drop in the magnitude of the ζ potential of mica with increasing salt is illustrated in Fig. V-7; here ψ is reduced in the immobile layer by ion adsorption and specific ion effects are evident. In Fig. V-8, the pH is potential determining and alters the electrophoretic mobility. Carbon blacks are industrially important materials having various acid–base surface impurities depending on their source and heat treatment.

Figure V-8 illustrates that there can be a pH of zero ζ potential; interpreted as the point of zero charge at the shear plane; this is called the *isoelectric point* (iep). Because of specific ion and Stern layer adsorption, the iep is not necessarily the *point of zero surface charge* (pzc) at the particle surface. An example of this occurs in a recent study of zircon ($ZrSiO_4$), where the pzc measured by titration of natural zircon is 5.9 ± 0.1

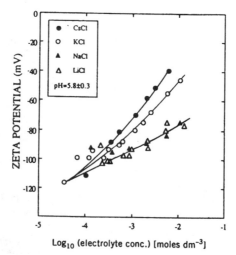

Fig. V-7. ζ potential of muscovite mica versus electrolyte concentration at pH 5.8 ± 0.3. (From Ref. 76.)

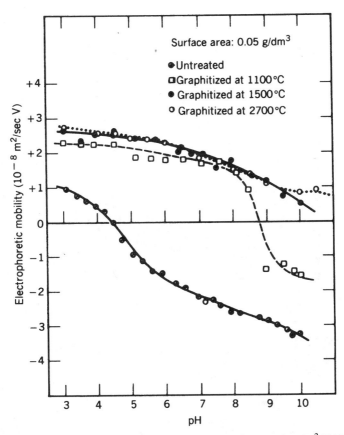

Fig. V-8. Electrophoretic mobility of carbon black dispersions in $10^{-3}M$ KNO$_3$ as a function of pH. (From Ref. 93.)

while the iep measured by electrophoresis is 5.7 ± 0.1; in commercial zircon containing a kaolinite impurity the difference is more striking with pzc = 6.1 ± 0.1 and iep = 5.5 ± 0.1 [37]. The pzc of oxides may be determined by the titration illustrated in Fig. V-9. The net adsorption of an acid or base gives a surface charge differing from the actual one by some constant, and this constant will vary with pH. One can thus obtain a plot of apparent surface charge versus pH. If a series of such plots are made for various concentrations of indifferent (nonspecifically adsorbing) electrolyte, there will tend to be a common intersection at the pH marking the pzc. There is some variation in the terminology in the literature, with pzc sometimes actually meaning iep.

pH effects can be complicated by the adsorption of hydrolyzable metal ions [94] and by surface conductivity [9, 95, 96]. Various investigators have addressed problems in electrical double-layer treatments [97–99]. van Oss and co-workers studied electrophoretic mobility and contact angles (see Chapter XIII) to show that in suspensions where hydrophilicity is determined by the electrical surface potential, reduction of ζ produces a more hydrophobic surface [100].

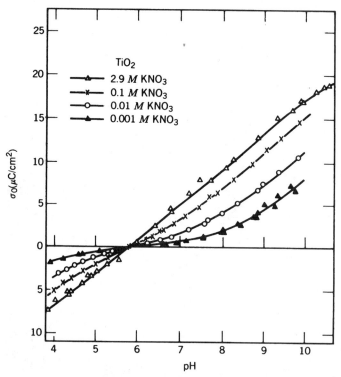

Fig. V-9. Variation of the surface charge density, σ of TiO$_2$ (rutile) as a function of pH in aqueous KNO$_3$: $\triangle$, 2.9M, $\times$, 0.1M; $\bigcirc$, 0.01M; $\triangle$, 10^{-3}M. (From Ref. 91.)

Relaxations in the double layers between two interacting particles can retard aggregation rates and cause them to be independent of particle size [101–103]. Discrepancies between theoretical predictions and experimental observations of heterocoagulation between polymer latices, silica particles, and ceria particles [104] have prompted Matijević and co-workers to propose that the charge on these particles may not be uniformly distributed over the surface [105, 106]. Similar behavior has been seen in the heterocoagulation of cationic and anionic polymer latices [107].

7. Electrocapillarity

It has long been known that the form of a curved surface of mercury in contact with an electrolyte solution depends on its state of electrification [108, 109], and the earliest comprehensive investigation of the electrocapillary effect was made by Lippmann in 1875 [110]. A sketch of his apparatus is shown in Fig. V-10.

Qualitatively, it is observed that the mercury surface initially is positively charged, and on reducing this charge by means of an applied potential, it is found that the height of the mercury column and hence the interfacial tension

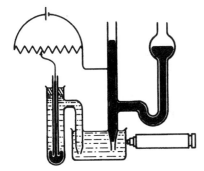

Fig. V-10. The Lippmann apparatus for observing the electrocapillary effect.

of the mercury increase, go through maxima, and then decrease. This roughly parabolic curve of surface tension versus applied potential, illustrated in Fig. V-11, is known as the *electrocapillary curve*. For mercury, the maximum occurs at -0.48 V applied potential E_{max} [in the presence of an "inert" electrolyte, such as potassium carbonate, and referred to the mercury in a normal calomel electrode (NCE)]. Vos and Vos [111] give the value of E_{max} as -0.5084 V versus NCE for $0.116M$ KCl at $25°C$, the difference from -0.48 V being presumably due to Cl^- adsorption at the Hg-solution interface.

The left-hand half of the electrocapillary curve is known as the ascending or anodic branch, and the right-hand half is known as the descending or cathodic branch. As is discussed in the following section, the left-hand branch is in general sensitive to the nature of the anion of the electrolyte, whereas the right-hand branch is sensitive to the nature of the cations present. A material change in solvent may affect the general shape and location of the entire curve and, with nonionic solutes, the curve is affected mainly in the central region.

It is necessary that the mercury or other metallic surface be polarized, that is, that there be essentially no current flow across the interface. In this way no chemical changes occur, and the electrocapillary effect is entirely associated with potential changes at the interface and corresponding changes in the adsorbed layer and diffuse layer.

Recalling the earlier portions of this chapter on Stern and diffuse layer theories, it can be appreciated that varying the potential at the mercury-electrolyte solution interface will vary the extent to which various ions adsorb. The great advantage in electrocapillarity studies is that the free energy of the interface is directly measurable; through the Gibbs equation, surface excesses can be calculated, and from the variation of surface tension with applied potential, the surface charge density can be calculated. Thus major parameters of double-layer theory are directly determinable. This transforms electrocapillarity from what would otherwise be an obscure phenomenon to a powerful method for gaining insight into the structure of the interfacial region at all metal-solution interfaces.

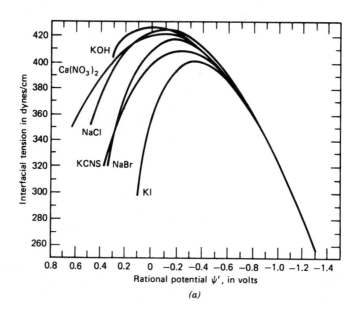

(a)

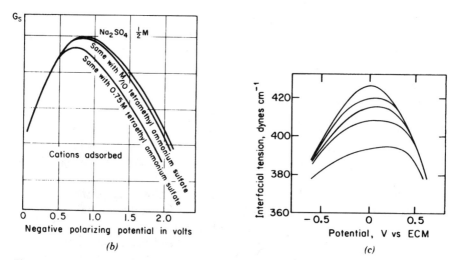

(b)   *(c)*

Fig. V-11. Electrocapillary curves: (*a*) adsorption of anions (from Ref. 113); (*b*) absorption of cations (from Ref. 6); (*c*) electrocapillary curves for *n*-pentanoic acid in 0.1*N* HClO$_4$. Solute activities from top to bottom are 0, 0.04761, 0.09096, 0.1666, and 0.500 (from Ref. 112).

A. Thermodynamics of the Electrocapillary Effect

The basic equations of electrocapillarity are the Lippmann equation [110]

$$\left(\frac{\partial \gamma}{\partial \mathbf{E}} \right)_{\mu} = -\boldsymbol{\sigma} \tag{V-49}$$

and the related equation

$$\left(\frac{\partial^2 \gamma}{\partial \mathbf{E}^2} \right)_{\mu} = -\frac{\partial \boldsymbol{\sigma}}{\partial \mathbf{E}} = C \tag{V-50}$$

where $\mathbf{E}$ is the applied potential, $\boldsymbol{\sigma}$ is the amount of charge on the solution side of the interface per square centimeter, C is the differential capacity of the double layer, and the subscript means that the solution (and the metal phase) composition is held constant.

A number of more or less equivalent derivations of the electrocapillary Eq. V-49 have been given, and these have been reviewed by Grahame [113]. Lippmann based his derivation on the supposition that the interface was analogous to a parallel-plate condenser, so that the reversible work dG, associated with changes in area and in charge, was given by

$$dG = \gamma \, d\mathcal{A} + \Delta\phi \, dq \tag{V-51}$$

where q is the total charge and $\Delta\phi$ is the difference in potential, the second term on the right thus giving the work to increase the charge on a condenser. Integration of Eq. V-51, keeping γ and $\Delta\phi$ constant, gives

$$G = \gamma \mathcal{A} + \Delta\phi \, q \tag{V-52}$$

Equation V-52 may now be redifferentiated, and on comparing the results with Eq. V-51, one obtains

$$0 = \mathcal{A} \, d\gamma + q \, d(\Delta\phi) \tag{V-53}$$

Since $q/\mathcal{A} = \boldsymbol{\sigma}$, Eq. V-53 rearranges to give Eq. V-49.

The treatments that are concerned in more detail with the nature of the adsorbed layer make use of the general thermodynamic framework of the derivation of the Gibbs equation (Section III-5B) but differ in the handling of the electrochemical potential and the surface excess of the ionic species [114–117]. The derivation given here is after that of Grahame and Whitney [117]. Equation III-76 gives the combined first- and second-law statements for the surface excess quantities

$$dE^s = T \ dS^s + \sum_i \mu_i \ dm_i^s + \gamma \ d\mathcal{A} \qquad \text{(V-54)}$$

Also,

$$\mu_i = \left(\frac{\partial E}{\partial m_i} \right)_{S, V, m_j, \mathcal{A}} \qquad \text{(V-55)}$$

or, on a mole basis

$$\overline{\mu}_i = \frac{\partial E}{\partial n_i} \qquad \text{(V-56)}$$

The chemical potential μ_i has been generalized to the *electrochemical* potential $\overline{\mu}_i$ since we will be dealing with phases whose charge may be varied. The problem that now arises is that one desires to deal with individual ionic species and that these are not independently variable. In the present treatment, the difficulty is handled by regarding the electrons of the metallic phase as the dependent component whose amount varies with the addition or removal of charged components in such a way that electroneutrality is preserved. One then writes, for the ith charged species,

$$\frac{\partial E}{\partial n_i} = \overline{\mu}_i + z_i \overline{\mu}_e \qquad \text{(V-57)}$$

where $\overline{\mu}_e$ is the electrochemical potential of the electrons in the system. The Gibbs equation (Eq. III-80) then becomes

$$d\gamma = -S^s \ dT - \sum_i \Gamma_i \ d\overline{\mu}_i - \sum_i \Gamma_i z_i \ d\overline{\mu}_e \qquad \text{(V-58)}$$

The electrochemical potentials $\overline{\mu}_i$ may now be expressed in terms of the chemical potentials μ_i and the electrical potentials (see Section V-9):

$$d\overline{\mu}_i = d\mu_i + z_i \mathcal{F} \ d\phi \qquad \text{(V-59)}$$

where $d\phi$ is the change in the electrical potential of the phase (for uncharged species $z_i = 0$, so the chemical and electrochemical potentials are equal). Since the metal phase α is at some uniform potential ϕ^α and the solution phase β is likewise at some potential ϕ^β, the components are grouped according to the phase they are in. The relationship Eq. V-59 is introduced into Eq. V-58 with this discrimination, giving

$$d\gamma = -S^s \, dT - \sum_i^c \Gamma_i \, d\mu_i - \sum_i^c \Gamma_i z_i \, d\mu_e - \mathcal{F} \sum_i^\alpha \Gamma_i z_i \, d\phi^\alpha$$

$$- \mathcal{F} \sum_i^\beta \Gamma_i z_i \, d\phi^\beta + \mathcal{F} \sum_i^\alpha \Gamma_i z_i \, d\phi^\alpha + \mathcal{F} \sum_i^\beta \Gamma_i z_i \, d\phi^\alpha \quad \text{(V-60)}$$

In obtaining Eq. V-60, it must be remembered that $d\bar{\mu}_e = d\mu_e - \mathcal{F} \, d\phi^\alpha$, since electrons are confined to the metal phase. Canceling and combining terms,

$$d\gamma = -S^s \, dT - \mathcal{F} \sum_i \Gamma_i z_i \, d(\phi^\beta - \phi^\alpha) - \sum_i^c \Gamma_i \, d\mu_i - \sum_i^c \Gamma_i z_i \, d\mu_e \quad \text{(V-61)}$$

or, because of the electroneutrality requirement,

$$\sum_i^c \Gamma_i z_i = \Gamma_e \quad d\gamma = -S^s \, dT - \boldsymbol{\sigma} \, d(\phi^\beta - \phi^\alpha) - \sum_i^{c+1} \Gamma_i \, d\mu_i \quad \text{(V-62)}$$

In the preceding equations, the summation index c indicates summation over all components other than the electrons. The quantity $\mathcal{F}\Sigma_i \Gamma_i z_i$ in Eq. V-61 has been identified with $\boldsymbol{\sigma}$, the excess charge on the solution side of the interface. In general, however, this identification is not necessarily valid because the Γ's depend on the location of the dividing surface. For a *completely polarized* surface, where the term implies that no ionic species is to be found on *both* sides of the interface, the matter of locating the dividing surface becomes trivial and Eq. V-62 may be used. At constant temperature and composition, Eq. V-62 reduces to the Lippmann equation (Eq. V-49); it should be noted, however, that $\mathbf{E}$ in Eq. V-49 refers to the externally measured potential difference, whereas $\Delta\phi$ is the difference in potential between the two phases in question. These two quantities are not actually the same but generally are thought to differ by some constant involving the nature of the reference electrode and of other junctions in the system; however, changes in them are taken to be equal so that $\mathbf{dE}$ may be substituted for $d(\Delta\phi)$.

B. Experimental Methods

The various experimental methods associated with electrocapillarity may be divided into those designed to determine surface tension as a function of applied potential and those designed to measure directly either the charge or the capacity of the double layer.

The capillary electrometer, due to Lippmann [110] and illustrated in Fig. V-10, is the classic and still very important type of apparatus. It consists of a standing tube connected with a mercury reservoir and terminating in a fine capillary at the lower end; this capillary usually is slightly tapered and its diameter is on the order of 0.05 mm. There is an optical system, for example, a cathetometer, for viewing the meniscus in the capillary. The rest of the system then consists of the solution to be studied, into which the capillary dips, a reference electrode, and a potentiometer circuit by means of which a variable voltage may be applied. The measurement is that of the height of the mercury column as a function of E, the applied potential. Since the column height is to be related to the surface tension, it is essential that the solution wet the capillary so that the mercury-glass contact angle is 180°, and the meniscus can be assumed to be hemispherical, allowing Eq. II-11 to be used to obtain γ.

Usually one varies the head of mercury or applied gas pressure so as to bring the meniscus to a fixed reference point [118]. Grahame and co-workers [119], Hansen and co-workers [120] (see also Ref. 121), and Hills and Payne [122] have given more or less elaborate descriptions of the capillary electrometer apparatus. Nowadays, the capillary electrometer is customarily used in conjunction with capacitance measurements (see below). Vos and Vos [111] describe the use of sessile drop profiles (Section II-7B) for interfacial tension measurements, thus avoiding an assumption as to the solution-Hg-glass contact angle.

The surface charge density σ may be determined directly by means of an apparatus, described by Lippmann and by Grahame [123], in which a steady stream of mercury droplets is allowed to fall through a solution. Since the mercury drops are positively charged, electrons flow back to the reservoir and into an external circuit as each drop forms. The total charge passing through the circuit divided by the total surface area of the drops yields σ. One may also bias the dropping electrode with an applied voltage and observe the variation in current, and hence in σ, with E, allowing for the concomitant change in surface tension and droplet size. Grahame [123] and others [124] have described capacitance bridges whereby a direct determination of the double-layer capacity can be made.

It should be noted that the capacity as given by $C_i = \sigma/E^r$, where σ is obtained from the current flow at the dropping electrode or from Eq. V-49, is an integral capacity (E^r is the potential relative to the electrocapillary maximum (ecm), and an assumption is involved here in identifying this with the potential difference across the interface). The differential capacity C given by Eq. V-50 is also then given by

$$C = C_i + E^r \left(\frac{\partial C_i}{\partial E^r} \right)_\mu \tag{V-63}$$

C. Results for the Mercury–Aqueous Solution Interface

The shape of the electrocapillary curve is easily calculated if it is assumed that the double layer acts as a condenser of constant capacity C. In this case, double integration of Eq. V-50 gives

$$\gamma = \gamma_{max} - \tfrac{1}{2}C(\mathbf{E}^r)^2 \qquad\qquad (V\text{-}64)$$

Incidentally, a quantity called the *rational potential* ψ^r is defined as $\mathbf{E}^r$ for the mercury–water interface (no added electrolyte) so, in general, $\psi^r = \mathbf{E} + 0.480$ V if a normal calomel reference electrode is used.

Equation V-64 is that of a parabola, and electrocapillary curves are indeed approximately parabolic in shape. Because $\mathbf{E}_{max}$ and γ_{max} are very nearly the same for certain electrolytes, such as sodium sulfate and sodium carbonate, it is generally assumed that specific adsorption effects are absent, and $\mathbf{E}_{max}$ is taken as a constant $(-0.480$ V) characteristic of the mercury–water interface. For most other electrolytes there is a shift in the maximum voltage, and ψ^r_{max} is then taken to be $\mathbf{E}_{max} - 0.480$. Some values for the quantities are given in Table V-5 [113]. Much information of this type is due to Gouy [125], although additional results are to be found in most of the other references cited in this section.

TABLE V-5
Properties of the Electrical Double Layer at the Electrocapillary Maximum

Electrolyte	Concentration (M)	$\mathbf{E}_{max}$	ψ^r_{max}	Γ^{max}_{salt} ($\mu C/cm^2$)
NaF	1.0	−0.472	0.008	—
	0.001	−0.482	−0.002	—
NaCl	1.0	−0.556	−0.076	3.6
	0.1	−0.505	−0.025	1.1
KBr	1.0	−0.65	−0.17	10.6
	0.01	−0.54	−0.06	0.6
KI	1.0	−0.82	−0.34	15.2
	0.001	−0.59	−0.11	1.3
NaSCN	1.0	−0.72	−0.24	14.0
K_2CO_3	0.5	−0.48	0.00	−2.2
NaOH	1.0	−0.48	0.00	Small
Na_2SO_4	0.5	−0.48	0.00	Small

Note: Some of Grahame's values for Γ^{max}_{salt} and ψ^r_{max} are included in this table. For a common cation, the sequence of anions in order of increasing adsorption is similar to that of the Hofmeister series in coagulation studies, and it is evident that specific adsorption properties are involved.

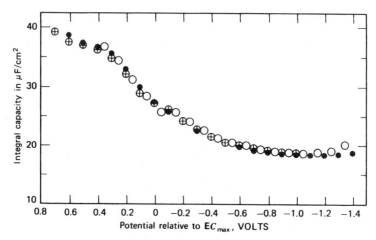

Fig. V-12. Variation of the integral capacity of the double layer with potential for 1 N sodium sulfate: ●, from differential capacity measurements; ⊕, from the electrocapillary curves; ○, from direct measurements. (From Ref. 113.)

The variation of the integral capacity with **E** is illustrated in Fig. V-12, as determined both by surface tension and by direct capacitance measurements; the agreement confirms the general correctness of the thermodynamic relationships. The differential capacity C shows a general decrease as **E** is made more negative but may include maxima and minima; the case of nonelectrolytes is mentioned in the next subsection.

At the electrocapillary maximum, $d\gamma/d\mathbf{E}$ is zero and hence σ is zero. There may still be adsorption of ions, but in equal amounts, that is, $\Gamma_+^{max} = \Gamma_-^{max}$ (for a 1 : 1 electrolyte).

It follows from Eq. V-62 that

$$\frac{d\gamma^{max}}{d\mu} = -\Gamma_{salt}^{max} \tag{V-65}$$

D. Effect of Uncharged Solutes and Changes of Solvent

The location and shape of the entire electrocapillary curve are affected if the general nature of the medium is changed. Fawcett and co-workers (see Ref. 126) have used nonaqueous media such as methanol, N-methylformamide, and propylene carbonate. In earlier studies, electrocapillary curves were obtained for 0.01M hydrochloric acid in mixed water-ethanol media of various compositions [117, 118]. The surface adsorption of methanol, obtained from

$$\Gamma = -\left(\frac{\partial \gamma}{\partial \mu}\right)_{\mathbf{E}} \tag{V-66}$$

was always positive and varied with the solution composition in a manner similar to the behavior at the solution-air interface.

Gouy [125] made a large number of determinations of the effect of added organic solutes, as have others, including Hansen and co-workers [127–129]. Figure V-11c shows data for n-pentanoic acid solutions; similar results are obtained for other acids and for various alcohols. The data may be worked up as follows. One first constructs plots of π versus ln a at constant potential, where a is the solute activity, defined as C/C_0, where C_0 is the solubility of the solute. For a given system, the plots were all of the same shape and superimposed if suitably shifted along the abscissa scale, that is, if a were multiplied by a constant (different for each voltage). The composite plot for 3-pentanol is shown in Fig. V-13 (see Refs. 112, 127–130). Differentiation and use of Eq. V-66 then yields data from which plots of Γ versus a may be constructed, that is, adsorption isotherms. The isotherms could be fit to the Frumkin equation

$$\frac{\theta}{(1-\theta)} = Bae^{2\alpha\theta} \qquad \text{(V-67)}$$

where $\theta = \Gamma/\Gamma_m$, in which Γ_m is Γ at the maximum coverage and the constants B and α describe the adsorption equilibrium and lateral interaction, respectively. A generalized Frumkin isotherm for two coadsorbing species contains an additional interaction parameter, α_3

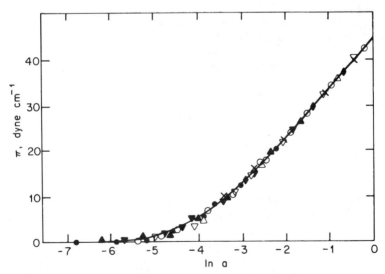

Fig. V-13. Composite $\pi/\ln a$ curve for 3-pentanol. The various data points are for different E values; each curve for a given E has been shifted horizontally to give the optimum match to a reference curve for an E near the electrocapillary maximum. (From Ref. 134.)

$$\frac{\theta_1}{n_1(1 - \theta_1 - \theta_2)^{n_1}} = B_1 a_1 \exp(2n_1\alpha_1\theta_1 + 2n_1\alpha_3\theta_2)$$

$$\frac{\theta_2}{n_2(1 - \theta_1 - \theta_2)^{n_2}} = B_2 a_2 \exp(2n_2\alpha_2\theta_2 + 2n_2\alpha_3\theta_1) \qquad \text{(V-68)}$$

characterizing the interactions between species 1 and 2 in the adsorbed layer. Here n_i = $\Gamma_{m,\text{water}}/\Gamma_{m,i}$ are coefficients accounting for the displacement of water molecules from the interface by the ith adsorbing species. These equations have been successfully applied to the adsorption of mixed surfactants [131–133] to illustrate the synergistic interaction between adsorbed surfactants, particularly those having very different chain lengths [131].

Evidence for two-dimensional condensation at the water–Hg interface is reviewed by de Levie [135]. Adsorption may also be studied *via* differential capacity data where the interface is modeled as parallel capacitors, one for the Hg–solvent interface and another for the Hg–adsorbate interface [136, 137].

E. Other Electrocapillary Systems

Scattered data are available on the electrocapillary effect when liquid metal phases other than pure mercury are involved. Frumkin and co-workers [138] have reported on the electrocapillary curves of amalgams with less noble metals than mercury, such as thallium or cadmium. The general effect is to shift the maximum to the right, and Koenig [116] has discussed the thermodynamic treatment for the adsorption of the metal solute at the interface. Liquid gallium give curves similar to those for mercury, again shifted to the right; this and other systems such as those involving molten salts as the electrolyte are reviewed by Delahay [139]. Narayan and Hackerman [140] studied adsorption at the In-Hg-electrolyte interface. Electrocapillary behavior may also be studied at the interface between two immiscible electrolyte solutions (see Ref. 141).

The equations of electrocapillarity become complicated in the case of the solid metal-electrolyte interface. The problem is that the work spent in a differential *stretching* of the interface is not equal to that in *forming* an infinitesimal amount of new surface, if the surface is under elastic strain. Couchman and co-workers [142, 143] and Mohliner and Beck [144] have, among others, discussed the thermodynamics of the situation, including some of the problems of terminology.

8. The Electrified Solid–Liquid Interface

In recent years, advances in experimental capabilities have fueled a great deal of activity in the study of the electrified solid–liquid interface. This has been the subject of a recent workshop and review article [145] discussing structural characterization, interfacial dynamics and electrode materials. The field of surface chemistry has also received significant attention due to many surface-sensitive means to interrogate the molecular processes occurring at the electrode surface. Reviews by Hubbard [146, 147] and others [148] detail the progress. In this and the following section, we present only a brief summary of selected aspects of this field.

A. Electrode–Solution Interface

Studies of the electrode–solution interface can be divided into electrical, physical, and chemical investigations. Of general interest from an electrical point of view is the applied potential needed to produce zero charge at the interface (the pzc). Bockris and co-workers [149, 150] compare three methods for determining the pzc (which is also the potential at the maximum of the electrocapillary curve from Eq. V-61). A neutral adsorbate should show a maximum or minimum in adsorption at the pzc and can be measured by radiolabeling of the adsorbate [151]. The interfacial capacitance should go through a minimum at the pzc and, interestingly, the *friction* at the metal–solution interface appears to go through a maximum at the pzc (an aspect of the Rehbinder effect—see Section VII-5). It might be thought that since σ is zero at the pzc the applied potential should correspond to the *absolute* potential difference between the metal and solution phases. This is not so since adsorption of both neutral solutes and solvent produces a potential at the metal–solution interface [152] (see Section V-9C).

The development of scanning probe microscopies and x-ray reflectivity (see Chapter VIII) has allowed molecular-level characterization of the structure of the electrode surface after electrochemical reactions [145]. In particular, the important role of adsorbates in determining the state of an electrode surface is illustrated by scanning tunneling microscopic (STM) images of gold (111) surfaces in the presence and absence of chloride ions [153]. Electrodeposition of one metal on another can also be measured via x-ray diffraction [154].

The understanding of the molecular processes occurring during electrochemical reactions has benefited from the development of spectroscopic techniques to interrogate surfaces (see Chapter VIII) [146]. In particular, low-energy electron diffraction (LEED), high-resolution electron-energy-loss spectroscopy (HREELS), and Auger electron spectroscopy (AES) have been applied to studies of surface coordination chemistry for silver and platinum electrodes [146]. Other vibrational spectroscopies such as reflection infrared spectroscopy [155] and Raman spectroscopy allow one to follow functional groups on the surfaces. Proximity to the silver surface produces a strong Raman signal, known as surface enhanced Raman spectroscopy [156].

The electrode surface may be modified by the attachment of molecules of desired redox properties [157] or by anchoring fluorinated hydrocarbon self-assembled monolayers (SAMs) (see Chapter XI) to inhibit corrosion of electrical contacts [158]. The field of electrocatalysis is an active one [156, 159] with much activity in the enhancement of reactions with surfactants (see Section V-8B). The creation of electrochemical sensors drives much research in electrode–solution interfaces. One approach is to attach ferrocene-terminated SAMs to an electrode; electron transfer occurs via tunneling and one can vary the distance between the ferrocene and the surface to engineer sensors of particular properties [160]. Biosensors involve immobilized protein molecules at an electrode surface. Rusling and co-workers have shown that myoglobin embedded in surfactant films [161, 162] or Langmuir–Blodgett lipid films [163] provided enhanced electron transfer and stability over myoglobin on the bare electrode surface.

B. Electrochemistry in Dispersed Phases

While much early work with dispersed electrochemical systems focused on silver halide sols [16], more recent studies by Rusling and co-workers and others exploited surfactant microstructures to influence electrochemical reactions [164, 165]. In the simplest of these systems, catalytic reactions of aryl halides can be enhanced in cationic surfactant micelles; a film of surfactant adsorbed on the electrode surface is thought to influence this process [166, 167]. In an important application to environmental remediation, they [167–170] have shown that surfactants or microemulsions are capable of dispersing nonpolar organohalides for subsequent electrochemical dehalogenation. Dispersion in microemulsions is preferable since micelles have a limited capacity and use added salts and mercury electrodes [171]. Microemulsions are stable suspensions of water, oil and surfactant having sizes in the 10–100-nm range (see Chapter XIV). The structures vary from spherical droplets to bicontinuous networks. In these systems, kinetic control and enhancement of electrochemical reactions is possible [165, 172] with improved mass transfer and higher surface area in the bicontinuous structures [173, 174]. In addition, composite films of surfactant and clay particles [175] or redox catalysts [176] can be used to create catalytic electrodes.

C. Photoelectrochemistry; Solar Energy Conversion

Light irradiation of the electrode-solution interface may produce *photovoltaic* or *photogalvanic* effects; in the former, a potential change is observed, and in the latter, current flow occurs. Semiconductors are usually involved, and the photon energy required is related to the band gap of the electrode material. These effects have been discussed by Williams and Nozik [177], and the field of organic photoelectrochemistry has been reviewed by Fox [178]. The electrodes may be modified or coated; gold coated with a phthalocyanine derivative shows light-modified voltammograms and H_2 evolution [179]. Polymerization of 1-vinylpyrene was achieved photoelectrochemically at an n-GaAs electrode [180], as an example of "photocatalysis."

There have been numerous reviews of photoelectrochemical cells for solar energy conversion; see Refs. 181–184 for examples. Figure V-14 shows a typical illustrative scheme for a cell consisting of an n-type semiconductor electrode as anode and an ordinary metal electrode as cathode separated by an electrolyte solution. The valence and conduction bands of the semiconductor are bent near the interface, and as a consequence, the electron-hole pair generated by illumination should separate, the electron going into the bulk semiconductor phase and thence around the external circuit to the metal electrode and the hole migrating to the interface to cause the opposite chemical reaction. Current flow, that is, electricity, is thus generated. As the diagram in Fig. V-14 indicates, there are a number of interrelations between the various potentials and energies. Note the approximate alignment of the solid-state and electrochemical energy scales, the former having vacuum as the reference point and the latter having the H^+/H_2 couple as the reference point.

There are difficulties in making such cells practical. High-band-gap semiconductors do not respond to visible light, while low-band-gap ones are prone to photocorrosion [182, 185]. In addition, both photochemical and entropy or thermodynamic factors limit the *ideal* efficiency with which sunlight can be converted to electrical energy [186].

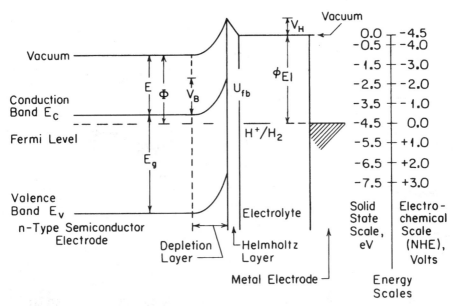

Fig. V-14. Energy level diagram and energy scales for an n-type semiconductor photoelectrochemical cell: E_g, band gap; E, electron affinity; Φ, work function; V_B, band bending; V_H, Helmholtz layer potential drop; ϕ_{El}, electrolyte work function; U_{fb}, flatband potential. (See Section V-9 for discussion of some of these quantities. (From Ref. 181.)

9. Types of Potentials and the Meaning of Potential Differences When Two Phases Are Involved

A. The Various Types of Potentials

Various kinds of potentials have been referred to in the course of this and the preceding chapter, and their interrelation is the subject of the present section. The chief problem is that certain types of potential differences are physically meaningful in the sense that they are operationally defined, whereas others that may be spoken of more vaguely are really conceptual in nature and may not be definable experimentally.

It is easy, for example, to define the potential difference between two points in a vacuum; this can be done by means of Coulomb's law in terms of the work required to transport unit charge from one point to the other. The potential of a point in space, similarly, is given by the work required to bring unit charge up from infinity. It is this type of definition that is involved in the term *Volta potential* ψ, which is the potential *just outside* or, practically speaking, about 10^{-3} cm from a phase. It is defined as $1/e$ times the work required to bring unit charge from infinity up to a point close to the surface. In like manner, one may define the *difference* in potential between two points, *both within a given phase*, in terms of the work required to transport unit charge from one point to the other. An example would be that of ζ potential or potential difference between bulk solution and the shear.layer of a charged particle.

The electrostatic potential *within* a phase, that is, $1/e$ times the electrical work of bringing unit charge from vacuum at infinity into the phase, is called the *Galvani*, or *inner*, *potential* ϕ. Similarly, the electrostatic potential difference between two phases is $\Delta\phi$. This quantity presents some subtleties, however. If one were to attempt to measure $\Delta\phi$ for a substance by bringing a charge from infinity in vacuum into the phase, the work would consist of two parts, the first being the work to bring the charge to a point just outside the surface, giving the Volta potential, and the second being the work to take the charge across the phase boundary into the interior, giving the *surface potential jump* χ. The situation is illustrated in Fig. V-15. The relationship between ϕ, ψ, and χ is simply

$$\phi = \chi + \psi \tag{V-69}$$

and the distinctions above between them are those introduced by Lange [187].

The problem is that ϕ, although apparently defined, is not susceptible to absolute experimental determination. The difficulty is that the unit charge involved must in practice be some physical entity such as an electron or an ion. When some actual charged species is transported across an interface, work will be involved, but there will be, in addition to the electrostatic work, a chemical work term that may be thought of as involving van der Waals forces, exchange forces, image forces, and so on. Some theoretical estimates of χ have been made, and Verwey [188] (see also Ref. 150) gives χ for the water-vacuum interface as -0.5 V, which has the physical implication that surface molecules are oriented with the hydrogens directed outward.

The *electrochemical potential* $\bar{\mu}_i^\alpha$ is defined as the total work of bringing a species i from vacuum into a phase α and is thus experimentally defined. It may be divided into a chemical work μ_i^α, the *chemical potential*, and the electrostatic work $z_i e \phi^\alpha$:

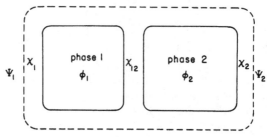

Fig. V-15. Volta potentials ψ, Galvani potentials ϕ, and surface potential jumps χ in a two-phase system. (From Ref. 187.)

$$\bar{\mu}_i^\alpha = \mu_i^\alpha + z_i e \phi^\alpha$$
$$\bar{\mu}_i^\alpha = \mu_i^\alpha + z_i e (\chi^\alpha + \psi^\alpha) \tag{V-70}$$

Since ψ^α is experimentally measurable, it is convenient to define another potential, the *real potential* α_i^α:

$$\alpha_i^\alpha = \mu_i^\alpha + z_i e \psi^\alpha \tag{V-71}$$

or

$$\bar{\mu}_i^\alpha = \alpha_i^\alpha + z_i e \chi^\alpha \tag{V-72}$$

Thus $\bar{\mu}_i$, α_i, and ψ are experimentally definable, while the surface potential jump χ, the chemical potential μ, and the Galvani potential difference between two phases $\Delta\phi = \phi^\beta - \phi^\alpha$ are not. While $\bar{\mu}_i$ is defined, there is a practical difficulty in carrying out an experiment in which the *only* change is the transfer of a charged species from one phase to another; electroneutrality generally is maintained in a reversible process so that it is a *pair* of species that is transferred, for example, a positive ion and a negative ion, and it is the sum of their electrochemical potential changes that is then determined. As a special case of some importance, the difference in chemical potential for a species i between two phases of *identical* composition is given by

$$\bar{\mu}_i^\beta - \bar{\mu}_i^\alpha = z_i e (\phi^\beta - \phi^\alpha) = z_i e (\psi^\beta - \psi^\alpha) \tag{V-73}$$

since now $\mu_i^\beta = \mu_i^\alpha$ and $\chi^\beta = \chi^\alpha$.†
    Finally, the difference in Volta potential between two phases is the *surface potential* ΔV discussed in Chapter III:

$$\Delta V = \psi^\beta - \psi^\alpha \tag{V-74}$$

Another potential, mentioned in the next section, is the *thermionic work function* Φ, where $e\Phi$ gives the work necessary to remove an electron from the highest populated level in a metal to a point outside. We can write

$$\Phi = \mu + e\chi \tag{V-75}$$

---

†Two phases at a different potential cannot have the identical chemical composition, strictly speaking. However, it would only take about 10^{-17}-mol electrons or ions to change the electrostatic potential of the 1-cm^3 portion of matter by 1 V, so that little error is involved.

Although certain potentials, such as the Galvani potential difference between two phases, are not experimentally well defined, *changes* in them may sometimes be related to a definite experimental quantity. Thus in the case of electrocapillarity, the imposition of a potential $\mathbf{E}$ on the phase boundary is taken to imply that a corresponding change in $\Delta\phi$ occurs, that is, $d\mathbf{E} = d(\Delta\phi)$. It might also be noted that the practice of dividing the electrochemical potential of a charged species into a chemical and an electrostatic part is actually an arbitrary one not strictly necessary to thermodynamic treatments. It has been used by Brønsted [189] and Guggenheim [190] as a useful way of separating out the feature common to charged species that their free energy does depend on the value of the electric field present. For further discussions of these various potentials, see Butler [191], Adam [192], and de Boer [193]. Case et al. [194] have reported on the determination of real potentials and on estimates of χ.

B. Volta Potentials, Surface Potential Differences, and the Thermionic Work Function

The thermionic work function for metals may be measured fairly accurately, and an extensive literature exists on the subject. A metal, for example, will spontaneously emit electrons, since their escaping tendency into space is enormously greater than that of the positive ions of the metal. The equilibrium state that is finally reached is such that the accumulated positive charge on the metal is sufficient to prevent further electrons from leaving. Alternatively, if a metal filament is negatively charged, electrons will flow from it to the anode; the rate of emission is highly temperature-dependent, and Φ may be calculated from this temperature dependence [195]. The function Φ may also be obtained from the temperature coefficient of the photoelectric emission of electrons or from the long-wavelength limit of the emission. As the nature of these measurements indicates, Φ represents an energy rather than a free energy.

When two dissimilar metals are connected, as illustrated in Fig. V-16, there is a momentary flow of electrons from the metal with the smaller work function to the other so that the electrochemical potential of the electrons becomes the same. For the two metals α and β

$$\bar{\mu}_e = -e(\Phi^\alpha + \psi^\alpha) = -e(\Phi^\beta + \psi^\beta) \tag{V-76}$$

The difference in Volta potential ΔV, which has been called the *surface* (or *contact*) *potential* in this book, is then given by

$$\Delta V = \psi^\alpha - \psi^\beta = \Phi^\beta - \Phi^\alpha \tag{V-77}$$

It is this potential difference that is discussed in Chapter IV in connection with monomolecular films. Since it is developed in the space between the phases, none of the uncertainties of phase boundary potentials is involved.

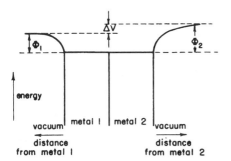

Fig. V-16. The surface potential difference ΔV.

In the case of films on liquids, ΔV was measured directly rather than through determinations of thermionic work functions. However, just as an adsorbed film greatly affects ΔV, so must it affect Φ. In the case of metals, the effect of adsorbed gases on ΔV is easily studied by determining changes in the work function, and this approach is widely used today (see Section VIII-2C). Much of Langmuir's early work on chemisorption on tungsten may have been stimulated by a practical appreciation of the important role of adsorbed gases in the behavior of vacuum tubes.

C. Electrode Potentials

We now consider briefly the matter of electrode potentials. The familiar Nernst equation was at one time treated in terms of the "solution pressure" of the metal in the electrode, but it is better to consider directly the net chemical change accompanying the flow of 1 faraday ($\mathcal{F}$), and to equate the electrical work to the free energy change. Thus, for the cell

$$Zn'/\text{solution containing } Zn^{2+} \text{ and } Ag^+/Ag/Zn'' \tag{V-78}$$

the net cell reaction is

$$\tfrac{1}{2}Zn + Ag^+ = \tfrac{1}{2}Zn^{2+} + Ag \tag{V-79}$$

and

$$-\mathcal{F}\mathcal{E} = \mu_{Ag}^{Ag} - \tfrac{1}{2}\mu_{Zn}^{Zn} - \mu_{Ag^+}^{S} + \tfrac{1}{2}\mu_{Zn^{2+}}^{S} \tag{V-80}$$

where the superscripts denote the phase involved. Equation V-80 may be abbreviated

$$\mathcal{F}\mathcal{E} = \Delta\mu_0 - RT\ln\frac{a_{Zn^{2+}}^{1/2}}{a_{Ag^+}} \tag{V-81}$$

Equation V-81 may be derived in a more detailed fashion by considering that the difference between the electrochemical potentials of the electrons at the Zn' and Zn" terminals, the chemical potentials being equal, gives $\Delta\phi$ and hence $\mathcal{E}$. Thus

$$\bar{\mu}_e'' - \bar{\mu}_e' = -(\phi'' - \phi') = -\mathcal{F}\mathcal{E} \tag{V-82}$$

Since the electrons are in equilibrium,

$$\bar{\mu}_e' = \bar{\mu}_e^{Zn} \qquad \bar{\mu}_e'' = \bar{\mu}_e^{Ag} \tag{V-83}$$

Also, there is equilibrium between electrons, metallic ions, and metal atoms within each electrode:

$$\tfrac{1}{2}\bar{\mu}_{Zn^{2+}}^{Zn} + \bar{\mu}_e^{Zn} = \tfrac{1}{2}\mu_{Zn}^{Zn} \tag{V-84}$$

$$\bar{\mu}_{Ag^+}^{Ag} + \bar{\mu}_e^{Ag} = \bar{\mu}_{Ag}^{Ag} \tag{V-85}$$

Then

$$-\mathcal{F}\mathcal{E} = \mu_{Ag}^{Ag} - \tfrac{1}{2}\mu_{Zn}^{Zn} - \bar{\mu}_{Ag^+}^{Ag} + \tfrac{1}{2}\bar{\mu}_{Zn^{2+}}^{Zn} \tag{V-86}$$

In addition, since there is equilibrium between the metal ions in the two phases,

$$\bar{\mu}_{Zn^{2+}}^{Zn} = \bar{\mu}_{Zn^{2+}}^{S} \qquad \bar{\mu}_{Ag^+}^{Ag} = \bar{\mu}_{Ag^+}^{S} \tag{V-87}$$

Substitution of Eq. V-87 into Eq. V-86 gives Eq. V-80, since

$$\tfrac{1}{2}\bar{\mu}_{Zn^{2+}}^{S} - \bar{\mu}_{Ag^+}^{S} = \tfrac{1}{2}\mu_{Zn^{2+}}^{S} + \mu_{Ag^+}^{S}$$

This approach, much used by Guggenheim [190], seems elaborate, but in the case of more complex situations than the above, it can be a very powerful one.

A problem that has fascinated surface chemists is whether, through suitable measurements, one can determine *absolute* half-cell potentials. If some one standard half-cell potential can be determined on an absolute basis, then all others are known through the table of standard potentials. Thus, if we know $\mathcal{E}^0$ for

$$Ag = Ag^+(aq) + e^- \tag{V-88}$$

we can immediately obtain $\mathcal{E}_{H_2/H^+}^0$ since $\mathcal{E}_{Ag/Ag^+}^0 - \mathcal{E}_{H_2/H^+}^0$ is -0.800 V at 25°C.

The standard states of Ag and of $Ag^+(aq)$ have the conventional definitions, but there is an ambiguity in the definition of the standard state of e^-. Suppose that a reference electrode R is positioned above a solution of $AgNO_3$, which in turn is in contact with an Ag electrode. The Ag electrode and R are connected by a wire. Per Faraday, the processes are

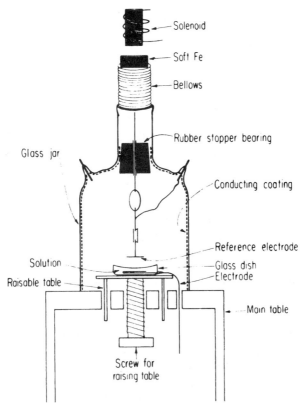

Fig. V-17. Schematic diagram for the apparatus for measurement of V_{obs} (see text). The vibrating reference electrode is positioned close to the surface of a $AgNO_3$ solution in which there is an Ag electrode, which, in turn, is in electrical contact with the reference electrode. (From Ref. 196.)

$$e^-(\text{in } R) = e^-(\text{in air})$$
$$e^-(\text{in air}) = e^-(\text{in solution})$$
$$\underline{e^-[\text{in solution} + Ag^+(aq)] = Ag}$$
$$e^-(\text{in } R) + Ag^+(aq) = Ag \qquad \text{(V-89)}$$

The potential corresponding to the reversible overall process is the measurable quantity V_{obs}. If we know the work function for R, that is, the potential Φ_R for $e^-(\text{in } R) = e^-(\text{in}$ air), then $V_{obs} - \Phi_R$ is $\mathcal{E}$ for the process

$$e^-(\text{in air}) + Ag^+(aq) = Ag \qquad \text{(V-90)}$$

On correcting to unit activity $Ag^+(aq)$, we can obtain $\mathcal{E}^0_{Ag/Ag^+}$. Electron solvation energy is neglected in this definition.

Figure V-17 shows the apparatus used by Gomer and Tryson [196] for measurements of V_{obs}. While their analysis (and literature review) is much more detailed than the foregoing discussion, they concluded that $\mathcal{E}^{0}_{H_2/H^+}$ is -4.73 ± 0.05 V at 25°C.

D. Irreversible Electrode Phenomena

An important and common phenomenon is that of overvoltage, or the observation that in order to pass appreciable current through an electrochemical cell, it is frequently necessary to apply a larger than reversible potential to the electrodes. A simple method for studying overvoltages is illustrated in Fig. V-18. The desired current is passed through the electrode to be studied E_x by means of the circuit $E_1 - E_x$, and the potential variation at E_x is measured by means of a potentiometer circuit $E_x - E_h$. One may either make the potential measurement while current is being passed through the circuit $E_1 - E_x$ or, by means of a rotating commutator or thyratron switching, make the measurement just after turning off the current.

Overvoltage effects may be divided roughly into three main classes with respect to causes. First, whenever current is flowing, there will be chemical change at the electrode and a corresponding local accumulation or depletion of material in the adjacent solution. This effect, known as *concentration polarization*, can be serious. Usually, however, it is possible to eliminate it by suitable stirring. Second, overvoltage arises from the Ohm's law potential drop when appreciable current is flowing. This effect is not very important when only small currents are involved and may be further minimized by the use of the circuit $E_x - E_h$ with the electrode E_h in the form of a probe positioned close to E_x

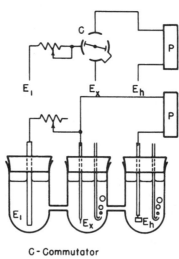

C - Commutator
P - Potentiometer

Fig. V-18. Apparatus for measuring overvoltages.

and on the opposite side to the one facing E_1. The commutator method also is designed to eliminate ohmic overvoltage. Finally, the most interesting class of overvoltage effect is that where some essential step in the electrode reaction itself is slow, presumably due to some activated chemical process being involved. In general, a further distinction between these three classes of overvoltage phenomena is that, on shutting off the polarizing current, ohmic currents cease instantly and concentration polarization decays slowly and in a complex way, whereas activation polarization often decays exponentially.

Considering now only the third type of overvoltage, most electrodes involving a metal and its ion in solution are fairly reversible, and rather high current densities are required to produce appreciable overvoltages. Actually, the effect has been studied primarily in connection with gas electrodes, particularly with respect to hydrogen overvoltage. In this case rather small current flow ($\mu A/cm^2$) can produce a sizable effect in terms of the overvoltage to cause visible hydrogen evolution; values range from essentially zero for platinized platinum (and 0.09 for smooth platinum) to 0.78 V for mercury.

The first law of electrode kinetics, observed by Tafel in 1905 [197], is that overvoltage η varies with current density i according to the equation

$$\eta = \alpha - b \ln i \qquad (V\text{-}91)$$

where η is negative for a cathodic process and positive for an anodic process.

This law may be accounted for in a simple way as follows:

$$O(\text{oxidized state}) + e^- = R(\text{reduced state}) \qquad (V\text{-}92)$$

that is activated. Then the forward rate is given by

$$R_f = i_f = \frac{kT}{h}\, e^{-\Delta G_f^{0\ddagger}/RT}(O) \qquad (V\text{-}93)$$

where $\Delta G_f^{0\ddagger}$ is the standard free energy of activation. Similarly, for the reverse rate,

$$R_b = i_b = \frac{kT}{h}\, e^{-\Delta G_b^{0\ddagger}/RT}(R) \qquad (V\text{-}94)$$

It is now assumed that $\Delta G_f^{0\ddagger}$ consists of a chemical component and an electrical component and that it is only the latter that is affected by changing the electrode potential. The specific assumption is that

$$\Delta G_f^{0\ddagger} = (\Delta G_f^{0\ddagger})_{\text{chem}} + \alpha \mathcal{F} (\phi_M - \phi_S) \qquad (V\text{-}95)$$

that is, that a certain fraction α of the potential difference between the metal and solution phases contributes to the activation energy. This coefficient α is known as the *transfer coefficient* (see Ref. 150). Then

$$i_f = \frac{kT}{h} e^{-(\Delta G^{0\ddagger}_{f\text{chem}}/RT)} e^{-(\alpha \mathcal{F}/RT)(\phi_M - \phi_S)}(O) \tag{V-96}$$

Equations such as V-96 are known as Butler-Volmer equations [150]. At equilibrium, there will be equal and opposite currents in both directions, $i_f^0 = i_b^0 = i^0$. By Eq. V-96 the apparent *exchange* current i^0 will be

$$i^0 = \frac{kT}{h} e^{-(\Delta G^{0\ddagger}_{f\text{chem}}/RT)} e^{-(\alpha \mathcal{F}/RT)(\phi_M^0 - \phi_S)}(O) \tag{V-97}$$

where $(\phi_M^0 - \phi_S)$ is now the equilibrium potential difference at the electrode. Combination of Eqs. V-96 and V-97 give

$$i_f = i^0 e^{-(\alpha \mathcal{F}/RT)(\phi_M - \phi_M^0)} = i^0 e^{-(\alpha \mathcal{F}/RT)\eta} \tag{V-98}$$

Equation V-98 may be written in the form

$$\ln i = \ln i^0 - \frac{\alpha \mathcal{F}}{RT} \eta \tag{V-99}$$

which, in turn, rearranges to the Tafel equation.

The treatment may be made more detailed by supposing that the rate-determining step is actually from species O in the OHP (at potential ϕ_2 relative to the solution) to species R similarly located. The effect is to make i^0 dependent on the value of ϕ_2 and hence on any changes in the electrical double layer. This type of analysis has permitted some detailed interpretations to be made of kinetic schemes for electrode reactions and also connects that subject to the general one of this chapter.

The measurement of α from the experimental slope of the Tafel equation may help to decide between rate-determining steps in an electrode process. Thus in the reduction water to evolve H_2 gas, if the slow step is the reaction of H_3O^+ with the metal M to form surface hydrogen atoms, M—H, α is expected to be about $\frac{1}{2}$. If, on the other hand, the slow step is the surface combination of two hydrogen atoms to form H_2, a second-order process, then α should be 2 (see Ref. 150).

10. Problems

1. Assume ψ_0 is -25 mV for a certain silica surface in contact with $0.001M$ aqueous NaCl at $25°C$. Calculate, assuming simple Gouy-Chapman theory (a) ψ at 200 Å from the surface, (b) the concentrations of Na^+ and of Cl^- ions 10 Å from the surface, and (c) the surface charge density in electronic charges per unit area.

2. Repeat the calculation of Problem 1, assuming all conditions to be the same except that the electrolyte is di-divalent (e.g., $MgSO_4$).

3. Using the Gouy-Chapman equations calculate and plot ψ at 5 Å from a surface as a function of ψ_0 (from 0 to 100 mV at $25°C$) for $0.002M$ and for $0.02M$ 1 : 1 electrolyte.

4. Show that Eq. V-12 can be written in the equivalent forms

$$y = 2\ln \frac{1 + e^{-\kappa x}\tanh(y_0/4)}{1 - e^{-\kappa x}\tanh(y_0/4)}$$

and

$$\kappa x = \ln \frac{(e^{y/2} + 1)(e^{y_0/2} - 1)}{(e^{y/2} - 1)(e^{y_0/2} + 1)}$$

5. Derive Eq. V-14.

6. Derive Eq. V-18 for small values of ψ_0.

7. Show what Eq. V-7 (for κ^2) becomes when written in the SI system. Calculate κ for $0.05M$, 2 : 2 electrolyte at $25°C$ using SI units; repeat the calculation in the cgs/esu system and show that the result is equivalent.

8. Show what Eq. V-16 (for σ) becomes when written in the SI system. Calculate σ for $0.02M$ 1 : 1 electrolyte at $25°C$ and $\psi_0 = -40$ mV.

9. Derive the general equation for the differential capacity of the diffuse double layer from the Gouy-Chapman equations. Make a plot of surface charge density σ versus this capacity. Show under what conditions your expressions reduce to the simple Helmholtz formula of Eq. V-17.

10. A circular metal plate of 10-cm^2 area is held parallel to and at a distance d above a solution, as in a surface potential measurement. The temperature is $25°C$.

(a) First, assuming ψ_0 for the solution interface is 40 mV, calculate ψ as a function of x, the distance normal to the surface on the air side. Assume that the air has 10^7 ions/cm^3. (Ignore the plate for this part.) Plot your result.

(b) Calculate the repulsive potential between the plate and the solution if $d = 2$ cm, assuming ψ_0 for the plate also to be 40 mV and using Eq. V-29. Are the assumptions of the equation good in this instance? Explain.

(c) Calculate the contribution of the solution-air interface to the surface potential if the plate is now used as a probe in a surface potential measurement and is 1 mm from the surface.

11. Using the conditions of the Langmuir approximation for the double-layer repulsion, calculate for what size particles in water at $25°C$ the double-layer repulsion energy should equal kT if the particles are 40 Å apart.

12. For water at $25°C$, Eq. V-36 can be written

$$\zeta = 12.9v/\mathbf{F}$$

Show what the units of ζ, v, and $\mathbf{F}$ must be.

13. Show what form Eq. V-36 takes if written in the SI system. Referring to the equation of Problem 12, show what number replaces 12.9 if ζ is in volts, v in meters per second, and $\mathbf{F}$ in volts per meter.

14. Verify the calculation of v in the paragraph preceding Eq. V-37. (Watch the units!)

15. Calculate ζ for the example in the second paragraph following Eq. V-46. Note the caution in Problem 16. Also, Eq. V-46 can be expressed in SI units: $E = \zeta P \epsilon_0/\eta k$, where ϵ_0 is the permittivity of vacuum, 8.85419×10^{-12} C^2 J^{-1} m^{-1}.

16. It was mentioned that streaming potentials could be a problem in jet aircraft. Suppose that a hydrocarbon fuel of dielectric constant 7 and viscosity 0.02 poise (P) is being pumped under a driving pressure of 25 atm. The potential between the pipe and the fuel is, say, 130 mV, and the fuel has a low ion concentration in it, equivalent to 10^{-7} M NaCl. Making and stating any necessary and reasonable assumptions, calculate the streaming potential that should be developed. *Note:* Consider carefully the handling of units.

17. Calculate the zeta potential for the system represented by the first open square point (for pH 3) in Fig. V-8.

18. Make an approximate calculation of ψ 90 Å from the surface of a 1-μm spherical particle of $\psi_0 = 30$ mV and in a 0.10 M 1:1 electrolyte at 25°C.

19. Make a calculation to confirm the numerical illustration following Eq. V-42. Show what form the equation takes in the SI system, and repeat the calculation in SI units. Show that the result corresponds to that obtained using cgs/esu units.

20. (a) Calculate the expected streaming potential $\mathbf{E}$ for pure water at 25°C flowing down a quartz tube under 8 atm pressure. Take ζ to be 150 mV. (b) Show what form Eq. V-46 takes in the SI system and repeat the calculation of part (a) in SI units.

21. Streaming potential measurements are to be made using a glass capillary tube and a particular electrolyte solution, for example, 0.01M sodium acetate in water. Discuss whether the streaming potential should or should not vary appreciably with temperature.

22. While the result should not have very exact physical meaning, as an exercise, calculating the ζ potential of lithium ion, knowing that its equivalent conductivity is 39 cm^2/(eq)(ohm) in water at 25°C.

23. The points in Fig. V-12 come from three types of experimental measurements. Explain clearly what the data are and what is done with the data, in each case, to get the σ-versus-ψ^r plot. What does the agreement between the three types of measurement confirm? Explain whether it confirms that ψ^r is indeed the correct absolute interfacial potential difference.

24. The following data (for 25°C) were obtained at the pzc for the Hg-aqueous NaF interface. Estimate Γ_{salt}^{max} and plot it as a function of the mole fraction of salt in solution. In the table, $f_\pm$ is mean activity coefficient such that $a_\pm = f_\pm m_\pm$, where $m_\pm$ is mean molality.

Concentration (M)	Density (g/cm³)	$f_\pm$	γ (ergs/cm²)
0.1	1.001	0.773	426.5
0.25	1.008	0.710	426.9
0.5	1.020	0.670	427.3
1.0	1.045	0.645	427.7
4.0	1.189	0.792	430.3
6.25	1.285	1.098	432.8
8.0	1.360	1.480	436.7
10.0	1.440	2.132	443.0

25. Assume that a salt, MX (1 : 1 type), adsorbs at the mercury-water interface according to the Langmuir equation:

$$\frac{x}{x_m} = \frac{bc}{1 + bc}$$

where x_m is the maximum possible amount, and x/x_m is 0.5 at $c = 0.1$ M. Neglect activity coefficient effects and estimate the value of the mercury-water interfacial tension at 25°C at the electrocapillary maximum where 0.01 M salt solution is used.

26. What is the magnitude of the attainable sedimentation potential given by Eq. V-47 for a 0.1% by volume aqueous suspension of 300 nm polystyrene latex spheres (density = 1.059 g/cm³) having a zeta potential of 100 mV? Assume you are using a typical swinging basket centrifuge with a 20-cm rotor radius and 6 cm centrifuge tubes filled 2/3 full. Plot the measurable potential versus the angular velocity of the centrifuge.

27. Referring to Eq. V-69, calculate the value of C for a 150-Å film of ethanol of dielectric constant 26. *Optional:* Repeat the calculation in the SI system.

28. Make a calculation to confirm the statement made in the footnote following Eq. V-74.

General References

J. O'M. Bockris and A. K. N. Reddy, *Modern Electrochemistry*, Plenum, New York, 1970.

P. Delahay, *Double Layer and Electrode Kinetics*, Interscience, New York, 1965.

D. C. Grahame, *Chem. Rev.*, **41**, 441 (1947).

E. A. Guggenheim, *Thermodynamics*, Interscience, New York, 1949.

H. S. Harned and B. B. Owen, *The Physical Chemistry of Electrolyte Solutions*, Reinhold, New York, 1950.

R. J. Hunter, *Zeta Potential in Colloid Science: Principles and Applications*, Academic, Orlando, FL, 1981.

J. N. Israelachvili, *Intermolecular and Surface Forces*, Academic, Orlando, FL, 1985.

G. Kortüm, *Treatise on Electrochemistry*, Elsevier, New York, 1965.

J. Lyklema, *Fundamentals of Interface and Colloid Science, Vol. 1 and 2*, Academic Press, San Diego, 1995.

H. R. Kruyt, *Colloid Science*, Elsevier, New York, 1952.

J. Th. G. Overbeek, *Advances in Colloid Science*, Vol. 3, Interscience, New York, 1950.

W. B. Russel, D. A. Saville, and W. R. Schowalter, *Colloidal Dispersions*, Cambridge University Press, Cambridge, UK, 1989.

M. J. Sparnaay, *The Electrical Double Layer*, Pergamon, New York, 1972.

G. Sposito, *The Thermodynamics of Soil Solutions*, Oxford University Press (Clarendon), Oxford, 1981.

H. van Olphen and K. J. Mysels, *Physical Chemistry: Enriching Topics from Colloid and Surface Chemistry*, Theorex (8327 La Jolla Scenic Drive), La Jolla, CA, 1975.

E. J. W. Verwey and J. Th. G. Overbeek, *Theory of the Stability of Lyophobic Colloids*, Elsevier, New York, 1948.

R. D. Vold and M. J. Vold, *Colloid and Interface Chemistry*, Addison-Wesley, Reading, MA, 1983.

Textual References

1. A. W. Adamson, *J. Chem. Ed.*, **55,** 634 (1978).

2. G. Gouy, *J. Phys.*, **9**(4), 457 (1910); *Ann. Phys.*, **7**(9), 129 (1917).

3. D. L. Chapman, *Phil. Mag.*, **25**(6), 475 (1913).

4. P. Debye and E. Hückel, *Phys. Z.*, **24,** 185 (1923); P. Debye, *Phys. Z.*, **25,** 93 (1924).

5. E. J. W. Verwey and J. Th. G. Overbeek, *Theory of the Stability of Lyophobic Colloids*, Elsevier, New York, 1948.

6. H. R. Kruyt, *Colloid Science*, Elsevier, New York, 1952.

7. R. O. James and G. A. Parks, in *Surface and Colloid Science*, Vol. 12, E. Matijevic, ed., Plenum, New York, 1982.

8. L. Blum, in *Fluid Interfacial Phenomena*, C. A. Croxton, ed., Wiley, New York, 1986.

9. W. B. Russel, D. A. Saville, and W. R. Schowalter, *Colloidal Dispersions*, Cambridge University Press, 1989.

10. H. S. Harned and B. B. Owen, *The Physical Chemistry of Electrolyte Solutions*, Reinhold, New York, 1950.

11. K. J. Mysels, *An Introduction to Colloid Chemistry*, Interscience, New York, 1959.

12. A. Marmur, *J. Colloid Interface Sci.*, **71,** 610 (1979).

13. A. L. Loeb, J. Th. G. Overbeek, and P. H. Wiersema, *The Electrical Double Layer Around a Spherical Particle*, MIT Press, Cambridge, MA, 1961.

13a. R. Natarajan and R. S. Schechter, *J. Colloid Interface Sci.*, **99,** 50 (1984).

14. J. Th. G. Overbeek and B. H. Bijsterbosch, in *Electrochemical Separation Methods*, P. G. Righetti, C. J. van Oss, and J. W. Vanderhoff, eds., Elsevier/North Holland Biomedical Press, New York, 1979.

15. S. Carnie, D. Chan, J. Mitchell, and B. Ninham, *J. Chem. Phys.*, **74,** 1472 (1981).

16. M. J. Sparnaay, *The Electrical Double Layer*, Pergamon, New York, 1972.

17. C. Outhwaite, B. Bhuiyan, and S. Levine, *J. Chem. Soc., Faraday Trans. 2*, **76**, 1388 (1980).

18. G. M. Torrie, J. P. Valleau, and G. N. Patey, *J. Chem. Phys.*, **76**, 4615 (1982).

19. V. Kralj-Iglič and A. Iglič, *J. Phys. II France* **6**, 477 (1996).

20. A. P. Winiski, A. C. McLaughlin, R. V. McDaniel, M. Eisenberg and S. McLaughlin, *Biochemistry*, **25**, 8206 (1986).

21. A. P. Winiski, M. Eisenberg, and S. McLaughlin, *Biochemistry*, **27**, 386 (1988).

22. M. P. Hentschel, M. Mischel, R. C. Oberthur, and G. Buldt, *FEBS Lett.*, **193**, 236 (1985).

23. O. Stern, *Z. Elektrochem.*, **30**, 508 (1924).

24. J. Lyklema and J. Th. G. Overbeek, *J. Colloid Sci.*, **16**, 595 (1961).

25. J. A. Davis, R. O. James, and J. O. Leckie, *J. Colloid Interface Sci.*, **63**, 480 (1978).

26. R. O. James, J. A. Davis, and J. O. Leckie, *J. Colloid Interface Sci.*, **65**, 331 (1978).

27. J. B. Sweeney, L. E. Scriven, and H. T. Davis, *J. Chem. Phys.*, **87**, 6120 (1987).

28. D. Y. C. Chan and D. J. Mitchell, *J. Colloid Interface Sci.*, **95**, 193 (1983); see also D. G. Hall, *ibid.*, **108**, 411 (1985).

29. B. Derjaguin and L. Landau, *Acta Physicochim. URSS*, **14**, 633 (1941).

30. B. V. Derjaguin, *Prog. Colloid Polym. Sci.*, **74**, 17 (1987).

31. I. Langmuir, *J. Chem. Phys.*, **6**, 873 (1938); R. Defay and A. Sanfeld, *J. Chim. Phys.*, **60**, 634 (1963).

32. A. Dunning, J. Mingins, B. A. Pethica, and P. Richmond, *J. Chem. Soc., Faraday Trans. 1*, **74**, 2617 (1978).

33. W. Olivares and D. A. McQuarrie, *J. Phys. Chem.*, **84**, 863 (1980).

34. G. Wilemski, *J. Colloid Interface Sci.*, **88**, 111 (1982).

35. P. L. Levine and G. M. Bell, *J. Colloid Interface Sci.*, **74**, 530 (1980).

36. V. M. Muller and B. V. Derjaguin, *Colloids & Surfaces*, **6**, 205 (1983).

37. E. Ruckenstein and D. Schiby, *Langmuir*, **1**, 612 (1985).

38. B. Duplantier, R. E. Goldstein, V. Romero-Ronchín, and A. I. Pesci, *Phys. Rev. Lett.*, **65**, 508 (1990).

39. R. E. Goldstein, A. I. Pesci, and V. Romero-Ronchín, *Phys. Rev. A*, **41**, 5504 (1990).

40. R. E. Goldstein, T. C. Halsey, and M. Leibig, *Phys. Rev. Lett.*, **66**, 1551 (1991).

41. M. C. Herman and K. D. Papadopoulos, *J. Colloid Interface Sci.*, **142**, 331 (1991).

42. A. K. Sengupta and K. D. Papadopoulos, *J. Colloid Interface Sci.*, **149**, 135 (1992).

43. K. D. Papadopoulos, A. Yato, and H. Nguyen, *J. Theor. Biol.*, **113**, 545 (1985).

44. S. L. Carnie, D. Y. C. Chan, and J. Stankovich, *J. Colloid Interface Sci.*, **165**, 116 (1994).

45. J. E. Sader, S. L. Carnie, and D. Y. C. Chan, *J. Colloid Interface Sci.*, **171**, 46 (1995).

46. R. A. Ring, *J. Chem. Soc. Faraday Trans. 2*, **81**, 1193 (1985).

47. E. Barouch, E. Matijevic, and T. H. Wright, *J. Chem. Soc. Faraday Trans. I*, **81**, 1819 (1985).

48. S. L. Carnie, D. Y. C. Chan, and J. S. Gunning, *Langmuir*, **10**, 2993 (1994).

49. L. K. Koopal and S. S. Dukhin, *Colloids & Surfaces*, **A73**, 201 (1993).

50. J. N. Israelachvili and G. E. Adams, *J. Chem. Soc. Faraday Trans. 1*, **74**, 975 (1978).

51. H. K. Christenson and R. G. Horn, *Chem. Phys. Lett.*, **98**, 45 (1983).

52. R. M. Pashley and B. W. Ninham, *J. Phys. Chem.*, **91**, 2902 (1987).

53. I. Larson, C. J. Drummond, D. Y. C. Chan, and F. Grieser, *J. Phys. Chem.*, **99**, 2114 (1995).

54. D. C. Prieve and B. M. Alexander, *Science*, **231**, 1269 (1986).

55. J. Penfold and J. D. F. Ramsay, *J. Chem. Soc. Faraday Trans. 1*, **81**, 117 (1985).

56. D. C. Prieve and E. Ruckenstein, *J. Colloid Interface Sci.*, **63**, 317 (1978).

57. J. L. Anderson, *Annu. Rev. Fluid Mech.*, **21**, 61 (1989).

58. See J. Th. Overbeek, *Advances in Colloid Science*, Vol. 3, Interscience, New York, 1950, p. 97.

59. R. Hidalgo-Alvarez, F. J. de las Nieves, A. J. Van der Linde, and B. H. Bijsterbosch, *Colloids & Surfaces*, **21**, 259 (1986).

60. D. Bastos and F. J. de las Nieves, *Colloid Polym. Sci.*, **271**, 860 (1993).

61. D. Bastos and F. J. de las Nieves, *Colloid Polym. Sci.*, **272**, 592 (1994).

62. A. A. Morfesis and R. L. Rowell, *Langmuir*, **6**, 1088 (1990).

63. B. J. Marlow and R. L. Rowell, *Langmuir*, **7**, 2970 (1991).

64. J. H. Prescott, S. Shiau, and R. L. Rowell, *Langmuir*, **9**, 2071 (1993).

65. R. W. O'Brien, D. W. Cannon, and W. N. Rowlands, *J. Colloid Interface Sci.*, **173**, 406 (1995).

66. M. C. Fair and J. L. Anderson, *J. Colloid Interface Sci.*, **127**, 388 (1989).

67. D. Velegol, J. L. Anderson, and S. Garoff, *Langmuir*, **12**, 4103 (1996).

68. J. P. Ebel, J. L. Anderson, and D. C. Prieve, *Langmuir*, **4**, 396 (1988).

69. N. von Smoluchowski, *Bull. Int. Acad. Polon. Sci., Classe Sci. Math. Nat.*, **1903**, 184.

70. S. McLaughlin and R. T. Mathias, *J. Gen. Physiol.*, **85**, 699 (1985).

71. P. B. Lorenz, *J. Phys. Chem.*, **57**, 430 (1953).

72. J. T. Davies and E. K. Rideal, *Interfacial Phenomena*, Academic, New York, 1963.

73. R. A. Van Wagenen and J. D. Andrade, *J. Colloid Interface Sci.*, **76**, 305 (1980).

74. R. Ramachandran and P. Somasundaran, *Colloids & Surfaces*, **21**, 355 (1986).

75. J. S. Lyons, D. N. Furlong, A. Homola, and T. W. Healy, *Aust. J. Chem.*, **34**, 1167 (1981); J. S. Lyons, D. N. Furlong, and T. W. Healy, ibid., **34**, 1177 (1981).

76. P. J. Scales, F. Grieser, and T. W. Healy, *Langmuir*, **6**, 582 (1990).

77. A. J. Rutgers and P. Nagels, *Nature*, **171**, 568 (1953).

78. B. J. Marlow and R. L. Rowell, *Langmuir*, **1**, 83 (1985).

79. U. Saxén, *Wied Ann.*, **47**, 46 (1892).

80. P. Mazur and J. Th. G. Overbeek, *Rec. Trav. Chim.*, **70**, 83 (1951).

81. P. B. Lorenz, *J. Phys. Chem.*, **57**, 430 (1953).

82. R. Hidalgo-Alvarez, F. J. de las Nieves, and G. Pardo, *J. Chem. Soc. Faraday Trans. 1*, **81**, 609 (1985).

83. A. J. Rutgers, M. de Smet, and W. Rigole, in *Physical Chemistry: Enriching Topics from Colloid and Surface Chemistry*, H. Van Olphen and K. J. Mysels, eds., Theorex (8327 La Jolla Scenic Drive), La Jolla, CA, 1975.

84. J. Kujlstra, H. P. van Leeuwen, and J. Lyklema, *J. Chem. Soc. Faraday Trans. 1*, **88**, 3441 (1992).

85. F. J. de las Nieves, F. J. Rubio-Hernandez and R. Hidalgo-Alvarez, *J. Non-Equilib. Thermodyn.*, **13**, 373 (1988).

86. R. Hidalgo-Alvarez, J. A. Moleon, F. J. de las Nieves, and B. H. Bijsterbosch, *J. Colloid Interface Sci.*, **149**, 23 (1992).

87. A. Fernández-Barbero, A. Martín-Rodríguez, J. Callejas-Fernández, and R. Hidalgo-Alvarez, *J. Colloid Interface Sci.*, **162**, 257 (1994).

88. J. F. Miller, O. Velev, S. C. C. Wu, and H. J. Ploehn, *J. Colloid Interface Sci.*, **174**, 490 (1995).

89. E. F. Burton, *Phil. Mag.*, **11**, 425 (1906); **17**, 583 (1909).

90. R. J. Pugh, "Dispersion and Stability of Ceramic Powders," in *Surface and Colloid Chemistry in Advanced Ceramics Processing*, Marcel Dekker, New York, 1994, Chapter 4.

91. D. E. Yates and T. W. Healy, *J. Chem. Soc., Faraday 1*, **76**, 9 (1980).

92. M. Mao, D. Fornasiero, J. Ralston, R. S. C. Smart, and S. Sobieraj, *Colloids & Surfaces*, **A85**, 37 (1994).

93. A. C. Lau, D. N. Furlong, T. W. Healy, and F. Grieser, *Colloids & Surfaces*, **18**, 93 (1986).

94. R. O. James, G. R. Wiese, and T. W. Healy, *J. Colloid Interface Sci.*, **59**, 381 (1977).

95. C. F. Zukoski IV and D. A. Saville, *J. Colloid Interface Sci.*, **114**, 45 (1986).

96. T. Morimoto and S. Kittaka, *J. Colloid Interface Sci.*, **44**, 289 (1973).

97. D. G. Hall, *J. Chem. Soc. Faraday II*, **76**, 1254 (1980); D. G. Hall and H. M. Rendall, *J. Chem. Soc. Faraday I*, **76**, 2575 (1980).

98. R. W. O'Brien and R. J. Hunter, *Can. J. Chem.*, **59**, 1878 (1981).

99. H. Ohshima, T. W. Healy, and L. R. White, *J. Chem. Soc. Faraday Trans. 2*, **79**, 1613 (1983).

100. W. Wu, R. F. Giese, Jr., and C. J. van Oss, *Colloids & Surfaces*, **A89**, 241 (1994).

101. S. S. Dukhin and J. Lyklema, *Faraday Disc. Chem. Soc.*, **90**, 261 (1990).

102. J. Kijlstra and H. van Leeuwen, *J. Colloid Interface Sci.*, **160**, 424 (1993).

103. S. Y. Shulepov, S. S. Dukhin, and J. Lyklema, *J. Colloid Interface Sci.*, **171**, 340 (1995).

104. H. Kihira, N. Ryde, and E. Matijević, *Colloids & Surfaces*, **64**, 317 (1992).

105. H. Kihira, N. Ryde, and E. Matijević, *J. Chem. Soc. Faraday Trans. 1*, **88**, 2379 (1992).

106. H. Kihira and E. Matijević, *Langmuir*, **8**, 2855 (1992).

107. J. A. Maroto and F. J. de las Nieves, *Prog. Colloid Polym. Sci.*, **98**, 89 (1995).

108. W. Henry, *Nicolson's J.*, **4**, 224 (1801).

109. See also H. R. Kruyt, ed., *Colloid Science*, Vol. 1, Elsevier, New York, 1952.

110. G. Lippmann, *Ann. Chim. Phys.*, **5**, 494 (1875).

111. H. Vos and J. M. Vos, *J. Colloid Interface Sci.*, **74**, 369 (1980).

112. K. G. Baikerikar and R. S. Hansen, *Surf. Sci.*, **50**, 527 (1975).

113. D. C. Grahame, *Chem. Rev.*, **41**, 441 (1947).

114. See S. Levine and G. M. Bell, *J. Colloid Sci.*, **17**, 838 (1962); *J. Phys. Chem.*, **67**, 1408 (1963).

115. A. Frumkin, *Z. Phys. Chem.*, **103**, 55 (1923).

116. F. O. Koenig, *J. Phys. Chem.*, **38**, 111 and 339 (1934).

117. D. C. Grahame and R. B. Whitney, *J. Am. Chem. Soc.*, **64**, 1548 (1942).

118. R. Parsons and M. A. V. Devanathan, *Trans. Faraday Soc.*, **49**, 673 (1953).

119. D. C. Grahame, R. P. Larsen, and M. A. Poth, *J. Am. Chem. Soc.*, **71**, 2978 (1949).

120. L. A. Hansen and J. W. Williams, *J. Phys. Chem.*, **39**, 439 (1935).

121. K. G. Baikerikar, R. S. Hansen, and G. L. Zweerink, *J. Electroanal. Chem.*, **129**, 285 (1981).

122. G. J. Hills and R. Payne, *Trans. Faraday Soc.*, **61**, 317 (1963).

123. D. C. Grahame, *J. Am. Chem. Soc.*, **63**, 1207 (1941).

124. Z. Borkowska, R. M. Denobriga, and W. R. Fawcett, *J. Electroanal. Chem.*, **124**, 263 (1981).

125. G. Gouy, *Ann. Phys.*, **6**, 5 (1916); **7**, 129 (1917).

126. W. R. Fawcett and Z. Borkowska, *J. Phys. Chem.*, **87**, 4861 (1983).

127. R. S. Hansen and K. G. Baikerikar, *J. Electroanal. Chem.*, **82**, 403 (1977).

128. D. E. Broadhead, K. G. Baikerikar, *J. Phys. Chem.*, **80**, 370 (1976).

129. K. G. Baikerikar and R. S. Hansen, *J. Colloid Interface Sci.*, **52**, 277 (1975).

130. M. Hamdi, D. Schuhmann, P. Vanel, and E. Tronel-Peyroz, *Langmuir*, **2**, 342 (1986).

131. R. Wüstneck, R. Miller, and J. Kriwanek, *Colloids & Surfaces*, **A81**, 1 (1993).

132. R. Wüstneck, H. Fiedler, K. Haage, and R. Miller, *Langmuir*, **10**, 3966 (1994).

133. R. Wüstneck, R. Miller, J. Kriwanek, and H. R. Holzbauer, *Langmuir*, **10**, 3738 (1994).

134. D. E. Broadhead, R. S. Hansen, and G. W. Potter, Jr., *J. Colloid Interface Sci.*, **31**, 61 (1969).

135. R. de Levie, *Chem. Rev.*, **88**, 599 (1988).

136. M. K. Kaisheva and V. K. Kaishev, *Langmuir*, **1**, 760 (1985).

137. K. G. Baikerikar and R. S. Hansen, *J. Colloid Interface Sci.*, **115**, 339 (1987).

138. A. Frumkin and A. Gorodetzkaja, *Z. Phys. Chem.*, **136**, 451 (1928); A. Frumkin and F. J. Cirves, *J. Phys. Chem.*, **34**, 74 (1930).

139. P. Delahay, *Double Layer and Electrode Kinetics*, Interscience, New York, 1965.

140. R. Narayan and N. Hackerman, *J. Electrochem. Soc.*, **118**, 1426 (1971).

141. L. Cunningham and H. Freiser, *Langmuir*, **1**, 537 (1985).

142. P. R. Couchman, D. H. Everett, and W. A. Jesser, *J. Colloid Interface Sci.*, **52**, 410 (1975).

143. P. R. Couchman and C. R. Davidson, *J. Electroanal. Chem.*, **85**, 407 (1977).

144. D. M. Mohliner and T. R. Beck, *J. Phys. Chem.*, **83**, 1169 (1979).

145. A. J. Bard et al., *J. Phys. Chem.*, **97**, 7147 (1993).

146. A. T. Hubbard, *Heterogen. Chem. Rev.*, **1**, 3 (1994).

147. A. T. Hubbard, *Chem. Rev.*, **88**, 633 (1988).

148. W. Schmickler and D. Henderson, *Prog. Surf. Sci.*, **22**, 323 (1986).

149. J. O'M. Bockris, S. D. Argade, and E. Gilead, *Electrochim. Acta*, **14**, 1259 (1969).

150. J. O'M. Bockris and A. K. N. Reddy, *Modern Electrochemistry*, Plenum, New York, 1970.

151. E. K. Krauskopf, K. Chan, and A. Wieckowski, *J. Phys. Chem.*, **91**, 2327 (1987).

152. H. Reiss and A. Heller, *J. Phys. Chem.*, **89**, 4207 (1985).

153. D. J. Trevor, C. E. D. Chidsey, and D. N. Loiacono, *Phys. Rev. Lett.*, **62**, 929 (1989).

154. M. G. Samant, M. F. Toney, G. L. Borges, L. Blum, and O. W. Melroy, *J. Phys. Chem.*, **92**, 220 (1988).

155. K. Kunimatsu, H. Seki, W. G. Golden, J. G. Gordon II, and M. R. Philpott, *Langmuir*, **2**, 464 (1986).

156. E. Yeager, *J. Electrochem. Soc.*, **128**, 160C (1981).

157. M. S. Wrighton, *Science*, **231**, 32 (1986).

158. H. H. Law, J. Sapjeta, C. E. D. Chidsey, and T. M. Putvinski, *J. Electrochem. Soc.*, **141**, 1977 (1994).

159. G. E. Cabaniss, A. A. Diamantis, W. R. Murphy, Jr., R. W. Linton, and T. J. Meyer, *J. Am. Chem. Soc.*, **107**, 1845 (1985).

160. C. E. D. Chidsey, *Science*, **251**, 919 (1991).

161. J. F. Rusling and A. F. Nassar, *J. Am. Chem. Soc.*, **115**, 11891 (1993).

162. A. F. Nassar, W. S. Willis, and J. F. Rusling, *Anal. Chem.*, **67**, 2386 (1995).

163. A. F. Nassar, Y. Narikiyo, T. Sagara, N. Nakashima, and J. F. Rusling, *J. Chem. Soc. Faraday Trans. 1*, **91**, 1775 (1995).

164. J. F. Rusling, in "Electrochemistry in Micelles, Microemulsions and Related Microheterogeneous Fluids," *Electroanalytical Chemistry, A Series of Advances*, Marcel Dekker, New York, 1994.

165. J. F. Rusling, *Electrochemistry and Electrochemical Catalysis in Microemulsions*, Plenum Press, 1994, Chapter 2.

166. J. F. Rusling, *Acc. Chem. Res.*, **24**, 75 (1991).

167. A. Sucheta, I. U. Haque, and J. F. Rusling, *Langmuir*, **8**, 1633 (1992).

168. S. Zhang and J. F. Rusling, *Environ. Sci. Technol.*, **29**, 1195 (1995).

169. Q. Huang and J. F. Rusling, *Environ. Sci. Technol.*, **29**, 98 (1995).

170. J. F. Rusling, S. Schweizer, S. Zhang, and G. N. Kamau, *Colloids & Surfaces*, **A88**, 41 (1994).

171. S. Zhang and J. F. Rusling, *Environ. Sci. Technol.*, **27**, 1375 (1993).

172. G. N. Kamau, N. Hu, and J. F. Rusling, *Langmuir*, **8**, 1042 (1992).

173. M. O. Iwunze, A. Sucheta, and J. F. Rusling, *Anal. Chem.*, **62**, 644 (1990).

174. D. Zhou, J. Gao, and J. F. Rusling, *J. Am. Chem. Soc.*, **117**, 1127 (1995).

175. N. Hu and J. F. Rusling, *Anal. Chem.*, **63**, 2163 (1991).

176. N. Hu, D. J. Howe, M. F. Ahmadi, and J. F. Rusling, *Anal. Chem.*, **64**, 3180 (1992).

177. F. Williams and A. J. Nozik, *Nature*, **311**, 21 (1984).

178. M. A. Fox, *Topics of Organic Electrochemistry*, Vol. 4, A. Fry, ed., Plenum, New York, 1985, p. 177.

179. P. C. Ricke and N. R. Armstrong, *J. Am. Chem. Soc.*, **106**, 47 (1984).

180. P. V. Kamat, R. Basheer, and M. A. Fox, *Macromolecules*, **18**, 1366 (1985).

181. A. J. Nozik, in *Photovoltaic and Photoelectrochemical Solar Energy Conversion*, F. Cardon, W. P. Gomes, and W. Dekeyser, eds., Plenum, New York, 1981.

182. B. Parkinson, *J. Chem. Ed.*, **60**, 338 (1983).

183. M. S. Wrighton, ACS Symposium Series No. 211, M. H. Chisholm, ed., American Chemical Society, Washington, DC, 1983.

184. M. Grätzel, *Acc. Chem. Res.*, **14**, 376 (1981); N. Vlachopoulos, P. Liska, J. Augustynski, and M. Grätzel, *J. Am. Chem. Soc.*, **110**, 1216 (1988).

185. R. M. Benito and A. J. Nozik, *J. Phys. Chem.*, **89**, 3429 (1985).

186. A. W. Adamson, J. Namnath, V. J. Shastry, and V. Slawson, *J. Chem. Ed.*, **61**, 221 (1984).

187. See E. Lange and F. O. Koenig, *Handbuch der Experimentalphysik*, Vol. 12, Part 2, Leipzig, 1933, p. 263.

188. E. J. W. Verwey, *Rec. Trav. Chim.*, **61**, 564 (1942).

189. J. N. Brønsted, *Z. Phys. Chem.*, **A143**, 301 (1929).

190. E. A. Guggenheim, *Thermodynamics*, Interscience, New York, 1949; *J. Phys. Chem.*, **33**, 842 (1929); **34**, 1540, 1758 (1930).

191. J. A. V. Butler, *Electrical Phenomena at Interfaces*, Methuen, London, 1951.

192. N. K. Adam, *The Physics and Chemistry of Surfaces*, 3rd ed., Oxford University Press, London, 1941.

193. J. H. de Boer, *Electron Emission and Adsorption Phenomena*, Macmillan, New York, 1935.

194. B. Case, N. S. Hush, R. Parsons, and M. E. Peover, *J. Electroanal. Chem.*, **10**, 360 (1965).

195. N. K. Adam, *Physical Chemistry*, Oxford University Press, London, 1958; see also Ref. 118.

196. R. Gomer and G. Tryson, *J. Chem. Phys.*, **66**, 4413 (1977).

197. J. Tafel, *Z. Phys. Chem.*, **50**, 641 (1905).

CHAPTER VI

Long-Range Forces

1. Introduction

One fascinating feature of the physical chemistry of surfaces is the direct influence of intermolecular forces on interfacial phenomena. The calculation of surface tension in section III-2B, for example, is based on the Lennard–Jones potential function illustrated in Fig. III-6. The wide use of this model potential is based in physical analysis of intermolecular forces that we summarize in this chapter. In this chapter, we briefly discuss the fundamental electromagnetic forces. The electrostatic forces between charged species are covered in Chapter V.

Generally speaking, intermolecular forces act over a short range. Were this not the case, the specific energy of a portion of matter would depend on its size; quantities such as molar enthalpies of formation would be *extensive* variables! On the other hand, the cumulative effects of these forces between macroscopic bodies extend over a rather long range and the discussion of such situations constitutes the chief subject of this chapter.

Although we know of gravitational, nuclear, and electromagnetic interactions, it is only the latter that are usually important in chemistry. Electromagnetic forces include repulsive and attractive forces. Two types of repulsive forces have been considered in preceding chapters: the Coulomb repulsion between like-charged ions and the repulsion between atoms brought too close together. The latter occurs over a very short range, hence the form of the Lennard–Jones potential illustrated in Fig. III-6, where the respulsion decays as the inverse twelfth power of distance. This repulsion is caused by the strong interaction between electron clouds as they are made to overlap and defines an atomic or molecular "diameter." Since the radial portion of the electron wave function is exponential, this repulsion is often represented by an exponential function.

Much of chemistry is concerned with the short-range wave-mechanical force responsible for the chemical bond. Our emphasis here is on the less chemically specific attractions, often called *van der Waals* forces, that cause condensation of a vapor to a liquid. An important component of such forces is the *dispersion* force, another wave-mechanical force acting between both polar and nonpolar materials. Recent developments in this area include the ability to measure dispersion interactions with a surface force apparatus along with increasingly sophisticated theoretical analyses of the force expressions.

One complication is the presence of two systems of units: cgs/esu and SI. As in Chapter V, we again present the equations in the cgs/esu system; we show alternative forms appropriate to the SI system in Tables VI-2 and VI-3.

2. Forces between Atoms and Molecules

We begin with the force between two point charges, q_1 and q_2, separated by a distance x in a vacuum from Coulomb's law

$$f = \frac{q_1 q_2}{x^2} \qquad\qquad \text{(VI-1)}$$

The potential energy of interaction $U = -\int f dx$ is then

$$U = \frac{q_1 q_2}{x} \qquad\qquad \text{(VI-2)}$$

where U is in ergs if q is in electrostatic units and x in centimeters. The electric field a distance x from a charge is

$$E = \frac{q}{x^2} \qquad\qquad \text{(VI-3)}$$

such that the force is given by the product qE. The sign of E follows that of q; the interaction energy is negative if attractive and positive if repulsive.

We next consider a molecule having a dipole moment $\mu = qd$, that is, one in which charges q^+ and q^- are separated by a distance d. A dipole aligned with a field experiences a potential energy, $U = \mu E$, where again U is in ergs if μ is in esu/cm. The conventional unit of a dipole moment is the *Debye*, 1 D = 1×10^{-18} esu/cm or 3.336×10^{-30} C/m, corresponding to unit electronic charges 0.21 Å apart.

At distances far from the dipole, the length d becomes unimportant and the dipole appears as a *point dipole*. The potential energy for a point dipole in the field produced by a charge (Eq. VI-3) is

$$U = \frac{\mu q}{x^2}. \qquad\qquad \text{(VI-4)}$$

The field produced far from a dipole ($x \gg d$) is (see Problem VI-1)

$$E = \frac{\mu}{x^3} (3 \cos^2 \theta + 1)^{1/2} \qquad \text{(VI-5)}$$

where θ is the angle between the position vector x and the dipole direction. Along the dipole direction ($\theta = 0$) this field becomes simply $E = 2\mu/x^3$. A dipole interacts with the field of a second dipole to give an interaction potential energy

$$U = \frac{\mu_1 \mu_2}{x^3} [2 \cos \theta_1 \cos \theta_2 - \sin \theta_1 \sin \theta_2 \cos \phi] \qquad \text{(VI-6)}$$

where the θ_i are the angles between the center-to-center line x and the dipoles and ϕ is the azimuthal angle as illustrated in Fig. VI-1. The maximum attraction occurs with aligned dipoles $\theta_1 = \theta_2 = 0$; thus for identical dipoles

$$U_{max} = -\frac{2\mu^2}{x^3} \qquad \text{(VI-7)}$$

whereas the maximum repulsion will be of the same magnitude when the dipoles are aligned in the opposite direction ($\theta_1 = \theta_2 = 180$). In a gas or liquid, thermal agitation tends to rotate the dipoles into random orientations while the interaction potential energy favors alignment. The resulting net interaction potential energy (due to Keesom in 1912) is

$$U_{av} = -\frac{2\mu^4}{3kTx^6} \qquad \text{(VI-8)}$$

This *orientation* interaction thus varies inversely with the sixth power of the distance between dipoles. Remember, however, that the derivation has assumed separations large compared with d.

Another interaction involving dipoles is that between a dipole and a polar-

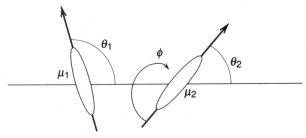

Fig. VI-1. Geometry for the interaction between two dipoles.

izable molecule. A field induces a dipole moment in a polarizable molecule or atom

$$\mu_{ind} = \alpha E \tag{VI-9}$$

where α is the polarizability and has units of volume in the cgs system. It follows from $U = \mu E$ that

$$U(\alpha E) = \mu_{ind}E = -\frac{\alpha E^2}{2} \tag{VI-10}$$

where the negative sign implies attraction and the factor of 0.5 arises because we integrate from zero field or infinite separation to the field or position of interest $\int_0^E \mu_{ind}\, dE$. The induced dipole is instantaneous on the time scale of molecular motions and the potential energy is independent of temperature and is averaged over all orientations to give

$$U(\alpha\mu) = -\frac{\alpha\mu^2}{x^6} \tag{VI-11}$$

a result worked out by Debye in 1920 and referred to as the Debye or induction interaction.

As an exercise, it is not difficult to show that the interaction of a polarizable molecule with a charge q is

$$U(\alpha q) = -\frac{\alpha q^2}{2x^4} \tag{VI-12}$$

We have two interaction potential energies between uncharged molecules that vary with distance to the minus sixth power as found in the Lennard–Jones potential. Thus far, none of these interactions accounts for the general attraction between atoms and molecules that are neither charged nor possess a dipole moment. After all, CO and N_2 are similarly sized, and have roughly comparable heats of vaporization and hence molecular attraction, although only the former has a dipole moment.

In 1930, London [1,2] showed the existence of an additional type of electromagnetic force between atoms having the required characteristics. This is known as the *dispersion* or London–van der Waals force. It is always attractive and arises from the fluctuating electron clouds in all atoms that appear as oscillating dipoles created by the positive nucleus and negative electrons. The derivation is described in detail in several books [1,3] and we will outline it briefly here.

The energy of a polarizable atom 1 in a field E is again given by Eq. VI-10,

$U(x) = -\alpha_1 E^2/2$, where now E is

$$E = \frac{2\bar{\mu}_2}{x^3} \tag{VI-13}$$

where $\bar{\mu}_2$ is the root-mean-square (rms) average dipole moment for the fluctuating electron–nucleus system of atom 2. The polarizability of an atom is in
general frequency-dependent and can be expressed as a sum over all excited
states of the transition moment squared divided by the energy, and if approximated by the largest term, then for atom 2

$$\alpha_2 \approx \frac{(\overline{ed})^2}{h\nu_0} \tag{VI-14}$$

where $(\overline{ed})^2$ is the electron charge times the displacement and averages to $\bar{\mu}_2^2$;
$h\nu_0$ is approximately the ionization energy. Thus the energy of atom 1 in the
average dipole field of atom 2 becomes

$$U(x) = -\frac{3h\nu_0\alpha_1\alpha_2}{4x^6} \tag{VI-15}$$

where the factor of 0.75 comes from the detailed derivation [2].

At the same level of approximation, the corresponding interaction between
two different atoms is [1]

$$U(x) = -\frac{3\alpha_1\alpha_2}{2x^6[(1/h\nu_1) + (1/h\nu_2)]} \tag{VI-16}$$

where $h\nu_1$ and $h\nu_2$ denote the characteristic energies for atoms 1 and 2 (roughly
the ionization energies). A useful approximation [4] is

$$U(x) = -\frac{363\alpha_1\alpha_2}{x^6[(\alpha_1/n_1)^{1/2} + (\alpha_2/n_2)^{1/2}]} \tag{VI-17}$$

where n_1 and n_2 are the numbers of electrons in the outer shells, x is in
angstroms, $U(x)$ is in kilocalories per mole, and the α are in cubic angstroms.
A widely used alternative, known as the Kirkwood–Müller form is

$$U(x) = -\frac{6mc^2\alpha_1\alpha_2}{x^6[(\alpha_1/\chi_1) + (\alpha_2/\chi_2)]} \tag{VI-18}$$

where m is the electronic mass (9.109×10^{-28} g), c is the velocity of light (2.998×10^{10} cm/sec), and χ_1 and χ_2 are the diamagnetic susceptibilities [5].

At small separations or for molecules with more complicated charge distributions, the more general form

$$u(x) = -C_1 x^{-6} - C_2 x^{-8} - C_3 x^{-10} - \cdots \qquad \text{(VI-19)}$$

includes dipole quadrupole x^{-8} and dipole–octupole and quadrupole–quadrupole x^{-10} interactions along with the dipole–dipole term. An evaluation by Fontana [6] gives C_1, C_2, and C_3, as 1400, 3000, and 7900 for Ar, respectively, and 1750, 5800 and 24,000 for Na with x expressed in angstroms and $u(x)$ in units of kT at 25°C. Because of the dependence of the van der Waals interaction on the absorption of electromagnetic radiation, spectroscopy of van der Waals molecules such as Ar_2, O_2-O_2, and $Ar-N_2$ have contributed to the study of van der Waals interactions [7].

Table VI-1 shows the approximate values for the Keesom ($\mu-\mu$), Debye, or induction ($\mu-\alpha$), and London or dispersion ($\alpha-\alpha$) interactions for several molecules. Even for highly polar molecules, the last is very important. The first two interactions are difficult to handle in condensed systems since they are sensitive to the microscopic structure through the molecular orientation. We will call all these interactions giving rise to an attraction varying with the inverse sixth power of the intermolecular distance *van der Waals interactions*. This is the dependence indicated by the a/V^2 terms in the van der Waals equation of state for a nonideal gas:

$$\left(P + \frac{a}{V^2}\right)(V - b) = RT \qquad \text{(VI-20)}$$

where V is the volume per mole, and a and b are constants, the former giving a measure of the attractive potential and the latter the actual volume of a mole of molecules.

TABLE VI-1

Contributions to van der Waals's Interactions between Neutral Molecules

Molecule	$10^{18}\mu$ (esu/cm)	$10^{24}\alpha$ (cm^3)	$h\nu_0$ (eV)	10^{60} $\mu-\mu$	C_1 $\mu-\alpha$	(ergs/cm^6) $\alpha-\alpha$
He	0	0.2	24.7	0	0	1.2
Ar	0	1.6	15.8	0	0	48
CO	0.12	1.99	14.3	0.003	0.057	67.5
HCl	1.03	2.63	13.7	18.6	5.4	105
NH_3	1.5	2.21	16	84	10	93
H_2O	1.84	1.48	18	190	10	47

TABLE VI-2
Conversions between cgs/esu and SI Constants and Their Units

Function	cgs/esu	SI
Potential	Volt$_{esu}$	300 volts (V)
Ionization energy $h\nu_0$	eV = 1.6×10^{-12} erg	eV = 1.6×10^{-19} joule (J)
Charge	$q[=]$esu	$\dfrac{q}{\sqrt{4\pi\epsilon_0}}$ [=]coulombs (C)
Polarizability	$\alpha[=]$cm^3	$\dfrac{\alpha}{4\pi\epsilon_0}$ [=]m^3
Dipole moment	$\mu[=]$D = 10^{-18} esu/cm	$\dfrac{\mu}{\sqrt{4\pi\epsilon_0}}$ [=]D = 3.336 $\times 10^{-30}$ C/m
Electronic charge	$e = 4.803 \times 10^{-10}$ esu	$e = 1.602 \times 10^{-19}$ C
Permittivity	1	$\epsilon_0 = 8.854 \times 10^{-12}$ C^2 J^{-1} m^{-1}
Boltzmann constant	$k = 1.38 \times 10^{-16}$ erg/K	$k = 1.38 \times 10^{-23}$ J/K

TABLE VI-3
Interaction Potential Energies in cgs/esu and SI

Function	cgs/esu	SI
Coulomb's law	$U = \dfrac{q_1 q_2}{x}$	$U = \dfrac{q_1 q_2}{4\pi\epsilon_0 x}$
Keesom interaction	$U = -\dfrac{2\mu^4}{3kTx^6}$	$U = -\dfrac{\mu^4}{3kT(4\pi\epsilon_0)^2 x^6}$
Debye, induction interaction	$U = -\dfrac{\mu^2\alpha}{x^6}$	$U = -\dfrac{\mu^2\alpha}{(4\pi\epsilon_0)^2 x^6}$
London, dispersion interaction	$U = -\dfrac{3h\nu_0\alpha_1\alpha_2}{4x^6}$	$U = -\dfrac{3h\nu_0\alpha_1\alpha_2}{4(4\pi\epsilon_0)^2 x^6}$

To first order, the dispersion (α–α) interaction is independent of the structure in a condensed medium and should be approximately pairwise additive. Qualitatively, this is because the dispersion interaction results from a small perturbation of electronic motions so that many such perturbations can add without serious mutual interaction. Because of this simplification and its ubiquity in colloid and surface science, dispersion forces have received the most significant attention in the past half-century. The way dispersion forces lead to long-range interactions is discussed in Section VI-3 below. Before we present this discussion, it is useful to recast the key equations in cgs/esu units and SI units in Tables VI-2 and VI-3.

3. Long-Range Forces

In this section we consider electromagnetic dispersion forces between macroscopic objects. There are two approaches to this problem; in the first, microscopic model, one assumes pairwise additivity of the dispersion attraction between molecules from Eq. VI-15. This is best for surfaces that are near one another. The macroscopic approach considers the objects as continuous media having a dielectric response to electromagnetic radiation that can be measured through spectroscopic evaluation of the material. In this analysis, the retardation of the electromagnetic response from surfaces that are not in close proximity can be addressed. A more detailed derivation of these expressions is given in references such as the treatise by Russel et al. [3]; here we limit ourselves to a brief physical description of the phenomenon.

A. The Microscopic Approach

The total interaction between two slabs of infinite extent and depth can be obtained by a summation over all atom–atom interactions if pairwise additivity of forces can be assumed. While definitely not exact for a condensed phase, this conventional approach is quite useful for many purposes [1,3]. This summation, expressed as an integral, has been done by de Boer [8] using the simple dispersion formula, Eq. VI-15, and following the nomenclature in Eq. VI-19:

$$U(x) = -C_1 n^2 \int_{V_1} \int_{V_2} \frac{dx_1 dx_2}{x^6} \qquad \text{(VI-21)}$$

where C_1 is $3h\nu_0\alpha^2/4$. Since ionization potentials are in the range of 10–20 eV and polarizabilities about 1–2×10^{-24} cm^3, C_1 will have a value in the range of 10^{-57} to 10^{-58} erg/cm^6 per atom.

The two volume integrals change the dependence on x and introduce the number density of atoms n (number/cm^3) such that the energy now varies as

$$U(x)_{\text{slab–slab}} = -\frac{\pi n^2 C_1}{12x^2} = -\frac{A}{12\pi x^2} \qquad \text{(VI-22)}$$

where U is now in energy per square centimeter and A is known as the *Hamaker constant*, after its developer [9], and is equal to $\pi^2 n^2 C_1$. The term A contains the quantity $(n\alpha)^2$ or $(\alpha/V)^2$, where V is the molecular volume. Since the polarizability scales as the molecular volume, the ratio (α/V) is nearly constant at about 0.1 for solids; Tabor and Winterton [10,11] give the approximation $A \sim h\nu_0/8$, or $A \sim 10^{-12}$ erg.

A more detailed description of the interaction accounts for the variation of the polarizability of the material with frequency. Then, the Hamaker constant across a vacuum becomes

$$A = \frac{3}{8} kT \left([\epsilon(0) - 1]^2/2 + \sum_{j=1}^{\infty} [\epsilon(i\nu_j) - 1]^2 \right) \qquad \text{(VI-23)}$$

where $\epsilon(i\nu_j)$ is the dielectric permittivity of the material at the imaginary frequency $i\nu_j$. Useful approximations to A using the first term in Eq. VI-23 are given in Section VI-4 for interactions across a medium. Values of A are tabulated in several texts [1,3,12]. If the two objects are of different materials, a common and not unreasonable approach is to use the geometric mean of the C_1 values; more approximately, the geometric mean of the A values has been used.

The integration in Eq. VI-21 may be carried out for various macroscopic shapes. An important situation in colloid science, two spheres of radius a yields

$$U(x)_{\text{sphere-sphere}} = -\frac{A}{6} \left(\frac{2a^2}{r^2 - 4a^2} + \frac{2a^2}{r^2 + \ln \dfrac{r^2 - 4a^2}{r^2}} \right) \qquad \text{(VI-24)}$$

for center-to-center separations r. For small surface-to-surface separations $x = r - 2a$, the interaction potential becomes $U(x) = -aA/12x$. In the direct experimental measurement of A (see Section VI-3D), it is easier to use a sphere and a flat plate or two crossed cylinders that have the same simple form

$$U(x)_{\text{sphere-slab}} = U(x)_{\text{crossed-cylinders}} = -\frac{aA}{6x} \qquad \text{(VI-25)}$$

If the cylinders are of different radii, then $a = (a_1 a_2)^{1/2}$. In these cases, $U(x)$ has a simple inverse dependence on x, so that the attraction is, indeed, long-range.

Short-range interactions between two spheres can be usefully approximated in terms of flat-plate interactions. In this approach, known as the Derjaguin approximation, for two large spheres of radii a_1 and a_2 at a surface-to-surface separation of $x = r - a_1 - a_2$ such that $x \ll a$, we can integrate the integration potential between small opposing circular regions of area $2\pi y dy$ as illustrated in Fig. VI-2. The surface is assumed to be locally flat on the length scale of dy. The energy can be written

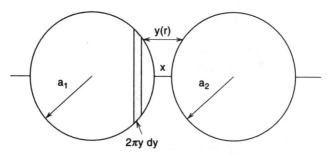

Fig. VI-2. Schematic representation of the geometry used for the Derjaguin approximation.

$$U(x)_{\text{sphere–sphere}} \approx 2\pi \int_0^\infty U_{\text{slab–slab}} r \; dr$$

$$\approx 2\pi \frac{a_1 a_2}{a_1 + a_2} \int_0^\infty U_{\text{slab–slab}} y \; dy \qquad \text{(VI-26)}$$

and the force is just

$$F = -\frac{\partial U}{\partial x} \approx 2\pi \frac{a_1 a_2}{a_1 + a_2} U_{\text{slab–slab}}(x) \qquad \text{(VI-27)}$$

Thus for equal-size spheres the force between them is just $\pi a U_{\text{slab–slab}}(x)$ and is directly related to the potential energy between two slabs [13]. This point is examined further in the problems at the end of the chapter.

Microscopic analyses of the van der Waals interaction have been made for many geometries, including, a spherical colloid in a cylindrical pore [14] and in a spherical cavity [15] and for flat plates with conical or spherical asperities [16,17].

B. The Retarded Dispersion Interaction

The dispersion force arises from the mutual interaction between fluctuating dipoles and is thus propagated at the speed of light. Surfaces more than $c/\nu_0 2\pi$ apart will thus become out-of-phase; that is, the propagation time for the field becomes comparable to the oscillating dipole period. The field then interacts with an instantaneous dipole in a different phase, thus diminishing the attraction, an effect known as *retardation*. This effect was recognized as early as 1948, by Casimir and Polder [18]. It can be derived from several points of view. Here we highlight the physical aspects of retardation between macroscopic surfaces.

The continuum treatment of dispersion forces due to Lifshitz [19,20] provides the appropriate analysis of retardation through quantum field theory. More recent analyses are more tractable and are described in some detail in several references [1,3,12,21,22].

A common approach to treating retardation in dispersion forces is to define an effective Hamaker constant that is not constant but depends on separation distance. Looking back at Eq. VI-22, this defines the effective Hamaker constant

$$A_{\text{eff}}(x) = -12\pi x^2 U_{\text{slab} - \text{slab}}(x) \qquad \text{(VI-28)}$$

in terms of the actual, retarded interaction potential energy. The frequency dependence of the Hamaker constant embodied in Eq. VI-23 reminds us that retardation will diminish the low-frequency (long-wavelength) parts of the interaction at the longest distances while the higher-frequency contributions will be supressed at smaller separations (even down to 5 nm). This effect will be most pronounced in a continuous medium; thus we restrict our attention here to two planar surfaces interacting across a liquid.

Application of the exact continuum analysis of dispersion forces requires significant calculations and the knowledge of the frequency spectrum of the material dielectric response over wavelengths $\lambda = 2\pi c/\nu$ around 10–10^4 nm. Because of these complications, it is common to assume that a primary absorption peak at one frequency in the ultraviolet, ν_{UV}, dominates the dielectric spectrum of most materials. This leads to an expression for the dielectric response

$$\epsilon(i\nu_j) = 1 + \frac{n^2 - 1}{1 + (\nu_j/\nu_{\text{UV}})^2} \qquad \text{(VI-29)}$$

where n is the low-frequency limit of the refractive index in the visible. This simplification and mathematical manipulations lead to an analytic representation of the effective Hamaker constant for two planar surfaces of medium 1 interacting through medium 2 [3]

$$A_{\text{eff}}(x) = \frac{3}{4} kT \left(\frac{\epsilon_1 - \epsilon_2}{\epsilon_1 + \epsilon_2} \right)^2 + \frac{3h\nu_{\text{UV}}}{16\sqrt{2}} \frac{(n_1^2 - n_2^2)^2}{(n_1^2 + n_2^2)^{3/2}} G(X) \qquad \text{(VI-30)}$$

where the dielectric constants, ϵ_1 and ϵ_2, are evaluated at zero frequency and the distance parameter X is given by

$$X = n_2(n_1^2 + n_2^2)^{1/2} \frac{x\nu_{\text{UV}}}{c}. \qquad \text{(VI-31)}$$

The distance dependence enters through $G(X)$, which goes to 1 as $X \to 0$ and goes to $4\sqrt{2}/\pi X$ as $X \to \infty$. An interpolating function is available for $G(X)$ at intermediate separations

$$G(X) = \left[1 + \left(\frac{\pi X}{4\sqrt{2}} \right)^{3/2} \right]^{-2/3}$$ (VI-32)

In general, the adsorption frequency ν_{UV} is not equal to the ionization frequency for isolated molecules; they are related through the Clausius–Mossotti–Lorentz–Lorentz equation as $\nu_{UV} = \nu_0 \sqrt{3/(n_1^2 + 2)}$ [1]. The difference between the frequencies is often not significant to the result for A_{eff} [3]. The values of ν_{UV} have been tabulated for various materials and are generally in the range of 3×10^{15} sec^{-1} [1,3]. Often, the second term in Eq. VI-30 dominates over the first, and in an electrolyte, the ions screen the interaction, causing the first term to be negligible [3].

C. Experimental Measurements

Experiments to directly measure dispersion and other surface forces date back to the 1950s, when force measurements between parallel glass or quartz plates were attempted [23–26] as well as measurements between a flat plate and a spherical lens [23, 24]. These are difficult measurements since forces for the latter geometry are on the order of 0.1 dyn for a separation distance of 100 nm. Surfaces must be quite smooth, dust-free, and devoid of any electrical charge, and since the force changes quite rapidly with distance, the balancing system must be very stiff as well as very sensitive. Despite these formidable experimental difficulties, the retarded dispersion interaction was confirmed by Derjaguin and co-workers [23, 24], Overbeek and co-workers [25, 27], and Kitchener and Prosser [26]. An example of these measurements from Overbeek's laboratory, shown in Fig. VI-3, gives the correct distance dependence for the retarded force between a sphere and a flat plate. Taking the derivative of Eq. VI-25 and combining it with Eq. VI-32 for the retarded Hamaker constant, we anticipate the force to decay as x^{-3} as seen in Fig. VI-3. The magnitudes of the Hamaker constants from these measurements are too high, and the roughness of the surface precludes measurement of the nonretarded regime.

A major advance in force measurement was the development by Tabor, Winterton and Israelachvili of a surface force apparatus (SFA) involving crossed cylinders coated with molecularly smooth cleaved mica sheets [11, 28]. A current version of an apparatus is shown in Fig. VI-4 from Ref. 29. The separation between surfaces is measured interferometrically to a precision of 0.1 nm; the surfaces are driven together with piezoelectric transducers. The combination of a stiff double-cantilever spring with one of a number of measuring leaf springs provides force resolution down to 10^{-3} dyn (10^{-8} N). Since its development, several groups have used the SFA to measure the retarded and unretarded dispersion forces, electrostatic repulsions in a variety of electrolytes, structural and solvation forces (see below), and numerous studies of polymeric and biological systems.

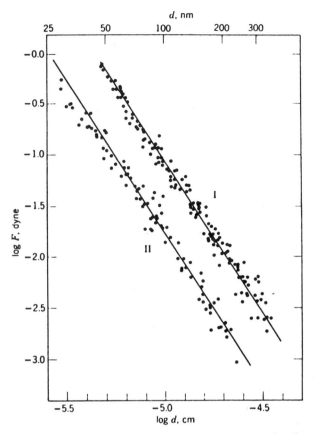

Fig. VI-3. Attraction between fused-silica flat plates and spheres of radius 413.5 cm (I) or 83.75 cm (II). The lines are drawn with a slope of −3.00. (From Ref. 27.)

Direct force measurements have been extended to other materials through a variety of techniques. Derjaguin and co-workers [30] used crossed filaments of radius in the 0.1–0.5-mm range to measure the transition from unretarded to retarded dispersion forces. A new surface forces apparatus uses bimorph force sensors instead of springs [31]. Smooth 2-mm glass spheres are mounted to the force sensor and to a piezoelectric actuator with a distance resolution of 0.1 nm. Atomic force microscopy (AFM) (see Section VIII-2C) provides another means to measure the force between surfaces. While the molecular tip of an AFM may be too small to provide meaningful surface force measurements, attaching a particle to the cantilever spring of the AFM provides measurements of surface forces between like and unlike surfaces with similar accuracy and resolution as in the SFA [32, 33]. In an elegant and very different experiment, Shih and Parsegian [34] were able to observe the van der Waals force on heavy alkali metal atoms in a molecular beam passing close to a gold surface.

The forces between colloidal surfaces in solution can be measured by a number of clever techniques. One important method involves x-ray diffraction from an ordered

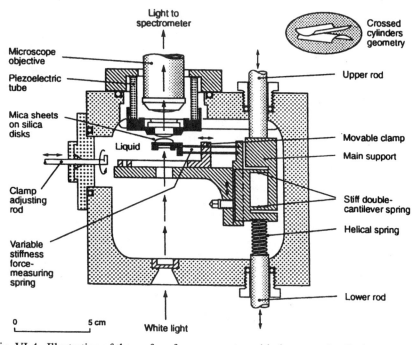

Fig. VI-4. Illustration of the surface force apparatus with the crossed-cylinder geometry shown as an inset. The surface separations are determined from the interference fringes from white light travelling vertically through the apparatus. At each separation, the force is determined from the deflection in the force measuring spring. For solution studies, the entire chamber is filled with liquid. (From Ref. 29.)

array of particles, viruses, bilayers, or proteins in equilibrium with a solution of known osmotic pressure [35, 36]. Variation of the osmotic pressure of the external solution imposes an *osmotic stress* on the suspension forcing the particles together. By mapping the pressure against interparticle separation, the force–distance profile can be determined. In a second approach with total internal reflection microscopy, Prieve and co-workers [37] developed a means to measure the weak forces between a colloidal particle and a planar surface in a liquid. Illuminating the solid–liquid interface with a laser beam undergoing total internal reflection provides a means to microscopically track the position of a particle near the surface with 1-nm resolution. A histogram of the equilibrium separations between the particle and the wall provides a measure of the force between them; forces down to 10^{-8} dyn (10^{-13} N) can be measured. In a variation of this technique, Brown and Staples added radiation pressure from another laser to manipulate particle positions and probe additional regimes of the force curve [38]. A third method, colloidal particle scattering, is a sensitive way to measure forces between two colloidal particles based on monitoring two-particle collision trajectories in shear flow. Introduced by van de Ven and co-workers [39], this technique is 3–4 orders of magnitude more sensitive than SFA or AFM.

The Hamaker constant can also be related to physical properties. The dispersion

TABLE VI-4
Hamaker Constants[a]

Substance	A $(10^{-13}$ erg$)$	Source (Reference No.)	Substance	A $(10^{-13}$ erg$)$	Source (Reference No.)
Water	0.6–0.7	41	Teflon	2.75–3.8	42
	4	22		3.8	1
	3.70	42	Polystyrene	6.37–6.58	42
	3.7–4.0	1		6.57	12
	4.35	12		6.6–7.9	1
n-Octane	4.26–5.02	40		7.8–9.8	40
	5.02	12	Silica[b]	6.55	42
	4.5	1		6.5	1
n-Hexadecane	6.2–6.31	40	Mica[c]	9.7	11
	6.31	12		10	1
	5.2	1	Gold	20–40	22
Benzene	5.0	1		43.3	12

[a]Calculated from various theoretical equations; some of the sources are themselves compilations rather than original.
[b]Experimental value: 5–6 [1].
[c]Experimental value: 1.07 [11]; 13.5 [1].

contribution to the surface free energy of a liquid can be obtained by integrating Eq. VI-22 from $x = d$ to $x = \infty$ to obtain

$$\gamma = \frac{A}{24\pi d^2} \qquad (VI\text{-}33)$$

where d is an effective molecular size [1]. Agreement with experiment is good for nonpolar liquids but not, as might be expected, in the case of polar ones such as water where forces other than dispersion are important. Also, the Hamaker constant for a liquid can be related to its equation of state [12, 40].

An assortment of values of the Hamaker constant A is collected in Table VI-4. These are a mixture of theoretical and experimental values; there is reasonable agreement between theory and experiment in the cases of silica, mica, and polystyrene.

4. Long-Range Forces in Solution

While the confirmation of the predicted long-range dispersion attraction between surfaces in air has been a major experimental triumph, the forces between particles in solution are of more general interest in colloid and surface chemistry. The presence of a condensed medium between the surfaces presents some theoretical complications as evident in Eq. VI-30 for the effective Hamaker constant. In addition, in aqueous solutions we must consider the effect of the electrostatic repulsion due to the electric double layer. The combined treatment of dispersion forces and electric double-layer repulsions is known as the *DLVO* (Derjaguin–Landau–Verwey–Overbeek) theory after its independent originators [43, 44].

A. Dispersion Forces in Condensed Media

The potential energy of two planar surfaces immersed in a medium is again given by Eq. VI-22 with Eq. VI-30. The dispersion interaction is reduced by the presence of an intervening fluid; that is, in Eq. VI-30 both ϵ_2 and n_2 are greater than unity so that A_{eff} is smaller than for surfaces interacting across a vacuum. One problem is the lack of measurements of Hamaker constants in various media. Two arbitrary planar surfaces 1 and 2 immersed in medium 3 will have a net Hamaker constant, A_{132}. This can be approximated in terms of A_{ij} for surfaces i and j interacting across a vacuum as

$$A_{132} = A_{12} - A_{13} - A_{23} + A_{33} \qquad (\text{VI-34})$$

or in terms of A_{ii} for two like surfaces i interacting across a vacuum

$$A_{132} = (\sqrt{A_{11}} - \sqrt{A_{33}})(\sqrt{A_{22}} - \sqrt{A_{33}}) \qquad (\text{VI-35})$$

These equations imply that A_{132} will exceed A_{12} if A_{33} is larger than $A_{13} + A_{23}$. This effect, termed *lyophobic bonding*, occurs if the solvent–surface interaction is weaker than that between the solvent molecules. More interestingly, the dispersion interaction will be *repulsive* ($A_{132} < 0$) when A_{13} and/or A_{23} are sufficiently large. Israelachvili [1] tabulates a number of A_{131} values; $A_{WHW} \approx A_{HWH} \approx 0.4 \times 10^{-13}$ erg, $A_{PWP} \approx 1 \times 10^{-13}$ erg, and $A_{QWQ} \approx 0.8 \times 10^{-13}$ erg, where W, H, P, and Q denote water, hydrocarbon, polystyrene and quartz respectively.

B. Electric Double-Layer Repulsion

Often the van der Waals attraction is balanced by electric double-layer repulsion. An important example occurs in the flocculation of aqueous colloids. A suspension of charged particles experiences both the double-layer repulsion and dispersion attraction, and the balance between these determines the ease and hence the rate with which particles aggregate. Verwey and Overbeek [44, 45] considered the case of two colloidal spheres and calculated the net potential energy versus distance curves of the type illustrated in Fig. VI-5 for the case of $\psi_0 = 25.6$ mV (i.e., $\psi_0 = kT/e$ at 25°C). At low ionic strength, as measured by κ (see Section V-2), the double-layer repulsion is overwhelming except at very small separations, but as κ is increased, a net attraction at all distances

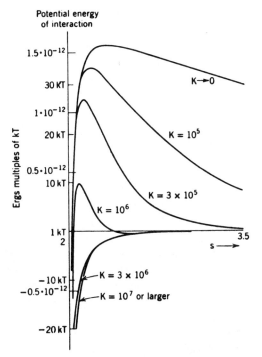

Fig. VI-5. The effect of electrolyte concentration on the interaction potential energy between two spheres where K is κ in cm^{-1}. (From Ref. 44.)

is finally attained. There is a critical region of κ such that a small potential minimum of about $kT/2$ occurs at a distance of separation s about equal to a particle diameter. This minimum is known as the *secondary minimum* and can lead to weak, reversible aggregation under certain conditions of particle size, surface potential, and Hamaker constant [3].

The presence of the large repulsive potential barrier between the secondary minimum and contact prevents flocculation. One can thus see why increasing ionic strength of a solution promotes flocculation. The net potential per unit area between two planar surfaces is given approximately by the combination of Eqs. V-31 and VI-22:

$$U(x)_{net} = \frac{64n_0kT}{\kappa} \left(\frac{e^{y_0/2} - 1}{e^{y_0/2} + 1} \right)^2 e^{-2\kappa x} - \frac{A}{12\pi x^2} \qquad (\text{VI-36})$$

where $y_0 = ze\psi_0/kT$. If we assume that flocculation will occur when no barrier exists, we require that $U(x) = 0$ and $dU(x)/dx = 0$ at some value of x. This defines a critical electrolyte concentration (after substituting for κ by means of Eq. V-7)

$$n_0 = \frac{1152}{\exp(4)} \frac{\epsilon^3 (kT)^5}{e^6 A^2 z^6} \left(\frac{e^{y_0/2} - 1}{e^{y_0/2} + 1} \right)^4 \tag{VI-37}$$

which varies as $1/z^6$ as embodied in the *Schulze–Hardy* rule discussed in Chapter V. Thus, for a z–z electrolyte, equivalent flocculation concentrations would scale as $1:(\frac{1}{2})^6:(\frac{1}{3})^6$ or $100:1.6:0.13$ for a 1–1, 2–2, and 3–3 electrolyte, respectively. Actually, the higher-valence ions have an increased tendency for specific adsorption, so the flocculation effectiveness becomes a matter of reduction of ψ_0 as well as reduction of the double layer thickness [46].

Quantitative measurements of flocculation rates have provided estimates of Hamaker constants in qualitative agreement with theory. One assumes diffusion-limited flocculation where the probability to aggregate decreases with the exponential of the potential energy barrier height of the type illustrated in Fig. VI-5. The barrier height is estimated from the measured flocculation rate; other measurements (see Section V-6) give the surface (or zeta) potential leaving the Hamaker constant to be determined from Equations like VI-36 [47–49]. Complications arise from the assumption of constant surface potential during aggregation, double-layer relaxation during aggregation [50–52], and nonuniform charge distribution on the particles [53–55]. In studies of the stability of ZnS sols in NaCl and $CaCl_2$ Durán and co-workers [56] found they had to add the Lewis acid–base interactions developed by van Oss [57] to the DLVO potential to model their measurements. Alternatively, the initial flocculation rate may be measured at an ionic strength such that no barrier exists. By this means A_{PWP} was found to be about 0.7×10^{-13} erg for aqueous suspensions of polystyrene latex [58]. The hydrodynamic resistance between particles in a viscous fluid must generally be recognized to obtain the correct flocculation rates [3].

Direct force measurements on the SFA by Israelachvili and co-workers and others also confirm DLVO theory for many cases [59–62]. An example of a force measurement is shown as a plot of force over radius, F/R, versus surface separation in Fig. VI-6 for lipid bilayer-coated surfaces in two salt solutions at two ionic strengths. The inset shows the effect of the van der Waals force at small separations. Note that the ordinate scale may be converted to erg/cm^2 or N/m^2 for two effective planar surfaces using the Derjaguin approximation, Eq. VI-27 (see Problem VI-7).† Interestingly, good agreement was also found with surfactant solutions well above the critical micelle concentration (see Section XII-5B) [63], although the micelles and their counterions do not contribute to the electrostatic screening [64]. A complete range of behavior was found as neat liquid ethylammonium nitrate was diluted with water [65]. Generally the DLVO potential works well until one gets to separations on the order of the Stern layer (see Section V-3), where the hydrated ions can eliminate the primary minimum

---

†This approximation may not be reliable if the force varies rapidly with distance. Also, at small separations the mica surfaces may become appreciably flattened, compromising the meaning of R, the cylinder radius.

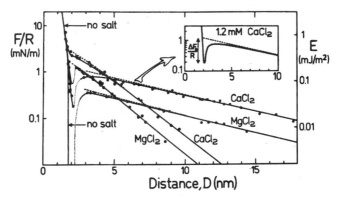

Fig. VI-6. The force between two crossed cylinders coated with mica and carrying adsorbed bilayers of phosphatidylcholine lipids at 22°C. The solid symbols are for 1.2 mM salt while the open circles are for 10.9 mM salt. The solid curves are the DLVO theoretical calculations. The inset shows the effect of the van der Waals force at small separations; the Hamaker constant is estimated from this to be $7 \pm 1 \times 10^{-14}$ erg. In the absence of salt there is no double-layer force and the adhesive force is -1.0 mN/m. (From Ref. 66.)

[1]. Derjaguin and co-workers used crossed filaments of various materials to measure DLVO potentials [30].

C. Forces Due to Solvent Structure

The discussion thus far has centered on continuum descriptions of the media; however, when distances on the order of molecular dimensions are involved, the structure of the fluid molecules near the solid surface can play a role. Correlations or molecular crowding in a liquid produce oscillatory forces very near surfaces. While these forces are largely geometric in origin, there are also monotonic repulsive or attractive forces between the surfaces depending on the affinity of the solvent molecules for the surface. A surface having ions or other surface groups that interact very strongly with water will have a net repulsion with another like surface, due to the energetic cost of removing the bound water layers. This is termed the *hydration* force; an analogous force in nonaqueous liquids is the *solvation* force. Likewise, if the surface repels water, there will be an energetic gain in removing water layers from between two surfaces producing the *hydrophobic* force whose counterpart in nonaqueous media is called the *solvophobic* force.

While evidence for hydration forces date back to early work on clays [1], the understanding of these solvent-induced forces was revolutionized by Horn and Israelachvili using the modern surface force apparatus. Here, for the first time, one had a direct measurement of the oscillatory forces between crossed mica cylinders immersed in a solvent, octamethylcyclotetrasiloxane (OMCTS) [67].

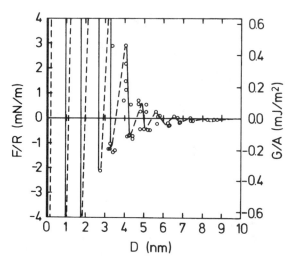

Fig. VI-7. The force between two crossed mica cylinders in dry OMCTS. The cylinder radii R were about 1 cm. The dashed lines show the presumed, experimentally inaccessible, transition between a repulsive maximum and an attractive minimum. (From Ref. 68.)

The damped oscillations with a period of about 1 nm corresponded well to the size of the OMCST molecules and extended to about 5 nm, or about five solvent layers. An example of these forces for the same system from Christenson and Blom [68] is shown in Fig. VI-7.

In aqueous electrolyte solutions, the oscillations are superimposed on the DLVO potential and additional hydration or hydrophobic forces [69–71]. The repulsive hydration force is generally exponential in form with a decay length of 1–2 nm attributed to the adsorption of hydrated ions to the mica surface [61, 72, 73]. The extensive early literature on this subject has been reviewed by Derjaguin and Churaev [74] and more recently by Israelachvili [1]. A variety of observations including infrared studies and thermal expansion measurements on clays, quartz, and other systems led numerous researchers to conclude that water at an interface may be structurally perturbed to a depth of tens of nanometers [75–81]. Dielectric constant and nuclear magnetic resonance (NMR) measurements have indicated that water adsorbed to alumina and other solids is structurally perturbed to a considerable depth [82] (see Section XVI-4). The structurally perturbed water at an interface was termed "vicinal" water by Drost-Hansen and suggestions about its behavior include a lower density than bulk water [79], a liquid crystalline state [80], and polarization layers [83]. Molecular dynamics studies find a perturbation in the dipole moment orientation of vicinal water [84]; however, there is some question as to the magnitude of the effect [85]. A statistical mechanical lattice theory, developed by Besseling and Scheutjens [86] with orientation-dependent interactions, can explain many anomalous features of water, including its interfacial tension and the thickness of the interfacial layers [87].

While there is little doubt that the presence of a solid surface significantly perturbs the structure of the water adjacent to the surface, there remains significant controversy about the origin of numerous measured long range forces. The hydration and hydrophobic forces have been attributed to Lewis acid–base interactions by van Oss [57] and the hydrophobic force has been cast in terms of the hydrogen bonding energy of cohesion surrounding a hydrophobic surface [88]. Very long-range hydration forces (10 nm) measured by SFA have been attributed to surfactants coming from millipore filters used to purify water [89]. The recent suggestion by Attard and co-workers [90, 91] that the very long-range attraction measured between hydrophobic surfaces in the SFA arises from the formation of submicroscopic bubbles or cavities between the surfaces has received much discussion [92]. Such vapor inclusions or cavities produce a capillary force (see Chapter II), which could account for many measurements of "hydrophobic" forces, as shown in recent AFM measurements of bubbles [33]. The stability of submicroscopic cavities has been questioned [92], and the issue is under debate. Another group discounts cavitation in their measurement of unusually long-range (300-nm) "hydrophobic" attractions between amphiphile layers; they attribute these forces to hydrogen bonding of vicinal water near the hydrophobic surface [88].

Nonpolar and anisotropic molecules are also significantly perturbed near surfaces. Electron and x-ray diffraction studies of films of long chain molecules on metal surfaces indicated the presence of structure well above the melting point in layers 100 molecules thick. There have been many observations [93] that the melting points of relatively thick films may differ considerably from the corresponding bulk characteristics. An extreme example of the long-range structural perturbation in a liquid adjacent to an interface arises in liquid crystals where the anchoring mechanism and bulk orientation are extremely sensitive to simple surface treatments such as rubbing [94]. In addition to the influence of surface orientation on static forces, this perturbation can have a profound effect on layer dynamics discussed in the next section.

The structural perturbations of a fluid near a surface have implications for the properties of thin films. Adsorbed films must be multimolecular in thickness before their surface approaches bulk phase properties (see Chapter XVI). Especially striking are the data on films adsorbed from the vapor phase at pressures approaching P^0, the condensation pressure. In a now classic paper, Derjaguin and Zorin [95] found that certain polar liquids reached a *limiting* thickness of about 10 nm when adsorbed on glass at P^0. The presence of the glass substrate caused a thick film to become an alternative state capable of coexisting in equilibrium with bulk fluid. Additional studies [82, 96, 97] report measurements of similarly thick films and of the film–bulk liquid–solid contact angle. As discussed further in Section X-6C, such films *must* be structurally perturbed relative to the bulk phase.

D.   Thin-Film Viscosity

In the context of the structural perturbations at fluid–solid interfaces, it is interesting to investigate the *viscosity* of thin liquid films. Early work on thin-film viscosity by Derjaguin and co-workers used a "blow off" technique to cause a liquid film to thin. This work showed elevated viscosities for some materials [98] and thin film viscosities lower than the bulk for others [99, 100]. Some controversial issues were raised particularly regarding surface roughness and contact angles in the experiments [101–103]. Entirely different types of data on clays caused Low [104] to conclude that the viscosity of interlayer water in clays is greater than that of bulk water.

The modification of the surface force apparatus (see Fig. VI-4) to measure viscosities between crossed mica cylinders has alleviated concerns about surface roughness. In dynamic mode, a slow, small-amplitude periodic oscillation was imposed on one of the cylinders such that the separation x varied by approximately 10% or less. In the limit of low shear rates, a simple equation defines the viscosity as a function of separation

$$\eta(x) = \frac{Kx}{12\pi^2 R^2 \nu} \left[\left(\frac{A_0}{A} \right)^2 - 1 \right]^{1/2} \qquad \text{(VI-38)}$$

where K is the spring constant ($\approx 3 \times 10^5$ dyn/cm), R is the radius of the mica cylinders (≈ 1 cm), ν is the oscillation frequency (≈ 1 sec^{-1}) and A is the amplitude of the oscillation of $x(A \ll x)$ and A_0 is that of the driven cylinder [105]. Surprisingly, the viscosity of tetradecane, aqueous NaCl or KCl, films was within 10% of the bulk value down to films as thin as 1 nm! The slip plane appeared to be within one molecule of the mica surface. The same observation was made for hexadecane with or without dissolved polymer [106]; however, the plane of slip moved out to the expected radius of gyration of adsorbed polymer when present.

Other SFA studies complicate the picture. Chan and Horn [107] and Horn and Israelachvili [108] could explain anomalous viscosities in thin layers if the first layer or two of molecules were immobile and the remaining intervening liquid were of normal viscosity. Other interpretations are possible and the hydrodynamics not clear, since as Granick points out [109] the measurements average over a wide range of surface separations, thus confusing the definition of a layer thickness. McKenna and co-workers [110] point out that compliance effects can introduce serious corrections in constrained geometry systems.

5.   Forces in Biological Systems

All the long-range forces discussed in this chapter play a role in biological processes. Interactions between membranes, proteins, ligands, antibodies and antigens, and enzymes and substrates involve electrostatic, van der Waals, steric, hydrophobic, and hydration forces. Many of the osmotic stress experiments have elucidated the interactions between proteins, membranes, and viruses [35]. Wiggins [111] has reviewed the large body of evidence for structural perturbations of water in membranes. In analogy to the colloidal aggregation studies discussed in Sections V-5 and VI-3B, phase transitions in proteins can be used to investigate their interactions [112, 113]. The surface force apparatus (SFA) discussed in Section VI-3C has been a valuable tool for understanding these complex systems [29].

In an extensive SFA study of protein receptor–ligand interactions, Leckband and co-workers [114] showed the importance of electrostatic, dispersion, steric, and hydrophobic forces in mediating the strong streptavidin–biotin interaction. Israelachvili and co-workers [66, 115] have measured the Hamaker constant for the dispersion interaction between phospholipid bilayers and find $A = 7.5 \pm 1.5 \times 10^{-13}$ erg in water.

6. The Disjoining Pressure

While most presentations of interactions between macroscopic bodies center on energies or forces, in 1936 Derjaguin introduced a different point of view with the concept of the *disjoining pressure* [116]. He defines the disjoining pressure, Π [117]: "The disjoining pressure is equal to a difference between the pressure of the interlayer on the surfaces confining it, and the pressure in the bulk phase, the interlayer being part of this phase and/or in equilibrium with it." For two plates immersed in a fluid, Π is the pressure in excess of the external pressure that must be applied to the fluid between the plates to maintain a given separation, that is, the force of attraction or repulsion between the plates per unit area. A mathematical definition in terms of the Gibbs energy is

$$\Pi = -\frac{1}{\mathcal{A}} \left(\frac{\partial G}{\partial x} \right)_{\mathcal{A}, T,} \qquad \text{(VI-39)}$$

where x is the thickness of the film or interlayer region and $\mathcal{A}$ is the plate area. Figure V-4 provides another illustration of double-layer repulsion, treating it as an excess osmotic pressure. The concept may be applied to soap films (note Problem VI-19) and to the adsorbed layer at the solid–vapor interface (see Section X-3). In this last case one can think of Π as the mechanical pressure that would have to be applied to a bulk substance to bring it into equilibrium with a film of the same substance.

Just as with interaction energies, Π can be regarded as the sum of several components. These include Π_m due to dispersion interaction, Π_e due to electrostatic interactions between charged surfaces, Π_a due to overlapping adsorbed layers of neutral molecules, and Π_s due to solvent structural effects [30, 118]. These last two contributions arise from the solvent structural effects discussed in Section VI-4C.

7. Anomalous Water

Some mention should be made of perhaps the major topic of conversation among surface and colloid chemists during the period 1966–1973. Some initial observations were made by Shereshefsky and co-workers on the vapor pressure of water in small capillaries (anomalously low) [119] but especially by Fedyakin in 1962, followed closely by a series of papers by Derjaguin and co-workers (see Ref. 120 for a detailed bibliography up to 1970–1971).

The reports were that water condensed from the vapor phase into 10–100-μm quartz or pyrex capillaries had physical properties distinctly different from those of bulk liquid water. Confirmations came from a variety of laboratories around the world (see the August 1971 issue of *Journal of Colloid Interface Science*), and it was proposed that a new *phase* of water had been found; many called this water *polywater* rather than the original Derjaguin term, *anomalous water*. There were confirming theoretical calculations (see Refs. 121, 122)! Eventually, however, it was determined that the microamounts of water that could be isolated from small capillaries was always contaminated by salts and other impurities leached from the walls. The nonexistence of anomalous or polywater as a new, pure phase of water was acknowledged in 1974 by Derjaguin and co-workers [123]. There is a mass of fascinating anecdotal history omitted here for lack of space but told very well by Frank [124].

8. Dipole-Induced Dipole Propagation

The discussion of forces in Sections VI-3 and VI-4 emphasized relatively nonspecific long-range dispersion or Lifshitz attractions and electrical double-layer repulsion. The material of Section VI-4C makes it clear, however, that structured interactions must exist and can be important. One model for such interactions, leading to attraction, is that of the propagation of polarization, that is, of the dipole-induced dipole effect. This may be described in terms of the situation in which there is an adsorbed layer of atoms on a surface. If the surface is considered to contain surface charges of individual value ze (where z might be the formal charge on a polar surface atom), then by Eq. VI-12 the interaction energy with a polarizable atom absorbed on the surface is

$$U_{01} = -\frac{(ze)^2\alpha}{2d^4} = -\frac{(ze)^2}{2d}\frac{\alpha}{d^3} \qquad \text{(VI-40)}$$

where d is a distance of separation; the induced dipole moment for the atom is $ze\alpha/d^2$, which corresponds to equal and opposite charge $ze\alpha/d^3$ having been induced in the atom along an axis normal to the surface. If a second layer of atoms is present, the field of this charge $ze\alpha/d^3$ induces a dipole moment in

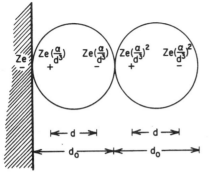

Fig. VI-8. Long-range interaction through propagated polarization.

an adjacent second-layer atom, with corresponding interaction energy

$$U_{12} = -\frac{(ze)^2}{2d} \left(\frac{\alpha}{d^3} \right)^3 \qquad \text{(VI-41)}$$

The situation is illustrated in Fig. VI-8. Continued step-by-step propagation leads to the general term

$$U_{i,(i+1)} = U_{(i-1),i} \left(\frac{\alpha}{d^3} \right)^2 \qquad \text{(VI-42)}$$

The locus of successive U values can be represented by an exponential relationship

$$U(x) = U_0 e^{-ax} \qquad \text{(VI-43)}$$

where x denotes distance from the surface, and a is given by

$$a = -\frac{1}{d_0} \ln \left(\frac{\alpha}{d^3} \right)^2 \qquad \text{(VI-44)}$$

where d_0 is the atomic diameter. The distance d may be regarded as an effective value smaller than d_0.

It is thus seen that the dipole-induced dipole propagation gives an exponential rather than an inverse x cube dependence of $U(x)$ with x. As with the dispersion potential, the interaction depends on the polarizability, but unlike the dispersion case, it is only the polarizability of the adsorbed species that is involved. The application of Eq. VI-43 to physical adsorption is considered in Section XVII-7D. For the moment, the treatment illustrates how a "long-range" interaction can arise as a propagation of short-range interactions.

9. Problems

1. Derive the expression for the electric field around a point dipole, Eq. VI-5, by treating the dipole as two charges separated by a distance d, then moving to distances $x \gg d$.

2. Calculate $U(\alpha, \alpha)$ for two Ar atoms 8 Å apart and compare to kT at 77 K (around the boiling point of Ar). Do the problem in both SI and cgs/esu units.

3. The atomic unit (AU) of dipole moment is that of a proton and electron separated by a distance equal to the first Bohr orbit, a_0. Similarly, the au of polarizability is a_0^3 [125]. Express μ and α for NH_3 using both the cgs/esu and SI approach.

4. Derive Eq. VI-12 for the interaction between a polarizable molecule and a charge.

5. Show how $U(x) = -aA/12x$ can be derived from Eq. VI-24 for small x.

6. Derive Eq. VI-22 from Eq. VI-15 by carrying out the required integrations.

7. From the discussion about the Derjaguin approximation for spheres and Eq. VI-26, show that the approximation for two crossed cylinders of radius R is

$$U(x)_{slab-slab} = -\frac{F(x)}{2\pi R} \tag{VI-45}$$

8. Verify the numerical prefactor coefficients in Eqs. VI-24 and VI-25.

9. Verify Eq. VI-33 and determine d for water and n-octane.

10. Explain qualitatively why Eq. VI-25 applies to the case of crossed cylinders as well as to a sphere and a flat plate.

11. The Hamaker constant for the case of a substance having two kinds of atoms is obtained by replacing the quantity $n^2\alpha^2$ by $n_1^2\alpha_1^2 + n_2^2\alpha_2^2 + 2n_1n_2\alpha_1\alpha_2$. Explain why this formulation should be correct and calculate the A appropriate for ice, taking the polarizabilities of H and of O to be 0.67×10^{-24} and 3.0×10^{-24} cm^3, respectively. Calculate the expected force between two thick slabs of ice that are 150 Å apart. Do the same for two spheres of ice of 2 μm radius and again 150 Å apart. Assume for $h\nu_0$ ice to be $\sim 10^{-11}$ erg.

12. The long-range van der Waals interaction provides a cohesive pressure for a thin film that is equal to the mutual attractive force per square centimeter of two slabs of the same material as the film and separated by a thickness equal to that of the film. Consider a long column of the material of unit cross section. Let it be cut in the middle and the two halves separated by d, the film thickness. Then, from one outside end of one of each half, slice off a layer of thickness d; insert one of these into the gap. The system now differs from the starting point by the presence of an isolated thin layer. Show by suitable analysis of this sequence that the opening statement is correct. *Note:* About the only assumptions needed are that interactions are superimposable and that they are finite in range.

13. Using Eqs. VI-30–VI-32 and data from the General References or handbooks, plot the *retarded* Hamaker constant for quartz interacting through water and through n-decane. Comment on the relative importance of the zero frequency contribution and that from the ν_{UV} peak.

14. From the data in Problem 13, calculate the interaction potential energy between

two slabs of quartz in water and in n-decane. Plot these on a log–log plot to visualize the power-law behavior.

15. A thin film of hydrocarbon spread on a horizontal surface of quartz will experience a negative dispersion interaction. Treating these as 1 = quartz, 2 = n-decane, 3 = vacuum, determine the Hamaker constant A_{123} for the interaction. Balance the negative dispersion force (nonretarded) against the gravitational force to find the equilibrium film thickness.

16. Determine the net DLVO interaction (electrostatic plus dispersion forces) for two *large* colloidal spheres having a surface potential ψ_0 = 51.4 mV and a Hamaker constant of 3×10^{-14} erg in a 0.002M solution of 1 : 1 electrolyte at 25°C. Plot $U(x)$ as a function of x for the individual electrostatic and dispersion interactions as well as the net interaction.

17. Show the reduction of Eq. VI-34 for the case of two identical materials 1 interacting through a medium 3. An analysis similar to that in Problem 12 provides a proof of your equation. Formulate this proof; it is due to Hamaker [44].

18. The interfacial free energy per unit area is given by the adhesion force $F_0/2\pi R$; estimate the Hamaker constant responsible for the adhesion force in the crossed-cylinder geometry illustrated in the inset to Fig. VI-6.

19. Consider the situation illustrated in Fig. VI-9, in which two air bubbles, formed in a liquid, are pressed against each other so that a liquid film is present between them. Relate the disjoining pressure of the film to the Laplace pressure P in the air bubbles.

20. Derjaguin and Zorin report that at 25°C, water at 0.98 of the saturation vapor pressure adsorbs on quartz to give a film 40 Å thick. Calculate the value of the disjoining pressure of this film and give its sign.

21. Show that Eq. VI-44 indeed follows from Eq. VI-42.

22. Calculate A/A_0 of Eq. VI-38 assuming that the mica cylinders are immersed in a dilute aqueous solution at 25°C and taking the parameters to have the indicated typical values.

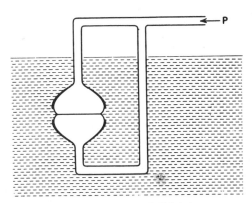

Fig. VI-9. A measurement of disjoining pressure.

General References

B. V. Derjaguin, N. V. Churaev, and V. M. Muller, *Surface Forces*, translation from Russian by V. I. Kisin, J. A. Kitchener, Eds., Consultants Bureau, New York, 1987.

R. J. Hunter, *Foundations of Colloid Science*, Vol. 1, Clarendon Press, Oxford, 1987.

J. Israelachvili, *Intermolecular and Surface Forces*, 2nd ed., Academic, San Diego, CA, 1992.

J. Lyklema, *Fundamentals of Interface and Colloid Science, Vol. 1 and 2*, Academic Press, San Diego, 1995.

J. Mahanty and B. W. Ninham, *Dispersion Forces*, Academic, New York, 1976.

S. Ross and I. D. Morrison, *Colloidal Systems and Interfaces*, Wiley, New York, 1988.

W. B. Russel, D. A. Saville, and W. R. Schowalter, *Colloidal Dispersions*, Cambridge University Press, Cambridge, UK, 1989.

H. van Olphen and K. J. Mysels, eds., *Physical Chemistry: Enriching Topics from Colloid and Surface Science*, Theorex (8327 La Jolla Scenic Drive, La Jolla, CA), 1975.

R. D. Vold and M. J. Vold, *Colloid and Interface Chemistry*, Addison-Wesley, Reading, MA, 1983.

Textual References

1. J. N. Israelachvili, *Intermolecular and Surface Forces*, Academic, San Diego, CA, 1992.
2. F. London, *Trans. Faraday Soc.*, **33**, 8 (1937).
3. W. B. Russel, D. A. Saville, and W. R. Schowalter, *Colloid Dispersions*, Cambridge University Press, 1989.
4. J. C. Slater and J. G. Kirkwood, *Phys. Rev.*, **37**, 682 (1931).
5. A. Müller, *Proc. Roy. Soc.* (London), **A154**, 624 (1936).
6. P. R. Fontana, *Phys. Rev.*, **123**, 1865 (1961).
7. G. E. Ewing, *Acc. Chem. Res.*, **8**, 185 (1975).
8. J. H. de Boer, *Trans. Faraday Soc.* **32**, 10 (1936).
9. H. C. Hamaker, *Physica*, **4**, 1058 (1937).
10. D. Tabor and R. H. S. Winterton, *Nature*, **219**, 1120 (1968).
11. D. Tabor and R. H. S. Winterton, *Proc. Roy. Soc.* (London) **A312**, 435 (1969).
12. S. Ross and I. Morrison, *Colloidal Systems and Interfaces*, Wiley, New York, 1988.
13. B. V. Derjaguin, *Kolloid-Z.*, **69**, 155 (1934).
14. K. D. Papadopoulos and C.-C. Kuo, *Colloids & Surfaces*, **46**, 115 (1990).
15. A. K. Sengupta and K. D. Papadopoulos, *J. Colloid Interface Sci.*, **152**, 534 (1992).
16. M. C. Herman and K. D. Papadopoulos, *J. Colloid Interface Sci.*, **136**, 285 (1990).
17. M. C. Herman and K. D. Papadopoulos, *J. Colloid Interface Sci.*, **142**, 331 (1991).
18. H. B. G. Casimir and D. Polder, *Phys. Rev.*, **73**, 360 (1948).
19. E. M. Lifshitz, *Soviet Physics JETP*, **2**, 73 (1956).
20. I. E. Dzyaloshinskii, E. M. Lifshitz, and L. P. Pitaevskii, *Adv. Phys.*, **10**, 165 (1961).
21. V. A. Parsegian and B. W. Ninham, *Nature*, **224**, 1197 (1969).

22. V. A. Parsegian and G. H. Weiss, *J. Colloid Interface Sci.*, **81**, 285 (1981).

23. B. V. Derjaguin, A. S. Titijevskaia, I. I. Abrikossova, and A. D. Malkina, *Disc. Faraday Soc.*, **18**, 24 (1954).

24. B. V. Derjaguin, I. I. Abrikossova, and E. M. Lifshitz, *Q. Rev. Chem. Soc.* (London), **10**, 292 (1956).

25. W. Black, J. G. V. de Jongh, J. T. G. Overbeek, and M. J. Sparnaay, *J. Chem. Soc., Faraday Trans. I*, **56**, 1597 (1960).

26. J. A. Kitchener and A. P. Prosser, *Proc. Roy. Soc.* (London), **A242**, 403 (1957).

27. G. C. J. Rouweler and J. T. G. Overbeek, *J. Chem. Soc., Faraday Trans. I*, **67**, 2117 (1971).

28. J. N. Israelachvili and D. Tabor, *Prog. Surf. Membr. Sci.*, **7**, 1 (1973).

29. D. Leckband, *Nature*, **376**, 617 (1995).

30. B. V. Derjaguin, Y. I. Rabinovich, and N. V. Churaev, *Nature*, **272**, 313 (1978).

31. P. M. Claesson, J. L. Parker, and J. C. Fröberg, *J. Disp. Sci. Technol.*, **15**, 375 (1994).

32. I. Larson, C. J. Drummond, D. Y. C. Chan, and F. Grieser, *J. Am. Chem. Soc.*, **115**, 11885 (1993).

33. W. A. Ducker, Z. Xu, and J. N. Israelachvili, *Langmuir*, **10**, 3279 (1994).

34. A. Shih and V. A. Parsegian, *Phys. Rev. A*, **12**, 835 (1975).

35. D. M. LeNeveu, R. P. Rand, and V. A. Parsegian, *Nature*, **259**, 601 (1976).

36. V. A. Parsegian, R. P. Rand, N. L. Fuller, and D. C. Rau, in *Methods in Enzymology: Biomembranes Part O*, Volume 127, L. Packer, ed. Academic, Orlando, FL, 1986.

37. D. C. Prieve, S. G. Bike, and N. A. Frej, *Disc. Faraday Soc.*, **90**, 209 (1990).

38. M. A. Brown and E. J. Staples, *Disc. Faraday Soc.*, **90**, 193 (1990).

39. T. G. M. van de Ven, P. Warszynski, X. Wu, and T. Dabros, *Langmuir*, **10**, 3046 (1994).

40. M. D. Croucher and M. L. Hair, *J. Phys. Chem.*, **81**, 1631 (1977).

41. B. W. Ninham and V. A. Parsegian, *Biophys. J.*, **10**, 646 (1970).

42. D. B. Hough and L. R. White, *Adv. Colloid Interface Sci.*, **14**, 3 (1980).

43. B. V. Derjaguin and L. Landau, *Acta Physicochem. URSS*, **14**, 633 (1941).

44. E. J. W. Verwey and J. T. G. Overbeek, *Theory of the Stability of Lyophobic Colloids*, Elsevier, Amsterdam, 1948.

45. E. J. W. Verwey and J. T. G. Overbeek, *J. Chem. Soc., Faraday Trans. I*, **42B**, 117 (1946).

46. J. T. G. Overbeek, *Pure Appl. Chem.*, **52**, 1151 (1980).

47. R. H. Ottewill and A. Watanabe, *Kolloid-Z.*, **173**, 7 (1960).

48. G. D. Parfitt and N. H. Picton, *J. Chem. Soc., Faraday Trans. I*, **64**, 1955 (1968).

49. C. G. Force and E. Matijević, *Kolloid-Z.*, **224**, 51 (1968).

50. S. S. Dukhin and J. Lyklema, *Faraday Disc. Chem. Soc.*, **90**, 261 (1990).

51. J. Kijlstra and H. van Leeuwen, *J. Colloid Interface Sci.*, **160**, 424 (1993).

52. S. Y. Shulepov, S. S. Dukhin, and J. Lyklema, *J. Colloid Interface Sci.*, **171**, 340 (1995).

53. H. Kihira, N. Ryde, and E. Matijević, *J. Chem. Soc., Faraday Trans. 1*, **88**, 2379 (1992).

54. H. Kihira and E. Matijević, *Langmuir*, **8**, 2855 (1992).

55. J. A. Maroto and F. J. de las Nieves, *Prog. Colloid Polym. Sci.*, **98**, 89 (1995).

56. J. D. G. Durán, M. C. Guindo, A. V. Delgado, and F. González-Caballero, *Langmuir*, **11**, 3648 (1995).

57. C. J. van Oss, *Colloids & Surfaces*, **A78**, 1 (1993).

58. J. W. T. Lichtenbelt, C. Pathmamanoharan, and P. H. Wiersma, *J. Colloid Interface Sci.*, **49**, 281 (1974).

59. J. N. Israelachvili and B. W. Ninham, *J. Colloid Interface Sci.*, **58**, 14 (1977).

60. J. N. Israelachvili and G. E. Adams, *Nature*, **262**, 774 (1976).

61. R. M. Pashley, *J. Colloid Interface Sci.*, **83**, 531 (1981).

62. J. Marra, *J. Phys. Chem.*, **90**, 2145 (1986).

63. R. M. Pashley and B. W. Ninham, *J. Phys. Chem.*, **91**, 2902 (1987).

64. J. Marra and M. L. Hair, *J. Phys. Chem.*, **92**, 6044 (1988).

65. R. G. Horn, D. F. Evans, and B. W. Ninham, *J. Chem. Phys.*, **92**, 3531 (1988).

66. J. Marra and J. N. Israelachvili, *Biochemistry*, **24**, 4608 (1985).

67. R. G. Horn and J. N. Israelachvili, *J. Chem. Phys.*, **75**, 1400 (1981).

68. H. K. Christenson and C. E. Blom, *J. Chem. Phys.*, **86**, 419 (1987).

69. J. N. Israelachvili, *Acc. Chem. Res.*, **20**, 415 (1987).

70. R. M. Pashley, P. M. McGuiggan, B. W. Ninham, and D. F. Evans, *Science*, **229**, 1088 (1985).

71. P. M. Claesson and H. K. Christenson, *J. Phys. Chem.*, **92**, 1650 (1988).

72. E. Ruckenstein and D. Schiby, *Chem. Phys. Lett.*, **95**, 439 (1983).

73. R. M. Pashley, *J. Colloid Interface Sci.*, **80**, 153 (1981).

74. B. V. Derjaguin and N. V. Churaev, in *Fluid Interfacial Phenomena*, C. A. Croxton, ed., Wiley, New York, 1986.

75. N. K. Roberts and G. Zundel, *J. Phys. Chem.*, **85**, 2706 (1981).

76. W. Drost-Hansen, C. V. Braun, R. Hochstim, and G. W. Crowther, "Colloidal and Interfacial Phenomena," in *Particulate and Multiphase Processes*, Vol. 3, T. Ariman and T. N. Veziroğlu, eds., Hemisphere Publishing, New York, 1987.

77. Y. Sun, H. Lin, and P. F. Low, *J. Colloid Interface Sci.*, **112**, 556 (1986).

78. D. J. Mulla and P. F. Low, *J. Colloid Interface Sci.*, **95**, 51 (1983).

79. P. M. Wiggins and R. T. van Ryn, *J. Macromol. Sci.-Chem.*, **A23**, 875 (1987).

80. B. V. Derjaguin, in *Fluid Interfacial Phenomena*, C. A. Croxton, ed., Wiley, New York, 1986.

81. B. V. Derjaguin, V. V. Karasev, and N. B. Ur'ev, *Colloids & Surfaces*, **25**, 397 (1987).

82. J. Tse and A. W. Adamson, *J. Colloid Interface Sci.*, **72**, 515 (1979).

83. E. Ruckenstein and D. Schiby, *Chem. Phys. Lett.*, **95**, 435 (1983).

84. P. F. Low, J. H. Cushman, and D. J. Diestler, *J. Colloid Interface Sci.*, **100**, 576 (1984).

85. P. Ahlström, O. Teleman, and B. Jönsson, *J. Am. Chem. Soc.*, **110**, 4198 (1988).

86. N. A. M. Besseling and J. M. H. M. Scheutjens, *J. Phys. Chem.*, **98**, 11597 (1994).

87. N. A. M. Besseling and J. Lyklema, *J. Phys. Chem.*, **98**, 11610 (1994).

88. C. J. van Oss, *Colloids & Surfaces B: Biointerfaces* (in press).

89. C. Toprakcioglu, J. Klein, and P. F. Luckham, *J. Chem. Soc., Faraday Trans. I*, **83**, 1703 (1987).

90. J. L. Parker, P. M. Claesson, and P. Attard, *J. Phys. Chem.*, **98**, 8468 (1994).

91. P. Attard, *Langmuir*, **12**, 1693 (1996).

92. J. C. Eriksson and S. Lunggren, *Langmuir*, **11**, 3325 (1995).

93. G. Karagounis, *Helv. Chim. Acta*, **37**, 805 (1954).

94. S. Chandrasekhar, *Liquid Crystals*, 2nd ed., Cambridge University Press, 1992.

95. B. V. Derjaguin and Z. M. Zorin, *Proc. 2nd Int. Congr. Surface Activity* (London), **2**, 145 (1957).

96. M. E. Tadros, P. Hu, and A. W. Adamson, *J. Colloid Interface Sci.*, **49**, 184 (1974).

97. A. W. Adamson and R. Massoudi, *Fundamentals of Adsorption*, Engineering Foundation, New York, 1983.

98. B. V. Derjaguin and V. V. Karasev, *Proc. 2nd Int. Conf. Surface Activity* **3**, 531 (1957).

99. B. V. Derjaguin, V. V. Karasev, I. A. Lavygin, I. I. Skorokhodov, and E. N. Khromova, *Special Disc. Faraday Soc.*, **1**, 98 (1970).

100. B. V. Derjaguin, V. V. Karasev, V. M. Starov, and E. N. Khromova, *J. Colloid Interface Sci.*, **67**, 465 (1968).

101. K. J. Mysels, *J. Phys. Chem.*, **68**, 3441 (1964).

102. W. D. Bascom and C. R. Singleterry, *J. Colloid Interface Sci.*, **66**, 559 (1978).

103. B. V. Derjaguin, V. V. Karasev, and E. N. Khromova, *J. Colloid Interface Sci.*, **66**, 573 (1978).

104. P. F. Low, *Soil Sci. Soc. Am. J.*, **40**, 500 (1976).

105. J. N. Israelachvili, *J. Colloid Interface Sci.*, **110**, 263 (1986).

106. J. N. Israelachvili, *Colloid Polym. Sci.*, **264**, 1060 (1986).

107. D. Y. C. Chan and R. G. Horn, *J. Chem. Phys.*, **83**, 5311 (1985).

108. R. G. Horn and J. N. Israelachvili, *Macromolecules*, **21**, 2836 (1988).

109. J. V. Alsten and S. Granick, *J. Colloid Interface Sci.*, **125**, 739 (1988).

110. L. J. Zapas and G. B. McKenna, *J. Rheol.*, **33**, 69 (1989).

111. P. M. Wiggins, *Prog. Polym. Sci.*, **13**, 1 (1987).

112. M. L. Broide, T. Tominic, and M. Saxowsky, *Phys. Rev. E*, **53**, 6325 (1996).

113. D. Rosenbaum, P. C. Zamora, and C. F. Zukoski, *Phys. Rev. Lett.*, **76**, 150 (1996).

114. D. Leckband, F. J. Schmitt, W. Knoll, and J. Israelachvili, *Biochemistry*, **33**, 4611 (1994).

115. J. N. Israelachvili, *Langmuir*, **10**, 3369 (1994).

116. B. V. Derjaguin and N. V. Churaev, *J. Colloid Interface Sci.*, **49**, 249 (1974).

117. B. V. Derjaguin, private communication.

118. N. V. Churaev and B. V. Derjaguin, *J. Colloid Interface Sci.*, **103**, 542 (1985).

119. See M. Folman and J. L. Shereshefsky, *J. Phys. Chem.*, **59**, 607 (1955).

120. L. C. Allen, *J. Colloid Interface Sci.*, **36,** 554 (1971).

121. L. C. Allen and P. A. Kollman, *Science*, **167,** 1443 (1970); note, however, *Nature*, **233,** 550 (1971).

122. J. W. Linnett, *Science*, **167,** 1719 (1970).

123. B. V. Derjaguin, Z. M. Zorin, Ya. I. Rabinovich, and N. V. Churaev, *J. Colloid Interface Sci.*, **46,** 437 (1974).

124. F. Frank, *Polywater*, MIT Press, Cambridge, MA, 1981.

125. J. Gready, G. B. Bacskay, and N. S. Hush, *Chem. Phys.*, **31,** 467 (1978).

CHAPTER VII

Surfaces of Solids

1. Introduction

The interface between a solid and its vapor (or an inert gas) is discussed in this chapter from an essentially phenomenological point of view. We are interested in surface energies and free energies and in how they may be measured or estimated theoretically. The study of solid surfaces at the molecular level, through the methods of spectroscopy and diffraction, is taken up in Chapter VIII.

A. The Surface Mobility of Solids

A *solid*, by definition, is a portion of matter that is rigid and resists stress. Although the surface of a solid must, in principle, be characterized by surface free energy, it is evident that the usual methods of capillarity are not very useful since they depend on measurements of equilibrium surface properties given by Laplace's equation (Eq. II-7). Since a solid deforms in an elastic manner, its shape will be determined more by its past history than by surface tension forces.

Practically speaking, however, solids may possess sufficient plasticity to flow, at least slowly, such that the methods of capillarity may be applied. An early experiment on thin copper wires provided a measure of the surface tension of copper near its melting point of 1370 dyn/cm [1]. The sintering process, driven by a reduction of surface energy through minimization of surface area, occurs because of bulk and surface mobility in metals and other solids. Surface scratches on silver heal [2], and copper or silver spheres fuse with flat surfaces of the same metal [3] below the melting point. The distance from the melting point is variable; small ice spheres develop a connecting neck if touching at $-10°C$, and a detailed study of the kinetics of the process is needed to determine the relative importance of surface and bulk diffusion [4, 5]. Some systems are sensitive to other conditions; sintering rates for MgO increase with ambient humidity [6]. Sintering is, of course, very important in the field of catalysis, where consolidation of small supported metal particles reduces surface area and the number of active sites. The rate law may be of the form

$$\Delta V = V_0 C \, \frac{t^n}{r^m} \qquad \text{(VII-1)}$$

257

where V_0 is the volume of a particle undergoing sintering, t is time, and r is the particle radius [7–9]

It is instructive to consider just how mobile the surface atoms of a solid might be expected to be. Following the approach in Section III-2, one may first consider the evaporation–condensation equilibrium. The number of molecules hitting a 1-cm^2 surface per second is from kinetic theory

$$Z = P \left(\frac{1}{2\pi MRT} \right)^{1/2} \tag{VII-2}$$

or for saturated water vapor at room temperature $Z = 1 \times 10^{22}$ molecules cm^{-2} sec^{-1}. When balanced with an equal average evaporation rate, the average lifetime of a molecule at the water–vapor interface is about 1 μsec. By contrast, a nonvolatile metal such as tungsten has a room-temperature vapor pressure of 10^{-40} atm and $Z = 1 \times 10^{-17}$ molecules cm^{-2} sec^{-1} such that the average lifetime becomes about 10^{32} sec! Even more volatile metals and most salts have very long lifetimes at room temperature. Nearer to melting, this process can become significant; copper at 725°C has an estimated vapor pressure of 10^{-8} mm Hg [10], which leads to an estimated surface lifetime of 1 hr. Low-melting solids such as ice, iodine and organic solids can have surface lifetimes comparable to those for liquids at moderate temperatures.

Not all molecules striking a surface necessarily condense, and Z in Eq. VII-2 gives an upper limit to the rate of condensation and hence to the rate of evaporation. Alternatively, actual measurement of the evaporation rate gives, through Eq. VII-2, an effective vapor pressure P_e that may be less than the actual vapor pressure P^0. The ratio P_e/P^0 is called the vaporization coefficient α. As a perhaps extreme example, α is only 8.3 $\times 10^{-5}$ for (111) surfaces of arsenic [11].

The diffusion time gives the same general picture. The bulk self-diffusion coefficient of copper is 10^{-11} cm^2/sec at 725°C [12]; the Einstein equation

$$\mathcal{D} = \frac{x^2}{2t} \tag{VII-3}$$

provides the Brownian displacement x in time t of 0.1 sec for 10 nm. At room temperature the time would be about 10^{27} sec since the apparent activation energy for copper diffusion is about 54 kcal/mol. Surface diffusion is often more important than bulk diffusion in surface chemistry. Surface diffusion on copper becomes noticeable around 700°C [13] and, because of its lower activation energy, is the dominant transport process over an important temperature range. Surface migration becomes measurable by field ion microscopy (Section VIII-2) as low as half the evaporation temperature. The surface region of solids near their melting point is often liquidlike [14, 15].

It is obvious that there is a very great range in the nature of solid surfaces. Those most often studied at room temperature are refractory solids, that is, are far below their melting points. Here surface atoms are relatively immobile, although vibrating around lattice positions; the surface is highly conditioned by its past history and cannot be studied by the methods of capillarity. Solids near their melting points show surface mobility in the form of exchange with the vapor phase and especially in the form of lateral mobility.

B. Effect of Processing on the Condition of Solid Surfaces

The immobility of the surface atoms of a refractory solid has the consequence that the surface energy and other physical properties depend greatly on the immediate history of the material. A clean cleavage surface of a crystal will have a different (and probably lower) surface energy than a ground, abraded, heat-treated or polished surface of the same material.

Grinding may induce changes in general physical properties such as density; in the case of quartz, a deep amorphous layer appears to form [16]. Electron diffraction studies of the surface region indicated that grinding removes surface material without changing its molecular crystallinity, while polishing leaves a fairly deep amorphous surface layer [17, 18] known as the *Beilby layer* [19, 20]. Apparently the Beilby layer is formed through a softening, if not melting, of the metal surface [21] and this layer can persist 2–10 nm into a metal such as gold or nickel [22]. Cold working of metals also affects the nature of the surface region [23].

Surface defects (Section VII-4C) are also influenced by the history of the sample. Such imperfections may to some extent be reversibly affected by processes such as adsorption so that it is not safe to regard even a refractory solid as having fixed surface actions. Finally, solid surfaces are very easily contaminated; detection of contamination is aided by ultra-high-vacuum techniques and associated cleaning protocols [24].

2. Thermodynamics of Crystals

A. Surface Tension and Surface Free Energy

Unlike the situation with liquids, in the case of a solid, the surface tension is not necessarily equal to the surface stress. As Gibbs [25] pointed out, the former is the work spent in *forming* unit area of surface (and may alternatively be called the *surface free energy*; see Sections II-1 and III-2), while the latter involves the work spent in *stretching* the surface. It is helpful to imagine that the process of forming a fresh surface of a monatomic substance is divided into two steps: first, the solid or liquid is cleaved so as to expose a new surface, keeping the atoms fixed in the same positions that they occupied when in the bulk phase; second, the atoms in the surface region are allowed to rearrange to their final equilibrium positions. In the case of the liquid, these two steps occur as one, but with solids the second step may occur only slowly because of the immobility of the surface region. Thus, with a solid it may be possible to stretch or

Fig. VII-1

to compress the surface region without changing the number of atoms in it, only their distances apart.

In Chapter III, surface free energy and surface stress were treated as equivalent, and both were discussed in terms of the energy to form unit additional surface. It is now desirable to consider an independent, more mechanical definition of surface stress. If a surface is cut by a plane normal to it, then, in order that the atoms on either side of the cut remain in equilibrium, it will be necessary to apply some external force to them. The total such force per unit length is the surface stress, and half the sum of the two surface stresses along mutually perpendicular cuts is equal to the surface tension. (Similarly, one-third of the sum of the three principal stresses in the body of a liquid is equal to its hydrostatic pressure.) In the case of a liquid or isotropic solid the two surface stresses are equal, but for a nonisotropic solid or crystal, this will not be true. In such a case the partial surface stresses or stretching tensions may be denoted as τ_1 and τ_2.

Shuttleworth [26] (see also Ref. 27) gives a relation between surface free energy and stretching tension as follows. For an anisotropic solid, if the area is increased in two directions by $d\mathcal{A}_1$ and $d\mathcal{A}_2$, as illustrated in Fig. VII-1, then the total increase in free energy is given by the reversible work against the surface stresses, that is,

$$\tau_1 = G^s + \mathcal{A}_1 \frac{dG^s}{d\mathcal{A}_1} \quad \text{and} \quad \tau_2 = G^s + \mathcal{A}_2 \frac{dG^s}{d\mathcal{A}_2} \qquad \text{(VII-4)}$$

where G^s is the free energy per unit area. If the solid is isotropic, Eq. VII-4 reduces to

$$\tau = \frac{d(\mathcal{A} G^s)}{d\mathcal{A}} = G^s + \frac{\mathcal{A} dG^s}{d\mathcal{A}} \qquad \text{(VII-5)}$$

For liquids, the last term in Eq. VII-5 is zero, so that $\tau = G^s$ (or $\tau = \gamma$, since we will use G^s and γ interchangeably); the same would be true of a solid if the change in area $d\mathcal{A}$ were to occur in such a way that an equilibrium surface configuration was always maintained. Thus the stretching of a wire under reversible conditions would imply that interior atoms would move into the surface as needed so that the increased surface area was not accompanied by any change in specific surface properties. If, however, the stretching were done under conditions such that full equilibrium did not prevail, a surface stress would be present whose value would differ from γ by an amount that could be time-dependent and would depend on the term $\mathcal{A} \, dG^s / d\mathcal{A}$.

B. The Equilibrium Shape of a Crystal

The problem of the equilibrium shape for a crystal is an interesting one. Since the surface free energy for different faces is usually different, the question becomes one of how to construct a shape of specified volume such that the total surface free energy is a minimum. A general, quasi-geometric solution has been given by Wulff [28] and may be described in terms of the following geometric construction.

Given the set of surface free energies for the various crystal planes, draw a set of vectors from a common point of length proportional to the surface free energy and of direction normal to that of the crystal plane. Construct the set of planes normal to each vector and positioned at its end. It will be possible to find a geometric figure whose sides are made up entirely from a particular set of such planes that do not intersect any of the other planes. The procedure is illustrated in Fig. VII-2 for a two-dimensional crystal for which γ_{10} is 250 ergs/cm and γ_{11} is 225 ergs/cm. Note that the optimum shape makes use of both types of planes; for these surface tensions there is a free energy gain in truncating the corners of a (11) sided square crystal to reduce its perimeter (note Problems VII-2–VII-6). A statement of Wulff's theorem is that for an equilibrium crystal there exists a point in the interior such that its perpendicular distance h_i from the ith face is proportional to γ_i. This is, of course, the basis of the construction in Fig. VII-2.

There are many studies involving the Wulff construction, including an analytical proof of the theorem [29], a determination of the shapes of face-centered cubic (FCC) and body-centered cubic (BCC) crystals [30], and investigations of the finite radius of curvature of edges [31] and the possibility of cusp formation [32]. Minimum free energy polyhedra do exist in nature, in rock salt [33], metals [34], and in other small crystals [35]. Wulff constructions have been worked out for crystals in a gravitational field [36], on a surface, or in a corner [37]. Combination of the Wulff construction with density functional theory of the solid (see Section IX-2) shows the lack of faceting in systems having a very short-range attraction [38].

Since the crystal shape, or *habit*, can be determined by kinetic and other nonequilibrium effects, an actual crystal may have faces that differ from those of the Wulff construction. For example, if a (100) plane is a stable or singular plane but by processing one produces a plane at a small angle to this, describable as an (x00) plane, where x is a large number, the surface may decompose into a set of (100) steps and (010) risers [39].

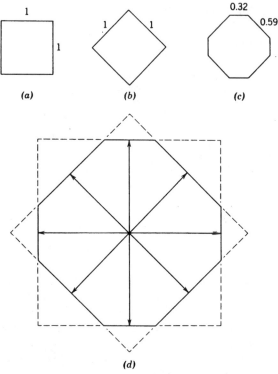

(a)   (b)   (c)

(d)

Fig. VII-2. Conformation for a hypothetical two-dimensional crystal. (a) (10)-type planes only. For a crystal of 1 cm^2 area, the total surface free energy is $4 \times 1 \times 250 = 1000$ ergs. (b) (11)-type planes only. For a crystal of 1-cm^2 area, the total surface free energy is $4 \times 1 \times 225 = 900$ ergs. (c) For the shape given by the Wulff construction, the total surface free energy of a 1-cm^2 crystal is $(4 \times 0.32 \times 250) + (4 \times 0.59 \times 225) = 851$ ergs. (d) Wulff construction considering only (10)- and (11)-type planes.

C. The Kelvin Equation

The Kelvin equation (Eq. III-18), which gives the increase in vapor pressure for a curved surface and hence of small liquid drops, should also apply to crystals. Thus

$$RT \ln \frac{P}{P^0} = \frac{2\gamma \overline{V}}{r} \qquad \text{(VII-6)}$$

Since an actual crystal will be polyhedral in shape and may well expose faces of different surface tension, the question is what value of γ and of r should be used. As noted in connection with Fig. VII-2, the Wulff theorem states that γ_i/r_i is invariant for all faces of an equilibrium crystal. In Fig. VII-2, r_{10} is the radius of the circle inscribed to the set of (10) faces and r_{11} is the radius of the circle inscribed to the set of (11) faces. (See also Ref. 40.) Equation VII-6 may also be applied to the solubility of small crystals (see Section X-2). In the case of relatively symmetrical crystals, Eq. VII-6 can be used as an approximate equation with r and γ regarded as mean values.

3.   Theoretical Estimates of Surface Energies and Free Energies

A.   Covalently Bonded Crystals

The nature of the theoretical approach to the calculation of surface energy quantities necessarily varies with the type of solid considered. Perhaps the simplest case is that of a covalently bonded crystal whose sites are occupied by atoms; in this instance no long-range interactions need be considered. The example of this type of calculation, par excellence, is that for the surface energy of diamond. Harkins [41] considered the surface energy at 0 K to be simply one-half of the energy to rupture that number of bonds passing through 1 cm^2, that is, $E^s = \frac{1}{2}E_{\text{cohesion}}$.

The unit cell for diamond is shown in Fig. VII-3, and it is seen that three bonds would be broken by a cleavage plane parallel to (111) planes. From the (111) interplanar distance of 2.32 Å, and the density of diamond 3.51 g/cm^3, one computes that 1.83×10^{15} bonds/cm^2 are involved, and using 90 kcal/mol as the bond energy, the resulting value for the surface energy is 5650 ergs/cm^2. For (100) planes, the value is 9820 ergs/cm^2. Since these figures are for 0 K, they are also equal to the surface free energy at that temperature.

Harkins then estimated T_c for diamond to be about 6700 K and, using Eq. III-10, found the entropy correction at 25°C to be negligible so that the preceding values also approximate the room temperature surface free energies. These

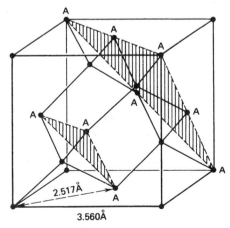

Fig. VII-3. Diamond structure. (From Ref. 41.)

values cannot be strictly correct, however, since no allowance has been made for surface distortion (see Sections VII-3B,C). Such distortion or *reconstruction* can be substantial. In the case of silicon (a material of great importance to the electronics industry), a molecular dynamics calculation predicts surface dimer formation [42].

B. Rare-Gas Crystals

Face-centered cubic crystals of rare gases are a useful model system due to the simplicity of their interactions. Lattice sites are occupied by atoms interacting via a simple van der Waals potential with no orientation effects. The principal problem is to calculate the net energy of interaction across a plane, such as the one indicated by the dotted line in Fig. VII-4. In other words, as was the case with diamond, the surface energy at 0 K is essentially the excess potential energy of the molecules near the surface.

The calculation is made by determining the primary contribution $E^{s'}$ to the surface energy, that of the two separate parts, holding all the atoms in fixed positions. The total energy is reduced by the rearrangement of the surface layer to its equilibrium position as

$$E^s = E^{s'} - E^{s''} \tag{VII-7}$$

Examination of Fig. VII-4 shows that the mutual interaction between a pair of planes occurs once if the planes are a distance a apart, twice for those $2a$ apart and so on. If the planes are labeled by the index l, as shown in the figure, the mutual potential energy is

$$2u' = \sum_{l \geq 1} l U(r) \tag{VII-8}$$

where u' is the surface energy per atom in the surface plane and $U(r)$ is the potential energy function in terms of interatomic spacing r.

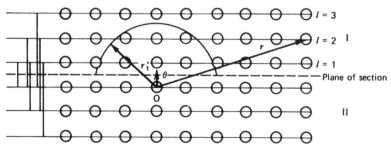

Fig. VII-4. Interactions across a dividing surface for a rare-gas crystal. (From Ref. 43.)

Various functions for $U(r)$ have been used; a classic form

$$U(r) = \lambda r^{-s} - \mu r^{-t} \qquad \text{(VII-9)}$$

recovers the Lennard-Jones potential if t is 12 and s is 6 corresponding to the van der Waals attraction. Experimental values for s and t can be obtained for rare gases from their virial coefficients. The summation indicated in Eq. VII-8 must be carried out over all interatomic distances. The distance from the origin to a point in the lattice is

$$d = (x^2 + y^2 + z^2)^{1/2}$$

where x, y, and z are given by $m_1 a$, $m_2 a$, and $m_3 a$; a is the side of the simple cubic unit cell; and the m's are integers. Thus

$$d = a(m_1^2 + m_2^2 + m_3^2)^{1/2} \qquad \text{(VII-10)}$$

Equation VII-8 then becomes

$$-2u' = \lambda a^{-s} \sum_{(m_1+m_2+m_3)\text{even},\, l \geq 1} \frac{l}{(m_1^2 + m_2^2 + m_3^2)^{s/2}}$$
$$- \mu a^{-t} \sum_{(m_1+m_2+m_3)\text{even},\, l \geq 1} \frac{l}{(m_1^2 + m_2^2 + m_3^2)^{t/2}} \qquad \text{(VII-11)}$$

Only even values of $m_1 + m_2 + m_3$ are used for the FCC lattice. The numerical values of these lattice sums are dependent on the exponents used for $U(r)$, and Eq. VII-11 may be written

$$-2u' = B_s \lambda a^{-s} - B_t \mu a^{-t} \qquad \text{(VII-12)}$$

Similarly, the energy of evaporation U_0 is given by

$$-2U_0 = \lambda a^{-s} \sum_{(m_1+m_2+m_3)\text{even}} \frac{1}{(m_1^2 + m_2^2 + m_3^2)^{s/2}}$$
$$- \mu a^{-t} \sum_{(m_1+m_2+m_3)\text{even}} \frac{1}{(m_1^2 + m_2^2 + m_3^2)^{t/2}} \qquad \text{(VII-13)}$$

or

$$-2U_0 = A_s\lambda a^{-s} - A_t\mu a^{-t} \tag{VII-14}$$

In these equations, the sums give twice the desired quantity because each atom is counted twice. In addition, the condition that the atoms be at their equilibrium distances gives, from differentiation of Eq. VII-14,

$$sA_s\lambda a^{-s} = tA_t\mu a^{-t} \tag{VII-15}$$

Eliminating λa^{-s} and μa^{-t} from Eqs. VII-2, VII-4, and VII-5,

$$u' = \frac{U_0(sB_t/A_t - tB_s/A_s)}{s - t} \tag{VII-16}$$

If the interaction between atoms that are not nearest neighbors is neglected, then the ratios B/A are each equal to the ratio of the number of nearest neighbors to a surface atom (across the dividing plane) to the number of nearest neighbors for an interior atom. The calculation then reduces to that given by Eq. III-15.

Returning to the complete calculation, $E^{s\prime}$ is then given by u' multiplied by the number of atoms per unit area in the particular crystal plane.

Benson and Claxton [44] calculated the lattice distortion term $E^{s\prime\prime}$ by allowing the first five layers to move outward until positions of minimum energy were found. The distortion dropped off rapidly; it was 3.5% for the first layer and only 0.04% for the fifth. Also the total correction to $E^{s\prime}$ was only about 1%. Their results using the Lennard-Jones 6-12 potential, Eq. VII-9, are given in Table VII-1. Shuttleworth [43] also calculated surface stresses using Eq. VII-5, obtaining negative values of about one-tenth of those for the surface energies.

Molecular dynamics calculations have been made on atomic crystals using a Lennard-Jones potential. These have to be done near the melting point in order for the iterations not to be too lengthy and have yielded density functions as one passes through the solid-vapor interface (see Ref. 45). The calculations showed considerable mobility in the surface region, amounting to the presence of a

TABLE VII-1
Surface Energies at 0 K for Rare-Gas Crystals

Rare Gas	$2a$ (Å)	E_{vap} (ergs/atom)	E^s (ergs/cm^2)			E^s_{liq} (ergs/cm^2)
			(100)	(110)	(111)	
Ne	4.52	4.08×10^{-14}	21.3	20.3	19.7	15.1
Ar	5.43	13.89	46.8	44.6	43.2	36.3
Kr	5.59	19.23	57.2	54.5	52.8	—
Xe	6.18	26.87	67.3	64.1	62.1	—

liquidlike surface layer. Surface free energies and stresses could be calculated as a function of temperature. In another type of molecular dynamics calculation, the outer layers of a Lennard-Jones crystal were pulse heated and the growth of a quasi-liquid layer tracked [46].

Molecular dynamics and density functional theory studies (see Section IX-2) of the Lennard-Jones 6-12 system determine the interfacial tension for the solid–liquid and solid–vapor interfaces [47–49]. The dimensionless interfacial tension $\gamma \sigma^2 / kT$, where σ is the Lennard-Jones molecular size, increases from about 0.83 for the solid–liquid interface to 2.38 for the solid–vapor at the triple point [49], reflecting the large energy associated with a solid–vapor interface.

C. Ionic Crystals

Surface Energies at 0 K. The complete history of the successive attempts to calculate the surface energy of simple ionic crystals is too long and complex to give here, and the following constitutes only a brief summary. The classic procedure entirely resembles that described for the rare-gas crystals, except that charged atoms occupy the sites, and a more complicated potential energy function must be used. For the face-centered-cubic alkali halide crystals, if some particular ion is selected as the origin, all ions with the same charge have coordinates for which $m_1 + m_2 + m_3$ is even, while ions with opposite charges have coordinates for which the sum is odd. As with the rare gas crystals, the first step is to calculate the mutual potential energy across two halves of a cleavage plane, and very similar lattice sums are involved, using, however, the appropriate potential energy function for like or unlike ions, as the case may be. Early calculations of this nature were made by Born and co-workers [50].

Refinements were made by Lennard-Jones, Taylor, and Dent [51–53], including an allowance for surface distortion. Their value of E^s for (100) planes of sodium chloride at 0 K was 77 ergs/cm^2. Subsequently, Shuttleworth obtained a value of 155 ergs/cm^2 [43].

Among several important developments, the potential function has been refined and one commonly used function takes the form [54]

$$U_{ij}(r) = \frac{z_i z_j e^2}{r} - \frac{C_{ij}}{r^6} - \frac{d_{ij}}{r^8} + b_i b_j e^{r/\rho} \qquad \text{(VII-17)}$$

where the first term on the right is the Coulomb energy between ions i and j, the next two terms are the dipole–dipole and dipole–quadrupole contributions to the van der Waals attraction and the last term is the short-range electronic repulsion. Although the van der Waals terms are unimportant contributors to the *total lattice energy*, the dipole-dipole (inverse r^6) term makes a 20–30% contribution to the *surface energy* [43, 55]. In the case of multivalent ion crystals, such as CaF$_2$, CaO, and so on, the results appear to be quite sensitive to the form of the potential function used. Benson and McIntosh [56] obtained values for the surface energy of (100) planes of MgO ranging from -298 to 1362 ergs/cm^2 depending on this variable!

Dynamic models for ionic lattices recognize explicitly the force constants between ions and their polarization. In "shell" models, the ions are represented as a shell and a core, coupled by a spring (see Refs. 57–59), and parameters are evaluated by matching bulk elastic and dielectric properties. Application of these models to the surface region has allowed calculation of surface vibrational modes [60] and LEED patterns [61–63] (see Section VIII-2).

It is important to evaluate the surface distortion associated with the assymetric field at the surface, a difficult task often simplified by assuming that distortion is limited to the direction normal to the plane [64, 65]. Benson and co-workers [65] calculated displacements for the first five planes in the (100) face of sodium chloride and found the distortion correction to $E^{s\prime}$ of about 100 ergs/cm^2 or about half of $E^{s\prime}$ itself. The displacements show a tendency toward ion pair formation, suggesting that lateral displacements to produce ion doublets should be considered [66]; however, other calculations yielded much smaller displacements [67].

The uncertainties in choice of potential function and in how to approximate the surface distortion contribution combine to make the calculated surface energies of ionic crystals rather uncertain. Some results are given in Table VII-2, but comparison between the various references cited will yield major discrepancies. Experimental verification is difficult (see Section VII-5). Qualitatively, one expects the surface energy of a solid to be distinctly higher than the surface tension of the liquid and, for example, the value of 212 ergs/cm^2 for (100)

TABLE VII-2

Calculated Surface Energies and Surface Stresses (ergs/cm^2)

Crystal[a]	(100) Planes		(110) Planes	
	E^s	τ	E^s	τ
LiF	480	1530	1047	407
LiCl	294	647	542	252
NaF	338	918	741	442
NaCl	212	415	425	256
NaBr	187	326	362	221
NaI	165	231	294	182
KCl	170	295	350	401
CsCl[b]	—	—	219	—
CaO[b]	509–879	—	—	—
MgO[b]	360–924	—	—	—
CaF$_2$[b]	—	—	1082	—

[a]From Ref. 67 unless otherwise indicated.
[b]From 68; see also Refs. 69 and 70.

planes of NaCl given in the table is, indeed, higher than the surface tension of the molten salt, 190 ergs/cm^2 [43]. Practically speaking, an extrapolation to room temperature of the value for the molten salt using semiempirical theory such as that of Ref. 71 is probably still the most reliable procedure.

Surface Stresses and Edge Energies. Some surface tension values, that is, values of the surface stress τ, are included in Table VII-2. These are obtained by applying Eq. VII-5 to the appropriate lattice sums. The calculation is very sensitive to the form of the lattice potential. Earlier calculations have given widely different results, including negative τ [43, 51, 52].

It is possible to set up the lattice sums appropriate for obtaining the edge energy, that is, the mutual potential energy between two cubes having a common edge. The dimensions of the edge energy k are ergs per centimeter, and calculations by Lennard-Jones and Taylor [51] gave a value of k of about 10^{-5} for the various alkali halides. A more recent estimate is 3×10^{-6} for the edge energy between (100) planes of sodium chloride [72]. As with surface stress values, these figures can only be considered as tentative estimates.

D. Molecular Crystals

A molecular crystal is one whose lattice sites are occupied by molecules, as opposed to atoms or monatomic ions, and the great majority of solids belong in this category. Included would be numerous salts, such as $BaSO_4$, involving a molecular ion most nonionic inorganic compounds, as, for example, CO_2 and H_2O and all organic compounds. Theoretical treatment is difficult. Relatively long-range van der Waals forces (see Chapter VI) are involved; surface distortion now includes surface reorientation effects. Surface states, called *polaritons*, have been studied by Rice and co-workers in the case of anthracene [73].

One molecular solid to which a great deal of attention has been given is ice. A review by Fletcher [74] cites calculated surface tension values of 100–120 ergs/cm^2 (see Ref. 75) as compared to an experimental measurement of 109 ergs/cm^2 [76]. There is much evidence that a liquidlike layer develops at the ice-vapor interface, beginning around $-35°C$ and thickening with increasing temperature [45, 74, 77, 78].

E. Metals

The calculation of the surface energy of metals has been along two rather different lines. The first has been that of Skapski, outlined in Section III-1B. In its simplest form, the procedure involves simply prorating the surface energy to the energy of vaporization on the basis of the ratio of the number of nearest neighbors for a surface atom to that for an interior atom. The effect is to bypass the theoretical question of the exact calculation of the cohesional forces of a metal and, of course, to ignore the matter of surface distortion.

Empirically, however, the results are reasonably accurate, and the approach is a very useful one. An application of it to various Miller index planes is given by

MacKenzie and co-workers [79]. Related is a statistical mechanical treatment by Reiss and co-workers [80] (see also Schonhorn [81]).

A related approach carries out lattice sums using a suitable interatomic potential, much as has been done for rare gas crystals [82]. One may also obtain the dispersion component to E^s by estimating the Hamaker constant A by means of the Lifshitz theory (Eq. VI-30), but again using lattice sums [83]. Thus for a FCC crystal the dispersion contributions are

$$E^s(100) = \frac{0.09184A}{a^2} \quad \text{and} \quad E^s(110) = \frac{0.09632A}{a^2} \tag{VII-18}$$

where a is the nearest-neighbor distance. The value so calculated for mercury is about half of the surface tension of the liquid and close to the γ^d of Fowkes (84; see Section IV-2).

The broken bond approach has been extended by Nason and co-workers (see Ref. 85) to calculate E^s as a function of surface composition for alloys. The surface free energy follows on adding an entropy of mixing term, and the free energy is then minimized.

The second model is a quantum mechanical one where free electrons are contained in a box whose sides correspond to the surfaces of the metal. The wave functions for the standing waves inside the box yield permissible states essentially independent of the lattice type. The kinetic energy corresponding to the rejected states leads to the surface energy in fair agreement with experimental estimates [86, 87].

Small metal clusters are also of interest because of their importance in catalysis. Despite the fact that small clusters should consist of mostly surface atoms, measurement of the photon ionization threshold for Hg clusters suggest that a transition from van der Waals to metallic properties occurs in the range of 20–70 atoms per cluster [88] and near-bulk magnetic properties are expected for Ni, Pd, and Pt clusters of only 13 atoms [89]! Theoretical calculations on Si_n and other semiconductors predict that the structure reflects the bulk lattice for 1000 atoms but the bulk electronic wave functions are not obtained [90]. Bartell and co-workers [91] study beams of molecular clusters with electron diffraction and molecular dynamics simulations and find new phases not observed in the bulk. Bulk models appear to be valid for their clusters of several thousand atoms (see Section IX-3).

As with rare gas and ionic crystals, there should be surface distortion in the case of metals. Burton and Jura [92] have estimated theoretically the increased interplanar spacing expected at the surface of various metals, and Abraham and Brundle [93] have reviewed models for surface segregation in alloys. Metals and other solids undergo surface melting at temperatures below the bulk melting point. One very good example of this occurs in the Pb(110) surface, where disorder appears as low as 75% of the melting temperature. Recent scanning tunneling microscopy and ion scattering studies (see Chapter VIII) show the melting of some surfaces and incomplete melting or faceting of other orientations [14]. The primary criteria for surface melting is the difference in free energy between a dry surface and one wetted by a thick liquid film. This can be expressed as the difference in surface tensions $\Delta\gamma = \gamma_{sv} - \gamma_{sl} - \gamma_{vl}$, where $\Delta\gamma < 0$ implies no surface melting while positive $\Delta\gamma$ indicates the formation of a liquid surface layer whose thickness diverges as the melting point is approached.

4. Factors Affecting the Surface Energies
and Surface Tensions of Actual Crystals

A. State of Subdivision

Surface chemists are very often interested in finely divided solids having high specific surface area, and it is worthwhile to consider briefly an illustrative numerical example of how surface properties should vary with particle size. We will refer the calculation to a 1-g sample of sodium chloride of density 2.2 g/cm^3 and with assumed surface energy of 200 ergs/cm^2 and edge energy 3×10^{-6} erg/cm. The original 1-g cube is now considered to be successively divided into smaller cubes, and the number of such cubes, their area and surface energies, and edge lengths and edge energies are summarized in Table VII-3.

It might be noted that only for particles smaller than about 1 μg or of surface area greater than a few square meters per gram does the surface energy become significant. Only for very small particles does the edge energy become important, at least with the assumption of perfect cubes.

B. Deviations from Ideality

The numerical illustration given above is so highly idealized that any agreement with experiment could hardly be more than coincidental. A number of *caveats* have been mentioned, including the inadequacies of theoretical calculations, the various surface distortions and reconstructions expected, surface

TABLE VII-3
Variation of Specific Surface Energy with Particle Size[a]

Side (cm)	Total Area (cm^2)	Total Edge (cm)	Surface Energy (ergs/g)	Edge Energy (ergs/g)
0.77	3.6	9.3	720	2.8×10^{-5}
0.1	28	550	5.6×10^3	1.7×10^{-3}
0.01	280	5.5×10^4	5.6×10^4	0.17
0.001	2.8×10^3	5.5×10^6	5.6×10^5	17
10^{-4} (1 μm)	2.8×10^4	5.5×10^8	5.6×10^6 (0.1 cal)	1.7×10^3
10^{-6} (100 Å)	2.8×10^6	5.5×10^{12}	5.6×10^8 (13 cal)	1.7×10^7 (0.4 cal)

[a] 1 g NaCl, $E^s = 200$ ergs/cm^2, $k = 3 \times 10^{-6}$ erg/cm.

melting, and effects of grinding and polishing on the surface condition. Some additional considerations should be noted when investigating crystal surfaces.

The *equilibrium* surface for a given crystal plane may not be smooth [94]. Surfaces undergo a transition from faceted to "rough" surfaces, known as the *roughening* transition. Below the roughening temperature a surface is faceted, and above this temperature thermal motion overcomes the interfacial energy and causes a faceted crystal to roughen. Above the roughening transition the interfacial width should diverge, and there should be a negligible dependence of the interfacial energy on lattice direction. This is analogous to thermal surface waves in liquids (see Sections II-8B and IV-3) [95].

Even if perfect cleavage planes are present, those that actually predominate may be determined by experimental conditions. Selective adsorption of some constituent may change the crystal habit by retarding the growth of certain planes (see Section IX-4); for example, sodium chloride crystallizes from urea solution in the form of octahedra instead of cubes and the shape of silver crystals in air is different from those annealed in nitrogen, due to the influence of adsorbed oxygen [39]. In addition to the well-known effects of oxygen on crystal growth, a recent study by Madey and co-workers shows the influence of thin metallic films on W(111) and Mo(111) crystals [96, 97]. The adsorption of a monolayer of a metal, such as Pd and Pt, having electronegativities greater than some critical value, causes W(111) and Mo(111) to restructure to a highly faceted pyramidal surface with the primary facets in the $\langle 211 \rangle$ direction. The dependence on electronegativity indicates the influence of electronic effects on the surface energies.

Actual crystal planes tend to be incomplete and imperfect in many ways. Nonequilibrium surface stresses may be relieved by surface imperfections such as overgrowths, incomplete planes, steps, and dislocations (see below) as illustrated in Fig. VII-5 [98, 99]. The distribution of such features depends on the past history of the material, including the presence of adsorbing impurities [100]. Finally, for sufficiently small crystals (1–10 nm in dimension), quantum-mechanical effects may alter various physical (e.g., optical) properties [101].

It is because of these complications, both theoretical and practical, that it is doubtful that calculated surface energies for solids will ever serve as more than a guide as to what to expect experimentally. Corollaries are that different preparations of the same substance may give different E^s values and that widely different experimental methods may yield different apparent E^s values for a given preparation. In this last connection, see Section VII-5 especially.

C. Fractal Surfaces

In the case of powders formed by grinding and particles formed by aggregation, surface roughness can be so extreme that, curiously, it can be treated by mathematical geometry (see Mandelbrot, Ref. 102; also Ref. 103). We can

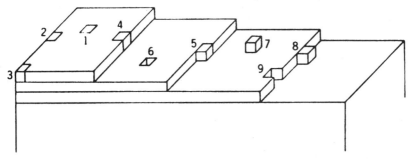

Fig. VII-5. Types of surface dislocations. (From Ref. 98.)

approach the subject in terms of a surface area measurement in which uniform spheres of radius r are packed as closely as possible on the surface. If the surface were a perfect plane, the number of spheres needed would be

$$N(r) = \frac{\mathcal{A}}{4\pi r^2} = Cr^{-2} \qquad \text{(VII-19)}$$

That is, the number of spheres needed would increase as r^{-2}. If, however, the surface is irregular, there will be pits and depressions into which only the smaller spheres can fit. The effect would be to make $N(r)$ increase more rapidly than as r^{-2}.

The situation can be illustrated in one dimension by considering the case of a line, Fig. VII-6a. If a ruler of length r is used, then $N(r)$ would vary as r^{-1}. We now make the line irregular in successive steps. The recipe for each step is to divide every line segment into three equal parts and then form an apex in the middle, consisting of two sides of the equilateral triangle whose base is the middle third of the original line segment. We now use a ruler to measure the length of the jagged line. Only 2 lengths of a ruler of length r_1 can be used, as shown in Fig. VII-6b, c, or d, and our measured "length" L_1 would be just $2r_1$. If, however, the ruler were of length $r_2 = \frac{1}{3}r_1$, a total of 8 lengths would be needed in the case of Fig. VII-6c or d, and the total length L_2 would now be $8r_2$ or $(\frac{8}{3})r_1$ or $(\frac{4}{3})L_1$. Turning to Fig. VII-6d, it would take 32 lengths of a ruler of length $r_3 = \frac{1}{3}r_2 = \frac{1}{9}r_1$, and L_3 would be $(\frac{16}{9})L_1$. Thus L increases as r decreases and $N(r)$ increases more rapidly than as r^{-1}.

The above recipe can be repeated indefinitely, and the mathematical result would be what is called a *self-similar* profile. That is, successive magnifications of a section (in this case by factors of 3) would give magnified "jagged" lines which could be superimposed exactly on the original one. In this limit, one finds for this case that

$$N(r) = Cr^{-D} \qquad \text{(VII-20)}$$

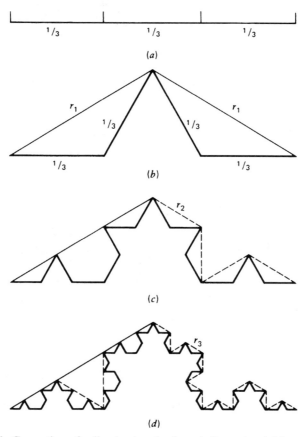

Fig. VII-6. Generation of a line having the fractal dimension 1.26 ... (see text).

where $D = \ln 4/\ln 3 = 1.262. \ldots$ Our jagged line is then said to have a *fractal* dimension of 1.262. ...

An analogous procedure can be applied to a plane surface. The surface can be "roughened" by the successive application of one or another recipe, just as was done for the line in Fig. VII-6. One now has a fractal or self-similar surface, and in the limit Eq. VII-20 again applies, or

$$N(\sigma) = C\sigma^{-D/2} \tag{VII-21}$$

where σ is the area of the object (e.g., cross-sectional area of the molecule) used to measure the apparent surface area and C is again a constant. The fractal dimension D will now be some number between 2 and 3.†

---

†Fractal lines and surfaces present a mathematical paradox. They are continuous and single valued yet have no definable slope or tangent plane!

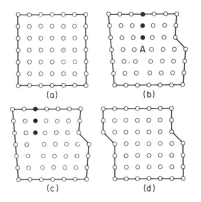

(a)   (b)

(c)   (d)

Fig. VII-7. Motion of an edge dislocation in a crystal undergoing slip deformation: (*a*) the undeformed crystal; (*b*, *c*) successive stages in the motion of the dislocation from right to left; (*d*) the undeformed crystal. (From Ref. 113 with permission.)

It turns out that many surfaces (and many line patterns such as shown in Fig. XV-7) conform empirically to Eq. VII-20 (or Eq. VII-21) over a significant range of r (or σ). Fractal surfaces thus constitute an extreme departure from ideal plane surfaces yet are amenable to mathematical analysis. There is a considerable literature on the subject, but Refs. 104–109 are representative. The fractal approach to adsorption phenomena is discussed in Section XVI-13.

D. Dislocations

Dislocation theory as a portion of the subject of solid-state physics is somewhat beyond the scope of this book, but it is desirable to examine the subject briefly in terms of its implications in surface chemistry. Perhaps the most elementary type of defect is that of an extra or interstitial atom—Frenkel defect [110]—or a missing atom or vacancy—Schottky defect [111]. Such point defects play an important role in the treatment of diffusion and electrical conductivities in solids and the solubility of a salt in the host lattice of another or different valence type [112]. Point defects have a thermodynamic basis for their existence; in terms of the energy and entropy of their formation, the situation is similar to the formation of isolated holes and erratic atoms on a surface. Dislocations, on the other hand, may be viewed as an organized concentration of point defects; they are lattice defects and play an important role in the mechanism of the plastic deformation of solids. Lattice defects or dislocations are not thermodynamic in the sense of the point defects; their formation is intimately connected with the mechanism of nucleation and crystal growth (see Section IX-4), and they constitute an important source of surface imperfection.

One type of dislocation is the *edge dislocation*, illustrated in Fig. VII-7. We imagine that the upper half of the crystal is pushed relative to the lower half, and the sequence shown is that of successive positions of the dislocation. An extra plane, marked as full circles, moves through the crystal until it emerges at the left. The process is much like moving a rug by pushing a crease in it.

The dislocation may be characterized by tracing a counterclockwise circuit around point A in Fig. VII-7b, counting the same number of lattice points in the plus and minus directions along each axis or row. Such a circuit would close if the crystal were perfect, but if a dislocation is present, it will not, as illustrated in the figure. This circuit is known as a *Burgers* circuit [114]; its failure to close distinguishes a dislocation from a point imperfection. The ends of the circuit define a vector, the *Burgers vector b*, and the magnitude and angle of the Burgers vector are used to define the magnitude and type of a dislocation.

The second type of dislocation is the *screw dislocation*, illustrated in Fig. VII-8a (from Frank [115]) and in Fig. VII-8b; each cube represents an atom or lattice site. The geometry of this may be imagined by supposing that a block of rubber has been sliced part way through and one section bent up relative to the other one. The crystal has a single plane, in the form of a spiral ramp; the screw dislocation can be produced by a slip on *any* plane containing the dislocation line AB—Fig. VII-8b. The distortion around a screw dislocation is mostly shear in nature, as suggested in Fig. VII-8a by showing unit cells as cubes displaced relative to one another in the direction of the slip vector. Combinations of screw and edge dislocations may also occur, of course. A photomicrograph of a carborundum crystal is shown in Fig. VII-9, illustrating spiral growth patterns [116].

The density of dislocations is usually stated in terms of the number of dislocation lines intersecting unit area in the crystal; it ranges from 10^8 cm^{-2} for "good" crystals to 10^{12} cm^{-2} in cold-worked metals. Thus, dislocations are separated by 10^2–10^4 Å, or every crystal grain larger than about 100 Å will have dislocations on its surface; one surface atom in a thousand is apt to be near a dislocation. By elastic theory, the increased potential energy of the lattice near

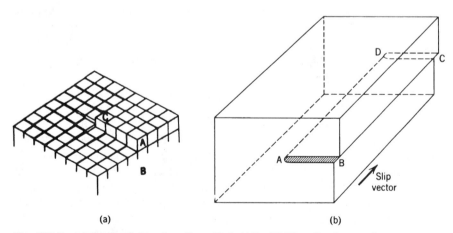

(a)   (b)

Fig. VII-8. (*a*) Screw dislocation (from Ref. 115). (*b*) The slip that produces a screw-type dislocation. Unit slip has occurred over *ABCD*. The screw dislocation *AD* is parallel to the slip vector. (From W. T. Read, Jr., *Dislocations in Crystals*, McGraw-Hill, New York, 1953, p. 15.)

Fig. VII-9. Screw dislocation in a carborundum crystal. (From Ref. 116.)

a dislocation is proportional to $|b|^2$. The core or dislocation line is severely strained, and the chemical potential of the material in it may be sufficiently high to leave it hollow. Frank [117] related the rigidity modulus μ, the surface tension, and the Burgers vector b by

$$r = \frac{\mu b^2}{8\pi^2 \gamma} \qquad \text{(VII-22)}$$

where r is the radius of the hollow cylinder. A crystal grown in a solvent medium may be especially prone to have hollow dislocations because of the probably relatively low solid-liquid interfacial tension. Adsorption of a gas may sufficiently lower γ to change a dislocation that emerges flush with the surface into a pit [100]. Etching of crystals tends to remove material preferentially from dislocation sites, producing etch pits, and the production of etch pits is in fact a way of seeing and counting dislocations by microscopy. In general, then, the surface sites of dislocations can be numerous enough to mar seriously the uniformity of a surface and in a way that is sensitive to past history and that may interact with the very surface phenomenon being studied, such as adsorption; in addition, surface dislocation appears to play a major role in surface kinetic processes such as crystal growth and catalyzed reactions.

The discovery of perfect geodesic dome closed structures of carbon, such as C_{60} has led to numerous studies of so-called Buckminster fullerene. Dislocations are important features of the structures of nested fullerenes also called onion skin, multilayered or Russian doll fullerenes. A recent theoretical study [118] shows that these defects serve to relieve large inherent strains in thick-walled nested fullerenes such that they can show faceted shapes.

5. Experimental Estimates of Surface Energies and Free Energies

There is a rather limited number of methods for obtaining experimental surface energy and free energy values, and many of them are peculiar to special solids or situations. The only general procedure is the rather empirical one of estimating a solid surface tension from that of the liquid. Evidence from a few direct measurements (see Section VII-1A) and from nucleation studies (Section IX-3) suggests that a solid near its melting point generally has a surface tension 10–20% higher than the liquid–about in the proportion of the heat of sublimation to that of liquid vaporization—and a value estimated at the melting point can then be extrapolated to another temperature by means of an equation such as III-10.

A. Methods Depending on the Direct
· Manifestation of Surface Tensional Forces

It was mentioned during the discussion on sintering (Section VII-1A) that some bulk flow can occur with solids near their melting point; solids may act like viscous liquids, in that their strain rate is proportional to the applied stress. As shown in Fig. VII-10, the strain rate (fractional elongation per unit time) versus stress (applied load per unit area) for 1-mil gold wires near 1000°C is nearly linear at small loads [119]. The negative strain rate at zero load is considered to be due to the surface tension of the metal and, therefore, the stress such as to give zero strain rate must just balance the surface tensional force along the circumference of the wire. Copper near its melting point has a surface tension of 1370 dyn/cm, while gold at 1000°C has values of 1300–1700 by this technique [1, 119]. Wire heated for long times at low stresses develops interesting waists and bulges analogous to the instability in liquid columns described in Section II-3. Solid paraffin has a surface tension of about 50 dyn/cm at 50°C, compared to that of about 25 dyn/cm for the liquid state [120].

An indirect estimate of surface tension may be obtained from the change in lattice parameters of small crystals such as magnesium oxide and sodium chloride owing to surface tensional compression [121]; however, these may represent nonequilibrium surface stress rather than surface tension [68]. Surface stresses may produce wrinkling in harder materials [122].

A direct measurement of surface tension is sometimes possible from the work of cleaving a crystal. Mica, in particular, has such a well-defined cleavage plane that it can be split into large sheets of fractional millimeter thickness. Orowan

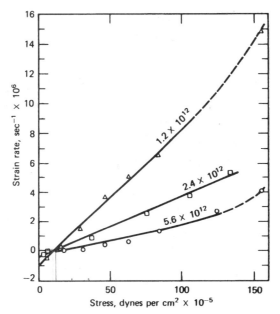

Fig. VII-10. Strain rate versus stress for 1-mil gold wire at various temperatures: △, 1020°C; □, 970°C; ○, 920°C. (From Ref. 119.)

[123], in reviewing the properties of mica, gives the equation

$$2\gamma = \frac{T^2 x}{2E} \qquad \text{(VII-23)}$$

where T is the tension in the sheet of thickness x that is being split and E is the modulus of elasticity. Essentially, one balances the surface tension energy for the two surfaces being created against the elastic energy per square centimeter. The process is not entirely reversible, so the values obtained, again, are only approximate. Interestingly, they were 375 ergs/cm^2 in air and 4500 ergs/cm^2 in a vacuum, the former apparently representing the work of cohesion of two surfaces with adsorbed water on them.

Gilman [124] and Westwood and Hitch [135] have applied the cleavage technique to a variety of crystals. The salts studied (with cleavage plane and best surface tension value in parentheses) were LiF (100, 340), MgO (100, 1200), CaF$_2$ (111, 450), BaF$_2$ (111, 280), CaCO$_3$ (001, 230), Si (111, 1240), Zn (0001, 105), Fe (3% Si) (100, about 1360), and NaCl (100, 110). Both authors note that their values are in much better agreement with a very simple estimate of surface energy by Born and Stern in 1919, which used only Coulomb terms and a hard-sphere repulsion. In more recent work, however, Becher and Freiman [126] have reported distinctly higher values of γ, the "critical fracture energy."

Westwood and Hitch suggest, incidentally, that the cleavage experiment, not being fully reversible, may give only a bond-breaking or nearest-neighbor type of surface energy with little contribution from surface distortion.

B. Surface Energies from Heats of Solution

The illustrative data presented in Table VII-3 indicate that the total surface energy may amount to a few tenths of a calorie per gram for particles on the order of 1 μm in size. When the solid interface is destroyed, as by dissolving, the surface energy appears as an extra heat of solution, and with accurate calorimetry it is possible to measure the small difference between the heat of solution of coarse and of finely crystalline material.

An excellent example of work of this type is given by the investigations of Benson and co-workers [127, 128]. They found, for example, a value of E^s = 276 ergs/cm^2 for sodium chloride. Accurate calorimetry is required since there is only a few calories per mole difference between the heats of solution of coarse and finely divided material. The surface area of the latter may be determined by means of the BET gas adsorption method (see Section XVII-5).

Brunauer and co-workers [129, 130] found values of E^s of 1310, 1180, and 386 ergs/cm^2 for CaO, Ca(OH)$_2$ and tobermorite (a calcium silicate hydrate). Jura and Garland [131] reported a value of 1040 ergs/cm^2 for magnesium oxide. Patterson and co-workers [132] used fractionated sodium chloride particles prepared by a volatilization method to find that the surface contribution to the low-temperature heat capacity varied approximately in proportion to the area determined by gas adsorption. Questions of equilibrium arise in these and adsorption studies on finely divided surfaces as discussed in Section X-3.

C. Relative Surface Tensions from Equilibrium Crystal Shapes

We noted in Section VII-2B that, given the set of surface tension values for various crystal planes, the Wulff theorem allowed the construction of the equilibrium or minimum free energy shape. This concept may be applied in reverse; small crystals will gradually take on their equilibrium shape upon annealing near their melting point and likewise, small air pockets in a crystal will form equilibrium-shaped voids. The latter phenomenon offers the possible advantage that adventitious contamination of the solid–air interface is less likely.

Nelson et al. [34] determined from void shapes that the ratio $\gamma_{100}/\gamma_{110}$ was 1.2, 0.98 and 1.14 for copper at 600°C, aluminum at 550°C, and molybdenum at 2000°C, respectively, and 1.03 for $\gamma_{100}/\gamma_{111}$ for aluminum at 450°C. Metal tips in field emission studies (see Section VIII-2C) tend to take on an equilibrium faceting into shapes agreeing fairly well with calculations [133].

Sundquist [35], studying small crystals of metals, noted a great tendency for rather rounded shapes and concluded that for such metals as silver, gold, copper, and iron there was not more than about 15% variation in surface tension between different crystal planes, indicating that these metals were solidified at a temperature above their roughening transition (see Section VII-4B) or where surface melting was occurring (Section VII-3E).

D. Dependence of Other Physical Properties on Surface Energy Changes at a Solid Interface

A few additional phenomena are worth mentioning. First, there are several that are not included here because they are discussed in more detail elsewhere. Thus solid-liquid interfacial tensions can be estimated from solubility measurements (Section X-2) and also from nucleation studies (Section IX-3). In addition, changes or differences in the interfacial free energy at a solid-vapor interface can be obtained from gas adsorption studies (Section X-3) and the corresponding differences in surface energies from heat of immersion data (Section X-3). Contact angle measurements give a difference between a solid-liquid and a solid-vapor interfacial free energy (Section X-4), and Section V-6 shows how to estimate the effect of potential on the free energy of a solid-electrolyte solution interface.

Many solids show marked swelling as a result of the uptake of a gas or a liquid. In certain cases involving the adsorption of a vapor by a porous solid, a linear relationship exists between the percentage of linear expansion of the solid and the film pressure of the adsorbed material [134, 135].

Most solid surfaces are marred by small cracks, and it appears clear that it is often because of the presence of such surface imperfections that observed tensile strengths fall below the theoretical ones. For sodium chloride, the theoretical tensile strength is about 200 kg/mm^2 [136], while that calculated from the work of cohesion would be 40 kg/mm^2 [137], and actual breaking stresses are a hundreth or a thousandth of this, depending on the surface condition and crystal size. Coating the salt crystals with a saturated solution, causing surface deposition of small crystals to occur, resulted in a much lower tensile strength but not if the solution contained some urea.

Of course, it is common knowledge (but often more easily said than done) that glass cracks easily at a scratch line, but in addition, a superficially flawless glass surface is marred by many fine cracks [138, 139], and very freshly drawn glass shows a high initial tensile strength that decreases rapidly with time. In general, any reduction in the solid interfacial tension should also reduce the observed tensile strength. Thus the tensile strength of glass in various liquids decreases steadily with decreasing polarity of the liquid [140]. The effect of adsorbed species on fracture strength has been discussed by Whitesides from a bonding point of view [141], and the general subject has been reviewed by El-Shall and Somasundaran [142].

Rehbinder and co-workers were pioneers in the study of environmental effects on the strength of solids [144]. As discussed by Frumkin and others [143–145], the measured hardness of a metal immersed in an electrolyte solution varies with applied potential in the manner of an electrocapillary curve (see Section V-7). A dramatic demonstration of this so-called Rehbinder effect is the easy deformation of single crystals of tin and of zinc if the surface is coated with an oleic acid monolayer [144].

6. Reactions of Solid Surfaces

Perhaps the simplest case of reaction of a solid surface is that where the reaction product is continuously removed, as in the dissolving of a soluble salt in water or that of a metal or metal oxide in an acidic solution. This situation is discussed in Section XVII-2 in connection with surface area determination.

More complex in its kinetics is the reaction of a solid with a liquid or gas to give a second solid as, for example,

$$CuSO_4 \cdot 5H_2O = CuSO_4 \cdot H_2O + 4H_2O \qquad \text{(VII-24)}$$

$$CaCO_3 = CaO + CO_2 \qquad \text{(VII-25)}$$

$$2\,Cu + COS\,(g) = Cu_2S + CO \qquad \text{(VII-26)}$$

$$Ag_2O + CO_2 = Ag_2CO_3 \qquad \text{(VII-27)}$$

$$6\,Fe_2O_3 = 4\,Fe_3O_4 + O_2 \qquad \text{(VII-28)}$$

The usual situation, true for the first three cases, is that in which the reactant and product solids are mutually insoluble. Langmuir [146] pointed out that such reactions undoubtedly occur at the linear interface between the two solid phases. The rate of reaction will thus be small when either solid phase is practically absent. Moreover, since both forward and reverse rates will depend on the amount of this common solid-solid interface, its extent cancels out at equilibrium, in harmony with the thermodynamic conclusion that for the reactions such as Eqs. VII-24 to VII-27 the equilibrium constant is given simply by the gas pressure and does not involve the amounts of the two solid phases.

Qualitative examples abound. Perfect crystals of sodium carbonate, sulfate, or phosphate may be kept for years without efflorescing, although if scratched, they begin to do so immediately. Too strongly heated or "burned" lime or plaster of Paris takes up the first traces of water only with difficulty. Reactions of this type tend to be autocatalytic. The initial rate is slow, due to the absence of the necessary linear interface, but the rate accelerates as more and more product is formed. See Refs. 147–153 for other examples. Ruckenstein [154] has discussed a kinetic model based on nucleation theory. There is certainly evidence that patches of product may be present, as in the oxidation of Mo(100) surfaces [155], and that surface defects are important [156]. There may be catalysis; thus reaction VII-27 is catalyzed by water vapor [157]. A *topotactic* reaction is one where the product or products retain the external crystalline shape of the reactant crystal [158]. More often, however, there is a complicated morphology with pitting, cracking, and pore formation, as with calcium carbonate [159].

In the case of reaction VII-28, the reactant and product are mutually soluble. Langmuir argued that in this case, escape of oxygen is easier from bulk Fe_2O_3

than from such units as an Fe_2O_3–Fe_3O_4 interface. The reaction therefore proceeds by a gradual escape of oxygen randomly from all portions of the Fe_2O_3, thus producing a solid solution of the two oxides.

The kinetics of reactions in which a new phase is formed may be complicated by the interference of that phase with the ease of access of the reactants to each other. This is the situation in corrosion and tarnishing reactions. Thus in the corrosion of a metal by oxygen the increasingly thick coating of oxide that builds up may offer more and more impedance to the reaction. Typical rate expressions are the logarithmic law,

$$y = k_1 \log(k_2 t + k_3) \qquad \text{(VII-29)}$$

where y denotes the thickness of the film, the parabolic law,

$$y^2 = k_1 t + k_2 \qquad \text{(VII-30)}$$

and of course, the simple law for an unimpeded reaction,

$$y = kt \quad \text{or} \quad y = k_1 + k_2 t \qquad \text{(VII-31)}$$

Thus the oxidation of light metals such as sodium, calcium, or magnesium follows Eq. VII-31, the low-temperature oxidation of iron follows Eq. VII-29, and the high-temperature oxidation follows Eq. VII-30. The controlling factor seems to be the degree of protection offered by the coating of oxide [160]. If, as in the case of the light metals, the volume of the oxide produced is less than that of the metal consumed, then the oxide tends to be porous and nonprotective, and the rate, consequently, is constant. Evans [160] suggests that the logarithmic equation results when there is discrete mechanical breakdown of the film of product. In the case of the heavier metals, the volume of the oxide produced is greater than that of the metal consumed, and although this tends to give a dense protective coating, if the volume difference is too great, flaking or other forms of mechanical breakdown may occur as a result of the compressional stress produced. There may be more complex behavior. In the case of brass alloyed with tin, corrosion appears first to remove surface Zn, but the accumulated Sn then forms a protective layer [161].

Studies have been made on the rate of growth of oxide films on different crystal faces of a metal using ellipsometric methods. The rate was indeed different for (100), (101), (110), and (311) faces of copper [162]; moreover, the film on a (311) surface was anisotropic in that its apparent thickness varied with the angle of rotation about the film normal.

The rate law may change with temperature. Thus for reaction VII-30 the rate was paralinear (i.e., linear after an initial curvature) below about 470°C and parabolic above this temperature [163], presumably because the CuS_2 product was now adherent. Non-simple rate laws are reported for the oxidation of iron coated with an iron-silicon solid solution [164] and that of zirconium coated with carbonitride [165]; several layers of material are present during these reactions.

The parabolic law may be derived on the basis of a uniform coating of product being formed through which a rate-controlling diffusion of either one or both of the reactants must occur [166]. For example, in the case of the reaction between liquid sulfur and silver, it was determined that silver diffused through the Ag_2S formed and that the reaction occurred at the Ag_2S–S rather than at the Ag–Ag_2S boundary [167]. The rate constant k_1 in the parabolic rate law generally shows an exponential dependence on temperature related to that of the rate-determining diffusion process. As emphasized by Gomes [168], however, activation energies computed from the temperature dependence of rate constants are meaningless except in terms of a specific mechanism.

The foregoing discussion indicates how corrosion products may form a protective coating over a metal and thus protect it from extensive reaction. Perhaps the best illustration of this is the fact that aluminum, which should react vigorously with water, may be used to make cooking utensils. If, however, the coating is impaired, then the expected reactivity of the metal is observed. Thus an amalgamated aluminum surface does react with water, and aluminum foil wet with mercurous nitrate or chloride solution and rolled up tightly may actually reach incandescence as a result of the heat of the reaction. For some representative studies in the field of passivation and corrosion inhibition see Refs. 169 and 170 by Hackerman and co-workers; a cation monolayer model is discussed by Griffin [171]. The protective film need not derive from the corrosion products but may be provided by added adsorbates, as in the use of imidazoles in the inhibition of copper corrosion [172]. Complicated electrical phenomenon may occur in a corroding system, such as current oscillations in the case of an iron anode [173]. Photoinduced corrosion, especially of semiconductors, is of concern in connection with solar energy conversion systems; see Refs. 174 and 175, and Section V-8C.

An important problem involving reactions of solid surfaces is the deterioration of wall paintings and stone monuments. A significant source of damage to *frescoes* is the presence of water, which can degrade the binding media, alter pigments, and promote microbial growth and the collection of dirt and pollutants. While environmental factors such as relative humidity and surface temperature can be controlled by conservators, the presence of hygroscopic salts facilitate water absorption. Deliquescent salts, those that absorb water to produce a saturated solution at the surface of the painting, are especially problematic. Ferroni and co-workers have studied the cycles of deliquescence and recrystallization in calcium nitrate, $Ca(NO_3)_2 \cdot 4H_2O$ [176, 177]. Their understanding of the reactions at solid surfaces of paintings has helped define conditions to minimize their wetting by water vapor and subsequent deterioration. Stone sculptures and buildings decay through a process where the calcium carbonate reacts with sulfur oxides found in atmospheric pollutants to produce a more soluble salt. Gypsum (calcium sulfate dihydrate) produces black scabs on stone surfaces which not only ruin the appearance of the object but weaken the material and eventually flake off to expose a fresh surface to continue the cycle of decay (see Ref. [178]).

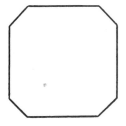

Fig. VII-11. Wulff construction of a two-dimensional crystal.

7. Problems

1. The exponents n and m in Eq. VII-1 are expected to be 1 and 3, respectively, under some conditions. Assuming spheres and letting $V_r = V/V_0$, derive the rate law $dV_r/dt = f(V_r)$. Expound on the peculiar nature of this rate law.

2. The surface tensions for a certain cubic crystalline substance are $\gamma_{100} = 160$ ergs/cm^2, $\gamma_{110} = 140$ ergs/cm^2, and $\gamma_{210} = \gamma_{120} = 140$ ergs/cm^2. Make a Wulff construction and determine the equilibrium shape of the crystal in the xy plane. (If the plane of the paper is the xy plane, then all the ones given are perpendicular to the paper, and the Wulff plot reduces to a two-dimensional one. Also, $\gamma_{100} = \gamma_{010}$, etc.)

3. Consider a hypothetical two-dimensional crystal having a simple square unit cell. Given that γ_{10} is 300 ergs/cm and γ_{11} is 100 ergs/cm, make a Wulff construction to show whether the equilibrium crystal should consist of (10)-type or (11)-type edges. Calculate directly the edge energy for both cases to verify your conclusion; base your calculation on a crystal of unit area.

4. Referring to Problem 3, what should the ratio γ_{10}/γ_{11} be if the equilibrium crystal is to be a regular octahedron, that is, to have (10) and (11) edges of equal length?

5. Referring to Fig. VII-2, assume the surface tension of (10) type planes to be 400 ergs/cm. (a) For what surface tension value of (11) type planes should the stable crystal habit just be that of Fig. VII-2a and (b) for what surface tension value of (11) type planes should the stable crystal habit be just that of Fig. VII-2b? Explain your work.

6. An enlarged view of a crystal is shown in Fig. VII-11; assume for simplicity that the crystal is two-dimensional. Assuming equilibrium shape, calculate γ_{11} if γ_{10} is 275 dyn/cm. Crystal habit may be changed by selective adsorption. What percentage of reduction in the value of γ_{10} must be effected (by, say, dye adsorption selective to the face) in order that the equilibrium crystal exhibit only (10) faces? Show your calculation.

7. Bikerman [179] has argued that the Kelvin equation should not apply to crystals, that is, in terms of increased vapor pressure or solubility of small crystals. The reasoning is that perfect crystals of whatever size will consist of plane facets whose radius of curvature is therefore infinite. On a molecular scale, it is argued that local condensation-evaporation equilibrium on a crystal plane should not be affected by the extent of the plane, that is, the crystal size, since molecular forces are short range. This conclusion is contrary to that in Section VII-2C. Discuss the situation. The derivation of the Kelvin equation in Ref. 180 is helpful.

8. According to Beamer and Maxwell [181], the element Po has a simple cubic

structure with $a = 3.34$ Å. Estimate by the Harkins' method the surface energy for (100) and (111) planes. Take the energy of vaporization to be 50 kcal/mol.

9. Make the following approximate calculations for the surface energy per square centimeter of solid krypton (nearest-neighbor distance 3.97 Å), and compare your results with those of Table VII-1. (*a*) Make the calculations for (100), (110), and (111) planes, considering only nearest-neighbor interactions. (*b*) Make the calculation for (100) planes, considering all interactions within a radius defined by the sum $m_1^2 + m_2^2 + m_3^2$ being 8 or less.

10. Calculate the surface energy at 0 K of (100) planes of radon, given that its energy of vaporization is 35×10^{-14} erg/atom and that the crystal radius of the radon atom is 2.5 Å. The crystal structure may be taken to be the same as for other rare gases. You may draw on the results of calculations for other rare gases.

11. Calculate the percentage of atoms that would be surface atoms in a particle containing 125 atoms; 1000 atoms. Assume simple cubic geometry.

12. Using the nearest-neighbor approach, and assuming a face-centered-cubic structure for the element, calculate the surface energy per atom for (*a*) an atom sitting on top of a (111) plane and (*b*) against a step of a partially formed (111) plane (e.g., position 8 in Fig. VII-5). Assume the energy of vaporization to be 60 kcal/mol.

13. Taking into account only nearest-neighbor interactions, calculate the value for the line or edge tension k for solid argon at 0 K. The units of k should be in ergs per centimeter.

14. Calculate the Hamaker constant for Ar crystal, using Eq. VII-18. Compare your value with the one that you can estimate from the data and equations of Chapter VI.

15. Metals A and B form an alloy or solid solution. To take a hypothetical case, suppose that the structure is simple cubic, so that each interior atom has six nearest neighbors and each surface atom has five. A particular alloy has a bulk mole fraction $x_A = 0.50$, the side of the unit cell is 4.0 Å, and the energies of vaporization E_A and E_B are 30 and 35 kcal/mol for the respective pure metals. The A—A bond energy is E_{AA} and the B—B bond energy is E_{BB}; assume that $E_{AB} = \frac{1}{2}(E_{AA} + E_{BB})$. Calculate the surface energy E^s as a function of surface composition. What should the surface composition be at 0 K? In what direction should it change on heating, and why?

16. Show that D is indeed $\ln 4/\ln 3$ for the self-similar profile of Fig. VII-6.

17. A fractal surface of dimension $D = 2.5$ would show an apparent area $\mathcal{A}_{app}$ that varies with the cross-sectional area σ of the adsorbate molecules used to cover it. Derive the equation relating $\mathcal{A}_{app}$ and σ. Calculate the value of the constant in this equation for $\mathcal{A}_{app}$ in cm²/g and σ in Å²/molecule if 1 μmol of molecules of 18 Å² cross section will cover the surface. What would $\mathcal{A}_{app}$ be if molecules of 40 Å² were used?

18. Calculate the surface tension of gold at 970°C. Take your data from Fig. VII-10.

19. The excess heat of solution of sample A of finely divided sodium chloride is 18 cal/g, and that of sample B is 12 cal/g. The area is estimated by making a microscopic count of the number of particles in a known weight of sample, and it is found that sample A contains 22 times more particles per gram than does sample B. Are the specific surface energies the same for the two samples? If not, calculate their ratio.

20. In a study of tarnishing the parabolic law, Eq. VII-30, is obeyed, with $k_2 = 0$. The film thickness y, measured after a given constant elapsed time, is determined in a

series of experiments, each at a different temperature. It is found that y so measured varies exponentially with temperature, and from $d \ln y / d(1/T)$ and apparent activation energy of 10 kcal/mol is found. If k_1 in Eq. VII-30 is in fact proportional to a diffusion coefficient, show what is the activation energy for the diffusion process.

General References

D. Avnir, ed., *The Fractal Approach to Heterogeneous Chemistry*, Wiley, New York, 1989.

D. D. Eley, ed., *Adhesion*, The Clarendon Press, Oxford, 1961. E. Passaglia, R. R. Stromberg, and J. Kruger, eds., *Ellipsometry in the Measurement of Surfaces and Thin Films*, National Bureau of Standards Miscellaneous Publication 256, Washington, DC, 1964.

R. C. Evans, *An Introduction to Crystal Chemistry*, Cambridge University Press, Cambridge, UK, 1966.

P. P. Ewald and H. Juretschke, *Structure and Properties of Solid Surfaces*, University of Chicago Press, Chicago, 1953.

R. S. Gould and S. J. Gregg, *The Surface Chemistry of Solids*, Reinhold, New York, 1951.

W. T. Read, Jr., *Dislocations in Crystals*, McGraw-Hill, New York, 1953. *Solid Surfaces*, ACS Symposium Series No. 33, American Chemical Society, Washington, DC, 1961.

G. A. Somorjai, *Principles of Surface Chemistry*, Prentice-Hall, Englewood Cliffs, NJ, 1972.

Textual References

1. H. Udin, A. J. Shaler, and J. Wulff, *J. Met.*, **1**(2); *Trans. AIME*, 186 (1949).

2. B. Chalmers, R. King, and R. Shuttleworth, *Proc. Roy. Soc.* (London), **A193,** 465 (1948).

3. G. C. Kuczynski, *J. Met.*, **1,** 96 (1949).

4. P. V. Hobbs and B. J. Mason, *Phil. Mag.*, **9,** 181 (1964).

5. H. H. G. Jellinek and S. H. Ibrahim, *J. Colloid Interface Sci.*, **25,** 245 (1967).

6. O. J. Whittemore and J. A. Varela, *Advances in Ceramics*, Vol. 10, American Ceramic Society, Washington, DC, 1985, p. 583.

7. J. S. Chappel, T. A. Ring, and J. D. Birchall, *J. Appl. Phys.*, **60,** 383 (1986).

8. E. Ruckenstein and D. B. Dadyburjor, *Rev. in Chem. Eng.*, **1,** 251 (1983).

9. A. Bellare and D. B. Dadyburjor, *AIChE J.*, **33,** 867 (1987).

10. H. N. Hersh, *J. Am. Chem. Soc.*, **75,** 1529 (1953).

11. G. M. Rosenblatt, *Acc. Chem. Res.*, **9,** 169 (1976).

12. G. Cohen and G. C. Kuczynski, *J. Appl. Phys.*, **21,** 1339 (1950).

13. E. Menzel, *Z. Phys.*, **132,** 508 (1952).

14. J. W. M. Frenken, H. M. van Pinxteren, and L. Kuipers, *Surf. Sci.*, **283,** 283 (1993).

15. N. H. Fletcher, *Phil. Mag.*, **7**(8), 255 (1962).

16. I. J. Lin and P. Somasundaran, *Powder Techn.*, **6**, 171 (1972).

17. L. H. Germer, in *Frontiers in Chemistry*, Vol. 4, R. E. Burk and O. Grummitt, eds., Interscience, New York, 1945.

18. See L. H. Germer and A. U. MacRae, *J. Appl. Phys.*, **33**, 2923 (1962).

19. G. Beilby, *Aggregation and Flow of Solids*, Macmillan, New York, 1921.

20. R. C. French, *Proc. Roy. Soc.* (London), **A140**, 637 (1933).

21. F. P. Bowden and D. Tabor, *The Friction and Lubrication of Solids*, The Clarendon Press, Oxford, 1950.

22. S. J. Gregg, *The Surface Chemistry of Solids*, Reinhold, New York, 1951.

23. J. J. Trillat, *CR*, **224**, 1102 (1947).

24. M. L. White, in *Clean Surfaces, Their Preparation and Characterization for Interfacial Studies*, G. Goldfinger, ed., Marcel Dekker, New York, 1970.

25. J. W. Gibbs, *The Collected Works of J. W. Gibbs*, Longmans, Green, New York, 1931, p. 315.

26. R. Shuttleworth, *Proc. Phys. Soc.* (London), **63A**, 444 (1950).

27. J. W. Cahn, *Acta Met.*, **28**, 1333 (1980).

28. G. Wulff, *Z. Krist.*, **34**, 449 (1901).

29. G. C. Benson and D. Patterson, *J. Chem. Phys.*, **23**, 670 (1955).

30. M. Drechsler and J. F. Nicholas, *J. Phys. Chem. Solids*, **28**, 2609 (1967).

31. C. Herring, *Phys. Rev.*, **82**, 87 (1951).

32. J. E. Taylor and J. W. Cahn, *Science*, **233**, 548 (1986).

33. H. G. Müller, *Z. Phys.*, **96**, 307 (1935).

34. R. S. Nelson, D. J. Mazey, and R. S. Barnes, *Phil. Mag.*, **11**, 91 (1965).

35. B. E. Sundquist, *Acta Met.*, **12**, 67, 585 (1964).

36. J. E. Avron, J. E. Taylor, and R. K. P. Zia, *J. Stat. Phys.*, **33**, 493 (1983).

37. R. K. P. Zia, J. E. Avron, and J. E. Taylor, *J. Stat. Phys.*, **50**, 727 (1988).

38. D. W. M. Marr and A. P. Gast, *Phys. Rev. E*, **52**, 4058 (1995).

39. W. M. Mullins, *Phil. Mag.*, **6**, 1313 (1961).

40. W. J. Dunning, in *Structure of Surfaces*, D. Fox, M. M. Labes, and A. Weissberg, eds., Wiley-Interscience, New York, 1963.

41. W. D. Harkins, *J. Chem. Phys.*, **10**, 268 (1942).

42. F. F. Abraham and I. P. Batra, *Surf. Sci.*, **163**, L752 (1985).

43. R. Shuttleworth, *Proc. Phys. Soc.* (London), **A62**, 167 (1949); **A63**, 444 (1950).

44. G. C. Benson and T. A. Claxton, *Phys. Chem. Solids*, **25**, 367 (1964).

45. J. Q. Broughton and G. H. Gilmer, *J. Phys. Chem.*, **91**, 6347 (1987).

46. F. F. Abraham and J. Q. Broughton, *Phys. Rev. Lett.*, **56**, 734 (1986).

47. W. A. Curtin, *Phys. Rev. B*, **39**, 6775 (1989).

48. J. Q. Broughton and G. H. Gilmer, *J. Chem. Phys.*, **84**, 5759 (1986).

49. D. W. Marr and A. P. Gast, *Phys. Rev. E*, **47**, 1212 (1993).

50. See M. Born and W. Heisenberg, *Z. Phys.*, **23**, 388 (1924); M. Born and J. E. Mayer, *Z. Phys.*, **75**, 1 (1932).

51. J. E. Lennard-Jones and P. A. Taylor, *Proc. Roy. Soc.* (London), **A109,** 476 (1925).

52. J. E. Lennard-Jones and B. M. Dent, *Proc. Roy. Soc.* (London), **A121,** 247 (1928).

53. B. M. Dent, *Phil. Mag.*, **8**(7), 530 (1929).

54. M. L. Huggins and J. E. Mayer, *J. Chem. Phys.*, **1,** 643 (1933).

55. F. van Zeggeren and G. C. Benson, *J. Chem. Phys.*, **26,** 1077 (1957).

56. G. C. Benson and R. McIntosh, *Can. J. Chem.*, **33,** 1677 (1955).

57. B. G. Dick and A. W. Overhauser, *Phys. Rev.*, **112,** 90 (1958).

58. M. J. L. Sangster, G. Peckham, and D. H. Saunderson, *J. Phys. C*, **3,** 1026 (1970).

59. C. R. A. Catlow, K. M. Diller, and M. J. Norgett, *J. Phys. C*, **10,** 1395 (1977).

60. T. S. Chen and F. W. de Wette, *Surf. Sci.*, **74,** 373 (1978).

61. M. R. Welton-Cook and M. Prutton, *Surf. Sci.*, **74,** 276 (1978).

62. A. J. Martin and H. Bilz, *Phys. Rev. B*, **19,** 6593 (1979).

63. C. B. Duke, R. J. Meyer, A. Paton, and P. Mark, *Phys. Rev. B*, **8,** 4225 (1978).

64. E. J. W. Verwey, *Rec. Trav. Chim.*, **65,** 521 (1946).

65. G. C. Benson, P. I. Freeman, and E. Dempsey, *ACS Symposium Series*, No. 33, American Chemical Society, 1961, p. 26.

66. K. Molière, W. Rathje, and I. N. Stranski, *Disc. Faraday Soc.*, **5,** 21 (1949); K. Molière and I. N. Stranski, *Z. Phys.*, **124,** 429 (1948).

67. P. W. Tasker, *Phil. Mag. A*, **39,** 119 (1979).

68. G. C. Benson and K. S. Yun, in *The Solid-Gas Interface*, E. A. Flood, ed., Marcel Dekker, New York, 1967. See also G. C. Benson and T. A. Claxton, *J. Chem. Phys.*, **48,** 1356 (1968).

69. G. C. Benson, *J. Chem. Phys.*, **35,** 2113 (1961).

70. G. C. Benson and T. A. Claxton, *Can. J. Phys.*, **41,** 1287 (1963).

71. H. Reiss and S. W. Mayer, *J. Chem. Phys.*, **34,** 2001 (1961).

72. H. P. Schreiber and G. C. Benson, *Can. J. Phys.*, **33,** 534 (1955).

73. K. Tomioka, M. G. Sceats, and S. A. Rice, *J. Chem. Phys.*, **66,** 2984 (1977).

74. N. H. Fletcher, *Rep. Prog. Phys.*, **34,** 913 (1971).

75. A. U. S. de Reuck, *Nature*, **179,** 1119 (1957).

76. W. M. Ketcham and P. V. Hobbs, *Phil. Mag.*, **19,** No. 162, 1161 (1969).

77. H. H. G. Jellinek, *J. Colloid Interface Sci.*, **25,** 192 (1967).

78. M. W. Orem and A. W. Adamson, *J. Colloid Interface Sci.*, **31,** 278 (1969).

79. J. K. MacKenzie, A. J. W. Moore, and J. F. Nicholas, *J. Phys. Chem. Solids*, **23,** 185 (1962).

80. H. Reiss, H. L. Frisch, E. Hefland, and J. L. Lebowitz, *J. Chem. Phys.* **32,** 119 (1960).

81. H. Schonhorn, *J. Phys. Chem.*, **71,** 4878 (1967).

82. J. F. Nicholas, *Australian J. Phys.*, **21,** 21 (1968).

83. T. Matsunaga and Y. Tamai, *Surf. Sci.*, **57,** 431 (1976).

84. F. M. Fowkes, *Ind. Eng. Chem.*, **56,** 40 (1964).

85. F. W. Williams and D. Nason, *Surf. Sci.*, **45,** 377 (1974).

86. N. D. Land and W. Kohn, *Phys. Rev. B*, **1,** 4555 (1970).

87. J. H. Rose and J. F. Dobson, *Solid State Commun.*, **37,** 91 (1981).

88. K. Rademann, B. Kaiser, U. Even, and F. Hensel, *Phys. Rev. Lett.*, **58,** 2319 (1987).

89. R. P. Messmer, *Surf. Sci.*, **106,** 225 (1981).

90. L. Brus, *J. Phys. Chem.*, **90,** 2555 (1986).

91. L. S. Bartell, *J. Phys. Chem.*, **99,** 1080 (1995).

92. J. J. Burton and G. Jura, *J. Phys. Chem.*, **71,** 1937 (1967).

93. F. F. Abraham and C. R. Brundle, *J. Vac. Sci. Technol.*, **18,** 506 (1981).

94. H. N. V. Temperley, *Proc. Cambridge Phil. Soc.*, **48,** 683 (1952).

95. W. K. Burton and N. Cabrera, *Disc. Faraday Soc.*, **5,** 33 (1949).

96. T. E. Madey, J. Guan, C.-Z. Dong, and S. M. Shivaprasad, *Surf. Sci.*, **287/288,** 826 (1993).

97. J. Guan, R. A. Campbell, and T. E. Madey, *J. Vac. Sci. Technol.*, **A13,** 1484 (1995).

98. I. N. Stranski, *Z. Phys. Chem.*, **136,** 259 (1928).

99. See also W. D. Harkins, *The Physical Chemistry of Surface Films*, Reinhold, New York, 1952.

100. W. J. Dunning, *J. Phys. Chem.*, **67,** 2023 (1963).

101. A. Henglein, *Prog. Colloid Polymer Sci.*, **73,** 1 (1987).

102. B. B. Mandelbrot, *The Fractal Geometry of Nature*, Freeman, New York, 1982; *Fractals: Form, Chance, and Dimension*, Freeman, New York, 1977.

103. D. Avnir, ed., *The Fractal Approach to Heterogeneous Chemistry*, Wiley, New York, 1989.

104. P. Pfeifer and D. Avnir, *J. Chem. Phys.*, **79,** 3558 (1983).

105. D. Farin, S. Peleg, D. Yavin, and D. Avnir, *Langmuir*, **1,** 399 (1985).

106. D. Avnir, D. Farin, and P. Pfeifer, *J. Chem. Phys.*, **79,** 3566 (1983).

107. J. J. Fripiat, L. Gatineau, and H. Van Damme, *Langmuir*, **2,** 562 (1986).

108. P. Meakin, "Fractal Aggregates and Their Fractal Measure," in *Phase Transitions and Critical Phenomena*, Vol. 12, C. Domb and J. L. Lebowitz, eds., Academic, New York, 1988.

109. P. Meakin, A. Coniglio, H. E. Stanley, and T. A. Witten, *Phys. Rev. A*, **34,** 3325 (1986).

110. J. Frenkel, *Z. Phys.*, **35,** 652 (1926).

111. C. Wagner and W. Schottky, *Z. Phys. Chem.*, **11B,** 163 (1930).

112. See F. A. Kruger and H. J. Vink, *Phys. Chem. Solids*, **5,** 208 (1958).

113. R. C. Evans, *An Introduction to Crystal Chemistry*, Cambridge University Press, Cambridge, UK, 1964.

114. J. M. Burgers, *Proc. Phys. Soc.* (London), **52,** 23 (1940).

115. F. C. Frank, *Dis. Faraday Soc.*, **5,** 48 (1949).

116. A. R. Verma, *Phil. Mag.*, **42,** 1005 (1951).

117. F. C. Frank, *Acta Cryst.*, **4,** 497 (1951).

118. D. J. Srolovitz, S. A. Safran, M. Homyonfer, and R. Tenne, *Phys. Rev. Lett.*, **74,** 1779 (1995).

119. B. H. Alexander, M. H. Dawson, and H. P. Kling, *J. Appl. Phys.*, **22,** 439 (1951).

120. E. B. Greenhill and S. R. McDonald, *Nature*, **171,** 37 (1953).

121. M. M. Nicolson, *Proc. Roy. Soc.* (London), **A228,** 507 (1955).

122. A. I. Murdoch, *Int. J. Eng. Sci.*, **16,** 131 (1978).

123. E. Orowan, *Z. Phys.*, **82,** 235 (1933); see also A. I. Bailey, *Proc. Int. Congr. Surf. Act. 2nd*, **3,** 406 (1957).

124. J. J. Gilman, *J. Appl. Phys.*, **31,** 2208 (1960).

125. A. R. C. Westwood and T. T. Hitch, *J. Appl. Phys.*, **34,** 3085 (1963).

126. P. F. Becher and S. W. Freiman, *J. Appl. Phys.*, **49,** 3779 (1978).

127. For a description of the apparatus, see G. C. Benson and G. W. Benson, *Rev. Sci. Instr.*, **26,** 477 (1955).

128. G. C. Benson, H. P. Schreiber, and F. van Zeggeren, *Can. J. Chem.*, **34,** 1553 (1956).

129. S. Brunauer, D. L. Kantro, and C. H. Weise, *Can. J. Chem.*, **34,** 729 (1956).

130. S. Brunauer, D. L. Kantro, and C. H. Weise, *Can. J. Chem.*, **37,** 714 (1959).

131. G. Jura and C. W. Garland, *J. Am. Chem. Soc.*, **74,** 6033 (1952); **75,** 1006 (1953).

132. D. Patterson, J. A. Morrison, and F. W. Thompson, *Can. J. Chem.*, **33,** 240 (1955).

133. M. Drechsler and J. F. Nicholas, *J. Phys. Chem. Solids*, **28,** 2609 (1967).

134. D. H. Bangham and N. Fakhoury, *J. Chem. Soc.*, **1931,** 1324; see also *Proc. Roy. Soc.* (London), **A147,** 152, 175 (1934).

135. D. J. C. Yates, *Proc. Roy. Soc.* (London), **A224,** 526 (1954).

136. I. N. Stranski, *Bericht.*, **75B,** 1667 (1942).

137. M. Polanyi, *Z. Phys.*, **7,** 323 (1921).

138. A. A. Griffith, *Phil. Trans. Roy. Soc.* (London), **A221,** 163 (1920).

139. W. C. Hynd, *Sci. J. Roy. Coll. Sci.*, **17,** 80 (1947).

140. D. McCammond, A. W. Neumann, and N. Natarajan, *J. Am. Ceram. Soc.*, **58,** 15 (1975).

141. G. M. Whitesides, *Atomistics of Fracture*, R. M. Latanision and J. R. Pickens, eds., Plenum, New York, 1983.

142. H. El-Shall and P. Somasundaran, *Powder Technol.*, **38,** 275 (1984).

143. A. Frumkin, *J. Colloid Sci.*, **1,** 277 (1946).

144. P. A. Rehbinder, V. I. Likhtman, and V. M. Maslinnikov, *CR Acad. Sci. URSS*, **32,** 125 (1941).

145. P. de Gennes, *CR* **304,** 547 (1987).

146. See I. Langmuir, *J. Am. Chem. Soc.*, **38,** 2221 (1916).

147. See W. D. Harkins, *The Physical Chemistry of Surface Films*, Reinhold, New York, 1952.

148. T. S. Renzema, *J. Appl. Phys.*, **23,** 1412 (1952).

149. B. Reitzner, *J. Phys. Chem.*, **65,** 948 (1961).

150. B. Reitzner, J. V. R. Kaufman, and F. E. Bartell, *J. Phys. Chem.*, **66,** 421 (1962).

151. F. P. Bowden and H. M. Montague-Pollock, *Nature*, **191,** 556 (1961).

152. P. W. M. Jacobs, F. C. Tompkins, and V. R. P. Verneker, *J. Phys. Chem.*, **66,** 1113 (1962).

153. F. L. Hirshfeld and G. M. J. Schmidt, *J. Polym. Sci.*, **2(A),** 2181 (1964).

154. E. Ruckenstein and T. Vavanellos, *AIChE J.*, **21,** 756 (1975).

155. T. B. Fryberger and P. C. Stair, *Chem. Phys. Lett.*, **93,** 151 (1982).

156. A. L. Testoni and P. C. Stair, *J. Vac. Sci. Technol.*, **A4,** 1430 (1986).

157. T. Morimoto and K. Aoki, *Langmuir*, **2,** 525 (1986).

158. L. S. D. Glasser, F. P. Glasser, and H. F. W. Taylor, *Q. Rev.*, **16,** 343 (1962).

159. R. Sh. Mikhail, S. Hanafi, S. A. Abo-el-enein, R. J. Good, and J. Irani, *J. Colloid Interface Sci.*, **75,** 74 (1980).

160. U. R. Evans, *J. Chem. Soc.*, **1946,** 207.

161. A. J. Bevolo, K. G. Baikerikar, and R. S. Hansen, *J. Vac. Sci. Technol.*, **A2,** 784 (1984).

162. J. V. Cathcart, J. E. Epperson, and G. F. Peterson, *Acta Met.*, **10,** 699 (1962); see also *Ellipsometry in the Measurement of Surfaces and Thin Films*, E. Passaglia, R. R. Stromberg, and J. Kruger, eds., *Natl. Bur. Stand. (U.S.)*, Misc. Publ., **256** (1964).

163. P. Hadjisavas, M. Caillet, A. Galerie, and J. Besson, *Rev. Chim. Miner.*, **14,** 572 (1977).

164. A. Abba, A. Galerie, and M. Caillet, *Mater. Chem.*, **5,** 147 (1980).

165. M. Caillet, H. F. Ayedi, and J. Besson, *J. Less-Common Met.*, **51,** 323 (1977).

166. G. Cohn, *Chem. Rev.*, **42,** 527 (1948).

167. N. F. Mott and R. W. Gurney, *Electronic Processes in Ionic Crystals*, Clarendon Press, Oxford, 1940.

168. W. Gomes, *Nature*, **192,** 865 (1961).

169. N. Hackerman, D. D. Justice, and E. McCafferty, *Corrosion*, **31,** 240 (1975).

170. F. M. Delnick and N. Hackerman, *J. Electrochem. Soc.*, **126,** 732 (1979).

171. G. L. Griffin, *J. Electrochem. Soc.*, **133,** 1315 (1986).

172. S. Yoshida and H. Ishida, *J. Chem. Phys.*, **78,** 6960 (1983).

173. O. Teschke, F. Galembeck, and M. A. Tenan, *Langmuir*, **3,** 400 (1987).

174. K. W. Frese, Jr., M. J. Madou, and S. R. Morrison, *J. Phys. Chem.*, **84,** 3172 (1980).

175. M. C. Madou, K. W. Frese, Jr., and S. R. Morrison, *J. Phys. Chem.*, **84,** 3423 (1980).

176. F. Piqué, P. Baglioni, L. Dei, and E. Ferroni, *Sci. Technol. Cult. Heritage*, **3,** 155 (1994).

177. F. Piqué, L. Dei, and E. Ferroni, *Stud. Conserv.*, **37,** 217 (1992).

178. G. G. Amoroso and V. Fassina, *Stone Decay and Conservation*, Materials Science Monograph 11, Elsevier, 1983.

179. J. J. Bikerman, *Phys. Stat. Sol.*, **10,** 3 (1965).

180. G. N. Lewis and M. Randall, *Thermodynamics*, 2nd ed., revised by K. S. Pitzer and L. Brewer, McGraw-Hill, New York, 1961.

181. W. H. Beamer and C. R. Maxwell, *J. Chem. Phys.*, **17,** 1293 (1949).

CHAPTER VIII

Surfaces of Solids: Microscopy and Spectroscopy

1. Introduction

A number of methods that provide information about the structure of a solid surface, its composition, and the oxidation states present have come into use. The recent explosion of activity in scanning probe microscopy has resulted in investigation of a wide variety of surface structures under a range of conditions. In addition, spectroscopic interrogation of the solid–high-vacuum interface elucidates structure and other atomic processes.

The function of this chapter is to review these methods with emphasis on the types of phenomenology involved and information obtained. Many of the effects are complicated, and full theoretical descriptions are still lacking. The wide variety of methods and derivative techniques has resulted in a veritable alphabet soup of acronyms. A short list is given in Table VIII-1 (see pp. 313–318); the IUPAC recommendations for the abbreviations are found in Ref. 1.

A few of the most frequently used techniques are discussed briefly in this chapter; references to those not covered are given in the table. Useful reviews are Refs. 2–5 and 6 and, for organic surfaces, Refs. 7–9 and 10. Also, many of the various measurements have found use in the study of the adsorbed state, and further examples of their use are to be found in Chapters VII, XVI, and XVII.

2. The Microscopy of Surfaces

A. Optical and Electron Microscopy

Conventional optical microscopy can resolve features down to about the wavelength of visible light or about 500 nm. Surface faceting and dislocations may be seen as in Fig. VII-10. Confocal microscopy adds the ability to make three-dimensional scans in solutions, biological membranes, and polymeric systems. [11]. A new approach, *photon tunneling microscopy* (PTM), improves resolution and contrast down to 1 nm vertically and one quarter of the wavelength of light in the lateral directions [12]. The striking improvement in resolution of surface features is due to the unique properties of the evanescent wave created when light tunnels from a medium of higher index of refraction into one of lower index. PTM has been applied to polymeric surfaces in air [12]. Many other optical microscopy techniques were discussed in Chapter IV.

Transmission electron microscopy (TEM) can resolve features down to about 1 nm and allows the use of electron diffraction to characterize the structure. Since electrons must pass through the sample; however, the technique is limited to thin films. One cryoelectron microscopic study of fatty-acid Langmuir films on vitrified water [13] showed faceted crystals. The application of TEM to Langmuir–Blodgett films is discussed in Chapter XV.

Scanning electron microscopy (SEM) is widely used to examine surfaces down to a resolution of 50 nm or so. The surface is scanned by a focused electron beam, and the intensity of *secondary* electrons is monitored. The secondary electrons are viewed on a cathode-ray tube (CRT) [14]. SEM images characteristically have a wide range in contrast and great depth of focus. Thus even quite rough surfaces show in startling clarity and depth (note Refs. 14–16). This depth of focus still makes SEM the method of choice when faced with a surface that is too rough to conveniently image with scanning probe microscopy.

B. Scanning Probe Microscopies

The ability to control the position of a fine tip in order to scan surfaces with subatomic resolution has brought scanning probe microscopies to the forefront in surface imaging techniques. We discuss the two primary techniques, *scanning tunneling microscopy* (STM) and *atomic force microscopy* (AFM); the interested reader is referred to comprehensive reviews [9, 17, 18].

The first method, STM, relies on measurement of the exponentially decaying tunneling current between a metal tip and conducting substrate. Since its development in the early 1980s and the recognition for its inventors with the 1986 Nobel prize [19], STM has found wide use in studies of inorganic materials [20–25], organic materials [9, 26–28], and dynamic processes, including reactions [9].

The two essential technologies enabling microscopy by electron tunneling are the formation of a small metallic tip and the fine position control of piezoelements. A tungsten wire is sharpened by electrochemical etching in an NaOH solution or a platinum/iridium wire is cut to form the probe [9]. The effective tip is formed by the few atoms closest to the scanned surface. Both vertical and lateral motion is controlled by piezoelectric transducers. Typically an STM image is gathered by scanning the tip in a plane less than a nanometer above the surface in a constant current feedback control mode as illustrated in Fig. VIII-1. Here a fixed tunneling current of pico- to nanoamperes is maintained by allowing the tip to move vertically; this motion is recorded as a surface contour. Striking images are obtained in this manner as shown in Fig. VIII-2.

Since scanning tunneling microscopy requires flat conducting surfaces, it is not surprising that most of its early application was to study inorganic materials [17, 19, 20, 29–34]. These studies include investigations of catalytic metal surfaces [24, 35–37], silicon and other oxides [21], superconductors [38], gold

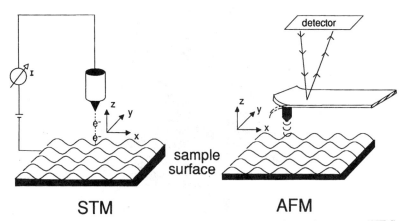

Fig. VIII-1. Schematic illustration of the scanning tunneling microscope (STM) and atomic force microscope (AFM). (From Ref. 9.)

[20, 23, 39], and alloys [22]. An example of an STM image of a silicon surface is shown in Fig. VIII-2a.

One of the most novel uses of STM has been to monitor the dynamics of surface processes including phase transitions, surface diffusion, epitaxial growth and corrosion. An example of this type of work is shown in Fig. VIII-2b, where the selective dissolution of Ag from an Ag–Au alloy is studied over a period of 42 min [22]. Corrosion first roughens the surface followed by vacancy migration to form smooth terraces. Similar surface motion is observed in electrochemically roughened gold films [23] and in gold deposited from vapor on mica at sufficient temperatures [20]. In a related application, scanning electrochemical microscopy (SECM), the STM tip serves as an ultramicroelectrode to reduce electroactive species, allowing the study of electrochemical reactions on the 100-nm scale [40]. A novel atom-tracking technique has allowed the direct measurement of Si dimers on Si(001) [25].

In addition to monitoring surface dynamics, the STM can be used to manipulate atoms and molecules into desired configurations on a surface. In parallel processes, atoms or molecules are moved along the surface without breaking the surface–adsorbate bond. Sufficient energy to cross the energy barrier to lateral diffusion, or approximately 10–33% of the adsorption energy (see Chapter XVI), must be supplied [41]. This is accomplished by either field-assisted diffusion caused by the action of the spatially inhomogeneous electric field near the probe tip on the dipole moment of the adsorbed atom, or by a sliding process where the tip–adsorbate attraction is used to reposition atoms [41]. The latter approach was used to produce the island of CO molecules on Pt shown in Fig. VIII-2c. Field-assisted diffusion can be used to measure the dipole moment and polarizabilities of adsorbates as done in field ion and field emission microscopy (see Section VIII-2C). Another means to manipulate atoms is via-perpendicular processes whereby an atom is transferred to and from the tip by the application of a voltage pulse, causing evaporation or electromigration [41]. This approach has been used to move Si atoms and clusters on silicon surfaces [42].

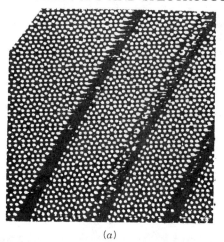

(a)

(b)

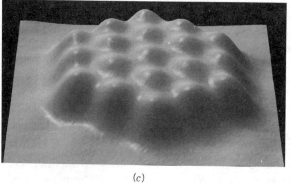

(c)

Organic molecules on conducting surfaces can be studied with STM; however, there remains some uncertainty about the tunneling mechanism through insulators [9]. Numerous materials have been imaged with STM, including ordered monolayers [18, 43], fullerenes [44], liquid crystals [9,18], alkanes [45], and biological molecules [26–28,46,47]. The most readily studied organic molecules possess some order on the surface as found in the self-assembling thioalkylmonolayers [43], Langmuir–Blodgett films [48], liquid crystals, and physically adsorbed alkanes.

A novel modification of the STM supplements images with those due to the thermopower signal across the tip–sample temperature gradient [49]. Images of guanine on graphite illustrate the potential of this technique.

Mounting a fine tip to a piezoelectrically positioned cantilever spring provides the means to measure surface forces in the range from 10^{-13} to 10^{-6} N. The atomic force microscope (AFM) illustrated in Fig. VIII-1b measures deflections in the cantilever due to capillary (Chapter II), electrostatic (Chapter V), van der Waals (Chapter VI), and frictional (Chapter XII) forces between the tip and the surface. Not restricted to conducting surfaces, AFM measurements can be made on organic and inorganic surfaces and immersed in liquids [9,50]. Conical tips of silicon have points of 5–50 nm (radius of curvature); however, numerous probes have been used including attaching a several micrometer colloidal particle to a cantilever as described by Butt and co-workers [50] and Ducker and co-workers [51].

In materials science, AFM has provided a useful tool to measure adhesion (see Chapter XII) and delamination [9,50], surface elasticity [50], and friction [52]. The measurement of colloidal forces reconfirms the DLVO prediction for electrostatic repulsion (see Chapter V) and reveals the importance of anion adsorption in the stability of gold sols (see Refs. 53, 54). Organic materials studied with AFM again include Langmuir–Blodgett films [9,52,55], organic crystals and liquid crystals [9], and latex particles [56]. A nice example of the complementary information given by TEM and AFM is shown in Fig. VIII-3, where Yang and co-workers studied the growth of lead selenide crystals under monolayers [57].

Modification of an AFM to operate in a dynamic mode aids the study of soft biological materials [58]. Here a stiff cantilever is oscillated near its resonant frequency with an amplitude of about 0.5 nm; forces are detected as a shift to a new frequency

Fig. VIII-2. Scanning tunneling microscopy images illustrating the capabilities of the technique: (a) a 10-nm-square scan of a silicon(111) crystal showing defects and terraces from Ref. 21; (b) the surface of an Ag–Au alloy electrode being electrochemically roughened at 0.2 V and 2 and 42 min after reaching 0.70 V (from Ref. 22); (c) an island of CO molecules on a platinum surface formed by sliding the molecules along the surface with the STM tip (from Ref. 41).

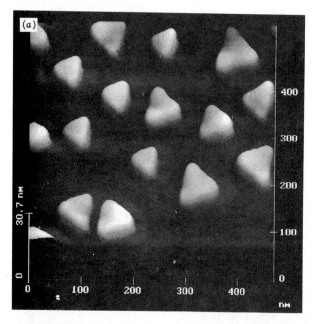

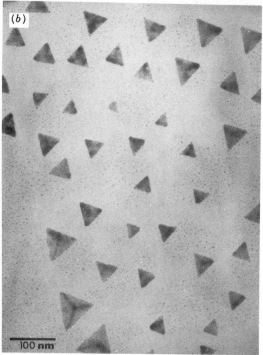

Fig. VIII-3. (a) Atomic force microscope (AFM) and (b) transmission electron microscope (TEM) images of lead selenide particles grown under arachidic acid monolayers.
(From Ref. 57.)

$$\omega = \sqrt{\frac{C + \partial F/\partial z}{m_{\text{eff}}}} \qquad\qquad \text{(VIII-1)}$$

where C and m_{eff} are the cantilever spring constant and effective mass and $\partial F/\partial z$ is the vertical force gradient. With this technique, Anselmetti and co-workers were able to measure attractive forces over surfaces of DNA and tobacco mosaic virus down to 4×10^{-12} N [58].

C. Field Emission and Field Ion Microscopies

The *field emission microscope* (FEM), invented in 1936 by Müller [59, 60], has provided major advances in the structural study of surfaces. The subject is highly developed and has been reviewed by several groups [2, 61, 62], and only a selective, introductory presentation is given here. Some aspects related to chemisorption are discussed in Chapter XVII.

The basic device is very simple. A tip of refractory metal, such as tungsten, is electrically heat-polished to yield a nearly hemispherical end of about 10^{-5} cm radius. A potential of about 10 kV is applied between the tip and a hemispherical fluorescent screen. The field, F, falls off with distance as kr^{-2}, and if the two radii of curvature are a and b, the total potential difference V is then

$$V = \int_a^b F\,dr = -k\left(\frac{1}{b} - \frac{1}{a}\right) \qquad\qquad \text{(VIII-2)}$$

If $1/b$ is neglected in comparison to $1/a$, then $k = aV$ and the field at the tip $F(a) = V/a$ or 10^9 V/cm in the present example. This very high value, equivalent to 100 V/nm, is sufficient to pull electrons from the metal, and these now accelerate along radial lines to hit the fluorescent screen. Individual atoms serve as emitting centers, and different crystal faces emit with different intensities, depending on the packing density and work function. Since the magnification factor b/a is enormous (about 10^6), it might be imagined that individual atoms could be seen, but the resolution is limited by the kinetic energy of the electrons in the metal at right angles to the emission line and obtainable resolutions are about 3–5 nm. Nevertheless, the technique produces miraculous pictures showing the various crystal planes that form the tip in patterns of light and dark. Adsorbates, such as N_2, may increase the emission intensity providing a means to follow the surface migration of adsorbed species.

An ingenious modification of the field emission microscope by Müller [23], known as the *field ion microscope* (FIM), makes use of the fact that a positively charged tip will strip nearby gas molecules of an electron, causing the positive ion to accelerate out radially and to hit the fluorescent screen. Helium is the most commonly used gas, although other gases have been used. This is a highly developed subject; for details, see the monograph by Müller and Tsong [63].

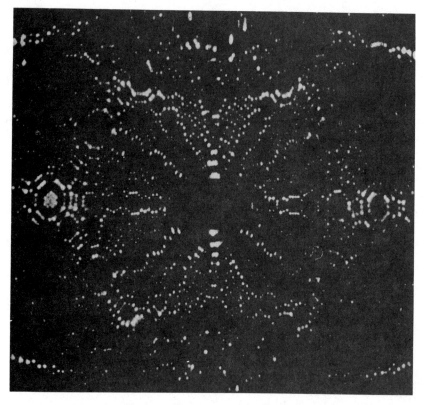

Fig. VIII-4. Field ion emission from clean tungsten. (From Ref. 64.)

Since field ion microscopes can be operated at cryogenic temperatures, atomic resolution is possible. A good example of a field ion emission photograph for a tungsten tip is shown in Fig. VIII-4. The spots represent individual tungsten atoms, and the patterns arise from the geometries of different crystallographic planes. Models of the arrangement of tungsten atoms on a hemispherical tip (Ref. 64) show the variety of crystal planes present, each with a different ability to ionize helium producing patterns such as those in Fig. VIII-4.

While field ion microscopy has provided an effective means to visualize surface atoms and adsorbates, field emission is the preferred technique for measurement of the energetic properties of the surface. The effect of an applied field on the rate of electron emission was described by Fowler and Nordheim [65] and is shown schematically in Fig. VIII-5. In the absence of a field, a barrier corresponding to the thermionic work function, Φ, prevents electrons from escaping from the Fermi level. An applied field, reduces this barrier to $\Phi - V$, where the potential V decreases linearly with distance according to $V = xF$. Quantum-mechanical tunneling is now possible through this finite barrier, and the solution for an electron in a finite potential box gives

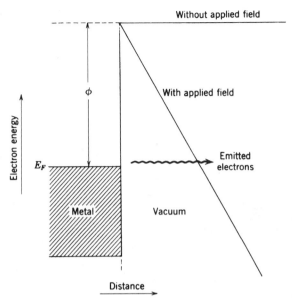

Fig. VIII-5. Schematic potential energy diagram for electrons in a metal with and without an applied field: Φ, work function; E_F, depth of the Fermi level. (From Ref. 62.)

$$P - \text{const} \exp\left[-\frac{2^{2/3}m^{1/2}}{\hbar} \int_0^l (\Phi - V)dx \right] \qquad \text{(VIII-3)}$$

where P is the probability of escape, m is the mass of the electron, and l is the width of the barrier, Φ/F. The integral can be evaluated and put into a convenient experimental form

$$\frac{i}{V^2} = A \exp\left(-\frac{B\Phi^{3/2}}{V} \right) \qquad \text{(VIII-4)}$$

where i is the total emission current, V is the applied voltage, and A and B are constants that depend on Φ and geometric factors. The V^2 dependence of i arises from more detailed considerations, including the rate of arrival of electrons at the surface. The important aspect of Eq. VIII-4 is that the emission current depends exponentially on $1/V$ with a coefficient that allows evaluation of the work function.

Measuring the electron emission intensity from a particular atom as a function of V provides the work function for that atom; its change in the presence of an adsorbate can also be measured. For example, the work function for the (100) plane of tungsten decreases from 4.71 to 4.21 V on adsorption of nitrogen. For more details, see Refs. 66 and 67 and Chapter XVII. Information about the surface tensions of various crystal planes can also be obtained by observing the development of facets in field ion microscopy [68].

D. Low-Energy Electron Diffraction (LEED)

It was recognized early in the history of diffraction studies that just as x-rays and high-energy electrons (say, 50 keV) gave information about the bulk periodicity of a crystal, low-energy electrons (around 100 eV) whose penetrating power is only a few atomic diameters should provide information about the surface structure of the solid. The first reported experiment of Davisson and Germer in 1927 [69] was hampered by difficulties in generating a monoenergetic beam of electrons and detecting its scattering without ultra-high-vacuum techniques. Even at a pressure of 10^{-6} torr, a surface becomes covered with a monolayer of adsorbed gas in about 1 sec; to study clean surfaces, pressures down to 10^{-10} are needed.

Much of the experimental development was carried out by MacRae and coworkers [70]. An experimental arrangement is shown in Fig. VIII-6 [62, 71, 72]. Electrons from a hot filament are given a uniform acceleration, striking the crystal normal to its surface. The scattered electrons may have been either elastically or inelastically scattered; only the former are used in the diffraction experiment. Of the several grids shown, the first is at the potential of the crystal, and the second is a repelling grid that allows only electrons of original energy to pass—those inelastically scattered, and hence of lower energy, are stopped.

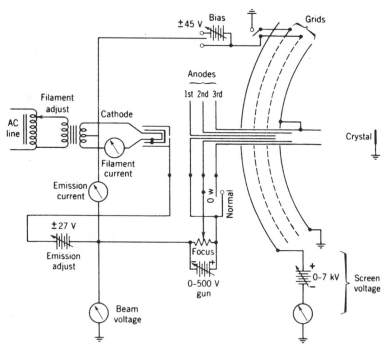

Fig. VIII-6. Low-energy electron diffraction (LEED) apparatus. (From Ref. 8.)

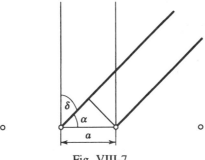

Fig. VIII-7.

The final grid is positively charged to accelerate the accepted electrons onto the fluorescent screen. The diffraction pattern may then be photographed.

The diffraction is essentially that of a grating. As illustrated in Fig. VIII-7, the Laue condition for incidence normal to the surface is

$$a \cos \alpha = n_1 \lambda \tag{VIII-5}$$

where a is the repeat distance in one direction, n_1 is an integer, and λ is the wavelength of the electrons. For a second direction, in the usual case of a two-dimensional grid of atoms,

$$b \cos \beta = n_2 \lambda \tag{VIII-6}$$

The diffraction pattern consists of a small number of spots whose symmetry of arrangement is that of the surface grid of atoms (see Fig. IV-10). The pattern is due primarily to the first layer of atoms because of the small penetrating power of the low-energy electrons (or, in HEED, because of the grazing angle of incidence used); there may, however, be weak indications of scattering from a second or third layer.

LEED angles must be corrected for refraction by the surface potential barrier [73]. Also, the intensity of a diffraction spot is temperature dependent because of the vibration of the surface atoms. As an approximation,

$$\frac{d \ln I}{dT} = \frac{12 \, h^2 (\cos^2 \phi)}{mk\lambda^2 \theta_D^2} \tag{VIII-7}$$

where I is the intensity of a given spot, ϕ is the angle between the direction of the incident beam and the diffracted beam (usually small), λ is the wavelength of the electron, and θ_D is the Debye temperature of the lattice [74].

It might be imagined that the structure of a clean surface of, say, a metal single crystal would simply be that expected from the bulk structure. This is not necessarily so; recall that in the case of the alkali metal halides the ions of the surface layer are displaced differently. The surface structure, however, should "fit" on the bulk structure, and it has become customary to describe the former in terms of the latter. A (1×1) surface structure has the same periodicity or "mesh" as the corresponding crystallographic plane of the bulk structure, as illustrated in Fig. VIII-8a. If alternate rows of atoms are shifted, the surface periodicity is reduced to a $C(2 \times 1)$ structure, the C denoting the presence of a center atom, as shown in Fig. VIII-8b. Somorjai has detailed the mathematical procedure for reducing a LEED pattern to a surface structure [62, 75].

LEED and its relatives have become standard procedures for the characterization of the structure of both clean surfaces and those having an adsorbed layer. Somorjai and co-workers have tabulated thousands of LEED structures [75], for example. If an adsorbate is present, the substrate surface structure may be altered, or "reconstructed," as illustrated in Fig. VIII-9 for the case of H atoms on a Ni(110) surface. Beginning with the (experimentally) hypothetical case of (100) Ar surfaces, Burton and Jura [76] estimated theoretically the free energy for a surface transition from a (1×1) to a $C(2 \times 1)$ structure as given by

$$\Delta G = 4.83 - 0.0592T \ (\text{ergs/cm}^2) \qquad \text{(VIII-8)}$$

Above 81.5 K the $C(2 \times 1)$ structure becomes the more stable. Two important points are, first, that a change from one surface structure to another can occur without any bulk phase change being required and, second, that the energy difference between alternative surface structures may not be very large, and the free energy difference can be quite temperature-dependent.

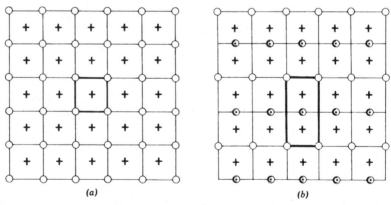

(a)   (b)

Fig. VIII-8. Surface structures: (a) (1×1) structure on the (100) surface of a FCC crystal (from Ref. 76); (b) $C(2 \times 1)$ surface structure on the (100) surface of a FCC crystal (from Ref. 76). In both cases the unit cell is indicated with heavy lines, and the atoms in the second layer with pluses. In (b) the shaded circles mark shifted atoms. (See also Ref. 77.)

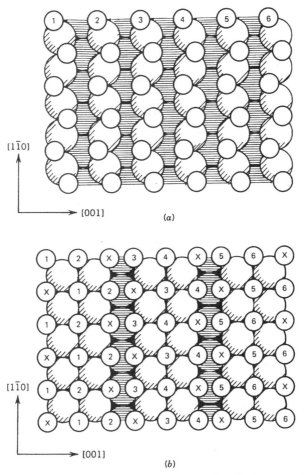

Fig. VIII-9. Alteration of the structure of a Ni(110) surface by H-atom adsorption: (a) structure for $\theta \simeq 1$; (b) structure for $\theta > 1$. [Reprinted with permission from G. Ertl, *Langmuir*, **3**, 4 (1987). Copyright 1987, American Chemical Society.]

It is also possible to determine the structure of the adsorbed layer itself, as illustrated in Fig. VIII-9. Even for relatively low coverages, the adsorbed layer may undergo a series of structural or "phase" changes (see Ref. 78). Much of the research of this type has been aimed at understanding the detailed mechanisms of catalyzed reactions (see Chapter XVIII), and an important development has been apparatus design that allows one to take a sample from atmospheric pressure to ultrahigh vacuum in a few seconds time (see Ref. 79). This makes it possible to observe surface structural and compositional changes at intervals during the reaction.

Finally, it has been possible to obtain LEED patterns from films of molecular solids deposited on a metal-backing. Examples include ice and naphthalene [80] and various phthalocyanines [81]. (The metal backing helps to prevent surface charging.)

3. Spectroscopic Methods

If a surface, typically a metal surface, is irradiated with a probe beam of photons, electrons, or ions (usually positive ions), one generally finds that photons, electrons, and ions are produced in various combinations. A particular method consists of using a particular type of probe beam and detecting a particular type of produced species. The method becomes a spectroscopic one if the intensity or efficiency of the phenomenon is studied as a function of the energy of the produced species at constant probe beam energy, or vice versa. Quite a few combinations are possible, as is evident from the listing in Table VIII-1, and only a few are considered here.

The same basic apparatus may allow more than one kind of measurement. This is illustrated in Fig. VIII-10, which shows the energy spectrum of electrons scattered from an incident beam, where E_p is the energy of the probe beam. As discussed above, the diffraction of elastically scattered electrons produces a LEED pattern, indicated in the inset in the figure. The incident beam may, however, be scattered inelastically by inducing vibrational excitations in adsorbate species. As shown in Fig. VIII-10b, this gives rise to an energy-loss spectrum (EELS or HREELS) characteristic of the species and its environment. If the incident beam energy is high enough, core electrons may be ejected, giving rise to the Auger effect discussed below and illustrated in Fig. VIII-10c.

The various spectroscopic methods do have in common that they typically allow analysis of the surface composition. Some also allow an estimation of the chemical state of the system and even of the location of nearest neighbors.

A. Auger Electron Spectroscopy (AES)

The physics of the method is as follows. A probe electron (2–3 keV usually) ionizes an inner electron of a surface atom, creating a vacancy. Suppose that a K electron has been ejected by the incident electron (or x-ray) beam. A more outward-lying electron may now drop into this vacancy. If, say, it is an L_1 electron that does so, the energy $E_K - E_{L_1}$ is now available and may appear as a characteristic x-ray. Alternatively, however, this energy may be given to an outer electron, expelling it from the atom. This *Auger electron* might, for example, come from the L_{III} shell, in which case its kinetic energy would be $(E_K - E_{L_1}) - E_{L_{III}}$. The $L_1 - K$ transition is the probable one, and an Auger spectrum is mainly that of the series $(E_K - E_{L_1}) - E_i$, where the ith type of outer electron is ejected.

The principal use of Auger spectroscopy is in the determination of surface composition, although peak positions are secondarily sensitive to the valence state of the atom. See Refs. 2, 82, and 83 for reviews.

Experimentally, it is common for LEED and Auger capabilities to be combined; the basic equipment is the same. For Auger measurements, a grazing angle of incident electrons is needed to maximize the contribution of surface

ENERGY DISTRIBUTION OF SCATTERED ELECTRONS FROM
A c(4×2) MONOLAYER OF C₂H₃ ON Rh(III) AT 300K

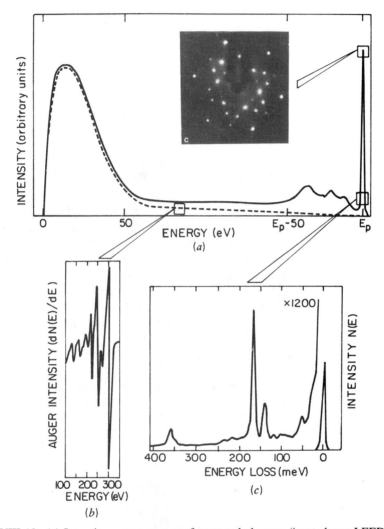

Fig. VIII-10. (*a*) Intensity versus energy of scattered electron (inset shows LEED pattern) for a Rh(111) surface covered with a monolayer of ethylidyne (CCH₃), the structure of chemisorbed ethylene. (*b*) Auger electron spectrum. (*c*) High-resolution electron energy loss spectrum. [Reprinted with permission from G. A. Somorjai and B. E. Bent, *Prog. Colloid Polym. Sci.*, **70,** 38 (1985) (Ref. 6). Copyright 1985, Pergamon Press.]

atoms. The voltage on the retarding grid (Fig. VIII-6) may be modulated to make the detector signal responsive to electrons of just that energy. As shown in Fig. VIII-10*b*, one obtains a *derivative* plot.

Hubbard and co-workers have developed a surface imaging approach, angular-distribution Auger microscopy (ADAM), based on resolution of the angular distribution of Auger electrons [84–86]. In this method, the sample is irradiated by a beam of electrons held at a fixed angle from the surface. The energy analyzer scans the angular distribution with a resolution of 1° in about three hours [85]. Surface atoms are illuminated by Auger electrons emitted by atoms deeper in the solid. The surface structure thus appears as silhouettes or minima in Auger intensity. This technique has been used to image single crystals [84], but is best applied to monatomic layers such as silver on platinum with and without an iodine overlayer [85, 86]. The structures found from ADAM have been verified with STM and LEED.

B. Photoelectron Spectroscopy (XPS, ESCA)

In photoelectron spectroscopy monoenergetic x-radiation ejects inner ($1s$, $2s$, $2p$, etc.) electrons. The electron energy is then $E_0 - E_i$, where E_0 is the x-ray quantum energy and E_i is that of the ith type of electron. The energy of the ejected electrons is determined by means of an electron spectrometer, thus obtaining a spectrum of both the primary photoelectrons and Auger electrons. The method is more accurate than Auger spectroscopy, and because of this, one can observe that a given type of electron has an energy that is dependent on the valence state of the atom. Thus for $1s$ sulfur electrons, there is a *chemical shift* of over 5 V, the ionization energy increasing as the valence state of sulfur varies from −2 to +6. The effect is illustrated in Fig. VIII-11 for the case of aluminum, showing how it is possible to analyze for oxidized aluminum on the surface.

Madey and co-workers followed the reduction of titanium with XPS during the deposition of metal overlayers on TiO_2 [87]. This shows the reduction of surface TiO_x molecules on adsorption of reactive metals. Film growth is readily monitored by the disappearance of the XPS signal from the underlying surface [88, 89]. This approach can be applied to polymer surfaces [90] and to determine the thickness of polymer layers on metals [91]. Because it is often used for chemical analysis, the method is sometimes referred to as *electron spectroscopy for chemical analysis* (ESCA). Since x-rays are very penetrating, a grazing incidence angle is often used to emphasize the contribution from the surface atoms.

C. Ion Scattering (ISS, LEIS)

If a beam of monoenergetic ions of mass M_i is elastically scattered at an angle θ by surface atoms of mass M_a, conservation of momentum and energy requires that

$$E_s = \left[\frac{\cos \theta + (r^2 - \sin^2 \theta)^{1/2}}{1 + r} \right]^2 E_i \qquad \text{(VIII-9)}$$

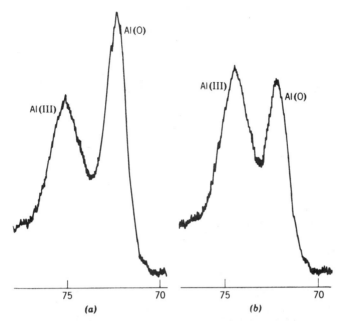

Fig. VIII-11. ESCA spectrum of Al surface showing peaks for the metal, Al(0), and for surface oxidized aluminum, Al(III): (a) freshly abraided sample; (b) sample after five days of ambient temperature air exposure showing increased Al(III)/Al(0) ratio due to surface oxidation. (From Instrument Products Division, E. I. du Pont de Nemours, Co., Inc.)

where E_s is the energy of the scattered ion, E_i its initial energy, and $r = M_a/M_i$. For the case of $\theta = 90°$, Eq. VIII-9 reduces to

$$E_s = \frac{M_a - M_i}{M_a + M_i} E_i \qquad \text{(VIII-10)}$$

These equations indicate that the energy of the scattered ions is sensitive to the mass of the scattering atom s in the surface. By scanning the energy of the scattered ions, one obtains a kind of mass spectrometric analysis of the surface composition. Figure VIII-12 shows an example of such a spectrum. Neutral, that is, molecular, as well as ion beams may be used, although for the former a velocity selector is now needed to define E_i.

Equations VIII-9 and VIII-10 are useful in relating scattering to M_a. It is also of interest to study the variation of scattering intensity with scattering angle. Often, the scattering is essentially specular, the intensity peaking sharply for an angle of reflection equal to the angle of incidence.

A useful complication is that if kinetic energy is not conserved; that is, if

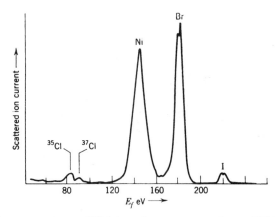

Fig. VIII-12. Energy spectrum of Ne$^+$ ions that are scattered over 90° by a halogenated nickel surface. The incident energy of the ions is 300 V. (From Ref. 92.)

the collision is inelastic, there should be a quite different angular scattering distribution. As an extreme, if the impinging molecule sticks to the surface for awhile before evaporating from it, memory of the incident direction is lost, and the most probable "scattering" is now normal to the surface, the probability decreasing with the cosine of the angle to the normal. In intermediate cases, some but not complete exchange of rotational and vibrational energy may occur between the molecule and the surface. Thus D_2 and HD but not H_2 were able to exchange rotational energy with the Ag(111) surface [62]; the inability of H_2 to exchange rotational energy efficiently is probably due to the relatively large energy separation of rotational energy states in this case.

Studies of inelastic scattering are of considerable interest in heterogeneous catalysis. The degree to which molecules are scattered specularly gives information about their residence time on the surface. Often new chemical species appear, whose trajectory from the surface correlates to some degree with that of the incident beam of molecules. The study of such *reactive* scattering gives mechanistic information about surface reactions.

An interesting development is that of *electron-stimulated desorption ion angular distribution* (ESDIAD). The equipment is essentially that for LEED—a focused electron beam (20–1000 eV) impinging on a surface results in the desorption of atomic and molecular ions, which are accelerated linearly from the surface by means of suitably charged grids. The direction cones are largely determined by the *orientation* of the bonds that are ruptured, and the result is a somewhat diffuse but rather LEED-like pattern that may be seen on a fluorescent screen and photographed. If necessary, the ions may be mass selected, but often this is not necessary as some single kind of ion predominates. Figure VIII-13 shows a H$^+$ ESDIAD pattern of NH_3 on an oxygen pretreated Ni(111) surface [93]. The technique has been used to study hindered rotation of PF_3 on the same surface [94].

NH₃ on Ni (111)

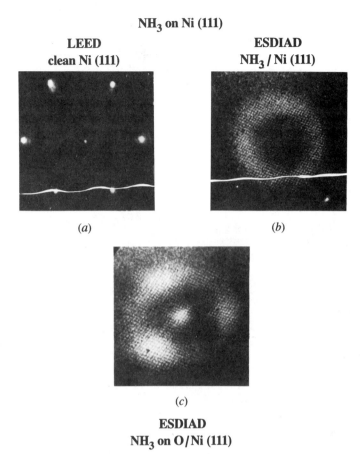

LEED
clean Ni (111)

ESDIAD
NH₃ / Ni (111)

(*a*)   (*b*)

(*c*)

ESDIAD
NH₃ on O / Ni (111)

Fig. VIII-13. LEED and ESDIAD on clean and oxygen-dosed Ni(111): (*a*) LEED, clean surface; (*b*) H⁺ ESDIAD of NH₃ on Ni(111), the halo suggesting free rotation of the surface NH₃ groups; (*c*) H⁺ ESDIAD after predosing with oxygen, then heated to 600 K and cooled before dosing with NH₃—only well-ordered chemisorbed NH₃ is now present. (From Ref. 93.)

4. Other Techniques

Optical second-harmonic generation (SHG) has recently emerged as a powerful surface probe [95, 96]. Second harmonic generation has long been used to produce frequency doublers from noncentrosymmetric crystals. As a surface probe, SHG can be caused by the break in symmetry at the interface between two centrosymmetric media. A high-powered pulsed laser is focused at an angle of incidence from 30° to 70° onto the sample at a power density of 10^5 to 10^8W/cm^2. The harmonic is observed in reflection or transmission at twice the incident frequency with a photomultiplier tube.

SHG measurements have been made on a large number of interfacial systems, usually to monitor adsorption and surface coverage [95]. In addition to adsorption of vapors onto metals, SHG has been used to study the electrochemical interface [96] and liquid–liquid interfaces [97, 98]. An important feature of SHG is its rapid time response, allowing time-resolved measurements down to picoseconds of surface processes such as band flattening at a semiconductor surface [95, 96].

Additional techniques are listed in Table VIII-1 and discussed in Chapters XVII and XVIII.

5. Problems

1. A LEED pattern is obtained for the (111) surface of an element that crystallizes in the face-centered close-packed system. Show what the pattern should look like in symmetry appearance. Consider only first-order nearest-neighbor diffractions.

2. The listing of techniques in Table VIII-1 is not a static one. It is expanded over what it was a few years ago and is continuing to expand. Try, in an imaginative yet serious manner, to suggest techniques not listed in the table. Explain what their values might be and, of course, propose a suitable acronym for each.

3. Discuss briefly which techniques listed in Table VIII-1 give information that is averaged over, that is, is representative of a macroscopic region of surface and which ones give information characteristic of a particular microscopic region. Take the dividing line of macroscopic versus microscopic to be about 1000 atoms in size.

4. Use Equation VIII-1 to determine the effective mass of the cantilever if the cantilever has a spring constant $C = 20$ N/m, the minimum detectable force gradient is $\partial F/\partial z = 4 \times 10^{-3}$ N/m, and the frequency shift is 200 kHz. How does the frequency shift depend on distance from the surface if the force has a $1/z^2$ distance dependence?

5. Derive Eqs. VIII-5 and VIII-6.

6. Derive Eq. VIII-10.

7. Ehrlich and co-workers [61] found that in a study of a tungsten surface, plots of log (i/V^2) versus $10^4/V$ gave straight lines, as expected from the Fowler-Nordheim equation. Their slope for a clean tungsten surface was -2.50, from which they calculated Φ to be 4.50 V. Six minutes after admitting a low pressure of nitrogen gas, the slope of the same type of plot was found to be -2.35. Calculate Φ for the partially nitrogen covered surface.

8. The relative intensity of a certain LEED diffraction spot is 0.25 at 300 K and 0.050 at 570 K using 390-eV electrons. Calculate the Debye temperature of the crystalline surface (in this case of Ru metal).

9. Calculate the energies of x-rays, electrons, $^4He^+$ ions, and H_2 molecules that correspond to a wavelength of 1.4 Å.

TABLE VIII-1
Techniques for Studying Surface Structure and Composition

Acronym	Technique and References	Description	Information
	Microscopy		
ADAM*	Angular-distribution Auger microscopy [85]	Surface atoms silhouetted by Auger electrons from atoms in bulk	Surface structure
AFM*	Atomic force microscopy [9, 47, 99]	Force measured by cantilever deflection as probe scans the surface	Surface structure
AM	Acoustic microscopy [100]	High-frequency acoustic waves are rastered across sample	Surface and below-surface structure
FEM*	Field emission microscopy [62, 101, 102]	Electrons are emitted from a tip in a high field	Surface structure
FIM*	Field ion microscopy [63, 62, 103]	He ions are formed in a high field at a metal tip	Surface structure
NSOM	Near-field scanning optical microscopy [103a]	Light from a sharp tip scatters off sample	Surface structure to 3 nm
PTM*	Photon tunneling microscopy [12]	An interface is probed with an evanescent wave produced by internal reflection of the illuminating light	Surface structure
SECM*	Scanning electrochemical microscopy [40]	An STM serves as microelectrode to reduce electroactive species	Electrochemical reactions on surfaces
SEM*	Scanning electron microscopy [7, 10, 14]	A beam of electrons scattered from a surface is focused	Surface morphology
SFM	Scanning force microscopy	Another name for AFM	Surface structure
STM*	Scanning tunneling microscopy [9, 19, 31]	Tunneling current from probe scans a conducting surface	Surface structure
SIAM*	Scanning interferometric apertureless microscopy [103b]	Laser light is reflected off the substrate, and scattering between an AFM tip and sample is processed interferometrically	Surface structure
	Diffraction		
HEED*	High-energy electron diffraction [104]	Diffraction of elastically back-scattered electrons (~20 keV, grazing incidence)	Surface structure
LEED*	Low-energy electron diffraction [62, 75, 105]	Elastic backscattering of electrons (10–200 eV)	Surface structure
RHEED	Reflection high-energy electron diffraction [78, 106]	Similar to HEED	Surface structure, composition
SHEED	Scanning high-energy electron diffraction [106]	Scanning version of HEED	Surface heterogeneity

TABLE VIII-1   (*Continued*)

Acronym	Technique and References	Description	Information
PED	Photoelectron diffraction [107–109]	x-rays (40–1500 eV) eject photoelectrons; intensity measured as a function of energy and angle	Surface structure
PhD, NPD, XPD	Other names for PED	Same as PED	Same as PED
APD, ARPEFS	Azimuthal PED, angle-resolved photoemission extended fine structure	Same as PED	Same as PED
Spectroscopy of Emitted X-rays or Photons			
IRE	Infrared emission [110]	Infrared emission from a metal surface is affected in angular distribution by adsorbed species	Orientation of adsorbed molecules
XES, SXES*	Soft x-ray emission spectroscopy [111]	An x-ray or electron beam ejects K electrons and the spectrum of the resulting x-rays is measured	Energy levels and chemical state of adsorbed molecules; surface composition
Spectroscopy of Emitted Electrons			
AES*	Auger electron spectroscopy [77, 112–114, 117]	An incident high-energy electron ejects an inner electron from an atom; an outer electron (e.g., L) falls into the vacancy and the released energy is given to an ejected Auger electron	Surface composition
ARAES	Angle-resolved AES [85, 115]		
PI, PIS	Penning ionization [116, 118]	Auger deexcitation of metastable noble-gas atoms	
MDS	Metastable deexcitation spectroscopy [119]	Same as PI	Surface valence–electron states
EELS*	Electron-energy-loss spectroscopy [120–127]	Incident electrons are scattered inelastically	Surface energy states; surface composition
CELS, EIS	Characteristic-energy-loss spectroscopy, electron-impact spectroscopy [128]	Same as EELS	Same as EELS
HREELS*	High-resolution electron energy-loss spectroscopy [129, 130]	Same as EELS	Identification of adsorbed species through their vibrational energy spectrum

TABLE VIII-1 *(Continued)*

Acronym	Technique and References	Description	Information
XPS*	X-ray photoelectron spectroscopy [131–137]	Monoenergetic x-rays eject electrons from various atomic levels; the electron energy spectrum is measured	Surface composition, oxidation state
ESCA*	Electron spectroscopy for chemical analysis [106, 138–142]	Same as XPS	Same as XPS
OSEE	Optically stimulated exoelectron emission [143]	Light falling on a surface in a potential field produces electron emission	Presence and nature of adsorbates
PES	Photoelectron spectroscopy	Same as XPS with UV light	Similar to XPS
UPS	Ultraviolet photoemission spectroscopy	Same as PES	Same as PES
ARPS	Angle-resolved PES [144–146]	Same as XPS, UPS	Same as XPS, UPS
INS	Ion neutralization spectroscopy [147]	An inert gas hitting surface is neutralized with the ejection of an Auger electron from a surface atom	Kinetics of surface reactions; chemisorption

Spectroscopy of Emitted Ions or Molecules

Acronym	Technique and References	Description	Information
DIET	Desorption induced by electronic transitions [147a]	General class of desorption and reaction phenomena induced by electron or photon bombardment	Same as ESD and PSD
ESD	Electron-stimulated (impact) desorption [148, 149]	An electron beam (100–200) eV ejects ions from a surface	Surface sites and adsorbed species
ESDIAD*	Electron-stimulated desorption ion angular distribution [150–152]	A LEED-like pattern of ejected ions is observed	Orientation of adsorbed species
ISS*	Ion scattering spectroscopy [153, 154]	Inelastic backscattering of ions (~1 keV ion beam)	Surface composition
LEIS*	Low-energy ion scattering [155–157]	A monoenergetic beam of rare-gas ions is scattered elastically by surface atoms	Surface composition
MBRS	Molecular beam spectroscopy [158]	A modulated molecular beam hits the surface and the time lag for reaction products is measured	Kinetics of surface reactions; chemisorption
MSS	Molecule surface scattering [159–161]	Translational and rotational energy distribution of a scattered molecular beam	Quantum mechanics of scattering processes

TABLE VIII-1   (*Continued*)

Acronym	Technique and References	Description	Information
PSD	Photon-stimulated desorption [149, 162–165]	Incident photons eject adsorbed molecules	Desorption mechanisms and dynamics
SIMS	Secondary-ion mass spectroscopy [106, 166–168] (L-SIMS: liquids) [169, 170]	Ionized surface atoms are ejected by impact of ~1 keV ions and analyzed by mass spectroscopy	Surface composition
TPD	Temperature-programmed desorption [171, 172]	The surface is heated and chemisorbed species desorb at characteristic temperatures	Characterization of surface sites and desorption kinetics
TDS, FDS	Thermal desorption spectroscopy, Flash desorption spectroscopy [173]	Similar to TPD	Similar to TPD
TPRS	Temperature-programmed reaction spectroscopy [174]	Same as TPD	Same as TPD

Spectroscopy of Emitted Neutrons

SANS	Small-angle neutron scattering [175, 176]	Thermal or cold neutrons are scattered elastically or inelastically	Surface vibrational states, pore size distribution; suspension structure

Incident-Beam Spectroscopy

APS	Appearance potential spectroscopy (see AES)	Intensity of emitted x-ray or Auger electrons is measured as a function of incident electron energy	Surface composition
AEAPS, SXAPS	Auger electron APS, Soft x-ray APS	Same as APS	Same as APS
EXAFS	Extended x-ray absorption fine structure [177, 178]	Variation of x-ray absorption as a function of x-ray energy beyond an absorption edge; the probability is affected by backscattering of the emitted electron from adjacent atoms	Number and interatomic distance of surface atoms
NEXAFS	Near-edge EXAFS	Variant of EXAFS	Variant of EXAFS
SEXAFS	Surface EXAFS	Same as EXAFS	Same as EXAFS
XANES	X-ray adsorption near-edge spectroscopy [178a]	Same as NEXAFS	Same as NEXAFS

TABLE VIII-1   (*Continued*)

Acronym	Technique and References	Description	Information
	Photophysical and Photochemical Effects		
CDAD	Circular dichroism photo-electron angular distribution [179]	Uses circularly polarized light	Orientation of adsorbed species
LIF	Laser-induced fluorescence [180–184]	Incident laser beam excites adsorbed species and surface atoms, leading to emission	Excited-state processes; surface analysis
LISC	Laser-induced surface chemistry [185, 186]	Similar to LIF; chemical changes occur	Photochemistry, desorption
REMPI	Resonance-enhanced multi-photon ionization [182]	Laser-induced ionization	
PAS	Photoacoustic spectros-copy; optoacoustic spectroscopy [187–191]	Modulated incident infrared radiation is absorbed, producing acoustic signal in surrounding gas	Vibrational states of surface and adsorbed species
	Incident UV-Visible Absorption, Reflection		
LA	Light absorption [192, 193]	UV-visible adsorption by reflection or transmission	Nature of adsorbed species
ELL	Ellipsometry [194, 195]	Depolarization of reflected light	Thickness of adsorbed film
	Infrared Spectroscopy		
IR, FTIR	Infrared absorption; Fourier transform IR [196–199]	Transmission absorption spectra	Analysis; state of adsorbed species
ATR	Attenuated total reflection [200, 201]	Reflected beam attenuated by absorption	Similar to IR
EIR	Enhanced IR [202]	Reflection from backside of metal film exposed to sample	Enhanced sensitivity
DRIFTS	Diffuse reflectance infrared Fourier-transform spectroscopy	Same as IR	Same as IR
IRE	Infrared emission [208, 209]	IR emission spectrum observed	High resolution, long wavelength
IRRAS	Infrared reflection-absorption spectros-copy [203–207]	Grazing-incidence polarized IR beam	Enhanced sensitivity
PMIRRAS	Polarization modulation IRRAS [203]		

TABLE VIII-1   (*Continued*)

Acronym	Technique and References	Description	Information
RS	Raman spectroscopy [210, 211]	Scattered monochromatic visible light shows frequency shifts corresponding to vibrational states of surface material	Can observe IR-forbidden absorptions; low sensitivity
RRS	Resonance Raman spectroscopy [212, 213]	Incident light is of wave length corresponding to an absorption band	Enhanced sensitivity
SERS	Surface-enhanced Raman spectroscopy [214–217]	Same as RS but with rough-ened metal (usually silver) substrate	Greatly enhanced intensity
SERRS	Surface-enhanced RRS [214, 217]	Same as SERS but using a wavelength corresponding to an absorption band	Same as SERS
Magnetic Spectroscopies			
ESR* (EPR)	Electron spin (paramag-netic) resonance [218–222]	Chemical shift of splitting of electron spin states in a magnetic field	Chemical state of adsorbed species
NMR*	Nuclear magnetic resonance [223, 224]	Chemical shift of splitting of nuclear spin states in a magnetic field ^{1}H [225], ^{13}C [226, 227], ^{15}N [228], ^{19}F [229], ^{129}Xe [230]	Chemical state; diffusion of adsorbed species
Other Techniques			
IETS, IET	Inelastic electron tunneling spectroscopy [231–232]	Current is measured as applied voltage is varied across a metal–film–metal sandwich	Vibrational spectroscopy of thin films
MS	Mössbauer Spectroscopy [233–236]	Chemical shift of nuclear energy states, usually of iron	Chemical state of atoms
SHG	Optical second-harmonic generation [95, 96]	A high-powered pulsed laser generates frequency-doubled response due to the asymmetry of the interface	Adsorption and surface coverage; rapid surface changes

*Covered in text.

General References

P. Echlin, ed., *Analysis of Organic and Biological Surfaces*, Wiley, New York, 1984.

C. S. Fadley, in *Electron Spectroscopy, Theory, Techniques, and Applications*, Vol. 2, C. R. Brundle and A. D. Baker, eds., Pergamon, New York, 1978.

R. Gomer, *Field Emission and Field Ionization*, Harvard University Press, Cambridge, MA, 1961.

N. B. Hannay, *Treatise on Solid State Chemistry*, Vol. 6A, *Surfaces I*, Plenum, New York, 1976.

P. K. Hansma, ed., *Tunneling Spectroscopy*, Plenum, New York, 1982.

D. M. Hercules, "Surface Analysis: An Overview," *Crit. Rev. Surf. Chem.*, **1**, 243 (1992).

A. T. Hubbard, ed., *Surface Imaging and Visualization*, CRC Press, Boca Raton, FL.

R. E. Lee, *Scanning Electron Microscopy and X-Ray Microanalysis*, PTR Prentice Hall, Englewood Cliffs, NJ, 1993.

S. N. Magonov and M.-H. Whangbo, *Surface Analysis with STM and AFM*, VCH Publishers, New York, 1996.

O. Marti and M. Amrein, eds., *STM and AFM in Biology*, Academic, San Diego, CA, 1993.

E. W. Müller and T. T. Tsong, *Field Ion Microscopy*, American Elsevier, New York, 1969.

G. A. Somorjai, *Science*, **201**, 489 (1978).

G. A. Somorjai, *Introduction to Surface Chemistry and Catalysis*, Wiley, New York, 1994.

J. P. Thomas and A. Cachard, eds., *Material Characterization Using Ion Beams*, Plenum, New York, 1976.

P. R. Thornton, *Scanning Electron Microscopy*, Chapman and Hall, 1968. See also *Scanning Electron Microscopy: Systems and Applications*, The Institute of Physics, London, 1973.

T. Wolfram, ed., *Inelastic Electron Tunneling Spectroscopy*, Springer-Verlag, New York, 1978.

Textual References

1. N. Sheppard, *Pure Appl. Chem.*, **63**, 887 (1991).

2. G. A. Somorjai, *Introduction to Surface Chemistry and Catalysis*, Wiley, New York, 1994.

3. J. H. Block, A. M. Bradshaw, P. C. Gravelle, J. Haber, R. S. Hansen, M. W. Roberts, N. Sheppard, and K. Tamaru, *Pure Appl. Chem.*, **62**, 2297 (1990).

4. G. S. Selwyn and M. C. Lin, *Lasers as Reactants and Probes in Chemistry*, Howard University Press, Washington, DC, 1985.

5. M. C. Lin and G. Ertl, *Ann. Ref. Phys. Chem.*, **37**, 587 (1986).

6. G. A. Somorjai and B. E. Bent, *Prog. Colloid Polym. Sci.*, **70**, 38 (1985).

7. J. J. A. Gardella, *Trends Anal. Chem.*, **3**, 129 (1984).

8. I. C. Sanchez, *Physics of Polymer Surfaces and Interfaces*, Butterworth-Heinemann, 1992.

9. J. E. Frommer, *Angew. Chem., Int. Ed. Engl.*, **31**, 1298 (1992).

10. J. D. Andrade, *Surface and Interfacial Aspects of Biomedical Polymers*, Plenum, New York, 1985.

11. C. J. R. Sheppard, *Advances in Optical and Electron Microscopy*, Vol. 10, Academic, London, 1987, pp. 1–98.

12. J. M. Guerra, M. Srinivasarao, and R. S. Stein, *Science*, **262**, 1395 (1993).

13. J. Majewski, L. Margulis, D. Jacquemain, F. Leveiller, C. Böhm, T. Arad, Y. Talmon, M. Lahav, and L. Leiserowitz, *Science*, **261**, 899 (1993).

14. R. E. Lee, *Scanning Electron Microscopy and X-Ray Microanalysis*, Prentice-Hall, Englewood Cliffs, NJ, 1993.

15. F. L. Baker and L. H. Princen, *Encyclopedia of Polymer Science*, Vol. 15, Wiley, New York, 1971, p. 498.

16. L. H. Princen, *Treatise on Coatings*, Vol. 2, Marcel Dekker, New York, 1976, Chapter 7.

17. X. L. Wu and C. M. Lieber, *Prog. Inorg. Chem.*, **39**, 431 (1991).

18. J. E. Frommer, *Imaging Chemical Bonds by SPM*, Kluwer Academic Publishers, 1995, pp. 551–566.

19. G. Binnig and H. Rohrer, *Angew. Chem., Int. Ed. Engl.* **26**, 606 (1987).

20. C. E. D. Chidsey, D. N. Loiacono, T. Sleator, and S. Nakahara, *Surf. Sci.*, **200**, 45 (1988).

21. R. Wiesendanger, G. Tarrach, D. Bürgler, and H. J. Güntherodt, *Europhys. Lett.*, **12**, 57 (1990).

22. I. C. Oppenheim, D. J. Trevor, C. E. D. Chidsey, P. L. Trevor, and K. Sieradzki, *Science*, **254**, 687 (1991).

23. D. J. Trevor and C. E. D. Chidsey, *J. Vac. Sci. Technol. B*, **9**, 964 (1991).

24. X. Xu, S. M. Vesecky, and D. W. Goodman, *Science*, **258**, 788 (1992).

25. B. S. Swartzentruber, *Phys. Rev. Lett.*, **76**, 459 (1996).

26. H. Fuchs, W. Schrepp, and H. Rohrer, *Surf. Sci.*, **181**, 39 (1987).

27. A. Stemmer, R. Reichelt, A. Engel, J. P. Rosenbusch, M. Ringger, H. R. Hidber, and H. J. Güntherodt, *Surf. Sci.*, **181**, 393 (1987).

28. G. Travaglini, H. Rohrer, M. Amrein, and H. Gross, *Surf. Sci.*, **181**, 380 (1987).

29. G. Binnig and H. Rohrer, *Surf. Sci.*, **126**, 236 (1983).

30. C. F. Quate, *Physics Today*, 26 (Aug. 1986).

31. R. Sonnenfeld and P. K. Hansma, *Appl. Phys. Lett.*, **232**, 211 (1986).

32. J. Nogami, S. Park, and C. F. Quate, *Phys. Rev. B*, **36**, 6221 (1987).

33. D. G. Frank, N. Batina, J. W. McCarger, and A. T. Hubbard, *Langmuir*, **5**, 1141 (1989).

34. D. G. Frank and A. T. Hubbard, *Langmuir*, **6**, 1430 (1990).

35. T. Gritsch, D. J. Coulman, R. J. Behm, and G. Ertl, *Phys. Rev. Lett.*, **63**, 1086 (1989).

36. F. M. Chua, Y. Kuk, and P. J. Silverman, *Phys. Rev. Lett.*, **63**, 386 (1989).

37. D. J. Coulman, J. Wintterlin, R. J. Behm, and G. Ertl, *Phys. Rev. Lett.*, **64**, 1761 (1990).

38. M. D. Kirk, J. Nogami, A. A. Baski, D. B. Mitzi, A. Kapitulnik, T. H. Geballe, and C. F. Quate, *Science*, **242**, 1673 (1988).

39. J. V. Barth, H. Brune, G. Ertl, and R. J. Behm, *Phys. Rev. B*, **42**, 9307 (1990).

40. A. J. Bard, F. R. Fan, D. T. Pierce, P. R. Unwin, D. O. Wipf, and F. Zhou, *Science*, **254**, 68 (1991).

41. J. A. Stroscio and D. M. Eigler, *Science*, **254**, 1319 (1991).

42. I. W. Lyo and P. Avourix, *Science*, **253**, 173 (1991).

43. C. A. Widrig, C. A. Alves, and M. D. Porter, *J. Am. Chem. Soc.*, **113**, 2805 (1991).

44. Y. Z. Li, M. Chander, J. C. Patrin, J. H. Weaver, L. P. F. Chibante, and R. E. Smalley, *Science*, **253**, 429 (1991).

45. J. P. Rabe and S. Buchholz, *Science*, **253**, 424 (1991).

46. L. Häussling, B. Michel, H. Ringsdorf, and H. Rohrer, *Angew. Chem., Int. Ed. Engl.*, **30**, 569 (1991).

47. O. Marti and M. Amrein, *STM and SFM in Biology*, Academic, San Diego, CA, 1993.

48. D. P. E. Smith, A. Bryant, C. F. Quate, J. P. Rabe, C. Gerber, and J. D. Swalen, *Proc. Natl. Acad. Sci.* (USA), **84**, 969 (1987).

49. J. C. Poler, R. M. Zimmerman, and E. C. Cox, *Langmuir*, **11**, 2689 (1995).

50. H. J. Butt, M. Jaschke, and W. Ducker, *Bioelectrochem. Bioenerget.* **38**, 191 (1995).

51. W. A. Ducker, Z. Xu, and J. N. Israelachvili, *Langmuir*, **10**, 3279 (1994).

52. R. M. Overney et al., *Nature*, **359**, 133 (1992).

53. S. Biggs, M. K. Chow, C. F. Zukoski, and F. Greiser, *J. Colloid Interface Sci.*, **160**, 511 (1993).

54. S. Biggs, P. Mulvaney, C. F. Zukoski, and F. Greiser, *J. Am. Chem. Soc.*, **116**, 9150 (1994).

55. O. Mori and T. Imae, *Langmuir*, **11**, 4779 (1995).

56. H. J. Butt and B. Gerharz, *Langmuir*, **11**, 4735 (1995).

57. J. Yang, J. H. Fendler, T. C. Jao, and T. Laurion, *Microsc. Res. Tech.*, **27**, 402 (1994).

58. D. Anselmetti, R. Lüthi, E. Meyer, T. Richmond, M. Dreier, J. E. Frommer, and H. J. Güntherodt, *Nanotechnology*, **5**, 87 (1994).

59. E. W. Müller, *Z. Phys.*, **37**, 838 (1936).

60. E. W. Müller, *Z. Phys.*, **120**, 261 (1942).

61. G. Ehrlich and F. G. Hudda, *J. Chem. Phys.*, **35**, 1421 (1961).

62. G. A. Somorjai, *Principles of Surface Chemistry*, Prentice-Hall, Englewood Cliffs, NJ, 1972.

63. E. W. Müller and T. T. Tsong, *Field Ion Microscopy*, American Elsevier, New York, 1969.

64. G. Ehrlich, *J. Appl. Phys.*, **15**, 349 (1964).

65. R. H. Fowler and L. W. Nordheim, *Proc. Roy. Soc.* (London), **A112**, 173 (1928).

66. R. Gomer, *Field Emission and Field Ionization*, Harvard University Press, Cambridge, MA, 1961.

67. R. C. Mowrey and D. J. Kouri, *J. Chem. Phys.*, **84**, 6466 (1986).

68. S. S. Brenner, *Surf. Sci.*, **2**, 496 (1964).

69. C. J. Davisson and L. H. Germer, *Phys. Rev.*, **30**, 705 (1927).

70. A. U. MacRae, *Science*, **139**, 379 (1963); see also L. H. Germer and A. U. MacRae, *J. Appl. Phys.*, **33**, 2923 (1962).

71. G. A. Somorjai, *Angew. Chem., Int. Ed. Engl.*, **16**, 92 (1977).

72. L. K. Kesmodel and G. A. Somarjai, *Acc. Chem. Res.*, **9**, 392 (1976).

73. See M. Scheffler, K. Kambe, and F. Forstmann, *Solid State Commun.*, **25**, 93 (1978).

74. T. W. Orent and R. S. Hansen, *Surf. Sci.*, **67**, 325 (1977).

75. H. Ohtani, C. T. Kao, M. A. Van Hoeve, and G. A. Somorjai, *Prog. Surf. Sci.*, **23**, 155 (1987).

76. J. J. Burton and G. Jura, *Structure and Chemistry of Solid Surfaces*, G. Somorjai, ed., Wiley, New York, 1969.

77. G. A. Somorjai and F. J. Szalkowski, *Adv. High Temp. Chem.*, **4**, 137 (1971).

78. G. Ertl, *Langmuir*, **3**, 4 (1987).

79. A. L. Cabrera, N. D. Spencer, E. Kozak, P. W. Davies, and G. A. Somorjai, *Rev. Sci. Instrum.*, **53**, 1888 (1982).

80. L. E. Firment and G. A. Somorjai, *J. Chem. Phys.*, **63**, 1037 (1975).

81. J. C. Buchholz and G. A. Somorjai, *J. Chem. Phys.*, **66**, 573 (1977).

82. G. A. Somorjai and F. J. Szalkowski, *Adv. High Temp. Chem.*, **4**, 137 (1971).

83. T. Smith, *J. Appl. Phys.*, **43**, 2964 (1972).

84. N. Batina, O. M. R. Chyan, D. G. Frank, T. Golden, and A. T. Hubbard, *Naturwissenschaften*, **77**, 557 (1990).

85. D. G. Frank, O. M. R. Chyan, T. Golden, and A. T. Hubbard, *J. Phys. Chem.*, **97**, 3829 (1993).

86. D. G. Frank, O. M. R. Chyan, T. Golden, and A. T. Hubbard, *J. Phys. Chem.*, **98**, 1895 (1994).

87. U. Diebold, J. M. Pan, and T. E. Madey, *Surf. Sci.*, **331–333**, 845 (1995).

88. H. P. Steinrück, F. Pesty, L. Zhang, and T. E. Madey, *Phys. Rev. B*, **51**, 2427 (1995).

89. J. M. Pan and T. E. Madey, *Catal. Lett.*, **20**, 269 (1993).

90. D. W. Dwight and W. M. Riggs, *J. Colloid Interface Sci.*, **47**, 650 (1974).

91. D. C. Muddiman, A. H. Brockman, A. Proctor, M. Houalla, and D. M. Hercules, *J. Phys. Chem.*, **98**, 11570 (1994).

92. H. H. Brongersma and P. M. Mul, *Surf. Sci.*, **35**, 393 (1973).

93. F. P. Netzer and T. E. Madey, *Surf. Sci.*, **119**, 422 (1982).

94. M. D. Alvey, J. T. Yates, Jr., and K. J. Uram, *J. Chem. Phys.*, **87**, 7221 (1987).

95. R. M. Corn and D. A. Higgins, *Chem. Rev.*, **94**, 107 (1994).

96. J. M. Lantz and R. M. Corn, *J. Phys. Chem.*, **98**, 9387 (1994).

97. D. A. Higgins, R. R. Naujok, and R. M. Corn, *Chem. Phys. Lett.*, **213**, 485 (1993).

98. R. R. Naujok, D. A. Higgins, D. G. Hanken, and R. M. Corn, *J. Chem. Soc. Faraday Trans.*, **91**, 1411 (1995).

99. O. Marti, H. O. Ribi, B. Drake, T. R. Albrecht, C. F. Quate, and P. K. Hansma, *Science*, **239**, 50 (1988).

100. C. F. Quate, *Physics Today*, 34 (Aug. 1985).

101. B. E. Koel and G. A. Somorjai, *Catalysis: Science and Technology*, Vol. 38, Springer-Verlag, New York, 1983.

102. M. Blaszczyszyn, R. Blaszczyszyn, R. Meclewski, A. J. Melmed, and T. E. Madey, *Surf. Sci.*, **131**, 433 (1983).

103. G. Ehrlich, *Adv. Catal.*, **1**, 255 (1963).

103a. F. Zenhausern, M. P. O'Boyle, and H. K. Wickramasinghe, *Appl. Phys. Lett.*, **65**, 1623 (1994).

103b. F. Zenhausern, Y. Martin, and H. K. Wickramasinghe, *Science*, **269**, 1083 (1995).

104. M. R. Leggett and R. A. Armstrong, *Surf. Sci.*, **24**, 404 (1971).

105. J. W. Anderegg and P. A. Thiel, *J. Vac. Sci. Technol.*, **A4**, 1367 (1986).

106. E. G. MacRae and H. D. Hagstrom, *Treatise on Solid State Chemistry*, Vol. 6A, Part I, Plenum, New York, 1976.

107. S. D. Kevan, D. H. Rosenblatt, D. R. Denley, B. C. Lu, and D. A. Shirley, *Phys. Rev. B*, **20**, 4133 (1979).

108. L. G. Petersson, S. Kono, N. F. T. Hall, S. Goldberg, J. T. Lloyd, and C. S. Fadley, *Mater. Sci. Eng.*, **42**, 111 (1980).

109. M. Sagurton, E. L. Bullock, and G. S. Fadley, *Surf. Sci.*, **182**, 287 (1987).

110. R. G. Greenler, *Surf. Sci.*, **69**, 647 (1977).

111. S. Evans, J. Pielaszek, and J. M. Thomas, *Surf. Sci.*, **55**, 644 (1976).

112. E. W. Plummer, C. T. Chen, and W. K. Ford, *Surf. Sci.*, **158**, 58 (1985).

113. M. C. Burrell and N. R. Armstrong, *Langmuir*, **2**, 37 (1986).

114. G. D. Stucky, R. R. Rye, D. R. Jennison, and J. A. Archer, *J. Am. Chem. Soc.*, **104**, 5951 (1982).

115. T. Matusudara and M. Onchi, *Surf. Sci.*, **72**, 53 (1978).

116. W. Sesselmann, B. Woratschek, G. Ertl, and J. Küppers, *Surf. Sci.*, **146**, 17 (1984).

117. M. C. Burrell and N. R. Armstrong, *Langmuir*, **2**, 30 (1986).

118. B. Woratschek, W. Sesselman, J. Küppers, G. Ertl, and H. Haberland, *Surf. Sci.*, **180**, 187 (1987).

119. W. Sesselman, B. Woratschek, J. Küppers, G. Ertl, and H. Haberland, *Phys. Rev. B*, **35**, 1547 (1987).

120. Y. Sakisaka, K. Akimoto, M. Nishijima, and M. Onchi, *Solid State Commun.*, **29**, 121 (1979).

121. R. J. Gorte, L. D. Schmidt, and B. A. Sexton, *J. Catal.*, **67**, 387 (1981).

122. M. C. Burrell and N. R. Armstrong, *J. Vac. Sci. Technol.*, **A1**, 1831 (1983).

123. E. M. Stuve, R. J. Madix, and B. A. Sexton, *Chem. Phys. Lett.*, **89**, 48 (1982).

124. B. A. Sexton, *Surf. Sci.*, **163**, 99 (1985).

125. N. D. Shinn and T. E. Madey, *J. Chem. Phys.*, **83**, 5928 (1985).

126. H. J. Freund, R. P. Messmer, W. Spiess, H. Behner, G. Wedler, and C. M. Kao, *Phys. Rev. B*, **33**, 5228 (1986).

127. B. A. Sexton, *Appl. Phys.*, **A26,** 1 (1981).

128. S. Trajmar, *Acc. Chem. Res.*, **13,** 14 (1980).

129. N. D. Shinn and T. F. Madey, *Surf. Sci.*, **173,** 379 (1986).

130. G. E. Mitchell, P. L. Radloff, C. M. Greenlief, M. A. Henderson, and J. M. White, *Surf. Sci.*, **183,** 403 (1987).

131. W. F. Egelhoff, Jr., *Surf. Sci. Rep.*, **6,** 253 (1987).

132. B. A. Sexton, *Mat. Forum*, **10,** 134 (1987).

133. R. P. Messmer, S. H. Lamson, and D. R. Salahub, *Phys. Rev. B*, **25,** 3576 (1982).

134. S. K. Suib, A. M. Winiecki, and A. Kostapapas, *Langmuir*, **3,** 483 (1987).

135. K. W. Nebesny, K. Zavadil, B. Burrow, and N. R. Armstrong, *Surf. Sci.*, **162,** 292 (1985).

136. N. Kakuta, K. H. Park, M. F. Finlayson, A. J. Bard, A. Campion, M. A. Fox, S. E. Webber, and J. M. White, *J. Phys. Chem.*, **89,** 5028 (1985).

137. D. E. Amory, M. J. Genet, and P. G. Rouxhet, *Surf. Interface Anal.*, **11,** 478 (1988).

138. G. L. Grobe III, J. A. Gardella, Jr., W. L. Hopson, W. P. McKenna, and E. M. Eyring, *J. Biomed. Mat. Res.*, **21,** 211 (1987).

139. J. L. Grant, T. B. Fryberger, and P. C. Stair, *Appl. Surf. Sci.*, **26,** 472 (1986).

140. T. J. Hook, J. A. Gardella, Jr., and L. Salvati, Jr., *J. Mat. Res.*, **2,** 132 (1987).

141. T. E. Madey, C. D. Wagner, and A. Joshi, *J. Electron. Spectrosc. & Related Phenomena*, **10,** 359 (1977).

142. T. J. Hook, J. A. Gardella, Jr., and L. Salvati, Jr., *J. Mat. Res.*, **2,** 117 (1987).

143. Y. Momose, T. Ishll, and T. Namekawa, *J. Phys. Chem.*, **84,** 2908 (1980).

144. W. F. Egelhoff, Jr., *J. Vac. Sci. Technol.*, **A3,** 1511 (1985).

145. R. C. White, C. F. Fadley, M. Sagurton, and Z. Hussain, *Phys. Rev. B*, **34,** 5226 (1986).

146. R. C. White, C. S. Fadley, M. Sagurton, P. Roubin, D. Chandesris, J. Lecante, C. Guillot, and Z. Hussain, *Solid State Comm.*, **59,** 633 (1986).

147. G. A. Somorjai, in *Treatise on Solid State Chemistry*, N. B. Hannay, ed., Vol. 6A, Part I, Plenum, New York, 1976.

147a. A. R. Burns, E. B. Stechel, and D. R. Jennison, eds., *Desorption Induced by Electronic Transitions, DIET V*, Springer Verlag, Heidelberg, 1994.

148. T. E. Madey, *Surf. Sci.*, **94,** 483 (1980).

149. R. L. Kurtz, R. Stockbauer, and T. E. Madey, *Nuc. Instru. Meth. Phys. Res.*, **B13,** 518 (1986).

150. T. E. Madey, J. M. Yates, Jr., A. M. Bradshaw, and F. M. Hoffmann, *Surf. Sci.*, **89,** 370 (1979).

151. N. D. Shinn and T. E. Madey, *J. Chem. Phys.*, **83,** 5928 (1985).

152. K. Bange, T. E. Madey, J. K. Sass, and E. M. Stuve, *Surf. Sci.*, **183,** 334 (1987).

153. J. C. Carver, M. Davis, and D. A. Goetsch, ACS Symposium Series No. 288, M. L. Deviney and J. L. Gland, eds., American Chemical Society, Washington, DC, 1985.

154. K. J. Hook and J. A. Gardella, Jr., *Macromolecules*, **20,** 2112 (1987).

155. S. H. Oberbury, P. C. Stair, and P. A. Agron, *Surf. Sci.*, **125,** 377 (1983).

156. S. M. Davis, J. C. Carber, and A. Wold, *Surf. Sci.*, **124,** L12 (1983).

157. T. J. Hook, R. L. Schmitt, J. A. Gardella, Jr., L. S. Salvati, Jr., R. L. Chin, *Anal. Chem.*, **58,** 1285 (1986).

158. H. J. Robota, W. Vielhaber, M. C. Lin, J. Segner, and G. Ertl, *Surf. Sci.*, **155,** 101 (1985).

159. R. B. Gerber, A. T. Yinnon, Y. Shimoni, and D. J. Kouri, *J. Chem. Phys.*, **73,** 4397 (1980).

160. A. Mödl, T. Gritsch, F. Budde, T. J. Chuang, and G. Ertl, *Phys. Rev. Lett.*, **57,** 384 (1986).

161. R. C. Mowrey and D. J. Kouri, *J. Chem. Phys.*, **84,** 6466 (1986).

162. P. C. Stair and E. Weitz, *Optical Soc. Am. B*, **4,** 255 (1987).

163. T. J. Chuang, H. Seki, and I. Hussla, *Surf. Sci.*, **158,** 525 (1985).

164. T. E. Madey, *Science*, **234,** 316 (1986).

165. S. van Smaalen, H. F. Arnoldus, and T. F. George, *Phys. Rev. B*, **35,** 1142 (1987).

166. P. L. Radloff and J. M. White, *Accts. Chem. Res.*, **19,** 287 (1986).

167. P. L. Radloff, G. E. Mitchell, C. M. Greenlief, J. M. White, and C. A. Mims, *Surf. Sci.*, **183,** 377 (1987).

168. K. J. Hook, T. J. Hook, J. W. Wandass, and J. A. Gardella, Jr., *Appl. Surf. Sci.*, **44,** 29 (1990).

169. W. V. Ligon, Jr. and S. Dorn, *Int. J. Mass Spectrom. Ion Proc.*, **63,** 315 (1985).

170. W. V. Ligon, Jr., "Evaluating the Composition of Liquid Surfaces Using Mass Spectrometry," in *Biological Mass Spectrometry*, Elsevier, Amsterdam, 1990.

171. T. M. Apple and C. Dybowski, *Surf. Sci.*, **121,** 243 (1982).

172. L. L. Lauderback and W. N. Delgass, ACS Symposium Series No. 248, T. E. Whyte, Jr., R. A. Dalla Betta, E. G. Derouane, and R. T. K. Baker, eds., American Chemical Society, Washington, DC, 1984.

173. L. W. Anders and R. S. Hansen, *J. Chem. Phys.*, **62,** 4652 (1975).

174. B. A. Sexton and R. J. Madix, *Surf. Sci.*, **105,** 177 (1981).

175. K. H. Rieder, *Surf. Sci.*, **26,** 637 (1971).

176. C. J. Glinka, L. C. Sander, S. A. Wise, M. L. Hunnicutt, and C. H. Lochmuller, *Anal. Chem.*, **57,** 2079 (1985).

177. J. H. Sinfelt, G. H. Via, and F. W. Lytle, *Catal. Rev.-Sci. Eng.*, **26,** 81 (1984).

178. O. R. Melroy, M. G. Samant, G. L. Borges, J. G. Gordon II, L. Blum, J. H. White, M. J. Albarelli, M. McMillan, and H. D. Abruna, *Langmuir*, **4,** 728 (1988).

178a. M. J. Fay, A. Proctor, D. P. Hoffmann, M. Houalla, and D. M. Hercules, *Mikrochim. Acta*, **109,** 281 (1992).

179. R. L. Dubs, S. N. Dixit, and V. McKoy, *Phys. Rev. Lett.*, **54,** 1249 (1985).

180. S. Garoff, D. A. Weitz, M. S. Alvarez, and J. I. Gersten, *J. Chem. Phys.*, **81,** 5189 (1984).

181. J. T. Lin, Xi-Yi Huang, and T. F. George, *J. Opt. Soc. Am. B*, **4,** 219 (1987).

182. M. C. Lin and G. Ertl, *Ann. Rev. Phys. Chem.*, **37,** 587 (1986).

183. P. de Mayo, L. V. Natarajan, and W. R. Ware, ACS Symposium Series 278, M. A. Fox, ed., American Chemical Society, Washington, DC, 1985.

184. J. K. Thomas, *J. Phys. Chem.*, **91,** 267 (1987).

185. Xi-Yi Huang, T. F. George, and Jian-Min Yuan, *J. Opt. Soc. Am. B*, **2,** 985 (1985).

186. H. Van Damme, and W. K. Hall, *J. Am. Chem. Soc.*, **101,** 4373 (1979).

187. E. G. Chatzi, M. W. Urban, H. Ishida, J. L. Loenig, A. Laschewski, and H. Rings-dorf, *Langmuir*, **4,** 846 (1988).

188. L. Rothberg, *J. Phys. Chem.*, **91,** 3467 (1987).

189. N. D. Spencer, *Chemtech*, June, 1986, p. 378.

190. J. A. Gardella, Jr., G. L. Grobe III, W. L. Hopson, and E. M. Eyring, *Anal. Chem.*, **56,** 1169 (1984).

191. J. B. Kinney, R. H. Staley, C. L. Reichel, and M. S. Wrighton, *J. Am. Chem. Soc.*, **103,** 4273 (1981).

192. S. Garoff, R. B. Stephens, C. D. Hanson, and G. K. Sorenson, *Opt. Commun.*, **41,** 257 (1982).

193. S. Garoff, D. A. Weitz, T. J. Gramila, and C. D. Hanson, *Opt. Lett.*, **6,** 245 (1981).

194. S. Engström and K. Bäckström, *Langmuir*, **3,** 568 (1987).

195. C. G. Gölander, and E. Kiss, *J. Colloid Interface Sci.*, **121,** 240 (1988).

196. T. Ishikawa, S. Nitta, and S. Kondo, *J. Chem. Soc., Faraday Trans. I*, **82,** 2401 (1986).

197. M. L. Hair, *Vibrational Spectroscopies for Adsorbed Species*, ACS Symposium Series No. 137, A. T. Bell and M. L. Hair, eds., American Chemical Society, Washington, DC, 1980.

198. J. L. Robbins and E. Marucchi-Soos, *J. Phys. Chem.*, **91,** 2026 (1987).

199. J. Sarkany, M. Bartok, and R. M. Gonzalez, *J. Phys. Chem.*, **91,** 4301 (1987).

200. M. I. Tejedor-Tejedor and M. A. Anderson, *Langmuir*, **2,** 203 (1986).

201. K. Wagatsuma, A. Hatta, and W. Suetaka, *Mol. Cryst. Liq. Cryst.*, **55,** 179 (1979).

202. A. Hatta, Y. Chiba, and W. Suetaka, *Surf. Sci.*, **158,** 616 (1985).

203. T. Wadayama, W. Suetaka, and Atushi Sekiguchi, *Jpn J. App. Phys.*, **27,** 501 (1988).

204. W. Tornquist, F. Guillaume, and G. L. Griffin, *Langmuir*, **3,** 477 (1987).

205. D. D. Saperstein and W. G. Golden, ACS Symposium Series 288, M. L. Deviney and J. L. Gland, eds., American Chemical Society, Washington, DC, 1985.

206. W. G. Golden, C. D. Snyder, and B. Smith, *J. Phys. Chem.*, **86,** 4675 (1982).

207. W. G. Golden, D. D. Saperstein, M. W. Severson, and J. Overend, *J. Phys. Chem.*, **88,** 574 (1984).

208. R. G. Tobin, S. Chiang, P. A. Thiel, and P. L. Richards, *Surf. Sci.*, **140,** 393 (1984).

209. K. Wagatsume, K. Monma, and W. Suetaka, *Appl. Surf. Sci.*, **7,** 281 (1981).

210. M. Ohsawa, K. Hashima, and W. Suetaka, *Appl. Surf. Sci.*, **20,** 109 (1984).

211. A. Campion, *Ann. Rev. Phys. Chem.*, **36,** 549 (1985).

212. T. Takenaka and K. Yamasaki, *J. Colloid Interface Sci.*, **78,** 37 (1980).

213. R. Rossetti and L. E. Brus, *J. Phys. Chem.*, **90,** 558 (1986).

214. O. Siiman, R. Smith, C. Blatchford, and M. Kerker, *Langmuir*, **1,** 90 (1985).

215. M. Kerker, *Accts. Chem. Res.*, **17,** 271 (1984).

216. D. A. Weitz, S. Garoff, J. I. Gersten, and A. Nitzan, *J. Chem. Phys.*, **78,** 5324 (1983).

217. A. Lepp and O. Siiman, *J. Phys. Chem.*, **89,** 3494 (1985).

218. L. Kevan, *Rev. Chem. Intermediates*, **8,** 53 (1967).

219. G. Martini, *J. Colloid Interface Sci.*, **80,** 39 (1981).

220. S. J. DeCanio, T. M. Apple, and C. R. Bybowski, *J. Phys. Chem.*, **87,** 194 (1983).

221. T. J. Pinnavaia, *Advanced Techniques for Clay Mineral Analysis*, J. J. Fripiat, ed., Elsevier, New York, 1981.

222. M. Che and L. Bonneviot, *Z. Physikalische Chemie Neue Folge*, **152,** 113 (1987).

223. T. M. Duncan and C. Bybowski, *Surf. Sci. Rep.*, **1,** 157 (1981).

224. F. D. Blum, *Spectroscopy*, **1,** 32 (1986).

225. T. W. Root and T. M. Duncan, *J. Cataly.*, **102,** 109 (1986).

226. R. K. Shoemaker and T. M. Apple, *J. Phys. Chem.*, **89,** 3185 (1985).

227. H. A. Resing, D. Slotfeldt-Ellingsen, A. N. Garroway, T. J. Pinnavaia, and K. Unger, *Magnetic Resonance in Colloid and Interface Science*, J. P. Fraissard and H. A. Resing, eds., D. Reidel, Hingham, MA, 1980.

228. J. A. Ripmeester, *J. Am. Chem. Soc.*, **105,** 2925 (1983).

229. A. R. Siedle and R. A. Newmark, *J. Am. Chem. Soc.*, **103,** 1240 (1981).

230. J. A. Ripmeester, *J. Mag. Reson.*, **56,** 247 (1984).

231. G. J. Gajda, R. H. Bruggs, and W. H. Weinberg, *J. Am. Chem. Soc.*, **109,** 66 (1987).

232. E. L. Wolf, ed., *Principles of Electron Tunneling Spectroscopy*, Oxford, New York, 1985.

233. R. R. Gatte and J. Phillips, *J. Catal.*, **104,** 365 (1987).

234. J. D. Brown, D. L. Williamson, and A. J. Smith, *J. Phys. Chem.*, **89,** 3076 (1985).

235. J. W. Niemantsverdriet, *J. Mol. Catal.*, **25,** 285 (1984).

236. F. Hong, B. L. Yang, L. H. Schwartz, and H. H. Kung, *J. Phys. Chem.*, **88,** 2525 (1984).

CHAPTER IX

The Formation of a New Phase—Nucleation and Crystal Growth

1. Introduction

The transition from one phase to another is an important and interesting phenomenon closely linked to surface physical chemistry. The topic covers a wide range of situations from the ubiquitous process of rain formation in clouds to the formation of opals from monodisperse colloidal silica sediments. Quantitative description of nucleation and growth is challenging and has been attempted for many years in many contexts, such as the condensation of a vapor, the crystallization of a liquid and the precipitation of a solute from solution. The general formulation of the model for nucleation relies on application of bulk material properties to small clusters or droplets of material. Newer simulations and statistical mechanical density functional models improve on this assumption, but are not yet widely applicable and do not provide extensive improvements in the qualitative understanding of the process. In this chapter we focus on the nucleation and growth of solid materials, yet we formulate the general model for vapor–liquid nucleations as well.

In the absence of participating foreign surfaces, small clusters of molecules form and grow by accretion to the point of becoming recognizable droplets or crystallites that may finally coalesce or grow to yield large amounts of the new phase. This sequence generally does not occur if the system is just beyond the saturation pressure, concentration, or crystallization temperature. Instead, the concentration or vapor pressure must be increased well beyond the equilibrium value before nucleation in the form of a fog or suspension occurs. Similarly, pure liquids must be considerably supercooled prior to crystallization. Very pure liquid water can be cooled to $-40°C$ before spontaneous freezing occurs; early observations of supercooling were made by Fahrenheit in 1714 [1].

The resistance to nucleation is associated with the surface energy of forming small clusters. Once beyond a critical size, the growth proceeds with the considerable driving force due to the supersaturation or subcooling. It is the definition of this *critical nucleus size* that has consumed much theoretical and experimental research. We present a brief description of the classic nucleation theory along with some examples of crystal nucleation and growth studies.

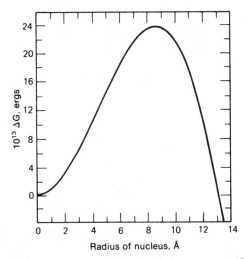

Fig. IX-1. Variation of ΔG with droplet size for water vapor at $0°C$ at four times the saturation pressure. (From Ref. 2.)

2. Classic Nucleation Theory

In the classic nucleation theory, the free energy of forming a cluster of radius r containing n atoms or molecules is the sum of two terms:

$$\Delta G = -n \, \Delta\mu + 4\pi r^2 \gamma \qquad (\text{IX-1})$$

the driving force for the phase transition, $n\Delta\mu$, or energy to transfer n molecules from the one phase to another and the surface energy $4\pi r^2 \gamma$ where γ is the interfacial tension. Since these terms are opposite in sign beyond the phase boundary, a plot of ΔG versus r goes through a maximum as illustrated in Fig. IX-1. For a condensation process at pressure P, $nA(\text{gas}) = A_n(\text{liq})$ the driving force becomes

$$\Delta G_{\text{cond}} = -n\Delta\mu = -nkT\ln \frac{P}{P^0} = -nkT\ln x \qquad (\text{IX-2})$$

where P^0 is the pressure or activity of the liquid phase. The number of molecules in the cluster is related to its size

$$n = \frac{4\pi}{3} r^3 \hat{\rho} \qquad (\text{IX-3})$$

through the number density of the cluster, $\hat{\rho} = \rho N_A/M$ molecules/volume, where N_A is Avagadro's number and M is molecular weight. Combining Eqs. IX-1 and IX-3 and taking the derivative of the free energy with respect to radius or number and setting it equal to zero provides the value of n or r at the maximum, we obtain

$$n_c = \frac{32\pi\gamma^3}{3\hat{\rho}^2 \, \Delta\mu^3} \tag{IX-4}$$

and

$$r_c = \frac{2\gamma}{\hat{\rho} \, \Delta\mu} \tag{IX-5}$$

defining the size of the critical nucleus. In the example presented in Fig. IX-1, the critical nucleus is about 8 Å in radius and contains about 90 molecules of water. For a liquid–vapor system, $r_c = 2\gamma/\hat{\rho}kT \ln x$ an equivalent of the Kelvin equation (Eq. III-21), giving the vapor pressure for a drop of radius r_c.

Combining Eqs. IX-1 and IX-5 gives the height of the energy barrier

$$\Delta G_{max} = \frac{16\pi\gamma^3}{3\hat{\rho}^2 \, \Delta\mu^2} = \frac{4\pi r_c^2 \gamma}{3} \tag{IX-6}$$

which equals one-third of the surface energy for the formation of the nucleus as stated long ago by Gibbs [3]. When applied to crystals, Eq. IX-6 may have slight differences in the numerical factor since nonspherical shapes may be formed.

The dynamic picture of a vapor at a pressure near P^0 is then somewhat as follows. If P is less than P^0, then ΔG for a cluster increases steadily with size, and although in principle all sizes would exist, all but the smallest would be very rare, and their numbers would be subject to random fluctuations. Similarly, there will be fluctuations in the number of embryonic nuclei of size less than r_c, in the case of P greater than P^0. Once a nucleus reaches the critical dimension, however, a favorable fluctuation will cause it to grow indefinitely. The experimental maximum supersaturation pressure is such that a large traffic of nuclei moving past the critical size develops with the result that a fog of liquid droplets is produced.

It follows from the foregoing discussion that the essence of the problem is that of estimating the rate of formation of nuclei of critical size, and a semirigorous treatment has been given by Becker and Doring (see Refs. 4–6). The simplifying device that was employed, which made the treatment possible, was to consider the case of a steady-state situation such that the average number of nuclei consisting of 2, 3, 4, ... , N molecules, although different in each case, did not change with time. A detailed balancing of evaporation and condensation rates was then set up for each size of nucleus, and by a clever integration procedure, the flux I, or the rate of formation of nuclei containing n molecules from ones containing $n - 1$ molecules, was estimated. The detailed treatment is rather lengthy, although not difficult, and the reader is referred to Refs. 1, 2, and 4 through 7 for details.

The final equation obtained by Becker and Doring may be written down immediately by means of the following qualitative argument. Since the flux I is taken to be the same for any size nucleus, it follows that it is related to the rate of formation of a cluster of two molecules, that is, to Z, the gas kinetic collision frequency (collisions per cubic centimeter-second).

For the steady-state case, Z should also give the forward rate of formation or flux of critical nuclei, except that the positive free energy of their formation amounts to a free energy of activation. If one correspondingly modifies the rate Z by the term $e^{-\Delta G_{max}/kT}$, an approximate value for I results:

$$I = Z \exp \frac{-\Delta G_{max}}{kT} \qquad \text{(IX-7)}$$

While Becker and Doring obtained a more complex function in place of Z, its numerical value is about equal to Z, and it turns out that the exponential term, which is the same, is the most important one. Thus the complete expression is

$$I = \frac{Z}{n_c} \left(\frac{\Delta G_{max}}{3\pi kT} \right)^{1/2} \exp \frac{-\Delta G_{max}}{kT} \qquad \text{(IX-8)}$$

where n_c is the number of molecules in the critical nucleus. The quantity $(1/n_c)(\Delta G_{max}/3\pi kT)^{1/2}$ has been called the *Zeldovich factor* [8, 9].

The full equation for I is obtained by substituting into Eq. IX-8 the expression for ΔG_{max} and the gas kinetic expression for Z:

$$I = 2n^2\sigma^2 \left(\frac{RT}{M} \right)^{1/2} \exp \left[\frac{-16\pi\gamma^3 M^2}{3kT\rho^2(RT\ln x)^2} \right] \qquad \text{(IX-9)}$$

where σ is the collision cross section and n is the number of molecules of vapor per cubic centimeter. Since Z is roughly $10^{23}P^2$, where P is given in millimeters of mercury, Eq. IX-9 may be simplified as

$$I = 10^{23}P^2 \exp \frac{-17.5\overline{V}^2\gamma^3}{T^3(\ln x)^2} \quad \text{(in nuclei/cm}^3 \cdot \text{sec)} \qquad \text{(IX-10)}$$

By way of illustration, the various terms in Eq. IX-10 are evaluated for water at 0°C in Table IX-1. Taking $\overline{V}$ as 20 cm^3/mole, γ as 72 ergs/cm^2, and P^0 as 4.6 mm, Eq. IX-10 becomes

TABLE IX-1
Evaluation of Equation IX-11 for Various Values of x

x	$\ln x$	$118/(\ln x)^2 = A$	e^{-A}	I, nuclei cm^{-3} sec^{-1}
1.0	0	—	0	0
1.1	0.095	1.31×10^4	10^{-5700}	10^{-5680}
1.5	0.405	720	10^{-310}	10^{-286}
2.0	0.69	246	10^{-107}	10^{-82}
3.0	1.1	95.5	10^{-42}	10^{-17}
3.5	1.25	75.5	10^{-33}	2×10^{-8}
4.0	1.39	61.5	$10^{-26.7}$	0.15
4.5	1.51	51.8	$10^{-22.5}$	10^3

$$I = 2 \times 10^{24} x^2 \exp \left[-\frac{118}{(\ln x)^2} \right] \tag{IX-11}$$

The figures in the table show clearly how rapidly I increases with x, and it is generally sufficient to define the critical supersaturation pressure such that $\ln I$ is some arbitrary value such as unity.

Frequently, vapor-phase supersaturation is studied not by varying the vapor pressure P directly but rather by cooling the vapor and thus changing P^0. If T_0 is the temperature at which the saturation pressure is equal to the actual pressure P, then at any temperature T, $P/P^0 = x$ is given by

$$\ln x = \frac{\Delta H_v}{R} \left(\frac{1}{T} - \frac{1}{T_0} \right) \tag{IX-12}$$

where ΔH_v is the latent heat of vaporization. Then, by Eq. IX-5, it follows that

$$\Delta T = T_0 - T = \frac{2\gamma M T_0}{r_c \Delta H_v \rho} = \frac{2\gamma \overline{V} T_0}{r_c \Delta H_v} \tag{IX-13}$$

[An interesting point is that ΔH_v itself varies with r [10].] As is the case when P is varied, the rate of nucleation increases so strongly with the degree of supercooling that a fairly sharp critical value exists for T. Analogous equations can be written for the supercooling of a melt, where the heat of fusion ΔH_f replaces ΔH_v.

At a sufficiently low temperature, the phase nucleated will be crystalline rather than liquid. The theory is reviewed in Refs. 1 and 7. It is similar to that for the nucleation of liquid drops. The kinetics of the heterogeneous nucleation to form crystalline films on solid substrates has been simulated by means of Monte Carlo calculations [11, 12].

The case of nucleation from a condensed phase, usually that of a melt, may be treated similarly. The chief modification to Eq. IX-7 that ensues is in the frequency factor; instead of free collisions between vapor molecules, one now has a closely packed liquid phase. The rate of accretion of clusters is therefore related to the diffusion process, and the situation was treated by Turnbull and Fisher [13]. Again, the reader is referred to the original literature for the detailed derivation, and the final equation is justified here only in terms of a qualitative argument. If one considers a crystalline nucleus that has formed in a supercooled melt, then the rate at which an additional molecule may add can be regarded as determined by the frequency with which a molecule may jump from one position in the liquid to another just at the surface of the solid. Such a jump is akin to those involved in diffusion, and the frequency may be approximated by means of absolute rate theory as being equal to the frequency factor kT/h times an exponential factor containing the free energy of activation for diffusion. The total rate of such occurrences per cubic centimeter of liquid is

$$Z = n \, \frac{kT}{h} \exp\left(-\frac{\Delta G_D}{kT}\right) \tag{IX-14}$$

where n is the number of molecules of liquid per cubic centimeter. The steady-state treatment is again used, and the final result is analogous to Eq. IX-7:

$$I = n \, \frac{kT}{h} \exp\left(-\frac{\Delta G_D}{kT}\right) \exp\left(-\frac{\Delta G_{max}}{kT}\right) \tag{IX-15}$$

For liquids that are reasonably fluid around their melting points, the kinetic factors in Eq. IX-15 come out about $10^{33}/cm^3 \cdot sec$, so that Eq. IX-15 becomes

$$I = 10^{33} \exp\left(-\frac{\Delta G_{max}}{kT}\right) \tag{IX-16}$$

where ΔG_{max} is given by Eq. IX-6, or by a minor modification of it, which allows for a nonspherical shape for the crystals.

A more recent model for the preexponential factor including viscous flow across the solid–liquid interface is [14]

$$Z = \frac{2(\gamma_{sl}kT)^{1/2}}{v_m^{5/3} \eta} \tag{IX-17}$$

where γ_{sl} is the interfacial energy per unit area of the solid–liquid interface, v_m is the molecule volume, and $\eta(T)$ is the liquid viscosity. Other preexponential factors have been proposed for solid-solid phase transitions [14].

Another important point in connection with the rate of nuclei formation in the case of melts or of solutions is that the rate reaches a maximum with degree of supercooling. To see how this comes about, r_c is eliminated between Eq. IX-6 and the one for liquids analogous to Eq. IX-13, giving

$$I = \frac{nkT}{h} \, e^{-A/kT} \qquad\qquad (IX\text{-}18)$$

where

$$A = \Delta G_D + \frac{4\pi\gamma}{3} \left(\frac{2\bar{V}\gamma T_0}{\Delta H_f \Delta T} \right)^2$$

On setting dI/dT equal to zero, one obtains for y at the maximum rate the expression

$$\frac{(y-1)^3}{y^2(3-y)} = \frac{4\pi\gamma}{3\Delta H_D} \left(\frac{2\gamma\bar{V}}{\Delta H_f} \right)^2 \qquad\qquad (IX\text{-}19)$$

where ΔH_D is the activation energy for diffusion, and $y = T_0/T$. Qualitatively, while the concentration of nuclei increases with increasing supercooling, their rate of formation decreases, due to the activation energy for diffusion or, essentially, due to the increasing viscosity of the medium. Glasses, then, result from a cooling so rapid that the temperature region of appreciably rapid nucleation is passed before much actual nucleation occurs (note Ref. 15).

Still another situation is that of a supersaturated or supercooled solution, and straightforward modifications can be made in the preceding equations. Thus in Eq. IX-2, x now denotes the ratio of the actual solute activity to that of the saturated solution. In the case of a nonelectrolyte, $x = S/S_0$, where S denotes the concentration. Equation IX-13 now contains ΔH_s, the molar heat of solution.

The classic nucleation theory is an excellent qualitative foundation for the understanding of nucleation. It is not, however, appropriate to treat small clusters as bulk materials and to ignore the sometimes significant and diffuse interface region. This was pointed out some years ago by Cahn and Hilliard [16] and is reflected in their model for interfacial tension (see Section III-2B).

Density functional theory from statistical mechanics is a means to describe the thermodynamics of the solid phase with information about the fluid [17–19]. In density functional theory, one makes an *ansatz* about the structure of the solid, usually describing the particle positions by Gaussian distributions around their lattice sites. The free energy is then a functional of the local density in the lattice, which is approximated as a spatially varying *effective liquid*. Once this correspondence is made, the energy is easily integrated over the solid local density for the total free energy. The energy is minimized with respect to the Gaussian width to find the equilibrium structure and energy of the solid phase. Several averaging techniques have allowed its implementation to more complex problems of the interfacial tension between solids and liquids [20, 21]. Simple systems such as hard spheres [22], adhesive spheres [23, 24], and Lennard-Jones [21] molecules have been addressed; however, its widespread application to problems in crystal growth has been limited thus far.

The interfacial tension also depends on curvature (see Section III-1C) [25–27]. This alters Eq. IX-1 by adding a radius-dependent surface tension

$$\gamma(r) = \gamma^0 \left[1 - \frac{2\delta}{r} + O\left(\frac{1}{r^2}\right) \right] \tag{IX-20}$$

where now the Tolman length, δ can be expressed as $\delta = 2c_0 K/\gamma^0$ in terms of the spontaneous curvature c_0 and the bending elastic modulus K and the second term is a complicated expression depending inversely on the square of the curvature. In practice, however, the first-order term may be negligible leaving only the $1/r^2$ dependence [27].

LaMer and Pound [28] noted that standard nucleation theory requires the use of the macroscopic interfacial tension down to critical nucleus sizes of about 1 nm. The exponential dependence on γ in Eq. IX-10 makes this a significant defect. An extreme case arises in condensation of mercury vapor because of the high liquid surface tension. The classical theory predicts a critical nucleus of only 1–10 *atoms* and underpredicts the critical supersaturation by a factor of 1000; a difference attributable to a 40% difference in the surface tension for the critical nuclei [29]. The surface tension may not vary strongly with size for more common materials. Benson and Shuttleworth [30] found that even for a crystallite containing 13 molecules the surface energy was only 15% less than that for a planar surface. Walton [31], with more allowance for distortion, found a 35% increase in the surface energy of a KCl crystal four ions in length. Condensation from a pure supersaturated gas, with no inert gas present, does not agree with theory perhaps due to a lack of a thermal steady state [32].

In principle, nucleation should occur for any supersaturation given enough time. The critical supersaturation ratio is often defined in terms of the condition needed to observe nucleation on a convenient time scale. As illustrated in Table IX-1, the nucleation rate changes so rapidly with degree of supersaturation that, fortunately, even a few powers of 10 error in the preexponential term make little difference. There has been some controversy surrounding the preexponential term and some detailed analyses are available [33–35].

Classic nucleation theory must be modified for nucleation near a critical point. Observed supercooling and superheating far exceeds that predicted by conventional theory and McGraw and Reiss [36] pointed out that if a usually neglected excluded volume term is retained the free energy of the critical nucleus increases considerably. As noted by Derjaguin [37], a similar problem occurs in the theory of cavitation. In binary systems the composition of the nuclei will differ from that of the bulk vapor [38–40]; a controversy over this composition has been examined by Mirabel and Reiss [41].

3. Experimental Nucleation Studies

A. One-Component Systems

Many studies have been made on nucleation from a vapor, and generally the experimental findings are consistent with the Becker–Doring theory. Water has been a popular medium to study; Volmer and Flood [42] found the critical value of x to be about 5.03 for water vapor at 261 K while the theoretical value for $\ln I \approx 1$ is 5.14. Sander and Damköhler [43] found that the critical value of x for water and its temperature dependence were in reasonable agreement with theory. Some aspects of nucleation theory and experiments are reviewed by Higuchi and O'Konski [44] and some results are summarized in Table IX-2. In general, the agreement between theory and experiment is quite good as also supported by more recent work [45–48].

An important approach to the study of nucleation of solids is the investigation of small droplets of large molecular clusters. Years ago, Turnbull showed that by studying small droplets one could eliminate impurities in all except a few droplets and study homogeneous nucleation at significant undercoolings [13].

One remarkably simple yet seemingly robust outcome of Turnbull's experiments was his empirical finding that the solid–liquid interfacial free energy was

TABLE IX-2
Critical Supersaturation Pressures for Vapors

Substance	Temperature (K)	γ (erg/cm^2)	Critical value of x	
			Observed	Calculated
Water	275	—	4.2	4.2
	264	77.0	4.85	4.85
	261	—	5.0	5.1
Methanol	270	24.8	3.0, 3.2	1.8
Ethanol	273	24.0	2.3	2.3
n-Propanol	270	25.4	3.0	3.2
Isopropyl alcohol	265	23.1	2.8	2.9
n-Butyl alcohol	270	26.1	4.6	4.5
Nitromethane	252	40.6	6.0	6.2
Ethyl acetate	242	30.6	8.6–12	10.0
Dibutyl phthalate	332	29.4	26–29[a]	—
Triethylene glycol	324	42.8	27–37[a]	—

[a]Values of interfacial tension of nucleus from turbulent jet measurements, by various equations [44].

directly related to the enthalpy of fusion per molecule as [49]

$$\gamma_{sl} = C_T \, \Delta H_{\text{fus}} \hat{\rho}^{2/3} \tag{IX-21}$$

where C_T, the Turnbull coefficient, equals 0.45 for metals and 0.32 for other nonmetallic solids. This result has been confirmed for freezing of CCl_4 and $CHCl_3$ and has been extended to solid-solid transitions where ΔH for the transition is known and $C_T = 0.2$–0.3 [50]. It also agrees with density functional theory analysis (see Section IX-2A) of freezing in adhesive spheres where $C_T = 0.47$ for the entire phase diagram [23, 24].

Bartell and co-workers have made significant progress by combining electron diffraction studies from beams of molecular clusters with molecular dynamics simulations [14, 51, 52]. Due to their small volumes, deep supercoolings can be attained in cluster beams; however, the temperature is not easily controlled. The rapid nucleation that ensues can produce new phases not observed in the bulk [14]. Despite the concern about the appropriateness of the classic model for small clusters, its application appears to be valid in several cases [51].

The assumption that macroscopic interfacial tensions are valid down to critical nucleus size has led many researchers to use nucleation data to calculate interfacial tension. Staveley and co-workers [53, 54] used Eq. IX-16 to calculate the interfacial tensions for various organic and inorganic substances. Mason [55] estimated from the supercooling of water drops that the interfacial tension between ice and water was about 22 ergs/cm^2, in agreement with the 26 ergs/cm^2 from melting-point change in a capillary (Section X-2) and within range of Gurney's theoretical estimate of 10 ergs/cm^2 [56]. Turnbull and Cormia [57] give values of 7.2, 9.6, and 8.3 ergs/cm^2 for the solid–liquid interfacial tensions of n-heptadecane, n-octadecane, and n-tetracosane, respectively.

The entropically driven disorder–order transition in hard-sphere fluids was originally discovered in computer simulations [58, 59]. The development of colloidal suspensions behaving as hard spheres (i.e., having negligible Hamaker constants, see Section VI-3) provided the means to experimentally verify the transition. Experimental data on the nucleation of hard-sphere colloidal crystals [60] allows one to extract the hard-sphere solid–liquid interfacial tension, $\gamma = 0.55 \pm 0.02kT/\sigma^2$, where σ is the hard-sphere diameter [61]. This value agrees well with that found from density functional theory, $\gamma = 0.6 \pm 0.02kT/\sigma^2$ [21] (Section IX-2A).

An interesting experimental technique is "heat development" of nuclei. The liquid is held at the desired temperature for a prescribed time, while nuclei accumulate; they are then made visible as crystallites by quickly warming the solution to a temperature just below T_0, where no new nuclei form but existing ones grow rapidly.

Nucleation in a cloud chamber is an important experimental tool to understand nucleation processes. Such nucleation by ions can arise in atmospheric physics; theoretical analysis has been made [62, 63] and there are interesting differences in the nucleating ability of positive and negative ions [64]. In water vapor, it appears that the full heat of solvation of an ion is approached after only 5–10 water molecules have associated with it. In a recent study using the common indoor pollutant radon to produce ions, He and Hopke [65] studied ion-induced particle formation that may have serious health effects if occurring in households.

B. Binary Systems and Solutions

As mentioned in Section IX-2A, binary systems are more complicated since the composition of the nuclei differ from that of the bulk. In the case of sulfuric acid and water vapor mixtures only some 10^{10} *molecules* of sulfuric acid are needed for water droplet nucleation that may occur at less than 100% relative humidity [38]. A rather different effect is that of "passivation" of water nuclei by long-chain alcohols [66] (which would inhibit condensation; note Section IV-6). A recent theoretical treatment by Bar-Ziv and Safran [67] of the effect of surface active monolayers, such as alcohols, on surface nucleation of ice shows the link between the inhibition of subcooling (enhanced nucleation) and the strength of the interaction between the monolayer and water.

The nucleation of a system by means of foreign bodies is, of course, a well-known phenomenon. Most chemists are acquainted with the practice of scratching the side of a beaker to induce crystallization. Of special interest, in connection with artificial rainmaking, is the nucleation of ice. Silver iodide, whose crystal structure is the same as ice and whose cell dimensions are very close to it, will nucleate water below $-4°C$ [68]. Other agents have been found. A fluorophlogopite (a mica) does somewhat better than AgI [69] (see also Ref. 70). Fletcher [71] comments that the free energy barrier to the growth of an ice cluster on the surface of a foreign particle should be minimized if the particle-ice-water contact angle is small, which implies that the surface should be hydrophobic to water. A nucleating agent need not be crystalline. Thus ice nucleation can occur at a liquid interface [72].

A very different nucleation scheme by Grieser and co-workers employs ultrasonic irradiation of salt solutions to create H· and OH· radicals in solution [73]. These radicals proceed to nucleate growth of quantum-sized (Q-state) particles of cadmium sulfide. Similar initiation has been used for polymer latices [74].

Two nucleation processes important to many people (including some surface scientists!) occur in the formation of gallstones in human bile and kidney stones in urine. Cholesterol crystallization in bile causes the formation of gallstones. Cryotransmission microscopy (Chapter VIII) studies of human bile reveal vesicles, micelles, and potential early crystallites indicating that the cholesterol crystallization in bile is not cooperative and the true nucleation time may be much shorter than that found by standard clinical analysis by light microscopy [75]. Kidney stones often form from crystals of calcium oxalates in urine. Inhibitors can prevent nucleation and influence the solid phase and intercrystallite interactions [76, 77]. Citrate, for example, is an important physiological inhibitor to the formation of calcium renal stones. Electrokinetic studies (see Section V-6) have shown the effect of various inhibitors on the surface potential and colloidal stability of micrometer-sized dispersions of calcium oxalate crystals formed in synthetic urine [78, 79].

Supersaturation phenomena in solutions are, of course, very important, but, unfortunately, data in this field tend to be not too reliable. Not only is it difficult to avoid accidental nucleation by impurities, but also solutions, as well as pure liquids, can exhibit memory effects whereby the attainable supersaturation depends on the past history, especially the thermal history, of the solution. In the case of slightly soluble salts, where precipitation is brought about by the mixing of reagents, it is difficult to know the effective degree of supersaturation.

If x is replaced by S/S_0, then Eq. IX-15 takes on the form

$$\ln t = k_1 - \frac{k_2 \gamma^3}{T^3 (\ln S/S_0)^2} \tag{IX-22}$$

and Stauff [80] found fair agreement with this equation in the case of potassium chlorate solutions. Gindt and Kern [81] took (N/t), the average rate of nucleation, as proportional to I, where N is the number of crystals produced, and from the slope of plots of log (N/t) versus $1/(\ln S/S_0)^2$ they deduced the rather low solid-solution interfacial tension value of 2–3 ergs/cm^2 for KCl and other alkali halides. Alternatively, one may make estimates from the critical supersaturation ratio, that is, the value of S/S_0 extrapolated to infinite rate of precipitation or zero induction time. Enüstün and Turkevich [82] found a value of 7.5 for SrSO$_4$, and from their solubility estimated interfacial tension (see Section X-2) deduced a critical nucleus of size of about 18 Å radius. More recently, Ring and co-workers gave nucleation rates of TiO$_2$ from alcohol solution [83, 84]. Uhler and Helz [85] studied the kinetics of PbS precipitating in the presence of a chelating agent.

4. Crystal Growth

The visible crystals that develop during a crystallization procedure are built up as a result of growth either on nuclei of the material itself or surfaces of foreign material serving the same purpose. Neglecting for the moment the matter of impurities, nucleation theory provides an explanation for certain qualitative observations in the case of solutions.

Once nuclei form in a supersaturated solution, they begin to grow by accretion and, as a result, the concentration of the remaining material drops. There is thus a competition for material between the processes of nucleation and of crystal growth. The more rapid the nucleation, the larger the number of nuclei formed before relief of the supersaturation occurs and the smaller the final crystal size. This, qualitatively, is the basis of what is known as von Weimarn's law [86]:

$$\frac{1}{d} = k \frac{S}{S_0} \tag{IX-23}$$

where d is some measure of the particle size. Although essentially empirical, the law appears to hold approximately.

A beautiful illustration of the balance between nucleation and crystal growth arises in the formation of "monodisperse" (i.e., narrow size distribution) colloids. One well-known example of such colloids occurs in natural opals formed from sediments of monodisperse silica particles [87]. Early work by LaMer and co-workers [88] showed that a short nucleation time followed by steady diffusion-limited growth produced monodisperse sulfur particles. More recent and broader studies by Matijevic̀ and co-workers have encompassed a myriad of inorganic colloids [89–91]. This science involves the chemical processes resulting in an insoluble precipitate as well as the physical understanding of the particle nucleation and growth. An interesting feature of this work is that the final shape of the particles may not reflect the crystal structure (see Chapter VII) and may vary with experimental conditions. A good example of this occurs in ZnO (zincite) particles, all having the same crystal structure, prepared at different pH as shown in Figure IX-2.

The mechanism of crystal growth has been a topic of considerable interest. In the case of a perfect crystal, the starting of a new layer involves a kind of nucleation since the first few atoms added must occupy energy-rich positions. Becker and Doring [4],

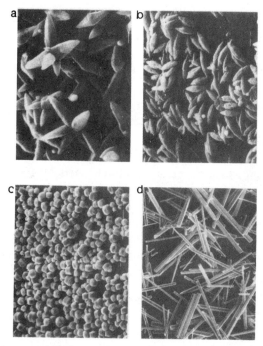

Fig. IX-2. SEM images of ZnO particles obtained by aging the following solutions: (a) $5.0 \times 10^{-3} M$ Zn(NO$_3$)$_2$ and $1.9 \times 10^{-2} M$ NH$_4$OH (pH 8.8) at 90°C for 3 hr; (b) $1.0 \times 10^{-4} M$ Zn(NO$_3$)$_2$ and $3.2 \times 10^{-4} M$ NH$_4$OH (pH 7.7) at 90°C for 1 hr; (c) $3.2 \times 10^{-3} M$ Zn(NO$_3$)$_2$ and 0.1 M triethanolamine (TEA, pH 8.9) at 90°C for 1 hr; (d) $4.0 \times 10^{-2} M$ Zn(NO$_3$)$_2$ and 0.2M TEA, and 1.2M NH$_4$OH (pH 12.1) at 150°C for 2 hr. The particles in (a), (b), and (c) average about 5 μm in size; those in (d) are much larger. (From Ref. 90.)

in fact, have treated crystal growth in terms of such surface nucleation processes; see also Ref. 60. Dislocations may also serve as surface nucleation sites, and, in particular, Frank [92] suggested that crystal growth might occur at the step of a screw dislocation (see Section VII-4C) so that the surface would advance in spiral form. However, while crystallization phenomena now occupy a rather large section of the literature, it is still by no means clear what mechanisms of crystal growth predominate. Buckley [93] comments that spiral patterns are somewhat uncommon and, moreover, occur on well-developed and hence *slowly* growing faces. Some interferometric studies of the concentration gradients around a growing crystal [94, 95] showed that, depending on the crystal, the maximum gradient may occur either near the center of a face or near the edges, and the pattern of fringes around a given crystal may change considerably from time to time and without necessarily any direct correlation with local growth rates. Clearly, the possibility of surface deposition at one point, followed by surface migration to a final site, must be considered. On the other hand, the Frank mechanism is widely accepted, and in individual cases it has been possible to observe the slow turning of a spiral pattern as crystal growth occurred [96]. The mechanism has been suggested in the case of some calcium phosphates [97].

The kinetics of crystal growth has been much studied; Refs. 98–102 are representative. Often there is a time lag before crystallization starts, whose parametric dependence may be indicative of the nucleation mechanism. The crystal growth that follows may be controlled by diffusion or by surface or solution chemistry (see also Section XVI-2C).

At equilibrium, crystal growth and dissolving rates become equal, and the process of *Ostwald ripening* may now appear, in which the larger crystals grow at the expense of the smaller ones. The kinetics of the process has been studied (see Ref. 103).

5. Epitaxial Growth and Surface Nucleation

The oriented overgrowth of a crystalline phase on the surface of a substrate that is also crystalline is called *epitaxial growth* [104]. Usually it is required that the lattices of the two crystalline phases match, and it can be a rather complicated process [105]. Some new applications enlist amorphous substrates or grow new phases on a surface with a rather poor lattice match.

In molecular beam epitaxy, evaporated atoms are directed toward a crystalline surface in an ultra-high-vacuum chamber, where in situ surface analysis may be performed (Chapter VIII). While epitaxial growth often requires lattice matching between the surface and the growing crystal phase, van der Waals interactions (see Chapter VI) dominate the growth of semiconductors and appear to relax the lattice matching constraint [106, 107]. Armstrong and co-workers have grown semiconductor GaSe and $MoSe_2$ films on GaP(111) after passivating the surface in a solution containing sulfur. The sulfur passivation creates a chemically unreactive smooth surface for the growth of semiconductor films. Molecular beam epitaxy of organic molecules creates ordered ultrathin films of large aromatic molecules on metals, insulators, and semiconductors [108] wherein the organic molecules are packed more closely than in films grown by Langmuir–Blodgett or self-assembly techniques (see Chapters XI and XV). Higher densities may lead to improved optical properties in phthalocyanine dye thin films [109]. Again, van der Waals interactions seem to dominate the structures formed in organic films [110].

A very different type of epitaxial growth may be pursued in solutions. The desire to produce nanometer-sized semiconducting, metallic, or magnetic particles has fueled much recent research in surface chemistry. One name for this pursuit is "band-gap engineering." Fendler and co-workers have had success growing monodisperse crystals under Langmuir monolayers [111]. Infusing H_2S through a monolayer nucleates PbS crystals in a lead nitrate solution (see Figure VIII-3). With the appropriate choice of amphiphile, equilateral triangular crystallites of PbS can be grown with certain faces [such as (111)] oriented parallel to the monolayer [112]. Similar results are obtained with cadmium sulfide [113] and other semiconductors [111]. Electrocrystallization under monolayers is another means to control nanocrystal formation such as in the case of silver crystallization under various surfactants [114].

6. Problems

1. Calculate the value of the Zeldovich factor for water at $20°C$ if the vapor is 5% supersaturated.

2. Because of the large surface tension of liquid mercury, extremely large supersaturation ratios are needed for nucleation to occur at a measurable rate. Calculate r_c and n_c at 400 K assuming that the critical supersaturation is $x = 40,000$. Take the surface tension of mercury to be 486.5 ergs/cm^2.

3. Calculate ΔG of Eqs. IX-1 and IX-2 for the case of n-octane and plot against r over the range $r = 5$ to 300 Å and for the two cases of $x = 2$ and $x = 250$. Assume $20°C$.

4. Assuming that for water ΔG_D is 7 kcal/mol, calculate the rate of nucleation for ice nuclei for several temperatures and locate the temperature of maximum rate. Discuss in terms of this result why glassy water might be difficult to obtain.

5. Calculate what the critical supersaturation ratio should be for water if the frequency factor in Eq. IX-10 were indeed too low by a factor of 10^{20}. Alternatively, taking the observed value of the critical supersaturation ratio as 4.2, what value for the surface tension of water would the "corrected" theory give?

6. Verify Fig. IX-1.

7. Given the expression IX-20 for the interfacial tension dependence on radius, derive the form of the dependence of the critical radius on the density and chemical potential, given that the tension is proportional to (a) $1/r$ and (b) $1/r^2$.

8. Using the temperature dependence of γ from Eq. III-11 with $n = \frac{11}{9}$ and the chemical potential difference $\Delta\mu$ from Eq. IX-2, sketch how you expect a curve like that in Fig. IX-1 to vary with temperature (assume ideal-gas behavior).

9. As a follow-up to Problem 2, the observed nucleation rate for mercury vapor at 400 K is 1000-fold less than predicted by Eq. IX-9. The effect may be attributed to a lowered surface tension of the critical nuclei involved. Calculate this surface tension.

General References

F. F. Abraham, *Homogeneous Nucleation Theory*, Academic, New York, 1974.

H. E. Buckley, *Crystal Growth*, Wiley, New York, 1951.

P. P. Ewald and H. Juretschke, *Structure and Properties of Solid Surfaces*. University of Chicago Press, Chicago, 1953.

N. H. Fletcher, *The Physics of Rainclouds*, Cambridge University Press, Cambridge, England, 1962.

W. E. Garner, *Chemistry of the Solid State*, Academic, New York, 1955.

J. M. McBride, *Science*, **256,** 814 (1993).

K. Nishioka and G. M. Pound, *Surface and Colloid Science*, E. Matijevic, ed., Wiley, New York, 1976.

A. C. Zettlemoyer, ed., *Nucleation*, Marcel Dekker, New York, 1969.

Textual References

1. W. J. Dunning, *Nucleation*, Marcel Dekker, New York, 1969.

2. R. S. Bradley, *Q. Rev.* (London), **5,** 315 (1951).

3. J. W. Gibbs, *Collected Works of J. W. Gibbs*, Longmans, Green, New York, 1931.

4. R. Becker and W. Doring, *Ann. Phys.*, **24,** 719 (1935).

5. See M. Volmer, *Kinetik der Phasenbildung*, Edwards Brothers, Ann Arbor, MI, 1945.

6. F. F. Abraham, *Homogeneous Nucleation Theory*, Chapter V, Academic, New York, 1974.

7. J. Frenkel, *Kinetic Theory of Liquids*, Clarendon Press, Oxford, 1946.

8. P. P. Wegener, *J. Phys. Chem.*, **91,** 2479 (1987).

9. P. P. Wegener, *Naturwissenschaften*, **74,** 111 (1987).

10. A. W. Adamson and M. Manes, *J. Chem. Ed.*, **61,** 590 (1984).

11. A. I. Michaels, G. M. Pound, and F. F. Abraham, *J. Appl. Phys.*, **45,** 9 (1974).

12. G. H. Gilmer, H. J. Leamy, and K. A. Jackson, *J. Crystal Growth*, **24/25,** 495 (1974).

13. D. Turnbull and J. C. Fisher, *J. Chem. Phys.*, **17,** 71 (1949).

14. L. S. Bartell, *J. Phys. Chem.*, **99,** 1080 (1995).

15. E. Ruckenstein and S. K. Ihm, *J. Chem. Soc., Faraday Trans. I*, **72,** 764 (1976).

16. J. W. Cahn and J. E. Hilliard, *J. Chem. Phys.*, **31,** 688 (1959).

17. D. W. Oxtoby, "Liquids, Freezing and the Glass Transition," in *Les Houches Session 51*, J. P. Hansen, D. Levesque, and J. Zinn-Justin, eds., Elsevier, New York, 1990.

18. J. F. Lutsko, *Phys. Rev. A*, **43,** 4124 (1991).

19. Y. Singh, *Phys. Rep.*, **207,** 351 (1991).

20. A. R. Denton and N. W. Ashcroft, *Phys. Rev. A*, **39,** 4701 (1989).

21. D. W. Marr and A. P. Gast, *Phys. Rev. E*, **47,** 1212 (1993).

22. W. A. Curtin, *Phys. Rev. B*, **39,** 6775 (1989).

23. D. W. Marr and A. P. Gast, *J. Chem. Phys.*, **99**, 2024 (1993).

24. D. W. M. Marr and A. P. Gast, *Phys. Rev. E*, **52**, 4058 (1995).

25. See I. W. Plesner, *J. Chem. Phys.*, **40**, 1510 (1964).

26. R. A. Oriani and B. E. Sundquist, *J. Chem. Phys.*, **38**, 2082 (1963).

27. W. K. Kegel, *J. Chem. Phys.*, **102**, 1094 (1995).

28. V. K. LaMer and G. M. Pound, *J. Chem. Phys.*, **17**, 1337 (1949).

29. J. Martens, H. Uchtmann, and F. Hensel, *J. Phys. Chem.*, **91**, 2489 (1987).

30. G. C. Benson and R. Shuttleworth, *J. Chem. Phys.*, **19**, 130 (1951).

31. A. G. Walton, *J. Chem. Phys.*, **36**, 3162 (1963).

32. B. Barschdorff, W. J. Dunning, P. P. Wegener, and B. J. C. Wu, *Nat. Phys. Sci.*, **240**, 166 (1972).

33. H. Reiss, in *Nucleation II*, A. C. Zettlemoyer, ed., Marcel Dekker, New York, 1976.

34. H. Reiss, *Adv. Colloid Interface Sci.*, **7**, 1 (1977).

35. G. Wilemski, *J. Chem. Phys.*, **88**, 5134 (1988).

36. R. McGraw and H. Reiss, *J. Stat. Phys.*, **20**, 385 (1979).

37. B. V. Derjaguin, *J. Colloid Interface Sci.*, **38**, 517 (1972).

38. A. Jaecker-Voirol and P. Mirabel, *J. Phys. Chem.*, **92**, 3518 (1988).

39. G. Wilemski, *J. Phys. Chem.*, **91**, 2492 (1987).

40. C. Flageollet-Daniel, J. P. Garnier, and P. Mirabel, *J. Chem. Phys.*, **78**, 2600 (1983).

41. P. Mirabel and H. Reiss, *Langmuir*, **3**, 228 (1987).

42. M. Volmer and H. Flood, *Z. Phys. Chem.*, **A170**, 273 (1934).

43. A. Sander and G. Damköhler, *Naturwissenschaften*, **31**, 460 (1943).

44. W. L. Higuchi and C. T. O'Konski, *J. Colloid Sci.*, **15**, 14 (1960).

45. M. A. Sharaf and R. A. Dobbins, *J. Chem. Phys.*, **77**, 1517 (1982).

46. P. P. Wegener and C. F. Lee, *J. Aerosol Sci.*, **14**, 29 (1983).

47. P. P. Wegener, *J. Phys. Chem.*, **91**, 2479 (1987).

48. J. P. Garner, Ph. Ehrhard, and Ph. Mirabel, *Atmos. Res.*, **21**, 41 (1987).

49. D. Turnbull, *J. Appl. Phys.*, **21**, 1022 (1950).

50. T. S. Dibble and L. S. Bartell, *J. Phys. Chem.*, **96**, 8603 (1992).

51. J. Huang and L. S. Bartell, *J. Phys. Chem.*, **99**, 3924 (1995).

52. L. S. Bartell and J. Chen, *J. Phys. Chem.*, **99**, 12444 (1995).

53. H. J. de Nordwall and L. A. K. Staveley, *J. Chem. Soc.*, **1954**, 224.

54. D. G. Thomas and L. A. K. Staveley, *J. Chem. Soc.*, **1952**, 4569.

55. B. J. Mason, *Proc. Roy. Soc.* (London), **A215**, 65 (1952).

56. R. Shuttleworth, *Proc. Phys. Soc.* (London), **63A**, 444 (1950).

57. D. Turnbull and R. L. Cormia, *J. Chem. Phys.*, **34**, 820 (1961).

58. W. G. Hoover and F. H. Ree, *J. Chem. Phys.*, **49**, 3609 (1968).

59. B. Alder, W. Hoover, and D. Young, *J. Chem. Phys.*, **49**, 3688 (1968).

60. J. K. G. Dhont, C. Smits, and H. N. W. Lekkerkerker, *J. Colloid Interface Sci.*, **152**, 386 (1992).

61. D. W. Marr and A. P. Gast, *Langmuir*, **10**, 1348 (1994).

62. A. W. Castleman, Jr., P. M. Holland, and R. G. Keese, *J. Chem. Phys.*, **68,** 1760 (1978).

63. A. I. Rusanov, *J. Colloid Interface Sci.*, **68,** 32 (1979).

64. H. Rabeony and P. Mirabel, *J. Phys. Chem.*, **91,** 1815 (1987).

65. F. He and P. K. Hopke, *J. Chem. Phys.*, **99,** 9972 (1993).

66. B. V. Derjaguin, Yu. S. Kurghin, S. P. Bakanov, and K. M. Merzhanov, *Langmuir*, **1,** 278 (1985).

67. R. Bar-Ziv and S. A. Safran, *Phys. Rev. E*, **49,** 4306 (1994).

68. B. Vonnegut, *J. Appl. Phys.*, **18,** 593 (1947).

69. J. H. Shen, K. Klier and A. C. Zettlemoyer, *J. Atmos. Sci.*, **34,** 957 (1977).

70. V. A. Garten and R. B. Head, *Nature*, **205,** 160 (1965).

71. N. H. Fletcher, private communication.

72. J. Rosinski, *J. Phys. Chem.*, **84,** 1829 (1980).

73. R. A. Hobson, P. Mulvaney, and F. Grieser, *J. Chem. Soc., Chem. Commun.*, **1994,** 823.

74. S. Biggs and F. Grieser, *Macromolecules*, **28,** 4877 (1995).

75. A. Kapluna, Y. Talmon, F. M. Konikoff, M. Rubin, A. Eitan, M. Tadmor, and D. Lichtenberg, *Fed. Eur. Biochem. Soc. Lett.*, **340,** 78 (1994).

76. J. Callejas-Fernández, R. Martínez-García, F. J. de las Nieves López, and R. Hidalgo-Alvarez, *Solid State Ionics*, **63–65,** 791 (1993).

77. C. L. Erwin and G. H. Nancollas, *J. Crystal Growth*, **53,** 215 (1981).

78. J. Callejas-Fernández, F. J. de las Nieves, R. Martínez-García, and R. Hidalgo-Alvarez, *Prog. Colloid Polym. Sci.*, **84,** 327 (1991).

79. J. Callejas, R. Martínez, F. J. de las Nieves, and R. Hidalgo-Alvarez, *J. Surf. Sci. Technol.*, **8,** 105 (1992).

80. J. Stauff, *Z. Phys. Chem.*, **A187,** 107, 119 (1940).

81. R. Gindt and R. Kern, *CR*, **256,** 4186 (1963).

82. B. V. Enüstün and J. Turkevich, *J. Am. Chem. Soc.*, **82,** 4502 (1960).

83. M. D. Lamey and T. A. Ring, *Chem. Eng. Science*, **41,** 1213 (1986).

84. J. H. Jean and T. A. Ring, *Langmuir*, **2,** 251 (1986).

85. A. D. Uhler and G. R. Helz, *J. Crystal Growth*, **66,** 401 (1984).

86. P. P. von Weimarn, *Chem. Rev.*, **2,** 217 (1925).

87. P. J. Daragh, A. J. Gaskin, and J. V. Sanders, *Sci. Am.*, **234,** 84 (1976).

88. V. K. LaMer and R. H. Dinegar, *J. Am. Chem. Soc.*, **72,** 4847 (1950).

89. E. Matijević, *Langmuir*, **10,** 8 (1994).

90. E. Matijević, *Chem. Mat.*, **5,** 412 (1993).

91. E. Matijević, *Langmuir*, **2,** 12 (1986).

92. F. C. Frank, *Discuss. Faraday Soc.*, **5,** 48 (1949).

93. H. E. Buckley, in *Structure and Properties of Solid Surfaces*, R. Gomer and C. S. Smith, eds., University of Chicago Press, Chicago, 1952, p. 271.

94. G. C. Krueger and C. W. Miller, *J. Chem. Phys.*, **21,** 2018 (1953).

95. S. P. Goldsztaub and R. Kern, *Acta Cryst.*, **6,** 842 (1953).

96. W. J. Dunning, private communication.

97. G. H. Nancollas, *J. Crystal Growth*, **42**, 185 (1977).

98. G. L. Gardner and G. H. Nancollas, *J. Phys. Chem.*, **87**, 4699 (1983).

99. Z. Amjad, P. G. Koutsoukos, and G. H. Nancollas, *J. Colloid Interface Sci.*, **101**, 250 (1984).

100. C. Chieng and G. H. Nancollas, *Desalination*, **42**, 209 (1982).

101. Lj. Maksimović, D. Babić, and N. Kallay, *J. Phys. Chem.*, **89**, 2405 (1985).

102. D. Shea and G. R. Helz, *J. Colloid Interface Sci.*, **116**, 373 (1987).

103. D. B. Dadyburjor and E. Ruckenstein, *J. Crystal Growth*, **40**, 279, 285 (1977).

104. G. S. Swei, J. B. Lando, S. E. Rickert, and K. A. Mauritz, *Encyclopedia of Polymer Science and Technology*, Vol. 6, 2nd ed., Wiley, New York, 1986.

105. J. Heughebaert, S. J. Zawacki, and G. H. Nancollas, *J. Crystal Growth*, **63**, 83 (1983).

106. C. Hammond, A. Back, M. Lawrence, K. W. Nebesny, P. A. Lee, R. Schlaf, and N. R. Armstrong, *J. Vac. Sci. Technol. A*, **13**, 1768 (1995).

107. R. Schlaf, D. Louder, O. Lang, C. Pettenkofer, W. Jaegermann, K. W. Nebesny, P. A. Lee, B. A. Parkinson, and N. R. Armstrong, *J. Vac. Sci. Technol. A*, **13**, 1761 (1995).

108. T. J. Schuerlein, A. Schmidt, P. A. Lee, K. W. Nebesny, B. A. Parkinson, and N. R. Armstrong, *Jpn. J. Appl. Phys.*, **34**, 3837 (1995).

109. A. Schmidt, L. K. Chau, A. Back, and N. R. Armstrong, "Epitaxial Phthalocyanine Ultrathin Films Grown by Organic Molecular Beam Epitaxy (OMBE)," in *Phthalocyanines*, Vol. 4, C. Leznof and A. P. B. Lever, eds., VCH Publications, 1996.

110. C. D. England, G. E. Collins, T. J. Schuerlein, and N. R. Armstrong, *Langmuir*, **10**, 2748 (1994).

111. J. H. Fendler and F. C. Meldrum, *Adv. Mat.*, **7**, 607 (1995).

112. J. Yang and J. H. Fendler, *J. Phys. Chem.*, **99**, 5505 (1995).

113. J. Yang, F. C. Meldrum, and J. H. Fendler, *J. Phys. Chem.*, **99**, 5500 (1995).

114. N. A. Kotov, N. E. D. Zaniquelli, F. C. Meldrum, and J. H. Fendler, *Langmuir*, **9**, 3710 (1993).

CHAPTER X

The Solid–Liquid Interface—Contact Angle

1. Introduction

The importance of the solid–liquid interface in a host of applications has led to extensive study over the past 50 years. Certainly, the study of the solid–liquid interface is no easier than that of the solid–gas interface, and all the complexities noted in Section VII-4 are present. The surface structural and spectroscopic techniques presented in Chapter VIII are not generally applicable to liquids (note, however, Ref. 1). There is, perforce, some retreat to phenomenology, empirical rules, and semiempirical models. The central importance of the Young equation is evident even in its modification to treat surface heterogeneity or roughness .

We begin by discussing the methods for estimating the solid–liquid interfacial properties via thermodynamic measurements on bulk systems. We then discuss the contact angle on a uniform perfect solid surface where the Young equation is applicable. The observation that most surfaces are neither smooth nor uniform is addressed in Section X-5, where contact angle hysteresis is covered. Then we describe briefly the techniques used to measure contact angles and some results of these measurements. Section X-7 is taken up with a discussion of theories of contact angles and some practical empirical relationships.

2. Surface Energies from Solubility Changes

This section represents a continuation of Section VII-5, which dealt primarily with the direct estimation of surface quantities at a solid-gas interface. Although in principle some of the methods described there could be applied at a solid-liquid interface, very little has been done apart from the study of the following Kelvin effect and nucleation studies, discussed in Chapter IX.

The Kelvin equation may be written

$$RT\ln \frac{a}{a_0} = \frac{2\gamma \overline{V}}{r} \tag{X-1}$$

in the case of a crystal, where γ is the interfacial tension of a given face and r is the radius of the inscribed sphere (see Section VII-2C); a denotes the activity, as measured, for example, by the solubility of the solid.

347

In principle, then, small crystals should show a higher solubility in a given solvent than should large ones. A corollary is that a mass of small crystals should eventually recrystallize to a single crystal (see Ostwald ripening, Section IX-4).

In the case of a sparingly soluble salt that dissociates into v^+ positive ions M and v^- negative ions A, the solubility S is given by

$$S = \frac{(M)}{v^+} = \frac{(A)}{v^-}$$
(X-2)

and the activity of the solute is given by

$$a = (M)^{v^+}(A)^{v^-} = S^{(v^+ + v^-)}(v^+)^{v^+}(v^-)^{v^-}$$
(X-3)

if activity coefficients are neglected. On substituting into Eq. X-1,

$$RT(v^+ + v^-)\ln \frac{S}{S_0} = \frac{2\gamma \overline{V}}{r}$$
(X-4)

Most studies of the Kelvin effect have been made with salts—see Refs. 2–4. A complicating factor is that of the electrical double layer presumably present; Knapp [5] (see also Ref. 6) gives the equation

$$RT\ln \frac{S}{S_0} = \frac{2\gamma \overline{V}}{r} - \frac{q^2 \overline{V}}{8\pi \epsilon r^4}$$
(X-5)

supposing the particle to possess a fixed double layer of charge q. Other potential difficulties are in the estimation of solution activity coefficients and the presence of defects affecting the chemical potential of the small crystals. See Refs. 7 and 8.

3. Surface Energies from Immersion, Adsorption, and Engulfment Studies

A. Enthalpy of Immersion

Immersion of a solid in a liquid generally liberates heat, and the enthalpy of immersion may be written

$$h_i = \gamma_{SL} - \gamma_S - T\left(\frac{\partial \gamma_{SL}}{\partial T} - \frac{\partial \gamma_S}{\partial T}\right)$$
(X-6)

where γ denotes the interfacial tension of the designated interface. This enthalpy can be related to the contact angle, θ through application of Young's equation (Section X-4A) [9] (see also Ref. 10)

TABLE X-1
Heats of Immersion at 25°C

	h_i (ergs/cm^2)^a				
Solid	H$_2$O	C$_2$H$_5$OH	n-Butylamine	CCl$_4$	n-C$_6$H$_{14}$
TiO$_2$ (rutile)	550	400	330	240^b	135
Al$_2$O$_3$^{c}	400–600	—	—	—	100
SiO$_2$	400–600^c	—	—	220^b	—
Graphon	32.2	110	106	—	103
Teflon	6	—	—	—	47^c

^{a}Ref. 18.
^{b}Ref. 16.
^{c}Ref. 17.

$$h_i = -\gamma_{LV} \cos\theta - (\gamma_S - \gamma_{SV}) - T\left(\frac{\partial\gamma_{SL}}{\partial T} - \frac{\partial\gamma_S}{\partial T}\right) \qquad \text{(X-7)}$$

In practice, $\gamma_S - \gamma_{SV}$ is negligible as is $\partial\gamma_S/\partial T$ for systems having large contact angles. Also, low energy surfaces have a relatively constant value of $\partial\gamma_{SL}/\partial T$ = 0.07 ±0.02 erg cm^{-2} K^{-1}, leaving

$$\cos\theta = \frac{-0.07T - h_i}{\gamma_{LV}} \qquad \text{(X-8)}$$

a simple relationship between the contact angle and enthalpy of immersion. Equation X-8 works well for nonpolar solids such as graphon where the calculated $\theta = 81°$ corresponds well to that measured (82°) [9].

The heat of immersion is measured calorimetrically with finely divided powders as described by several authors [9, 11–14] and also in Section XVI-4. Some h_i data are given in Table X-1. Polar solids show large heats of immersion in polar liquids and smaller ones in nonpolar liquids. Zettlemoyer [15] noted that for a given solid, h_i was essentially a linear function of the dipole moment of the wetting liquid.

There are complexities. The wetting of carbon blacks is very dependent on the degree of surface oxidation; Healey et al. [19] found that q_{imm} in water varied with the fraction of hydrophilic sites as determined by water adsorption isotherms. In the case of oxides such as TiO$_2$ and SiO$_2$, q_{imm} can vary considerably with pretreatment and with the specific surface area [17, 20, 21]. Morimoto and co-workers report a considerable variation in q_{imm} of ZnO with the degree of heat treatment (see Ref. 22).

One may obtain the difference between the heat of immersion of a clean surface and one with a preadsorbed film of the same liquid into which immersion is carried out. This difference can now be related to the heat of adsorption of the film, and this aspect of immersion data is discussed further in Section XVII-4.

B. Surface Energy and Free Energy Changes from Adsorption Studies

It turns out to be considerably easier to obtain fairly precise measurements of a change in the surface free energy of a solid than it is to get an absolute experimental value. The procedures and methods may now be clear-cut, and the calculation has a thermodynamic basis, but there remain some questions about the physical meaning of the change. This point is discussed further in the following material and in Section X-6.

There is always some degree of adsorption of a gas or vapor at the solid-gas interface; for vapors at pressures approaching the saturation pressure, the amount of adsorption can be quite large and may approach or exceed the point of monolayer formation. This type of adsorption, that of vapors near their saturation pressure, is called *physical adsorption*; the forces responsible for it are similar in nature to those acting in condensation processes in general and may be somewhat loosely termed "van der Waals" forces, discussed in Chapter VII. The very large volume of literature associated with this subject is covered in some detail in Chapter XVII.

The present discussion is restricted to an introductory demonstration of how, in principle, adsorption data may be employed to determine changes in the solid-gas interfacial free energy. A typical adsorption isotherm (of the physical adsorption type) is shown in Fig. X-1. In this figure, the amount adsorbed per gram of powdered quartz is plotted against P/P^0, where P is the pressure of the adsorbate vapor and P^0 is the vapor pressure of the pure liquid adsorbate.

The surface excess per square centimeter Γ is just n/Σ, where n is the moles adsorbed per gram and Σ is the specific surface area. By means of the Gibbs equation (III-80), one can write the relationship

$$d\gamma = -\Gamma RT \, d\ln a \qquad (\text{X-9})$$

For an ideal gas, the activity can be replaced by the pressure, and

$$d\gamma = -\Gamma RT \, d\ln P \qquad (\text{X-10})$$

or

$$\pi = -\int d\gamma = RT \int \Gamma \, d\ln P \qquad (\text{X-11})$$

where π is the film pressure. It then follows that

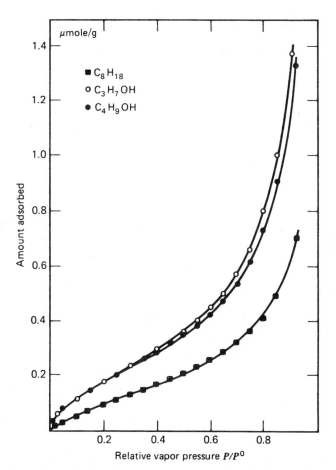

Fig. X-1. Adsorption isotherms for *n*-octane, *n*-propanol, and *n*-butanol on a powdered quartz of specific surface area 0.033 m^2/g at 30°C. (From Ref. 23.)

$$\pi = \gamma_S - \gamma_{SV} = \frac{RT}{\Sigma} \int_0^P n \, d\ln P \qquad \text{(X-12)}$$

$$\pi^0 = \gamma_S - \gamma_{SV^0} = \frac{RT}{\Sigma} \int_0^{P^0} n \, d\ln P \qquad \text{(X-13)}$$

where *n* is the moles adsorbed.

Equations X-12 and X-13 thus provide a thermodynamic evaluation of the change in interfacial free energy accompanying adsorption. As discussed further in Section X-5C, typical values of π for adsorbed films on solids range up to 100 ergs/cm^2.

A somewhat subtle point of difficulty is the following. Adsorption isotherms are quite often entirely reversible in that adsorption and desorption curves are identical. On the other hand, the solid will not generally be an equilibrium crystal and, in fact, will often have quite a heterogeneous surface. The quantities γ_S and γ_{SV} are therefore not very well defined as separate quantities. It seems preferable to regard π, which *is* well defined in the case of reversible adsorption, as simply the change in interfacial free energy and to leave its further identification to treatments accepted as modelistic.

The film pressure can be subjected to further thermodynamic manipulation, as discussed in Section XVII-13. Thus

$$\pi - T \, \frac{d\pi}{dT} = H_S - H_{SV} \tag{X-14}$$

where the subscript SV denotes a solid having an adsorbed film (in equilibrium with some vapor pressure P of the gaseous adsorbate). Equation X-14 is analogous to Eq. III-8 for a one-component system.

A heat of immersion may refer to the immersion of a clean solid surface, $q_{S,\,imm}$, or to the immersion of a solid having an adsorbed film on the surface. If the immersion of this last is into liquid adsorbate, we then report $q_{SV,\,imm}$; if the adsorbed film is in equilibrium with the saturated vapor pressure of the adsorbate (i.e., the vapor pressure of the liquid adsorbate P^0), we will write $q_{SV^0,\,imm}$. It follows from these definitions that

$$q_{S,\,imm} - q_{SV,\,imm} = \pi - T \, \frac{d\pi}{dT} + \Gamma \, \Delta H_v \tag{X-15}$$

where the last term is the number of moles adsorbed (per square centimeter of surface) times the molar enthalpy of vaporization of the liquid adsorbate (see Ref. 24).

This discussion of gas adsorption applies in similar manner to adsorption from solution, and this topic is taken up in more detail in Chapter XII.

C. Engulfment

Neumann and co-workers have used the term *engulfment* to describe what can happen when a foreign particle is overtaken by an advancing interface such as that between a freezing solid and its melt. This effect arises in floatation processes described in Section XIII-4A. Experiments studying engulfment have been useful to test semiempirical theories for interfacial tensions [25–27] and have been used to estimate the surface tension of cells [28] and the interfacial tension between ice and water [29].

4. Contact Angle

A. Young's Equation

It is observed that in most instances a liquid placed on a solid will not wet it but remains as a drop having a definite angle of contact between the liquid and solid phases. The situation, illustrated in Fig. X-2, is similar to that for a

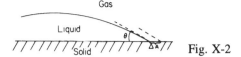

Fig. X-2

lens on a liquid (Section IV-2E), and a simple derivation leads to a very useful relationship.

The change in surface free energy, ΔG^s, accompanying a small displacement of the liquid such that the change in area of solid covered, $\Delta\mathcal{A}$, is

$$\Delta G^s = \Delta\mathcal{A}\,(\gamma_{SL} - \gamma_{SV^0}) + \Delta\mathcal{A}\,\gamma_{LV}\cos(\theta - \Delta\theta) \qquad (\text{X-16})$$

At equilibrium

$$\lim_{\Delta\mathcal{A}\to 0}\frac{\Delta G^s}{\Delta\mathcal{A}} = 0$$

and

$$\gamma_{SL} - \gamma_{SV^0} + \gamma_{LV}\cos\theta = 0\dagger \qquad (\text{X-17})$$

or

$$\gamma_{LV}\cos\theta = \gamma_{SV^0} - \gamma_{SL} \qquad (\text{X-18})$$

Alternatively, in combination with the definition of work of adhesion (Eq. XII-20),

$$w_{SLV} = \gamma_{LV}(1 + \cos\theta) \qquad (\text{X-19})$$

Equation X-17 was stated in qualitative form by Young in 1805 [30], and we will follow its designation as Young's equation. The equivalent equation, Eq. X-19, was stated in algebraic form by Dupré in 1869 [31], along with the definition of work of adhesion. An alternative designation for both equations, which are really the same, is that of the Young and Dupré equation (see Ref. 32 for an emphatic dissent).

It is important to keep in mind that the phases are mutually in equilibrium. In particular, the designation γ_{SV^0} is a reminder that the solid surface must be in equilibrium with the saturated vapor pressure P^0 and that there must therefore be an adsorbed film of film pressure π^0 (see Section X-3B). Thus

---

†It can be shown that regardless of the macroscopic geometry of the system, $\Delta\theta/\Delta\mathcal{A}$ behaves as a second-order differential and drops out in taking the limit of $\Delta\mathcal{A}\to 0$.

$$\gamma_{LV} \cos \theta = \gamma_S - \gamma_{SL} - \pi^0 \qquad (\text{X-20})$$

This distinction between γ_S and γ_{SV^0} seems first to have been made by Bangham and Razouk [33]; it was also stressed by Harkins and Livingstone [34]. Another quantity, introduced by Bartell and co-workers [35] is the *adhesion tension* **A**, which will be defined here as

$$A_{SLV} = \gamma_{SV^0} - \gamma_{SL} = \gamma_{LV} \cos \theta \qquad (\text{X-21})$$

Both here and in Eq. X-19 the subscript SLV serves as a reminder that the work of adhesion and the adhesion tension involve γ_{SV^0} rather than γ_S.

For practical purposes, if the contact angle is greater than 90°, the liquid is said not to wet the solid—in such a case drops of liquid tend to move about easily on the surface and not to enter capillary pores. On the other hand, a liquid is considered to wet a solid only if the contact angle is zero. It must be understood that this last is a limiting extreme only in a geometric sense. If θ is zero, Eq. X-17 ceases to hold, and the imbalance of surface free energies is now given by a spreading coefficient (see Section IV-2A).

$$S_{L/S(V)} = \gamma_{SV^0} - \gamma_{LV} - \gamma_{SL} \qquad (\text{X-22})$$

Alternatively, the adhesion tension now exceeds γ_{LV} and may be written as K_{SLV}.

The preceding definitions have been directed toward the treatment of the solid-liquid-gas contact angle. It is also quite possible to have a solid-liquid-liquid contact angle where two mutually immiscible liquids are involved. The same relationships apply, only now more care must be taken to specify the extent of mutual saturations. Thus for a solid and liquids A and B, Young's equation becomes

$$\gamma_{AB} \cos \theta_{SAB} = \gamma_{SB(A)} - \gamma_{SA(B)} = A_{SAB} \qquad (\text{X-23})$$

Here, θ_{SAB} denotes the angle as measured in liquid A, and the phases in parentheses have saturated the immediately preceding phase. A strictly rigorous nomenclature would be yet more complicated; we simply assume that A and B are saturated by the solid and further take it for granted that the two phases at a particular interface are mutually saturated. *If* mutual saturation effects are neglected, then the combination of Eqs. X-23 and X-21 gives

$$A_{SA} - A_{SB} = \gamma_{SB} - \gamma_{SA} = A_{SAB} \qquad (\text{X-24})$$

If in applying Eq. X-24 in adhesion tension corresponding to $\theta = 0$ is obtained, then spreading occurs, and the result should be expressed by K or by a spreading coefficient. Thus adhesion tension is a unifying parameter covering both contact angle and spreading behavior. A formal, rigorous form of Eq. X-24 would be

$$A_{SAV} - A_{SBV} = A_{SAB} + (\pi_{SV(B)} - \pi_{SV(A)}) + (\pi_{SB(A)} - \pi_{SA(B)}) \tag{X-25}$$

where, for example, $\pi_{SB(A)}$ is the film pressure or interfacial free energy lowering at the solid-liquid B interface due to saturation with respect to A.

There are some subtleties with respect to the physicochemical meaning of the contact angle equation, and these are taken up in Section X-7. The preceding, however, serves to introduce the conventional definitions to permit discussion of the experimental observations.

5. Contact Angle Hysteresis

The phenomenon of contact angle hysteresis has been recognized and studied for a long time. Numerous researchers have studied its origin experimentally and theoretically [36, 37]. The general observation is that the contact angle measured for a liquid *advancing* across a surface exceeds that of one *receding* from the surface. An everyday example is found in the appearance of a raindrop moving down a windowpane or an inclined surface [38–40]. This difference, known as *contact angle hysteresis*, can be quite large, as much as 50° for water on mineral surfaces. This can be quite important in coating processes.

Although not fully understood, contact line hysteresis is generally attributed to surface roughness, surface heterogeneity, solution impurities adsorbing on the surface, or swelling, rearrangement or alteration of the surface by the solvent. The local tilting of a rough surface or the local variation in interfacial energies on a heterogeneous surface can cause the contact angle to vary. It is not yet clear whether, like other hysteretic phenomena (such as found in magnetism), contact angle hysteresis can be described by irreversible transitions or "jumps" between domains of equilibrium states [41]. Here we review some of the main features of heterogeneous or rough surfaces and their effect on contact angle measurements.

A. Heterogeneous Surfaces

Surfaces having heterogeneities due to impurities or polycrystallinity over small length scales can be modeled with the Cassie equation [42]

$$\gamma_{LV} \cos \theta_c = f_1(\gamma_{S1V} - \gamma_{S1L}) + f_2(\gamma_{S2V} - \gamma_{S2L}) \tag{X-26}$$

or

$$\cos \theta_c = f_1 \cos \theta_1 + f_2 \cos \theta_2 \tag{X-27}$$

where f_1 and f_2 are the fractions of the surface occupied by surface types having contact angles θ_1 and θ_2. Another averaging expression has been proposed [43]

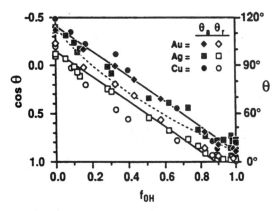

Fig. X-3. Variation of contact angle with f_{OH}, the fraction of the surface covered by $HS(CH_2)_{11}OH$ in a mixture with $HS(CH_2)_{11}CH_3$. Solid line is comparison with Eq. X-27, and dashed line is from Eq. X-28. (From Ref. 44.)

$$(1 + \cos \theta_c)^2 = f_1(1 + \cos \theta_1)^2 + f_2(1 + \cos \theta_2)^2 \qquad \text{(X-28)}$$

wherein forces have been averaged rather than surface tensions. While to these authors Eq. X-28 seems physically more correct than Eq. X-27, experiments on mixed monolayers of alkylthiols by Laibinis and Whitesides [44] showed that the Cassie equation fit the variation of angle with surface fraction much better than Eq. X-28, as illustrated in Fig. X-3.

An interesting application of the Cassie equation (X-27) arises in the case of a screen or woven material where now f_2 is the fraction of open area, γ_{S2V} is zero and γ_{S2L} is simply γ_{LV}. The relationship then becomes

$$\cos \theta_c = f_1 \cos \theta_1 - f_2 \qquad \text{(X-29)}$$

Several groups found good agreement with the apparent contact angle of water drops on screens and textiles [45–47]; this is important in processes for waterproofing fabrics. A natural example occurs in the structure of feathers where an interlocking set of barbules, the fine support fibers on feather barbs, produce a highly porous ($f_2 = 0.5$) resilient structure. The apparent contact typical of "water on a duck's back" is about 150° (receding), whereas the true contact angle of the feather material is about 100°.

There have been numerous studies of heterogeneous surfaces [48–54]. Johnson and Dettre [48] and later Neumann and Good [49] showed that heterogeneous surfaces cause metastable equilibrium states of the system, allowing for multiple contact angles. An example of the magnitude and variability of contact angle hysteresis is shown in Fig. X-4 for titania slides coated with varying amounts of a surfactant. Qualitatively, the advancing angle is determined by the

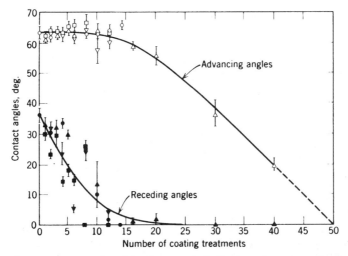

Fig. X-4. Water contact angle of titania-coated glass after treatment with trimethyloc-tadecylammonium chloride as a function of the number of coating treatments with 1.1% polydibutyl titanate. (From Ref. 51.)

less polar portions of the plate where the water will adhere and the receding angle is influenced by the more polar portions.

Generally, a contact line traversing a heterogeneous surface will become pinned to the patches, producing a lower contact angle (for a review of the theories of pinning, see Ref. 55). There are energy barriers to surpass as the contact line moves across these pinning points. The ability to cross these energy barriers and hence the magnitude of the contact angle hysteresis is very sensitive to ambient vibrations [56, 57], hence the variability in the literature. As a test of the forces pinning the contact line to heterogeneities, Nadkarni and Garoff [58] measured the distortion of a contact line traversing a single 20–100-μm defect produced by melting a polymethylmethacrylate ($\theta = 90°$) particle on a polystyrene ($\theta = 83°$) surface. The contact line on a vertical Wilhelmy plate follows

$$y(x) = \frac{f}{\pi\gamma_{LV}\theta_0^2} \ln\left\{\frac{L}{|x - x_d|}\right\} \tag{X-30}$$

where x and y are the horizontal and vertical directions, f is the force exerted by the heterogeneity on the contact line, L is a long-length-scale cutoff, ≈ 2 mm, and θ_0 is the contact angle on the uniform surface. A fit to this expression provided an assessment of f, which follows Hooke's law independent of the defect structure. Marsh and Cazabat [59] found the same contact line structure; however, they found a larger than expected velocity for the increase in width due to dynamic contact angle effects (see Section X-5D).

There is a great need for model heterogeneous surfaces to study contact angle hysteresis. One completely random heterogeneous surface has been created very recently by Decker and Garoff [57]. By coating a surface with a fluorinated surfactant then degrading it with ultraviolet illumination in the presence of water vapor, they produce a surface having randomly heterogeneous surface energy. This is an important advance in the ability to study hysteresis in a consistent manner. Self-assembled monolayers [44] represent another system of variable heterogeneity.

Another approach to the systematic study of contact angle hysteresis is to produce surfaces of known patterning. Studying capillary rise theoretically on a surface having vertical strips of alternating surface energy, Gaydos and Neumann [60] determined that the minimum patch size necessary to produce contact angle hysteresis was surprisingly large, approximately one micrometer. In measurements on a surface having vertical line defects, Marsh and Cazabat [61] found that the healing length from the defect to the unperturbed contact line varied linearly with the width of the vertical defect up to a saturation value when the defect is approximately 0.75 mm in width. When the strips are turned horizontally, Li and Neumann [62] conclude that the advancing angle will approach the minimum free energy state while the receding angle will be subject to vibrations and can attain one of many metastable states. This helps to explain the relative reproducibility of the advancing contact angle as compared to the receding angle. These studies illustrate the usefulness of models and experiments on surfaces having carefully controlled geometries.

B. Surface Roughness

The effect of surface roughness on contact angle was modeled by several authors about 50 years ago [42, 45, 63, 64]. The basic idea was to account for roughness through r, the ratio of the actual to projected area. Thus $\Delta \mathcal{A}_{SL} = r\Delta \mathcal{A}_{SL \text{ apparent}}$ and similarly for $\Delta \mathcal{A}_{SV}$, such that the Young equation (Eq. X-18) becomes

$$\cos \theta_{\text{rough}} = r \cos \theta_{\text{true}} \tag{X-31}$$

hence angles less than 90° are decreased by roughness, while if θ is greater than 90°, the angle increases. If the contact angle is large and the surface sufficiently rough, the liquid may trap air so as to give a composite surface effect as illustrated in Fig. X-5 and an apparent contact angle from Eq. X-29:

$$\cos \theta_{\text{apparent}} = r f_1 \cos \theta_1 - f_2 \tag{X-32}$$

Dettre and Johnson [37] (see also Good [64]) made calculations on a mathematical model consisting of sinusoidal grooves such as those in Fig. X-5 concentric with a spherical drop (i.e., gravity was neglected). Minimization of the surface free energy with the local contact angle defined by the Young equation led to a drop configuration such that the apparent angle θ in Fig. X-5 was that given by Eq. X-31. The free energy contained barriers as the drop was deformed such that the liquid front moved over successive ridges. The presence of these small energy barriers suggested that hysteresis arose when the drop had insufficient vibrational energy to surmount them. Similar anal-

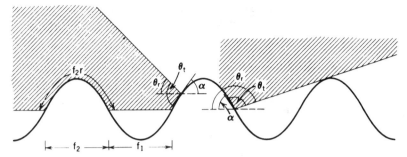

Fig. X-5. Drop edge on a rough surface.

yses have been made by other authors [63, 65, 66]. Clearly, roughness is not adequately defined by r but is also a matter of topology: the same roughness in the form of parallel grooves gives an entirely different behavior than one in the form of pits [67].

A modification of Eq. X-31 has been made for fractal surfaces (see Section VII-4C). Contact angles near $90°$ on fractal surfaces are well described by [68]

$$\cos \theta_{\text{fractal}} = \left(\frac{L}{l}\right)^{D-2} \cos \theta_{\text{true}} \qquad (\text{X-33})$$

where L and l are the upper and lower limits of fractal behavior, respectively, and D is the fractal dimension. While a contact angle of $180°$ is fundamentally impossible, Tsujii and co-workers [68] have attained an angle of $174°$ on a fractal surface of an alkylketene dimer (a paper sizing agent). By contrast, a flattened surface of the same material has a contact angle of $109°$ as illustrated in Fig. X-6.

Roughness has important implications in wetting applications. While the eutectic solder, SnPb, normally forms a contact angle of $15–20°$ with copper, it completely wets the surface of rough electroplated copper and forms a fractal spreading front [69].

Lin et al. [70, 71] have modeled the effect of surface roughness on the dependence of contact angles on drop size. Using two geometric models, concentric rings of cones and concentric conical crevices, they find that the effects of roughness may obscure the influence of line tension on the drop size variation of contact angle. Conversely, the presence of line tension may account for some of the drop size dependence of measured hysteresis.

C. Liquid–Surface Interactions: Surface Changes and Autophobicity

The third set of causes of contact angle hysteresis arise as a result of specific interactions between the liquid and solid phases. When a liquid traverses a solid, it may alter the molecular structure of the solid, thereby producing contact angle hysteresis. This occurs on polymer substrates where the solvent swells, solvates, or otherwise perturbs the surface [72, 73]. One example is the large hysteresis of methylene iodide on agar gels ($66–30°$) [74]. This effect has been examined in some detail by Neumann and co-workers studying the effect of alkane swelling

Fig. X-6. Drop of water on an alkylketene dimer surface: (*a*) fractal surface $D = 2.29$; (*b*) flat surface. (From Ref. 68.)

of a fluoropolymer surface on contact angle [75]. The contact angle hysteresis increases with the amount of time the alkane is in contact with the surface, as shown in Fig. X-7. Other studies have used stretched (and hence anisotropic) polymer films [76].

Surfactant-coated surfaces may also rearrange on contact with a liquid as shown by Israelachvili and co-workers [77]. This mechanism helps to explain hysteresis occurring on otherwise smooth and homogeneous surfaces.

When a surface-active agent is present in a liquid droplet, it can adsorb to the surface, lower the surface energy, and cause the liquid contact angle to increase. This phenomenon, known as *autophobicity*, was postulated by Zisman many years ago [78, 79]. Autophobicity is quite striking in wetting films on clean

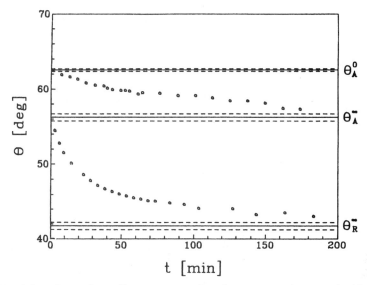

Fig. X-7. Advancing and receding contact angles of octane on mica coated with a fluoropolymer FC 722 (3M) versus the duration of the solid–liquid contact. The solid lines
represent the initial advancing and infinite time advancing and receding contact lines
and the dashed lines are 95% confidence limits. (From Ref. 75.)

surfaces since they first advance across the surface then retract via a stick-jump
motion since they no longer wet the interface [80]. Novotny and Marmur [81]
have shown that this process can occur by adsorption from the vapor ahead
of the contact line. Thus, as found in wetting films, a thin film, known as a
precursor film, exists on the surface around the liquid drop [82, 83] (see Section
X-7).

Ruch and Bartell [84], studying the aqueous decylamine–platinum system, combined
direct estimates of the adsorption at the platinum–solution interface with contact angle
data and the Young equation to determine a solid–vapor interfacial energy change of up
to 40 ergs/cm^2 due to decylamine adsorption. Healy (85) discusses an adsorption model
for the contact angle in surfactant solutions and these aspects are discussed further in
Ref. 86.

D. Dynamic Contact Angles

Since wetting films and coating processes involve the flow of fluid over the solid, the
contact angle cannot be viewed as simply a static quantity, it also depends on the speed
with which the three-phase line advances or recedes. An understanding of the dynamic
contact angle requires detailed hydrodynamic analysis outside the scope of this book;
hence, we simply summarize the basic findings and refer to the reader to several good
references [87–90]. The key element of the analysis is the proper definition of the boundary condition at the three-phase line. The presence of a singularity at the moving contact

line has fueled much investigation of the fluid mechanical model. Two main approaches have arisen to remove the singularity: to treat the fluid-solid interface with a slip boundary condition near the contact line [91], or to treat the case where a thin immobilized liquid film precedes the contact line and the contact angle is thus defined between the macroscopic layer and the thin film [87].

The fundamental dimensionless parameter characterizing the speed of contact line motion is the capillary number, $Ca = U\mu/\gamma_{LV}$, where U is the velocity and μ is the fluid viscosity. Ca represents the relative importance of viscous forces to interfacial tension in determining the shape of the free surface near the contact line. Recent experimental work by Garoff and co-workers [89, 92, 93] has provided a measurement of the liquid–vapor profile in the vicinity of the contact line. They found that the profiles were consistent with either the slip or thin-film boundary conditions except at the highest Ca (9×10^{-3}). In an analysis of dewetting situations, Brochard-Wyart and deGennes [90] found that hydrodynamic effects dominate at low contact angles and low velocities but molecular effects become important when the contact angle is large and the velocity is high, a situation frequently encountered in practical applications. Neumann and co-workers have developed a high-precision capillary rise technique to measure contact angles at low velocities [94].

6. Experimental Methods and Measurements of Contact Angle

A. Experimental Methods to Measure Contact Angle

The classic techniques for measuring contact angle have been reviewed in detail by Neumann and Good [95]. The most commonly used method involves directly measuring the contact angle for a drop of liquid resting on a horizontal solid surface (a *sessile* drop) as illustrated in Fig. X-8. Commercial contact angle goniometers employ a microscope objective to view the angle directly much as Zisman and co-workers did 50 years ago [96]. More sophisticated approaches involve a photograph or digital image of the droplet [46, 97–100]. An entirely analogous measurement can be made on a sessile bubble captured at a solid–liquid interface as illustrated in Fig. X-8 [101, 102]. The use of bub-

Sessile Drops

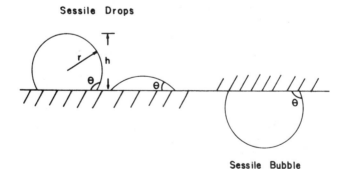

Sessile Bubble

Fig. X-8. Use of sessile drops or bubbles for contact angle determination.

bles has the advantage of minimizing the effect of adventitious contamination.

The axisymmetric drop shape analysis (see Section II-7B) developed by Neumann and co-workers has been applied to the evaluation of sessile drops or bubbles to determine contact angles between 50° and 180° [98]. In two such studies, Li, Neumann, and co-workers [99, 100] deduced the line tension from the drop size dependence of the contact angle and a modified Young equation

$$\cos \theta = \cos \theta_\infty - \frac{\sigma}{\gamma_{LV} R} \qquad \text{(X-34)}$$

where σ is the line tension, R is the drop radius, and θ_∞ is the contact angle at $R = \infty$. This size dependence thus implies a line tension; the magnitude and even the sign of the line tension were the subject of controversy in earlier studies [103–106]. A similar analysis can be made on variable size bubbles [101, 107].

The drop shape analysis technique can also be applied to images taken from above sessile drops to determine contact angles [108]. In one approach, the diameter of the drop in contact with the surface is measured from above and the contact angle is determined from integration of the Laplace equation (II-7) given the surface tension and drop volume. This approach is applicable only to systems where the contact angle is less than 90° since the observed diameter will exceed the contact diameter when $\theta > 90°$. For these nonwetting cases, the maximum equatorial diameter can be measured to determine contact angle [109]. The contact angle can also be calculated from interferometric measurement of the meniscus profile of a horizontal film in a capillary [110].

The capillary rise on a Wilhelmy plate (Section II-6C) is a nice means to obtain contact angles by measurement of the height, h, of the meniscus on a partially immersed plate (see Fig. II-14) [111, 112]. Neumann has automated this technique to replace manual measurement of h with digital image analysis to obtain an accuracy of ±0.06° (and a repeatability to 95%, in practice, of ±0.01°) [108]. The contact angle is obtained directly from the height through

$$\sin \theta = 1 - \frac{\rho g h^2}{2\gamma} = 1 - \frac{1}{2} \left(\frac{h}{a} \right)^2 \qquad \text{(X-35)}$$

where a is the capillary length (Eq. II-10). The automated measurement of h can be used to determine the dynamic contact angle at low speeds around 0.5 mm/min [113]. The Wilhelmy slide method has been modified to accept pulse displacements to measure the response of the contact angle and the energetics of contact line hysteresis on random heterogeneous surfaces (see Section X-4B) [57].

Yaminsky and Yaminskaya [114] have used a Wilhelmy plate to directly measure the interfacial tension (and hence infer the contact angle) for a surfactant solution on silica. An elegant (and underused) method due to Langmuir and Schaeffer [115] employs reflection of light from the interface to precisely determine the contact angle at the edge of a meniscus of a vertical plate or capillary tube (see Ref. 116).

If the solid in question is available only as a finely divided powder, it may be compressed into a porous plug so that the capillary pressure required to pass a nonwetting liquid can be measured [117]. If the porous plug can be regarded as a bundle of capillaries of average radius r, then from the Laplace equation (II-7) it follows that

$$\Delta P = \frac{2\gamma_{LV} \cos \theta}{r} \tag{X-36}$$

where, depending on the value of θ, ΔP is the pressure required to force or restrain entry of the liquid. For a wetting liquid, $\Delta P_0 = 2\gamma'_{LV}/r$, such that measurement of the pressure required to prevent a wetting liquid from entering provides the capillary radius r. Repeating the measurement with the nonwetting liquid, and eliminating r from Eq. X-36 yields

$$\cos \theta = \frac{\Delta P}{\Delta P_0} \frac{\gamma'_{LV}}{\gamma_{LV}} \tag{X-37}$$

B. Results of Contact Angle Measurements

It is clear from our discussion of contact angle hysteresis that there is some degree of variability in reported contact angle values. The data collected in Table X-2, therefore, are intended mainly as a guide to the type of behavior to be expected. The older data comprise mainly results for refractory and relatively polar solids, while newer data are for polymeric surfaces.

Bartell and co-workers [118], employing mainly solids such as graphite and stibnite, observed a number of regularities in contact angle behavior but were hampered by the fact that most organic liquids appear to wet such solids. Zisman and co-workers [78] were able to take advantage of the appearance of low-surface-energy polymers to make extensive contact angle studies with homologous series of organic liquids. Table X-2 includes some data on the temperature dependence of contact angle and some values for π^0, the film pressure of adsorbed vapor at its saturation pressure P^0.

There is appreciable contact angle hysteresis for many of the systems reported in Table X-2; the customary practice of reporting advancing angles has been followed.

As a somewhat anecdotal aside, there has been an interesting question as to whether gold is or is not wet by water, with many publications on either side. This history has been reviewed by Smith [119]. The present consensus seems to be that absolutely pure gold is water-wet and that the reports of nonwetting are a documentation of the ease with which gold surface becomes contaminated (see Ref. 120, but also 121). The detection and control of surface contaminants has been discussed by White [121]; see also Gaines [122].

TABLE X-2
Selected Contact Angle Data

Liquid γ (ergs/cm^2)	Solid	θ (deg)	$d\theta/dT$ (deg/K)	π^0 (ergs/cm^2)	Ref.
		Advancing Contact Angle, 20–25°C			
Mercury (484)	PTFE[a]	150	—	—	78
	Glass	128–148	—	—	116, 124, 125
Water (72)	n-H[b]	111	—	—	123
	Paraffin	110	—	—	126
	PTFE[a]	108, 112	—	—	78, 126
		98	—	8.8	127
	FEP[c]	108	−0.05	—	128
		111.59 ± 0.19	—	—	129
	PET[j]	79.09 ± 0.12	—	—	129
	FC721[k]	119.05 ± 0.16	—	—	129
	Butyl rubber	110.8–113.3	—	—	130
	PMMA[l]	59.3	—	—	29
	Polypropylene	108	−0.02	—	131
	Human skin	90, 75[d]	—	—	136, 137
	Talc	78.3	—	—	138
	Polyethylene	103	−0.01	—	132
		96	−0.11	—	128
		94	—	—	133
		93	—	—	134
		88	—	14	135
	Naphthalene	88[e]	−0.13	—	139
	Graphite	86	—	19	140
			—	59	141
	Graphon	82	—	—	133
	Stearic acid[f]	80	—	98	135
	Sulfur	78	—	—	142
	Pyrolytic carbon	72	—	228	135
	Platinum	40	—	—	133
	Silver iodide	17	—	—	143
	Glass	Small	—	~20[g]	144
	Gold	0	—	—	119
CH$_2$I$_2$	PTFE	85, 88	—	—	137, 145
(67, 50.8[h])	Paraffin	61	—	—	126
		60	—	—	137
CH$_2$I$_2$	Talc	53, 64.1	—	—	138, 146
(67, 50.8[h])	Polyethylene	46, 51.9	—	—	137, 145
		40[e]	—	—	134
Formamide (58)	FEP[c]	95.38 ± 0.20	—	—	129
		92	−0.06	—	128
	Polyethylene	75	−0.01	—	128
	PET[j]	61.50 ± 0.37	—	—	129

TABLE X-2   (*Continued*)

Liquid γ (ergs/cm^2)	Solid	θ (deg)	$d\theta/dT$ (deg/K)	π^0 (ergs/cm^2)	Ref.
	PMMA	50.0	—	—	29
	Talc	67.1	—	—	138
CS$_2$ (~35[h])	Ice[i]	35	0.35	—	111
Benzene (28)	PTFE[a]	46	—	—	78
	n-H[b]	42	—	—	125
	Paraffin	0	—	—	79
	Graphite	0	—	—	79
n-Propanol	PTFE[a]	43	—	8.8	139
(23)	Paraffin	22	—	—	147
	Polyethylene	7	—	5	148
n-Decane	Graphite	120	—	—	149
(23)	PTFE[a]	40	−0.11	~1.0	150, 151
		35	—	—	152
		32	−0.12	—	153
Decane	FEP[c]	43.70 ± 0.15	—	—	129
n-Octane	PTFE[a]	30	−0.12	—	153
(21.6)		26	—	1.8, 3.0	127, 154
		26	—	—	152

Solid	Liquid 1	Liquid 2	Contact angle, θ_{S12}	Ref.
Solid-Liquid-Liquid Contact Angles, 20–25°C				
Stibnite	Water	Benzene	130	118
Aluminum oxide	Water	Benzene	22	118
PTFE[a]	Water	*n*-Decane	~180	151
	Benzyl alcohol	Water	30	155
PE[b]	Water	*n*-Decane	~180	151
	Paraffin oil	Water	30	155
Mercury	Water	Benzene	~100[d]	156
Glass	Mercury	Gallium	~0	157

[a]Polytetrafluoroethylene (Teflon).

[b]*n*-Hexatriacontane.

[c]Polytetrafluoroethylene-*co*-hexafluoropropylene.

[d]Not cleaned of natural oils.

[e]Single crystal.

[f]Langmuir-Blodgett film deposited on copper.

[g]From graphic integration of the data of Ref. 144.

[h]See Ref. 158.

[i]At approximately −10°C.

[j]Polyethylene terephthalate.

[k]Fluoropolymer (3M).

[l]Polymethylmethacrylate.

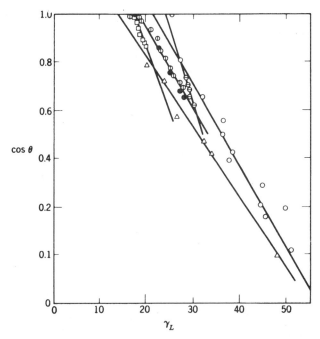

Fig. X-9. Zisman plots of the contact angles of various homologous series on Teflon; ○, RX; ⊖, alkylbenzenes; ⏀, *n*-alkanes; ●, dialkyl ethers; □, siloxanes; △, miscellaneous polar liquids. (Data from Ref. 78.)

A major contribution to the rational organization of contact angle data was made by Zisman and co-workers. They observed that cos θ (advancing angle) is usually a monotonic function of γ_L for a homologous series of liquids. The proposed function was

$$\cos \theta_{SLV} = a - b\gamma_L = 1 - \beta(\gamma_L - \gamma_c) \tag{X-38}$$

Figure X-9 shows plots of cos θ versus γ_L for various series of liquids on Teflon (polytetrafluoroethylene) [78]. Each line extrapolates to zero θ at a certain γ_L value, which Zisman has called the critical surface tension γ_c; since various series extrapolated to about the same value, he proposed that γ_c was a quantity characteristic of a given solid. For Teflon, the representative γ_c was taken to be about 18 and was regarded as characteristic of a surface consisting of —CF$_2$— groups.

The critical surface tension concept has provided a useful means of summarizing wetting behavior and allowing predictions of an interpolative nature. A schematic summary of γ_c values is given in Fig. X-10 [123]. In addition, actual contact angles for various systems can be estimated since β in Eq. X-38 usually has a value of about 0.03–0.04.

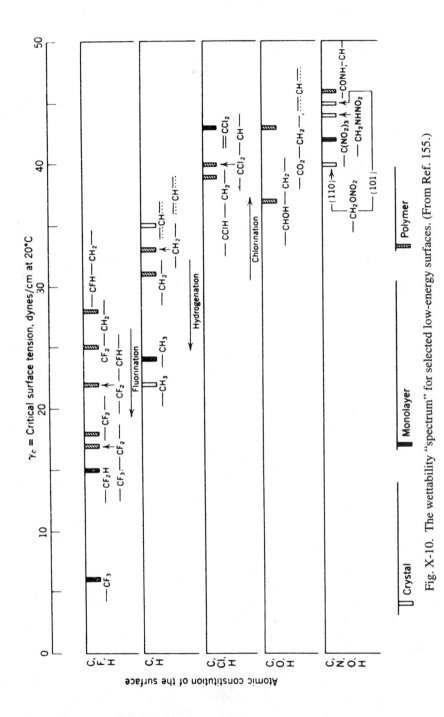

Fig. X-10. The wettability "spectrum" for selected low-energy surfaces. (From Ref. 155.)

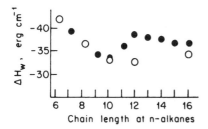

Fig. X-11. Heats of wetting from θ (●) and calorimetric heats of immersion (○) of PTFE in n-alkanes. (From Ref. 67.)

An important and expensive problem in surface science occurs in the prevention of the attachment of marine animals such as barnacles to ship surfaces, a process known as *biofouling*. Baier and Meyer [159] have shown that the Zisman plot can be used to predict biofouling, thus avoiding costly field tests to find a successful coating to prevent biofouling.

The effect of temperature on contact angle is seldom very great, as a practical observation. Some values of $d\theta/dT$ are included in Table X-2; a common figure is about -0.1 degrees/K (but note the case of CS_2 on ice; also rather large temperature changes may occur in L_1-L_2-S systems (see Ref. 160).

Knowledge of the temperature coefficient of θ provides a means of calculating the heat of immersion. Differentiation of Eq. X-18 yields

$$h_i = q_{imm} = E_{SV} - E_{SL} = E_L \cos \theta - T\gamma_L \; \frac{d \cos \theta}{dT} \qquad \text{(X-39)}$$

Since both sides of Eq. X-39 can be determined experimentally, from heat of immersion measurements on the one hand and contact angle data, on the other hand, a test of the thermodynamic status of Young's equation is possible. A comparison of calorimetric data for n-alkanes [18] with contact angle data [95] is shown in Fig. X-11. The agreement is certainly encouraging.

The measurement of contact angle has found several biomedical applications. In order to understand susceptibility to infection. Sherman and co-workers [161] measured the contact angle of water on various parts of the rabbit intestine at varying ages. Increasing hydrophobicity with age (from suckling to adult rabbits) may partially account for age-related differences in susceptibility to bacteria. A similar study focused on the lining of the stomach [162]. Contact angle measurements made in vivo on human teeth and various dental restorative materials showed invariant angles due to the presence of a surface biofilm that made the surfaces essentially indistinguishable [163]. The oral film even removed the effects of surface roughness. Finally, contact angles play a central role in the barrier function of condoms, a topic receiving enhanced attention due to the small size ($\approx 0.1\mu$m) of the HIV virus [164].

Values for π^0, the film pressure of the adsorbed film of the vapor (of the liquid whose contact angle is measured), are scarce. Vapor phase adsorption data, required by Eq. X-13, cannot be obtained in this case by the usual volumetric method (see Chapter

XVII). This method requires the use of a powdered sample (to have sufficient adsorption to measure), and interparticle capillary condensation grossly distorts the isotherms at pressures above about $0.9P^0$ [165]. Also, of course, contact angle data are generally for a smooth, macroscopic surface of a solid, and there is no assurance that the surface properties of the solid remain the same when it is in powdered form. More useful and more reliable results may be obtained by the gravimetric method (see Chapter XVII) in which the amount adsorbed is determined by a direct weighing procedure. In this case it has been possible to use stacks of 200–300 thin sheets of material, thus obtaining sufficient adsorbent surface (see Ref. 166). Probably the most satisfactory method, however, is that using ellipsometry to measure adsorbed film thickness (Section IV-3D); there is no possibility of capillary condensation effects, and the same smooth and macroscopic surface can be used for contact angle measurements. Most of the π^0 data in Table X-2 were obtained by this last procedure; the corresponding contact angles are for the identical surface. In addition to the values in Table X-2, Whalen and Hu quote 5–7 ergs/cm^2 for n-octane, carbon tetrachloride, and benzene on Teflon [167]; Tamai and co-workers [168] find values of 5–11 ergs/cm^2 for various hydrocarbons on Teflon.

Another approach has been to measure a contact angle with and without the presence of vapor of a third component, one insoluble in the contact angle forming liquid [145]. The π^0 value is then inferred from any change in contact angle. By this means Fowkes and co-workers obtained a value of 7 ergs/cm^2 for cyclohexane on polyethylene but zero for water and methylene iodide. The discrepancy with the value for water in Table X-2 could be due to differences in the polyethylene samples. It was suggested [145], moreover, that the vapor adsorption method reflects polar sites while contact angle is determined primarily by the nonpolar regions. A possibility, however, is that the interpretation of the effect of the third component vapor on contact angle is in error. The method balances the absorbing ability of the third component against that of the contact angle forming liquid and so does not actually give a simple π^0 for a single species.

Contact angle measurements can be used to follow adsorption processes as in Neumann and co-worker's study of human albumin adsorption onto hydrophobic and hydrophilic surfaces. Using drop shape analysis they could track θ and hence γ every 10 sec during the initial rapid adsorption [169]. The contact angle can also be used to infer the degree of conformational change in a protein on adsorption. The idea rests on the hydrophobic nature of the interior of most folded proteins; when they unfold or denature, this hydrophobic region becomes exposed to the solution. This is illustrated by a study showing that protein adsorbed onto polystyrene underwent more conformational change than one bound to an amine functionalized surface of plasma treated polystyrene [170].

Contact angles are also commonly used to characterize self-assembled monolayers (SAMs). Investigating SAMs of HS(CH$_2$)$_{16}$OR, Whitesides and co-workers [171] show the effect of burying the polar group in the monolayer by changing the length of the R group as illustrated in Fig. X-12. Contact angles for SAMs on different surfaces are independent of the underlying metal [44, 171].

Contact angle will vary with liquid composition, often in a regular way as illustrated in Fig. X-13 (see also Ref. 136). Li, Ng, and Neumann have studied the contact angles of binary liquid mixtures on teflon and found that the equation of state that describes

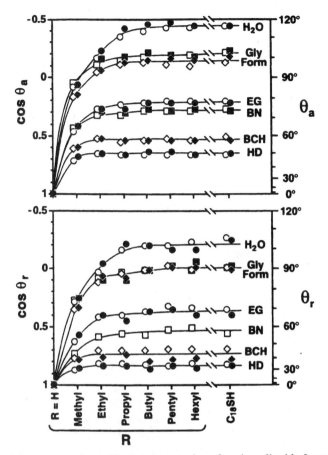

Fig. X-12. Advancing and receding contact angles of various liquids [water (circles), Gly = glycerol (squares), Form = formamide (diamonds), EG = ethylene glycol (circles), BN = αbromonapthalene (squares), BCH = bicyclohexyl (diamonds), HD = hexadecane (circles)] on monolayers of $HS(CH_2)_{16}OR$ having a range of R groups adsorbed on gold and silver (open and filled symbols respectively). (From Ref. 171.)

single-component interfacial tensions (see Section X-7) does not work for binary liquid mixtures [173]. From phase rule considerations, one would need an equation that has three independent variables such as γ_{LV}, γ_{SV}, and x, the mole fraction of the mixture.

Contact angle behavior in liquid mixtures is more complicated when one of the components is volatile such as found in *wine tears* along the edge of a glass of a hearty wine or "digestif." Since water has a much higher surface tension than alcohol, evaporation of alcohol produces a surface tension gradient driving a thin film up along the side of a wine glass, where the liquid accumulates and forms drops or tears. This effect has been quantitatively studied (appropriately in France) by Fournier and Cazabat [174].

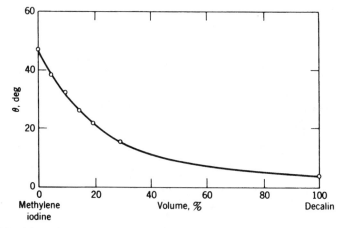

Fig. X-13. Advancing contact angles for methylene iodide–decalin mixtures on polyethylene. (From Ref. 172.)

7. Theories of Contact Angle Phenomena

A. Thermodynamics of the Young Equation

The extensive use of the Young equation (Eq. X-18) reflects its general acceptance. Curiously, however, the equation has never been verified experimentally since surface tensions of solids are rather difficult to measure. While Fowkes and Sawyer [140] claimed verification for liquids on a fluorocarbon polymer, it is not clear that their assumptions are valid. Nucleation studies indicate that the interfacial tension between a solid and its liquid is appreciable (see Section IX-3) and may not be ignored. Indirect experimental tests involve comparing the variation of the contact angle with solute concentration with separate adsorption studies [175].

It is, in fact, possible to consider at least *three* different contact angles for a given system! Let us call them θ_m, θ_{th}, and θ_{app}. The first, θ_m, is the microscopic angle between the liquid and the ridge of solid; it is the angle determined by the balance of surface stresses taking into account local deformations. The second, θ_{th}, is the thermodynamic or intrinsic angle obtained in the derivation of Eq. X-18 and in general is the angle obtained in a free energy minimization for the overall system. Finally, for a variety of reasons outlined in the preceding sections, one will measure experimentally an apparent angle θ_{app}, which is some kind of average such as that in Eq. X-27.

The microscopic complexity of the contact angle is illustrated in Fig. X-14, which shows the edge of a solidified drop of glass—note the foot that spreads out from the drop. Ruckenstein [176] discusses some aspects of this, and de Gennes [87] has explained the independence of the spreading rate on the nature of the substrate as due to a precursor film present also surrounding a nonspread-

Fig. X-14. SEM picture of a drop of cooled glass on Fernico metal (which has the same coefficient of thermal expansion), ×130. (From Ref. 183.)

ing drop [177]. The precursor film in a volatile system is produced by vapor condensation on the solid. In nonvolatile polymer liquids the film forms by flow [83, 178] and can have a profound effect on the dynamics [179].

Returning to θ_{th}, there is no dearth of thermodynamic proofs that the Young equation represents a condition of thermodynamic equilibrium at a three-phase boundary involving a smooth, homogeneous, and incompressible solid [104, 180]. Global-system free energy minimizations are mathematically difficult because of the problem of analytical representation of the free surface of a deformed drop; however, the case of an infinite drop is not hard, and the Young equation again results [10]. Widom and co-workers [181] have done numerical calculations and an analytical functional differentiation of the free energy of droplet shape in the presence of gravity and find that Young's equation remains valid when gravity is acting on the droplet.

Bikerman [182] criticized the derivation of Eq. X-18 out of concern for the ignored vertical component of γ_L. On soft surfaces a circular ridge is raised at the periphery of a drop (see Ref. 67); on harder solids there is no visible effect, but the stress is there. It has been suggested that the contact angle is determined by the balance of surface stresses rather than one of surface free energies, the two not necessarily being the same for a

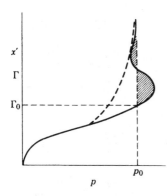

Fig. X-15. Variation of adsorbed film thickness with pressure. (From Ref. 10.)

nonequilibrium solid (see Section VII-2). Marmur [184] has set forth the dependence of the intrinsic contact angle on colloidal interactions near the contact line. He called into question the validity of the Young equation.

A somewhat different point of view is the following. Since γ_{SV} and γ_{SL} always occur as a difference, it is possible that it is this *difference* (the adhesion tension) that is the fundamental parameter. The adsorption isotherm for a vapor on a solid may be of the form shown in Fig. X-1, and the asymptotic approach to infinite adsorption as saturation pressure is approached means that at P^0 the solid is in equilibrium with bulk liquid. As Derjaguin and Zorin [144] note (see also Refs. 10 and 185), in a contact angle system, the adsorption isotherm must cross the P^0 line and have an unstable region, as illustrated in Fig. X-15. See Section XVII-12. A Gibbs integration to the first crossing gives

$$\pi^0 = \gamma_S - \gamma_{SV^0} = RT \int_{\Gamma=0}^{\Gamma^0} \Gamma \, d\ln P \qquad (X\text{-}40)$$

while that to the limiting condition of an infinitely thick and hence duplex film gives

$$I = \gamma_S - \gamma_{SL} - \gamma_L = RT \int_{\Gamma=0}^{\infty} \Gamma \, d\ln P \qquad (X\text{-}41)$$

From the difference in Eqs. X-40 and X-41, it follows that

$$\gamma_{SV^0} - (\gamma_{SL} + \gamma_L) = RT \int_{\Gamma=\Gamma^0}^{\infty} \Gamma \, d\ln P = \Delta I \qquad (X\text{-}42)$$

where ΔI is given by the net shaded area in the figure. Also, ΔI is just the spreading coefficient S_{LSV}, or

$$\Delta I = \gamma_L(\cos\theta - 1) \tag{X-43}$$

The integral ΔI, while expressible in terms of surface free energy differences, is defined independently of such individual quantities. A contact angle situation may thus be viewed as a consequence of the ability of two states to coexist: bulk liquid and thin film.

Koopal and co-workers [186] have extended this thermodynamic analysis to investigate the competitive wetting of a solid by two relatively immiscible liquids. They illustrate the tendency of silica to be preferentially wet by water over octane, a phenomenon of importance in oil reservoirs.

Equation X-43 may, alternatively, be given in terms of a disjoining pressure (Section VI-6) integral [187]:

$$\gamma_L \cos\theta = \gamma_L + \int_{x^0}^{\infty} \Pi \, dx \tag{X-44}$$

Here, x denotes film thickness and x^0 is that corresponding to Γ^0. An equation similar to Eq. X-42 is given by Zorin et al. [188]. Also, film pressure may be estimated from ζ potential changes [189]. Equation X-43 has been used to calculate contact angles in dilute electrolyte solutions on quartz; results are in accord with DLVO theory (see Section VI-4B) [190]. Finally, the π^0 term may be especially important in the case of liquid–liquid–solid systems [191].

B. Semiempirical Models: The Girifalco–Good–Fowkes–Young Equation

A model that has proved to be very stimulating to research on contact angle phenomena begins with a proposal by Girifalco and Good [192] (see Section IV-2B). If it is assumed that the two phases are mutually entirely immiscible and interact only through additive dispersion forces whose constants obey a geometric mean law, that is, $C_{1,AB} = (C_{1,AA}C_{1,BB})^{1/2}$ in Eq. VI-22; then the interfacial free energies should obey the equation

$$\gamma_{AB} = \gamma_A + \gamma_B - 2\Phi(\gamma_A\gamma_B)^{1/2} \tag{X-45}$$

The parameter Φ should be unity if molecular diameters also obey a geometric mean law [193] and is often omitted. Equation X-44, if applied to the Young equation with omission of Φ, leads to the relationship [192]

$$\cos\theta = -1 + 2\left(\frac{\gamma_S}{\gamma_L}\right)^{1/2} - \frac{\pi^0}{\gamma_L} \tag{X-46}$$

The term $-\pi^0/\gamma_L$ corrects for the adsorption of vapor on the solid. (To the extent that contact angle is an equilibrium property, the solid *must* be in equilibrium with the saturated vapor pressure P^0 of the bulk liquid.) The term has generally been neglected, mainly for lack of data; also, however, there has been some tendency to affirm that the term is always negligible if θ is large (e.g., Ref. 95). Recently Li, Xie and Neumann found that vapor adsorption on a hydrophobic fluorocarbon surface has a negligible effect on contact angle [194]. Neglecting the π^0 term in Eq. X-45, $\cos\theta$ should be a linear function of $1/\sqrt{\gamma_L}$ and data for various liquids on PTFE do cluster reasonably well along such a line [195].

In order to include other interactions such as dipolar or hydrogen bonding, many semiempirical approaches have been tried [196, 197, 200], including adding terms to Eq. X-45 [198, 201] or modifying the definition of Φ [202, 199]. Perhaps the most well-known of these approaches comes from Fowkes' [203, 204] suggestion that the interactions across a water–hydrocarbon interface are dominated by dispersion forces such that Eq. X-45 could be modified as

$$\gamma_{AB} = \gamma_A + \gamma_B - 2(\gamma_A^d \gamma_B^d)^{1/2} \tag{X-47}$$

where γ^d denotes an effective surface tension due to the dispersion or general van der Waals component of the interfacial tension [205–207]. For a nonpolar solid or liquid, Eq. X-46 now becomes (neglecting the term in π_{SV}^0)

$$\cos\theta = -1 + \frac{2(\gamma_S^d \gamma_L^d)^{1/2}}{\gamma_L} \tag{X-48}$$

an expression called the *Girifalco–Good–Fowkes–Young* equation. For liquids whose γ_L^d = γ_L, Eq. X-48 suggests that Zisman's critical surface tension corresponds to γ_S^d. This treatment provides, for example, an interpretation of why water, a polar liquid, behaves toward hydrocarbons as though the surface were nonpolar (see, however, Section XVI-4 for an indication that polar interactions *are* important between water and hydrocarbons).

Good, van Oss, and Caudhury [208–210] generalized this approach to include three different surface tension components from Lifshitz–van der Waals (dispersion) and electron-donor/electron-acceptor polar interactions. They have tested this model on several materials to find these surface tension components [29, 138, 211, 212]. These approaches have recently been disputed on thermodynamic grounds [213] and based on experimental measurements [214, 215].

The matter of π^0 is of some importance to the estimation of solid interfacial tensions and has yet to be resolved. On one hand, it has been shown that the Giralfco and Good model carries the implication of π^0 being generally significant [10], and on the other hand, both Good [193] and Fowkes [145] propose equations that predict π^0

to be zero for water on low-energy polymers. It is certainly desirable to have accurate and uncontested values of π^0 established. Needed also is a better understanding of the thermodynamic and structural properties of adsorbed films as well as of the boundary layer at the liquid–solid interface.

Li and Neumann sought an equation of state of interfacial tensions of the form $\gamma_{SL} = f(\gamma_{LV}, \gamma_{SV})$. Based on a series of measurements of contact angles on polymeric surfaces, they revised an older empirical law (see Refs. 216, 217) to produce a numerically robust expression [129, 218]

$$\gamma_{SL} = \gamma_{LV} + \gamma_{SV} - 2\sqrt{\gamma_{LV}\,\gamma_{SV}}\, \exp -0.0001247(\gamma_{LV} - \gamma_{SV})^2 \qquad \text{(X-49)}$$

or when combined with the Young equation

$$\cos\theta = -1 + 2\sqrt{\frac{\gamma_{SV}}{\gamma_{LV}}}\, \exp -0.0001247(\gamma_{LV} - \gamma_{SV})^2 \qquad \text{(X-50)}$$

that holds for a variety of solid–liquid systems [129]. Some controversy has surrounded the equation of state approach [219–223].

C. Potential-Distortion Model

A rather different approach is to investigate possible adsorption isotherm forms for use with Eq. X-43. As is discussed more fully in Section XVII-7, in about 1914 Polanyi proposed that adsorption be treated as a compression of a vapor in the potential field $U(x)$ of the solid; with sufficient compression, condensation to liquid adsorbate would occur. If $U_0(x)$ denotes the field necessary for this, then

$$U_0 = kT\ln\frac{P^{0\prime}}{P}$$

where $P^{0\prime}$ is the vapor pressure of the liquid film. If the bulk of the observed adsorption is attributed to the condensed liquidlike layer, then $\Gamma = x/d^3$, where d is a molecular dimension such that d^3 equals the molecular volume. Halsey and coworkers [224] took $U(x)$ to be given by Eq. VI-22, but for the present purpose it is convenient to use an exponential form $U_0 = U^0 e^{-ax}$. Equation X-48 then becomes

$$kT\ln\frac{P^{0\prime}}{P} = U^0 e^{-ax} \qquad \Gamma = \frac{x}{d^3} \qquad \text{(X-51)}$$

and insertion of these relationships into Eq. X-11 gives an analytical expression for π (see Ref. 10).

The adsorption isotherm corresponding to Eq. X-51 is of the shape shown in Fig. X-1, that is, it cannot explain contact angle phenomena. The ability of a liquid film to coexist with bulk liquid in a contact angle situation suggests that the film structure has been modified by the solid and is different from that of the liquid, and in an empirical way, this modified structure corresponds to an effective vapor pressure $P^{0\prime}$, $P^{0\prime}$ representing the vapor pressure that bulk liquid would have were its structure that of the film. Such a structural perturbation should relax with increasing film thickness, and this can be represented in exponential form:

$$kT\ln \frac{P^{0\prime}}{P^0} = \beta e^{-\alpha x} \tag{X-52}$$

where P^0 is now the vapor pressure of normal liquid adsorbate.

Combination of Eqs. X-51 and X-52 gives

$$kT\ln \frac{P^0}{P} = U^0 e^{-ax} - \beta e^{-\alpha x} \tag{X-53}$$

This isotherm is now of the shape shown in Fig. X-14, with x_0 determined by the condition $U^0 e^{-ax_0} = \beta e^{-\alpha x_0}$. The integral of Eq. X-42 then becomes

$$\Delta I = \frac{U^0}{d^3} e^{-ax_0} \left(\frac{1}{a} - \frac{1}{\alpha} \right) \tag{X-54}$$

The condition for a finite contact angle is then that $\Delta I/\gamma_L < 1$, or that the adsorption potential field decay more rapidly than the structural perturbation.

Equation X-53 has been found to fit data on the adsorption of various vapors on low-energy solids, the parameters a and α being such as to predict the observed θ [135, 148].

There is no reason why the distortion parameter β should not contain an entropy as well as an energy component, and one may therefore write $\beta = \beta_0 - sT$. The entropy of adsorption, relative to bulk liquid, becomes $\Delta S^0 = s \exp(-\alpha x)$. A critical temperature is now implied, $T_c = \beta_0/s$, at which the contact angle goes to zero [151]. For example, T_c was calculated to be 174°C by fitting adsorption and contact angle data for the n-octane-PTFE system.

An interesting question that arises is what happens when a thick adsorbed film (such as reported at P^0 for various liquids on glass [144] and for water on pyrolytic carbon [135]) is layered over with bulk liquid. That is, if the solid is immersed in the liquid adsorbate, is the same distinct and relatively thick interfacial film still present, forming some kind of discontinuity or interface with bulk liquid, or is there now a smooth gradation in properties from the surface to the bulk region? This type of question seems not to have been studied, although the answer should be of importance in fluid flow problems and in formulating better models for adsorption phenomena from solution (see Section XI-1).

D. The Microscopic Meniscus Profile

The microscopic contour of a meniscus or a drop is a matter that presents some mathematical problems even with the simplifying assumption of a uniform, rigid solid. Since bulk liquid is present, the system must be in equilibrium with the local vapor pressure so that an equilibrium adsorbed film must also be present. The likely picture for the case of a nonwetting drop on a flat surface is

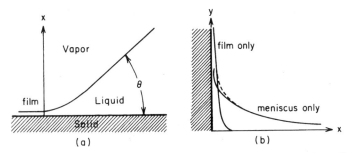

Fig. X-16. (a) Microscopic appearance of the three-phase contact region. (b) Wetting meniscus against a vertical plate showing the meniscus only, adsorbed film only, and joined profile. (From Ref. 226 with permission. Copyright 1980 American Chemical Society.)

shown in Fig. X-16a. There is a region of negative curvature as the drop profile joins the plane of the solid. Analyses of the situation have been reviewed by Good [200]. In a mathematical analysis by Wayner [225] a characteristic distance x_0 appears that is identified with the adsorbed film thickness and found to be 1–2 Å. Adsorption was taken as due to dispersion forces only, however, and some approximations were made in the analysis.

The preceding x_0 estimate contrasts with the values of 5–60 Å found by direct ellipsometric measurements of adsorbed film thicknesses for various organic vapors on PTFE on polyethylene [127, 148]. The matter needs more attention. On the theoretical side, the adsorption potential function used is of the type of Eq. VI-22, which cannot be correct for distances on the order of an atomic diameter, and a rigorous analysis is needed. On the experimental side, the observed adsorbed film thicknesses could reflect "lakes" filling in irregularities in the surface or on polar sites, in which case the appropriate contact angle to use might be closer to the receding than to the advancing one.

A detailed mathematical analysis has been possible for a second situation, of a wetting meniscus against a flat plate, illustrated in Fig. X-16b. The relevant equation is [226]

$$y''(1 + y'^2)^{-1.5} = \frac{2y}{a^2} - \frac{U}{\gamma V}$$
(X-55)

where a is the capillary length, V the molar volume of the liquid, and U the adsorption potential function. Equation X-55 has been solved by a numerical method for the cases of $U = U_0/x^n$, $n = 2, 3$. Again, a characteristic distance x_0 is involved, but on the order of 1000 Å if the U function is to match observed adsorption isotherms. Less complete analyses have been made for related situations, such as a wetting meniscus between horizontal flat plates, the upper plate extending beyond the lower [227, 228]. Meniscus profiles have, however, been calculated for the case of a nearly horizontal plate dipping into a liquid and found to agree with measured film thickness in the transition region, as determined ellipsometrically [229, 230].

7. Problems

1. Microcrystals of $SrSO_4$ of 30 Å diameter have a solubility product at 25°C which is 6.4 times that for large crystals. Calculate the surface tension of the $SrSO_4$-H_2O interface. Equating surface tension and surface energy, calculate the increase in heat of solution of this $SrSO_4$ powder in joules per mole.

2. Reference 3 gives the equation log $(a/a_0) = 16/x$, where a is the solubility activity of a crystal, a_0 is the normal value, and x is the crystal size measured in angstroms. Derive this equation.

3. Given the following data for the water-graphite system, calculate for 25°C (a) the energy of immersion of graphite in water, (b) the adhesion tension of water on graphite, (c) the work of adhesion of water to graphite, and (d) the spreading coefficient of water on graphite. Energy of adhesion: 280 mJ/m^2 at 25°C; surface tension of water at 20, 25, and 30°C is 72.8, 72.0, and 71.2 mJ/m^2, respectively; contact angle at 25°C is 90°.

4. Calculate the heat of immersion of talc in water around 25°C. Comment on the value.

5. Calculate from the data available in the text the value of q_{imm} for Teflon in benzene at around 20°C.

6. Estimate the specific surface area of the quartz powder used in Fig. X-1. Assume that a monolayer of C_4H_9OH is present at $P/P^0 = 0.2$ and that the molecule is effectively spherical in shape.

7. From the variation of contact angle with SAM composition in Fig. X-3, what do you think the measuring liquid is? What would the plot in Fig. X-3 look like if, say, n-decane were used instead?

8. Why do you think the Cassie equation Eq. X-27 might work better than Eq. X-28 for predicting the contact angle as a function of surface polarity?

9. It has been estimated that for the n-decane-PTFE system, π^0 is 0.82 mJ/m^2 at 15°C and 1.54 mJ/m^2 at 70°C [152]. Make a calculation of the difference in heat of immersion in n-decane of 1 m^2 of clean PTFE surface and of 1 m^2 of surface having an adsorbed film in equilibrium with P^0. Assume 25°C.

10. Suppose that a composite surface consists of patches of $\theta_1 = 20°$ and $\theta_2 = 70°$. Compare Eqs. X-27 and X-28 by plotting θ_c versus f_2 as calculated by each equation.

11. Derive Eq. X-35. Hint: The algebraic procedure used in obtaining Eq. II-15 is suggested as a guide.

12. As an extension of Problem 11, integrate a second time to obtain the equation for the meniscus profile in the Neumann method. Plot this profile as y/a versus x/a, where y is the vertical elevation of a point on the meniscus (above the flat liquid surface), x is the distance of the point from the slide, and a is the capillary constant. (All meniscus profiles, regardless of contact angle, can be located on this plot.)

13. Bartell and co-workers report the following capillary pressure data in porous plug experiments using powdered carbon. Benzene, which wets carbon, showed a capillary pressure of 6200 g/cm^2. For water, the pressure was 12,000 g/cm^2, and for ben-

zene displacing water in the plug, the entry pressure was 5770 g/cm^2. Calculate the water-carbon and carbon-benzene-water contact angles and the adhesion tension at the benzene-carbon interface.

14. Using the data of Table X-2, estimate the contact angle for benzene on aluminum oxide and the corresponding adhesion tension.

15. Using the data of Table X-2, estimate the contact angle for gallium on glass and the corresponding adhesion tension. The gallium-mercury interfacial tension is 37 mJ/m^2 at 25°C, and the surface tension of gallium is about 700 mJ/m^2.

16. Water at 20°C rests on solid naphthalene with a contact angle of 90°, while a water-ethanol solution of surface tension 35 dyn/cm shows an angle of 30°. Calculate (a) the work of adhesion of water to naphthalene, (b) the critical surface tension of naphthalene, and (c) γ^d for naphthalene.

17. (a) Estimate the contact angle for ethanol on Teflon. Make an educated guess, that is, with explanation, as to whether the presence of 10% hexane in the ethanol should appreciably affect θ and, if so, in what direction. (b) Answer the same questions for the case where nylon rather than Teflon is the substrate [nylon is essentially (CONH-CH$_2$)$_x$]. (c) Discuss in which of these cases the hexane should be more strongly adsorbed at the solution-solid interface.

18. Fowkes and Harkins reported that the contact angle of water on paraffin is 111° at 25°C. For a 0.1M solution of butylamine of surface tension 56.3 mJ/m^2, the contact angle was 92°. Calculate the film pressure of the butylamine absorbed at the paraffin-water interface. State any assumptions that are made.

19. The surface tension of liquid sodium at 100°C is 220 ergs/cm^2, and its contact angle on glass is 66°. (a) Estimate, with explanation of your procedure, the surface tension of glass at 100°C. (b) If given further that the surface tension of mercury at 100°C is 460 ergs/cm^2 and that its contact angle on glass is 143°, estimate by a different means than in (a) the surface tension of the glass. By "different" is meant a different conceptual procedure.

20. Using appropriate data from Table II-9, calculate the water-mercury interfacial tension using the simple Girifalco and Good equation and then using Fowkes' modification of it.

21. Suppose that the line tension for a given three-phase line is 1 × 10^{-2} dyn. Calculate θ for drops of radius 0.1, 0.01, and 0.001 cm if the value for a large drop is 56. Assume water at 20°C.

22. The contact angle of ethylene glycol on paraffin is 83° at 25°C, and γ_L^d for ethylene glycol is 28.6 mJ/m^2. For water γ_L^d is 22.1 mJ/m^2, and the surface tensions of ethylene glycol and of water are 48.3 and 72.8 mJ/m^2, respectively. Neglecting the π^0's, calculate from the above data what the contact angle should be for water on paraffin.

23. An astronaut team has, as one of its assigned experiments, the measurement of contact angles for several systems (to test the possibility that these may be different in gravity-free space). Discuss some methods that would be appropriate and some that would not be appropriate to use.

24. What is the critical surface tension for human skin? Look up any necessary data and make a Zisman plot of contact angle on skin versus surface tension of water-alcohol mixtures. (Note Ref. 136.)

25. Calculate γ_L^d for n-propanol from its contact angle on PTFE using $\gamma = 19.4$ mJ/m^2 for PTFE and (a) including and (b) neglecting π_{SL}^0.

26. Calculate γ^d for naphthalene; assume that it interacts with water only with dispersion forces.

27. Derive Eq. X-50 from Eq. X-49.

28. (a) Estimate the contact angle for butyl acetate on Teflon (essentially a CF$_2$ surface) at 20°C using data and semiempirical relationships in the text. (b) Use the Zisman relationship to obtain an expression for the spreading coefficient of a liquid on a solid, $S_{L/S}$, involving only γ_L as the variable. Show what $S_{L/S}$ should be when $\gamma_L = \gamma_c$ (neglect complications involving π^0).

29. Take the data from Fig. X-12 on the propyl monolayers and make a Zisman plot to determine the critical surface tension for the surface.

30. Where do you stand on the controversy between Neumann and co-workers [221–223] and those who have criticized this approach [219, 220]?

31. Estimate the interfacial tension gradient formed in alcohol–water mixtures as a function of alcohol content. Determine the minimum alcohol content necessary to form wine tears on a vertical glass wall [174] (experimental verification is possible).

32. Discuss the paradox in the wettability of a fractal surface (Eq. X-33). A true fractal surface is infinite in extent and a liquid of a finite contact angle will trap air at some length scale. How will this influence the contact angle measured for a fractal surface?

33. Plot the shape of the contact line pinned to a defect using Eq. X-30 for water on polyethylene, stearic acid, and platinum. Assume that the upper cutoff length is 2 mm. How does the shape of the pinned contact line compare with your observations of raindrops on dirty windows?

General References

Adv. Chem. Ser, No. 43, American Chemical Society, Washington, DC, 1964.

J. D. Andrade, ed., *Surface and Interfacial Aspects of Biomedical Polymers*, Plenum, New York, 1985.

R. Defay and I. Prigogine, *Surface Tension and Adsorption*, Wiley, New York, 1966.

R. E. Johnson, Jr., and R. H. Dettre, in *Surface and Colloid Science*, Vol. 2, E. Matijević, ed., Wiley-Interscience, New York, 1969.

J. Mahanty and B. W. Ninham, *Dispersion Forces*, Academic, New York, 1976.

A. W. Neumann and J. K. Spelt, eds., *Applied Surface Thermodynamics*, Marcel Dekker Inc., New York, 1996.

S. Ross and I. D. Morrison, *Colloidal Systems and Interfaces*, Wiley, New York, 1988.

J. S. Rowlinson and B. Wisdom, *Molecular Theory of Capillarity*, Clarendon Press, Oxford, 1982.

K. L. Sutherland and I. W. Wark, *Principles of Flotation*, Australian Institute of Mining and Technology, Inc., Melbourne, 1955.

Textual References

1. A. T. Hubbard, *Acc. Chem. Res.*, **13**, 177 (1980).
2. B. V. Enüstün and J. Turkevich, *J. Am. Chem. Soc.*, **82**, 4502 (1960).
3. B. V. Enüstün, M. Enuysal, and M. Dösemeci, *J. Colloid Interface Sci.*, **57**, 143 (1976).
4. F. van Zeggeren and G. C. Benson, *Can. J. Chem.*, **35**, 1150 (1957).
5. L. F. Knapp, *Trans. Faraday Soc.*, **18**, 457 (1922).
6. H. E. Buckley, *Crystal Growth*, Wiley, New York, 1951.
7. R. I. Stearns and A. F. Berndt, *J. Phys. Chem.*, **80**, 1060 (1976).
8. See. *J. Phys. Chem.*, **80**, 2707–2709 for comments on Ref. 5.
9. D. A. Spagnolo, Y. Maham, and K. T. Chuang, *J. Phys. Chem.*, **100**, 6626 (1996).
10. A. W. Adamson and I. Ling, *Adv. Chem.*, **43**, 57 (1964).
11. J. W. Whalen, *Adv. Chem. Ser.*, No. 33, 281 (1961); *J. Phys. Chem.*, **65**, 1676 (1961).
12. S. Partya, F. Rouquerol, and J. Rouquerol, *J. Colloid Interface Sci.*, **68**, 21 (1979).
13. A. C. Zettlemoyer and J. J. Chessick, *Adv. Chem. Ser.*, No. 43, 88 (1964).
14. M. Topic, F. J. Micale, H. Leidheiser, Jr., and A. C. Zettlemoyer, *Rev. Sci. Instr.*, **45**, 487 (1974).
15. A. C. Zettlemoyer, *Ind. Eng. Chem.*, **57**, 27 (1965).
16. W. D. Harkins, *The Physical Chemistry of Surfaces*, Reinhold, New York, 1952.
17. W. H. Wade and N. Hackerman, *Adv. Chem. Ser.*, No. 43, 222 (1964).
18. J. R. Whalen and W. H. Wade, *J. Colloid Interface Sci.*, **24**, 372 (1967).
19. F. H. Healey, Yung-Fang Yu, and J. J. Chessick, *J. Phys. Chem.*, **59**, 399 (1955).
20. See R. L. Venable, W. H. Wade, and N. Hackerman, *J. Phys. Chem.*, **69**, 317 (1965).
21. R. V. Siriwardane and J. P. Wightman, *J. Adhes.*, **15**, 225 (1983).
22. T. Morimoto, Y. Suda, and M. Nagao, *J. Phys. Chem.*, **89**, 4881 (1985).
23. B. Bilinski and E. Chibowski, *Powder Tech.*, **35**, 39 (1983).
24. J. C. Melrose, *J. Colloid Interface Sci.*, **24**, 416 (1967).
25. S. N. Omenyi and A. W. Neumann, *J. Appl. Phys.*, **47**, 3956 (1976).
26. S. N. Omenyi, A. W. Neumann, W. W. Martin, G. M. Lespinard, and R. P. Smith, *J. Appl. Phys.*, **52**, 796 (1981).
27. E. Moy and A. W. Neumann, *Colloids & Surfaces*, **43**, 349 (1990).
28. D. R. Absolom, Z. Policova, E. Moy, W. Zingg, and A. W. Neumann, *Cell Biophys.*, **7**, 267 (1985).
29. C. J. van Oss, R. F. Giese, R. Wentzek, J. Norris, and E. M. Chuvilin, *J. Adhes. Sci. Technol.*, **6**, 503 (1992).
30. T. Young, in *Miscellaneous Works*, Vol. 1, G. Peacock, ed., Murray, London, 1855, p. 418.
31. A. Dupré, *Theorie Mecanique de la Chaleur*, Paris, 1869, p. 368.
32. J. C. Melrose, *Adv. Chem. Ser.*, No. 43, 158 (1964).
33. D. H. Bangham and R. I. Razouk, *Trans. Faraday Soc.*, **33**, 1459 (1937).

34. W. D. Harkins and H. K. Livingston, *J. Chem. Phys.*, **10**, 342 (1942).

35. See F. E. Bartell and L. S. Bartell, *J. Am. Chem. Soc.*, **56**, 2205 (1934).

36. J. F. Oliver, C. Huh, and S. G. Mason, *Colloid Surf.*, **1**, 79 (1980).

37. R. H. Dettre and R. E. Johnson, Jr., *Adv. Chem. Ser.*, No. 43, 112, 136 (1964). See also Ref. 102.

38. H. Lomas, *J. Colloid Interface Sci.*, **33**, 548 (1970). See also H. M. Princen, *J. Colloid Interface Sci.*, **36**, 157 (1971) and H. Lomas, *J. Colloid Interface Sci.*, **37**, 247 (1971).

39. R. A. Brown, F. M. Orr, Jr., and L. E. Scriven, *J. Colloid Interface Sci.*, **73**, 76 (1980).

40. E. B. Dussan V, *J. Fluid Mech.*, **174**, 381 (1987).

41. A. Marmur, *Adv. Coll. Int. Sci.*, **50**, 121 (1994).

42. A. B. D. Cassie, *Disc. Faraday Soc.*, **3**, 11 (1948).

43. J. N. Israelachvili and M. L. Gee, *Langmuir*, **5**, 288 (1989).

44. P. E. Laibinis and G. M. Whitesides, *J. Am. Chem. Soc.*, **114**, 1990 (1992).

45. R. N. Wenzel, *Ind. Eng. Chem.*, **28**, 988 (1936); *J. Phys. Colloid Chem.*, **53**, 1466 (1949).

46. S. Baxter and A. B. D. Cassie, *J. Text. Inst.*, **36**, T67 (1945); A. B. D. Cassie and S. Baxter, *Trans. Faraday Soc.*, **40**, 546 (1944).

47. R. H. Dettre and R. E. Johnson, Jr., *Symp. Contact Angle, Bristol*, **1966.**

48. R. E. Johnson, Jr. and R. H. Dettre, *J. Phys. Chem.*, **68**, 1744 (1964).

49. A. W. Neumann and R. J. Good, *J. Colloid Interface Sci.*, **38**, 341 (1972).

50. D. C. Pease, *J. Phys. Chem.*, **49**, 107 (1945).

51. R. H. Dettre and R. E. Johnson, Jr., *J. Phys. Chem.*, **69**, 1507 (1965).

52. L. S. Penn and B. Miller, *J. Colloid Interface Sci.*, **78**, 238 (1980).

53. L. W. Schwartz and S. Garoff, *J. Colloid Interface Sci.*, **106**, 422 (1985).

54. L. W. Schwartz and S. Garoff, *Langmuir*, **1**, 219 (1985).

55. L. Lèger and J. F. Joanny, *Rep. Prog. Phys.*, **1992**, 431 (1992).

56. G. D. Nadkarni and S. Garoff, *Langmuir*, **10**, 1618 (1994).

57. E. L. Decker and S. Garoff, *Langmuir*, **12**, 2100 (1996).

58. G. D. Nadkarni and S. Garoff, *Europhys. Lett.*, **20**, 523 (1992).

59. J. A. Marsh and A. M. Cazabat, *Phys. Rev. Lett.*, **71**, 2433 (1993).

60. J. Gaydos and A. W. Neumann, *Adv. Coll. Int. Sci.*, **49**, 197 (1994).

61. J. A. Marsh and A. M. Cazabat, *Europhys. Lett.*, **23**, 45 (1993).

62. D. Li and A. W. Neumann, *Colloid Polym. Sci.*, **270**, 498 (1992).

63. R. Shuttleworth and G. L. J. Bailey, *Dis. Faraday Soc.*, **3**, 16 (1948).

64. R. J. Good, *J. Am. Chem. Soc.*, **74**, 5041 (1952); see also J. D. Eick, R. J. Good, and A. W. Neumann, *J. Colloid Interface Sci.*, **53**, 235 (1975).

65. J. J. Bikerman, *Ind. Eng. Chem., Anal. Ed.*, **13**, 443 (1941).

66. A. M. Schwartz and F. W. Minor, *J. Colloid Sci.*, **14**, 584 (1959).

67. A. W. Neumann, *Adv. Colloid Interface Sci.*, **4**, 105 (1974).

68. T. Onda, S. Shibuichi, N. Satoh, and K. Tsujii, *Langmuir*, **12**, 2125 (1996).

69. F. G. Yost, J. R. Michael, and E. T. Eisenmann, *Acta Met. Mat.*, **43**, 299 (1994).

70. D. Li, F. Y. H. Lin, and A. W. Neumann, *J. Colloid Interface Sci.*, **142**, 224 (1991).

71. D. Li, F. Y. H. Lin, and A. W. Neumann, *J. Colloid Interface Sci.*, **159,** 86 (1993).

72. J. D. Andrade, D. E. Gregonis, and L. M. Smith, in *Surface and Interfacial Aspects of Biomedical Polymers*, Vol. 1, J. D. Andrade, ed., Plenum, New York, 1985.

73. E. Ruckenstein and S. V. Gourisankar, *J. Colloid Interface Sci.*, **107,** 488 (1985).

74. A. S. Michaels and R. C. Lummis, private communication.

75. R. V. Sedev, J. G. Petrov, and A. W. Neumann, *J. Colloid Interface Sci.*, **180,** 36 (1996).

76. R. J. Good, J. A. Kvikstad, and W. O. Bailey, *J. Colloid Interface Sci.*, **35,** 314 (1971).

77. Y. L. Chen, C. A. Helm, and J. N. Israelachvili, *J. Phys. Chem.*, **95,** 10736 (1991).

78. W. A. Zisman, *Adv. Chem. Ser.*, No. 43 (1964).

79. H. W. Fox, E. F. Hare, and W. A. Zisman, *J. Phys. Chem.*, **59,** 1097 (1955); O. Levine and W. A. Zisman, *J. Phys. Chem.*, **61,** 1068, 1188 (1957).

80. B. Frank and S. Garoff, *Langmuir*, **11,** 87 (1995).

81. V. J. Novotny and A. Marmur, *J. Colloid Interface Sci.*, **145,** 355 (1991).

82. F. Heslot, A. M. Cazabat, and P. Levinson, *Phys. Rev. Lett.*, **62,** 1286 (1989).

83. D. Ausserre, A. M. Picard, and L. Lèger, *Phys. Rev. Lett.*, **57,** 2671 (1986).

84. R. J. Ruch and L. S. Bartell, *J. Phys. Chem.*, **64,** 513 (1960).

85. P. J. Scales, F. Grieser, D. N. Furlong, and T. W. Healy, *Colloids & Surfaces*, **21,** 55 (1986).

86. R. A. Pyter, G. Zografi, and P. Mukerjee, *J. Colloid Interface Sci.*, **89,** 144 (1982).

87. P. G. de Gennes, *Rev. Mod. Phys.*, **57,** 827 (1985).

88. J. F. Joanny and D. Andleman, *J. Colloid Interface Sci.*, **119,** 451 (1987).

89. E. B. Dussan V, E. Ramé, and S. Garoff, *J. Fluid Mech.*, **230,** 97 (1991).

90. F. Brochard-Wyart and P. G. de Gennes, *Adv. Colloid Interface Sci.*, **39,** 1 (1992).

91. E. B. Dussan V, *J. Fluid Mech.*, **77,** 665 (1976).

92. J. A. Marsh, S. Garoff, and E. B. Dussan V, *Phys. Rev. Lett.*, **70,** 2778 (1993).

93. E. Ramé and S. Garoff, *J. Colloid Interface Sci.*, **177,** 234 (1996).

94. R. V. Sedev, C. J. Budziak, J. G. Petrov, and A. W. Neumann, *J. Colloid Interface Sci.*, **159,** 392 (1993).

95. A. W. Neumann and R. J. Good, in *Techniques of Measuring Contact Angles, Surface and Colloid Science*, Vol. II, *Experimental Methods*, R. J. Good and R. R. Stromberg, ed., Plenum, New York, 1979.

96. W. C. Bigelow, D. L. Pickett, and W. A. Zisman, *J. Colloid Sci.*, **1,** 513 (1946); see also R. E. Johnson and R. H. Dettre, *J. Colloid Sci.*, **20,** 173 (1965).

97. J. K. Spelt, Y. Rotenberg, D. R. Absolom, and A. W. Neumann, *Colloids & Surfaces*, **24,** 127 (1987).

98. R. M. Prokop, O. I. del Rio, N. Niyakan, and A. W. Neumann, *Can. J. Chem. Eng.*, **74,** 534 (1996).

99. D. Li and A. W. Neumann, *Colloids & Surfaces*, **43,** 195 (1990).

100. D. Duncan, D. Li, J. Gaydos, and A. W. Neumann, *J. Colloid Interface Sci.*, **169,** 256 (1995).

101. R. H. Ottewill, private communication; see also A. M. Gaudin, *Flotation*, McGraw-Hill, New York, 1957, p. 163.

102. A. W. Adamson, F. P. Shirley, and K. T. Kunichika, *J. Colloid Interface Sci.*, **34,** 461 (1970).

103. R. J. Good and M. N. Koo, *J. Colloid Interface Sci.*, **71,** 283 (1979).

104. B. A. Pethica, *J. Colloid Interface Sci.*, **62,** 567 (1977).

105. S. Torza and S. G. Mason, *Kolloid-Z Z. Polym.*, **246,** 593 (1971).

106. F. P. Buff and H. J. Saltzburg, *J. Chem. Phys.*, **26,** 23 (1957).

107. D. Platikanov and M. Nedyalkov, "Contact Angles and Line Tension at Microscopic Three Phase Contacts," in *Microscopic Aspects of Adhesion and Lubrication*, J. M. Georges, ed., Elsevier, Amsterdam, 1982.

108. D. Y. Kwok, F. Y. H. Lin, and A. W. Neumann, "Contact Angle Studies on Perfect and Imperfect Solid Surfaces," in *Proc. 30th Int. Adhesion Symp., Yokohama Japan*, 1994.

109. E. Moy, P. Cheng, Z. Policova, S. Treppo, D. Kwok, D. R. Mack, P. M. Sherman, and A. W. Neumann, *Colloids & Surfaces*, **58,** 215 (1991).

110. A. S. Dimitrov, P. A. Kralchevsky, A. D. Nikolov, and D. T. Wasan, *Colloids & Surfaces*, **47,** 299 (1990).

111. A. W. Neumann and D. Renzow, *Z. Phys. Chemie Neue Folge*, **68,** 11 (1969); W. Funke, G. E. H. Hellweg, and A. W. Neumann, *Angew. Makromol. Chemie*, **8,** 185 (1969).

112. J. B. Cain, D. W. Francis, R. D. Venter, and A. W. Neumann, *J. Colloid Interface Sci.*, **94,** 123 (1983).

113. D. Y. Kwok, C. J. Budziak, and A. W. Neumann, *J. Colloid Interface Sci.*, **173,** 143 (1995).

114. V. V. Yaminsky and K. B. Yaminskaya, *Langmuir*, **11,** 936 (1995).

115. I. Langmuir and V. J. Schaeffer, *J. Am. Chem. Soc.*, **59,** 2405 (1937).

116. R. J. Good and J. K. Paschek, in *Wetting, Spreading, and Adhesion*, J. F. Padday, ed., Academic, 1978.

117. F. E. Bartell and C. W. Walton, Jr., *J. Phys. Chem.*, **38,** 503 (1934); F. E. Bartell and C. E. Whitney, *J. Phys. Chem.*, **36,** 3115 (1932).

118. F. E. Bartell and H. J. Osterhof, *Colloid Symposium Monograph*, The Chemical Catalog Company, New York, 1928, p. 113.

119. T. Smith, *J. Colloid Interface Sci.*, **75,** 51 (1980).

120. M. E. Schrader, *J. Phys. Chem.*, **74,** 2313 (1970).

121. M. L. White, in *Clean Surfaces, Their Preparation and Characterization for Interfacial Studies*, G. Goldfinger, ed., Marcel Decker, New York, 1970.

122. G. L. Gaines, Jr., *J. Colloid Interface Sci.*, **79,** 295 (1981).

123. H. W. Fox and W. A. Zisman, *J. Colloid Sci.*, **7,** 428 (1952).

124. *International Critical Tables*, Vol. 4, McGraw-Hill, New York, 1928, p. 434.

125. H. K. Livingston, *J. Phys. Chem.*, **48,** 120 (1944).

126. J. R. Dann, *J. Colloid Interface Sci.*, **32,** 302 (1970).

127. P. Hu and A. W. Adamson, *J. Colloid Interface Sci.*, **59,** 605 (1977).

128. F. D. Petke and B. R. Ray, *J. Colloid Interface Sci.*, **31,** 216 (1969).

129. D. Li and A. W. Neumann, *J. Colloid Interface Sci.*, **148,** 190 (1992).

130. C. J. Budziak, E. I. Vargha-Butler, and A. W. Neumann, *J. Appl. Polym. Sci.* **42,** 1959 (1991).

131. H. Schonhorn, *J. Phys. Chem.*, **70,** 4086 (1966).

132. H. Schonhorn, *Nature*, **210,** 896 (1966).

133. A. C. Zettlemoyer, *J. Colloid Interface Sci.*, **28,** 343 (1968).

134. H. Schonhorn and F. W. Ryan, *J. Phys. Chem.*, **70,** 3811 (1966).

135. M. E. Tadros, P. Hu, and A. W. Adamson, *J. Colloid Interface Sci.*, **49,** 184 (1974).

136. A. W. Adamson, K. Kunichika, F. Shirley, and M. Oren, *J. Chem. Ed.*, **45,** 702 (1968).

137. A. El-Shimi and E. D. Goddard, *J. Colloid Interface Sci.*, **48,** 242 (1974).

138. Z. Li, R. F. Giese, C. J. van Oss, J. Yvon, and J. Cases, *J. Colloid Interface Sci.*, **156,** 279 (1993).

139. J. B. Jones and A. W. Adamson, *J. Phys. Chem.*, **72,** 646 (1968).

140. F. M. Fowkes and W. M. Sawyer, *J. Chem. Phys.*, **20,** 1650 (1952).

141. G. E. Boyd and H. K. Livingston, *J. Am. Chem. Soc.*, **64,** 2383 (1942).

142. B. Janczuk, E. Chibowski, and W. Wojcik, *Powder Tech.*, **45,** 1 (1985).

143. J. A. Koutsky, A. G. Walton, and E. Baer, *Surf. Sci.*, **3,** 165 (1965).

144. B. V. Derjaguin and Z. M. Zorin, *Proc. 2nd Int. Congr. Surf. Act., London*, 1957, Vol. 2, p. 145.

145. F. M. Fowkes, D. C. McCarthy, and M. A. Mostafa, *J. Colloid Interface Sci.*, **78,** 200 (1980).

146. F. E. Bartell and H. H. Zuidema, *J. Am. Chem. Soc.*, **58,** 1449 (1936).

147. W. R. Good, *J. Colloid Interface Sci.*, **44,** 63 (1973).

148. J. Tse and A. W. Adamson, *J. Colloid Interface Sci.*, **72,** 515 (1979).

149. B. Janczuk, *Croatica Chem. Acta*, **58,** 245 (1985).

150. A. W. Neumann, G. Haage, and D. Renzow, *J. Colloid Interface Sci.*, **35,** 379 (1971).

151. A. W. Adamson, *J. Colloid Interface Sci.*, **44,** 273 (1973).

152. H. W. Fox and W. A. Zisman, *J. Colloid Interface Sci.*, **5,** 514 (1950).

153. C. L. Sutula, R. Hautala, R. A. Dalla Betta, and L. A. Michel, Abstracts, 153rd Meeting, American Chemical Society, April 1967.

154. J. W. Whalen, *Vacuum Microbalance Techniques*, Vol. 8, A. W. Czanderna, ed., Plenum, 1971.

155. E. G. Shafrin and W. A. Zisman, *J. Phys. Chem.*, **64,** 519 (1960).

156. W. D. Bascom and C. R. Singleterry, *J. Phys. Chem.*, **66,** 236 (1962).

157. H. Peper and J. Berch, *J. Phys. Chem.*, **68,** 1586 (1964).

158. F. M. Fowkes, D. C. McCarthy, and M. A. Mostafa, *J. Colloid Interface Sci.*, **78,** 200 (1980).

159. R. E. Baier and A. E. Meyer, *Biofouling*, **6,** 165 (1992).

160. M. C. Phillips and A. C. Riddiford, *Nature*, **205,** 1005 (1965).

161. D. R. Mack, A. W. Neumann, Z. Policova, and P. M. Sherman, *Am. J. Physiol.*, **262,** G171 (1992).

162. D. R. Mack, A. W. Neumann, Z. Policova, and P. M. Sherman, *Pediatric Res.*, **35,** 209 (1994).

163. P. O. Glantz, M. D. Jendresen, and R. E. Baier, *Surf. & Colloid Phenomena: Methodology*, **1982,** 119.

164. R. Schmukler, J. Casamento, R. E. Baier, and R. B. Beard, "Testing of the Barrier

Function of Condoms: An Overview," in *Winter Annual Meeting of the Am. Soc. Mech. Eng.*, 1989.

165. W. D. Wade and J. W. Whalen, *J. Phys. Chem.*, **72,** 2898 (1968).

166. T. D. Blake and W. H. Wade, *J. Phys. Chem.*, **75,** 1887 (1971).

167. J. W. Whalen and P. C. Hu, *J. Colloid Interface Sci.*, **65,** 460 (1978).

168. Y. Tamai, T. Matsunaga, and K. Horiuchi, *J. Colloid Interface Sci.*, **60,** 112 (1977).

169. R. Miller, S. Treppo, A. Voigt, and A. W. Neumann, *Colloids & Surfaces*, **69,** 203 (1993).

170. E. Moy, F. Y. H. Lin, J. W. Vogtle, Z. Policova, and A. W. Neumann, *Colloid Polym. Sci.*, **272,** 1245 (1994).

171. P. E. Laibinis, C. D. Bain, R. G. Nuzzo, and G. M. Whitesides, *J. Phys. Chem.*, **99,** 7663 (1995).

172. A. Baszkin and L. Ter-Minassian-Saraga, *J. Colloid Interface Sci.*, **43,** 190 (1973).

173. D. Li, C. Ng, and A. W. Neumann, *J. Adhes. Sci. Technol.*, **6,** 601 (1992).

174. J. B. Fournier and A. M. Cazabat, *Europhys. Lett.*, **20,** 517 (1992).

175. R. Williams, *J. Phys. Chem.*, **79,** 1274 (1975).

176. E. Ruckenstein, *J. Colloid Interface Sci.*, **86,** 573 (1982).

177. A. M. Cazabat, *Contemp. Phys.*, **28,** 347 (1987).

178. L. Lèger, A. M. Guinet-Picard, D. Ausserre, and C. Strazielle, *Phys. Rev. Lett.*, **60,** 2390 (1988).

179. P. Silberzan and L. Lèger, *Macromolecules*, **25,** 1267 (1992).

180. R. E. Johnson, Jr., *J. Phys. Chem.*, **63,** 1655 (1959).

181. E. M. Blokhuis, Y. Shilkrot, and B. Widom, *Mol. Phys.*, **86,** 891 (1995).

182. J. J. Bikerman, *Proc. 2nd Int. Congr. Surf. Act., London, 1957*, Vol. 3, p. 125.

183. W. Radigan, H. Ghiradella, H. L. Frisch, H. Schonhorn, and T. K. Kwei, *J. Colloid Interface Sci.*, **49,** 241 (1974).

184. A. Marmur, *J. Adhes. Sci. Technol.*, **6,** 689 (1992).

185. B. V. Derjaguin and N. V. Churaev, in *Fluid Interface Phenomena*, C. A. Croxton, ed., Wiley, New York, 1986.

186. L. J. M. Schlangen, L. K. Koopal, M. A. Cohen Stuart, and J. Lyklema, *Colloids & Surfaces*, **A89,** 157 (1994).

187. Z. M. Zorin, V. P. Romanov, and N. V. Churaev, *Colloid Polym. Sci.*, **257,** 968 (1979).

188. Z. M. Zorin, V. P. Romanov, and N. V. Churaev, *Colloid Polym. Sci.*, **267,** 968 (1979).

189. E. Chibowski and L. Holysz, *J. Colloid Interface Sci.*, **77,** 37 (1980).

190. N. V. Churaev and V. D. Sobolev, *Adv. Coll. Int. Sci.*, **61,** 1 (1995).

191. B. Janczuk and E. Chibowski, *J. Colloid Interface Sci.*, **95,** 268 (1983); B. Janczuk, T. Bialopiotrowicz, and E. Chibowski, *Mat. Chem. Phys.*, **15,** 489 (1987).

192. L. A. Girifalco and R. J. Good, *J. Phys. Chem.*, **61,** 904 (1957). See also R. J. Good, *Adv. Chem. Ser.*, No. 43, 74 (1964).

193. R. J. Good and E. Elbing, *J. Colloid Interface Sci.*, **59,** 398 (1977).

194. D. Li, M. Xie, and A. W. Neumann, *Colloid Polym. Sci.*, **271,** 573 (1993).

195. R. J. Good and L. A. Girifalco, *J. Phys. Chem.*, **64,** 561 (1960).

196. S. Wu, *J. Polym. Sci., Part C*, **34,** 19 (1971); *J. Adhes.*, **5,** 39 (1973).

197. A. El-Shimi and E. D. Goddard, *J. Colloid Interface Sci.*, **48**, 242 (1974).

198. Y. Tamai, T. Matsunaga, and K. Horiuchi, *J. Colloid Interface Sci.*, **60**, 112 (1977). See also Y. Tamai, *J. Phys. Chem.*, **79**, 965 (1975).

199. D. E. Sullivan, *J. Chem. Phys.*, **74**, 2604 (1981).

200. G. Körösi and E. Kovats, *Colloids & Surfaces*, **2**, 315 (1981).

201. F. M. Fowkes, *J. Phys. Chem.*, **72**, 3700 (1968); J. F. Padday and N. D. Uffindell, ibid., 3700 (1968).

202. R. J. Good, in *Surface and Colloid Science*, Vol. 11, R. J. Good and R. R. Stromberg, eds., Plenum, New York, 1979.

203. F. M. Fowkes, *J. Phys. Chem.*, **67**, 2538 (1963); *Adv. Chem. Ser.*, No. 43, 99 (1964).

204. F. M. Fowkes, in *Chemistry and Physics of Interfaces*, S. Ross, ed., American Chemical Society, 1971.

205. J. N. Israelachvili, *J. Chem. Soc., Faraday Trans.*, **69**, 1729 (1973).

206. Per M. Claesson, C. E. Blom, P. C. Herder, and B. W. Ninham, *J. Colloid Interface Sci.*, **114**, 234 (1986).

207. J. T. Koberstein, *Encyclopedia of Polymer Science and Engineering*, Vol. 8, 2nd ed., Wiley, New York, 1987.

208. R. J. Good, *Ind. Eng. Chem.*, **62**, 54 (1970).

209. C. J. van Oss, R. J. Good, and M. K. Chaudhury, *Separ. Sci. Tech.*, **22**, 1 (1987).

210. C. J. van Oss, R. J. Good, and M. K. Chaudhury, *Langmuir*, **4**, 884 (1988).

211. J. Norris, R. F. Giese, C. J. van Oss, and P. M. Costanzo, *Clays & Clay Miner.*, **40**, 327 (1992).

212. C. J. van Oss, R. J. Good, and R. F. Giese, *Langmuir*, **6**, 1711 (1990).

213. D. Y. Kwok, D. Li, and A. W. Neumann, *Colloids & Surfaces*, **89**, 181 (1994).

214. D. Y. Kwok, D. Li, and A. W. Neumann, *Langmuir*, **10**, 1323 (1994).

215. D. Y. Kwok and A. W. Neumann, *Can. J. Chem. Eng.*, **74**, 551 (1996).

216. A. W. Neumann, R. J. Good, C. J. Hope, and M. Sejpal, *J. Colloid Interface Sci.*, **49**, 291 (1974).

217. A. W. Adamson, *Physical Chemistry of Surfaces*, J. Wiley & Sons, New York, 5th edition, 1990.

218. D. Li and A. W. Neumann, *J. Colloid and Interface Sci.*, **137**, 304 (1990).

219. R. E. Johnson, Jr. and R. H. Dettre, *Langmuir*, **5**, 293 (1989).

220. I. D. Morrison, *Langmuir*, **5**, 540 (1989).

221. D. Li, E. Moy, and A. W. Neumann, *Langmuir*, **6**, 885 (1990).

222. J. Gaydos, E. Moy, and A. W. Neumann, *Langmuir*, **6**, 888 (1990).

223. D. Li and A. W. Neumann, *Langmuir*, **9**, 3728 (1993).

224. G. D. Halsey, *J. Chem. Phys.*, **16**, 931 (1948).

225. P. C. Wayner, Jr., *J. Colloid Interface Sci.*, **77**, 495 (1980); *ibid.*, **88**, 294 (1982).

226. A. W. Adamson and A. Zebib, *J. Phys. Chem.*, **84**, 2619 (1980).

227. F. Renk, P. C. Wayner, Jr., and B. M. Homsy, *J. Colloid Interface Sci.*, **67**, 408 (1978).

228. See B. V. Derjaguin, V. M. Starov, and N. V. Churaev, *Colloid J.*, **38**, 875 (1976).

229. J. G. Troung and P. C. Wayner, Jr., *J. Chem. Phys.*, **87**, 4180 (1987).

230. D. Beaglehole, *J. Phys. Chem.*, **93**, 893 (1989).

CHAPTER XI

The Solid–Liquid Interface—Adsorption from Solution

This chapter on adsorption from solution is intended to develop the more straightforward and important aspects of adsorption phenomena that prevail when a solvent is present. The general subject has a vast literature, and it is necessary to limit the presentation to the essential features and theory.

A logical division is made for the adsorption of nonelectrolytes according to whether they are in dilute or concentrated solution. In dilute solutions, the treatment is very similar to that for gas adsorption, whereas in concentrated binary mixtures the role of the solvent becomes more explicit. An important class of adsorbed materials, self-assembling monolayers, are briefly reviewed along with an overview of the essential features of polymer adsorption. The adsorption of electrolytes is treated briefly, mainly in terms of the exchange of components in an electrical double layer.

1. Adsorption of Nonelectrolytes from Dilute Solution

The adsorption of nonelectrolytes at the solid–solution interface may be viewed in terms of two somewhat different physical pictures. In the first, the adsorption is confined to a monolayer next to the surface, with the implication that succeeding layers are virtually normal bulk solution. The picture is similar to that for the chemisorption of gases (see Chapter XVIII) and arises under the assumption that solute–solid interactions decay very rapidly with distance. Unlike the chemisorption of gases, however, the heat of adsorption from solution is usually small; it is more comparable with heats of solution than with chemical bond energies.

In the second picture, an interfacial layer or region persists over several molecular diameters due to a more slowly decaying interaction potential with the solid (note Section X-7C). This situation would then be more like the physical adsorption of vapors (see Chapter XVII), which become multilayer near the saturation vapor pressure (e.g., Fig. X-15). Adsorption from solution, from this point of view, corresponds to a partition between bulk and interfacial phases; here the Polanyi potential concept may be used (see Sections X-7C, XI-1A, and XVII-7).

While both models find some experimental support, the monolayer has been much more amenable to simple analysis. As a consequence, most of the discussion in this chapter is in terms of the monolayer model, although occasional *caveats* are encountered. We begin with adsorption from dilute solution because the models take on a more accessible algebraic form and are easier to develop than those for concentrated solutions.

A. Isotherms

The moles of a solute species adsorbed per gram of adsorbent n_2^s can be expressed in terms of the mole fraction of the solute on the surface N_2^s and the moles of adsorption sites per gram n^s as

$$n_2^s = N_2^s n^s = f(C_2, T) \tag{XI-1}$$

and is generally a function of C_2, the equilibrium solute concentration; and T, the temperature. A useful discussion of definitions and terminology for adsorption has been given by Everett [1]; n_2^s has been called the *specific reduced surface excess*, for example. At constant temperature, $n_2^s = f_T(C_2)$, and this is called the *adsorption isotherm function*. The usual experimental approach is to determine this function by measuring the adsorbed amount as a function of concentration at a given temperature.

Various functional forms for f have been proposed either as a result of empirical observation or in terms of specific models. A particularly important example of the latter is that known as the *Langmuir* adsorption equation [2]. By analogy with the derivation for gas adsorption (see Section XVII-3), the Langmuir model assumes the surface to consist of adsorption sites, each having an area σ^0. All adsorbed species interact only with a site and not with each other, and adsorption is thus limited to a monolayer. Related lattice models reduce to the Langmuir model under these assumptions [3,4]. In the case of adsorption from solution, however, it seems more plausible to consider an alternative phrasing of the model. Adsorption is still limited to a monolayer, but this layer is now regarded as an ideal two-dimensional solution of equal-size solute and solvent molecules of area σ^0. Thus lateral interactions, absent in the site picture, cancel out in the ideal solution; however, in the first version σ^0 is a property of the solid lattice, while in the second it is a property of the adsorbed species. Both models attribute differences in adsorption behavior entirely to differences in adsorbate–solid interactions. Both present adsorption as a competition between solute and solvent.

It is perhaps fortunate that both versions lead to the same algebraic formulations, but we will imply a preference for the two-dimensional solution picture by expressing surface concentrations in terms of mole fractions. The adsorption process can be written as

$$N_2(\text{solute in solution}) + N_1^s(\text{adsorbed solvent})$$
$$= N_2^s(\text{adsorbed solute}) + N_1(\text{solvent in solution}) \tag{XI-2}$$

a chemical equilibrium between adsorbed and solution species. The equilibrium constant for this process is

$$K = \frac{N_2^s a_1}{N_1^s a_2}$$ (XI-3)

where a_1 and a_2 are the solvent and solute activities in solution, and by virtue of the model, the activities in the adsorbed layer are given by the respective mole fractions N_2^s and N_1^s. Since the treatment is restricted to dilute solutions, a_1 is constant, and we can write $b = K/a_1$; also, $N_1^s + N_2^s = 1$ so that Eq. XI-3 becomes

$$N_2^s = \frac{n_2^s}{n^s} = \theta = \frac{ba_2}{1 + ba_2}$$ (XI-4)

where θ is the fraction of surface occupied by solute. The moles of adsorption sites per gram of adsorbate can be related to the area per molecule $n^s = (\Sigma/N \sigma^0)$ through Σ, the surface area per gram. In sufficiently dilute solution, the activity coefficient, a_2, in Eq. XI-4 can be replaced by C_2.

The equilibrium constant K can be written

$$K = e^{(-\Delta G^0/RT)} = e^{(\Delta S^0/R)} e^{(-\Delta H^0/RT)}$$ (XI-5)

where ΔG^0 is the adsorption free energy and ΔH^0 is the enthalpy of adsorption, often denoted by $-Q$, where Q is the heat of adsorption. Thus the constant b can be written

$$b = b' e^{(Q/RT)}$$ (XI-6)

It is not necessary to limit the model to idealized sites; Everett [5] has extended the treatment by incorporating surface activity coefficients as corrections to N_1^s and N_2^s. The adsorption enthalpy can be calculated from the temperature dependence of the adsorption isotherm [6]. If the solution is taken to be ideal, then

$$\left(\frac{\partial \ln C_2}{\partial T}\right)_{n_2} = -\frac{\Delta H^0}{RT^2}$$ (XI-7)

a Clausius–Clapeyron-type equation results (note. Eq. XVII-106).

Returning to Eq. XI-4, with C_2 replacing a_2, at low concentrations n_2^s will be proportional to C_2 with a slope $n^s b$. At sufficiently high concentrations n_2^s approaches the limiting value n^s. Thus n^s is a measure of the capacity of the adsorbent and b of the intensity of the adsorption. In terms of the ideal model, n^s should not depend on temperature, while b should show an exponential dependence given in Eq. XI-6. The two constants are conveniently evaluated by putting Eq. XI-4 in the form

$$y = \left(\frac{C_2}{n_2^s}\right) = \frac{1}{n^s b} + \frac{C_2}{n^s} = x$$ (XI-8)

Thus a plot of C_2/n_2^s versus C_2 should give a straight line of slope $1/n^s$ and intercept $1/n^s b$.

An equation algebraically equivalent to Eq. XI-4 results if instead of site adsorption the surface region is regarded as an interfacial solution phase, much as in the treatment in Section III-7C. The condition is now that the (constant) volume of the interfacial solution is $v^s = N_1^s V_1 + N_2^s V_2$, where V_1 and V_2 are the molar volumes of the solvent and solute, respectively. If the activities of the two components in the interfacial phase are replaced by the volume fractions, the result is

$$\frac{v_2^s}{v^s} = \theta = \frac{ba_2}{1 + ba_2}$$ (XI-9)

where v_2^s is the volume of adsorbed solute [7].

Most surfaces are heterogeneous so that b in Eq. XI-6 will vary with θ. The experimentally observed adsorption isotherm may then be written

$$\Theta(C_2, T) = \int_0^\infty f(b)\theta(C_2, b, T)db$$ (XI-10)

where $f(b)$ is the distribution function for b and $\theta(C_2, b, T)$ is the adsorption isotherm function (e.g., Eq. XI-4). In a sense, this approach is an alternative to the use of surface-activity coefficients.

The solution to this integral equation is discussed in Section XVII-14, but one particular case is of interest here. If $\theta(C_2, b, T)$ is given by Eq. XI-4 and the variation in b and θ is attributed entirely to a variation in the heat of adsorption Q and $f(Q)$ is taken to be

$$f(Q) = \alpha e^{(-Q/nRT)}$$ (XI-11)

then the solution to Eq. XI-10 is of the form [8,9]

$$\Theta = \frac{n_2^s}{n^s} = \alpha RTnb' C_2^{1/n} = aC_2^{1/n}$$ (XI-12)

known as the *Freundlich adsorption isotherm* after its inventor [10]. See Refs. 11–13 for representative values of a and n for aqueous organic solutes.

The Freundlich equation, unlike the Langmuir one, does not become linear at low concentrations but remains convex to the concentration axis; nor does it show a satura-tion or limiting value. The constants an^s and n may be obtained from a plot of log n_2^s versus log C_2, and, roughly speaking, the intercept an^s gives a measure of the adsorbent capacity and the slope $1/n$ of the intensity of adsorption. As just mentioned, the shape of the isotherm is such that n is a number greater than unity. There is no assurance that the derivation of the Freundlich equation is unique; consequently, if the data fit the equation, it is only likely, but not proved, that the surface is heterogeneous.

Alternative approaches treat the adsorbed layer as an ideal solution or in terms of a Polanyi potential model (see Refs. 12–14 and Section XVII-7); a related approach has been presented by Myers and Sircar [15]. Adsorption *rates* have been modeled as diffusion controlled [16,17].

An important qualitative rule describing adsorption behavior is that a polar (non-polar) adsorbent will preferentially adsorb the more polar (nonpolar) component of a nonpolar (polar) solution. Polarity is used in the general sense of ability to engage in hydrogen bonding or dipole–dipole-type interactions as opposed to nonspecific disper-sion interactions. This rule is found in many areas of adsorption, particularly in surfac-tants and proteins where the concept of hydrophobic interactions describes adsorption from aqueous solution [18]. A semiquantitative extension of the foregoing is a ver-sion of Traube's rule (discussed in Section III-7E) [19], which—as given by Freundlich [10]—states: "The adsorption of organic substances from aqueous solutions increases strongly and regularly as we ascend the homologous series." This concept has been illustrated frequently in the recent surfactant literature; some useful citations are given in Refs. 20–23.

As discussed in Chapter III, the progression in adsorptivities along a homologous series can be understood in terms of a constant increment of work of adsorption with each additional CH_2 group. This is seen in self-assembling monolayers discussed in Section XI-1B. The film pressure π may be calculated from the adsorption isotherm by means of Eq. XI-7 as modified for adsorption from dilute solution:

$$\pi = \frac{RT}{\Sigma} \int_0^{C_2} n_2^s d \ln C_2 \qquad (XI\text{-}13)$$

If the Langmuir equation is obeyed, combination of Eqs. XI-4 and XI-13 gives the von *Szyszkowski* equation [24,25]

$$\pi = \frac{RT}{N\sigma^0} \ln(1 + bC_2) \qquad (XI\text{-}14)$$

so that equal values of bC_2 correspond to equal values of π.

B. Self-Assembling Monolayers

An important example of physicochemisorption is the creation of highly ordered self-assembling monolayers (SAMs). Sharing many features in com-mon with Langmuir–Blodgett (LB) films (discussed in Section XV-8), SAMs are created by the adsorption of long-chain acids, alkylthiols, or alkylchloro-silanes to coinage metals such as gold and oxides such as silica [26–29]. The

primary advantage of a thin-film SAM over an LB film is its stability, due
to the strong physical or chemical bonds with the surface. For a comparison of SAM and LB films, see Sagiv [30]. These layers have been studied
by contact angle measurement [27,31], ellipsometry [30,32–34], infrared spectroscopy [32,34–36], ultraviolet spectroscopy [37], electron diffraction [38],
x-ray diffraction and atomic force microscopy (AFM) [39], helium diffraction
[40], and electrochemistry [41].

The tendency to form organized monolayers improves with chain length.
This is illustrated in a study of adsorption kinetics in alkanoic acid monolayers on alumina by Chen and Frank [36]. They find that the Langmuir kinetic
equation, discussed in Section XVII-3, (see Problem XI-6)

$$\frac{\partial \theta}{\partial t} = k_a C_2 (1 - \theta) - k_d \theta \qquad \text{(XI-15)}$$

where k_a and k_d are the adsorption and desorption rate constants, fits adsorption
data for linear fatty acids having 12 or more carbons as illustrated in Fig. XI-1.
This reduces to the Langmuir isotherm, Eq. XI-4, at long times. The ratio of
rates depends on the free energy of adsorption as

$$b = \frac{k_a}{k_d} \propto e^{(-\Delta G)} \qquad \text{(XI-16)}$$

Using guest fluorophores, they show that the adsorption energy increases linearly with the alkyl chain length as

$$-\Delta G^0 = -\Delta G^{0,h} + N_c W \qquad \text{(XI-17)}$$

where N_c is the number of carbons in the chain and W is the energy per methylene group; for alkanoic acid adsorbing onto alumina from hexadecane W is
962 J/mol. Typical energies for other acids, alcohols, and amides onto various
surfaces vary from 400 to 4000 J/mol and are tabulated in Ref. 36.

The acid monolayers adsorb via physical forces [30]; however, the interactions between the head group and the surface are very strong [29]. While
chemisorption controls the SAMs created from alkylthiols or silanes, it is often
preceded by a physical adsorption step [42]. This has been shown quantitatively by FTIR for siloxane polymers chemisorbing to alumina illustrated in Fig.
XI-2. The fact that irreversible chemisorption is preceded by physical adsorption explains the utility of equilibrium adsorption models for these processes.

The silanization reaction has been used for some time to alter the wetting
characteristics of glass, metal oxides, and metals [44]. While it is known that
trichlorosilanes polymerize in solution, only very recent work has elucidated
the mechanism for surface reaction. A novel FTIR approach allowed Tripp and
Hair to prove that octadecyl trichlorosilane (OTS) does not react with dry silica,

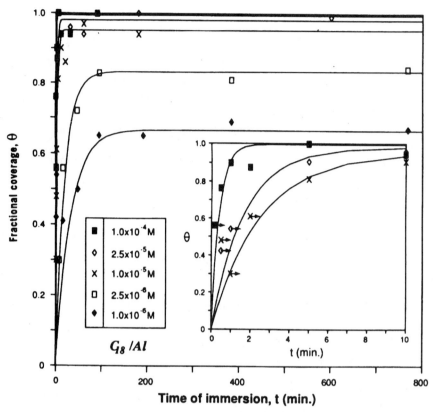

Fig. XI-1. Adsorption kinetics for C_{18} alkanoic acids adsorbing onto alumina for various solution concentrations from Ref. 36. Lines are the fit to Eq. XI-15.

produces low surface coverage with silica carrying 2.5-4.0 H_2O/nm^2, but has an order of magnitude greater coverage of OTS in the presence of more than a monolayer of water [35]. This evidence alters the previous picture of the reaction of the trichlorosilane directly with silanol groups on the surface (see, e.g., Ref. 29). Mathauer and Frank have shown that disordered partial monolayers of OTS can be solidified by "backfilling" with a second silane [37]. Their fluorescence spectroscopy indicates that binary mixtures of silanes are homogeneously distributed on the surface whether coadsorbed or sequentially adsorbed in either order. Sequential "backfilling" produces slightly better layers, presumably because the fresh solution is better able to fill vacancies than one already partially polymerized during the silanization.

Alkanethiols and other sulfur-bearing hydrocarbons covalently attach to metal surfaces; alkanethiol onto gold is the most widely studied of these systems [27–29,31,32,45]. These SAMs are ordered provided the alkane chain contains nine or more carbons [32]. Binary solutions of two alkanethiols also appear

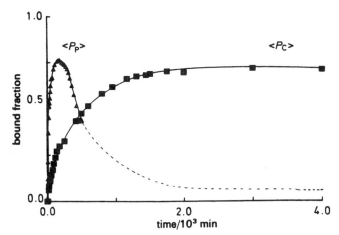

Fig. XI-2. Variation of physically adsorbed (P_P) and chemically adsorbed (P_C) segments as a function of time for cyclic polymethylsiloxane adsorbing from CCl$_4$ onto alumina (from Ref. 43). Note that the initial physisorption is overcome by chemical adsorption as the final state is reached. [T. Cosgrove, C. A. Prestidge, and B. Vincent, *J. Chem. Soc. Faraday Trans.*, **86**(9), 1377–1382 (1990). Reproduced by permission of The Royal Society of Chemistry.]

to adsorb as a homogeneous mixture [45]. These systems form a particularly robust surface coating resistant to acid solutions but susceptible to degradation by base [46]. Because of this stability, vinyl groups on the ends of the alkanes can be reacted in situ to produce acid, alcohol, or halogenated surfaces [46]. Another even more robust layer can be created by adsorption of polymers having alkylthiol side chains [34,47]. These layers show less crystallinity but better chemical stability than regular thiols. Side chains of 10 carbons are optimal for order and packing density.

The self-assembly process can be continued to form multilayer films of up to 25 layers [33,48,49]. The reliability of this process is illustrated in Fig. XI-3, where the thickness grows linearly with the number of reacted layers. These thick layers have many interesting applications.

SAMs are generating attention for numerous potential uses ranging from chromatography [50] to substrates for liquid crystal alignment [51]. Most attention has been focused on future application as nonlinear optical devices [49]; however, their use to control electron transfer at electrochemical surfaces has already been realized [52]. In addition, they provide ideal model surfaces for studies of protein adsorption [53].

C. Multilayer Adsorption

The Langmuir equation (Eq. XI-4) applies to many systems where adsorption occurs from dilute solution, but some interesting cases of sigmoid isotherms have been reported [54–56]. In several of these studies [54,55] the isotherms

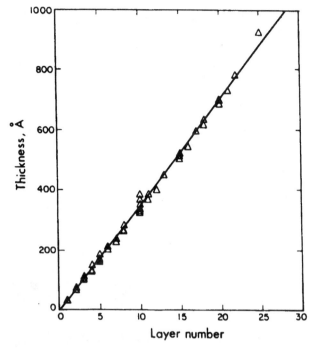

Fig. XI-3. Ellipsometric film thickness as a function of number of layers of methyl 23-(trichlorosilyl)tricosanoate on silicon wafers (Ref. 33).

showed a marked increase as the saturation concentration was approached in analogy to low-temperature gas adsorption near the saturation pressure (see Section XVII-5). There are instances [56] where the multilayer formation precedes the saturation condition indicative of a capillary-induced phase transition analogous to capillary condensation in vapors (see Section XVII-16B).

2. Polymer Adsorption

Since the study of polymer adsorption is a somewhat specialized topic, interwoven with the field of polymer chemistry, we present only a brief summary of its essential features. Several recent reviews provide more detailed information about this extensive subject [57–59]. The adsorption of polymers to solid surfaces is very important to many technologies, including colloidal stabilization, lubrication, adhesion, and separations. In these and other applications, the key feature of polymer adsorption is the range of configurations allowable at the solid–liquid interface. As macromolecules adhere to the solid surface, they leave large loops and tails dangling into solution to create the important properties for stability, adhesion, or lubrication. A typical chain configuration is illustrated in Fig. XI-4. It is the understanding of this distribution of segments

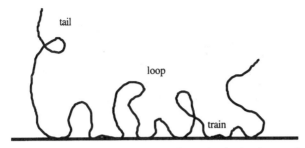

Fig. XI-4. Schematic diagram of the structure of an adsorbed polymer chain. Segments are distributed into trains directly attached to the surface and loops and tails extending into solution.

that has both fascinated and challenged the physical chemist studying polymer adsorption.

Upon adsorption, a polymer loses both translational and configurational entropy. This energetic cost is balanced by the gain in energy on segment–surface contacts. The interaction between segments and the surface is gauged by the parameter χ_s, which is the difference between the solvent and segment adsorption energies. This energy, typically on the order of kT, has been measured by solvent displacement [60]. Since individual segments are attracted to the surface by an energy of kT, they are free to exchange with one another while the whole polymer chain having hundreds or thousands of segments contacting the surface remains attached. These surface interactions are complicated by the polymer molecular weight, the solvent quality, the polydispersity, and the chain stiffness.

Polymers typically exhibit a *high-affinity* adsorption isotherm as shown in Fig. XI-5; here the adsorbed amount increases very rapidly with bulk concentration and then becomes practically independent of concentration.

While these isotherms can be modeled with the Langmuir equation (XI-4), their lack of interesting features and experimental difficulties caused by long equilibration times have lessened interest in their measurement. Those systems not exhibiting "high-affinity" isotherms are usually polydisperse or so low in molecular weight that they resemble small-molecule mixtures (see Section XI-5). Generally, the adsorbed amount increases with the molecular weight of the macromolecule, becoming independent of molecular weight for very long polymers.

Of particular interest has been the study of the polymer configurations at the solid–liquid interface. Beginning with lattice theories, early models of polymer adsorption captured most of the features of adsorption such as the loop, train, and tail structures and the influence of the surface interaction parameter (see Refs. 57, 58, 62 for reviews of older theories). These lattice models have been expanded on in recent years using modern computational methods [63,64] and have allowed the calculation of equilibrium partitioning between a poly-

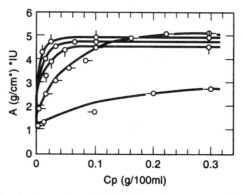

Fig. XI-5. Adsorption isotherm from Ref. 61 for polystyrene on chrome in cyclohexane at the polymer theta condition. The polymer molecular weights $\times 10^4$ are (-O) 11, (O-) 67, (ϕ) 242, (ϕ) 762, and (O) 1340. Note that all the isotherms have a high-affinity form except for the two lowest molecular weights.

mer solution and the surface. The fundamental idea behind these computations is to determine the segment distributions for polymers interacting with adsorbing walls. This is done by weighting the probability distribution function for polymer segments with a surface attraction term e^χ reflecting the Boltzmann weighting for a surface attachment. This probability distribution is then determined in a mean field due to interactions between segments. The lattice calculations are necessarily computationally intensive and beyond our scope here. The mean-field equations can be solved in the continuum limit, where the lattice size shrinks to zero. This amounts to solving the diffusion equation for the probability distribution function, G

$$\frac{\partial G}{\partial s} - \frac{b^2}{6} \nabla^2 G + \omega(\mathbf{z})G = \delta(s)\delta(\mathbf{z} - \mathbf{z}') \qquad \text{(XI-18)}$$

where ω is the mean-field potential due to the interaction of the chain segment with all other polymer segments. This potential depends on the segment concentration, $\phi(\mathbf{z})$ as

$$\omega(\mathbf{z}) = \{-\ln(1 - \phi(\mathbf{z})) + \chi[1 - 2\phi(\mathbf{z})]\} \qquad \text{(XI-19)}$$

where χ is the Flory–Huggins interaction parameter reflecting the solution conditions; $\chi = 0.5$ for an ideal or "theta" solution and $\chi = 0$ for a good solvent. Since the polymer volume fraction ϕ depends on the probability distribution function, Eqs. XI-18 and XI-19 must be solved self-consistently. When appropriate boundary conditions for the adsorbing wall are applied to Eq. XI-18, a concentration profile for loops, trains, and tails results as shown in Fig. XI-6 [59,65].

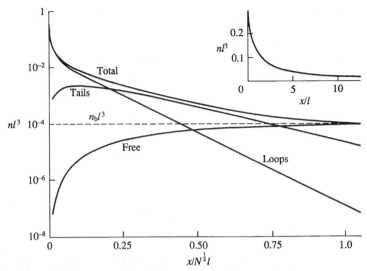

Fig. XI-6. Polymer segment volume fraction profiles for $N = 10^4$, $\chi = 0.5$, and $\chi_s \approx 1$, on a semilogarithmic plot against distance from the surface scaled on the polymer radius of gyration showing contributions from loops and tails. The inset shows the overall profile on a linear scale, from Ref. 65.

As evident from Fig. XI-6, the mean field produces concentration profiles that decay exponentially with distance from the surface [66]. A useful approximate solution to Eq. XI-18 captures the exponential character of the loop concentration profile [67]. Here a chain of length N at a bulk concentration of ϕ_b has a loop profile that can be estimated by

$$\phi(z) = A \exp\left[-\left(\frac{24B}{N}\right)^{1/2} z \right]$$ (XI-20)

with

$$A = \left(\frac{\chi_s - 1}{2 - 4\chi} \right)$$

and

$$B = \ln\left(\frac{AN}{6\phi_b \ln(1/\phi_b)} \right)$$

where z is normalized on the segment size. While limited to values of $A \leq 1$, this expression provides helpful insight into adsorbed polymer layer structure (see Problem XI-7). By contrast, scaling models predict a power-law decay for the concentration profile [68,69].

An early analytic theory by Hoeve accurately predicts the number of loops $n_l(s)$ and trains $n_t(s)$ having s segments

$$\frac{n_t(s)}{n_t} = \frac{1}{\langle t \rangle} \left(\frac{\langle t \rangle - 1}{\langle t \rangle} \right)^{s-1} \qquad \text{(XI-21)}$$

and

$$\frac{n_l(s)}{n_l} = \frac{s^{-3/2}}{2.612} \qquad \text{(XI-22)}$$

where n_l and n_t are the total number of segments in loops and trains; the train distribution is related to the average train length, which becomes $\langle t \rangle = 4.3$ for long chains.

The polymer concentration profile has been measured by small-angle neutron scattering from polymers adsorbed onto colloidal particles [70,71] or porous media [72] and from flat surfaces with neutron reflectivity [73] and optical reflectometry [74]. The fraction of segments bound to the solid surface is nicely revealed in NMR studies [75], infrared spectroscopy [76], and electron spin resonance [77]. An example of the concentration profile obtained by inverting neutron scattering measurements appears in Fig. XI-7, showing a typical surface volume fraction of 0.25 and layer thickness of 10–15 nm. The profile decays rapidly and monotonically but does not exhibit power-law scaling [70].

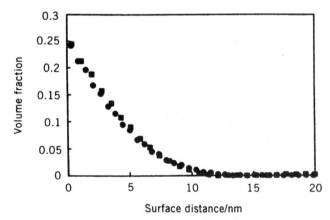

Fig. XI-7. Volume fraction profile of 280,000-molecular-weight poly(ethylene oxide) adsorbed onto deuterated polystyrene latex at a surface density of 1.21 mg/m^2 and suspended in D$_2$O, from Ref. 70.

The thickness of an adsorbed polymer layer is important for many applications. Thus many studies have centered on measuring a moment of the concentration profile. Ellipsometric techniques reveal an optical thickness that depends on the index of refraction difference between the polymer and the solvent [78,79]. This measure of thickness weights the loop and train contributions more heavily than the tails. The influence of the tails enters into hydrodynamic measurements of polymer layer thicknesses either through dynamic light scattering [80] or viscometry on colloidal particles [81] or permeability of pores carrying adsorbed polymer [82]. The adsorbed polymer thickness increases with the adsorbed amount and the polymer molecular weight, the thickness scaling as a power of the molecular weight with the exponent varying from 0.4 to 0.7. Generally, the hydrodynamic measures of thickness are larger owing to the significant drag exerted on the fluid by the dangling tails.

The high molecular weights of polymers and their many attachments with the surface render their equilibration quite slow. The attachment to the surface is often transport-limited and can be characterized in flow experiments [74,83–85]. Long polymers do not desorb into pure solvent on time scales accessible in the laboratory; however, they are readily exchanged with other polymers. Long polymers will displace short ones as the system gains entropy by doing so. The displacement is quite sensitive to the surface interaction parameter χ_s; shorter polymers will displace longer ones provided they have a higher affinity for the surface [86].

Many more complex polymers are receiving attention because of their technological importance. The adsorption of polyelectrolytes combines the interesting features of polymer configurational statistics with the added complexity of electrostatics [87–90], requiring characterization of such features as the charge on the polymer and the surface [91] and the effect of counterions on adsorption [92,93]. The adsorption of diblock copolymers is particularly interesting for colloidal stabilization as they produce polymers anchored to the surface by one end. These "tethered chains" have received much theoretical and experimental attention in recent years [94–99]. Many of the experiments have focused on the forces between surfaces bearing these polymer layers [91,93,100] using the surface forces apparatus described in Chapter VI. In contrast to adsorbed homopolymers, diblock copolymers show an adsorbed layer thickness increasing linearly with the molecular weight of the soluble block, making them attractive for colloidal stabilization [98,100].

Proteins adsorbing to solid surfaces are a ubiquitous feature of medicine, biotechnology, food processing, and environmental engineering and thus have received much attention in the past 10 years [101–103]. The fact that naturally occurring macromolecules adhere to solid surfaces impacts the blood clotting cascade, enzymatic reactions in detergents, and biological sensors. While some flexible proteins behave much the same as water-soluble polymers, most are globular in structure and exhibit different adsorption behavior. Globular proteins lose entropy on folding and are held in place by intramolecular forces between residues [104]. Many proteins unfold on adsorption to regain some configurational entropy while maximizing contacts with the surface [105]. Such conformation changes on adsorption have been implicated in adsorption irreversibility and immobility [106].

Protein adsorption has been studied with a variety of techniques such as ellipsometry [107,108], ESCA [109], surface forces measurements [102], total internal reflection fluorescence (TIRF) [103,110], electron microscopy [111], and electrokinetic measurement of latex particles [112,113] and capillaries [114]. The TIRF technique has recently been adapted to observe surface diffusion [106] and orientation [115] in adsorbed layers. These experiments point toward the significant influence of the protein-surface interaction on the adsorption characteristics [105,108,110]. A very important interaction is due to the hydrophobic interaction between parts of the protein and polymeric surfaces [18], although often electrostatic interactions are also influential [116]. Protein desorption can be affected by altering the pH [117] or by the introduction of a complexing agent [118].

Enzymes are important catalysts in biological organisms and are of increasing use in detergents and sensors. It is of interest to understand not only their adsorption characteristics but also their catalytic activity on the surface. The interplay between adsorption and deactivation has been clearly illustrated [119] as has the ability of a protein to cleave a surface-bound substrate [120].

3. Irreversible Adsorption

There are numerous references in the literature to irreversible adsorption from solution. *Irreversible adsorption* is defined as the lack of desorption from an adsorbed layer equilibrated with pure solvent. Often there is no evidence of strong surface–adsorbate bond formation, either in terms of the chemistry of the system or from direct calorimetric measurements of the heat of adsorption. It is also typical that if a better solvent is used, or a strongly competitive adsorbate, then desorption is rapid and complete. Adsorption irreversibility occurs quite frequently in polymers [4] and proteins [121–123] but has also been observed in small molecules and surfactants [124–128]. Each of these cases has a different explanation and discussion.

Fleer, Cohen Stuart, and co-workers attribute irreversibility in polymer adsorption to slight polydispersity in the molecular-weight distribution and the experimental inaccessibility of the extreme dilutions required to reach a sufficiently low region on the isotherm [4]. The "high-affinity" nature of the adsorption causes the isotherms to be so abrupt that one cannot dilute the system enough, to a volume fraction of approximately (10^{-10}), to reduce the surface coverage. Other groups attribute apparent irreversibility in polymer adsorption to kinetic limitations. While polymers do not desorb into solvent, they will readily exchange with other polymers in solution as clearly shown by Granick and co-workers [86].

Proteins often have the same high-affinity isotherms as do synthetic polymers and are also slow to equilibrate, due to many contacts with the surface. Proteins, however, have the additional complication that they can partially or completely unfold at the solid–liquid interface to expose their hydrophobic core units to a hydrophobic surface

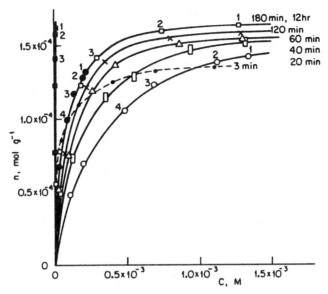

Fig. XI-8. Adsorption of BaDNNS on TiO_2 at 23°C from n-heptane solution. $\square$, $\times$, $\triangle$, [], $\bigcirc$, $\bullet$: adsorption points for indicated equilibration times. $\blacksquare$, $\bullet$: desorption points following 12-hr and 20-min equilibrations, respectively. (From Ref. 124.)

[105]. This type of behavior can cause irreversibility in the adsorption isotherm as well as immobility on the surface [106].

An example of the time effects in irreversible adsorption of a surfactant system is shown in Fig. XI-8 for barium dinonylnapthalene sulfonate (an oil additive) adsorbing on TiO_2 (anatase). Adsorption was irreversible for aged systems, but much less so for those equilibrating for a short time. The adsorption of aqueous methylene blue (note Section XI-4) on TiO_2 (anatase) was also irreversible [128]. In these situations it seems necessary to postulate at least a two-stage sequence, such as

$$A(\text{solution}) \rightleftharpoons A_1^s(\text{surface})$$
$$xA_1^s(\text{surface}) \rightarrow A_x^s(\text{surface}) \qquad (XI\text{-}23)$$

where A denotes adsorbate; A_1^s is the adsorbed state, which may be in rapid equilibrium with the solution; and A_x^s denotes some kind of denatured, polymeric or condensed phase state [124,129–131]. There remains, however, a fundamental paradox (see Problem XI-8).

4. Surface Area Determination

The estimation of surface area from solution adsorption is subject to many of the same considerations as in the case of gas adsorption discussed in Chapter XVII, but with the added complication that larger molecules are involved, whose surface orientation and pore penetrability may be uncertain. A first condition is that a definite adsorption model be obeyed, which in practice means that area determinations are limited to adsorption following the Langmuir equation (Eq. XI-4). The constant n^s is found, for example, from a plot of the data according to Eq. XI-8, and the specific surface area follows from $n^s = (\Sigma/N_A \sigma^0)$, leaving the choice of the correct value of σ^0. If chemisorption is occurring from solution σ^0 represents the spacing of adsorbent sites, while for physisorption σ^0 more likely reflects the area of the adsorbate molecule in its particular orientation. In addition, many powdered adsorbents are likely to be *fractal* in nature (see Section VII-4C), so that the apparent surface area depends on the size of the adsorbate molecule used [132].

The self-assembled monolayers described above are often well understood in terms of a close-packed area per molecule; however, the careful determination of tilt angle and surface density is necessary to confirm these values. Dyes have received much attention because of the ease of obtaining accurate concentrations colorimetrically. While dye adsorption generally follows the Langmuir equation, multilayer adsorption can occur. Additional problems that may arise due to dye association in solution are discussed by Padday [133] and Barton [134]. Rahman and Ghosh [135] used the Langmuir adsorption of pyridine (molecular area 24 Å^2) on various oxides to determine surface areas. Pugh [136] has used a number of acid and base probe molecules to identify chemical sites on several minerals used as polymer fillers. Surface areas may be estimated from the exclusion of like-charge ions from a charged interface [137]. This method is intriguing in that no estimation of site or molecular area is called for. Area determination with binary liquid systems (see next section) has been proposed by Everett [138] and discussed by Schay and Nagy [139].

5. Adsorption in Binary Liquid Systems

A. Adsorption at the Solid-Solution Interface

The discussion so far has been confined to systems in which the solute species are dilute, so that adsorption was not accompanied by any significant change in the activity of the solvent. In the case of adsorption from binary liquid mixtures, where the complete range of concentration, from pure liquid A to pure liquid B, is available, a more elaborate analysis is needed. The terms solute and solvent are no longer meaningful, but it is nonetheless convenient to cast the equations around one of the components, arbitrarily designated here as component 2.

We suppose that the Gibbs dividing surface (see Section III-5) is located at the surface of the solid (with the implication that the solid itself is not soluble). It follows that the surface excess Γ_2^s, according to this definition, is given by (see Problem XI-9)

$$\Gamma_2^s = \frac{n^s}{\Sigma}\,(N_2^s - N_2^l) \tag{XI-24}$$

Here, n^s denotes the total number of moles associated with the adsorbed layer, and N_1^s and N_1^l are the respective mole fractions in that layer and in solution at equilibrium. As before, it is assumed, for convenience, that mole numbers refer to that amount of system associated with one *gram of adsorbent*. Equation XI-24 may be written

$$\Gamma_2^s = \frac{n^s}{\Sigma}\left(\frac{n_2^s}{n^s} - \frac{n_2^l}{n^l}\right) \tag{XI-25}$$

where n_2^s and n_2^l are the moles of component 2 in the adsorbed layer and in solution. Since $n_2^s + n_2^l = n_2^0$, the total number of moles of component 2 present, and $n^s + n^l = n_0$, the total number of moles in the system, substitution into Eq. XI-25 yields

$$\Gamma_s^s = \frac{n_0}{\Sigma}\,(N_2^0 - N_2^l) = \frac{n_0 \Delta N_2^l}{\Sigma} \tag{XI-26}$$

where N_2^0 is the mole fraction of component 2 before adsorption.
   Another form of Eq. XI-26 may be obtained:

$$\Gamma_2^s = \frac{n_0 \Delta N_2^l}{\Sigma} = \frac{n_2^s N_1^l - n_1^s N_2^l}{\Sigma} \tag{XI-27}$$

It is important to note that the experimentally defined or *apparent adsorption* $n_0 \Delta N_2^l / \Sigma$, while it gives Γ_2^s, does *not* give the amount of component 2 in the adsorbed layer n_2^s. Only in dilute solution where $N_2^l \to 0$ and $N_1^l \simeq 1$ is this true. The adsorption isotherm, Γ_2^s plotted against N_2, is thus a *composite isotherm* or, as it is sometimes called, the *isotherm of composition change*.
   Equation XI-27 shows that Γ_2^s can be viewed as related to the difference between the individual adsorption isotherms of components 1 and 2. Figure XI-9 [140] shows the composite isotherms resulting from various combinations of individual ones. Note in particular Fig. XI-9a, which shows that even in the absence of adsorption of component 1, that of component 2 must go through a maximum (due to the N_1^l factor in Eq. XI-27), and that in all other cases the apparent adsorption of component 2 will be negative in concentrated solution.
   Everett and co-workers [141] describe an improved experimental procedure for obtaining Γ_i^s quantities. Some of their data are shown in Fig. XI-10. Note the negative region for n_1^s at the lower temperatures. More recent but similar data were obtained by Phillips and Wightman [142].

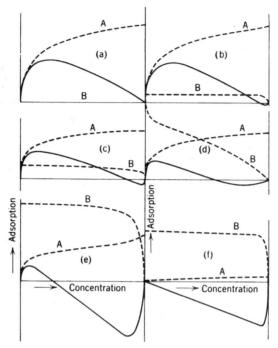

Fig. XI-9. Composite adsorption isotherms: ------, individual isotherms; ———, isotherms of composition change. (From Ref. 140.)

An elegant and very interesting approach was that of Kipling and Tester [143], who determined the separate adsorption isotherms for the vapors of benzene and of ethanol on charcoal, that is, the adsorbent was equilibrated with the vapor in equilibrium with a given solution, and from the gain in weight of the adsorbent and the change in solution composition, following adsorption, the amounts of each component in the adsorbed film could be calculated. These individual component isotherms could then be inserted in Eq. XI-27 to give a calculated apparent solution adsorption isotherm, which in fact, agreed well with the one determined directly. Their data are illustrated in Fig. XI-11. They also determined the separate adsorption isotherms for benzene and charcoal and ethanol and charcoal; these obeyed Langmuir equations for gas adsorption,

$$\theta_1 = \frac{b_1 P_1}{1 + b_1 P_1} \qquad \theta_2 = \frac{b_2 P_2}{1 + b_2 P_2} \qquad \text{(XI-28)}$$

from which the constants b_1 and b_2 were thus separately evaluated. The composite vapor adsorption isotherms of Fig. XI-11a were then calculated using these constants and the added assumption that in this case no bare surface was present. The Langmuir equation for the competitive adsorption of two gas-phase

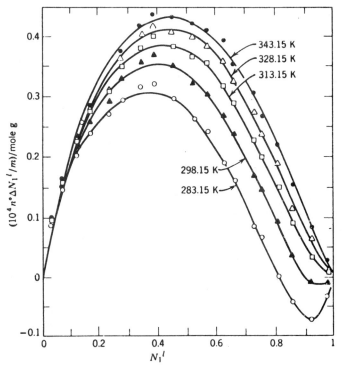

Fig. XI-10. Isotherm of composition change or surface excess isotherm for the adsorption of (1) benzene and (2) n-heptane on Graphon. (From Ref. 141.)

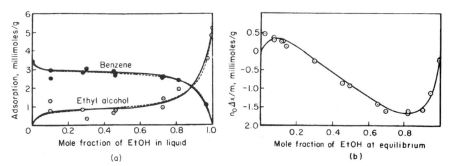

Fig. XI-11. Relation of adsorption from binary liquid mixtures to the separate vapor pressure adsorption isotherms, system: ethanol-benzene-charcoal: (a) separate mixed-vapor isotherms; (b) calculated and observed adsorption from liquid mixtures. (From Ref. 143.)

components is (see Section XVII-3)

$$\theta_2 = \frac{b_2 P_2}{1 + b_1 P_1 + b_2 P_2} \tag{XI-29}$$

and the effect of their assumptions was to put Eq. XI-29 in the form

$$\theta_2 = \frac{b_2 P_2}{b_1 P_1 + b_2 P_2} \simeq \frac{b_2 P_2^0 N_2^l}{b_1 P_1^0 N_1^l + b_2 P_2^0 N_2^l} \tag{XI-30}$$

and to identify b_2 and b_1 with the separately determined values from Eqs. XI-28. Again, good agreement was found, as shown by the dashed lines in Fig. XI-11a. Similar comparisons have been made by Myers and Sircar [144].

The Langmuir model as developed in Section XI-1 may be applied directly to Eq. XI-24 [5]. We replace a_1 and a_2 in Eq. XI-3 by N_1 and N_2 (omitting the superscript l as no longer necessary for clarity), and solve for N_2^s (with N_1 replaced by $1 - N_2$):

$$N_2^s = N_2 \frac{K}{1 + (K - 1)N_2} \tag{XI-31}$$

This is now substituted into Eq. XI-24, giving

$$\Gamma_2^s = \frac{n^s}{\Sigma} \frac{(K - 1)N_1 N_2}{1 + (K - 1)N_2} \tag{XI-32}$$

Or, using the apparent adsorption, $n^0 \Delta N_2$, Eq. XI-30 may be put in the linear form

$$\left(\frac{N_1 N_2}{n_0 \Delta N_2}\right) = \frac{1}{n^s(K - 1)} + \frac{1}{n^s}(N_2) \tag{XI-33}$$

which bears a close resemblance to that for the simple Langmuir equation, Eq. XI-4. Note that for $K = 1$, $\Gamma_2^s = 0$, that is, no fractionation occurs at the interface. Everett [5] found Eq. XI-33 to be obeyed by several systems, for example, that of benzene and cyclohexane on Spheron 6.

In general, one should allow for nonideality in the adsorbed phase (as well as in solution), and various authors have developed this topic [5,137,145–149]. Also, the adsorbent surface may be heterogeneous, and Sircar [150] has pointed out that a given set of data may equally well be represented by nonideality of the adsorbed layer on a uniform surface or by an ideal adsorbed layer on a heterogeneous surface.

Isotherms of type a in Fig. XI-9 are relatively linear for large N_2, that is,

$$n_0 \Delta N_2^l = a - b N_2^l \tag{XI-34}$$

Now, Eq. XI-27 can be written in the form

$$n_0 \Delta N_2^l = n_2^s - n^s N_2^l \tag{XI-35}$$

(since $n^s = n_1^s + n_2^s$ and $N_1 + N_2 = 1$). Comparing Eqs. XI-34 and XI-35, the slope b gives the monolayer capacity n^s. The surface area Σ follows if the molecular area can be estimated. The treatment assumes that in the linear region the surface is mostly occupied by species 2 so that n_2^s is nearly constant. See Refs. 151 and 152.

There is a number of very pleasing and instructive relationships between adsorption from a binary solution at the solid-solution interface and that at the solution-vapor and the solid-vapor interfaces. The subject is sufficiently specialized, however, that the reader is referred to the general references and, in particular, to Ref. 153. Finally, some studies on the effect of high pressure (up to several thousand atmospheres) on binary adsorption isotherms have been reported [154]. Quite appreciable effects were found, indicating that significant partial molal volume changes may occur on adsorption.

B. Heat of Adsorption at the Solid–Solution Interface

Rather little has been done on heats of wetting of a solid by a solution, but two examples suggest a fairly ideal type of behavior. Young et al. [152] studied the Graphon-aqueous butanol system, for which monolayer adsorption of butanol was completed at a fairly low concentration. From direct adsorption studies, they determined θ_b, the fraction of surface covered by butanol, as a function of concentration. They then prorated the heat of immersion of Graphon in butanol, 113 ergs/cm^2, and of Graphon in water, 32 ergs/cm^2, according to θ_b. That is, each component in the adsorbed film was considered to interact with its portion of the surface independently of the other,

$$q_{imm} = N_1^s q_1 + N_2^s q_2 \tag{XI-36}$$

where q_{imm} is the heat of immersion in the solution and q_1 and q_2 are the heats of immersion in the respective pure liquid components. To this prorated q_{imm} was added the heat effect due to concentrating butanol from its aqueous solution to the composition of the interfacial solution using bulk heat of solution data. They found that their heats of immersion calculated in this way agreed very well with the experimental values.

As a quite different and more fundamental approach, the isotherms of Fig. XI-10 allowed a calculation of K as a function of temperature. The plot of $\ln K$ versus $1/T$ gave an enthalpy quantity that should be just the difference between the heats of immersion of the Graphon in benzene and in n-heptane, or 2.6×10^{-3} cal/m^2 [141]. The experimental heat of immersion difference is 2.4×10^{-3} cal/m^2, or probably indistinguishable. The relationship between calorimetric and isosteric heats of adsorption has been examined further by Myers [155]

6. Adsorption of Electrolytes

The interaction of an electrolyte with an adsorbent may take one of several forms. Several of these are discussed, albeit briefly, in what follows. The electrolyte may be adsorbed in toto, in which case the situation is similar to that for molecular adsorption. It is more often true, however, that ions of one sign are held more strongly, with those of the opposite sign forming a diffuse or secondary layer. The surface may be polar, with a potential ψ, so that primary adsorption can be treated in terms of the Stern model (Section V-3), or the adsorption of interest may involve exchange of ions in the diffuse layer.

In the case of ion exchangers, the primary ions are chemically bonded into the framework of the polymer, and the exchange is between ions in the secondary layer. A few illustrations of these various types of processes follow.

A. Stern Layer Adsorption

Adsorption at a charged surface where both electrostatic and specific chemical forces are involved has been discussed to some extent in connection with various other topics. These examples are drawn together here for a brief review along with some more specific additional material. The Stern equation, Eq. V-19, may be put in a form more analogous to the Langmuir equation, Eq. XI-4:

$$\frac{\theta}{1 - \theta} = C_2 \exp \frac{ze\psi + \Phi}{kT} \tag{XI-37}$$

The effect is to write the adsorption free energy or, approximately, the energy of adsorption Q as a sum of electrostatic and chemical contributions. A review is provided by Ref. 156.

Stern layer adsorption was involved in the discussion of the effect of ions on ζ potentials (Section V-6), electrocapillary behavior (Section V-7), and electrode potentials (Section V-8) and enters into the effect of electrolytes on charged monolayers (Section XV-6). More specifically, this type of behavior occurs in the adsorption of electrolytes by ionic crystals. A large amount of work of this type has been done, partly because of the importance of such effects on the purity of precipitates of analytical interest and partly because of the role of such adsorption in coagulation and other colloid chemical processes. Early studies include those by Weiser [157], by Paneth, Hahn, and Fajans [158], and by Kolthoff and co-workers [159]. A recent calorimetric study of proton adsorption by Lyklema and co-workers [160] supports a new thermodynamic analysis of double-layer formation. A recent example of this is found in a study by Cerofolini and Boara [161] on the preferential adsorption of K^+ over Na^+ into hydrated silica gel.

It is possible for neutral species to be adsorbed even where specific chemical interaction is not important owing to a consequence of an electrical double layer. As discussed in connection with Fig. V-4, the osmotic pressure of solvent is reduced in the region between two charged plates. There is, therefore, an equilibrium with bulk solution that can be shifted by changing the external ionic strength. As a typical example of such an effect, the interlayer spacing of montmorillonite (a sheet or layer type aluminosilicate see Fig. XI-12) is very dependent on the external ionic strength. A spacing of about 19 Å in dilute electrolyte reduces to about 15 Å in 1–2M 1:1 electrolyte [162]. Phenomenologically, a strong water adsorption is repressed by added electrolyte.

If specific chemical interactions are not dominant, the adsorption of an ionic species is largely determined by its charge. This is the basis for an early conclusion by Bancroft [164] that the order of increasing adsorption of ions by sols is that of increasing charge. Within a given charge type, the sequence may be that of increasing hydration enthalpy [165]; thus: $Cs^+ > Rb^+ > NH_4^+$, K^+ $> Ag^+ > Na^+ > Li^+$. The extent of adsorption, in turn, determines the power of such ions to coagulate sols and accounts for the related statement that the coagulating ability of an ion will be greater the higher its charge. This rule,

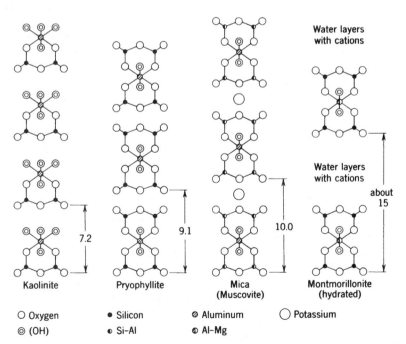

Fig. XI-12. End-on view of the layer structures of clays, pyrophillite, and mica. (From Ref. 163.)

TABLE XI-1

Potential-Determining Ion and Point of Zero Charge[a]

Material	Potential-Determining Ion	Point of Zero Charge
Fluorapatite, $Ca_5(PO_4)_3(F, OH)$	H^+	pH 6
Hydroxyapatite, $Ca_5(PO_4)_3(OH)$	H^+	pH 7
Alumina, Al_2O_3	H^+	pH 9
Calcite, $CaCO_3$	H^+	pH 9.5
Fluorite, CaF_2	Ca^{2+}	pCa 3
Barite (synthetic), $BaSO_4$	Ba^{2+}	pBa 6.7
Silver iodide	Ag^+	pAg 5.6
Silver chloride	Ag^+	pAg 4
Silver sulfide	Ag^+	pAg 10.2

[a]From Ref. 166.

the Schulze–Hardy rule, is discussed in Section VI-4B. Surface charge may be controlled or fixed by a potential-determining ion.

Table XI-1 (from Ref. 166) lists the potential-determining ion and its concentration giving zero charge on the mineral. There is a large family of minerals for which hydrogen (or hydroxide) ion is potential determining—oxides, silicates, phosphates, carbonates, and so on. For these, adsorption of surfactant ions is highly pH-dependent. An example is shown in Fig. XI-14. This type of behavior has important applications in flotation and is discussed further in Section XIII-4.

Electrolyte adsorption on metals is important in electrochemistry [167,168]. One study reports the adsorption of various anions an Ag, Au, Rh, and Ni electrodes using ellipsometry. Adsorbed film thicknesses now also depend on applied potential.

A detailed study by Grieser and co-workers [169] of the forces between a gold-coated colloidal silica sphere and a gold surface reveals the preferential adsorption of citrate ions over chloride to alter the electrostatic interaction.

B. Surfactant Adsorption

In the adsorption of long-chain surfactant molecules to charged surfaces, both chemical and electrical interactions may be important. These mechanisms are nicely reviewed by Berg [170], and we offer only a brief summary of important features here. The amphiphilic nature of surfactants causes them to follow Traube's rule (Section XI-1A), and numerous studies have focused on the effect of chain length on adsorption [20–23,171]. Rusling and co-workers have used flow voltammetry to study the adsorption and desorption dynamics of a series of amphiphilic ferrocenes [172]. They find adsorption free energies for their surfactants to increase with length by about 750 J/mol CH_2, in agreement with the findings for SAMs discussed in Section XI-1B. Ralston and co-workers

[173] have investigated the adsorption of amphiphilic napthalene derivatives onto graphite. They find adsorption to follow a Frumkin equation (see Section V-7D) with coefficients and exponents determined by the balance between chemical and electrostatic interactions. The hydrophobic interaction between long-chain surfactants can produce a second adsorption layer described by a new molecular theory by Lyklema [174].

The complexities of surfactant adsorption isotherms have been elucidated via the self-consistent field lattice calculations discussed in Section XI-2 [63,175]. Combining experimental studies of cationic and anionic surfactant adsorption with lattice calculations, Koopal and co-workers have distinguished four regions of the adsorption isotherm [175]. As illustrated in Fig. XI-13 the low concentration regime increases linearly with concentration as in Henry's law adsorption of gases (Section XVII-6). In region II, the slope is much larger than unity because of the combination of electrostatic interactions between the head groups and the surface and the lateral attraction between adjacent tails, as found in many other surfactant isotherms such as those by Conner and Ottewill [176]. Region III is characterized by a common intersection point in isotherms measured at different ionic strengths and indicates micellar adsorption (see Section XIII-5), where the surfactant counterions balance head-group charge. In region IV, the surfactant solution is above its critical micelle concentration (CMC) and the solution chemical potential ceases to change causing a plateau in adsorption.

In regime II Koopal and co-workers characterize adsorption in terms of

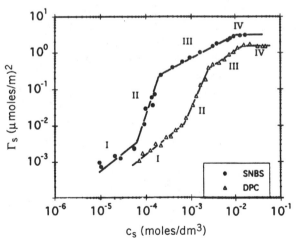

Fig. XI-13. Adsorption isotherms for SNBS (sodium *p*-3-nonylbenzene sulfonate) (pH 4.1) and DPC (dodecyl pyridinium chloride) (pH 8.0) on rutile at approximately the same surface potential and NaCl concentration of $0.01M$ showing the four regimes of surfactant adsorption behavior, from Ref. 175. [Reprinted with permission from Luuk K. Koopal, Ellen M. Lee, and Marcel R. Böhmer, *J. Colloid Interface Science*, **170**, 85–97 (1995). Copyright Academic Press.]

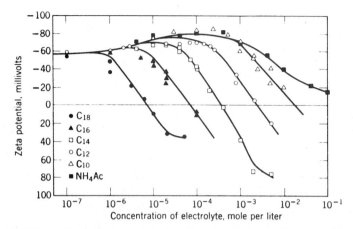

Fig. XI-14. Effect of hydrocarbon chain length on the ζ potential of quartz in solutions of alkylammonium acetates and in solutions of ammonium acetate. (From Ref. 183.)

hemimicelles having the charged head groups adjacent to the surface. The existence of surfactant *hemimicelles* at the solid surface, in analogy with the micelle formation in solution (see Section XIII-5), has been the subject of controversy for some time [177]. Recently AFM measurements by Manne and co-workers have proven the existence of ordered arrays of hemicylindrical hemimicelles on graphite [178]. In contrast to the Koopal experiments, however, these hemimicelles form on hydrophobic surfaces and thus have the surfactant tails closest to the surface.

Electrokinetic experiments elucidate the effect of surfactant adsorption on surface charge. This is possible for large flat surfaces via streaming potential measurements [179] or for colloidal particles via electrophoresis where one determines the ζ potential. An example is shown in Fig. XI-14, showing that adsorption of cations by quartz eventually reduces ζ to zero. If the concentration required to produce zero ζ potential is determined by Φ in Eq. XI-37, then the observation that log C_2 depends linearly on chain length follows Traube's rule. The slope of $(\partial \ln C_2/\partial N_c)_{\zeta=0}$, where N_c is the chain length, provides an energy increment of 2.5 kJ/CH$_2$, in agreement with that at the air–water interface (see Section III-7E) and behavior of some SAMs (Section XI-1B).

The surface forces apparatus (Section VI-3C) has revealed many features of surfactant adsorption and its effect on the forces between adsorbent surfaces [180,181]. A recent review of this work has been assembled by Parker [182].

C. Counterion Adsorption–Ion Exchange

A very important class of adsorbents consists of those having charged sites due to ions or ionic groups bound into the lattice. The montmorillonite clays, for example, consist of layers of tetrahedral SiO$_4$ units sharing corners with octahedral Al^{3+}, having coordinated oxygen and hydroxyl groups, as illustrated in Fig. XI-12 [163]. In not-too-acid solution, cation exchange with protons is possible, and because of the layer structure, swelling effects occur, which are understandable in terms of ionic strength effects on double-layer repulsion (see Section VI-4B and Ref. 184).

A tremendous variety of structures is known, and some of the three-dimensional network ones are porous enough to show the same type of swelling phenomena as the layer structures—and also ion exchange behavior. The zeolites fall in this last category and have been studied extensively, both as ion exchangers and as gas adsorbents (e.g., Refs. 185 and 186). As an example, Goulding and Talibudeen have reported on isotherms and calorimetric heats of Ca^{2+}-K^+ exchange for several aluminosilicates [187].

Organic ion exchangers were introduced in 1935, and a great variety is now available. The first ones consisted of phenol-formaldehyde polymers into which natural phenols had been incorporated, but now various polystyrene polymers are much more common. Here RSO_3^- groups, inserted by sulfonation of the polymer, are sufficiently acidic that ion exchange can occur even in quite acid solution. The properties can be controlled by varying the degree of sulfonation and of crosslinking. Other anionic groups, such as $RCOO^-$, may be introduced to vary the selectivity. Also, anion exchangers having RNH_2 groups are in wide use. Here addition of acid gives the $RNH_3^+ X^-$ function, and anions may now exchange with X^-.

Both the kinetics and the equilibrium aspects of ion exchange involve more than purely surface chemical considerations. Thus, the formal expression for the exchange

$$AR + B = BR + A \qquad (XI\text{-}38)$$

where R denotes the matrix and A and B the exchanging ions, suggests a simple mass action treatment. The AR and BR centers are distributed throughout the interior of the exchanger phase and can be viewed as forming a nonideal solution. One may represent their concentrations in terms of mole fractions or, for ions of differing charge, equivalent fractions (this point is discussed in Ref. 188); sizable activity coefficient corrections are generally needed. Alternatively, the exchanger phase may be treated as essentially a concentrated electrolyte solution so that volume concentrations are used, but again with activity coefficient corrections. The nonideality may be approached by considering the exchanger phase to act as a medium permeable to cations but not to the R or lattice ions, so that ion exchange appears as a Donnan equilibrium (see Section XV-7A), and specific recognition can then be given to the swelling effects that occur. Finally, detailed surface structural information may be obtained from surface spectroscopic techniques, as in ESCA (Section VIII-3B) of ion exchange on cleaved mica surfaces [189].

The rates of ion exchange are generally determined by diffusion processes; the rate-determining step may either be that of diffusion across a boundary film of solution or

that of diffusion in and through the exchanger base [190]. The whole matter is complicated by electroneutrality restrictions governing the flows of the various ions [191,192].

Stahlberg has presented models for ion-exchange chromatography combining
the Gouy–Chapman theory for the electrical double layer (see Section V-2) with
the Langmuir isotherm (Eq. XI-4) [193] and with a specific adsorption model
[194].

As may be gathered, the field of ion-exchange adsorption and chromatography is far too large to be treated here in more than this summary fashion. Refs.
195 and 196 are useful monographs.

7. Photophysics and Photochemistry of the Adsorbed State

There is a large volume of contemporary literature dealing with the structure and
chemical properties of species adsorbed at the solid-solution interface, making use of
various spectroscopic and laser excitation techniques. Much of it is phenomenologically
oriented and does not contribute in any clear way to the surface chemistry of the system;
included are many studies aimed at the eventual achievement of solar energy conversion. What follows here is a summary of a small fraction of this literature, consisting
of references which are representative and which also yield some specific information
about the adsorbed state.

A. Photophysics of Adsorbed Species

The typical study consists of adsorbing a species having a known photoexcited emission behavior and observing the emission spectrum, lifetime, and quenching by either
coadsorbed or solution species. Pyrene and surfactants containing a pyrene moiety have
been widely used, and also the complex ion Ru(2,2′-bipyridine)$_3^{2+}$, R. Pyrene, P, has the
property of *excimer* formation, that is, formation of the complex PP*, where P* denotes
excited-state pyrene, and the emission from PP* can give information about surface
mobility. Both pyrene emission and that from R* have spectral and lifetime characteristics that are sensitive to local environment. Emission from R* may be quenched by
electron-accepting or electron-donating species; in fact, it is the powerful oxidizing and
reducing nature of R* that has made it an attractive candidate for solar energy conversion schemes.

Surface heterogeneity may be inferred from emission studies such as those studies
by de Schrijver and co-workers on P and on R adsorbed on clay minerals [197,198]. In
the case of adsorbed pyrene and its derivatives, there is considerable evidence for surface mobility (on clays, metal oxides, sulfides), as from the work of Thomas [199], de
Mayo and co-workers [200], Singer [201] and Stahlberg et al. [202]. There has also been
evidence for ground-state bimolecular association of adsorbed pyrene [66,203]. The sensitivity of pyrene to the polarity of its environment allows its use as a probe of surface
polarity [204,205]. Pyrene or other emitters may be used as probes to study the structure
of an adsorbate film, as in the case of Triton X-100 on silica [206], sodium dodecyl sulfate at the alumina surface [207] and hexadecyltrimethylammonium chloride adsorbed
onto silver electrodes from water and dimethylformamide [208]. In all cases progressive structural changes were concluded to occur with increasing surfactant adsorption.

In the case of Ru(2,2′-bipyridine)$_3^{2+}$ adsorbed on porous Vycor glass, it was inferred that structural perturbation occurs in the excited state, R*, but not in the ground state [209].

Many of the adsorbents used have "rough" surfaces; they may consist of clusters of very small particles, for example. It appears that the concept of self-similarity or fractal geometry (see Section VII-4C) may be applicable [210,211]. In the case of quenching of emission by a coadsorbed species, Q, some fraction of Q may be "hidden" from the emitter if Q is a small molecule that can fit into surface regions not accessible to the emitter [211].

B. Photochemistry at the Solid–Solution Interface

The most abundant literature is that bearing on solar energy conversion, mainly centered on the use of Ru(2,2′-bipyridine)$_3^{2+}$ and its analogues. The excited state of the parent compound was found some years ago to be a powerful reducing agent [212], allowing the following spontaneous reactions to be written:

$$R \xrightarrow{h\nu} R^* \tag{XI-39}$$

$$R^* + H_3O^+ \rightarrow R^+ + \tfrac{1}{2}H_2 + H_2O \tag{XI-40}$$

$$R^+ + \tfrac{3}{2}H_2O \rightarrow R + \tfrac{1}{4}O_2 + H_3O^+ \tag{XI-41}$$

$$\tfrac{1}{2}H_2O \rightarrow \tfrac{1}{2}H_2 + \tfrac{1}{4}O_2 \tag{XI-42}$$

It is thus energetically feasible for R to catalyze the use of visible light to "split" water. The problem is that the reactions are multielectron ones and the actual individual steps are highly subject to reversal or to interception, leading to unproductive degradation of the excitation energy. Much use has been made of heterogeneous systems in an effort to avoid the problem. Thus, if R is adsorbed on a semiconductor such as TiO$_2$, R* might inject an electron into the conduction band of the adsorbent, which might then migrate away and thus not back-react with the R$^+$ formed. The subject has been extensively reviewed (see Refs. 213,214), along with the general photophysical behavior of TiO$_2$ [215].

Silica and silicates have been used as a support for Ru(II) complexes [216–219], as has porous Vycor glass [220] and clays [221]. Photoproduction of hydrogen using Pt-doped TiO$_2$ has been reported [222], as well as the photodecomposition of water using Pd and Ru doping [223]. Iron oxide (α-Fe$_2$O$_3$) is also effective [224], as is a mixture of supported CdS and Pt on oxide particles [225]. CdS, again a semiconductor [226], promotes both photoisomerization [227] and electron transfer [228]. Other surface photochemistries include that of ketones on various silicas [229] and zeolites [230], enone cycloaddition on silica gel [231], and surface-bound carbonyl complexes of Ru(0) [232]. Ru(2,2′-bipyridine)$_3^{2+}$ promotes the photoreduction of anthraquinones (on SiO$_2$ and ZrO$_2$) [233] and the hydrogenation of ethylene and acetylene (using colloidal Pt and Pd) [234], and Ru complexes can promote the photocleavage of C$_2$H$_2$ to give CH$_4$ [235].

Returning to the matter of water splitting, no commercially attractive system has been found—yields are too low, UV light is needed, and a sacrificial reagent is used whose cost is prohibitive. A great deal of interesting surface chemistry and photophysics has been learned in the process, however. Actually, reaction XI-42 is of dubious practicality even if it were feasible since an explosive and difficult to separate mixture of gases is produced. Of much greater potential value would be the photoreduction of CO_2 to CH_4 (by water), and this reaction has actually been reported to occur using a Ru colloid and a Ru(II) complex [236].

8. Problems

1. An adsorbed film obeys a modified Amagat equation of state, $\pi\sigma = qkT$ (see Eq. III-107). Show that this corresponds to a Freundlich adsorption isotherm (Eq. XI-12) and comment on the situation.

2. One hundred milliliters of an aqueous solution of methylene blue contains 3.0 mg dye per liter and has an optical density (or molar absorbancy) of 0.60 at a certain wavelength. After the solution is equilibrated with 25 mg of a charcoal the supernatant has an optical density of 0.20. Estimate the specific surface area of the charcoal assuming that the molecular area of methylene blue is 197 Å^2.

3. The adsorption of stearic acid from n-hexane solution on a sample of steel powder is measured with the following results:

Concentration (mM/l)	Adsorption (mg/g)	Concentration (mM/l)	Adsorption (mg/g)
0.01	0.786	0.15	1.47
0.02	0.864	0.20	1.60
0.04	1.00	0.25	1.70
0.07	1.17	0.30	1.78
0.10	1.30	0.50	1.99

Explain the behavior of this system, and calculate the specific surface area of the steel.

4. Dye adsorption from solution may be used to estimate the surface area of a powdered solid. Suppose that if 3.0 g of a bone charcoal is equilibrated with 100 ml of initially $10^{-4}M$ methylene blue, the final dye concentration is $0.3 \times 10^{-4}M$, while if 6.0 g of bone charcoal had been used, the final concentration would have been $0.1 \times 10^{-4}M$. Assuming that the dye adsorption obeys the Langmuir equation, calculate the specific surface area of the bone charcoal in square meters per gram. Assume that the molecular area of methylene blue is 197 Å^2.

5. The adsorption of the surfactant Aerosol OT onto Vulcan Rubber obeys the Langmuir equation [237]; the plot of C/x versus C is linear. For $C = 0.5$ mmol/l, C/x is 100 μmol/g, and the line goes essentially through the origin. Calculate the saturation adsorption in micromoles per gram.

6. Derive Eq. XI-15, assuming a Langmuir adsorption process described in Eq. XI-2, where k_a and k_d are the adsorption and desorption rate constants. Treat the solution

as sufficiently dilute that $a_1 = 1$ and a_2 can be replaced by C_2. What are the units for k_a and k_d?

7. Estimate the thickness of a polymer layer from the loop profile in Eq. XI-20. Assume $\chi = 0$, $\chi_s = 2$, $\phi = 0.01$, and $N = 10^4$. Calculate the second moment of this profile (this is often measured by ellipsometry) and compare this thickness to the radius of gyration of the coil $R_g = \sqrt{N/6}$.

8. Irreversible adsorption discussed in Section XI-3 poses a paradox. Consider, for example, curve 1 of Fig. XI-8, and for a particular system let the "equilibrium" concentration be 0.025 g/100 cm^3, corresponding to a coverage, θ of about 0.5. If the adsorption is irreversible, no desorption would occur on a small dilution; on the other hand, more adsorption would occur if the concentration were increased. If adsorption is possible but not desorption, why does the adsorption *stop* at $\theta = 0.5$ instead of continuing up to $\theta = 1$? Comment on this paradox and on possible explanations.

9. Derive Eq. XI-24.

10. In a study of the effect of pressure on adsorption, for a system obeying the Langmuir equation (and Eq. XI-3, assuming ideal solutions and $a_1 = 1$), the value of K is 2.75×10^4 at 1 atm pressure and 1.23×10^4 at 3000 bar and 25°C. Consult the appropriate thermodynamic texts and calculate ΔV, the volume change for the adsorption process of Eq. XI-2. Comment on the physical significance of ΔV.

11. For adsorption on Spheron 6 from benzene–cyclohexane solutions, the plot of $N_1 N_2 / n_0 \Delta N_2$ versus N_2 (cyclohexane being component 2) has a slope of 2.3 and an intercept of 0.4. (a) Calculate K. (b) Taking the area per molecule to be 40 Å^2, calculate the specific surface area of the spheron 6. (c) Plot the isotherm of composition change. *Note:* Assume that n^s is in millimoles per gram.

12. Equation XI-32

$$n_2^s(\text{apparent}) = n^s \, \frac{(K - 1)N_1 N_2}{1 + (K - 1)N_2}$$

will, under certain conditions, predict negative apparent adsorption. When such conditions prevail, explain which of the isotherms of composition change in Fig. XI-9 will most closely correspond to the calculated isotherm.

13. The example of Section XI-5B may be completed as follows. It is found that $\theta = 0.5$ at a butanol concentration of 0.3 g/100 cm^3. The heat of solution of butanol is 25 cal/g. The molecular area of adsorbed butanol is 40 Å^2. Show that the heat of adsorption of butanol at this concentration is about 50 ergs/cm^2.

14. Plot the number distribution of segments in loops and trains against segment number s from Eqs. XI-21 and 22. How do these profiles differ?

15. Estimate the slopes in region II of the isotherms in Fig. XI-13. Derive adsorption laws from these estimates and discuss the differences in the two systems responsible for the different isotherms.

16. Referring to Section XI-6B, the effect of the exclusion of coions (ions of like charge to that of the interface) results in an increase in solution concentration from n_0 to n_0'. Since the solution must remain electrically neutral, this means that the counterions (ions of charge opposite to that of the interface) must also increase in concentration from n_0 to n_0'. Yet Fig. V-1 shows the counterions to be positively adsorbed. Should not their concentration therefore *decrease* on adding the adsorbent to the solution? Explain.

17. The UV-visible absorption spectrum of $Ru(2,2'\text{-bipyridine})_3^{2+}$ has a maximum at about 450 nm, from which the energy in volts for process XI-39 may be estimated. The standard reduction potential for the R^+/R couple is about 1.26 V at 25°C. Estimate from this information (and standard reduction potentials) the potential in volts for processes XI-40 and XI-41. Repeat the calculation for alkaline solutions.

General References

R. Defay and I. Prigogine, *Surface Tension and Adsorption*, translated by D. H. Everett, Wiley, New York, 1966.

W. Eitel, *Silicate Science*, Vol. 1, Academic, New York, 1964.

G. J. Fleer, J. M. H. M. Scheutjens, M. A. Cohen Stuart, T. Cosgrove, and B. Vincent, *Polymers at Interfaces*, Elsevier, 1993.

J. X. Khym, *Analytical Ion-Exchange Procedures in Chemistry and Biology*, Prentice-Hall, Englewood Cliffs, NJ, 1974.

J. J. Kipling, *Adsorption from Solutions of Non-Electrolytes*, Academic, New York, 1965.

E. Lederer and M. Lederer, *Chromatography*, Elsevier, New York, 1955.

G. H. Osborn, *Synthetic Ion Exchangers*, 2nd ed., Chapman and Hall, London, 1961.

G. Schay, in *Surface and Colloid Science*, E. Matijevic, ed., Wiley-Interscience, New York, 1969.

A. Ulman, *An Introduction to Ultrathin Organic Films From Langmuir-Blodgett to Self-Assembly*, Academic, New York, 1991.

H. van Olphen and K. J. Mysels, eds., in *Physical Chemistry: Enriching Topics from Colloid and Surface Science*, Theorex, La Jolla, CA, 1975.

Textual References

1. D. H. Everett, *Pure Appl. Chem.*, **31,** 579 (1972).

2. I. Langmuir, *J. Am. Chem. Soc.*, **40,** 1361 (1918).

3. J. P. Badiali, L. Blum, and M. L. Rosinberg, *Chem. Phys. Lett.*, **129,** 149 (1986).

4. G. J. Fleer, J. M. H. M. Scheutjens, M. A. Cohen Stuart, T. Cosgrove, and B. Vincent, *Polymers at Interfaces*, Elsevier, 1993.

5. S. G. Ash, D. H. Everett, and G. H. Findenegg, *Trans. Faraday Soc.*, **64,** 2645 (1968).

6. P. Somasundaran and D. W. Fuerstenau, *Trans. SME*, **252,** 275 (1972).

7. A. Klinkenberg, *Rec. Trav. Chim.*, **78,** 83 (1959).

8. J. Zeldowitsh, *Acta Physicochim.* (USSR), **1,** 961 (1934).

9. G. Halsey and H. S. Taylor, *J. Chem. Phys.*, **15,** 624 (1947).

10. H. Freundlich, *Colloid and Capillary Chemistry*, Methuen, London, 1926.

11. K. Urano, Y. Koichi, and Y. Nakazawa, *J. Coll. Int. Sci.*, **81,** 477 (1981).

12. M. Greenbank and M. Manes, *J. Phys. Chem.*, **86,** 4216 (1982).

13. R. S. Hansen and J. W. V. Fackler, *J. Phys. Chem.*, **57,** 634 (1953).

14. G. Belfort, *AIChE J.*, **27,** 1021 (1981).

15. A. L. Myers and S. Sircar, *Advances in Chemistry Series*, No. 202, American Chemical Society, Washington, DC, 1983.

16. H. L. Frisch and K. J. Mysels, *J. Phys. Chem.*, **87,** 3988 (1983).

17. K. J. Mysels and H. L. Frisch, *J. Colloid Interface Sci.*, **99,** 136 (1984).

18. R. D. Tilton, C. R. Robertson, and A. P. Gast, *Langmuir*, **7,** 2710 (1991).

19. I. Traube, *Annalen*, **265,** 27 (1891).

20. P. Somasundaran, R. Middleton, and K. V. Viswanathan, ACS Symposium Series, American Chemical Society, 1984, Chapter 17, pp. 269–290, 253.

21. B. Kronberg and P. Stenius, *J. Colloid Interface Sci.* **102,** 410 (1984).

22. E. Tronel-Peyroz, D. Schuhmann, H. Raous, and C. Bertrand, *J. Colloid Interface Sci.*, **97,** 541 (1984).

23. G. Belfort, G. L. Altshuler, K. K. Thallam, C. P. Feerick, and K. L. Woodfield, *AIChE J.*, **30,** 197 (1984).

24. B. von Szyszkowski, *Z. Phys. Chem.*, **64,** 385 (1908).

25. H. P. Meissner and A. S. Michaels, *Ind. Eng. Chem.*, **41,** 2782 (1949).

26. D. L. Allara and R. G. Nuzzo, *Langmuir*, **1,** 45 (1985).

27. C. D. Bain and G. M. Whitesides, *Angew. Chem., Int. Ed. Engl.*, **28,** 506 (1989).

28. G. M. Whitesides and P. E. Laibinis, *Langmuir*, **6,** 87 (1990).

29. A. Ulman, *An Introduction to Ultrathin Organic Films From Langmuir-Blodgett to Self-Assembly*, Academic, San Diego, CA, 1991.

30. J. Gun, R. Iscovici, and J. Sagiv, *J. Colloid Interface Sci.*, **101,** 201 (1984).

31. E. B. Troughton, C. D. Bain, G. M. Whitesides, R. G. Nuzzo, D. L. Allara, and M. D. Porter, *Langmuir*, **4,** 365 (1988).

32. M. D. Porter, T. B. Bright, D. L. Allara, and C. E. D. Chidsey, *J. Am. Chem. Soc.*, **109,** 3559 (1987).

33. N. Tillman, A. Ulman, and T. L. Penner, *Langmuir*, **5,** 101 (1989).

34. F. Sun, D. W. Grainger, and D. G. Castner, *Langmuir*, **9,** 3200 (1993).

35. C. P. Tripp and M. L. Hair, *Langmuir*, **8,** 1120 (1992).

36. S. H. Chen and C. W. Frank, *Langmuir*, **5,** 978 (1989).

37. K. Mathauer and C. W. Frank, *Langmuir*, **9,** 3446 (1993).

38. L. Strong and G. L. Whitesides, *Langmuir*, **4,** 546 (1988).

39. G. Liu, P. Fenter, C. E. D. Chidsey, D. F. Ogletree, P. Eisenberger, and M. Salmeron, *J. Chem. Phys.*, **101,** 4301 (1994).

40. N. Camillone III, C. E. D. Chidsey, G. Liu, and G. Scoles, *J. Chem. Phys.*, **98,** 4234 (1993).

41. C. E. D. Chidsey and D. N. Loiacono, *Langmuir*, **6,** 682 (1990).

42. P. Silberzan, L. Leger, D. Ausserre, and J. J. Benattar, *Langmuir*, **7,** 1647 (1991).

43. T. Cosgrove, C. A. Prestidge, and B. Vincent, *J. Chem. Soc. Faraday Trans.*, **86,** 1377 (1990).

44. W. D. Bascom, *J. Colloid Interface Sci.*, **27,** 789 (1968).

45. C. D. Bain and G. M. Whitesides, *Science*, **240,** 62 (1988).

46. S. R. Wasserman, Y.-T. Tao, and G. M. Whitesides, *Langmuir*, **5,** 1074 (1989).

47. F. Sun, D. W. Grainger, and D. G. Castner, *J. Vac. Sci. Technol.*, **A12**, 2499 (1994).

48. T. M. Putvinski, M. L. Schilling, H. E. Katz, C. E. D. Chidsey, A. M. Mujsce, and A. B. Emerson, *Langmuir*, **6**, 1567 (1990).

49. H. E. Katz, G. Scheller, T. M. Putvinski, M. L. Schilling, W. L. Wilson, and C. E. D. Chidsey, *Science* **254**, 1485 (1991).

50. J. G. Dorsey and K. A. Dill, *Chem. Rev.*, **89**, 331 (1989).

51. J. Y. Yang, K. Mathauer, and C. W. Frank, *Microchemistry*, Elsevier Science, 1994, Chapter 4, pp. 441–454.

52. C. E. D. Chidsey, C. R. Bertozzi, T. M. Putvinski, and A. M. Mujsce, *J. Am. Chem. Soc.*, **112**, 4301 (1990).

53. K. L. Prime and G. L. Whitesides, *Science*, **252**, 1164 (1991).

54. R. S. Hansen, Y. Fu, and F. E. Bartell, *J. Phys. Colloid Chem.*, **53**, 769 (1949).

55. O. M. Dzhigit, A. V. Kiselev, and K. G. Krasilnikov, *Dokl. Akad. Nauk SSSR*, **58**, 413 (1947).

56. F. E. Bartell and D. J. Donahue, *J. Phys. Chem.*, **56**, 665 (1952).

57. G. J. Fleer and J. Lyklema, *Adsorption from Solution at the Solid/Liquid Interface*, Academic Press, Orlando, FL, 1983, Chapter 4, pp. 153–220.

58. E. A. Di Marzio, *Physics of Polymer Surfaces and Interfaces*, Butterworth-Heinemann, London, 1992, Chapter 4, pp. 73–96.

59. H. J. Ploehn and W. B. Russel, *Adv. Chem. Eng.*, **15**, 137 (1990).

60. G. P. van der Beek, M. A. Cohen Stuart, G. J. Fleer, and J. E. Hofman, *Langmuir*, **5**, 1180 (1989).

61. A. Takahashi, M. Kawaguchi, H. Hirota, and T. Kato, *Macromolecules*, **13**, 884 (1980).

62. A. Silberberg, *Pure Appl. Chem.*, **26**, 583 (1971).

63. J. M. H. M. Scheutjens and G. J. Fleer, *J. Phys. Chem.*, **83**, 1619 (1979).

64. J. M. H. M. Scheutjens and G. J. Fleer, *J. Phys. Chem.*, **84**, 178 (1980).

65. H. J. Ploehn and W. B. Russel, *Macromolecules*, **22**, 266 (1989).

66. K. Char, C. W. Frank, and A. P. Gast, *Langmuir*, **5**, 1096 (1989).

67. W. B. Russel, private communication.

68. P. G. de Gennes, *Macromolecules*, **14**, 1637 (1981).

69. P. G. de Gennes and P. Pincus, *J. Physique Lett.*, **44**, L (1983).

70. T. Cosgrove, *J. Chem. Soc. Faraday Trans.*, **86**, 1323 (1990).

71. L. Auvray and P. G. de Gennes, *Europhys. Lett.*, **2**, 647 (1986).

72. P. Auroy, L. Auvray, and L. Léger, *Macromolecules*, **24**, 2523 (1991).

73. T. Cosgrove, T. G. Heath, J. S. Phipps, and R. M. Richardson, *Macromolecules*, **24**, 94 (1991).

74. F. Leermakers and A. Gast, *Macromolecules*, **24**, 718 (1991).

75. K. G. Barnett, T. Cosgrove, D. S. Sissons, M. A. Cohen Stuart, and B. Vincent, *Macromolecules*, **14**, 1018 (1981).

76. M. Kawaguchi, S. Yamagiwa, A. Takahashi, and T. Kato, *J. Chem. Soc., Faraday Trans.*, **86**, 1383 (1990).

77. H. Sakai and Y. Imamura, *Bull. Chem. Soc. Jpn.*, **53**, 1749 (1980).

78. J. J. Lee and G. G. Fuller, *Macromolecules*, **17**, 375 (1984).

79. M. Kawaguchi and A. Takahashi, *Macromolecules*, **16**, 1465 (1983).

80. M. A. Cohen Stuart, F. H. W. H. Waajen, T. Cosgrove, B. Vincent, and T. L. Crowley, *Macromolecules*, **17**, 1825 (1984).

81. F. Lafuma, K. Wong, and B. Cabane, *J. Colloid Interface Sci.*, **143**, 9 (1991).

82. P. F. McKenzie, V. Kapur, and J. L. Anderson, *Colloids & Surfaces*, **A86**, 263 (1994).

83. M. A. Cohen Stuart, *Polym. J.*, **23**, 669 (1991).

84. M. R. Munch and A. P. Gast, *J. Chem. Soc. Faraday Trans.*, **86**, 1341 (1990).

85. J. C. Dijt, M. A. Cohen Stuart, and G. J. Fleer, *Macromolecules*, **27**, 3207 (1994).

86. S. Granick, *Physics of Polymer Surfaces and Interfaces*, Butterworth-Heinemann, London, 1992, Chapter 10, pp. 227–244.

87. M. A. Cohen Stuart, *J. Phys. France*, **49**, 1001 (1988).

88. P. Berndt, K. Kurihara, and T. Kunitake, *Langmuir*, **8**, 2486 (1992).

89. S. Biggs and T. W. Healy, *J. Chem. Soc., Faraday Trans.*, **90**, 3415 (1994).

90. M. Trau, F. Grieser, T. W. Healy, and L. R. White, *J. Chem. Soc., Faraday Trans.*, **90**, 1251 (1994).

91. M. A. G. Dahlgren, P. M. Claesson, and R. Audebert, *J. Colloid Interface Sci.*, **166**, 343 (1994).

92. M. A. G. Dahlgren, *Langmuir*, **10**, 1580 (1994).

93. M. A. G. Dahlgren, A. Waltermo, E. Blomberg, L. Sjöström, T. B. Jönsson, and P. M. Claesson, *J. Phys. Chem.*, **97**, 11769 (1993).

94. P. G. de Gennes, *Macromolecules*, **13**, 1069 (1980).

95. S. T. Milner, T. A. Witten, and M. E. Cates, *Macromolecules*, **21**, 2610 (1988).

96. S. T. Milner, *Science*, **251**, 905 (1991).

97. A. Halperin, M. Tirrell, and T. P. Lodge, *Adv. Polym. Sci.*, **100**, 31 (1992).

98. P. F. McKenzie, R. M. Webber, and J. L. Anderson, *Langmuir*, **10**, 1539 (1994).

99. S. Weber, editor, *Solvent and Polymer Self-Organization*, NATO ASI, Kluwer, 1995.

100. M. Tirrell, S. Patel, and G. Hadziioannou, *Proc. Natl. Acad. Sci. (USA)*, **84**, 4725 (1987).

101. J. Lyklema, *Colloids & Surfaces*, **10**, 33 (1984).

102. P. M. Claesson, E. Blomberg, J. C. Fröberg, T. Nylander, and T. Arnebrandt, *Adv. Colloid Interface Sci.*, **57**, 161 (1995).

103. J. D. Andrade, *Surface and Interfacial Aspects of Biomedical Polymers*, Vol. 2, *Protein Adsorption*, Plenum, 1985.

104. K. A. Dill, S. Bromberg, K. Yue, K. M. Fiebig, D. P. Yee, P. D. Thomas, and H. S. Chan, *Protein Sci.*, **4**, 561 (1995).

105. W. Norde and J. Lyklema, *J. Colloid Interface Sci.*, **66**, 257 (1978).

106. R. D. Tilton, C. R. Robertson, and A. P. Gast, *J. Colloid Interface Sci.*, **137**, 192 (1990).

107. U. Jönsson, I. Lundström, and I. Rönnberg, *J. Colloid Interface Sci.*, **117**, 127 (1987).

108. G. T. Taylor, P. J. Troy, and S. K. Sharma, *Marine Chem.*, **45,** 15 (1994).

109. B. D. Ratner, T. A. Horbett, D. Shuttleworth, and H. R. Thomas, *J. Colloid Interface Sci.*, **83,** 630 (1981).

110. Y. L. Chen, S. A. Darst, and C. R. Robertson, *J. Colloid Interface Sci.*, **118,** 212 (1987).

111. J. Feder and I. Giaever, *J. Colloid Interface Sci.*, **78,** 144 (1980).

112. J. M. Peula, J. Callejas, and F. J. Nieves, *Surf. Properties Biomat.* 61 (1992).

113. J. L. Ortega-Vinuesa, A. Martin Rodriguez, and R. Hidalgo-Alvarez, *Colloids & Surfaces*, **A95,** 216 (1995).

114. W. Norde and E. Rouwendal, *J. Colloid Interface Sci.*, **139,** 169 (1990).

115. M. A. Bos and J. M. Kleijn, *Biophys. J.*, **68,** 2573 (1995).

116. J. Stahlberg, B. Jonsson, and C. Horvath, *Anal. Chem.*, **64,** 3118 (1992).

117. J. L. Ortega-Vinuesa and R. Hidalgo-Alvarez, *Colloids & Surfaces B: Biointerfaces*, **80,** 365 (1993).

118. C. J. van Oss, W. Wu, R. F. Giese, and J. O. Naim, *Colloids & Surfaces B*, **4,** 185 (1995).

119. P. F. Brode III, C. R. Erwin, D. S. Rauch, D. S. Lucas, and D. N. Rubingh, *J. Biol. Chem.*, **269,** 23538 (1994).

120. P. B. Gaspers, A. P. Gast, and C. R. Robertson, *J. Colloid Interface Sci.*, **172,** 518 (1995).

121. G. Penners, Z. Priel, and A. Silberberg, *J. Colloid Interface Sci.*, **80,** 437 (1981).

122. W. Norde, F. MacRitchie, G. Nowicka, and J. Lyklema, *J. Colloid Interface Sci.*, **112,** 447 (1986).

123. H. G. de Bruin, C. J. van Oss, and D. R. Assolom, *J. Colloid Interface Sci.*, **76** (1980).

124. M. E. Zawadski, Y. Harel, and A. W. Adamson, *Langmuir*, **3,** 363 (1987).

125. P. J. Dobson and P. Somasundaran, *J. Coll. Interface Sci.*, **97,** 481 (1984).

126. D. N. Misra, *J. Dental Res.*, **65,** 706 (1986).

127. J. Tamura, J. T. Tse, and A. W. Adamson, *J. Jpn. Petrol. Inst.*, **27,** 385 (1984).

128. M. E. Zawadski and A. W. Adamson, *Fundamentals of Adsorption*, Engineering Foundation, New York, 1987, p. 619.

129. M. E. Soderquist and A. G. Walton, *J. Colloid Interface Sci.*, **76,** 386 (1980).

130. C. Peterson and T. K. Kwei, *J. Phys. Chem.*, **35,** 1330 (1961).

131. F. Rowland, R. Bulas, E. Rothstein, and F. R. Eirich, *Ind. Eng. Chem.*, **57,** 46 (1965).

132. A. Y. Meyer, D. Farin, and D. Avnir, *J. Am. Chem. Soc.*, **108,** 7897 (1986).

133. J. F. Padday, *Pure and Applied Chemistry, Surface Area Determination*, Butterworths, London, 1969.

134. S. S. Barton, *Carbon*, **25,** 343 (1987).

135. M. A. Rahman and A. K. Ghosh, *J. Colloid Interface Sci.*, **77,** 50 (1980).

136. R. J. Pugh, *Min. Metallurg. Proc.*, 151 (1992).

137. H. J. Van den Hul and J. Lyklema, *J. Colloid Interface Sci.*, **23,** 500 (1967).

138. D. H. Everett, *Trans. Faraday Soc.*, **61,** 2478 (1965).

139. G. Schay and L. G. Nagy, *J. Colloid Interface Sci.*, **38**, 302 (1972).

140. J. J. Kipling, *Q. Rev.* (London), **5**, 60 (1951).

141. S. G. Ash, R. Bown, and D. H. Everett, *J. Chem. Thermodyn.*, **5**, 239 (1973).

142. K. M. Phillips and J. P. Wightman, *J. Colloid Interface Sci.*, **108**, 495 (1985).

143. J. J. Kipling and D. A. Tester, *J. Chem. Soc.*, **1952**, 4123.

144. A. L. Myers and S. Sircar, *J. Phys. Chem.*, **76**, 3415 (1972).

145. G. D. Parfitt and P. C. Thompson, *Trans. Faraday Soc.*, **67**, 3372 (1971).

146. G. Schay, *Surf. Colloid Sci.*, **2**, 155 (1969); *J. Colloid Interface Sci.*, **42**, 478 (1973).

147. A. Dabrowski, M. Jaroniec, and J. Oscik, *Ads. Sci. Tech.*, **3**, 221 (1986).

148. S. Sircar, *J. Chem. Soc., Faraday Trans. I*, **82**, 831 (1986).

149. D. Schiby and E. Ruckenstein, *Colloids & Surfaces*, **15**, 17 (1985).

150. S. Sircar, *J. Chem. Soc., Faraday Trans. I*, **82**, 843 (1986); *Langmuir*, **3**, 369 (1987).

151. J. J. Kipling, *Proc. Int. Congr. Surf. Act., 3rd, Mainz*, **1960**, Vol. 2, p. 77.

152. G. J. Young, J. J. Chessick, and F. H. Healey, *J. Phys. Chem.*, **60**, 394 (1956).

153. W. A. Schroeder, *J. Am. Chem. Soc.*, **73**, 1122 (1951).

154. S. Ozawa, M. Goto, K. Kimura, and Y. Ogino, *J. Chem. Soc., Faraday Trans. I*, **80**, 1049 (1984); S. Ozawa, K. Kawahara, M. Yamabe, H. Unno, and Y. Ogino, *ibid.*, **80**, 1059 (1984).

155. A. L. Myers, *Langmuir*, **3**, 121 (1987).

156. R. O. James, in *Polymer Colloids*, R. Buscall, T. Corner, and J. F. Stageman, eds., Elsevier Applied Science, New York, 1985.

157. H. B. Weiser, *Colloid Chemistry*, Wiley, New York, 1950.

158. See K. Fajans, *Radio Elements and Isotopes. Chemical Forces and Optical Properties of Substances*, McGraw-Hill, New York, 1931.

159. I. M. Kolthoff and R. C. Bowers, *J. Am. Chem. Soc.*, **76**, 1503 (1954).

160. A. De Keizer, L. G. J. Fokkink, and J. Lyklema, *Colloids & Surfaces*, **49**, 149 (1990).

161. G. F. Cerofolini and G. Boara, *J. Colloid Interface Sci.*, **161**, 232 (1993).

162. A. M. Posner and J. P. Quirk, *J. Colloid Sci.*, **19**, 798 (1964).

163. F. F. Aplan and D. W. Fuerstenau, in *Froth Flotation*, D. W. Fuerstenau, ed., American Institute of Mining and Metallurgical Engineering, New York, 1962.

164. W. D. Bancroft, *J. Phys. Chem.*, **19**, 363 (1915).

165. S. Kondo, T. Tamaki, and Y. Ozeki, *Langmuir*, **3**, 349 (1987).

166. D. W. Fuerstenau, *Chem. Biosurfaces*, **1**, 143 (1971).

167. W. Paik, M. A. Genshaw, and J. O'M. Bockris, *J. Phys. Chem.*, **74**, 4266 (1970).

168. M. Kawaguchi, K. Hayashi, and A. Takahashi, *Colloids & Surfaces*, **31**, 73 (1988).

169. S. Biggs, P. Mulvaney, C. F. Zukoski, and F. Grieser, *J. Am. Chem. Soc.*, **116**, 9150 (1994).

170. J. C. Berg, *Absorbency*, Elsevier Science, Amsterdam, 1985, Chapter V, pp. 149–195.

171. J. F. Scamehorn, R. S. Schechter, and W. H. Wade, *J. Colloid Interface Sci.*, **85,** 463 (1982).

172. W. Peng, D.-L. Zhou, and J. F. Rusling, *J. Phys. Chem.*, **99,** 6986 (1995).

173. B. S. Kim, R. A. Hayes, and J. Ralston, *Carbon*, **33,** 25 (1995).

174. J. Lyklema, *Progr. Colloid Polym. Sci.*, **95,** 91 (1994).

175. L. K. Koopal, E. M. Lee, and M. R. Bohmer, *J. Colloid Interface Sci.*, **170,** 85 (1995).

176. P. Conner and R. H. Ottewill, *J. Colloid Interface Sci.*, **37,** 642 (1971).

177. S. G. Dick, D. W. Fuerstenau, and T. W. Healy, *J. Colloid Interface Sci.*, **37,** 595 (1971).

178. S. Manne, J. P. Cleveland, H. E. Gaub, G. D. Stucky, and P. K. Hansma, *Langmuir*, **10,** 4409 (1994).

179. P. J. Scales, F. Grieser, T. W. Healy, and L. J. Magid, *Langmuir*, **8,** 277 (1992).

180. J. L. Parker, V. V. Yaminsky, and P. M. Claesson, *J. Phys. Chem.*, **97,** 7706 (1993).

181. M. W. Rutland and J. L. Parker, *Langmuir*, **10,** 1110 (1994).

182. J. L. Parker, *Prog. Surf. Sci.*, **47,** 205 (1994).

183. P. Somasundaran, T. W. Healy, and D. W. Fuerstenau, *J. Phys. Chem.*, **68,** 3562 (1964).

184. G. H. Bolt and R. D. Miller, *Soil Sci. Am. Proc.*, **19,** 285 (1955); A. V. Blackmore and R. D. Miller, ibid., **25,** 169 (1961). H. van Olphen, *J. Phys. Chem.*, **61,** 1276 (1957).

185. R. M. Barrer and R. M. Gibbons, *Trans. Faraday Soc.*, **59,** 2569 (1963), and preceding references.

186. G. L. Gaines, Jr., and H. C. Thomas, *J. Phys. Chem.*, **23,** 2322 (1955).

187. K. W. T. Goulding and O. Talibudeen, *J. Colloid Interface Sci.*, **78,** 15 (1980).

188. R. M. Barrer and R. P. Townsend, *Zeolites*, **5,** 287 (1985); *J. Chem. Soc., Faraday Trans. 2*, **80,** 629 (1984).

189. P. C. Herder, Per M. Claesson, and C. E. Blom, *J. Colloid Interface Sci.*, **119,** 155 (1987); P. M. Claesson, P. Herder, P. Stenius, J. C. Eriksson, and R. M. Pashley, ibid., **109,** 31 (1986).

190. G. E. Boyd, A. W. Adamson, and L. S. Myers, Jr., *J. Am. Chem. Soc.*, **69,** 2836 (1947).

191. J. J. Grossman and A. W. Adamson, *J. Phys. Chem.*, **56,** 97 (1952).

192. R. Schlögl and F. Helfferich, *J. Chem. Phys.*, **26,** 5 (1957).

193. J. Stahlberg, *J. Chromatogr.*, **356,** 231 (1986).

194. J. Stahlberg, *Anal. Chem.*, **66,** 440 (1994).

195. J. A. Marinsky and Y. Marcus, eds., *Ion Exchange and Solvent Extraction*, Marcel Dekker, New York, 1973.

196. R. J. Hunter, *Foundations of Colloid Science*, Vol. 1, Clarendon Press, Oxford, 1987.

197. K. Viaene, J. Caigui, R. A. Schoonheydt, and F. C. De Schrijver, *Langmuir*, **3,** 107 (1987).

198. R. A. Schoonheydt, P. De Pauw, D. Vliers, and F. C. De Schrijver, *J. Phys. Chem.*, **88,** 5113 (1984).

199. J. K. Thomas, *J. Phys. Chem.*, **91,** 267 (1987).

200. P. de Mayo, L. V. Natarajan, and W. R. Ware, *J. Phys. Chem.*, **89,** 3526 (1985).

201. C. Francis, J. Lin, and L. A. Singer, *Chem. Phys. Lett.*, **94,** 162 (1983).

202. J. Stahlberg, M. Almgren, and J. Alsins, *Anal. Chem.*, **60,** 2487 (1988).

203. R. K. Bauer, P. de Mayo, W. R. Ware, and K. C. Wu, *J. Phys. Chem.*, **86,** 3781 (1982).

204. J. Stahlberg and M. Almgren, *Anal. Chem.*, **57,** 817 (1985).

205. M. F. Ahmadi and J. F. Rusling, *Langmuir*, **11,** 94 (1995).

206. P. Levitz and H. Van Damme, *J. Phys. Chem.*, **90,** 1302 (1986).

207. P Chandar, P. Somasundaran, and N. J. Turro, *J. Colloid Interface Sci.*, **117,** 31 (1987).

208. M. F. Ahmadi and J. F. Rusling, *Langmuir*, **7,** 1529 (1991).

209. W. Shi, S. Wolfgang, T. C. Strekas, and H. D. Gafney, *J. Phys. Chem.*, **89,** 974 (1985).

210. D. Pines, D. Huppert, and D. Avnir, *J. Phys. Chem.*, **89,** 1177 (1988).

211. D. Avnir, *J. Am. Chem. Soc.*, **109,** 2931 (1987).

212. D. Gafney and A. W. Adamson, *J. Am. Chem. Soc.*, **94,** 8238 (1972); J. N. Demas and A. W. Adamson, ibid., **95,** 5159 (1973).

213. J. Kiwi, E. Borgarello, E. Pelizzetti, M. Visca, and M. Grätzel, in *Photogeneration of Hydrogen*, A. Harriman and M. A. West, eds., Academic, New York, 1982.

214. K. Kalyanasundaram and M. Grätzel, *Coord. Chem. Rev.*, **69,** 57 (1986).

215. A. Heller, Y. Degani, D. W. Johnson, Jr., and P. K. Gallagher, *J. Phys. Chem.*, **91,** 5987 (1987).

216. J. K. Thomas and J. Wheeler, *J. Photochem.*, **28,** 285 (1985).

217. E. P. Giannelis, D. G. Norcera, and T. J. Pinnavaia, *Inorg. Chem.*, **26,** 203 (1987).

218. A. J. Frank, I. Willner, Z. Goren, and Y. Degami, *J. Am. Chem. Soc.*, **109,** 3568 (1987).

219. S. Abdo, P. Canesson, M. Cruz, J. J. Fripiat, and H. Van Damme, *J. Phys. Chem.*, **85,** 797 (1981).

220. R. C. Simon, E. A. Mendoza, and H. D. Gafney, *Inorg. Chem.*, **27,** 2733 (1988); W. Shi and H. D. Gafney, *J. Am. Chem. Soc.*, **109,** 1582 (1987).

221. T. Nakamura and J. K. Thomas, *Langmuir*, **1,** 568 (1985).

222. J. Kiwi and M. Grätzel, *J. Phys. Chem.*, **90,** 637 (1986).

223. K. Yamaguti and S. Sato, *J. Chem. Soc., Faraday Trans. I*, **81,** 1237 (1985).

224. J. Kiwi and M. Grätzel, *J. Chem. Soc., Faraday Trans. I*, **83,** 1101 (1987).

225. A. Sobczynski, A. J. Bard, A. Campion, M. A. Fox, T. Mallouk, S. E. Webber, and J. M. White, *J. Phys. Chem.*, **91,** 3316 (1987).

226. L. Spanhel, M. Haase, H. Weller, and A. Henglein, *J. Am. Chem. Soc.*, **109,** 5649 (1987).

227. T. Hasegawa and P. de Mayo, *Langmuir*, **2,** 362 (1986).

228. J. Kuczynski and J. K. Thomas, *Langmuir* **1,** 158 (1985).

229. N. J. Turro, *Tetrahedron*, **43,** 1589 (1987).

230. N. J. Turro, C. Cheng, L. Abrams, and D. R. Corbin, *J. Am. Chem. Soc.*, **109,** 2449 (1987).

231. R. Farwaha, P. de Mayo, and Y. C. Toong, *J. Chem. Soc., Chem. Commun.*, 739 (1983).

232. D. K. Liu, M. S. Wrighton, D. R. McKay, and G. E. Maciel, *Inorg. Chem.*, **23,** 212 (1984).

233. I. Willner and Y. Degan, *Israel J. Chem.*, **22,** 163 (1982).

234. Y. Degani and I. Willner, *J. Chem. Soc., Perkin Trans. II*, 37 (1986).

235. I. Willner, ACS Symposium Series No. 278, M. A. Fox, ed., American Chemical Society, Washington, DC, 1985.

236. R. Maidan and I. Willner, *J. Am. Chem. Soc.*, **108,** 8100 (1986).

237. J. C. Abram and G. D. Parfitt, *Proc. 5th Conf. Carbon*, Pergamon, New York, 1962, p. 97.

CHAPTER XII

Friction, Lubrication, and Adhesion

1. Introduction

This chapter and the two that follow are introduced at this time to illustrate some of the many extensive areas in which there are important applications of surface chemistry. Friction and lubrication as topics properly deserve mention in a textbook on surface chemistry, partly because these subjects do involve surfaces directly and partly because many aspects of lubrication depend on the properties of surface films. The subject of adhesion is treated briefly in this chapter mainly because it, too, depends greatly on the behavior of surface films at a solid interface and also because friction and adhesion have some interrelations. Studies of the interaction between two solid surfaces, with or without an intervening liquid phase, have been stimulated in recent years by the development of equipment capable of the direct measurement of the forces between macroscopic bodies.

Friction can now be probed at the atomic scale by means of atomic force microscopy (AFM) (see Section VIII-2) and the surface forces apparatus (see Section VI-4); these approaches are leading to new interpretations of friction [1,1a,1b]. The subject of friction and its related aspects are known as *tribology*, the study of surfaces in relative motion, from the Greek root *tribos* meaning rubbing.

2. Friction between Unlubricated Surfaces

A. Amontons' Law

The coefficient of friction μ between two solids is defined as F/W, where F denotes the frictional force and W is the load or force normal to the surfaces, as illustrated in Fig. XII-1. There is a very simple law concerning the coefficient of friction μ, which is amazingly well obeyed. This law, known as *Amontons' law*, states that μ is independent of the apparent area of contact; it means that, as shown in the figure, with the same load W the frictional forces will be the same for a small sliding block as for a large one. A corollary is that μ is independent of load. Thus if $W_1 = W_2$, then $F_1 = F_2$.

Although friction between objects is a matter of everyday experience, it is curious that Amontons' law, although of fairly good general validity, seems

Fig. XII-1. Amontons' law.

quite contrary to intuitive thinking. Indeed, when Amontons, a French army engineer, presented his findings to the Royal Academy in 1699 (*Mem. Acad. Roy. Sci.*, **1699**, 206), they were met with some skepticism. Amontons anticipated some of this in his paper and attempted to explain his findings by qualitative reasoning. As is often the case, however, it is the law as given by the data that has stood up over the course of time rather than the associated explanations. Amontons was thinking partly of friction as due to work associated with vertical motions as a result of surface roughness and, as is shown, the actual explanation (or at least the one now accepted) is quite different. Amontons went on to assert that the coefficient of friction was always $\frac{1}{3}$; this is, indeed, about the usual value but, as discussed in Section XII-4, the frictional coefficient between really clean surfaces can be much larger; in the case of some plastics, it may be much smaller.

The basic law of friction has been known for some time. Amontons was, in fact, preceded by Leonardo da Vinci, whose notebook illustrates with sketches that the coefficient of friction is independent of the apparent area of contact (see Refs. 2 and 3). It is only relatively recently, however, that the probably correct explanation has become generally accepted.

B. Nature of the Contact between Two Solid Surfaces

There is abundant evidence that even very smooth-appearing surfaces are irregular on a molecular scale. The nature of such surface irregularities may be studied by electron microscopy or scanning probe microscopies (see Section VII-2). These imaging techniques provide a very direct appreciation of the presence of imperfections, steps and so forth, on a surface (see Section VII-4), as well as of grooves left by one surface sliding over another. Low-energy electron diffraction and field emission techniques have also provided much information about surface structures (see Section VII-3).

The results of such studies have made it clear that the surfaces of crystalline materials may have quite irregular steps of hundreds or thousands of angstroms in depth and that lapped or ground surfaces may be quite jagged on this length scale. Even polished metal surfaces are not really smooth. As discussed in Section VII-1B, polishing appears to bring about local melting, and the effect is that ragged asperities are smoothed out; but even so, the resulting surface has a waved appearance. In the case of metals, it is also apt to have an oxide coating [4].

As a result of the irregular nature of even the smoothest surfaces available, two surfaces brought into contact will touch only in isolated regions. In fact,

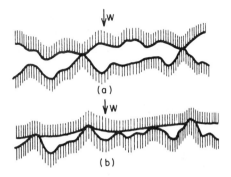

Fig. XII-2. Comparison of actual contact areas for (*a*) metal-on-metal and (*b*) plastic-on-metal. (From Ref. 2.)

on initial contact, one would expect to have only three points of touching, but even for minute total loads, the pressure at these points would be sufficient to cause deformation leading to multiple contacts. Actual contact regions might appear as in Fig. XII-2. Case *a* in the figure applies to metal-metal contacts where the local pressure is high enough to cause plastic flow of both metals, with local welding. Case *b* illustrates the situation if a soft material, such as a plastic, is pressed against a hard one, such as a metal.

The true area of contact is clearly much less than the apparent area. The former can be estimated directly from the resistance of two metals in contact. It may also be calculated if the statistical surface profiles are known from roughness measurements. As an example, the true area of contact, *A*, is about 0.01% of the apparent area in the case of two steel surfaces under a 10-kg load [4a].

Another indication that friction is confined to a real area that is much smaller than the apparent area is in the very high local temperatures that can prevail during a sliding motion. These can be estimated by rubbing a slider of one metal over the surface of another one and noting the potential difference that develops [4a]. An alternative procedure that has been used when one of the surfaces is transparent, such as glass, is to view the local hot spots with an infrared-sensitive cell and estimate the local temperature from the spectrum of the emitted infrared light [5]. Local temperature increases of hundreds of degrees have been recorded by these means; the final limitation may be the melting point of the lower melting material. In the case of plastics, microscopic examination of a cross section along the track of the slider reveals that melting has occurred to some depth. (See also Ref. 6.)

In summary, it has become quite clear that contact between two surfaces is limited to a small fraction of the apparent area, and, as one consequence of this, rather high local temperatures can develop during rubbing. Another consequence, discussed in more detail later, is that there are also rather high local pressures. Finally, there is direct evidence [7,8] that the two surfaces do not remain intact when sliding past each other. Microscopic examination of the track left by the slider shows gouges and irregular pits left in the softer metal by the harder ones. Similarly, a radioautograph obtained by pressing a photographic emulsion against the harder metal shows spots of darkening corresponding to specks of the radioactively labeled softer metal. This type of evidence is interpreted to mean that fairly strong seizure or actual welding occurs at the sites of contact.

C. Role of Shearing and Plowing—Explanation of Amontons' Law

As two surfaces are brought together, the pressure is extremely large at the initial few points of contact, and deformation immediately occurs to allow more and more to develop. This plastic flow continues until there is a total area of contact such that the local pressure has fallen to a characteristic yield pressure P_m of the softer material.

Normally, then, the actual contact area is determined by the yield pressure, so that

$$A = \frac{W}{P_m} \qquad \qquad \text{(XII-1)}$$

For most metals P_m values are in the range of 10 to 100 kg/mm^2. Thus in a friction experiment with a load of 10 kg, the true contact area would indeed be about the 10^{-3} cm^2 estimated from conductivity measurements. There are considerable uncertainties in the absolute values of A calculated in this way, since work hardening may increase P_m at the contact points. Nevertheless, the relationship given by Eq. XII-1 should still apply. For very light loads, such that the elastic limit is *not* exceeded, a hemispherical rider on a flat surface will have an area of contact, $A = kW^{2/3}$ that is no longer directly proportional to the load. If there are numerous asperities following a Gaussian distribution, A is again proportional to W [1].

In a typical friction measurement, a slider is pressed against a stationary block, and the force required to move it is measured. This force generally comprises two terms; the first is the force F required to shear the junctions at the points of actual contact given by

$$F = As_m \qquad \qquad \text{(XII-2)}$$

where s_m is the shear strength per unit area. With materials of different hardness, a harder slider will plow a track in the softer material and the second term, F', is the force associated with this plowing action. The force required to displace the softer material from the front of the slider is proportional to the cross section of the slider (see Fig. XII-3)

$$F' = kA' \qquad \qquad \text{(XII-3)}$$

where A' denotes the cross section of the plowed track.

The plowing contribution can be estimated by employing a rider that is very

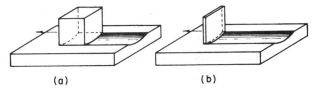

Fig. XII-3. Illustration of (a) shearing and (b) plowing actions.

thin, Fig. XII-3b, so that the shear contribution is minimized. Usually the plowing term is important only for the case of a hard material rubbing against a soft one; if both are hard, friction is due mostly to the shear term. As an approximation, then, A may be eliminated from Eqs. XII-1 and XII-2 to give

$$F = W \frac{s_m}{\mathbf{P}_m} \qquad \text{(XII-4)}$$

or

$$\mu = \frac{s_m}{\mathbf{P}_m} = \text{const} \qquad \text{(XII-5)}$$

This is Amontons' law as stated in the opening paragraphs of this chapter.

Another point in connection with Eq. XII-5 is that both the yielding and the shear will involve mainly the softer material, so that μ is given by a ratio of properties of the same substance. This ratio should be nearly independent of the nature of the metal itself since s_m and $\mathbf{P}_m$ tend to vary together in agreement with the observation that for most frictional situations, the coefficient of friction lies between about 0.5 and 1.0. Also, temperature should not have much effect on μ, as is observed.

As suggested by the preceding analysis, Amontons' law does, indeed, hold well if neither of the two surfaces is too soft. Bowden and Tabor [4a] cite data for the coefficient of friction of aluminum on aluminum giving a nearly constant value of μ for loads ranging from 0.037 to 4000 g. Friction measurements with the atomic force microscope (termed *friction force microscopy* [FFM]—see Section VIII-2) use the AFM tip as a single asperity to measure the load and friction simultaneously [9]. These microscopic measurements of friction can provide a map of the friction properties of a surface. In a remarkable example of Amontons' law, Mate found [9] that the friction on a carbon coated magnetic disk is the same, $\mu = 0.15$, with a tungsten FFM (friction force microscope) tip at a speed of 0.5 μm/sec and a load of nanonewtons as that measured with a conventional disk slider carrying a load of grams at 10 cm/sec. The case of polymers is not so simple, however, in that plastic deformation tends to be fairly important in determining A, as discussed in Section XII-5C.

The coefficient of friction may also depend on the relative velocity of the two surfaces. This will, for example, affect the local temperature, the extent of work hardening of metals, and the relative importance of the plowing and shearing terms. These facts work out such that the coefficient of friction tends to decrease with increasing sliding speeds [3,10], contrary to what is sometimes known as *Coulomb's law*, which holds that μ should be independent of sliding velocity. At the very low speeds commonly used, about 0.01 cm/sec, the effect is small and so is not usually involved in discussions of friction (see the next section, however).

D. Static and Stick-Slip Friction

In general it is necessary to distinguish static and kinetic coefficients of friction, μ_s and μ_k. The former is given by the force needed to initiate motion, and the latter, by the force to maintain a given sliding speed. There is a general argument to show that $\mu_s > \mu_k$. Suppose that μ_s were less than μ_k, then, for a load W, $F_s = W\mu_s$, and $F_k = W\mu_k$, that is, $F_s < F_k$. Now if a force were applied that lay between these two values, a paradoxical situation would exist. Since F would be greater than F_s, the rider *should* move, but since F is less than F_k, the rider *should not* move: the contradiction is inescapable and leads to the conclusion that the static coefficient must be equal to or greater than the sliding or kinetic coefficient. Note Problem XII-8 for a practical application of this point.

The static coefficient of friction tends to increase with time of contact, and it has been argued that μ_s for zero time should be equal to μ_k. Actually, μ_k is a function of sliding speed and appears to go through a maximum with increasing speed. This maximum occurs below 10^{-8} cm/sec for titanium, at about 10^{-4} cm/sec for indium, and at about 1 cm/sec for the very soft plastic, Teflon [13]. Thus most materials show a negative dependence of μ_k on sliding speeds in the usual range of 0.1 cm/sec. It might be noted that most junctions caused by the seizure of asperities seem to be about 0.001 cm in size so that it is necessary to traverse about this distance to obtain a meaningful value of μ_k. A practical limit is thus set on experiments at very small sliding speeds.

A phenomenon that may develop, especially at small sliding speeds, is known as *stick-slip friction* in which the slider moves in jumps that may be of quite high frequency. This effect is partly a consequence of having some play and lack of rigidity in the mechanism holding the slider. Once momentary sticking occurs, the slider is pushed back against the elastic restoring force of the holding mechanism as a result of the continued motion of the latter. When the restoring force exceeds that corresponding to μ_s, the slider moves forward rapidly, overshoots if there is sufficient play, and sticks again. The greater the difference between static and sliding frictional coefficients, the more prone is the system to stick–slip friction; with lubricated surfaces, there may be a fairly well defined temperature above which the phonomenon is observed. Friction force microscopy on smooth graphite surfaces illustrates molecular stick–slip friction [11]. Slip occurs when the derivative of the periodic force equals the spring constant of the AFM. Stick–slip friction is encouraged if μ_k *decreases* with sliding speed (why?) and with increasing W [12]. A very common, classroom illustration of stick–slip friction is, incidentally, the squeaking of chalk on a chalkboard. An important and interesting feature of current models of earthquakes is stick–slip motion of the tectonic plates.

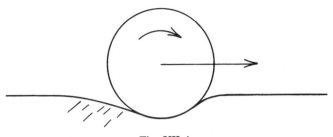

Fig. XII-4.

E. Rolling Friction

The slowing down of a wheel or sphere rolling on a surface is not, properly speaking, a frictional effect since there is no sliding of one surface against another. The effect, however, is as though there were a frictional force, and should therefore be mentioned. As illustrated in Fig. XII-4, a hard sphere rolling on a soft material travels in a moving depression. The material is compressed in front and rebounds at the rear; and were it perfectly elastic, the energy stored in compression would be returned to the sphere at its rear. Actual materials are not perfectly elastic, however, so energy dissipation occurs, with the result that kinetic energy of the rolling is converted into heat.

3. Two Special Cases of Friction

A. Use of Skid Marks to Estimate Vehicle Speeds

An interesting and very practical application of Amontons' law occurs in the calculation of the minimum speed of a vehicle from the length of its skid marks.

In the absence of skidding, the coefficient of *static* friction applies; at each instant, the portion of the tire that is in contact with the pavement has zero velocity. Rolling tire "friction" is more of the type discussed in Section XII-2E. If, however, skidding occurs, then since rubber is the softer material, the coefficient of friction as given by Eq. XII-5 is determined mainly by the properties of the rubber used and will be nearly the same for various types of pavement. Actual values of μ turn out to be about unity.

If μ is taken to be a constant during a skid, application of Amontons' law leads to a very simple relationship between the initial velocity of the vehicle and the length of the skid mark. The initial kinetic energy is $mv^2/2$, and this is to be entirely dissipated by the braking action, which amounts to a force F applied over the skid distance d. By Amontons' law,

$$F = \mu W = mg\mu \qquad \text{(XII-6)}$$

Then

$$\frac{mv^2}{2} = Fd = mg\mu d \qquad\qquad (\text{XII-7})$$

or

$$v = (2d\mu g)^{1/2} \qquad\qquad (\text{XII-8})$$

Thus if Amontons' law is obeyed, the initial velocity is determined entirely by the coefficient of friction and the length of the skid marks. The mass of the vehicle is not involved, neither is the size or width of the tire treads, nor how hard the brakes were applied, so long as the application is sufficient to maintain skidding.

The situation is complicated, however, because some of the drag on a skidding tire is due to the elastic hysteresis effect discussed in Section XII-2E. That is, asperities in the road surface produce a traveling depression in the tire with energy loss due to imperfect elasticity of the tire material. In fact, tires made of high-elastic hysteresis material will tend to show superior skid resistance and coefficient of friction.

As might be expected, this simple picture does not hold perfectly. The coefficient of friction tends to increase with increasing velocity and also is smaller if the pavement is wet [14]. On a wet road, μ may be as small as 0.2, and, in fact, one of the principal reasons for patterning the tread and sides of the tire is to prevent the confinement of a water layer between the tire and the road surface. Similarly, the texture of the road surface is important to the wet friction behavior. Properly applied, however, measurements of skid length provide a conservative estimate of the speed of the vehicle when the brakes are first applied, and it has become a routine matter for data of this kind to be obtained at the scene of a serious accident.

2. Ice and Snow

The coefficient of friction between two unlubricated solids is generally in the range of 0.5–1.0, and it has therefore been a matter of considerable interest that very low values, around 0.03, pertain to objects sliding on ice or snow. The first explanation, proposed by Reynolds in 1901, was that the local pressure caused melting, so that a thin film of water was present. Qualitatively, this explanation is supported by the observation that the coefficient of friction rises rapidly as the temperature falls, especially below about $-10°C$, if the sliding speed is small. Moreover, there is little doubt that formation of a water film is actually involved [3,4].

Although the theory of pressure melting is attractive, certain difficulties develop when it is given more quantitative consideration. Thus if the freezing point is to be lowered only to $-10°C$, a local pressure of about 1000 kg/cm^2 must prevail, corresponding to an actual contact area of a few hundredths of a square centimeter for a person of average weight. In the case of skis, this represents an unreasonably small fraction of the total area. In fact, at $-10°C$, μ is about 0.4 for small speeds. At about 5 m/sec sliding speed, however, μ was found to drop rapidly to about 0.04 [4].

An alternative explanation that has been advanced is that at the lower temperatures the water film that leads to low friction is produced as a result of local heating. Here, a trial calculation shows that the frictional energy is sufficient to melt an appreciable layer of the ice or snow without having to assume inordinately small areas of actual contact. Moreover, experiments with skis constructed of different materials or waxed with different preparations have indicated that smaller coefficients of friction are obtained if the surface of the ski has a low heat conductivity than if it has a high value [15]. Since the area of actual contact is still expected to be only a fraction of the apparent area, the frictional melting theory would require that the heat due to friction not be conducted away too rapidly by the body of the ski.

Another indication of the probable incorrectness of the pressure melting explanation is that the variation of the coefficient of friction with temperature for ice is much the same for other solids, such as solid krypton and carbon dioxide [16] and benzophenone and nitrobenzene [4]. In these cases the density of the solid is greater than that of the liquid, so the drop in μ as the melting point is approached cannot be due to pressure melting.

While pressure melting may be important for snow and ice near $0°C$, it is possible that even here an alternative explanation will prove important. Ice is a substance of unusual structural complexity, and it has been speculated that a liquidlike surface layer is present near the melting point [17,18]; if this is correct, the low μ values observed at low sliding speeds near $0°C$ may be due to a peculiarity of the surface nature of ice rather than to pressure melting.

4. Metallic Friction—Effect of Oxide Films

The study of the friction between metals turns out to be almost two subjects—that relating to truly clean metal surfaces and that involving metals with adsorbed gases or oxide coatings. If metal surfaces are freed of all surface contamination by electron bombardment of the heated surface in a vacuum, then, in the cases of tungsten, copper, nickel, and gold, the coefficients of friction can be quite large. Similarly, nickel surfaces so treated seized when placed in contact and rubbed slightly against each other. That is, a firm metal-meal weld occurred such that it was necessary to pry the pieces apart [19]; iron surfaces gave a μ value of 3.5 at room temperature, but seized at $300°C$. Machlin and Yankee [20] using fresh surfaces formed by machining in an inert atmosphere, again found that seizure usually occurred. With two dissimilar metals, however, this may not always happen if the two metals are mutually insoluble. Thus cadmium and iron, which are mutually soluble, did weld together on rubbing, but not silver and iron. The welds that form when seizure occurs appear to be of full strength, that is, for two like metals the strength of the weld is essentially that of the metal itself. An interesting correlation reported by Buckley [21] is that μ decreases with increasing d-bond character to the metal-metal bond.

The behavior in the presence of air is quite different. For example, Tingle [22] found that the friction between copper surfaces decreased from a μ value of 6.8 to one of 0.80 as progressive exposure of the clean surfaces led to increasingly thick oxide layers. As noted by Whitehead [23], several behavior patterns are possible. At very light loads, the oxide layer may be effective in preventing metal-metal contact (and electrical conductivity); the coefficient of friction will then tend to be in the range of 0.6–1.0. At heavier loads, depending on the metal, the film may break down, and μ increases (and the conductivity) as metal-metal contacts form. This was true for copper but not for aluminum or silver. In the case of aluminum, the oxide film is broken even with very small loads, perhaps because the substrate metal is softer than its oxide so that the latter cracks easily. In the case of silver, oxide formation is negligible, so again friction is relatively independent of load. In all cases, even though metal-metal contacts may be present, the coefficient of friction is less than that between really pure surfaces, either because of adsorbed gases or possibly patches and fragments of oxide material.

Bowden and Tabor [24] proposed, for such heterogeneous surfaces, that

$$F = A[\alpha s_m + (1 - \alpha)s_o] \tag{XII-9}$$

where α is the fraction of surface contact that is of the metal-to-metal type, and s_m and s_o are the shear strengths of the metal and the oxide, respectively. As may be gathered, the effect of oxide coatings on frictional properties is, in a way, a special case of the effect of films in general. Boundary lubrication, discussed in Section XII-7, extends the subject further.

5. Friction between Nonmetals

A. Relatively Isotropic Crystals

Substances in this category include Krypton, sodium chloride, and diamond, as examples, and it is not surprising that differences in detail as to frictional behavior do occur. The softer solids tend to obey Amontons' law with μ values in the "normal" range of 0.5–1.0, provided they are not too near their melting points. Ionic crystals, such as sodium chloride, tend to show irreversible surface damage, in the form of cracks, owing to their brittleness, but still tend to obey Amontons' law. This suggests that the area of contact is mainly determined by plastic flow rather than by elastic deformation.

Diamond behaves somewhat differently in that μ is low *in air*, about 0.1. It is dependent, however, on which crystal face is involved, and rises severalfold in vacuum (after heating) [1,2,25]. The behavior of sapphire is similar [24]. Diamond surfaces, incidentally, can have an oxide layer. Naturally occurring ones may be hydrophilic or hydrophobic, depending on whether they are found in formations exposed to air and water. The relation between surface wettability and friction seems not to have been studied.

B. Layer Crystals

A number of substances such as graphite, talc, and molybdenum disulfide have sheetlike crystal structures, and it might be supposed that the shear strength along such layers would be small and hence the coefficient of friction. It is true that μ for graphite is about 0.1 [26,27]. However, adsorbed gases play a role since on thorough outgassing, μ rises to about 0.6 [2,26,28]. Also, microscopic examination of slider tracks shows that strips of graphite layers have rolled up into miniature rollers about 0.05 μm in diameter [29], and it may well be that it is these that give the low μ values. The role of adsorbed gases may be to facilitate the detachment of layer fragments.

The structurally similar molybdenum disulfide also has a low coefficient of friction, but now not increased in vacuum [2,30]. The interlayer forces are, however, much weaker than for graphite, and the mechanism of friction may be different. With molecularly smooth mica surfaces, the coefficient of friction is very dependent on load and may rise to extremely high values at small loads [4]; at normal loads and in the presence of air, μ drops to a near normal level.

C. Polymers

A number of friction studies have been carried out on organic polymers in recent years. Coefficients of friction are for the most part in the normal range, with values about as expected from Eq. XII-5. The detailed results show some serious complications, however. First, μ is very dependent on load, as illustrated in Fig. XII-5, for a copolymer of hexafluoroethylene and hexafluoropropylene [31], and evidently the area of contact is determined more by elastic than by plastic deformation. The difference between static and kinetic coefficients of friction was attributed to transfer of an oriented film of polymer to the steel rider during sliding and to low adhesion between this film and the polymer surface. Tetrafluoroethylene (Telfon) has a low coefficient of friction, around 0.1, and in a detailed study, this lower coefficient and other differences were attributed to the rather smooth molecular profile of the Teflon molecule [32].

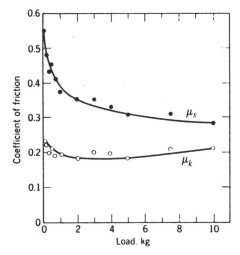

Fig. XII-5. Coefficient of friction of steel sliding on hexafluoropropylene as a function of load (first traverse). Velocity 0.01 cm/sec; 25°C. (From Ref. 31.)

More recently, emphasis has been given to adhesion between the polymer and substrate as a major explanation for friction [33], the other contribution being from elastic hysteresis (see Section XII-2E). The adhesion may be mostly due to van der Waals forces, but in some cases there is a contribution from electrostatic charging. Yethiraj and co-workers have carried out simulations of polymers combining Monte Carlo methods with density functional theory [34,35]. They show segmental depletions and enhancements near solid walls, which may help provide a molecular explanation for friction and adhesion in polymers. Brochard-Wyart and de Gennes have shown that the stretching transition polymers attached to a surface will experience at a critical shear rate will result in significant slippage and a consequently reduced friction [36].

6. Some Further Aspects of Friction

An interesting aspect of friction is the manner in which the area of contact changes as sliding occurs. This change may be measured either by conductivity, proportional to $A^{1/2}$ if, as in the case of metals, it is limited primarily by a number of small metal-to-metal junctions, or by the normal adhesion, that is, the force to separate the two substances. As an illustration of the latter, a steel ball pressed briefly against indium with a load of 15 g required about the same 15 g for its subsequent detachment [37]. If relative motion was set in, a μ value of 5 was observed and, on stopping, the normal force for separation had risen to 100 g. The ratio of 100:15 g may thus be taken as the ratio of junction areas in the two cases.

Even if no perceptible motion occurs (see later, however), application of a force leads to microdisplacements of one surface relative to the other and, again, often a large increase in area of contact. The ratio F/W in such an experiment will be called ϕ since it does not correspond to either the usual μ_s or μ_k; ϕ can be related semiempirically to the area change, as follows [38]. We assume that for two solids pressed against each other at rest the area of contact A_0 is given by Eq. XII-1, $A = W/P_m$. However, if shear as well as normal stress is present, then a more general relation for threshold plastic flow is

$$\mathbf{P}^2 + \alpha s^2 = \mathbf{P}_m^2 = \alpha s_0^2 \qquad \text{(XII-10)}$$

where s is the shear strength of the interfacial region, and s_0 is that of the bulk metal; the constant α is partly geometric, depending on the nature of the asperities. If we let $s = ms_0$, and write $\mu = s/\mathbf{P}$, we obtain [39]

$$\mu = \left(\frac{m^2}{\alpha(1 - m^2)} \right)^{1/2} \qquad \text{(XII-11)}$$

The constant α can be estimated by applying a load so as to establish an A_0 and then removing the load and measuring the force F_0 to cause motion. Under these conditions, $\mathbf{P} = 0$, $A_0 s = F_0$, and $A_0 \mathbf{P}_m = W$, so that from Eq. XII-10,

$$\alpha = \left(\frac{W}{F_0} \right)^2 \tag{XII-12}$$

For steel on indium, α was about 3, while for platinum against platinum, it was about 12 [4].

For very clean metal surfaces, m should approach unity, and μ becomes very large, as observed; with even a small decrease in m, μ falls to about unity, or to the type of value found for practically "clean" surfaces. And if a boundary film is present, making $m < 0.2$, Eq. XII-11 reduces to

$$\mu = \left(\frac{m^2}{\alpha} \right)^{1/2} = \left(\frac{m^2 s_0^2}{\alpha s_0^2} \right)^{1/2} = \frac{s}{\mathbf{P}_m} \tag{XII-13}$$

which is a more general form of Eq. XII-5.

Derjaguin and co-workers [40] have proposed the equation

$$F = \mu W + \mu A \mathbf{p}_0 \tag{XII-14}$$

where, in effect, $A\mathbf{p}_0$ adds to the external force W, owing to an internal or adhesion pressure $\mathbf{p}_0$. In fact, a number of the many systems showing deviations from Amontons' law would fit Eq. XII-14 approximately, in addition to the confirming behavior cited (paraffin on glass).

Finally, if the sliding surfaces are in contact with an electrolyte solution, an analysis indicates that the coefficient of friction should depend on the applied potential [41].

7. Friction between Lubricated Surfaces

A. Boundary Lubrication

Two limiting conditions exist where lubrication is used. In the first case, the oil film is thick enough so that the surface regions are essentially independent of each other, and the coefficient of friction depends on the hydrodynamic properties, especially the viscosity, of the oil. Amontons' law is not involved in this situation, nor is the specific nature of the solid surfaces.

As load is increased and relative speed is decreased, the film between the two surfaces becomes thinner, and increasing contact occurs between the surface regions. The coefficient of friction rises from the very low values possible for fluid friction to some value that usually is less than that for unlubricated surfaces. This type of lubrication, that is, where the nature of the surface region is

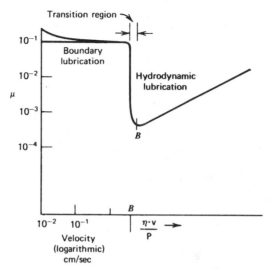

Fig. XII-6. Regions of hydrodynamic and boundary lubrications. (From Ref. 42.)

important, is known as *boundary lubrication*. A nice review of boundary lubrication along with a summary of recent studies has been presented by Spikes [42a]. The general picture is illustrated in Fig. XII-6 by what is known as the *Stribeck curve* [39]; the abscissa quantity, (viscosity) (speed)/(load), is known as the generalized Sommerfeld number [2]. Mate has recently made measurements with an AFM tungsten tip on silicon with perfluoropolyether lubricants that show that hydrodynamic lubrication (μ negligible, below the detection limit) persists down to separations of a few chain diameters [43]. Then, in a similar fashion to that shown in Fig. XII-6, there is a rapid transition to boundary lubrication where μ rises to about 0.5 to 0.8 and increases linearly with the load.

Much of the classic work with boundary lubrication was carried out by Sir William Hardy [44,45]. He showed that boundary lubrication could be explained in terms of adsorbed films of lubricants and proposed that the hydrocarbon surfaces of such films reduced the fields of force between the two parts.

Hardy studied a number of lubricants of the hydrocarbon type, such as long-chain paraffins, alcohols, and acids. An excess of lubricant was generally used, either as the pure liquid or as a solution of the solid in petroleum ether. In some cases the metal surface was then polished. The general observation was that μ values of 0.05 to 0.15 were obtained, that is, much lower than for unlubricated surfaces. Also, for a given homologous series, μ decreased nearly linearly with increasing molecular weight, although with the fatty acids, μ leveled off at about 0.05 at a molecular weight of 200. These relationships are shown in Fig. XII-7.

Levine and Zisman [46] confirmed and extended Hardy's results, using films on glass and on metal surfaces that were deposited by adsorption either from

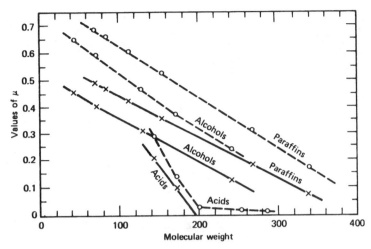

Fig. XII-7. Variations of μ with molecular weight of the boundary lubricant: ——, curve for spherical slider; ----, curve for plane slider. (From Ref. 44.)

solution or from the molten compound. With alcohols the leveling off occurred at a molecular weight of about 280, again at $\mu = 0.05$, so that the behavior is much the same as for the acids, but shifted toward higher molecular weights. With hydrocarbons on copper, some data by Russell and co-workers [47] indicate that the leveling off occurs at about a molecular weight of 400 at $\mu = 0.10$. These points of no further decrease in μ were considered by Levine and Zisman to correspond to the formation of a condensed monolayer. This conclusion was supported by the observation that the air-methylene iodide-film contact angles ceased to increase with molecular weight at about the same stage.

There is no doubt that the reduction of μ by such as the foregoing lubricants is due to a thin and perhaps monomolecular film of adsorbed material. Thus Frewing [48] found that, using a hemispherical steel rider, the coefficients of friction for various alcohols, esters, and acids were about the same whether an excess of lubricant was used or whether one or more monolayers were built up by means of the Langmuir-Blodgett technique (Section XV-7). Later, films of low-molecular-weight alcohols, aldehydes, paraffins, and water formed by vapor-phase adsorption on a steel surface were found to exhibit μ values that decreased to a minimum as the monolayer point was reached [49]. Similarly, monolayers of halogenated compounds adsorbed on metal or glass slides from solution in aqueous or organic solvents or from the vapor phase led to μ values typical for boundary lubrication [47].

It is evident that boundary lubrication is considerably dependent on the state of the monolayer. Frewing [48] found that, on heating, the value of μ rose sharply near the melting point sometimes accompanied by a change from smooth to stick–slip sliding. Very likely these points of change correspond to the transition between an expanded film and a condensed film in analogy with the transitions for films on water. The solidification of simple fluids confined in thin films between two solid walls has been well documented [50–52].

Israelachvili and co-workers have addressed the effect of the state of the monolayer by studying a series of surfactant monolayers of variable surface density [53,54] with the surface force apparatus (described in Chapter VI). The lubricating behavior of surfactant monolayers can be characterized by three regimes illustrated in the pseudo–phase diagram for friction in Fig. XII-8. Solid-like materials slide along a shear plane via stick–slip or smooth motion. The friction is medium to low. Liquidlike materials are always close to equilibrium and the molecules have a rapid response time when sheared and friction is again low. Intermediate densities produce "amorphous" layers that have significant interdigitation between layers and yet slow relaxation times. This interpenetration produces adhesion hysteresis (see Section XII-8) and significant friction. As illustrated in Fig. XII-8, these regimes can be reached through variation of the surfactant surface coverage, temperature, load, velocity, or exposure to organic vapors.

A further indication that condensed films are required for good boundary lubrication is in the behavior of fatty-acid monolayers [10,46,49]. On most surfaces the change in μ correlates with the bulk melting point in about the same manner as for other films, but for lauric acid on copper, the change in μ occurs at about 110°C [46] and for stearic acid on zinc at about 130°C [39,55]. While these temperatures greatly exceed the bulk melting points for the acids, they do correspond to the softening points of the corresponding *metal salts* of the fatty acids. In fact, it is known from adsorption studies that fatty acids chemisorb on the more electropositive metals by salt formation with the oxide coating (see Section XI-1B).

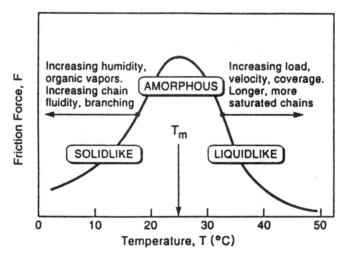

Fig. XII-8. A schematic "friction phase diagram" showing the trends found in the friction forces of surfactant monolayers. (From Ref. 53.)

Klein and co-workers have documented the remarkable lubricating attributes of polymer "brushes" tethered to surfaces by one end only [56]. Studying zwitterionic polystyrene-X attached to mica by the zwitterion end group in a surface forces apparatus, they found $\mu < 0.001$ for loads of 100 μN and speeds of 15–450 nm/sec. They attributed the low friction to strong repulsions existing between such polymer layers. At higher compression, stick–slip motion was observed. In a related study, they compared the friction between polymer brushes in toluene ($\mu < 0.005$) to that of mica in pure toluene $\mu = 0.7$ [57].

The lubricating properties of tears are an important feature in normal blinking. Kalachandra and Shah measured the coefficient of friction of ophthalmic solutions (artificial tears) on polymer surfaces and found no correlation with viscosity, surface tension or contact angle [58]. The coefficient of friction appears to depend on the structure of the polymer surfaces and decreases with increasing load and sliding speed.

B. The Mechanism of Boundary Lubrication

Hardy's explanation that the small coefficients of friction observed under boundary lubrication conditions were due to the reduction in the force fields between the surfaces as a result of adsorbed films is undoubtedly correct in a general way. The explanation leaves much to be desired, however, and it is of interest to consider more detailed proposals as to the mechanism of boundary lubrication.

It has been pointed out that the value of μ in boundary lubrication depends greatly on the state of the adsorbed film and that, generally speaking, the film must be in a condensed state to give a low coefficient of friction. In combination with Hardy's explanation, the picture of a contact region might then be supposed to look as shown schematically in Fig. XII-9. However, this cannot be correct for at least two reasons. First, it is generally found that with very light loads, the coefficient of friction under boundary lubrication conditions *rises* to near normal values [4,23]; such an effect cannot be accounted for in terms of Hardy's model. The effect is illustrated in Fig. XII-10. Parenthetically, and more understandably, boundary lubrication fails under very high loads [59] and with repeated traverses of the same track [60].

Second, it is found that metal–metal contacts are still present even under normal boundary lubrication conditions where μ is small. Very clear evidence

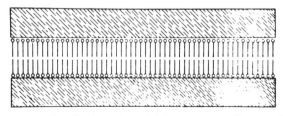

Fig. XII-9. Contact region in boundary lubrication according to Hardy. (From Ref. 45.)

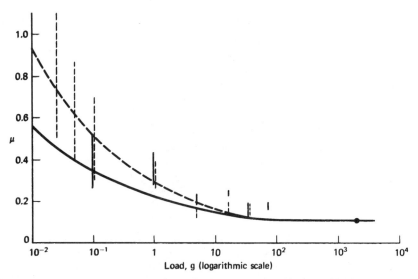

Fig. XII-10. Variation of μ with load. The data are the friction of copper lubricated with lauric acid (——) and with octacosanoic acid (----). (From Ref. 23.)

for this has been provided by radioautographic studies [55]. The transfer of metal from a radioactive slider is mapped by placing the surface against a photographic emulsion and noting where darkening occurs. The significant aspect of the results is that, while the total amount of transfer is greatly reduced when boundary lubrication is present over that in the absence of lubrication, the reduction is more in the *size* of each fragment than in the number of fragments. Bowden and Tabor [3] note that the number of spots may be reduced by a factor of 3 or 4 with a fatty-acid boundary lubricant while the total amount of metal transferred drops by a factor of perhaps 10^5. This observation bears on a matter of practical importance, namely, that boundary lubricants can give comparable friction coefficients yet differ enormously in the amount of wear allowed.

The radioautographic work suggests another model illustrated in Fig. XII-11. The load is supported over area A, with metal contacts of shear strength s_m over a portion of the area αA and film–film contacts of shear strength s_f over the rest of the area. In analogy to Eq. XII-9, one can write the total frictional force, F as

$$F = A[\alpha s_m + (1 - \alpha)s_f] = A_1 s_m + A_2 s_f \qquad \text{(XII-15)}$$

For boundary lubrication, α must be on the order of 10^{-4} to account for the great reduction in metal pickup, therefore, most of the friction must be due to film–film interactions.

This second picture, while an advance over Hardy's, again encounters difficulties. It does not suggest how A_2 could be so much greater than A_1 or why s_f

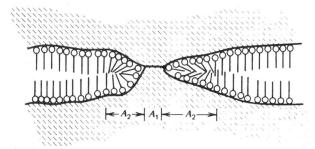

Fig. XII-11. Nature of the contact region in boundary lubrication according to Bowden and Tabor. (From Ref. 4a.)

should be higher with small loads. Also, it does not explain the behavior of the electrical conductivity. At small loads, the conductivity is very small, as would be expected in terms of the model, because of the small values of A_1 and the probably high electrical resistance of the film. However, at loads such that μ falls to its normal low value, the electrical conductivity rises to about the *same* as for unlubricated surfaces. There is still very little metal–metal contact, so the film must now have become conducting. Also, to explain why Amontons' law holds for boundary lubrication over the usual range of loads, it is necessary to suppose that A_2 is proportional to load. The model does not suggest how this could be true.

A third model, proposed by one of the authors 30 years ago [61], seems to meet these objections. It is supposed that as two surfaces are pressed together under increasing load, it is *only* for very light loads that the situation is that of Fig. XII-11. That is, with small loads, there will be regions of slight metal–metal contact with a surrounding fringe of essentially normal monolayer. The load is then being supported both by the metal–metal contact region and that of film–film contact. Since A_2 corresponds to the area of the surrounding fringes of film–film contact, the geometry of the situation requires that A_2/A_1 increase as A_1 decreases. Amontons' law should therefore not be obeyed at small loads. In addition, μ should be at near normal values for unlubricated surfaces to the extent that it arises from A_1. In addition, s_f might be fairly large itself as a steric consequence of the entangling of the hydrocarbon chains.

Turning now to the situation for normal loads, there can be no doubt that most of the load is being supported by the boundary film and that therefore the film itself is under mechanical pressure. This pressure will cause transport of film molecules from the thin high-pressure regions to adjacent wider gaps. Transport of fatty-acid films can occur rapidly over short distances, either through surface mobility or by way of vapor-phase hopping [62]. The pressurized film that remains might consist of molecules lying more or less flat, and would therefore resemble the L_1 state of monolayers (see Section IV-4).

The mechanism of boundary lubrication may then be pictured as follows. At the unusually prominent asperities, the local pressure exceeds the yield pressure of the metal, film material is completely displaced, and an area of metal–metal contact develops. A much smaller, and probably elastic, deformation of the metal is sufficient, however, to place a relatively large area of film under varying mechanical pressure, so that most of the load is "floating" on pressurized film. The area of the pressurized film is approximately proportional to the load, so that Amontons' law should be obeyed. In addition, the coefficient of shear for the film–film fraction now applies to film molecules that are lying flat and is thus much smaller than that of the film under low pressure, corresponding to the observed low μ values under boundary lubrication conditions with normal loads.

Bowden and Tabor [4] cite support for the general idea that film molecules are forced to a horizontal configuration in load-bearing regions and the general idea was proposed by Wilson in 1955 [63].

It must not be forgotten that any mechanism for boundary lubrication must be a dynamic one, since it is a kinetic rather than a static coefficient of friction that is generally involved. In the case of nonlubricated surfaces, it is supposed that contact regions form and are sheared as the surfaces move past each other, with new regions constantly forming as new asperities pass. There is thus a steady-state situation. Presumably, a similar situation is involved in the case of boundary lubrication, with regions forming and disappearing. It is doubtful that an equilibrium analysis can be very accurate at high sliding speeds since a finite time would be required for solid deformation and migration of film molecules. The molecular transport can be assessed with the following numerical example.

It is known that even condensed films must have surface diffusional mobility; Rideal and Tadayon [64] found that stearic acid films transferred from one surface to another by a process that seemed to involve surface diffusion to the occasional points of contact between the solids. Such transfer, of course, is observed in actual friction experiments in that an uncoated rider quickly acquires a layer of boundary lubricant from the surface over which it is passed [46]. However, there is little quantitative information available about actual surface diffusion coefficients. One value that may be relevant is that of Ross and Good [65] for butane on Spheron 6, which, for a monolayer, was about 5×10^{-3} cm^2/sec. If the average junction is about 10^{-3} cm in size, this would also be about the average distance that a film molecule would have to migrate, and the time required would be about 10^{-4} sec. This rate of junctions passing each other corresponds to a sliding speed of 100 cm/sec so that the usual speeds of 0.01 cm/sec should not be too fast for pressurized film formation. See Ref. 62 for a study of another mechanism for surface mobility, that of evaporative hopping.

There is a breakdown of boundary lubrication under extreme pressure conditions. The effect is considered to be related to that of increasing temperature [59]; this is not unreasonable since the amount of heat to be dissipated will increase with load and a parallel increase in the local temperature would be expected.

C. Forces and Friction between Smooth Surfaces

The surface forces apparatus of crossed mica cylinders (Section VI-4D) has provided a unique measurement of friction on molecular scales. The apparatus is depicted in Fig. VI-3, and the first experiments involved imposing a variation or pulsing in the separation between the cylinders and measuring the compliance or variation in the force between them. Then the technique was suitably modified to provide lateral motion and force measurement at fixed applied pressure or surface separation. At close approach, the glue holding the mica surfaces deforms resulting in plane circular areas of closest approach. As a result of viscous resistance, the actual lateral displacement is less than that expected, the relative motion of the cylinders thus having an out-of-phase component with respect to that otherwise expected. The result may be expressed as an apparent Newtonian viscosity.

Since we have parallel-plate geometry, the viscosity of the liquid layer is given by

$$F = \eta A \, \frac{v}{x} \qquad\qquad \text{(XII-16)}$$

where F is the force causing relative velocity v between parallel plates of area A separated by a distance x (note Section VI-3C). Alternatively, the viscosity η is given by the ratio of shearing stress F/A to the rate of strain, v/x. From the definition of μ and using Eq. XII-1, we obtain

$$\mu = \frac{\eta}{P_m} \, \frac{v}{x} \qquad\qquad \text{(XII-17)}$$

where P_m is the yield pressure of the glue-backed mica (see Problem 12, however).

Results have been intriguing. For hexadecane (effective molecular diameter 4 Å), the viscosity for a 24-Å-thick film was 20 times the bulk viscosity; at 8 Å thickness, it rose to 10^4 times the bulk viscosity [66], in contrast to findings using the pulsing method (Section VI-7). The normal pressure at this point was about 100 atm, and at higher pressures a *drop* in η occurred. Moreover, in this thickness region, the film appeared able to switch back and forth between a fluid and a plastic or solidlike state. Other work [67] found the shearing stress at a given x to be independent *both* of sliding velocity and of applied load. Such behavior is unexpected both in terms of Newtonian viscosity and of Amontons' law. See Section VI-7 for some additional aspects.

The traditional, essentially phenomenological modeling of boundary lubrication should retain its value. It seems clear, however, that newer results such as those discussed here will lead to spectacular modification of explanations at the molecular level. Note, incidentally, that the tenor of recent results was anticipated in much earlier work using the blow-off method for estimating the viscosity of thin films [68].

While friction does not correlate directly with adhesion (there are many examples of high-friction surfaces having low adhesion and vice versa), it appears to relate well to adhesion hysteresis. Adhesion hysteresis is the difference in adhesive energy measured on loading and unloading. Isrealachvili and co-workers have found a relationship between adhesion hysteresis and friction measured for model surfactant layers on molecularly smooth mica in a surface forces apparatus [53,54]. This is illustrated in Fig. XII-12, where the friction traces for two identical surfactant monolayers in air and decane vapors are shown with the adhesion load curves. In air there is significant adhesion hysteresis (unloading, $\gamma_R >$ loading, γ_A) corresponding to measurable friction. They correlate the friction force F to move a distance D

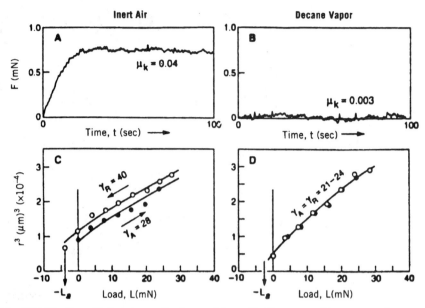

Fig. XII-12. Top: friction traces for two calcium alkylbenzenesulfonate monolayers on mica where the monolayers are in a liquidlike state. A—in inert air atmosphere; B—in saturated decane vapor. Bottom: contact radius–load curves showing adhesion energy measured under the same conditions as the friction traces. (From Ref. 53.)

$$F = \frac{A\Delta\gamma}{D} = \frac{A(\gamma_R - \gamma_A)}{D} \qquad (\text{XII-18})$$

with the adhesion hysteresis per unit area times the contact area A. In decane vapor, for this surfactant, both the friction and the adhesion hysteresis disappear.

8. Adhesion

A. Ideal Adhesion

Ideal adhesion simply means the adhesion expected under one or another model situation of uniform materials having intimate contact over a well-defined area. In these cases, the important quantity is the *work of adhesion* w_{AB} between two phases, which is given by

$$w_{AB} = \gamma_A + \gamma_B - \gamma_{AB} \qquad (\text{XII-19})$$

and represents the work necessary to separate a unit area of the interface AB into two liquid–vapor or solid–vapor interfaces A and B. Equation XII-19 may be written in a more complete form:

$$w_{A(B)B(A)} = \gamma_{A(B)} + \gamma_{B(A)} - \gamma_{AB} = -\pi_{A(B)} - \pi_{B(A)} \qquad \text{(XII-20)}$$

where a phase in parentheses saturates the one preceding it. Recalling the numerical example of Section VI-2A, for water A and benzene B, $w_{A(B)B(A)} = 56$ ergs/cm^2 and $w_{AB} = 76$ ergs/cm^2. These are fairly representative values for works of adhesion, and it is instructive to note that they correspond to fairly large separation forces. The reason is that these energies should be developed largely over the first few molecular diameters of separation, since intermolecular forces are rather short range. If the effective distance is taken to be 10 Å, for example, then the figure of 56 ergs/cm^2 becomes an average force of 5.6 $\times 10^8$ dyn/cm^2 or 560 kg/cm^2, which is much larger than most failing loads of adhesive joints.

Equation XII-20 may be combined with various semiempirical equations. Thus if Antonow's rule applies (Eq. IV-8), one obtains

$$w_{A(B)B(A)} = 2\gamma_{B(A)} \qquad \text{(XII-21)}$$

(which agrees well with the benzene-water example). The Girifalco–Good equation, Eq. IV-10,

$$w_{AB} = 2(\gamma_A \gamma_B)^{1/2} \qquad \text{(XII-22)}$$

does not agree as well. If substance B has a finite contact angle with solid substance A, then

$$w_{A(B)B(A)} = \gamma_{B(A)}(1 + \cos \theta_B) \qquad \text{(XII-23)}$$

and in combination with Zisman's relationship, Eq. X-37:

$$w_{A(B)B(A)} = \gamma_{B(A)}(2 + \beta\gamma_c) - \beta\gamma_{B(A)}^2 \qquad \text{(XII-24)}$$

The interesting implication of Eq. XII-24 is that for a given solid, the work of adhesion goes through a maximum as $\gamma_{B(A)}$ is varied [69]. For the low-energy surfaces Zisman and co-workers studied, β is about 0.04, and w_{max} is approximately equal to the critical surface tension γ_c itself; the liquid for this optimum adhesion has a fairly high contact angle.

There has been considerable elaboration of the simple Girifalco and Good relationship, Eq. XII-22. As noted in Sections IV-2A and X-6B, the surface free energies that appear under the square root sign may be supposed to be expressible as a sum of dispersion, polar, and so on, components. This type of approach has been developed by Dann [70] and Kaelble [71] as well as by Schonhorn and co-workers (see Ref. 72). Good (see Ref. 73) has preferred to introduce polar interactions into a detailed analysis of the meaning of Φ in Eq. IV-7. While there is no doubt that polar interactions are important, these are orientation dependent and hence structure sensitive.

These authors doubt that such interactions can be estimated other than empirically without fairly accurate knowledge of the structure in the interfacial region. Sophisticated scattering, surface force, and force microscopy measurements are contributing to this knowledge; however, a complete understanding is still a long way off. Even submonolayer amounts of adsorbed species can affect adhesion as found in metals and oxides [74].

Metal to ceramic (oxide) adhesion is very important to the microelectronics industry. An electron transfer model by Burlitch and co-workers [75] shows the importance of electron donating capability in enhancing adhesion. Their calculations are able to explain the enhancement in adhesion when a NiPt layer is added to a Pt–NiO interface.

The adhesion between two solid particles has been treated. In addition to van der Waals forces, there can be an important electrostatic contribution due to charging of the particles on separation [76]. The adhesion of hematite particles to stainless steel in aqueous media increased with increasing ionic strength, contrary to intuition for like-charged surfaces, but explainable in terms of electrical double-layer theory [77,78]. Hematite particles appear to form physical bonds with glass surfaces and chemical bonds when adhering to gelatin [79].

The adhesion of biological cells and microorganisms is a complicated yet important topic [80–84]. In particular, the adhesion of blood cells to injured blood vessels prevents blood loss while adhesion to medical implant devices causes deleterious clotting. Fundamental aspects of platelet adhesion have been studied by Zingg and Neumann [85] and adhesion of erythrocytes to various surfaces have been addressed [86,87]. The introduction of medical implants into the body is fraught with potential problems of a surface chemical nature and can result in costly litigation and liability issues (for a discussion, see Ref. 88). Baier and co-workers have studied the correlation between surface chemistry, cell adhesion, and biocompatibility, and several good references on this topic are available [89–91]. One topic of interest to all surface scientists (from at least a practical point of view!) is the adhesion of dental plaque to teeth and dental implants; Baier and co-workers provide a more scientific analysis [92]. Biofouling in marine environments is an expensive problem, thus several studies focus on the adhesive materials used by sea mussels [93,94]. A micropipet technique developed by Evans [95] has been used to measure forces of adhesion in endothelial cells [96] and the surface forces apparatus in Fig. VI-3 can be adapted to measurement of reversible adhesive forces [97].

A second ideal model for adhesion is that of a liquid wetting two plates, forming a circular meniscus, as illustrated in Fig. XII-13. Here a Laplace pressure $\mathbf{P} = 2\gamma_L/x$ draws the plates together and, for a given volume of liquid,

$$f = \frac{2\gamma_L V}{x^2}$$

(XII-25)

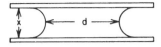

Fig. XII-13. Illustration of adhesion between two plates due to a meniscus of a wetting liquid.

so that the force f to separate the plates can be very large if the film is thin (note Problem XII-15). As an interesting example, if the medium is nonaqueous, small droplets of water bridging the two surfaces can give rise to a strong adhesion [97].

B. Polymer Adhesion

Many important adhesives are made from glassy or elastomeric polymers. There are familiar examples encountered in everyday life such as epoxy resins used to repair broken china and rubber cement and adhesive tape applied to paper. There are two primary mechanisms for polymer adhesion (for a good review, see Brown [98]). Elastomers are crosslinked polymers above their glass transition temperature. On fracture, these polymers dissipate energy through viscoelastic losses depending on the speed of crack propagation and the crosslink density [99]. A low crosslink density will produce a viscous material having insufficient elasticity to resist flow. Too much crosslinking will render the material too rigid for good adhesion. Thus an optimal crosslink density exists and is tailored to the end use such as removable adhesives in tape and note pads. Glassy polymers adhere by entanglements and specific interactions between chains. The final strength of a polymer–polymer interface depends on the rupture [100] or pull-out [101,102] of chains.

One way to promote adhesion between elastomers and solid surfaces is to chemically bond (graft) a layer of chains to the surface. When this is done, the attached chains can interpenetrate the elastomer and provide adhesion due to entanglements and other interactions. This is shown in theoretical studies of adhesion promoters by de Gennes and co-workers [103] and in experimental studies on siloxane polymers [104]. In both theory and experiment there is a maximum in adhesion with grafting density. This is due to the ability of a chain from a layer of moderate grafting density to interpenetrate the elastomeric network while at high-grafting-density interpenetration is reduced. Failure between adhesive tape and a solid surface shows flow instability patterns similar to those found in dewetting studies [105].

C. Practical Adhesion

The usual practical situation is that in which two solids are bonded by means of some kind of glue or cement. A relatively complex joint is illustrated in Fig. XII-14. The strength of a joint may be measured in various ways. A common standard method is the *peel test* in which the normal force to separate the joint

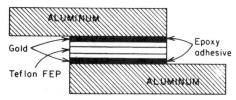

Fig. XII-14. A composite adhesive joint. (From Ref. 106.)

is measured, as illustrated in Fig. XII-15. (See Ref. 108, in which a more elaborate cantilever beam device is also used, and Ref. 109.) Alternatively, one may determine the shear adhesive strength as the force *parallel* to the surface that is needed to break the joint.

Two models for adhesive joint failure are the following. One is essentially a classic, mechanical picture, relating stored elastic energy and the work of crack propagation, known as the *Griffith-Irwin criterion* (see Ref. 110). Bikerman [111], however, has argued that the actual situation is usually one of a *weak boundary layer*. This was thought to be a thin layer (but of greater than molecular dimensions) of altered material whose mechanical strength was less than that of either bulk phase. While oxides, contamination, and so on, could constitute a weak boundary layer, it was also possible that this layer could be structurally altered but pure material. Interestingly, Schonhorn and Ryan [112] found that proper cleaning of a plastic greatly improved the ability to bond it to metal substrates. Good ([113], but see Ref. 110 also) has argued that the *molecular* interface may often be the weakest layer and that adhesive failure is at the true interface. A rather different idea, especially applicable to the peel test procedure, is due to Derjaguin (see Ref. 114). The peeling of a joint can produce static charging (sometimes visible as a glow or flashing of light), and much of the work may be due to the electrical work of, in effect, separating the plates of a charged condenser.

Returning to more surface chemical considerations, most literature discussions that relate adhesion to work of adhesion or to contact angle deal with surface free energy quantities. It has been pointed out that structural distortions are generally present in adsorbed layers and must be present if bulk liquid adsorbate forms a finite contact angle with the substrate (see Ref. 115). Thus both the entropy and the energy of adsorption are important (relative to bulk liquid). The

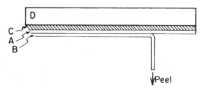

Fig. XII-15. Diagram of peel test. *A* and *B*, adhesive joint; *C*, double Scotch adhesive tape; and *D*, rigid support. (From Ref. 107. By permission of IBC Business Press, Ltd.)

same is likely true at a liquid-solid or a solid-solid interface, and it has been proposed that gradations of two extreme classes exist [116]. In class A systems there is little structural perturbation at the interface and therefore little entropy of adhesion. In class B systems structural changes in one or both phases are important in the interfacial region, and there is an adverse entropy of adhesion, compensated by a greater than expected energy of adhesion. The distinction between energy and entropy may be important in practical adhesion situations. It is possible that the practical adhesive strength of a joint is determined more by the *energy* than by the *free energy* of adhesion; this would be true if the crucial point of failure occurred as an irreversible rather than a reversible process. Other aspects being the same, including the free energy of adhesion, the practical adhesion should be less for a class A than for a class B system. See Section X-6 for related discussion.

Other aspects are often dominant in practical adhesion situations. In contrast to the situation in friction, quite large areas of actual interfacial contact are desired for good adhesion. A liquid advancing over a rough surface can trap air, however, so that only a fraction of the apparent area may be involved in good interfacial contact. A low contact angle assists in preventing such composite surface formation and, with this in mind, it has been advocated that one criterion for good adhesion is that the adhesive spread on the surface it is to bond [117, 118]—even though, as shown, this may not give the optimum ideal work of adhesion.

However, the actual strengths of "good" adhesive joints are only about a tenth of the ideal values [119], so that failure to wet completely does not seem to be an adequate explanation. Since adhesive joints are subject to shear as well as to normal force, there can be a "peeling" action in which the applied force concentrates at an edge to produce a crack that then propagates. Such stress concentrations can occur without deliberate intent simply because of imperfections and trapped bubbles in the interfacial region [120]. Zisman [69] remarks that adhesives may give stronger joints with rough surfaces simply because such imperfections, including trapped air bubbles, will not be coplanar, and the propagation of cracks will then be less easy. An adhesive joint may be weakened by penetration of water [72]; the effect is of great importance in the "stripping" of blacktop road surfaces where water penetrates the interface between the rock aggregate and the asphalt.

As mentioned, there are cases where lack of adhesion is desirable. The coined word for this attribute is *abhesion*. As a nonbiological example, good abhesion is often desired between ice and either a metal or a polymer surface (as on airplane wings, or in ice trays). *Ab*hesion of ice to metals is poor, and failure tends to occur in the ice itself. It is good with polytetrafluoroethylene (Teflon), apparently in part because in freezing against a surface that it wets poorly, air bubbles are produced at the ice-plastic interface. These allow stress concentration to lead to interfacial crack propagation. However, if water is cycled through several freezing and melting operations, dissolved air is removed and abhesion becomes poor (see Refs. 121–125). Fairly good abhesion has been reported for ice with polysiloxane-polycarbonate copolymers, adhesive strengths of as low as 200 g/cm^2 being found [126].

9. Problems

1. The coefficient of friction for copper on copper is about 0.9. Assuming that asperities or junctions can be represented by cones of base and height each about 5×10^{-4} cm, and taking the yield pressure of copper to be 30 kg/mm^2, calculate the local temperature that should be produced. Suppose the frictional heat to be confined to the asperity, and take the sliding speed to be 10 cm/sec and the load to be 20 kg.

2. The resistance due to a circular junction is given by $R = 1/2ak$, where a is the radius of the junction and k is specific conductivity of the metal. For the case of two steel plates, the measured resistance is $5 \times 10^{-5}\Omega$ for a load of 50 kg; the yield pressure of steel is 60 kg/mm^2, and the specific resistance is $5 \times 10^{-5}\Omega$/cm. Calculate the number of junctions, assuming that it is their combined resistance that is giving the measured value.

3. Deduce from Fig. XII-5, using the data for μ_s, how the contact area appears to be varying with load, and plot A versus W.

4. Calculate the angle of repose for a solid block on an inclined plane if the coefficient of friction is 0.52.

5. In a series of tests a car is brought to a certain speed, then braked by applying a certain force F on the brake pedal, and the deceleration a is measured. The pavement is dry concrete, and force F_0 is just sufficient to cause skidding. Sketch roughly how you think the plot of a versus F should look, up to F values well beyond F_0.

6. Derive the equation corresponding to Eq. XII-8 for the case where the road is inclined to the horizontal by an angle θ.

7. Discuss why stick-slip friction is favored if μ decreases with sliding speed.

8. Many modern automobiles are now equipped with an antilocking braking system. Explain in terms of μ_s and μ_k why this is advantageous.

9. The friction of steel on steel is being studied, using as boundary lubricants (a) $CH_3(CH_2)_8COOH$ and (b) $HOOC(CH_2)_8COOH$. Explain what your estimated value is for the coefficient of friction in case (a) and whether the value for case (b) should be about the same, or higher, or lower. The measurements are under conventional conditions, at 25°C. Discuss more generally what differences in frictional behavior might be found between the two cases.

10. Calculate the friction force between the surfactant layers in air in Fig. XII-12 using the relationship in Eq. XII-18. How does this compare with the friction shown in Fig. XII-12?

11. Discuss the dependence of the "friction phase diagram" on temperature, monolayer density, velocity, load and solvent vapor. Explain why each of these variables will drive one to the right or left in Fig. XII-8.

12. Derive Eq. XII-17. In an experiment using hexadecane and crossed mica cylinders, the circular flat contact area is about 10^{-3} cm in diameter and the two surfaces oscillate back and forth to the extent of 1% of their diameter per second. The separation distance is 10 Å and the yield pressure of the glue-backed mica is 0.1 kg/mm^2.

The reported apparent viscosity is 200 poise. Estimate the coefficient of friction that corresponds to these data. Discuss any assumptions and approximations.

13. The statement was made that the work of adhesion between two dissimilar substances should be larger than the work of cohesion of the weaker one. Demonstrate a basis on which this statement is correct and a basis on which it could be argued that the statement is incorrect.

14. Derive from Eq. XII-24 an expression for the maximum work of adhesion involving only β and γ_c. Calculate this maximum work for $\gamma_c = 22$ dyn/cm and $\beta = 0.030$, as well as γ_L for this case, and the contact angle.

15. As illustrated in Fig. XII-13, a drop of water is placed between two large parallel plates; it wets both surfaces. Both the capillary constant a and d in the figure are much greater than the plate separation x. Derive an equation for the force between the two plates and calculate the value for a 1-cm^3 drop of water at 20°C, with $x = 0.5$, 1, and 2 mm.

General References

J. J. Bikerman, *The Science of Adhesive Joints*, Academic, New York, 1961.

F. P. Bowden and D. Tabor, *The Friction and Lubrication of Solids*, Part I (1950) and Part II (1964), Clarendon Press, Oxford.

F. P. Bowden and D. Tabor, *Friction, An Introduction to Tribology*, Anchor Books, New York, 1973.

B. Bushan, ed., *Handbook of Micro/nanotribology*, CRC Press,1995.

D. D. Eley, ed., *Adhesion*, Clarendon Press, Oxford, 1961.

R. Houwink and G. Salomon, eds., *Adhesion and Adhesives*, Elsevier, New York, 1965.

J. Israelachvili, *Intermolecular & Surface Forces*, 2nd ed., Academic, San Diego, CA, 1992.

D. H. Kaelble, *Physical Chemistry of Adhesion*, Wiley-Interscience, New York, 1971.

D. F. Moore, *Principles and Applications of Tribology*, Pergamon, New York, 1975.

A. W. Newmann, in *Wetting, Spreading, and Adhesion*, J. F. Padday, ed., Academic Press, New York, 1978.

J. A. Schey, *Metal Deformation Processes*, Marcel Dekker, New York, 1970.

I. L. Singer and H. M. Pollock, eds., *Fundamentals of Friction: Macroscopic and Microscopic Processes*, Kluwer, 1996.

F. P. Tabor, *Surface and Colloid Science*, E. Matijevic, ed., Wiley-Interscience, New York, 1972.

Textual References

1. See *Chemical & Engineering News*, April 24, 1989, p. 37.

1a. W. N. Unertl and M. Grunze, *Physical and Chemical Mechanisms of Tribology*, special issue, *Langmuir*, **12**, 4481–4610 (1996).

1b. J. Krim, *Sci. Am.*, 74 (Oct. 1996).

2. D. F. Moore, *Principles and Applications of Tribology*, Pergamon, New York, 1975.

3. F. P. Bowden and D. Tabor, *Friction*, Anchor Books, New York, 1973.

4. F. P. Bowden and D. Tabor, *The Friction and Lubrication of Solids*, Part II, Clarendon Press, Oxford, 1964.

4a. F. P. Bowden and D. Tabor, *The Friction and Lubrication of Solids*, Oxford University Press, New York, 1950.

5. F. P. Bowden and P. H. Thomas, *Proc. Roy. Soc.* (London), **A223,** 29 (1954).

6. K. Tanaka and Y. Uchiyrma, in *Advances in Polymer Friction and Wear*, Lieng-Huang Lee, ed., *Polymer Science and Technology*, Vol. 5B, Plenum, New York, 1974.

7. See *A Discussion of Friction, Proc. Roy. Soc.* (London), **A212,** 439 (1952).

8. B. W. Sakmann, J. T. Burwell, and J. W. Irvine, *J. Appl. Phys.*, **15,** 459 (1944).

9. C. M. Mate, "Atomic Scale Friction," in *Handbook of Micro/Nano Tribology*, B. Bhushan, ed., CRC Press, 1995.

10. T. Fort, Jr., *J. Phys. Chem.*, **66,** 1136 (1962).

11. C. M. Mate, G. M. McClelland, R. Erlandsson, and S. Chiang, *Phys. Rev. Lett.*, **59,** 1942 (1987).

12. B. J. Briscoe, D. C. B. Evans, and D. Tabor, *J. Colloid Interface Sci.*, **61,** 9 (1977).

13. E. Rabinowicz, in *Friction and Wear*, R. Davies, ed., Elsevier, New York, 1959; see also J. T. Burwell and E. Rabinowicz, *J. Appl. Phys.*, **24,** 136 (1953).

14. C. E. O'Hara and J. W. Osterburg, *An Introduction to Criminalistics*, Macmillan, New York, 1959, p. 310.

15. F. P. Bowden, *Proc. Roy. Soc.* (London), **A217,** 462 (1953).

16. F. P. Bowden and G. W. Rowe, *Proc. Roy. Soc.* (London), **A228,** 1 (1955).

17. N. H. Fletcher, *Phil. Mag.*, **7**(8), 255 (1962).

18. A. W. Adamson, L. M. Dormant, and M. Orem, *J. Colloid Interface Sci.*, **25,** 206 (1967).

19. F. P. Bowden and J. E. Young, *Nature*, **164,** 1089 (1949).

20. E. S. Machlin and W. R. Yankee, *J. Appl. Phys.*, **25,** 576 (1954).

21. D. H. Buckley, *J. Colloid Interface Sci.*, **58,** 36 (1977).

22. E. D. Tingle, *Trans. Faraday Soc.*, **46,** 93 (1950).

23. J. R. Whitehead, *Proc. Roy. Soc.* (London), **A201,** 109 (1950).

24. F. P. Bowden and D. Tabor, *Ann. Rep.*, **42,** 20 (1945).

25. S. V. Pepper, *J. Vac. Sci. Tech.*, **20,** 643 (1982).

26. F. P. Bowden and J. E. Young, *Proc. Roy. Soc.* (London), **A208,** 444 (1951).

27. I. M. Feng, *Lub. Eng.*, **8,** 285 (1952).

28. F. P. Bowden, J. E. Young, and G. Rowe, *Proc. Roy. Soc.* (London), **A212,** 485 (1952).

29. J. Spreadborough, *Wear*, **5,** 18 (1962).

30. See G. W. Rowe, *Wear*, **3,** 274 (1960).

31. R. C. Bowers and W. A. Zisman, *Mod. Plast.*, **41,** (December 1963).

32. C. M. Pooley and D. Tabor, *Proc. Roy. Soc.* (London), **A329,** 251 (1972).

33. D. Tabor, in *Advances in Polymer Friction and Wear*, Lieng-Huang Lee, ed., *Polymer Science and Technology*, Vol. 5A, Plenum, New York, 1974.

34. A. Yethiraj and C. E. Woodward, *J. Chem. Phys.*, **102,** 5499 (1995).

35. A. Yethiraj, *J. Chem. Phys.*, **101,** 2489 (1994).

36. F. Brochard-Wyart and P. G. de Gennes, *Langmuir*, **8,** 3033 (1992).

37. J. S. McFarlane and D. Tabor, *Proc. Roy. Soc.* (London), **A202,** 244 (1950).

38. J. T. Burwell and E. Rabinowicz, *J. Appl. Phys.*, **24,** 136 (1953).

39. C. H. Riesz, in *Metal Deformation Processes*, J. A. Schey, ed., Marcel Dekker, New York, 1970.

40. B. V. Derjaguin, V. V. Karassev, N. N. Zakhavaeva, and V. P. Lazarev, *Wear*, **1,** 277 (1957–58).

41. J. O'M. Bockris and S. D. Argade, *J. Chem. Phys.*, **50,** 1622 (1969).

42. A. Bondi, *Physical Chemistry of Lubricating Oils*, Reinhold, New York, 1951.

42a. H. A. Spikes, *Langmuir*, **12,** 4567 (1996).

43. C. M. Mate, *Phys. Rev. Lett.*, **68,** 3323 (1992).

44. W. B. Hardy, *Collected Works*, Cambridge University Press, Cambridge, England, 1936.

45. W. B. Hardy and I. Bircumshaw, *Proc. Roy. Soc.* (London), **A108,** 1 (1925).

46. O. Levine and W. A. Zisman, *J. Phys. Chem.*, **61,** 1068, 1188 (1957).

47. J. A. Russell, W. E. Campbell, R. A. Burton, and P. M. Ku, *ASLE Trans.*, **8,** 48 (1958).

48. J. J. Frewing, *Proc. Roy. Soc.* (London), **A181,** 23 (1942).

49. A. V. Fraioli, F. H. Healey, A. C. Zettlemoyer, and J. J. Chessick, *Abstracts of Papers, 130th Meeting of the American Chemical Society*, Atlantic City, NJ, September 1956.

50. J. N. Israelachvili, *Intermolecular and Surface Forces*, Academic, 1992.

51. J. Klein and E. Kumacheva, *Science*, **269,** 816 (1995).

52. J. Israelachvili and A. Berman, *Israel J. Chem.*, **35,** 85 (1995).

53. H. Yoshizawa, Y. L. Chen, and J. Israelachvili, *J. Phys. Chem.*, **97,** 4128 (1993).

54. J. Israelachvili, Y. L. Chen, and H. Yoshizawa, *J. Adhes. Sci. Technol.*, **8,** 1231 (1994).

55. D. Tabor, *Proc. Roy. Soc.* (London), **A212,** 498 (1952).

56. J. Klein, E. Kumacheva, D. Mahalu, D. Perahia, and L. J. Fetters, *Nature*, **370,** 634 (1994).

57. J. Klein, E. Kumacheva, D. Perahia, D. Mahalu, and S. Warburg, *Faraday Disc.*, **98,** 173 (1994).

58. S. Kalachandra and D. O. Shah, *Ann. Ophthalmol.*, **17,** 708 (1985).

59. C. G. Williams, *Proc. Roy. Soc.* (London), **A212,** 512 (1952).

60. R. L. Cottington, E. G. Shafrin, and W. A. Zisman, *J. Phys. Chem.*, **62,** 513 (1958).

61. A. W. Adamson, in *Physical Chemistry of Surfaces*, 2nd ed., Wiley, New York, 1967.

62. A. W. Adamson and V. Slawson, *J. Phys. Chem.*, **85,** 116 (1981).

63. R. W. Wilson, *Proc. Phys. Soc.*, **68B,** 625 (1955).

64. E. Rideal and J. Tadayon, *Proc. Roy. Soc.* (London), **A225,** 346, 357 (1954).

65. J. W. Ross and R. J. Good, *J. Phys. Chem.*, **60,** 1167 (1956).

66. J. Van Alsten and S. Granick, *Phys. Rev. Lett.*, **28,** 2570 (1988).

67. J. N. Israelachvili, P. M. McGuiggan, and A. M. Homola, *Science*, **240,** 189 (1988).

68. B. V. Deryaguin, V. V. Karassev, N. N. Zakhavaeva, and V. P. Lazarev, *Wear*, **1,** 277 (1957–58).

69. W. A. Zisman, *Advances in Chemistry*, No. 43, American Chemical Society, Washington, DC, 1964, p. 1.

70. J. R. Dann, *J. Colloid Interface Sci.*, **32,** 302 (1970).

71. D. H. Kaelble, in *Physical Chemistry of Adhesion*, Wiley-Interscience, New York, 1971.

72. H. Schonhorn and H. L. Frisch, *J. Polym. Sci.*, **11,** 1005 (1973).

73. R. J. Good, in *Adhesion Science and Technology*, Vol. 9A, L. H. Lee, ed., Plenum, New York, 1975.

74. S. V. Pepper, *J. Appl. Phys.*, **50,** 8062 (1979).

75. R. S. Boorse, P. Alemany, J. M. Burlitch, and R. Hoffmann, *Chem. Mat.*, **5,** 459 (1993).

76. B. V. Derjaguin, V. M. Muller, N. S. Mikhovich, and Yu. P. Toporov, *J. Colloid Interface Sci.*, **118,** 553 (1987); B. V. Deryaguin, V. M. Muller, Yu. P. Toporov, and I. N. Aleinikova, *Powder Tech.*, **37,** 87 (1984).

77. N. Kallay, B. Biškup, M. Tomić, and E. Matijević, *J. Colloid Interface Sci.*, **114,** 357 (1986); G. Thompson, N. Kallay, and E. Matijević, *Chem. Eng. Sci.*, **39,** 1271 (1984).

78. N. Kallay, E. Barouch, and E. Matijević, *Adv. Colloid Interface Sci.*, **27,** 1 (1987).

79. N. Ryde and E. Matijević, *J. Colloid Interface Sci.*, **169,** 468 (1995).

80. D. J. L. McIver and S. Schürch, *J. Adhes.*, **22,** 253 (1987).

81. N. Mozes, F. Marchal, M. P. Hermesse, J. L. Van Haecht, L. Reuliaux, A. J. Leonard, and P. G. Rouxhet, *Biotech. Bioeng.*, **30,** 439 (1987).

82. D. R. Absolom, C. Thomson, G. Kruzyk, W. Zingg, and A. W. Neumann, *Colloids & Surfaces*, **21,** 447 (1986).

83. E. Ruckenstein and S. V. Gourisankar, *J. Colloid Interface Sci.*, **101,** 436 (1984).

84. R. E. Baier, A. E. Meyer, J. R. Natiella, R. R. Natiella, and J. M. Carter, *J. Biomed. Mat. Res.*, **18,** 337 (1984).

85. W. Zingg and A. W. Neumann, *Biofouling*, **4,** 293 (1991).

86. S. Lahooti and A. W. Neumann, *J. Adhes.*, **54,** 241 (1995).

87. S. Lahooti, H. K. Yueh, and A. W. Neumann, *Colloids & Surfaces*, **B3,** 333 (1995).

88. J. A. Gould et al., *J. Appl. Biomat.*, **4,** 355 (1993).

89. R. E. Baier, *J. Biomech. Eng.*, **104,** 257 (1982).

90. R. E. Baier, A. E. Meyer, J. R. Natiella, R. R. Natiella, and J. M. Carter, *J. Biomed. Mat. Res.*, **18,** 337 (1984).

91. R. E. Baier, K. H. Rittle, and A. E. Meyer, "Influences of Surface Chemical Factors

on Selective Cellular Retention," in *Principles of Cell Adhesion*, P. D. Richardson and M. Steiner, eds., CRC Press, 1995.

92. U. Nassar, A. E. Meyer, R. E. Ogle, and R. E. Baier, *Peridontology 2000*, **8**, 114 (1995).

93. M. P. Olivieri, R. E. Loomis, A. E. Meyer, and R. E. Baier, *J. Adhes. Sci. Technol.*, **4**, 197 (1990).

94. M. P. Olivieri, R. E. Baier, and R. E. Loomis, *Biomaterials*, **13**, 1000 (1992).

95. E. A. Evans, *Biophys. J.*, **30**, 265 (1980).

96. F. Moussy, F. Y. H. Lin, S. Lahooti, Z. Policova, W. Zingg, and A. W. Neumann, *Colloids & Surfaces*, **B2**, 493 (1994).

97. H. K. Christenson, R. G. Horn, and J. N. Israelachvili, *J. Colloid Interface Sci.*, **88**, 79 (1982).

98. H. R. Brown, *MRS Bull.*, **21**, 24 (1996).

99. P. G. de Gennes, *CR Acad. Sci. Paris*, **307**, 1949 (1988).

100. H. R. Brown, V. R. Deline, and P. F. Green, *Nature*, **341**, 221 (1989).

101. C. Creton, E. J. Kramer, C. Y. Hui, and H. R. Brown, *Macromolecules*, **25**, 3075 (1992).

102. A. R. C. Baljon and M. O. Robbins, *Science*, **271**, 482 (1996).

103. F. Brochard-Wyart, P. G. de Gennes, L. Lèger, and E. Raphael, *J. Phys. Chem.*, **98**, 9405 (1994).

104. M. Deruelle, M. Tirrell, Y. Marciano, H. Hervet, and L. Lèger, *Faraday Disc.*, **98**, 55 (1994).

105. B. Zhang Newby, M. K. Chaudhury, and H. R. Brown, *Science*, **269**, 1407 (1995).

106. R. F. Roberts, F. W. Ryan, H. Schonhorn, G. M. Sessler, and J. E. West, *J. Appl. Polym. Sci.*, **20**, 255 (1976).

107. A. Baszkin and L. Ter-Minassian-Saraga, *Polymer*, **19**, 1083 (1978).

108. W. D. Bascom, P. F. Becher, J. I. Bitner, and J. S. Murday, Special Technical Publication 640, American Society for Testing and Materials, Philadelphia, 1978.

109. D. Maugis and M. Barquins, *J. Phys. D; Appl. Phys.*, **11**, 1989 (1978).

110. R. J. Good, Special Technical Publication 640, American Society for Testing and Materials, Philadelphia, 1978.

111. J. J. Bikerman, *The Science of Adhesive Joints*, 2nd ed., Academic, New York, 1968.

112. H. Schonhorn and F. W. Ryan, *J. Polym. Sci.*, **7**, 105 (1969).

113. R. J. Good, *J. Adhes.*, **4**, 133 (1972).

114. H. Schonhorn, in *Polymer Surfaces*, D. T. Clark and W. J. Frost, eds., Wiley-Interscience, New York, 1978.

115. A. W. Adamson, *J. Colloid Interface Sci.*, **44**, 273 (1973).

116. A. W. Adamson, F. P. Shirley, and K. T. Kunichika, *J. Colloid Interface Sci.*, **34**, 461 (1970).

117. J. R. Huntsberger, *Advances in Chemistry*, No. 43, American Chemical Society, Washington, D.C., 1964, p. 180; also, L. H. Sharpe and H. Schonhorn, ibid., p. 189.

118. T. A. Bush, M. E. Counts, and J. P. Wightman, in *Polymer Science and Technology*, L. Lee, ed., Plenum, New York, 1975.

119. See D. D. Eley, ed., *Adhesion*, Clarendon Press, Oxford, 1961.

120. D. W. Dwight, M. E. Counts, and J. P. Wightman, *Colloid and Interface Science*, Vol. III, *Adsorption, Catalysis, Solid Surfaces, Wetting, Surface Tension, and Water*, Academic, New York, 1976, p. 143.

121. W. A. Zisman, *Ind. Eng. Chem.*, **55**(10), 18 (1963).

122. H. H. G. Jellinek, *J. Colloid Interface Sci.*, **25**, 192 (1967).

123. W. D. Bascom, R. L. Cottington, and C. R. Singleterry, *J. Adhes.*, **1**, 246 (1969).

124. M. Landy and A. Freiberger, *J. Colloid Interface Sci.*, **25**, 231 (1967).

125. P. Barnes, D. Tabor, and J. C. F. Walker, *Proc. Roy. Soc.* (London), **A324**, 127 (1971).

126. H. H. G. Jellinek, H. Kachi, S. Kittaka, M. Lee, and R. Yokota, *Colloid Polym. Sci.*, **256**, 544 (1978).

CHAPTER XIII

Wetting, Flotation, and Detergency

1. Introduction

We continue, in this chapter, the discussion of areas of applied surface chemistry that was initiated in Chapter XII. The topics to be taken up here are so grouped because they have certain aspects in common. In each case contact angles are important and most of this work has involved the use of surface-active materials designed to adsorb at the interface and alter the interfacial tension. As was true in the case of friction and lubrication, a great deal of work has been done in the fields of wetting, flotation, and detergency, to a point that a certain body of special theory has developed as well as a very large literature of applied surface chemistry. It is not the intent here, however to cover these fields in any definitive way, but rather to use them to illustrate the role of surface chemistry in areas of practical importance. The topic of detergency is of great importance and interest and probably worthy of a chapter of its own. Because of its colloid chemistry nature, however, it is only summarized briefly in this chapter.

2. Wetting

A. Wetting as a Contact Angle Phenomenon

The terms *wetting and nonwetting* as employed in various practical situations tend to be defined in terms of the effect desired. Usually, however, wetting means that the contact angle between a liquid and a solid is zero or so close to zero that the liquid spreads over the solid easily, and nonwetting means that the angle is greater than $90°$ so that the liquid tends to ball up and run off the surface easily.

To review briefly, a contact angle situation is illustrated in Fig. XIII-1, and the central relationship is the Young equation (see Section X-4A):

$$\cos \theta = \frac{\gamma_{SV} - \gamma_{SL}}{\gamma_{LV}} \tag{XIII-1}$$

for the case of a finite contact angle, and the spreading coefficient $S_{L/S}$,

Fig. XIII-1.

$$S_{L/S} = \gamma_{SV} - \gamma_{LV} - \gamma_{SL} \tag{XIII-2}$$

should wetting occur.

Thus, to encourage wetting, γ_{SL} and γ_{LV} should be made as small as possible. This is done in practice by adding a surfactant to the liquid phase. The surfactant adsorbs to both the liquid–solid and liquid–vapor interfaces, lowering those interfacial tensions. Nonvolatile surfactants do not affect γ_{SV} appreciably (see, however, Section X-7). It might be thought that it would be sufficient merely to lower γ_{LV} and that a rather small variety of additives would suffice to meet all needs. Actually it is equally if not more important that the surfactant lower γ_{SL}, and each solid will make its own demands.

Much attention has been devoted in recent years to wetting, dewetting, and instabilities at the wetting front. As with the contact angle, recent theoretical work has focused on the importance of long-range forces [1] and hydrodynamics [2, 3] on wetting. The structural forces in simple liquids result in stratified flows of molecular dimension [4, 5]. The primary effect of long-range interactions is that liquids do not spread down to monolayer dimensions but rather retain a "pancake" structure of finite thickness. The pancake structure has been revealed by ellipsometric studies by Cazabat and co-workers [6] and is nicely summarized in a review [7]. The effect of the surface energy has been studied using Langmuir–Blodgett films or self-assembled monolayers to create surfaces of different energies [8, 9]. In the case of "dry wetting" where a nonvolatile material is spreading onto a surface where $S_{L/S}$ is nearly zero, Silberzan and Léger found wetting behavior between partial and total wetting [9]. In this regime, the spreading is highly sensitive to the surface polarizability.

The thermodynamics of the wetting transition as the critical point is approached has been treated for simple fluids [10–12] and polymer solutions [13], and the influence of vapor adsorption on wetting has been addressed [14]. Widom has modeled the prewetting transition with a van der Waals–like theory and studied the boundary tension between two coexisting surface phases [15]. The line tension at the three phase contact line may diverge as the wetting transition is approached [16]. The importance of the precursor film (see Section X-7A) on wetting behavior was first realized by Marmur and Lelah in 1980 in their discovery of the dependence of spreading rates on the size of the solid surface [17]. The existence of this film has since received considerable attention [1, 18, 19].

Wetting behavior on rough surfaces is an important topic for many practical applications. Analytical models provide an equilibrium interface shape as a function of the film thickness and interaction potential [20]. Harden and Andleman have investigated thermal fluctuations on the surface of films on rough solids and found, as one might expect, that for thick films the surface fluctuations were dominated by capillary waves (see Section IV-3C) while thin films had surface waves correlated with the solid surface roughness [21]. Experimental measurements of the thickness by x-ray reflectivity [18] are in general agreement with models. Fermigier and co-workers have studied wetting on heterogeneous surfaces in a narrow "Hele–Shaw" cell having defects made by ink droplets [21a]. Air bubbles remained trapped on the defect as a result of contact line pinning (see Section X-5A).

There are various situations where good contact is desired between a liquid, usually an aqueous one, and an oily, greasy, or waxy surface. Examples would include sprays of various kinds, such as insecticidal sprays, which should wet the waxy surface of leaves or the epidermis of insects; animal dips, where wetting of greasy hair is desired; inks, which should wet the paper properly; scouring of textile fibers, including the removal of unwanted natural oils and the subsequent wetting of the fibers by desirable lubricants; the laying of dust, where a fluid must penetrate between dust particles as on roads or in coal mines [22]. The eye is lubricated by a tear film, important both for the normal eye and for contact lens tolerance [23] since both surfaces should be wet.

Wetting agents or surfactant materials typically consist of polar–nonpolar-type molecules. The nonpolar portion is usually hydrocarbon in nature but may also be a fluorocarbon or a silicone; it may be aliphatic or aromatic. The polar portion may involve almost any of the functional groups of organic chemistry. The group may contain oxygen, as in carboxylic acids, esters, ethers or alcohols; it may have sulfur as in sulfonic acids sulfates, or their esters; it may contain phosphorous, nitrogen or a halogen; it may or may not be ionic.

Silicon surfactants enable water to wet hydrophobic surfaces while not inhibiting its tendency to wet hydrophilic surfaces. Such surfactants are called "superspreaders" if the addition of a small amount (~0.1%) allows water to spread into a thin wetting film on a hydrophobic surface within tens of seconds [24]. While most wetting studies involve microscopic observation of the spreading front, several groups have adapted the Wilhelmy plate (see Section II-6C) technique to measure the interfacial tension in wetting liquids or *wetting tension* [25, 26]. Ellipsometric studies [6, 8] and x-ray reflectivity [18] have provided more quantitative analysis of the layer structure. A problem in wetting arises due to contact angle hysteresis (see Section X-5). Usually it is the advancing angle that is important, but in the case of sheep dips and other such processes it is the receding angle as the animal is removed from the bath that impacts the degree of retention of the dip.

A drop of surfactant solution will, under certain conditions, undergo a fingering instability as it spreads on a surface [27, 28]. This instability is attributed to the Marongoni effect (Section IV-2D) where the process is driven by surface tension gradients. Pesach and Marmur have shown that Marongoni flow is also responsible for enhanced spreading

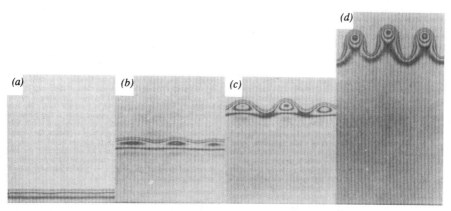

Fig. XIII-2. The development of the Marangoni fingering instability in a wetting thin film of a siloxane oil climbing a vertical surface with a temperature gradient ($d\gamma/dx =$ 0.18 Pa) at times (a) 1.5 min, (b) 6.5 min, (c) 10 min, and (d) 17 min. The tip-to-tip wavelength in (d) is 0.5 mm. (From Ref. 31.)

in binary mixtures of two wetting liquids [29]. Davis and co-workers found the Marongoni effect responsible for superspreading of water–silicone surfactant solutions on a hydrophobic surface in saturated water vapor [24]. A recent study indicates that the layering or formation of stratified structures in wetting layers of superspreaders may have a role in their unique properties [19]. Theoretical analysis by Carles and Cazabat describes the thickness of surface–tension–gradient-driven spreading flows [30].

A similar Marongoni instability can be provoked in a single component system by a temperature gradient [31] as illustrated in Fig. XIII-2. The wavelength of the instability is approximately

$$\text{Wavelength} = 25h_0(3\text{Ca})^{-1/3} \qquad \text{(XIII-3)}$$

where h_0 is the film thickness and Ca is the capillary number, which was defined in Chapter X as $\text{Ca} = V\mu/\gamma_{\text{LV}}$, where V is the velocity of the film and μ is the fluid viscosity. The film speed in this case is

$$V = \frac{h_0\, d\gamma/dx}{2\mu} \qquad \text{(XIII-4)}$$

where $d\gamma/dx = d\gamma/dT \times dT/dx$ is the vertical surface tension gradient produced by the temperature gradient imposed on the surface [31]. Substituting Eq. XIII-4 into Eq. XIII-3, one sees that the wavelength becomes independent of viscosity and varies slowly with the surface tension gradient.

The unstable situation caused when a spread film begins to *dewet* the surface has been studied [32, 33]. Dewetting generally proceeds from hole formation or retraction of the film edge [32] and hole formation can be a nucleation process or spinodal decomposition [34]. Brochart-Wyart and de Gennes provide a nice illustration of the importance of dewetting in hydroplaning of automobile tires on wet surfaces [34]. In this situation, a water film of approximately 1 μm must thin to form dry patches in timescales on the order of 0.005 sec! The attachment of polymers to the solid surface combined with their addition to the solution can stabilize nonwetting thin films [35,36].

The *speed* of wetting has been measured by running a tape of material that is wetted either downward through the liquid–air interface, or upward through the interface. For a polyester tape and a glycerol–water mixture, a wetting speed of about 20 cm/sec and a dewetting speed of about 0.6 cm/sec has been reported [37]. Conversely, the time of rupture of thin films can be important (see Ref. 38).

B. Wetting as a Capillary Action Phenomenon

For some types of wetting more than just the contact angle is involved in the basic mechanism of the action. This is true in the laying of dust and the wetting of a fabric since in these situations the liquid is required to penetrate between dust particles or between the fibers of the fabric. The phenomenon is related to that of capillary rise, where the driving force is the pressure difference across the curved surface of the meniscus. The relevant equation is then Eq. X-36,

$$\Delta P = \frac{2\gamma_{LV}\cos\theta}{r} \qquad\qquad (XIII-5)$$

where r denotes the radius (or equivalent radius) of the capillary. It is helpful to write Eq. XIII-5 in two separate forms. If θ is not zero, then

$$\Delta P = \frac{2(\gamma_{SV} - \gamma_{SL})}{r} \qquad\qquad (XIII-6)$$

so that the principal requirement for a large ΔP is that γ_{SL} be made as small as possible since for practical reasons it is not usually possible to choose γ_{SV}. On the other hand, if θ is zero, Eq. XIII-5 takes the form

$$\Delta P = \frac{2\gamma_{LV}}{r} \qquad\qquad (XIII-7)$$

and the requirement for a large ΔP is that γ_{LV} be large. The net goal is then to find a surfactant that reduces γ_{SL} without at the same time reducing γ_{LV}. Since any given surfactant affects both interfacial tensions, the best agent for producing such opposing effects can be expected to vary from one system to another even more than with ordinary wetting agents. Capillary phenomena in assemblies of parallel cylinders have been studied by Princen [39].

Capillary pressure gradients and Marongoni flow induce flow in porous media comprising glass beads or sand particles [40–42]. Wetting and spreading processes are an important consideration in the development of inkjet inks and paper or transparency media [43]; see the article by Marmur [44] for analysis of capillary penetration in this context.

In addition to ΔP being large, it is also desirable in promoting capillary penetration that the *rate* of entry be large. For the case of horizontal capillaries or, in general, where gravity can be neglected, Washburn [45] gives the following equation for the rate of entry of a liquid into a capillary radius r:

$$v = \frac{r\gamma_{L_1L_2} \cos \theta_{SL_1L_2}}{4(\eta_1 l_1 + \eta_2 l_2)} \tag{XIII-8}$$

This equation is for the general case of wetting liquid L_1 displacing liquid L_2 where the respective lengths of the liquid columns are l_1 and l_2 and the viscosities are η_1 and η_2. For a single liquid displacing air, the quantity $\gamma_L(\cos \theta)/\eta$ has the dimensions of velocity and thus gives a measure of the penetrating power of the liquid in a given situation.

The Washburn equation has most recently been confirmed for water and cyclohexane in glass capillaries ranging from 0.3 to 400 μm in radii [46]. The contact angle formed by a moving meniscus may differ, however, from the static one [46, 47]. Good and Lin [48] found a difference in penetration rate between an outgassed capillary and one with a vapor adsorbed film, and they propose that the driving force be modified by a film pressure term.

The Washburn model is consistent with recent studies by Rye and co-workers of liquid flow in V-shaped grooves [49]; however, the experiments are unable to distinguish between this and more sophisticated models. Equation XIII-8 is also used in studies of *wicking*. Wicking is the measurement of the rate of capillary rise in a porous medium to determine the average pore radius [50], surface area [51] or contact angle [52].

3. Water Repellency

Complementary to the matter of wetting is that of water repellency. Here, the desired goal is to make θ as large as possible. For example, in steam condensers, heat conductivity is improved if the condensed water does not wet the surfaces, but runs down in drops.

Fabrics may be made water-repellent by reversing the conditions previously discussed in the promotion of the wetting of fabrics. In other words, it is again a matter of capillary action, but now a large negative value of ΔP is desired. As illustrated in Fig. XIII-3a, if ΔP is negative (and hence if the contact angle is greater than 90°), the liquid will tend not to penetrate between the fibers, whereas if ΔP is positive, liquid will pass through easily. It should be noted

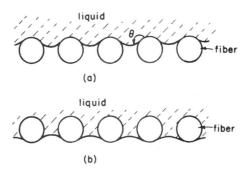

liquid

θ

fiber

(a)

liquid

fiber

(b)

Fig. XIII-3. Effect of contact angle in determining water repellency of fabrics.

that a fabric so treated that it acts as in Fig. XIII-3a is only water-repellent, not waterproof. The fabric remains porous, and water will run through it if sufficient hydrostatic pressure is applied.

Since a finite contact angle is present, Eq. XIII-6 shows that the surface tension of the liquid is not directly involved but rather the quantity ($\gamma_{SV} - \gamma_{SL}$) must be made negative. This can be done by coating the solids to reduce γ_{SV} such that its critical surface tension (Section X-5B) is less than about 40. The open mesh or screen structure leads to larger contact angles (Eq. X-29) and waterproofing agents can be designed to roughen the surface to increase the contact angle (Eqs. X-31 and X-33). An example was shown in Fig. X-6, where a superrepellent fractal surface of a sizing agent gained its repellency from surface roughness.

Apparently a negative ΔP with $\theta < 90°$ can be found for particular pore geometries [53]. A different type of water repellency is desired to prevent the deterioration of blacktop roads consisting of crushed rock coated with bituminous materials. Here the problem is that water tends to spread into the stone-oil interface, detaching the aggregate from its binder [54]. No entirely satisfactory solution has been found, although various detergent-type additives have been found to help. Much more study of the problem is needed.

Contemporary concern about pollution has made it important to dispose of oil slicks from spills. The suitable use of surfactants may reverse the spreading of the slick, thereby concentrating the slick for easier removal.

4. Flotation

A very important but rather complex application of surface chemistry is to the separation of various types of solid particles from each other by what is known as *flotation*. The general method is of enormous importance to the mining industry; it permits large-scale and economic processing of crushed ores whereby the desired mineral is separated from the *gangue* or non-mineral-containing material. Originally applied only to certain sulfide and oxide ores, flotation methods are now used not only for these but also in many other cases as well. A partial list of ores so treated commercially would include nickel and gold, as well as calcite, fluorite, barite (barium sulfate), scheelite (calcium tungstate), manganese carbonate and oxides, iron oxides, garnet, iron titanium oxides, silica and silicates, coal, graphite, sulfur, and soluble salts such as sylvite (potassium chloride). Flotation is also widely used to *remove* undesired minerals in water purification and in cleaning up industrial residues (see Refs. 55–57).

Prior to about 1920, flotation procedures were rather crude and rested primarily on the observation that copper and lead-zinc ore *pulps* (crushed ore mixed with water) could be *benefacted* (improved in mineral content) by treatment with large amounts of fatty and oily materials. The mineral particles collected in the oily layer and thus could be separated from the gangue and the water. Since then, oil flotation has been largely replaced by froth or foam flotation. Here, only minor amounts of oil or surfactant are used and a froth is formed by agitating or bubbling air through the suspension. The oily froth or foam is concentrated in mineral particles and can be skimmed off as shown schematically in Fig. XIII-4.

It was observed very early that rather minor variations in the composition of the oils could make great differences in performance and many secret recipes developed. The field is unusual in that empirical practice has been in the lead, with theory struggling to explain. Some basic aspects are fairly well under-

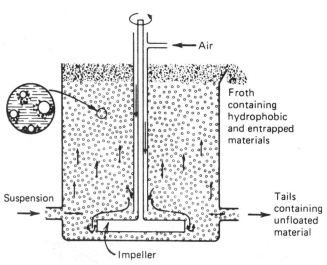

Fig. XIII-4. Schematic diagram of a froth flotation cell. Note the mineralized bubble shown in the inset. [Reprinted with permission from P. Somasumdaran, "Interfacial Chemistry of Particulate Flotation." *AIChE Symp. Ser.*, **71**(150), 2 (1975) (Ref. 58). Reproduced by permission of the American Institute of Chemical Engineers.]

stood and special additives have been designed. These include *collectors*, which adsorb on the mineral particles to alter the contact angle; *activators*, which enhance the selectivity of the collector; *depressants*, which selectively reduce the action of the collector; and *frothing agents*, to promote foam formation. A brief review by Wark [59] provides a good summary.

A. *The Role of Contact Angle in Flotation*

The basic phenomenon involved is that particles of ore are carried upward and held in the froth by virtue of their being attached to an air bubble, as illustrated in the inset to Fig. XIII-4. Consider, for example, the gravity-free situation indicated in Fig. XIII-5 for the case of a spherical particle. The particle may be entirely in phase A or entirely in phase B. Alternatively, it may be located in the interface, in which case both γ_{SA} and γ_{SB} contribute to the total surface free energy of the system. Also, however, some liquid–liquid interface has been eliminated. It may be shown (see Problem XIII-12) that if there is a finite contact angle, θ_{SAB}, the stable position of the particle is at the interface, as shown in Fig. XIII-5*b*. Actual measured detachment forces are in the range of 5 to 20 dyn [60].

It is helpful to consider qualitatively the numerical magnitude of the surface tensional stabilization of a particle at a liquid–liquid interface. For simplicity, we will assume θ = 90°, or that $\gamma_{SA} = \gamma_{SB}$. Also, with respect to the interfacial areas, $\mathcal{A}_{SA} = \mathcal{A}_{SB}$, since the particle will lie so as to be bisected by the plane of the liquid–liquid interface, and $\mathcal{A}_{AB} = \pi r^2$. The free energy to displace the particle from its stable position will then be just $\pi r^2 \gamma_{AB}$. For a particle of 1-mm radius, this would amount to about 1 erg, for $\gamma_{AB} = 40$ ergs/cm^2. This corresponds roughly to a restoring force of 10 dyn, since this work must be expended in moving the particle out of the interface, and this amounts to a displacement equal to the radius of the particle.

In the usual situation illustrated in Fig. XIII-6 the particle is supported at a liquid–air interface against gravitational attraction. As was seen, the restoring force

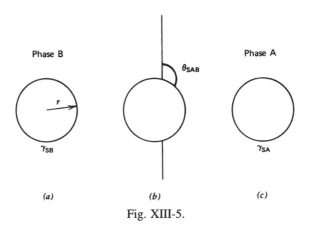

Fig. XIII-5.

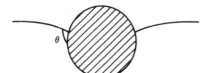

Fig. XIII-6.

stabilizing the particle at the interface varies approximately with the particle radius; for the particle to remain floating, this restoring force must be equal to or exceed that of gravity. Since the latter varies with the cube of the radius, it is clear that there will be a maximum size of particle that can remain at the interface. Thus referring to the preceding example, if the particle density is 3 g/cm^3, then the maximum radius is about 0.1 cm. The preceding analysis has been on an elementary basis; Rapacchietta and Neumann [61] have investigated the difficult problem of determining the actual surface profile of Fig. XIII-6.

In practice, it may be possible with care to float somewhat larger particles than those corresponding to the theoretical maximum. As illustrated in Fig. XIII-7, if the particle has an irregular shape, it will tend to float such that the three-phase contact occurs at an asperity since the particle would have to be depressed considerably for the line of contact to advance further. The resistance to rounding a sharp edge has been investigated by Mason and co-workers [62].

The preceding upper limit to particle size can be exceeded if more than one bubble is attached to the particle.† A matter relating to this and to the barrier that exists for a bubble to attach itself to a particle is discussed by Leja and Poling [63]; see also Refs. 64 and 65. The attachment of a bubble to a surface may be divided into steps, as illustrated in Figs. XIII-8a–c, in which the bubble is first distorted, then allowed to adhere to the surface. Step 1, the distortion step, is not actually unrealistic, as a bubble impacting a surface does distort, and only after the liquid film between it and the surface has sufficiently thinned does

Fig. XIII-7.

†An instructive and decorative illustration of this is found in the following parlor experiment. Some water is poured into a glass bowl and about 1% by weight of sodium bicarbonate is added and then some moth balls. About one-third of the volume of vinegar is then poured in carefully. The carbon dioxide that is slowly generated clings as bubbles to the moth balls, and each ball rises to the surface when a net buoyancy is reached. On reaching the surface, some of the bubbles break away, the ball sinks, and the process is repeated.

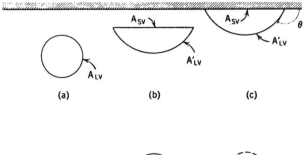

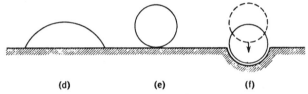

Fig. XIII-8. Bubble distortion and adhesion.

adhesion suddenly occur (see Refs. 64 and 66 to 68). For step 1, the surface free energy change is

$$\Delta G_1 = \Delta \mathcal{A}_{LV}\, \gamma_{LV} \qquad\qquad \text{(XIII-9)}$$

where $\Delta \mathcal{A}_{LV} = (\mathcal{A}'_{LV} + \mathcal{A}_{SV} - \mathcal{A}_{LV})$, and, for the second step,

$$\Delta G_2 = (\gamma_{SV} - \gamma_{SL} - \gamma_{LV})\mathcal{A}_{SV} = -w_a \mathcal{A}_{SV} \qquad\qquad \text{(XIII-10)}$$

so that for the overall process

$$\Delta G = -w_a \mathcal{A}_{SV} + \gamma_{LV}\Delta \mathcal{A}_{LV} = -w_{\text{pract}} \qquad\qquad \text{(XIII-11)}$$

where w_a is the work of adhesion (*not* the same as w_{SL}). The quantity $-\Delta G$ thus represents the practical work of adhesion of a bubble to a flat surface, while ΔG_1 is approximately the activation energy barrier to the adhesion.

As a numerical illustration, consider a bubble 0.1 cm in diameter in a system for which $w_a = 30$ ergs/cm^2, $\gamma_{LV} = 30$ ergs/cm^2, and $\theta = 90°$. The distorted bubble is just hemispherical, and we find $w_{\text{pract}} = 0.19$ erg, and $\Delta G_1 = 0.18$ erg. The gravitational energy available from the upward motion is only 0.066 erg, or less than the ΔG_1 barrier, so that adhesion might be difficult. Conversely, if the bubble were clinging to the upper side of a surface, as in Fig. XIII-8d, ΔG_2 gives a measure of the energy barrier for detachment and is 0.38 erg; again, the barrier is more than the gravitational energy available, this time from the upward motion of the center of gravity of the bubble in the reverse step 1 plus step 2. One can thus account qualitatively for the reluctance of bubbles to attach to a surface and, once attached, to leave it.

If the contact angle is zero, as in Fig. XIII-8e, there should be no tendency to adhere to a flat surface. Leja and Poling [63] point out, however, that, as shown in Fig. XIII-8f, if the surface is formed in a hemispherical cup of the same radius as the bubble, then for step 1a, the free energy change of attachment is

$$\Delta G_{1a} = -w_a \mathcal{A}_{SV} \qquad \text{(XIII-12)}$$

that is, there is now no distortion energy. In terms of the foregoing numerical example, the work of adhesion would be 0.47 erg, or greater than that for a contact angle of 90° and a flat surface.

The flotation of mica has been correlated to the adhesion force measured from surface force (SFA—see Section VI-4) experiments although, to these authors, it is clear that dynamic effects prevent an absolute comparison [69, 70].

These illustrations bear on the two principal mechanisms of flotation: nucleation and growth of the particle at the bubble surface [71] and collision between bubbles and particles [72]. The collision mechanism seems to be the more important process [59]. The efficiency of a bubble in collecting mineral particles is given as a product $E_{col} = E_c E_a E_s$, where E_c, E_a and E_s are the collision, attachment, and stability efficiencies, respectively. The collision efficiency is well understood by aggregation models and the stability of the particle–bubble complex is generally unity for systems of interest in flotation; however, the attachment efficiency was only recently measured by Ralston and co-workers for hydrophobized quartz particles [73]. E_a is higher for smaller bubbles and it decreases with increasing particle size.

The importance of the thin film between the mineral particle and the air bubble has been discussed in a review by Pugh and Manev [74]. In this paper, modern studies of thin films via SFA and interferometry are discussed. These film effects come into play in the stability of foams and froths. Johansson and Pugh have studied the stability of a froth with particles. Small (30-μm), moderately hydrophobic ($\theta_c \approx 65°$) quartz particles stabilized a froth, while more hydrophobic particles destabilized it and larger particles had less influence [75].

B. Flotation of Metallic Minerals

Clearly, it is important that there be a large contact angle at the solid particle-solution–air interface. Some minerals, such as graphite and sulfur, are naturally hydrophobic, but even with these it has been advantageous to add materials to the system that will adsorb to give a hydrophobic film on the solid surface. (Effects can be complicated—sulfur flotability oscillates with the number of preadsorbed monolayers of hydrocarbons such as n-heptane [76].) The use of surface modifiers or collectors is, of course, essential in the case of naturally hydrophilic minerals such as silica.

In the case of lead and copper ores, the use of xanthates,

$$S=C\begin{array}{c} {}^{O-R} \\ {}_{S^- K^+} \end{array}$$

has been widespread, and a reasonable explanation was that a reaction of the type

$$Pb(OH)_2 + 2ROCS_2^- = \left[Pb-S-C\begin{array}{c} {}^{S} \\ {}_{OR} \end{array} \right]_2 + 2OH^- \qquad (XIII-13)$$

occurred. However, early empirical observations that found dissolved oxygen played a role have now been recognized as correct. Thus, ethyl xanthate does not adsorb on copper in deaerated systems [77, 78]; not only does it appear that oxidation of the surface is important, but also the actual surfactant may be an oxidation product of the xanthate, dixanthogen,

$$(R-O-\underset{\underset{S}{\|}}{C}-S)_2$$

Perxanthate ion may also be implicated [59]. Even today, the exact nature of the surface reaction is clouded [59, 79–81], although Gaudin [82] notes that the role of oxygen is very determinative in the chemistry of the mineral–collector interaction.

Various chemical tricks are possible. Zinc ores are not well floated with xanthates, but a pretreatment with dilute copper sulfate rectifies the situation by electrodepositing a thin layer of copper on the mineral particles (note Ref. 83 for complexities). Chelating agents such as oximes may be used instead of xanthates [84]. Treatment of an ore containing a mixture of iron, zinc, and lead minerals with dilute cyanide solution will inhibit adsorption of the collector on the first two, but not on the last. In this case, cyanide is called a *depressant*. Depressants are also used to inhibit the undesired coflotation of talc, sulfate, graphite, and so on; organic polymers have been useful [85]. STM and AFM studies of galena (PbS) surfaces show the formation of 0.3–0.6-nm pits during the surface chemical reactions controlling flotation [86].

Very finely divided minerals may be difficult to purify by flotation since the particles may adhere to larger, undesired minerals—or vice versa, the fines may be an impurity to be removed. The latter is the case with TiO_2 (anatase) impurity in kaolin clay [87]. In *carrier flotation*, a coarser, separable mineral is added that will selectively pick up the fines [88,89]. The added mineral may be in the form of a floc (ferric hydroxide), and the process is called *adsorbing colloid flotation* [90]. The fines may be aggregated to reduce their loss, as in the addition of oil to agglomerate coal fines [91].

In addition to a large contact angle, it is also desirable that the mineral-laden bubbles not collapse when they reach the surface of the slurry, that is, that a stable (although not *too* stable) froth of such bubbles be possible. With this in mind, it is common practice to add various frothing agents to the system such as long-chain alcohols and pine oils. However, it turns out that the actions of the frothing agent and the collector are not independent of each other. The prime function of the frothing agent may therefore be to stabilize the attachment of the particle to the bubble, and its frothing ability, per se, may well be a matter of secondary importance. Also, as mentioned in the next chapter, well-mineralized bubbles themselves constitute a stable froth system.

C. Flotation of Nonmetallic Minerals

The examples in the preceding section, of the flotation of lead and copper ores by xanthates, was one in which chemical forces predominated in the adsorption of the collector. Flotation processes have been applied to a number of other minerals that are either ionic in type, such as potassium chloride, or are insoluble oxides such as quartz and iron oxide, or ink pigments [needed to be removed in waste paper processing [92]]. In the case of quartz, surfactants such as alkyl amines are used, and the situation is complicated by micelle formation (see next section), which can also occur in the adsorbed layer [93, 94].

As another example, sylvite (KCl) may be separated from halite (NaCl) by selective flotation in the saturated solutions, using long-chain amines such as a dodecyl ammonium salt. There has been some mystery as to why such similar salts should be separable by so straightforward a reagent. One suggestion has been that since the $R\!-\!NH_3^+$ ion is small enough to fit into a K^+ ion vacancy, but too large to replace the Na^+ ion, the strong surface adsorption in the former case may be due to a kind of isomorphous surface substitution (see Refs. 65, 95, and especially 96). Barytes ($BaSO_4$) may be separated from unwanted oxides by means of oleic acid as a collector; the same is true of calcite (CaF_2). Flotation is widely used to separate calcium phosphates from siliceous and carbonate minerals [97].

The flotation of the insoluble oxide minerals turns out to be best understood in terms of electrical double-layer theory. That is, the potential ψ of the mineral is as important as are specific chemical interactions. Hydrogen ion is potential determining (see Section V-3); for quartz, the pH of zero charge is about 3 while for, say, goethite, FeO(OH), it is 6.7 [99]. As illustrated in Fig. XIII-9, anionic surfactants are effective for goethite below pH 6.7 since the mineral is then positively charged, and cationic surfactants work at higher pH's since the charge is then negative.

The adsorption appears to be into the Stern layer, as was illustrated in Fig. V-3. That is, the adsorption itself reduces the ζ potential of such minerals; in fact, at higher surface coverages of surfactant, the ζ potential can be reversed, indicating that chemical forces are at least comparable to electrostatic ones. The rather sudden drop in ζ potential beyond a certain concentration suggested to

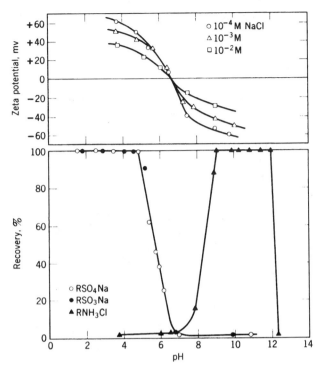

Fig. XIII-9. The dependence of the flotation properties of goethite on surface charge. Upper curves are ζ potential as a function of pH at different concentrations of sodium chloride: lower curves are the flotation recovery in $10^{-3}M$ solutions of dodecylammonium chloride, sodium dodecyl sulfate, or sodium dodecyl sulfonate. (From Ref. 99.)

Fuerstenau and Healy [100] and to Gaudin and Fuerstenau [101] that some type of near phase transition can occur in the adsorbed film of surfactant. They proposed, in fact, that surface micelle formation set in, reminiscent of Langmuir's explanation of intermediate type film on liquid substrates (Section IV-6).

In addition to the collector, polyvalent ions may show sufficiently strong adsorption on oxide, sulfide, and other minerals to act as potential-determining ions (see Ref. 98). Judicious addition of various salts, then, as well as pH control, can permit a considerable amount of selectivity.

5. Properties of Association Colloids—Micelles

The surface-active agents (surfactants) responsible for wetting, flotation and detergency exhibit rather special and interesting properties characteristic of what are called *association colloids* or, in the older literature, *colloidal electrolytes*. These properties play an important role in determining, at least indirectly, the detergency of a given surfactant and are therefore considered here briefly before discussing the mechanism of detergent action. The field is a vast one with many fascinating aspects beyond the scope of this book; the reader is referred to the General References for a more detailed review of the literature.

Many physical properties of surfactant solutions show consistent behavior with increasing concentration as illustrated in Fig. XIII-10 [102]. Sharp changes in properties occur in the region of the *critical micelle concentration* (CMC). The near-constancy of the osmotic pressure above the CMC suggests that something analogous to a phase transition is occurring and, although no macroscopic phase separation can be observed, the sudden increase in light scattering indicates the formation of colloidal species. The well-documented explanation is that in the region of the CMC aggregation of the surfactants into fairly large *micelles* occurs. Above the CMC, additional surfactants form additional micelles leaving a concentration of single chains constant at about the CMC. It is somewhat apart from the scope of this book to consider the physical chemistry of micellization in any detail, but the phenomenon is so ubiquitous in detergent solutions that a brief discussion is necessary.

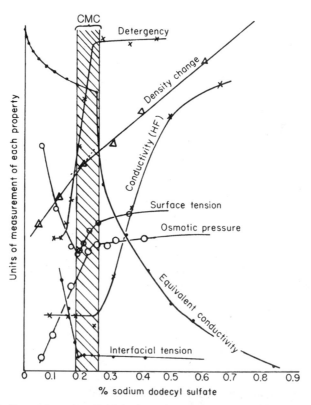

Fig. XIII-10. Properties of colloidal electrolyte solutions—sodium dodecyl sulfate. (From Ref. 102a.)

McBain [103] was one of the first to recognize the formation of micelles, and since then a voluminous literature has developed. See Refs. 104–110 for reviews. Micelles are often fairly narrowly dispersed in size, containing 50–100 monomer units (more for nonionic surfactants [111, 112]). Experimental evidence comes from both static and dynamic (Section IV-3C) light scattering [112–117], fluorescence quenching [118–123], small-angle neutron scattering [124–129], cryotransmission electron microscopy (Section VIII-2) [130–132], nuclear magnetic resonance (NMR) [133–134] dielectric spectroscopy [135], and electron spin resonance (ESR) [136, 137]. An important micellar property, particularly for detergency (see Section XIII-6) is *solubilization* where insoluble molecules such as benzene, dyes, or proteins are supported within micelles. Solubilization has been widely investigated [138–142] with recent experimental studies [143–145] complemented by theory [146–148]. The first measurements of micelle diffusion rates made use of a solubilized dye as a label [149, 150]; however, that method has now been supplanted by dynamic light scattering. Representative values of $\mathcal{D}$ are around 10^{-6} cm^2/sec [114–116, 151]. The lifetime of a micelle is very important to technological processes such as detergency [152], foamability [153], and wetting [154] and has been studied by pressure jump experiments by Shah and co-workers [152–155].

Surface active electrolytes produce charged micelles whose effective charge can be measured by electrophoretic mobility [117, 156]. The net charge is lower than the degree of aggregation, however, since some of the counterions remain associated with the micelle, presumably as part of a Stern layer (see Section V-3) [157]. Combination of self-diffusion with electrophoretic mobility measurements indicates that a typical micelle of a univalent surfactant contains about 100 monomer units and carries a net charge of 50–70. Additional colloidal characterization techniques are applicable to micelles such as ultrafiltration [158].

Micellar structure has been a subject of much discussion [104]. Early proposals for spherical [159] and lamellar [160] micelles may both have merit. A schematic of a spherical micelle and a unilamellar vesicle is shown in Fig. XIII-11. In addition to the most common spherical micelles, scattering and microscopy experiments have shown the existence of rodlike [161, 162], disklike [163], threadlike [132] and even quadruple-helix [164] structures. Lattice models (see Fig. XIII-12) by Leermakers and Scheutjens have confirmed and characterized the properties of spherical and membrane like micelles [165]. Similar analyses exist for micelles formed by diblock copolymers in a selective solvent [166]. Other shapes proposed include ellipsoidal [167] and a sphere-to-cylinder transition [168]. Fluorescence depolarization and NMR studies both point to a rather fluid micellar core consistent with the disorder implied by Fig. XIII-12.

The concentration of free surfactant, counterions, and micelles as a function of overall surfactant concentration is shown in Fig. XIII-13. Above the CMC, the concentration of free surfactant is essentially constant while the counterion concentration increases and

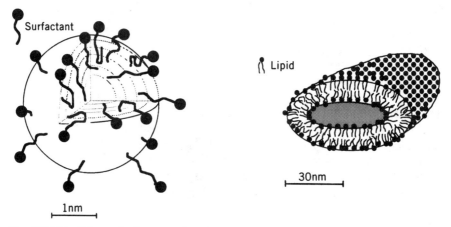

Fig. XIII-11. Schematic diagram of a spherical micelle and a unilamellar vesicle. (From Ref. 118.)

the micellar concentration increases approximately linearly. Numerous theories exist, including mass-action-based models [169–171] and those including various entropy estimates for the system [172]. Molecular dynamics simulations illustrate the importance of the hydrophobic attraction on micelle formation [173].

Amphiphilic diblock copolymers in a solvent that is selective for one of the blocks form micelles that have received extensive study [174]. Among the interesting block copolymer systems are block polyelectrolytes [175], fluorocarbon–hydrocarbon diblocks [176], and block copolymers compatibilizing immiscible polymer blends [177]. Many technological applications, including enhanced oil recovery, cosmetics, and paints, involve complex mixtures of polymers and surfactants (see Ref. 178 for a review of phase behavior). Numerous interesting properties exist including gelation [179, 180], threadlike structures [181] and rheological changes [182, 183].

A temperature known as the *Kraft temperature* denotes the point above which the

Fig. XIII-12. Lattice model of a spherical micelle. (From Ref. 160a.)

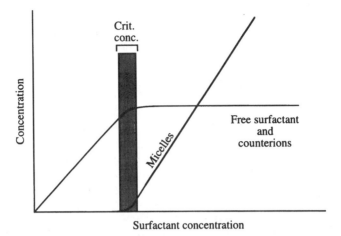

Fig. XIII-13. Concentrations of individual species in a surfactant solution.

solubility of a surfactant increases rapidly. A recent study provides a linear relation between the Kraft point and the logarithm of the CMC with a constant negative slope [184]. At temperatures below the Kraft temperature, surfactant molecules form a variety of liquid crystalline phases [185–187].

Other properties of association colloids that have been studied include calorimetric measurements of the heat of micelle formation (about 6 kcal/mol for a nonionic species, see Ref. 188) and the effect of high pressure (which decreases the aggregation number [189], but may raise the CMC [190]). Fast relaxation methods (rapid flow mixing, pressure-jump, temperature-jump) tend to reveal two relaxation times t_1 and t_2, the interpretation of which has been subject to much disagreement—see Ref. 191. A "fast" process of $t_1 \sim 1$ msec may represent the rate of addition to or dissociation from a micelle of individual monomer units, and a "slow" process of $t_2 < 100$ msec may represent the rate of total dissociation of a micelle (192; see also Refs. 193–195).

Practical systems will often have mixtures of surfactants. A useful rule is that the CMC of the mixture, C_m, is given by [196,197]

$$\frac{1}{c_m} = \sum_i \frac{N_i}{C_i} \qquad \text{(XIII-14)}$$

where N_i and C_i are the mole fraction and the CMC of the ith species, respectively.

The traditional association colloid is of the M^+R^- type where R^- is the surfactant ion, studied in aqueous solution. Such salts also form micelles in nonaqueous and nonpolar solvents. These structures, termed *inverse micelles*, have the polar groups inward if some water is present [198]; however, the presence of water may prevent the observation of a well-defined CMC [198,199]. Very complex structures may be formed in nearly anhydrous media (see Ref. 200).

There are other interesting surfactant structures including Gemini (dimeric) surfactants [130, 201], trimeric surfactants [132] and bola amphiphiles having two polar head groups attached to one hydrophobic tail [202]. Many nonionic detergents exist, in particular the ethylene oxide–alkane molecules are often studied. These also form micelles and can solubilize materials [203,204].

Surfactants have also been of interest for their ability to support reactions in normally inhospitable environments. Reactions such as hydrolysis, aminolysis, solvolysis, and, in inorganic chemistry, of aquation of complex ions, may be retarded, accelerated, or differently sensitive to catalysts relative to the behavior in ordinary solutions (see Refs. 205 and 206 for reviews). The acid–base chemistry in micellar solutions has been investigated by Drummond and co-workers [207]. A useful model has been the *pseudophase* model [206–209] in which reactants are either in solution or solubilized in micelles and partition between the two as though two distinct phases were involved. In inverse micelles in nonpolar media, water is concentrated in the micellar core and reactions in the micelle may be greatly accelerated [206, 210]. The confining environment of a solubilized reactant may lead to stereochemical consequences as in photodimerization reactions in micelles [211] or vesicles [212] or in the generation of radical pairs [213].

Much use has been made of micellar systems in the study of photophysical processes, such as in excited-state quenching by energy transfer or electron transfer (see Refs. 214–218 for examples). In the latter case, ions are involved, and their selective exclusion from the Stern and electrical double layer of charged micelles (see Ref. 219) can have dramatic effects, and ones of potential importance in solar energy conversion systems.

Finally, micellar systems are useful in separation methods. Micelles may bind heavy-metal ions, or, through solubilization, organic impurities. Ultrafiltration, chromatography, or solvent extraction may then be used to separate out such contaminants [220–222].

6. Detergency

Detergency may be defined as the removal of foreign material from solid surfaces by surface-active agents. This definition covers liquid and solid soil removal from fabrics, metal surfaces and so on, while excluding *purely mechanical* (e.g., abrading) and *purely chemical* (by reaction) cleansing. Common soap is the oldest and most familiar detergent and its use in laundering clothes is perhaps the best example of detergent action. Other important industrial examples include the removal of solder flux from printed-circuit boards [223, 224] and detergents added to lubricating oils to remove oxidation products from metal surfaces [225]. The field of detergency is a very important one for industries and consumers alike. The subject has received quite a bit of attention and is the subject of several excellent reviews [226, 227].

A. General Aspects of Soil Removal

The soil that accumulates on fabrics is generally of an oily nature and contains particles of dust, soot, and the like. Thus soil may be either a solid, a liquid, or a mixture of both. The oily material consists of animal fats and fatty acids, petroleum hydrocarbons, and a residuum of quite miscellaneous substances [227, 228].

The cleaning process proceeds by one of three primary mechanisms: solubilization, emulsification, and roll-up [229]. In *solubilization* the oily phase partitions into surfactant micelles that desorb from the solid surface and diffuse into the bulk. As mentioned above, there is a body of theoretical work on solubilization [146, 147] and numerous experimental studies by a variety of spectroscopic techniques [143–145,230]. *Emulsification* involves the formation and removal of an emulsion at the oil–water interface; the removal step may involve hydrodynamic as well as surface chemical forces. Emulsion formation is covered in Chapter XIV. In *roll-up* the surfactant reduces the contact angle of the liquid soil or the surface free energy of a solid particle aiding its detachment and subsequent removal by hydrodynamic forces. Adam and Stevenson's beautiful photographs illustrate roll-up of lanoline on wood fibers [231]. In order to achieve roll-up, one requires the surface free energies for soil detachment illustrated in Fig. XIII-14 to obey

$$\gamma_{SO} \geq \gamma_{OW} + \gamma_{SW} \qquad\qquad \text{(XIII-15)}$$

Thus, adding surfactants to minimize the oil–water and solid–water interfacial tensions causes removal to become spontaneous. On the other hand, a mere decrease in the surface tension of the water–air interface, as evidenced, say, by foam formation, is *not* a direct indication that the surfactant will function well as a detergent. The decrease in γ_{OW} or γ_{SW} implies, through the Gibb's equation (see Section III-5) adsorption of detergent.

In addition to lowering the interfacial tension between a soil and water, a surfactant can play an equally important role by partitioning into the oily phase carrying water with it [232]. This reverse solubilization process aids hydrodynamically controlled removal mechanisms. The partitioning of surface-active agents between oil and water has been the subject of fundamental studies by Grieser and co-workers [197, 233].

The experimental study of detergency ranges from the very practical to the molecular surface-sensitive approaches discussed in this book. In more practical studies, model

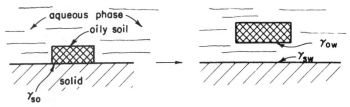

Fig. XIII-14. Surface tensional relationships in soil removal.

soils and soiling procedures have been developed. Standard soils generally comprise a mixture of carbon black and grease such as vaseline; however, many model soils are tailored to particular applications. The old launderometer was basically a washing machine having fairly reproducible conditions of temperature, agitation and so on. The *tergotometer* is widely used for studies of particulate and oily soil removal [226]. It has a two-liter temperature-controlled reservoir that holds test cloths that are subject to agitation by paddle. Another approach has a test surface on a cylindrical drum that is concentric with another drum having viewing windows to monitor detachment during washing (rotation) [226]. Soil removal is generally measured by reflectance, through an empirical relation between whiteness and amount of soil, or by radioactive labeling.

Many of the surface sensitive techniques described in Chapters II, IV, and VIII are being applied to problems in detergency. Carroll summarizes these studies in his review [226]. Microscopic measurements of droplet shape and contact angle are important tools to monitor detergency on the droplet length scale, and videomicroscopy allows study of the dynamics of these processes [227]. The substrate materials are inspected with the microscopic techniques of Chapter VIII, especially SEM and AFM. Dynamic surface tension measurements provide oil–water interfacial tension in analogy with the liquid–vapor tensions discussed in Chapter II. Ellipsometry, evanescent wave scattering, and electrophoresis studies have all been applied to the problem of soil detachment or deposition on surfaces and a novel optical-fiber technique monitors this process via total light transmitted through a fiber [226].

B. Factors in Detergent Action

The fact that successful detergents seem always to show the colloidal properties discussed has led to the thought that micelles must be directly involved in detergent action. McBain (see [234]), for example, proposed that solubilization was one factor in detergent action. Since micelles are able to solubilize dyes and other organic molecules, the suggestion was that oily soil might similarly be incorporated into detergent micelles. As illustrated in Fig. XIII-10, however, detergent ability rises before the CMC is reached and remains practically constant thereafter. Since the concentration of micelles rises steadily from the CMC on, there is thus no direct correlation between their concentration and detergent action. It therefore seems necessary to conclude that detergent action is associated with the long-chain monomer ion or molecule and that very likely the properties that make for good detergent action also lead to micelle formation as a *competing* rather than as a contributing process.

Much of ordinary soil involves particulate, more or less greasy matter, and an important attribute of detergents is their ability to keep such material suspended in solution once it is detached from the fabric and thus to prevent its redeposition. Were this type of action not present, washing would involve a redistribution rather than a removal of dirt. Detergents thus do possess *suspending power*. For example, carbon suspensions that otherwise would settle rapidly are stable indefinitely if detergent is present, and similarly such other solids as manganese dioxide [235, 236]. Evidently, detergent is adsorbed at the particle-solution interface, and suspending action apparently results partly through a resulting change in the charge of the particle and perhaps mainly through a deflocculation of particles that originally were agglomerates. Apart from charge repulsion, the presence of an adsorbed long-chain molecule prevents the coalescence of particles in much the same manner as it prevents the contact of surfaces in boundary lubrication (see Section XII-7B). A related effect, and one that is also important to good detergency is called *protective action*. This refers to the prevention of the particles of solid from adhering to the fabric.

Another component of detergent formulations is known as a *builder*—a substance with no detergent properties of its own but one that enhances the performance of a detergent (see Ref. 237). A typical builder would be pyrophosphate, $Na_4P_2O_7$; its effect may be partly as a sequestering agent for the metal ions that contribute to hardness [238]. It may play some role in affecting surface charges [239]. Recent concerns about excessive phosphate discharge into natural waters have led to a search for alternative builders. There has been some success with mixtures of electrolytes and sequestering agents [240].

The general picture of detergent action that has emerged is that of a balance of opposing forces (see Ref. 228). The soil tends to remain on the fabric either through surface tensional adhesion or mechanical entrapment and, on the other hand, tends to remain in suspension as a result of the suspending power and protective action of the detergent. See Ref. 241 for a discussion of optimum detergency conditions.

C. Adsorption of Detergents on Fabrics

The adsorption of detergent-type molecules on fabrics and at the solid–solution interface in general shows a complexity that might be mentioned briefly. Some fairly characteristic data are shown in Fig. XIII-15 [242]. There is a break at point *A*, marking a sudden increase in slope, followed by a *maximum* in the amount adsorbed. The problem is that if such data represent true equilibrium in a two-component system, it is possible to argue a second law violation (note Problem XIII-14) (although see Ref. 243).

The phenomenon occurs not only in fabric-aqueous detergent systems but also in the currently important problem of tertiary oil recovery. One means of displacing residual oil from a formation involves the use of detergent mixtures (usually petroleum sulfonates), and the degree of loss of detergent through adsorption by the oil-bearing formation is a matter of great importance. The presence of adsorption maxima (and even minima) in laboratory-determined isotherms is, to say the least, distressing!

After reviewing various earlier explanations for an adsorption maximum, Trogus, Schechter, and Wade [244] proposed perhaps the most satisfactory one so far (see also Ref. 245). Qualitatively, an adsorption maximum can occur if the surfactant consists of at least *two* species (which can be closely related); what is necessary is that species 2 (say) preferentially forms micelles (has a lower CMC) relative to species 1 and also adsorbs more strongly. The adsorbed state may also consist of aggregates or "hemi-micelles," and even for a pure component the situation can be complex (see Section XI-6 for recent AFM evidence of surface micelle formation and [246] for polymeric surface micelles). Similar adsorption maxima found in adsorption of nonionic surfactants can be attributed to polydispersity in the surfactant chain lengths [247]. Surface-active impuri-

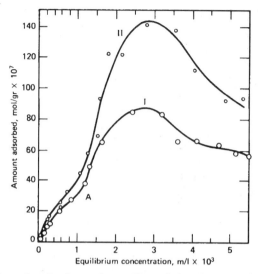

Fig. XIII-15. Adsorption isotherm for sodium dodecylbenzenesulfonate on cotton. Curve I, 30°C; curve II, 0°C. (From Ref. 64.)

ties play an important role in adsorption and a purification scheme for surfactants may aid systematic studies [248]. There are several models of surfactant adsorption (e.g., Ref. 249). A surface equation of state approach to the adsorption of sodium-n-alkyl sulfate surfactants at the air–water interface shows a pronounced odd/even-chain-length effect [250] and another theory of adhesion to surfaces shows why sodium dodecylsulfate adheres more strongly to "dirt" than to textiles [251].

Finally, adsorption dynamics can play an important role in detergency [232]. It is the subject of a model study [252] and a comprehensive review [253].

D. Detergents in Commercial Use

The reader is referred to monographs in the General References section for details but, very briefly, there are three main classes of detergents: anionic, cationic, and nonionic. The first group may be designated as MR, where R is the detergent anion, and includes the traditional soaps (fatty-acid salts) as well as a large number of synthetic detergents such as the sulfates and sulfonates (e.g., sodium dodecylbenzenesulfonate and sodium lauryl sulfate). Cationic detergents, RX, where R is now a detergent cation, include a variety of long-chain quaternary amines and amine salts such as Sapamines (e.g., $[(CH_3)_3CCH{=}CHNHCOC_{16}H_{33}]_2^+SO_4^{2-}$). Surfactants of this type find uses in connection with wetting, waterproofing, emulsion formation or breaking, dispersants for inks, and so on.

Nonionic detergents, as the name implies, are not electrolytes, although they do possess the general polar-nonpolar character typical of surfactants. Examples of common types would include polyether esters, for example, $RCO(OCH_2CH_2)_xCH_2CH_2OH$; alkyl-arylpolyether alcohols, for example, $RC_6H_5(OCH_2CH_2)_xCH_2CH_2OH$; and amides, for example, $RCON[(OCH_2CH_2)_xOH]_2$. Detergents of this type compare favorably with soaps and synthetic anionic detergents and find a considerable use in household products such as window- and car-washing preparations (so that incomplete rinsing, which allows for smooth draining, is possible without leaving a powdery residue on drying), insecticides, and detergents for automatic washers.

7. Problems

1. The following table lists some of the types of interfacial tensions that occur in systems of practical importance. Each row corresponds to a different system; not all of the types of interfacial tensions are necessarily present in a given system, and of those present there may be some that, from the nature of the situation, are not under control (i.e., from a practical point of view cannot be modified). For each system certain changes are indicated that constitute a desired goal for that situation, for example, wetting, and detergency. Thus inc (dec) means that it is desirable for good performance to introduce a surfactant that will increase (decrease) the particular surface tension involved. For each case state which practical situation is involved and discuss briefly why the indicated modifications in surface tension should be desired.

	γ_{SA}	γ_{SW}	γ_{SO}	γ_{OW}	γ_{WA}
(a)		dec	inc	dec	
(b)	dec	inc			
(c)		dec			dec
(d)		dec			inc

(S = solid, A = air, W = water or aqueous phase, O = oily or organic water-insoluble phase.)

2. A fabric is made of wool fibers of individual diameter 25 μm and density 1.5 g/cm^3. The advancing angle for water on a single fiber is 120°. Calculate (a) the contact angle on fabric so woven that its bulk density is 0.8 g/cm^3 and (b) the depth of a water layer that could rest on the fabric without running through. Make (and state) necessary simplifying assumptions.

3. Templeton obtained data of the following type for the rate of displacement of water in a 30-μm capillary by oil (n-cetane) (the capillary having previously been wet by water). The capillary was 10 cm long, and the driving pressure was 45 cm of water. When the meniscus was 2 cm from the oil end of the capillary, the velocity of motion of the meniscus was 3.6×10^{-2} cm/sec, and when the meniscus was 8 cm from the oil end, its velocity was 1×10^{-2} cm/sec. Water wet the capillary, and the water-oil interfacial tension was 30 dyn/cm. Calculate the apparent viscosities of the oil and the water. Assuming that both come out to be 0.9 of the actual bulk viscosities, calculate the thickness of the stagnant annular film of liquid in the capillary.

4. Show that for the case of a liquid–air interface Eq. XIII-8 predicts that the distance a liquid has penetrated into a capillary increases with the square root of the time.

5. Calculate ΔG_1, ΔG_2, and w_{pract} (Section XIII-4A) for a bubble in a flotation

system for which w_{SL} is 25 ergs/cm^2 and γ_L is 40 ergs/cm^2. Calculate also ΔG_{1a}, the free energy of adhesion of the bubble to a hemispherical cup in the surface. Take the bubble radius to be 0.2 cm.

6. Fuerstenau and co-workers observed in the adsorption of a long-chain ammonium ion RNH_3^+ on quartz that at a concentration of $10^{-3}M$ there was six-tenths of a monolayer adsorbed and the ζ potential was zero. At $10^{-5}M$ RNH_3^+, however, the ζ potential was -60 mV. Calculate what fraction of a monolayer should be adsorbed in equilibrium with the $10^{-5}M$ solution. Assume a simple Stern model.

7. A surfactant is known to lower the surface tension of water and also is known to adsorb at the water–oil interface but not to adsorb appreciably at the water–fabric interface. Explain briefly whether this detergent should be useful in (a) waterproofing of fabrics or (b) in detergency and the washing of fabrics.

8. Contact angle is proportional to $(\gamma_{SV} - \gamma_{SL})$, therefore addition of a surfactant that adsorbs at the solid–solution interface should decrease γ_{SL} and therefore increase the quantity above and make θ smaller. Yet such addition in flotation systems increases θ. Discuss what is incorrect or misleading about the opening statement.

9. Reference 115 gives the diffusion coefficient of DTAB (dodecyltrimethylammonium bromide) as 1.07×10^{-6} cm^2/sec. Estimate the micelle radius (use the Einstein equation relating diffusion coefficient and friction factor and the Stokes equation for the friction factor of a sphere) and compare with the value given in the reference. Estimate also the number of monomer units in the micelle. Assume 25°C.

10. Calculate the film thickness for the Marongoni instability shown in Fig. XIII-2 using the relationship in Eq. XIII-3, assuming that the interfacial tension is 20 mN/m.

11. Micelle formation can be treated as a mass action equilibrium, for example,

$$40Na^+ + 80R^- = (Na_{40}R_{80})^{-40}$$

The CMC for sodium dodecylbenzenesulfonate is about $10^{-3}M$ at 25°C. Calculate K for the preceding reaction, assuming that it is the only process that occurs in micelle formation. Calculate enough points to make your own quantitative plot corresponding to Fig. XIII-13. Include in your graph a plot of $(Na^+)(R^-)$. *Note:* It is worthwhile to invest the time for a little reflection on how to proceed before launching into your calculation!

12. Show that the stable position of a spherical particle is, indeed, that shown in Fig. XIII-5b if θ_{SAB} is nonzero. *Optional:* by what percent of its radius should the particle extend into phase A if $\gamma_{AB} \cos \theta_{SAB}$ is -34 erg/cm^2 and γ_{AB} is 40 ergs/cm^2.

13. A surfactant solution is a mixture of DTAC (dodecyltrimethylammonium chloride) and CPC (cetyl pyridinium chloride); the respective CMCs of the pure surfactants are $2 \times 10^{-3}M$ and $9 \times 10^{-4}M$ (Ref. 140). Make a plot of the CMC for mixtures of these surfactants versus the mole fraction of DTAC.

14. It was stated in Section XIII-6C that an adsorption maximum, as illustrated in Fig. XIII-15, implies a second law violation. Demonstrate this. Describe a specific set of operations or a "machine" that would put this violation into practice.

General References

A. N. Clark and D. J. Wilson, *Foam Flotation: Theory and Applications*, Marcel Dekker, New York, 1983.

D. H. Everett, ed., *Basic Principles of Colloid Science*, CRC Press, Boca Raton, FL, 1988.

D. W. Fuerstenau, *Pure Appl. Chem.*, **24,** 135 (1970).

R. J. Hunter, *Foundations of Colloid Science*, Vol. 1, Clarendon Press, Oxford, 1987.

E. Jungermann, *Cationic Surfactants*, Marcel Dekker, New York, 1970.

D. Karsa, ed., *Industrial Applications of Surfactants*, CRC Press, Boca Raton, FL, 1987. Part II, Royal Society of Chemistry, Letchworth, UK.

R. G. Laughlin, *The Aqueous Phase Behavior of Surfactants*, Academic, London, 1994.

R. Lemlich, ed., *Adsorptive Bubble Separation Techniques*, Academic, New York, 1972.

J. L. Moilliet, ed., *Waterproofing and Water Repellency*, Elsevier, New York, 1963.

S. R. Morrison, *The Chemical Physics of Surfaces*, Plenum, London, 1977.

A. W. Neumann and J. K. Spelt, eds., *Applied Surface Thermodynamics. Interfacial Tension and Contact Angles*, Marcel Dekker, New York, 1996.

L. I. Osipow, *Surface Chemistry: Theory and Industrial Applications*, Krieger, New York, 1977.

M. P. Pileni, ed., *Structure and Reactivity in Reverse Micelles*, Elsevier, New York, 1989.

M. J. Rosen and H. A. Goldsmith, *Systematic Analysis of Surface-Active Agents*, Wiley-Interscience, New York, 1972.

S. Ross and I. D. Morrison, *Colloidal Systems and Interfaces*, Wiley-Interscience, New York, 1988.

S. A. Safran, *Statistical Thermodynamics of Surfaces, Interfaces and Membranes*, Addison-Wesley, Reading, MA, 1994.

K. L. Sutherland and I. W. Wark, *Principles of Flotation*, Australasian Institute of Mining and Metallurgy, Melbourne, 1955.

R. D. Vold and M. J. Vold, *Colloid and Interface Chemistry*, Addison-Wesley, Reading, MA, 1983.

Textual References

1. P. G. de Gennes, *Rev. Mod. Phys.*, **57,** 827 (1985).

2. E. B. Dussan V, E. Ramé, and S. Garoff, *J. Fluid Mech.*, **230,** 97 (1991).

3. Q. Chen, E. Ramé, and S. Garoff, *Phys. Fluids*, **7,** 2631 (1995).

4. P. G. de Gennes and A.-M. Cazabat, *CR Acad. Sci. Paris, Serie II*, **310,** 1601 (1990).

5. N. Fraysse, M. P. Valignat, A.-M. Cazabat, F. Heslot, and P. Levinson, *J. Colloid Interface Sci.*, **158,** 27 (1993).

6. M. P. Valignat, N. Fraysse, and A.-M. Cazabat, *Langmuir*, **9,** 3255 (1993).

7. A.-M. Cazabat, N. Fraysse, F. Heslot, P. Levinson, J. Marsh, F. Tiberg, and M. P. Valignat, *Adv. Colloid Interface Sci.*, **48,** 1 (1994).

8. F. Heslot, A.-M. Cazabat, P. Levinson, and N. Fraysse, *Phys. Rev. Lett.*, **65,** 599 (1990).

9. P. Silberzan and L. Léger, *Phys. Rev. Lett.*, **66,** 185 (1991).

10. G. F. Teletzke, L. E. Scriven, and H. T. Davis, *J. Chem. Phys.*, **77,** 5794 (1982).

11. R. E. Brenner, Jr., G. F. Teletzke, L. E. Scriven, and H. T. Davis, *J. Chem. Phys.*, **80,** 589 (1984).

12. W. C. Kung, L. E. Scriven, and H. T. Davis, *Chem. Phys.*, **149,** 141 (1990).

13. H. Nakanishi and P. Pincus, *J. Chem. Phys.*, **79,** 997 (1983).

14. L. J. M. Schlangen, L. K. Koopal, M. A. Cohen Stuart, and J. Lyklema, *Colloids & Surfaces*, **A89,** 157 (1994).

15. S. Perković, E. M. Blokhuis, E. Tessler, and B. Widom, *J. Chem. Phys.*, **102,** 7584 (1995).

16. I. Szleifer and B. Widom, *Mol. Phys.*, **75,** 925 (1992).

17. A. Marmur and M. D. Lelah, *J. Colloid Interface Sci.*, **78,** 262 (1980).

18. S. Garoff, E. B. Sirota, S. K. Sinha, and H. B. Stanley, *J. Chem. Phys.*, **90,** 7505 (1989).

19. F. Tiberg and A.-M. Cazabat, *Langmuir*, **10,** 2301 (1994).

20. M. O. Robbins, D. Andelman, and J.-F. Joanny, *Phys. Rev. A*, **43,** 4344 (1991).

21. J. L. Harden and D. Andelman, *Langmuir*, **8,** 2547 (1992).

21a. A. Paterson, M. Fermigier, P. Jenffer, and L. Limat, *Phys. Rev. E*, **51,** 1291 (1995).

22. J. O. Glanville and J. P. Wightman, *Fuel*, **58,** 819 (1979).

23. A. Sharma and E. Ruckenstein, *J. Colloid Interface Sci.*, **111,** 8 (1986).

24. S. Zhu, W. G. Miller, L. E. Scriven, and H. T. Davis, *Colloids & Surfaces*, **A90,** 63 (1994).

25. V. V. Yaminsky, P. M. Claesson, and J. C. Eriksson, *J. Colloid Interface Sci.*, **161,** 91 (1993).

26. R. A. Hayes, A. C. Robinson, and J. Ralston, *Langmuir*, **10,** 2850 (1994).

27. A. Marmur and M. D. Lelah, *Chem. Eng. Commun.*, **13,** 133 (1981).

28. S. M. Troian, X. L. Wu, and S. A. Safran, *Phys. Rev. Lett.*, **62,** 1496 (1989).

29. D. Pesach and A. Marmur, *Langmuir*, **3,** 519 (1987).

30. P. Carles and A.-M. Cazabat, *J. Colloid Interface Sci.*, **157,** 196 (1993).

31. A.-M. Cazabat, F. Heslot, S. M. Troian, and P. Carles, *Nature*, **346,** 824 (1990).

32. H. S. Kheshgi and L. E. Scriven, *Chem. Eng. Sci.*, **46,** 519 (1991).

33. G. Elender and E. Sackmann, *J. Phys. II France* **4,** 455 (1994).

34. F. Brochard-Wyart and P. G. de Gennes, *J. Phys. Cond. Matter*, **6,** 23A(suppl.), A9–A12 (1994).

35. R. Yerushalmi-Rozen, J. Klein, and L. J. Fetters, *Science*, **263,** 793 (1994).

36. R. Yerushalmi-Rozen and J. Klein, *Langmuir*, **11,** 2806 (1995).

37. T. D. Blake and K. J. Ruschak, *Nature*, **282,** 489 (1979).

38. A. Sharma and E. Ruckenstein, *J. Colloid Interface Sci.*, **113,** 456 (1986).

39. H. M. Princen, *J. Colloid Interface Sci.*, **34,** 171 (1970).

40. M. V. Karkare and T. Fort, *Langmuir*, **9,** 2398 (1993).

41. M. V. Karkare, H. R. La, and T. Fort, *Langmuir*, **9,** 1684 (1993).

42. M. V. Karkare and T. Fort, *Langmuir*, **10,** 3701 (1994).

43. D. L. Briley, *Hewlett-Packard J.*, 28 (Feb. 1994).

44. A. Marmur, *J. Colloid Interface Sci.*, **122,** 209 (1988).

45. E. W. Washburn, *Phys. Rev. Ser. 2*, **17,** 273 (1921).

46. L. R. Fisher and P. D. Lark, *J. Colloid Interface Sci.*, **69,** 486 (1979).

47. G. E. P. Elliott and A. C. Riddiford, *Nature*, **195,** 795 (1962); also M. Haynes and T. Blake, private communication.

48. R. J. Good and N. J. Lin, *J. Colloid Interface Sci.*, **54,** 52 (1976).

49. R. R. Rye, J. A. Mann Jr., and F. G. Yost, *Langmuir*, **12,** 555 (1996).

50. R. F. Giese, Z. Li, C. J. van Oss, H. M. Kerch, and H. E. Burdette, *J. Am. Ceram. Soc.*, **77,** 2220 (1994).

51. C. J. van Oss, W. Wu, and R. F. Giese, *Particulate Sci. Technol.*, **11,** 193 (1993).

52. C. J. van Oss, R. F. Giese, Z. Li, K. Murphy, J. Norris, M. K. Chaudhury, and R. J. Good, *J. Adhes. Sci. Technol.*, **6,** 413 (1992).

53. B. Kim and P. Harriott, *J. Colloid Interface Sci.*, **115,** 1 (1987).

54. J. L. Moilliet, Ed., *Water Proofing and Water Repellency*, Elsevier, New York, 1963.

55. M. Sarker, M. Bettler, and D. J. Wilson, *Separ. Sci. Tech.*, **22,** 47 (1987).

56. R. Lemlich, ed., *Adsorptive Bubble Separation Techniques*, Academic, New York, 1972.

57. A. N. Clarke and D. J. Wilson, *Foam Flotation: Theory and Applications*, Marcel Dekker, New York, 1983.

58. P. Somasundaran, *AIChE Symp. Ser.*, **71,** 1 (1975).

59. I. W. Wark, *Chem. Aust.*, **46,** 511 (1979).

60. B. Jańczuk, *J. Colloid Interface Sci.*, **93,** 411 (1983).

61. A. V. Rapacchietta and A. W. Neumann, *J. Colloid Interface Sci.*, **59,** 555 (1977).

62. J. F. Oliver, C. Huh, and S. G. Mason, *J. Colloid Interface Sci.*, **59,** 568 (1977).

63. J. Leja and G. W. Poling, *Preprint, International Mineral Processing Congress*, London, April 1960.

64. V. I. Klassen and V. A. Mokrousov, eds., *An Introduction to the Theory of Flotation*, Butterworths, London, 1963.

65. A. M. Gaudin, *Flotation*, McGraw-Hill, New York, 1957.

66. S. P. Frankel and K. J. Mysels, *J. Phys. Chem.*, **66,** 190 (1962).

67. A. Vrij, *Discuss. Faraday Soc.*, **42,** 23 (1966).

68. I. B. Ivanov, B. Radoev, E. Manev, and A. Scheludko, *Disc. Faraday Soc.*, **66,** 1262 (1970).

69. R. J. Pugh, M. W. Rutland, E. Manev, and P. M. Claesson, *Prog. Colloid Polym. Sci.*, **98,** 284 (1995).

70. S. Tchaliovska, P. Herder, R. Pugh, P. Stenius, and J. C. Eriksson, *Langmuir*, **6,** 1535 (1990).

71. A. F. Taggart, *Handbook of Ore Dressing*, Wiley, New York, 1945.

72. A. M. Gaudin, *Flotation*, McGraw-Hill, New York, 1957.

73. D. Hewitt, D. Fornasiero, and J. Ralston, *J. Chem. Soc., Faraday Trans.*, **91,** 1997 (1995).

74. R. J. Pugh and E. Manev, "The Study of Thin Aqueous Films as Models for Froths and Flotation," in *Innovations in Flotation Technology*, P. Mavros and K. A. Matis, eds., Kluwer, 1992, pp. 1–24.

75. G. Johansson and R. J. Pugh, *Int. J. Min. Proc.*, **34,** 1 (1992).

76. E. Chibowski, L. Holysz, and P. Stanszczuk, *Polish J. Chem.*, **59,** 1167 (1985).

77. G. W. Poling and J. Leja, *J. Phys. Chem.*, **67,** 2121 (1963).

78. G. Guarnaschelli and J. Leja, *Sep. Sci.*, **1**(4), 413 (1966).

79. N. P. Finkelstein and G. W. Poling, *Min. Sci. Eng.*, **4**, 177 (1977).

80. R. N. Tipman and J. Leja, *Colloid Polym. Sci.*, **253**, 4 (1975).

81. S. C. Termes and P. E. Richardson, *Int. J. Min. Proc.*, **18**, 167 (1986).

82. A. M. Gaudin, *J. Colloid Interface Sci.*, **47**, 309 (1974).

83. J. Ralston and T. W. Healy, *Intern. J. Min. Proc.*, **7**, 175 (1980).

84. D. R. Nagaraj and P. Somasundaran, *Min. Eng.*, Sept. 1981, p. 1351.

85. R. J. Pugh, *Intern. J. Min. Proc.*, **25**, 101, 131 (1989).

86. B. S. Kim, R. A. Hayes, C. A. Prestidge, J. Ralston, and R. S. C. Smart, *Langmuir*, **11**, 2554 (1995).

87. Y. H. C. Wang and P. Somasundaran, *Trans. Soc. Min. Eng. of AIME*, **272**, 1970 (1969).

88. P. Somasundaran, *Mining Eng.*, Aug. 1984, p. 1177.

89. Y. H. Chia and P. Somasundaran, in *Ultrafine Grinding and Separation of Industrial Minerals*, S. G. Malghan, ed., AIME, New York, 1983.

90. K. Gannon and D. J. Wilson, *Separ. Sci. Tech.*, **21**, 475 (1986); G. McIntyre, J. J. Rodriguez, E. L. Thackston, and D. J. Wilson, ibid., **17**, 359 (1982).

91. W. Wojcik and A. M. Al Taweel, *Powder Tech.*, **40**, 179 (1984).

92. A. Larsson, Per Stenius, and L. Ödberg, *Svensk Papperstidning*, **88**, R2 (1985).

93. B. E. Novich and T. A. Ring, *Langmuir*, **1**, 701 (1985).

94. A. M. Gaudin and D. W. Fuerstenau, *Trans. AIME*, **202**, 958 (1955).

95. D. W. Fuerstenau and M. C. Fuerstenau, *Min. Eng.*, March 1956.

96. V. A. Arsentiev and J. Leja, *Colloid and Interface Science*, Vol. 5, Academic, New York, 1976.

97. D. J. Johnston and J. Leja, *Min. Proc. Extr. Metallurg.*, **87**, C237 (1978).

98. R. J. Pugh and K. Tjus, *J. Colloid Interface Sci.*, **117**, 231 (1987); R. J. Pugh and L. Bergström, ibid., **124**, 570 (1988).

99. F. F. Aplan and D. W. Fuerstenau, in *Froth Flotation*, 50th Anniversary Volume, D. W. Fuerstenau, ed., American Institute of Mining and Metallurgical Engineering, New York, 1962.

100. D. W. Fuerstenau and T. W. Healy, *Adsorptive Bubble Separation Techniques*, Academic Press, 1971, p. 92; D. W. Fuerstenau, *Pure Appl. Chem.*, **24**, 135 (1970).

101. A. M. Gaudin and D. W. Fuerstenau, *Min. Eng.* (Oct. 1955).

102. W. C. Preston, *J. Phys. Colloid Chem.*, **52**, 84 (1948).

102a. R. J. Williams, J. N. Phillips, and K. J. Mysels, *Trans. Faraday Soc.*, **51**, 728 (1955).

103. J. W. McBain, *Trans. Faraday Soc.*, **9**, 99 (1913).

104. R. M. Menger, *Acc. Chem. Res.*, **12**, 111 (1979).

105. B. Lindman and H. Wennerström, *Top. Curr. Chem.*, **87**, 1 (1980).

106. D. G. Hall and G. J. T. Tiddy, *Surf. Sci. Ser.*, **11**, 55 (1981).

107. K. Shinoda, *J. Phys. Chem.*, **89**, 2429 (1985).

108. R. Nagarajan, *Adv. Colloid Interface Sci.*, **26**, 205 (1986).

109. R. G. Laughlin, Current Opinion in *Colloid Interface Sci.*, **1**, 384 (1996).

110. W. Gelbart and A. Ben-Shaul, *J. Phys. Chem.*, **200**, 13169 (1996).

111. M. J. Schick, S. M. Atlas, and F. R. Eirich, *J. Phys. Chem.*, **66**, 1326 (1962).

112. J. P. Wilcoxon and E. W. Kaler, *J. Chem. Phys.*, **86,** 4684 (1987).

113. P. Debye, *J. Phys. Colloid Chem.*, **53,** 1 (1949).

114. G. Biresaw, D. C. McKenzie, C. A. Bunton, and D. F. Nicoli, *J. Phys. Chem.*, **89,** 5144 (1985).

115. D. Chatenay, W. Urbach, R. Messager, and D. Langevin, *J. Chem. Phys.*, **86,** 2343 (1987).

116. N. J. Chang and E. W. Kaler, *J. Phys. Chem.*, **89,** 2996 (1985).

117. T. Imae and T. Kohsaka, *J. Phys. Chem.*, **96,** 10030 (1992).

118. F. Grieser and C. J. Drummond, *J. Phys. Chem.*, **92,** 5580 (1988).

119. F. Grieser and R. Tausch-Treml, *J. Am. Chem. Soc.*, **102,** 7258 (1980).

120. M. Almgren, *Adv. Colloid Interface Sci.*, **41,** 9 (1992).

121. M. Almgren and J. E. Lofroth, *J. Colloid Interface Sci.*, **81,** 486 (1981).

122. G. G. Warr, F. Grieser, and D. F. Evans, *J. Chem. Soc., Faraday Trans. I*, **82,** 1829 (1986); G. G. Warr and F. Grieser, ibid., **82,** 1813 (1986).

123. F. C. De Schryver, Y. Croonen, E. Geladé, M. Van der Auweraer, J. C. Dederen, E. Roelants, and N. Boens, in *Surfactants in Solution*, K. L. Mittal and B. Lindman, eds., Plenum, New York, 1984.

124. L. J. Magid, *Colloids & Surfaces*, **19,** 129 (1986).

125. J. B. Hayter and J. Penfold, *J. Chem. Soc., Faraday Trans. I*, **77,** 1851 (1981).

126. T. L. Lin, S.-H. Chen, E. Gabriel, and M. F. Roberts, *J. Am. Chem. Soc.*, **108,** 3499 (1986).

127. T. L. Lin, M. Y. Tseng, S.-H. Chen, and M. F. Roberts, *J. Phys. Chem.*, **94,** 7239 (1990).

128. P. Lo Nostro and S.-H. Chen, *J. Phys. Chem.*, **97,** 6535 (1993).

129. Y. C. Liu, P. Baglioni, J. Teixeira, and S. H. Chen, *J. Phys. Chem.*, **98,** 10208 (1994).

130. R. Zana and Y. Talmon, *Nature*, **362,** 228 (1993).

131. D. P. Siegel, W. J. Green, and Y. Talmon, *Biophys. J.*, **66,** 402 (1994).

132. D. Danino, Y. Talmon, H. Levy, G. Beinert, and R. Zana, *Science*, **269,** 1420 (1995).

133. Y. S. Lee and K. W. Woo, *Bull. Kor. Chem. Soc.*, **14,** 392 (1993).

134. Y. S. Lee and K. W. Woo, *Bull. Kor. Chem. Soc.*, **14,** 599 (1993).

135. J. O. Saeten, J. Sjöblom, and B. Gestblom, *J. Phys. Chem.*, **95,** 1449 (1991).

136. P. Baglioni, M. F. Ottaviani, G. Martini, and E. Ferroni, "Esr Study of Spin Labels in Surfactant Micelles," in *Surfactants In Solution*, Vol. 1, K. L. Mittal and B. Lindman, eds., Plenum, 1984.

137. P. Baglioni and L. Kevan, *Heterogen. Chem. Rev.*, **2,** 1 (1995).

138. S. D. Christian, G. A. Smith, and E. E. Tucker, *Langmuir*, **1,** 564 (1985).

139. R. Bury and C. Treiner, *J. Colloid Interface Sci.*, **103,** 1 (1985).

140. M. A. Chaiko, R. Nagarajan, and E. Ruckenstein, *J. Colloid Interface Sci.*, **99,** 168 (1984).

141. R. Mallikurjun and D. B. Dadyburjor, *J. Colloid Interface Sci.*, **84,** 73 (1981).

142. P. Mukerjee, in *Solution Chemistry of Surfactants*, Vol. 1, K. L. Mittal, ed., Plenum, New York, 1979.

143. R. Nagarajan, M. Barry, and E. Ruckenstein, *Langmuir*, **2,** 210 (1986).

144. T. L. Calvert, R. J. Phillips, and S. R. Dungan, *AIChE J.*, **40**, 1449 (1994).

145. D. J. McClements and S. R. Dungan, *Colloids & Surfaces*, **A 104**, 127 (1995).

146. R. Nagarajan and E. Ruckenstein, *Langmuir*, **7**, 2934 (1991).

147. K. A. Cogan, F. A. M. Leermakers, and A. P. Gast, *Langmuir*, **8**, 429 (1992).

148. R. Nagarajan, *Curr. Opinion in Colloid Interface Sci.*, **1**, 391 (1996).

149. D. Stigter, R. J. Williams, and K. J. Mysels, *J. Phys. Chem.*, **59**, 330 (1955).

150. J. Clifford and B. A. Pethica, *J. Phys. Chem.*, **70**, 3345 (1966).

151. D. F. Evans, *J. Colloid Interface Sci.*, **101**, 292 (1984).

152. S. G. Oh and D. O. Shah, *J. Dispers. Sci. Technol.*, **15**, 297 (1994).

153. S. G. Oh and D. O. Shah, *Langmuir*, **7**, 1316 (1991).

154. S. G. Oh and D. O. Shah, *Langmuir*, **8**, 1232 (1992).

155. R. Leung and D. O. Shah, *J. Colloid Interface Sci.*, **113**, 484 (1986).

156. T. Imae and N. Hayashi, *Langmuir*, **9**, 3385 (1993).

157. D. Stigter, *J. Phys. Chem.*, **68**, 3603 (1964).

158. F. Grieser, T. W. Healey, W. P. Hsu, J. P. Kratohvil, and G. G. Warr, *Colloids & Surfaces*, **42**, 97 (1989), ibid p. 537–578.

159. G. S. Hartley, *Aqueous Solutions of Paraffin-Chain Salts*, Hermann et Cie, Paris, 1936.

160. See *Colloid Chemistry, Theoretical and Applied*, Reinhold, New York, 1944; also see Ref. 102.

160a. K. A. Dill and P. J. Flory, *Proc. Natl. Acad. Sci.* (USA), **78**, 676 (1981).

161. T. L. Lin, Y. Hu, S.-H. Chen, M. F. Roberts, J. Samseth, and K. Mortensen, *Prog. Colloid Polym. Sci.*, **97**, 128 (1994).

162. T. L. Lin, S.-H. Chen, E. Gabriel, and M. F. Roberts, *J. Phys. Chem.*, **94**, 855 (1990).

163. T. L. Lin, C. C. Liu, M. F. Roberts, and S.-H. Chen, *J. Phys. Chem.*, **95**, 6021 (1991).

164. J. Köning, C. Boettcher, H. Winkler, E. Zeitler, Y. Talmon, and J. H. Fuhrhop, *J. Am. Chem. Soc.*, **115**, 693 (1993).

165. F. A. M. Leermakers and J. M. H. M. Scheutjens, *J. Colloid Interface Sci.*, **136**, 231 (1990).

166. F. A. M. Leermakers, C. M. Wijmans, and G. J. Fleer, *Macromolecules*, **28**, 3434 (1995).

167. C. Tanford, *J. Phys. Chem.*, **78**, 2468 (1974).

168. R. Nagarajan, K. M. Shah, and S. Hammond, *Colloids & Surfaces*, **4**, 147 (1982).

169. R. Nagarajan and E. Ruckenstein, *J. Colloid Interface Sci.*, **60**, 221 (1977).

170. G. Kegeles, *J. Phys. Chem.*, **83**, 1728 (1979).

171. A. Ben-Nalm and F. H. Stillinger, *J. Phys. Chem.*, **84**, 2872 (1980).

172. R. Nagarajan, *Colloids & Surfaces* **A71**, 39 (1993).

173. J. F. Rusling and T. F. Kumosinski, *J. Phys. Chem.*, **99**, 9241 (1995).

174. A. P. Gast, *Langmuir*, **12**, 4060 (1996).

175. I. Astafieva, X. F. Zhong, and A. Eisenberg, *Macromolecules*, **26**, 7339 (1993).

176. P. Lo Nostro, C. Y. Ku, S.-H. Chen, and J. S. Lin, *J. Phys. Chem.*, **99**, 10858 (1995).

177. S. H. Anastasiadis, I. Gancarz, and H. T. Koberstein, *Macromolecules*, **22,** 1449 (1989).

178. L. Piculell and B. Lindman, *Adv. Colloid Interface Sci.*, **41,** 149 (1992).

179. K. Loyen, I. Iliopoulos, R. Audebert, and U. Olsson, *Langmuir*, **11,** 1053 (1995).

180. A. Sarrazin-Cartalas, I. Iliopoulos, R. Audebert, and U. Olsson, *Langmuir*, **10,** 1421 (1994).

181. N. Kemenka, A. Kaplun, Y. Talmon, and R. Zana, *Langmuir*, **10,** 2960 (1994).

182. I. Iliopoulos, T. K. Wang, and R. Audebert, *Langmuir*, **7,** 617 (1991).

183. B. Magny, I. Iliopoulos, R. Audebert, L. Piculell, and B. Lindman, *Prog. Colloid Polym. Sci.*, **89,** 118 (1992).

184. T. Gu and J. Sjöblom, *Colloids & Surfaces*, **64,** 39 (1992).

185. P. C. Schulz and J. E. Puig, *Colloids & Surfaces*, **A71,** 83 (1993).

186. G. J. T. Tiddy, in *Modern Trends of Colloid Science in Chemistry and Biology*, H. Eicke, ed., Birkhäuser-Verlag, Basel, 1985; *Phys. Rep.*, **57,** 1 (1980).

187. N. Moucharafieh and S. E. Friberg, *Molec. Cryst. Liq. Cryst.*, **49,** 23 (1979).

188. J. L. Woodhead, J. A. Lewis, G. N. Malcolm, and I. D. Watson, *J. Colloid Interface Sci.*, **79,** 454 (1981).

189. N. Nishikido, M. Shinozaki, G. Sugihara, M. Tanaka, and S. Kaneshina, *J. Colloid Interface Sci.*, **74,** 474 (1980).

190. G. Sugihara and P. Mukerjee, *J. Phys. Chem.*, **85,** 1612 (1981).

191. G. Kegels, *Arch. Biochem. Biophys.*, **200,** 279 (1980).

192. E. A. G. Aniansson, S. N. Wall, M. Almgren, H. Hoffmann, I. Kielmann, W. Ulbricht, B. Zana, J. Lang, and C. Tondre, *J. Phys. Chem.*, **80,** 905 (1976).

193. N. Muller, *J. Phys. Chem.*, **76,** 3017 (1972).

194. A. H. Colen, *J. Phys. Chem.*, **78,** 1676 (1974).

195. M. Grubic and R. Strey, quoted in M. Teubner, *J. Colloid Interface Sci.*, **80,** 453 (1981).

196. P. M. Holland and D. N. Rubingh, *J. Phys. Chem.*, **87,** 1984 (1983).

197. G. G. Warr, F. Grieser, and T. W. Healy, *J. Phys. Chem.*, **87,** 1220 (1983).

198. H. Eicke and H. Christen, *Helv. Chim. Acta*, **61,** 2258 (1978).

199. E. Ruckenstein and R. Nagarajan, *J. Phys. Chem.*, **84,** 1349 (1980).

200. N. Muller, *J. Phys. Chem.*, **79,** 287 (1975).

201. S. Karaborni, K. Esselink, P. A. J. Hilbers, B. Smit, J. Karthäuser, N. M. van Os, and R. Zana, *Science*, **266,** 254 (1994).

202. R. Nagarajan, *Chem. Eng. Commun.*, **55,** 251 (1987).

203. M. J. Schwuger, *Kolloid-Z. Z.-Polym.*, **240,** 872 (1970).

204. F. M. Fowkes, in *Solvent Properties of Surfactant Solutions*, K. Shinoda, ed., Marcel Dekker, New York, 1967.

205. J. H. Fendler, *Accs. Chem. Res.*, **9,** 153 (1976).

206. C. A. Bunton and G. Savelli, *Adv. Phys. Org. Chem.*, **22,** 213 (1986).

207. C. J. Drummond, F. Grieser, and T. W. Healey, *J. Chem. Soc., Faraday Trans. I*, **85,** 521 (1989).

208. S. J. Dougherty and J. C. Berg, *J. Colloid Interface Sci.*, **49,** 135 (1974).

209. D. G. Hall, *J. Phys. Chem.*, **91,** 4287 (1987).

210. S. Friberg and S. I. Ahmad, *J. Phys. Chem.*, **75,** 2001 (1971).

211. K. Takagi, B. R. Suddaby, S. L. Vadas, C. A. Backer, and D. G. Whitten, *J. Am. Chem. Soc.*, **108**, 7865 (1986).

212. T. Imae, T. Tsubota, H. Okamura, O. Mori, K. Takagi, M. Itoh, and Y. Sawaki, *J. Phys. Chem.*, **99**, 6046 (1995).

213. N. A. Porter, E. M. Arnett, W. J. Brittain, E. A. Johnson, and P. J. Krebs, *J. Am. Chem. Soc.*, **108**, 1014 (1986).

214. M. Almgren, P. Linse, M. Van der Auweraer, F. C. De Schryver, and Y. Croonen, *J. Phys. Chem.*, **88**, 289 (1984).

215. M. D. Hatlee, J. J. Kozak, G. Rothenberger, P. P. Infelta, and M. Grätzel, *J. Phys. Chem.*, **84**, 1508 (1980).

216. S. J. Atherton, J. H. Baxendale, and B. M. Hoey, *J. Chem. Soc., Faraday Trans. I*, **78**, 2167 (1982).

217. S. Hashimoto and J. K. Thomas, *J. Phys. Chem.*, **89**, 2771 (1985).

218. M. Van der Auweraer and F. C. De Schryver, *Chem. Phys.*, **111**, 105 (1987).

219. J. A. Beunen and E. Ruckenstein, *J. Colloid Interface Sci.*, **96**, 469 (1983).

220. R. O. Dunn, Jr. and J. F. Scamehorn, *Separ. Sci. Tech.*, **20**, 257 (1985).

221. W. L. Hinze, in ACS Symposium Series No. 342, W. L. Hinze and D. W. Armstrong, eds., American Chemical Society, Washington, DC, 1987.

222. R. Nagarajan and E. Ruckenstein, *Separ. Sci. Tech.*, **16**, 1429 (1981).

223. S. P. Beaudoin, C. S. Grant, and R. G. Carbonell, *Ind. Eng. Chem. Res.*, **34**, 3307 (1995).

224. S. P. Beaudoin, C. S. Grant, and R. G. Carbonell, *Ind. Eng. Chem. Res.*, **34**, 3318 (1995).

225. J. R. Ganc and R. Nagarajan, "Aggregation Behavior of Common Motor Oil Additives," in *International Fuels and Lubricants Meeting*, Toronto, Canada, 1991.

226. B. J. Carroll, *Colloids & Surfaces*, **A74**, 131 (1993).

227. C. A. Miller and K. H. Raney, *Colloids & Surfaces*, **A74**, 169 (1993).

228. A. M. Schwartz, in *Surface and Colloid Science*, E. Matijevic, ed., Wiley-Interscience, 1972, p. 195.

229. J. A. Kabin, A. E. Sáez, C. S. Grant, and R. G. Carbonell, *Ind. Eng. Chem. Res.* (in press).

230. M. Almgren, F. Grieser, and J. K. Thomas, *J. Am. Chem. Soc.*, **101**, 279 (1979).

231. N. K. Adam and D. G. Stevenson, *Endeavour*, **12**, 25 (1953).

232. A. E. Sáez, private communication.

233. G. C. Allan, J. R. Aston, F. Grieser, and T. W. Healy, *J. Colloid Interface Sci.*, **128**, 258 (1989).

234. *Advances in Colloid Science*, Vol. I, Interscience, New York, 1942.

235. J. W. McBain, R. S. Harborne, and A. M. King, *J. Soc. Chem. Ind.*, **42**, 373T (1923).

236. L. Greiner and R. D. Vold, *J. Phys. Colloid Chem.*, **53**, 67 (1949).

237. K. Durham, Ed., *Surface Activity and Detergency*, Macmillan, New York, 1961.

238. J. L. Moilliet, B. Collie, and W. Black, *Surface Activity*, E. & F. N. Spon, London, 1961.

239. K. Durham, *Proc. 2nd Int. Congr. Surf. Act.*, Vol. 4, 1957, p. 60.

240. P. Krings, M. J. Schwuger, and C. H. Krauch, *Naturwissenschaften*, **61**, 75 (1974);

P. Berth, G. Jakobi, E. Schmadel, M. J. Schwuger, and C. H. Krauch, *Angew. Chem.*, **87,** 115 (1975).

241. K. H. Raney, W. J. Benton, and C. A. Miller, *J. Colloid Interface Sci.*, **117,** 282 (1987).

242. A. Fava and H. Eyring, *J. Phys. Chem.*, **60,** 890 (1956).

243. D. G. Hall, *J. Chem. Soc., Faraday Trans. I*, **76,** 386 (1980).

244. F. J. Trogus, R. S. Schechter, and W. H. Wade, *J. Colloid Interface Sci.*, **70,** 293 (1979).

245. K. P. Ananthapadmanabhan and P. Somasundaran, *Colloids & Surfaces*, **77,** 105 (1983).

246. J. Zhu, A. Eisenberg, and R. B. Lennox, *J. Am. Chem. Soc.*, **113,** 5583 (1991).

247. P. J. Scales, F. Grieser, D. N. Furlong, and T. W. Healy, *Colloids & Surfaces*, **21,** 55 (1986).

248. K. Lunkenheimer, H. J. Pergande, and H. Krüger, *Rev. Sci. Instrum.*, **58,** 2313 (1987).

249. J. H. Harwell, J. C. Hoskins, R. S. Schechter, and W. H. Wade, *Langmuir*, **1,** 251 (1985).

250. K. Lunkenheimer, G. Czichocki, R. Hirte, and W. Barzyk, *Colloids & Surfaces*, **101,** 187 (1995).

251. C. J. van Oss and P. M. Costanzo, *J. Adhes. Sci. Technol.*, **6,** 477 (1992).

252. C. H. Chang, N. H. L. Wang, and E. I. Franses, *Colloids & Surfaces*, **62,** 321 (1992).

253. G. Kretzschmar and R. Miller, *Adv. Colloid Interface Sci.* **36,** 65 (1991).

CHAPTER XIV

Emulsions, Foams, and Aerosols

1. Introduction

In this chapter, we carry our discussion of applied surface chemistry into finely divided dispersions of liquids, vapor, or solids, Emulsions and foams are grouped together since they involve two partially miscible fluids and, usually, a surfactant. Both cases comprise a dispersion of one phase in another; if both phases are liquid, the system is called an emulsion, whereas if one fluid is a gas, the system may constitute a foam or an aerosol. The dispersed system usually consists of relatively large units, that is, the size of the droplets or gas bubbles will ordinarily range upward from a few tenths of a micrometer. The systems are generally unstable with respect to separation of the two fluid phases, that is, toward breaking in the case of emulsions and collapse in the case of foams, and their degree of practical stability is largely determined by the charges and surface films at the interfaces.

Although it is hard to draw a sharp distinction, emulsions and foams are somewhat different from systems normally referred to as colloidal. Thus, whereas ordinary cream is an oil-in-water emulsion, the very fine aqueous suspension of oil droplets that results from the condensation of oily steam is essentially colloidal and is called an oil *hydrosol*. In this case the oil occupies only a small fraction of the volume of the system, and the particles of oil are small enough that their natural sedimentation rate is so slow that even small thermal convection currents suffice to keep them suspended; for a cream, on the other hand, as also is the case for foams, the inner phase constitutes a sizable fraction of the total volume, and the system consists of a network of interfaces that are prevented from collapsing or coalescing by virtue of adsorbed films or electrical repulsions.

Microemulsions are treated in a separate section in this chapter. Unlike *macro-* or ordinary emulsions, microemulsions are generally thermodynamically stable. They constitute a distinctive type of phase, of structure unlike ordinary homogeneous bulk phases, and their study has been a source of fascination. Finally, *aerosols* are discussed briefly in this chapter, although the topic has major differences from those of emulsions and foams.

500

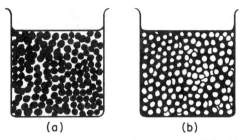

Fig. XIV-1. The two types of emulsion: (*a*) oil in water (O/W); (*b*) water in oil (W/O).

2. Emulsions—General Properties

An *emulsion* may be defined as a mixture of particles of one liquid with some second liquid. The two common types of emulsions are oil-in-water (O/W) and water-in-oil (W/O), where the term *oil* is used to denote the water-insoluble fluid. These two types are illustrated in Fig. XIV-1, where it is clear that the majority or "outer" phase is continuous, whereas the minority or "inner" phase is not. These two emulsion types are distinguished by their ability to disperse oil or water-soluble dyes, their dilution with oil or water, and their conductivity (O/W emulsions have much higher conductivity than do W/O ones; see Ref. 1 for reviews).

Apart from chemical composition, an important variable in the description of emulsions is the volume fraction, ϕ, or the ratio of the volume of the inner to outer phase. For spherical droplets, of radius a, the volume fraction is given by the number density, n, times the spherical volume, $\phi = 4\pi a^3 n/3$. It is easy to show that the maximum packing fraction of spheres is $\phi = 0.74$ (see Problem XIV-2). Many physical properties of emulsions can be characterized by their volume fraction. The viscosity of a dilute suspension of rigid spheres is an example where the Einstein limiting law is [2]

$$\eta = \eta_0(1 + 2.5\phi) \tag{XIV-1}$$

Since emulsion droplets are not rigid spheres, the coefficient of ϕ is around 3–6 for many emulsion systems [3–5]. More concentrated emulsions are non-Newtonian: η depends on shear rate and are thixotropic (η decreasing with shear rate) or rheopectic (η increasing with shear rate).

The conductivity of a dilute emulsion can be treated by classic theory (see Maxwell [6]) assuming spherical droplets

$$\frac{K - K_m}{K + 2K_m} = \frac{K_d - K_m}{K_d + 2K_m} \phi \tag{XIV-2}$$

where K, K_m, and K_d are the specific conductivities of the emulsion, the external medium, and the dispersed phase, respectively. More complex relationships are available for more concentrated emulsions [7].

Increasingly, dielectric measurements are being used to characterize the water content of emulsions. One model for the dielectric constant of a suspension, ϵ

$$\frac{(\epsilon_d - \epsilon_m)^3}{(\epsilon_d - \epsilon)^3} \left(\frac{\epsilon}{\epsilon_m}\right) = \frac{1}{(1 - \phi)^3} \qquad \text{(XIV-3)}$$

in terms of ϵ_d and ϵ_m, the permittivities of the dispersed and continuous phases, respectively, is adequate for O/W emulsions in the microwave frequency range [8] but fails with concentrated W/O emulsions. The frequency-dependent dielectric permittivity is, in general, complex as $\epsilon(\omega) = \epsilon'(\omega) + i\epsilon''(\omega)$ where $\epsilon'(\omega)$ is the real, storage, or polarization component and $\epsilon''(\omega)$, is the imaginary, loss contribution. Time-domain dielectric spectroscopy is useful for characterizing W/O emulsions in terms of a Cole–Cole modified Debye function [9, 10]

$$\epsilon(\omega) = \epsilon_\infty + \frac{\epsilon_s - \epsilon_\infty}{1 + (i\omega\tau)^{1-\alpha}} \qquad \text{(XIV-4)}$$

where ϵ_s and ϵ_∞ are the static and high-frequency permittivities, respectively, τ the relaxation time, and α the width of the distribution. An example shown in Fig. XIV-2 shows the characteristic decay in the storage component and a peak in the loss component at a frequency corresponding to the characteristic relaxation time of 1.6 nsec.

Other important characterization techniques include electrophoresis measurements of droplets [11, 12] (see Section XIV-3C), infrared absorption of the constituent species [13], and light or x-ray scattering. NMR self-diffusion measurements can be used to determine droplet sizes in W/O emulsions [14].

As an emulsion ages, the particle size distribution changes and the number of droplets decreases. On a global scale, the rate of *creaming*, as with milk, reflects the separation into concentrated and dilute emulsion phases. The droplet size distribution can be tabulated from cryoelectron microscopy (see Section VIII-2) photographs [15] or indicated by dynamic light scattering or NMR measurements. Larger particles are better characterized with a *Coulter counter*, where the electrical resistance is monitored as an emulsion flows through a small orifice. A size distribution is readily accumulated by monitoring the resistance as a series of particles pass through. Emulsion particle sizes often follow a log-normal distribution given by

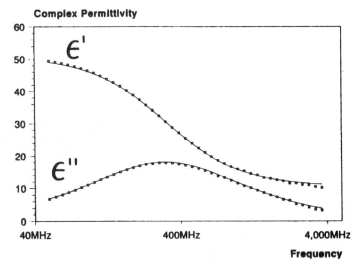

Fig. XIV-2. Dielectric relaxation spectrum of a water-in-oil emulsion containing 40% water in triglyceride with a salt concentration of 5 wt % at a temperature of 25°C. The squares are experimental points and the lines are fits to Eq. XIV-4. (From Ref. 9.)

$$p(x) = \frac{1}{\sigma \sqrt{2\pi}} \exp \left[-\frac{(\ln x - \ln x_m)^2}{2\sigma^2} \right] \tag{XIV-5}$$

where $p(x)$ is the probability of finding an emulsion drop of size x if x_m is the mean size and σ is the logarithmic standard deviation. Qualitatively, a log-normal distribution differs from the usual Gaussian curve in giving greater probability to the extremes in x [16].

The foregoing survey gives an indication of the complexity of emulsion systems and the wealth of experimental approaches available. We are limited here, however, to some selected aspects of a fairly straightforward nature.

3. Factors Determining Emulsion Stability

If two pure, immiscible liquids, such as benzene and water, are vigorously shaken together, they will form a dispersion, but it is doubtful that one phase or the other will be uniquely continuous or dispersed. On stopping the agitation, phase separation occurs so quickly that it is questionable whether the term emulsion really should be applied to the system. A surfactant component is generally needed to obtain a stable or reasonably stable emulsion. Thus, if a little soap is added to the benzene–water system, the result on shaking is a true emulsion that separates out only very slowly. Theories of emulsion stability have therefore been concerned with the nature of the interfacial films and interparticle forces that are important in preventing droplet coalescence.

A. *Macroscopic Theories of Emulsion Stabilization*

It is quite clear, first of all, that since emulsions present a large interfacial area, any reduction in interfacial tension must reduce the driving force toward coalescence and should promote stability. We have here, then, a simple thermodynamic basis for the role of emulsifying agents. Harkins [17] mentions, as an example, the case of the system paraffin oil-water. With pure liquids, the interfacial tension was 41 dyn/cm, and this was reduced to 31 dyn/cm on making the aqueous phase $0.001M$ in oleic acid, under which conditions a reasonably stable emulsion could be formed. On neutralization by $0.001M$ sodium hydroxide, the interfacial tension fell to 7.2 dyn/cm, and if also made $0.001M$ in sodium chloride, it became less than 0.01 dyn/cm. With olive oil in place of the paraffin oil, the final interfacial tension was 0.002 dyn/cm. These last systems emulsified spontaneously—that is, on combining the oil and water phases, no agitation was needed for emulsification to occur.

The surface tension criterion indicates that a stable emulsion should not exist with an inner phase volume fraction exceeding 0.74 (see Problem XIV-2). Actually, emulsions of some stability have been prepared with an inner phase volume fraction as high as 99%; see Lissant [18]. Two possible explanations are the following. If the emulsion is very heterogeneous in particle size (note Ref. 16), ϕ may exceed 0.74 by virtue of smaller drops occupying the spaces between larger ones, and so on with successively smaller droplets. Such a system would not be an equilibrium one, of course. It appears, however, that in some cases the thin film separating two emulsion drops may have a *lower* surface tension than that of the bulk interfacial tension [19]. As a consequence, two approaching drops will spontaneously deform to give a flat drop–drop contact area with the drops taking on a polyhedral shape such as found in foams.

One may rationalize emulsion type in terms of interfacial tensions. Bancroft [20] and later Clowes [21] proposed that the interfacial film of emulsion-stabilizing surfactant be regarded as duplex in nature, so that an inner and an outer interfacial tension could be discussed. On this basis, the type of emulsion formed (W/O vs. O/W) should be such that the inner surface is the one of higher surface tension. Thus sodium and other alkali metal soaps tend to stabilize O/W emulsions, and the explanation would be that, being more water- than oil-soluble, the film–water interfacial tension should be lower than the film–oil one. Conversely, with the relatively more oil-soluble metal soaps, the reverse should be true, and they should stabilize W/O emulsions, as in fact they do. An alternative statement, known as *Bancroft's rule*, is that the external phase will be that in which the emulsifying agent is the more soluble [20]. A related approach is discussed in Section XIV-5.

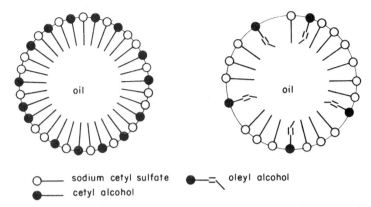

sodium cetyl sulfate   oleyl alcohol
cetyl alcohol

Fig. XIV-3. Steric effects in the penetration of sodium cetyl sulfate monolayers by cetyl alcohol and oleyl alcohol.

B. Specific Chemical and Structural Effects

The energetics and kinetics of film formation appear to be especially important when two or more solutes are present, since now the matter of monolayer penetration or complex formation enters the picture (see Section IV-7). Schulman and co-workers [77, 78], in particular, noted that especially stable emulsions result when the adsorbed film of surfactant material forms strong penetration complexes with a species present in the oil phase. The stabilizing effect of such mixed films may lie in their slow desorption or elevated viscosity. The dynamic effects of surfactant transport have been investigated by Shah and co-workers [22] who show the correlation between micellar lifetime and droplet size. More stable micelles are unable to rapidly transport surfactant from the bulk to the surface, and hence they support emulsions containing larger droplets.

The importance of steric factors in the formation of penetration complexes is made evident by the observation that although sodium cetyl sulfate plus cetyl alcohol gives an excellent emulsion, the use of oleyl alcohol instead of cetyl alcohol leads to very poor emulsions. As illustrated in Fig. XIV-3, the explanation may lie in the difficulty in accommodating the kinked oleyl alcohol chain in the film.

An important aspect of the stabilization of emulsions by adsorbed films is that of the role played by the film in resisting the coalescence of two droplets of inner phase. Such coalescence involves a local mechanical compression at the point of encounter that would be resisted (much as in the approach of two boundary lubricated surfaces discussed in Section XII-7B) and then, if coalescence is to occur, the discharge from the surface region of some of the surfactant material.

Proteins, glucosides, lipids, sterols, and so on, although generally very water soluble, are nonetheless frequently able to impart considerable stability to emulsions (and foams). Small globular proteins are less effective than more surface-active species; however, large hydrophobic proteins are not efficient because of their low solubility [23]. The widely studied milk protein, casein, is a highly surface-active example. The coating of liqid droplets by chitosan extracted from shrimp and crab shells may inhibit lipid adsorption in the intestinal tract and provide hypercholesterolemic qualities as a dietary supplement [11].

An interesting molecule, known as *Kemp's triacid* (*cis,cis*-1,3,5-trimethylcyclohexane-1.3.5-tricarboxylic acid), undergoes conformational changes as a function of pH and is thought to mimic some aspects of enzymatic recognition. The pH-induced conformational changes alter the surface activity, causing it to deteriorate with increasing pH, thus producing a pH-sensitive emulsion [24]. An unusual type of W/O emulsion is that in which the "water" phase is a molten salt hydrate [25] or salt [26]. If the salt is a nitrate, the emulsion may be explosive and have important industrial application.

Bibette [27] developed a fractionation procedure to produce relatively monodisperse O/W emulsions in the 100–1000-nm size range. The method relies on a weak, reversible aggregation produced by excess micelles of surfactant in solution with the emulsion. If properly fractionated, the emulsion droplets are not subject to Ostwald ripening and can remain stable for years. Interesting monodisperse emulsions of oil-based ferrofluids have been created in this way. Such droplets are paramagnetic and thus have induced dipolar moments in a magnetic field. The dipolar interaction can cause reversible aggregation into chains [28] or ellipsoidal structures [29], of interest for "smart fluid" applications as working fluids for robotic hands, clutches, and vibration isolation. Bibette and co-workers [30] have studied the forces between ferrofluid emulsion droplets by measuring the droplet separation as a function of the applied magnetic field, and hence dipole, strength.

C. Long-Range Forces as a Factor in Emulsion Stability

There appear to be two stages in the collapse of emulsions: flocculation, in which some clustering of emulsion droplets takes place, and coalescence, in which the number of distinct droplets decreases (see Refs. 31–33). Coalescence rates very likely depend primarily on the film-film surface chemical repulsion and on the degree of irreversibility of film desorption, as discussed. However, if emulsions are centrifuged, a compressed polyhedral structure similar to that of foams results [32–34]—see Section XIV-8—and coalescence may now take on mechanisms more related to those operative in the thinning of foams.

Flocculation involves droplet approach and should be sensitive to long-range forces. The oil droplets in an O/W emulsion are generally negatively charged—as may be determined from ζ-potential measurements described in Chapter V (see Ref. 2). Some mobility measurements of *n*-hexadecane drops in water of varying pH are shown in Fig. XIV-4. The oil droplets in this case acquire a charge due to preferential adsorption of the constituent ions [12]. In most emulsions, the surface charge will be determined by the charged surfactant.

The manner in which potential should vary across an oil-water interface is shown in Fig. XIV-5, after van den Tempel [35]. Here ΔV denotes the surface

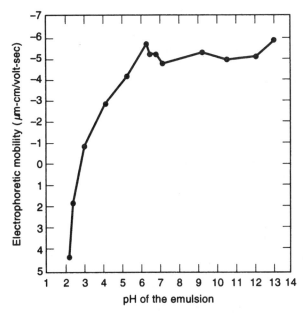

Fig. XIV-4. Electrophoretic mobility of n-hexadecane drops versus the pH of the emulsion. (From Ref. 12.)

potential difference between the two phases, and χ is the surface potential jump (see Section V-11). If some electrolyte is present, the solubility of the cations and the anions in general will be different in the two phases. Usually the anions will be somewhat more oil-soluble than the cations, so that, as illustrated in Fig. XIV-5b, there should be a net negative charge on the oil droplets.

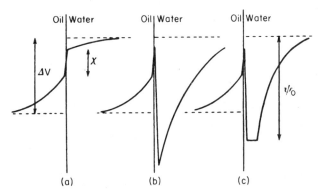

Fig. XIV-5. Variation in potential across an oil-water interface: (a) in the absence of electrolyte, (b) with electrolyte present, and (c) in the presence of soap ions and a large amount of salt. (From Ref. 35.)

The repulsion between oil droplets will be more effective in preventing floc-culation the greater the thickness of the diffuse layer and the greater the value of ψ_0, the surface potential. These two quantities depend oppositely on the elec-trolyte concentration, however. The total surface potential should increase with electrolyte concentration, since the absolute excess of anions over cations in the oil phase should increase. On the other hand, the half-thickness of the double layer decreases with increasing electrolyte concentration. The plot of emulsion stability versus electrolyte concentration may thus go through a maximum.

If an ionic surfactant is present, the potentials should vary as shown in Fig. XIV-5c, or similarly to the case with nonsurfactant electrolytes. In addi-tion, however, surfactant adsorption decreases the interfacial tension and thus contributes to the stability of the emulsion. As discussed in connection with charged monolayers (see Section XV-6), the mutual repulsion of the charged polar groups tends to make such films expanded and hence of relatively low π value. Added electrolyte reduces such repulsion by increasing the counterion concentration; the film becomes more condensed and its film pressure increases. It thus is possible to explain qualitatively the role of added electrolyte in reduc-ing the interfacial tension and thereby stabilizing emulsions.

The charge on a droplet surface produces a repulsive barrier to coales-cence into the London–van der Waals primary attractive minimum (see Section VI-4). If the droplet size is appropriate, a secondary minimum exists outside the repulsive barrier as illustrated by DLVO calculations shown in Fig. XIV-6 (see also Refs. 36–38). Here the influence of pH on the repulsive barrier between n-hexadecane drops is shown in Fig. XIV-6a, while the secondary minimum is enlarged in Fig. XIV-6b [39]. The inset to the figures contains t_c the coa-lescence time. Emulsion particles may flocculate into the secondary minimum without further coalescence.

Some studies have been made of W/O emulsions; the droplets are now aque-ous and positively charged [40, 41]. Albers and Overbeek [40] carried out calcu-lations of the interaction potential not just between two particles or droplets but between one and all nearest neighbors, thus obtaining the variation with particle density or ϕ. In their third paper, these authors also estimated the magnitude of the van der Waals long-range attraction from the shear gradient sufficient to detach flocculated droplets (see also Ref. 42).

An important industrial example of W/O emulsions arises in water-in-crude-oil emulsions that form during production. These emulsions must be broken to aid trans-portation and refining [43]. These suspensions have been extensively studied by Sjöblom and co-workers [10, 13, 14] and Wasan and co-workers [44]. Stabilization arises from combinations of surface-active components, asphaltenes, polymers, and particles; the composition depends on the source of the crude oil. Certain copolymers can mimic the emulsion stabilizing fractions of crude oil and have been studied in terms of their pressure–area behavior [45].

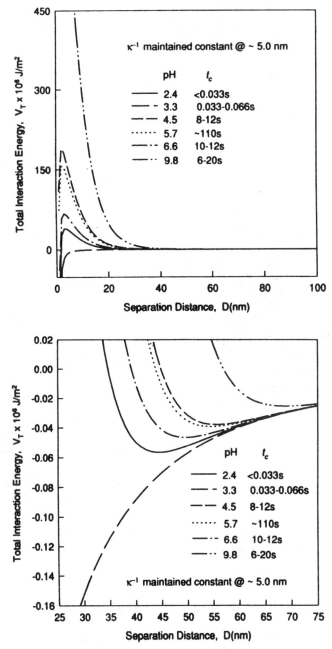

Fig. XIV-6. (a) The total interaction energy determined from DLVO theory for *n*-hexadecane drops for a constant ionic strength (κ^{-1} – 5.0 nm) at various emulsion pH; (b) enlargement of the secondary minimum region of (a). (From Ref. 39.)

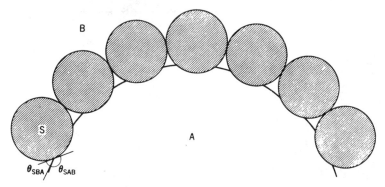

Fig. XIV-7. Stabilization of an emulsion by small particles.

D. Stabilization of Emulsions by Solid Particles

Powders constitute an interesting type of emulsion-stabilizing agent. For example, benzene–water emulsions are stabilized by calcium carbonate, the solid particles collecting at the oil–water interface and armoring the benzene droplets. Similarly, Scarlett and co-workers [46] report the stabilizing of toluene–water emulsions by pyrite as well as that of water–benzene emulsions by charcoal and by mercuric iodide; glycerol tristerate crystals stabilize water–paraffin oil emulsions [47]. Fat particles stabilize W/O emulsions such as those in margarine [9].

It was pointed out in Section XIII-4A that if the contact angle between a solid particle and two liquid phases is finite, a stable position for the particle is at the liquid–liquid interface. Coalescence is inhibited because it takes work to displace the particle from the interface. In addition, one can account for the type of emulsion that is formed, O/W or W/O, simply in terms of the contact angle value. As illustrated in Fig. XIV-7, the bulk of the particle will lie in that liquid that most nearly wets it, and by what seems to be a correct application of the early *"oriented wedge"* principle (see Ref. 48), this liquid should then constitute the outer phase. Furthermore, the action of surfactants should be predictable in terms of their effect on the contact angle. This was, indeed, found to be the case in a study by Schulman and Leja [49] on the stabilization of emulsions by barium sulfate.

4. The Aging and Inversion of Emulsions

As illustrated in Fig. XIV-8, at least three types of aging processes occur for emulsions. The inner phase droplets may undergo flocculation, that is, clustering together without losing their identity; if, as part of or subsequent to flocculation, the flocks undergo a gravity separation, the entire process is called *creaming*. If coalescence occurs, then eventual breaking of the emulsion must follow, giving two liquid layers—see Refs. 50 and 51 for a general discussion

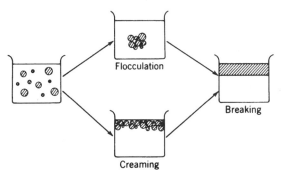

Fig. XIV-8. Types of emulsion instability.

of the process. Emulsion *inversion* involves very dynamic, complex events, as discussed further below.

A. Flocculation and Coagulation Kinetics

There are two approaches to the kinetics of emulsion flocculation. The first stems from a relationship due to Smoluchowski [52] for the rate of diffusional encounters, or flux:

$$J = 16\pi \mathcal{D} a n^2 \qquad \text{(XIV-6)}$$

where n is the number density of particles of radius a and $\mathcal{D}$ is their diffusion coefficient given by the Stokes–Einstein expression $\mathcal{D} = kT/6\pi\eta a$. If there is an energy barrier to aggregation, E^*, then the rate of effective encounters, dn/dt, becomes

$$\frac{dn}{dt} = -kn^2 \qquad \text{(XIV-7)}$$

where

$$k = \frac{8kT}{3\eta} e^{-E^*/kT}$$

The quantity dn/dt is essentially the rate of disappearance of primary particles and hence the rate of decrease in the total number of particles. Integration gives

$$\frac{1}{n} = \frac{1}{n_0} + kt \qquad \text{(XIV-8)}$$

Equation XIV-8 is rather approximate, and more detailed treatments are given

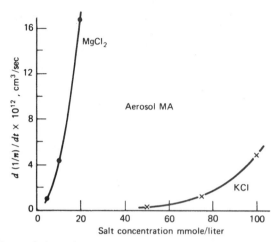

Fig. XIV-9. Effects of electrolyte on the rate of flocculation of Aerosol MA–stabilized emulsions. (From Ref. 35.)

by Kruyt [53] and Becher [3]; the problem is that of handling the formation of aggregates of more than two particles. As a result, the form of Eq. XIV-8 is not well obeyed, although $d(1/n)/dt$ is still used as a measure of the initial flocculation or coagulation rate.

For example, van den Tempel [35] reports the results shown in Fig. XIV-9 on the effect of electrolyte concentration on flocculation rates of an O/W emulsion. Note that $d(1/n)/dt$ (equal to k in the simple theory) increases rapidly with ionic strength, presumably due to the decrease in double-layer half-thickness and perhaps also due to some Stern layer adsorption of positive ions. The preexponential factor in Eq. XIV-7, $k_0 = (8kT/3\eta)$, should have the value of about 10^{-11} cm^3, but at low electrolyte concentration, the values in the figure are smaller by tenfold or a hundredfold. This reduction may be qualitatively ascribed to charged repulsion.

The preceding treatment relates primarily to flocculation rates, while the irreversible aging of emulsions involves the coalescence of droplets, the prelude to which is the thinning of the liquid film separating the droplets. Similar theories were developed by Spielman [54] and by Honig and co-workers [55], which added hydrodynamic considerations to basic DLVO theory. A successful experimental test of these equations was made by Bernstein and co-workers [56] (see also Ref. 57). Coalescence leads eventually to separation of bulk oil phase, and a practical measure of emulsion stability is the rate of increase of the volume of this phase, V, as a function of time. A useful equation is

$$\frac{t}{V} = \frac{1}{aV_\infty} + \frac{t}{V_\infty} \qquad (XIV-9)$$

(see Refs. 1, 58).

There have been some studies of the equilibrium shape of two droplets pressed against each other (see Ref. 59) and of the rate of film thinning [60, 61], but these are based on hydrodynamic equations and do not take into account film–film barriers to final rupture. It is at this point, surely, that the chemistry of emulsion stabilization plays an important role.

B. Inversion and Breaking of Emulsions

An interesting effect is that in which an A/B type of emulsion inverts to a B/A type. Generally speaking, the methods whereby inversion may be caused to take place involve introducing a condition such that the opposite type of emulsion would normally be the stable one. First, an emulsion would *have* to invert if ϕ exceeded 0.74, if the inner phase consisted of uniform rigid spheres; as noted in Section XIV-3A, this value of ϕ represents the point of close packing. Actual emulsion droplets are deformable, of course, and not monodisperse. Continued addition of inner phase may result in inversion, but the effect is not assured and emulsions can be made to exceed $\phi = 0.74$. Emulsions greatly surpassing this value resemble biliquid "foams" having polyhedral cells; Sebba [62, 63] has termed the unit cells of such emulsions as "aphrons". An *aphron* may be defined generally as a phase bounded by an encapsulating soapy film [64].

Deemulsification, or the breaking of an emulsion, can be accomplished by the judicious use of one of the preceding methods of emulsion inversion or by methods that accelerate the coalescence rate of droplets. Also, a *phase change* in one of the two liquid phases may be helpful; thus emulsions may be broken by heating to near the boiling point of the inner phase or by freezing and then rewarming. Absorption chromatography has been used as a means of removing the emulsifying agent and thus breaking the emulsion [65]. The application of a high electric field can induce coalescence and aid deemulsification [66].

5. The Hydrophile–Lipophile Balance

There is a very large technology that makes use of emulsions, and somewhat as in flotation, empirical observation still leads theory, in this case with respect to the prediction of the type and stability of emulsion that a given set of constituents will produce. A very useful numerical rating scheme, however, was introduced by Griffin [67] and is known as the *hydrophile–lipophile balance* (HLB) number. First, numbers are assigned on a one-dimensional scale of surfactant action, as given in Table XIV-1; note the correlation with Bancroft's rule (Section XIV-3A). Each surfactant is then rated according to this scale (see Refs. 1 and 3 for detailed listings and Ref. 68 for a bibliography). It is assumed that surfactant mixtures can be assigned an HLB number on a weight-prorated basis.

The use of the HLB system can be illustrated as follows. Suppose that a certain

TABLE XIV-1
The HLB Scale

Surficant Solubility Behavior in Water	HLB Number		Application
No dispersibility in water	$\left\{\begin{array}{c} 0 \\ 2 \end{array}\right.$		
Poor dispersibility	$\left.\begin{array}{c} 4 \\ \{6 \end{array}\right\}$		W/O emulsifier
Milky dispersion; unstable	$\{8$		
Milky dispersion; stable	$\{10\}$		Wetting agent
Translucent to clear solution	$\{12\}$	Detergent	
Clear solution	$\left\{\begin{array}{c} 14 \\ 16 \\ 18 \end{array}\right.$	Solubilizer	O/W emulsifier

desired O/W emulsion is formed using, say, various proportions of Span 65 (sorbitol tristearate, HLB 2.1) and Tween 60 (polyoxyethylene sorbitan monostearate, HLB 14.9). It is found that the optimum emulsion (smallest droplets) is obtained with 80% Tween 60 and 20% Span, average HLB = 12.3. The assumption is then that with any other mixture of surfactants, optimum performance for the particular system will again be at HLB = 12.3. The *absolute* performance of the two mixtures might differ, but each should be at *its optimum*. The next step, in practice, would be to make up a number of such optimum mixtures and find the one whose absolute performance was best.

Davies [69] (see also Ref. 70) carried the additivity principle further by developing a list of HLB functional group numbers, given in Table XIV-2. The empirical HLB number for a given surfactant is computed by adding 7 to the algebraic sum of the group numbers. Thus the calculated HLB number for cetyl alcohol, $C_{16}H_{33}OH$, would be $7 + 1.9 + 16(-0.475) = 1.3$.

The *phase-inversion temperature* (PIT) is defined as the temperature where, on heating, an oil—water—emulsifier mixture inverts from O/W to a W/O emulsion [23]. The PIT correlates very well with the HLB as illustrated in Fig. XIV-10 [72, 73]. The PIT can thus be used as a guide in emulsifier selection.

The HLB system has made it possible to organize a great deal of rather messy information and to plan fairly efficient systematic approaches to the optimization of emulsion preparation. If pursued too far, however, the system tends to lose itself in complexities [74]. It is not surprising that HLB numbers are not really additive: their effective value depends on what particular oil phase is involved and the emulsion depends on volume fraction. Finally, the host of physical characteristics needed to describe an emulsion cannot be encapsulated by a single HLB number (note Ref. 75).

TABLE XIV-2
Group HLB Numbers

Hydrophilic Groups	HLB	Lipophilic Groups	HLB
—SO$_4$Na	38.7	—CH— ⎤	
—COOK	21.1	—CH$_2$— ⎥	
—COONa	19.1	—CH$_3$— ⎬	−0.475
Sulfonate	~11.0	—CH= ⎦	
—N (tertiary amine)	9.4	—(CH$_2$—CH$_2$—CH$_2$—O—)	−0.15
Ester (sorbitan ring)	6.8		
Ester (free)	2.4		
—COOH	2.1		
—OH (free)	1.9		
—O—	1.3		
—OH (sorbitan ring)	0.5		

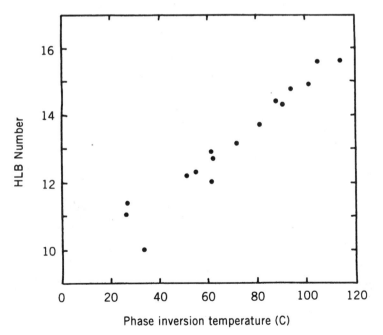

Fig. XIV-10. The correlation between the HLB number and the phase inversion temperature in cyclohexane of nonionic surfactants. (From Ref. 71.)

6. Microemulsions

A beautiful and elegant example of the intricacies of surface science is the formation of transparent, thermodynamically stable microemulsions. Discovered about 50 years ago by Winsor [76] and characterized by Schulman [77, 78], microemulsions display a variety of useful and interesting properties that have generated much interest in the past decade. Early formulations, still under study today, involve the use of a long-chain alcohol as a *cosurfactant* to stabilize oil droplets 10–50 nm in diameter. Although transparent to the naked eye, microemulsions are readily characterized by a variety of scattering, microscopic, and spectroscopic techniques, described below.

Microemulsions are equilibrium phases coexisting with excess oil, water, or both. The general phase behavior was classified by Winsor [76], and the phases are sometimes referred to as *Winsor I*, oil droplets in water coexisting with an excess oil phase, *Winsor II*, water droplets in oil in coexistence with water, and *Winsor III*, a bicontinuous, or "middle" phase, in equilibrium with both oil and water phases. Phase diagrams are complex, as might be imagined. A schematic pseudoternary phase diagram is shown in Fig. XIV-11, where the surfactant

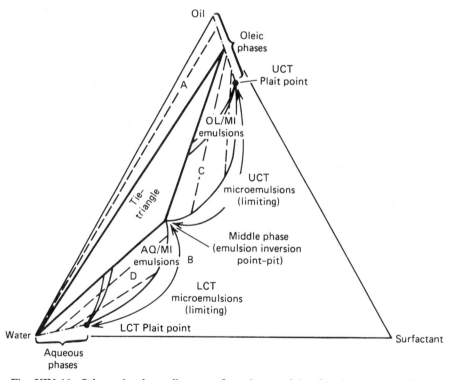

Fig. XIV-11. Schematic phase diagram of a microemulsion-forming system. (From Ref. 77.)

corner represents a fixed mixture of surfactant and cosurfactant [78a]. More generally, the phase diagram will be a stack of triangular diagrams at varying temperature (see Ref. 78b). The diagram in Fig. XIV-11 shows three two phase regions: A, containing oil and water surfactant solutions (perhaps unstable emulsions); and C and D, consisting of a microemulsion phase in equilibrium with oil-rich or water-rich phases, respectively. The single-phase region, B, is a microemulsion phase. Near the inversion point exists a middle or bicontinuous phase in coexistence with both oil and water phases. The various gel or liquid crystalline phases have been omitted from the surfactant corner of the diagram. Representative literature discussing microemulsions and their phase behavior is cited in Refs. 79–93.

Cationic surfactants may be used [94] and the effect of salinity and valence of electrolyte on charged systems has been investigated [95–98]. The phospholipid lecithin can also produce microemulsions when combined with an alcohol cosolvent [99]. Microemulsions formed with a double-tailed surfactant such as Aerosol OT (AOT) do not require a cosurfactant for stability (see, for instance, Refs. 100, 101). Morphological "hysteresis" has been observed in the inversion process and the formation of stable mixtures of microemulsion indicated [102].

The structure of microemulsions of composition away from area B in Fig. XIV-11 consists of swollen micelles of either oil or aqueous interior (note Refs. 103–105). Around the region B, however, changes in properties, such as electrical conductivity, suggest the formation of a mutually intertwined and extensive *bicontinuous structure* [81, 106–108]. In this region of the phase diagram, the surfactant-laden interfaces have little tendency for curvature in either direction [109]. Interpenetrating geometric figures can have this property, although it seems likely that the structures are rapidly fluctuating. It is possible to actually see the structure of a bicontinuous microemulsion through *freeze-fracture* electron microscopy. A platinum replica of a fracture surface from a rapidly frozen thin sample is viewed with transmission electron microscopy; the amazing interpenetrating structure is shown in Fig. XIV-12.

These fascinating bicontinuous or "sponge" phases have attracted considerable theoretical interest. *Percolation theory* [112] is an important component of such models as it can be used to describe conductivity and other physical properties of microemulsions. Topological analysis [113] and geometric models [114] are useful, as are thermodynamic analyses [115–118] balancing *curvature elasticity* and entropy. Similar elastic modulus considerations enter into models of the properties and stability of droplet phases [119–121] and phase behavior of microemulsions in general [97, 122].

The structure of microemulsions have been studied by a variety of experimental means. Scattering experiments yield the droplet size or *persistence length* (3–6 nm) for nonspherical phases. Small-angle neutron scattering (SANS) [123] and x-ray scattering [124] experiments are appropriate; however, the isotopic substitution of D_2O for H_2O

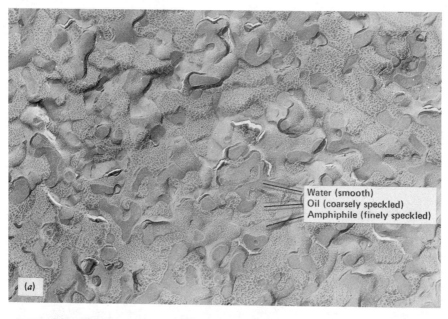

Water (smooth)
Oil (coarsely speckled)
Amphiphile (finely speckled)

(a)

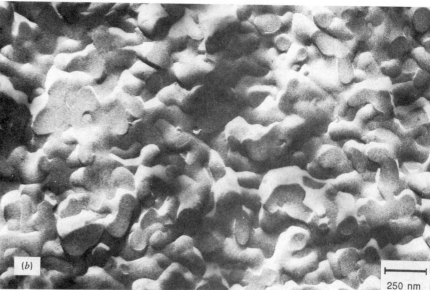

(b)

250 nm

Fig. XIV-12. Freeze-fracture transmission electron micrographs of a bicontinuous microemulsion consisting of 37.2% *n*-octane, 55.8% water, and the surfactant pentaethylene glycol dodecyl ether. In both cases 1 cm ≈ 2000 Å (for purposes of microscopy, a system producing relatively coarse structures has been chosen). [(*a*) Courtesy of P. K. Vinson, W. G. Miller, L. E. Scriven, and H. T. Davis—see Ref. 110; (*b*) courtesy of R. Strey—see Ref. 111.]

in SANS can alter the structure [96, 125]. Static and dynamic light scattering [100, 108, 126–129] reveals droplet sizes (around 5 nm) and interactions. Gradient NMR methods provide a measure of the diffusion coefficient of oil, water, or surfactant and allow demonstration of the presence of a bicontinuous phase [130–135]. Dielectric spectroscopy has been applied to microemulsions much in the same way as to emulsions as described in Section XIV-2 [136–138]. Photophysical probes ellucidate droplet exchange [101, 139], surface potential [140] and droplet size [141] and electron spin resonance (ESR) measurements characterize the lifetime of a probe radical in a droplet (10^{-6} sec) [142], the phase inversion point [143], and the interfacial region of microemulsions [144, 145]. More classic measurements include conductivity [146, 147], viscosity [148–150], calorimetry [151], and wetting studies [152].

Among other interesting properties of microemulsions are ultralow interfacial tensions between phases C and D in Fig. XIV-11. Values as low as 10^{-3} may be reached [153–156]. There are various theoretical explanations [154–158]. As might be inferred from the phase diagram, a middle-phase microemulsion is almost a universal solvent. It can solubilize water [159, 160] and hydrocarbons [161, 162], as well as polar organic species [163]. The oil solubilization is useful in drug delivery [164] and cosmetics [165] applications. This general property has made microemulsions interesting media for the study of bimolecular reactions, as in photophysical processes [166–168]. Both the low interfacial tensions and the solubilizing properties have encouraged the study of microemulsion systems in tertiary oil recovery [169–171]. Finally, microemulsions can serve as useful templates for nanoparticle synthesis when the synthetic reaction can be carried out in the droplet phase. Interesting nanoparticles made this way include silver halides for photographic emulsions [172], magnetic colloids [172, 173], and superconductors [172, 174].

7. Foam Structure

A foam can be considered as a type of emulsion in which the inner phase is a gas, and as with emulsions, it seems necessary to have some surfactant component present to give stability. The resemblance is particularly close in the case of foams consisting of nearly spherical bubbles separated by rather thick liquid films; such foams have been given the name *kugelschaum* by Manegold [175].

The second type of foam contains mostly gas phase separated by thin films or laminas. The cells are polyhedral in shape, and the foam can be thought of as a space-filling packing of more or less distorted polyhedra; such foams have been called *polyederschaum*. They may result from a sufficient drainage of a kugelschaum or be formed directly, as in the case of a liquid of low viscosity. Again, there is a parallel with emulsions—centrifuged emulsions can consist of polyhedral cells of inner liquid separated by thin films of outer liquid. Sebba [64, 176] reports on foams that range between both extremes, which he calls *microgas emulsions*, that is, clusters of encapsulated gas bubbles.

The discussion here is confined to the more common type of foam, the polyederschaum and their interesting geometric aspects. If three bubbles are joined, as in Fig. XIV-13, the three separating films or septa meet to form a small triangular column of liquid (perpendicular to the paper in the figure)

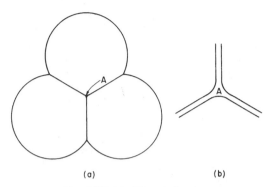

(a)   (b)

Fig. XIV-13. Plateau border.

known as the *Plateau border*. This equilibrium between three fluid laminas were studied in some detail by Plateau† [177] and by Gibbs [178] and the Plateau border plays an important role in film drainage. An enlarged drawing of it is shown in Fig. XIV-13*b*, and it is seen, first of all, that the high curvature of the boundaries of the area *A* must mean that a considerable pressure drop occurs between the gas and liquid phases. The resulting tendency for liquid to be sucked from the film into the border plays an important role in foam drainage, as will be seen.

With three bubbles, the septa must meet at 120° if the system is to be mechanically stable. A fourth bubble could now be added as shown in Fig. XIV-14, but this would not be stable. The slightest imbalance or disturbance would suffice to move the septa around until an arrangement such as in Fig. XIV-14*b* resulted. Thus a two-dimensional foam consists of a more or less uniform hexagonal type of network.

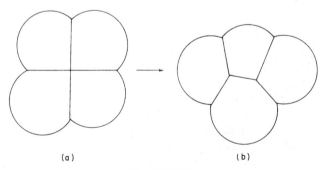

(a)   (b)

Fig. XIV-14.

†This ancient treatise makes fascinating reading!

The situation becomes more complex in the case of a three-dimensional foam. Since the septa should all be identical, again three should meet at 120° angles to form borders or lines, and four lines should meet at a point, at the tetrahedral angle of 109°28'. This was observed to be the case by Matzke [179] in his extensive statistical study of the geometric features of actual foams.

Intuitively, one would like the cells to consist of some single type of regular polyhedron, and one that closely meets the foregoing requirements is the pentagonal dodecahedron (a figure having 12 pentagonal sides). Matzke did find more than half of the faces in actual foams were five-sided, and about 10% of the polyhedra were pentagonal dodecahedrons. Earlier, Desch [180] reached similar conclusions.

However, the pentagonal dodecahedron gives angles that are slightly off (116°33' between faces and 108° between lines) and, moreover, cannot fill space exactly. This problem of space filling has been discussed by Gibbs [178] and Lord Kelvin [181]; the latter showed that a truncated octahedron (a figure having six square and eight hexagonal sides) would fill space exactly and, by suitable curving of the faces, give the required angles. However, experimentally, only 10% of the sides in Matzke's foams were four-sided.

The foregoing discussion leads to the question of whether actual foams do, in fact, satisfy the conditions of zero resultant force on each side, border, and corner without developing local variations in pressure in the liquid interiors of the laminas. Such pressure variations would affect the nature of foam drainage (see below) and might also have the consequence that films within a foam structure would, on draining, more quickly reach a point of instability than do isolated plane films.

8. Foam Drainage

A. Drainage of Single Films

Consider first the simple case of a plane soap film stretched across a rectangular frame, as illustrated in Fig. XIV-15. When first formed, by dipping the frame into a soap solution and raising it carefully (in a closed system to avoid evaporation), the initial film is relatively thick. Drainage begins immediately, however, and with soaps forming liquid (as opposed to solid) interfacial films, a pattern of interference fringes develops rather rapidly. This pattern, as indicated by the shadings in the figure, consists of repeated bands of rainbowlike colors resulting from the interference between light reflected from the front and the back surfaces of the film. As a result of draining, the film thickness decreases from bottom to top. The process of thinning is beautiful to watch; as it occurs, each colored band moves downward, and the spacings between them increase. The theoretical treatment of these interference effects is somewhat complicated, and the reader is referred to Rayleigh [182], Reinhold and Rucker [183], and Bikerman [184] (see Ref. 185 for a listing of colors for various film thicknesses).

Fig. XIV-15. Drainage of soap films. (Courtesy of K. J. Mysels.)

Eventually the top of a draining film becomes thinner than the wavelength of visible light and appears black in reflected light. Black films are of particular interest because their properties are governed by surface chemical interactions, mainly the electrostatic repulsion and van der Waals attraction (Section VI-3), between the surfactant layers [186, 187]. In fact, the study of horizontal, nondraining films provides one means of investigating intermolecular forces [188–190], or the resultant disjoining pressure [191, 192]. Early observations by Perrin [193] later confirmed by Mysels and co-workers [194] and by Overbeek [189] indicated that black films coexist in patches of varying thickness with stepwise transitions between regions. This stratification has been further quantified by Exerowa and co-workers [195], Platikanov and co-workers [196] and experiments by Wasan and co-workers on films containing colloidal particles showed the stratification [197] and illustrated its mechanism [198, 199]. The thinnest of these black films, the *Newton black film*, is on the order of 4 nm, or about the thickness of a soap bilayer. Recent x-ray scattering studies [196, 200] showed the inaccuracies in optical measurements of such thin films.

Exerowa and co-workers [201] suggest that surfactant association initiates black film formation; the growth of a black film is discussed theoretically by de Gennes [202]. A characteristic of thin films important for foam stability, their permeability to gas, has been studied in some depth by Platikanov and co-workers [203, 204]. A review of the stability and permeability of amphiphile films is available [205].

In an interesting medical application, the formation of a stable black foam film from amniotic fluid can be used as an assessment of fetal lung maturity [206].

The preceding discussion may have suggested that liquid soap films present a rather quiescent appearance as they thin. Quite the contrary is the case. On looking at such films, one sees a multicolored activity, particularly along the border between the film and its supporting frame. The appearance is one of rapid, fluid motion of a very complex character but consisting, roughly, of swirls of thinner (and hence differently colored) film rising from the edge and moving inward as well as upward.

This brief description of film drainage is given to emphasize the fact that the general thinning of soap films occurs by a mechanism that involves the formation of patches of thinner film at the border, the excess liquid presumably being discharged into the border channel. Drainage is thus determined by an edge effect, and its rate is governed by what happens at the film boundaries; as a consequence, the general rate of thinning varies inversely with the width of the film [207]. As discussed earlier, there must be a high curvature at the film edges so that the liquid in the border channel must be at a lower pressure than in the body of the film; it is presumably by virtue of this pressure difference that the thin patches are formed (see Ref. 189). Drainage by the gradual descent of liquid through the film interior is only important in very thick films.

B. Drainage of Foams

The very thick or kugelschaum type of foam probably does drain by a hydrodynamic mechanism, but the thin-film polyederschaum foam has septa similar to the individual soap films discussed and probably drains by the same border mechanism. As a result of such drainage, the film laminas become increasingly thin, and rupture begins to occur here and there. In some cases the uppermost films rupture first, so that the volume of the foam decreases steadily with time, whereas in other cases it is mostly interior laminas that rupture, so that the gas cells become increasingly large and the foam decreasingly dense. In addition, cell breakage as well as intercell gas diffusion changes the distribution of cell sizes and shapes with time. The local dynamics within the foam during drainage have been investigated by ESR [208]. By altering the electrolyte concentration, Khristov and Exerowa [209] create foams of common black (~8 nm) and Newton black (~4 nm) films. Then, by altering the pressure on the Plateau borders [210], they can regulate the rate of foam drainage to isolate other physicochemical effects.

The rupture process of a soap film is of some interest. In the case of a film spanning a frame, as in Fig. XIV-15, it is known that rupture tends to originate at the margin, as shown in the classic studies of Mysels [207, 211]. Rupture away from a border may occur spontaneously but is usually studied by using a spark [212] as a trigger (α-radiation will also initiate rupture [213]). An aureole or ridge of accumulated material may be seen on the rim of the growing hole [212, 214] (see also Refs. 215, 216). Theoretical analysis has been in the form of nucleation [217, 218] or thin-film instability [219].

9. Foam Stability

A notable characteristic of stable films is their resistance to mechanical disturbance. Gibbs [178] considered the important property to be the elasticity of the film E:

$$E = 2 \frac{d\gamma}{d \ln \mathcal{A}} \qquad \text{(XIV-10)}$$

where $\mathcal{A}$ is the area of the film (see Section III-7D). For the case of a two-component system, Eq. XIV-10 can be put in the form

$$E = 4(\Gamma_2^1)^2 \frac{d\mu_2}{dm_2} \qquad \text{(XIV-11)}$$

Here Γ_2^1 is the surface excess of component 2, μ_2 is the chemical potential of that component, and m_2 is its amount per unit area of film. Qualitatively, E gives a measure of the ability of a film to adjust its surface tension in an instant of stress. If the surface should be extended, the surface concentration of surfactant drops, and the local surface tension rises accordingly; the film is thus protected against rupture. For pure liquids, E as given would be zero, and this is in accord with the observation that the pure liquids do not give a stable foam. Conversely, Eq. XIV-11 indicates that in order for E to be large, both Γ_2^1 and $d\mu_2/dm_2$ should be large. In effect, this means that the surfactant concentration should be high, but not too high. The elasticity E may have values in the range of 10–40 dyn/cm [220]. The importance of elasticity, however, has been questioned [221].

The foregoing is an equilibrium analysis, yet some transient effects are probably important to film resilience. Rayleigh [182] noted that surface freshly formed by some insult to the film would have a greater than equilibrium surface tension (note Fig. II-15). A recent analysis [222] of the effect of surface elasticity on foam stability relates the *nonequilibrium* surfactant surface coverage to the foam retention time or time for a bubble to pass through a wet foam. The adsorption process is important in a new means of obtaining a foam by supplying vapor phase surfactants [223].

In addition to elasticity and resilience as properties giving stability to foams, a high surface viscosity appears also to be important. While several groups found a correlation between film viscosity and foam stability [224], the addition of water-soluble polymers as "thickeners" slow drainage but not in proportion to their viscosity [208]. It is thought that such additives interact with the surfactant molecules to stabilize films in addition to enhancing viscosity. The size and redistribution of foam cells may also correlate with foam stability [225, 226]; recent measurements of shaving cream use a dynamic light scattering technique to follow this coarsening process [227]. Reviews of foam stability and hydrophobic foams are found in Refs. 226 and 228.

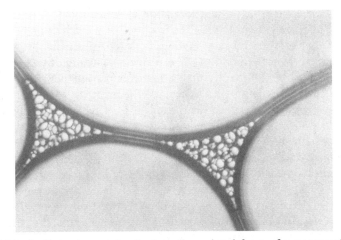

Fig. XIV-16. A photomicrograph of a two-dimensional foam of a commercial ethoxylated alcohol nonionic surfactant solution containing emulsified octane in which the oil drops have drained from the foam films into the Plateau borders. (From Ref. 234.)

Foam *rheology* has been a challenging area of research of interest for the yield behavior and stick–slip flow behavior (see the review by Kraynik [229]). Recent studies by Durian and co-workers combine simulations [230] and a dynamic light scattering technique suited to turbid systems [231], *diffusing wave spectroscopy* (DWS), to characterize *coarsening* and *shear-induced* rearrangements in foams. The dynamics follow stick–slip behavior similar to that found in earthquake faults and friction (see Section XII-2D).

The action of antifoam agents in preventing foaming can be analyzed qualitatively along lines similar to the preceding material as discussed by Ross and co-workers [232] and by Bikerman [233]. A recent study by Wasan and co-workers [234, 235] of the antifoaming properties of oils and hydrophobic particles suggests that the antifoam drops get trapped in the Plateau borders, as illustrated in Fig. XIV-16, and destabilize the film at the borders. Shah and co-workers [236] have shown that organic ions, on the other hand, interact with surfactant to decrease the CMC, increase the area per molecule of the surfactant, and decrease the surface viscosity to reduce foam stability.

An important application of foams arises in *foam displacement*, another means to aid enhanced oil recovery. The effectiveness of various foams in displacing oil from porous media has been studied by Shah and co-workers [237, 238]. The displacement efficiency depends on numerous physicochemical variables such as surfactant chain length and temperature with the surface properties of the foaming solution being an important determinant of performance.

10. Aerosols

Brief mention should be made of the important topic of *aerosols*, more or less stable suspensions of liquid or solid particles in a gas. The manufacture of fine particles, the behavior and formation of smokes and fly ash, and atmospheric pollution and atmospheric chemical reactions (such as involving depletion of the ozone layer) all may involve the properties of aerosols. Aerosols may be studied from the point of view of their generation by chemical reaction (e.g., Refs. 239, 240) or by nucleation of supersaturated vapors (Chapter IX) and of their coagulation and flocculation (e.g., Refs. 241, 242) or of their stabilization by charging effects (e.g., Ref. 243). Their size distributions may be determined by a variety of techniques [244–246], including a recent aerodynamic measurement [247]. Because of the importance of aerosol deposition in the respiratory tract, particularly for radon exposure, a study of aerosol hygroscopicity is designed to mimic the conditions in the human respiratory tract [248].

A relatively recent development has been that of techniques allowing measurements on *individual* aerosol particles. The classic method is that used by Millikan in 1909 in his famous oil droplet experiment. A single charged droplet could be held in position by means of an electrostatic field. Contemporary equipment is highly sophisticated, with detector/feedback devices that will automatically hold a droplet in place (see Ref. 249). There is usually insufficient restoring force, however, to permit gas flow past the particle. An important development has been the electrodynamic balance, whereby ac fields provide both radial and axial restoring forces [249, 250]. In the case of liquid droplets, stabilization can be obtained if there is a gradient in the pressure of the vapor [251]. *Photophoretic* forces, due to anisotropic heating by absorbed light, can provide levitation (see Ref. 252).

The importance of the particle levitation methods is that they allow the study of how a single particle responds to changes in environment. The infrared molecular spectroscopy of single particles is possible [253], as are photophysical studies using adsorbed or dissolved dyes.

11. Problems

1. Discuss briefly at least two reasons why two pure immiscible liquids do not form a stable emulsion.

2. Show that the maximum possible value for ϕ is 0.74 in the case of an emulsion consisting of uniform, rigid spheres.

3. Show what the maximum possible value of ϕ is for the case of a two-dimensional emulsion consisting of uniform, rigid circles (or, alternatively, of a stacking of right circular cylinders).

4. Suppose that the specific conductivities of an oil and a liquid phase are 2×10^{-3} and 2×10^{-4} Ω^{-1} cm^{-1}, respectively. Calculate and plot versus ϕ the specific conductivities of O/W and W/O emulsions formed from these phases.

5. Emulsion A has a droplet size distribution that obeys the ordinary Gaussian error curve. The most probable droplet size is 5 μm. Make a plot of $p/p(\text{max})$, where $p(\text{max})$ is the maximum probability, versus size if the width at $p/p(\text{max}) = \frac{1}{2}$ corresponds to

a standard deviation of 1.02 μm. Emulsion B has a droplet size distribution that obeys the log-normal distribution, Eq. XIV-5, with the same most probable droplet size as for emulsion A and with a logarithmic standard deviation of 1.02. Plot the distribution for emulsion B on the same plot as that for emulsion A.

6. The inset in Figure XIV-6 shows the coalescence time t_c for the droplets for the pH corresponding to each DLVO curve. Does DLVO theory adequately explain the variation of t_c with pH? What additional factors may play a role?

7. From the data in Fig. XIV-2 and Eq. XIV-4, see if you can assign ϵ_s, ϵ_∞, τ, and α.

8. Consider the case of an emulsion of 1 liter of oil in 1 liter of water having oil droplets of 0.6 μm diameter. If the oil–water interface contains a close-packed monolayer of surfactant of 18 Å^2 per molecule, calculate how many moles of surfactant are present.

9. A mixture of 70% Tween 60 and 30% Span 65 gives optimum behavior in a given emulsion system. What composition mixture of sodium lauryl sulfate and cetyl alcohol should also give optimum behavior in the same system?

10. A surfactant mixture having an HLB number of 8 should give a good W/O emulsion in which the oil phase is lanolin. Suggest two possible surfactant mixtures that you, an aspiring cosmetic chemist, might use; you have been told that your formulations must contain 10% cetyl alcohol.

11. Referring to Fig. XIV-11 and assuming the composition scale to be in weight fraction, describe the expected sequence of phase changes on adding surfactant to a system initially consisting of 50% water and 50% oil. Give the approximate system composition at which phase(s) appear or disappear and the approximate compositions of these phases.

12. Consider the case of two soap bubbles having a common septum. The bubbles have radii of curvature R_1 and R_2, and the radius of curvature of the common septum is R. Show under what conditions R would be zero and under what conditions it would be equal to R_2.

13. Derive Eq. XIV-11 from Eq. XIV-10. State the approximations involved. Explain whether the surface elasticity should be small or large for a surfactant film if the bulk surfactant concentration is about its CMC.

14. Make an estimate of the hydrostatic pressure that might be present in the Plateau border formed by the meeting of three thin black films. Make the assumptions of your calculation clear.

15. The "oil" droplets in a certain benzene-water emulsion are nearly uniform in size and show a diffusion coefficient of 3.75×10^{-8} cm^2/sec at 25°C. Estimate the number of benzene molecules in each droplet.

General References

P. Becher, ed., *Encyclopedia of Emulsion Technology*, Marcel Dekker, New York, 1983.

J. J. Bikerman, *Foams*, Springer-Verlag, New York, 1973.

J. T. Davies and E. K. Rideal, *Interfacial Phenomena*, Academic, New York, 1961.

S. Friberg, ed., *Food Emulsions*, Marcel Dekker, New York, 1976.

S. K. Friedlander, *Smoke, Dust, and Haze*, Wiley, New York, 1977.

W. M. Gelbart, A. Ven-Shaul, and D. Roux, eds., *Micelles, Membranes, Microemulsions and Monolayers*, Springer-Verlag, Berlin, 1994.

W. C. Griffin, in *Kirk-Othner Encyclopedia of Chemical Technology*, 3rd ed., Vol. 8, M. Grayson, ed., Wiley-Interscience, New York, 1979.

E. Manegold, *Schaum, Strassenbau, Chemie und Technik*, Heidelberg, 1953.

J. L. Moilliet, B. Collie, and W. Black, *Surface Activity*, E. & F. N. Spon, London, 1961.

D. H. Napper, *Polymeric Stabilization of Colloidal Dispersions*, Academic, New York, 1983.

L. I. Osipow, *Surface Chemistry*, Reinhold, New York, 1962.

L. H. Princen, in *Treatise on Coatings*, Vol. 1, Part 3, R. R. Myers and J. S. Long, eds., Marcel Dekker, New York, 1972, Chapter 2.

S. Ross, *Foams: Encyclopedia of Chemical Technology*, 3rd ed., Vol. 2, Wiley, New York, 1980.

S. Ross and I. D. Morrison, *Colloidal Systems and Interfaces*, Wiley, New York, 1988.

S. A. Safran, *Statistical Thermodynamics of Surfaces, Interfaces and Membranes*, Addison-Wesley, Reading, MA, 1994.

F. Sebba, *Foams and Biliquid Foams—Aphrons*, Wiley, New York, 1987.

P. Sherman, *Emulsion Science*, Academic, New York, 1968.

K. Shinoda, *Principles of Solution and Solubility*, Marcel Dekker, New York, 1978.

K. Shinoda and S. Friberg, *Emulsions and Solubilization*, Wiley, New York, 1986.

A. L. Smith, Ed., *Theory and Practice of Emulsion Technology*, Society of Chemical Industry, London, 1976.

R. D. Vold and M. Vold, *Colloid and Interface Chemistry*, Addison-Wesley, Reading, MA, 1983.

Textual References

1. P. Becher, in *Interfacial Phenomena in Apolar Media*, H. Eicke and G. D. Parfitt, eds., Marcel Dekker, New York, 1987; *Nonionic Surfactants Physical Chemistry*, M. J. Schick, ed., Marcel Dekker, New York, 1987.

2. A. Einstein, *Ann. Phys.*, **19**, 289 (1906); **34**, 591 (1911).

3. P. Becher, *Emulsions*, Reinhold, New York, 1965.

4. P. Sherman, *Emulsion Science*, Academic, New York, 1968, p. 131.

5. H. Eicke, R. Kubik, and H. Hammerich, *J. Colloid Interface Sci.*, **90**, 27 (1982).

6. J. C. Maxwell, *A Treatise on Electricity and Magnetism*, 2nd ed., Clarendon Press, Oxford, 1881.

7. R. E. Meredith and C. W. Tobias, *J. Electrochem. Soc.*, **108**, 286 (1961).

8. C. Thomas, J. P. Perl, and D. T. Wasan, *J. Colloid Interface Sci.*, **139**, 1 (1990).

9. T. Skodvin, J. Sjöblom, J. O. Saeten, T. Wärnheim, and B. Gestblom, *Colloids & Surfaces*, **A83**, 75 (1994).

10. T. Skodvin, J. Sjöblom, J. O. Saeten, O. Urdahl, and B. Gestblom, *J. Colloid Interface Sci.*, **166**, 43 (1994).

11. P. Fäldt, B. Bergenst, and P. M. Claesson, *Colloids & Surfaces*, **A71**, 187 (1993).

12. S. R. Deshiikan and K. D. Papadopoulos, *J. Colloid Interface Sci.*, **174,** 302 (1995).

13. J. Sjöblom, O. Urdahl, K. G. N. B., L. Mingyuan, J. O. Saeten, A. A. Christy, and T. Gu, *Adv. Colloid Interface Sci.*, **41,** 241 (1992).

14. B. Balinov, O. Urdahl, O. Söderman, and J. Sjöblom, *Colloids & Surfaces*, **A82,** 173 (1994).

15. K. J. Lissant and K. G. Mayhan, *J. Colloid Interface Sci.*, **42,** 201 (1973).

16. L. H. Princen, *Appl. Polym. Symp.*, No. 10, 159 (1969).

17. W. D. Harkins, *The Physical Chemistry of Surface Active Films*, Reinhold, New York, 1952, p. 90.

18. K. J. Lissant, *J. Soc. Cosmet. Chem.*, **21,** 141 (1970).

19. H. M. Princen, M. P. Aronson, and J. C. Moser, *J. Colloid Interface Sci.*, **75,** 246 (1980).

20. W. D. Bancroft, *J. Phys. Chem.*, **17,** 501 (1913); **19,** 275 (1915).

21. G. H. A. Clowes, *J. Phys. Chem.*, **20,** 407 (1916).

22. S. G. Oh, M. Jobalia, and D. O. Shah, *J. Colloid Interface Sci.*, **155,** 511 (1993).

23. B. Bergenstand, P. M. Claesson, "Surface Forces in Emulsions," in *Food Emulsions*, 2nd ed., K. Larsson and S. E. Friberg, eds., Marcel Dekker, New York, 1990.

24. T. Sengupta and K. D. Papadopoulos, *J. Colloid Interface Sci.*, **163,** 234 (1994).

25. D. R. MacFarlane and C. A. Angel, *J. Phys. Chem.*, **88,** 4779 (1984).

26. H. A. Bramfield, in *Encyclopedia of Emulsion Technology*, Vol. III, P. Becher, ed., Marcel Dekker, New York, 1987.

27. J. Bibette, *J. Colloid Interface Sci.*, **147,** 474 (1991).

28. J. H. E. Promislow, A. P. Gast, and M. Fermigier, *J. Chem. Phys.*, **102,** 5492 (1995).

29. J. H. E. Promislow and A. P. Gast, *Langmuir*, **12,** 4095 (1996).

30. F. L. Calderon, T. Stora, O. M. Monval, P. Poulin, and J. Bibette, *Phys. Rev. Lett.*, **72,** 2959 (1994).

31. R. D. Vold and R. C. Groot, *J. Soc. Cosmet. Chem.*, **14,** 233 (1963).

32. S. J. Rehfeld, *J. Phys. Chem.*, **66,** 1966 (1962).

33. S. R. Reddy and H. S. Fogler, *J. Colloid Interface Sci.*, **79,** 105 (1981).

34. R. D. Vold and R. C. Groot, *J. Colloid Sci.*, **19,** 384 (1964).

35. M. van den Tempel, *Rec. Trav. Chim.*, **72,** 419 (1953).

36. J. A. Kitchener and P. R. Musselwhite, in *Emulsion Science*, P. Sherman, ed., Academic, New York, 1968, p. 104.

37. F. Huisman and K. J. Mysels, *J. Phys. Chem.*, **73,** 489 (1969).

38. L. H. Princen, in *Treatise on Coatings*, Vol. 1, Part 3, R. R. Myers and J. S. Long, eds., Marcel Dekker, New York, 1972, Chapter 2.

39. S. R. Deshiikan and K. D. Papadopoulos, *J. Colloid Interface Sci.*, **174,** 313 (1995).

40. W. Albers and J. Th. G. Overbeek, *J. Colloid Sci.*, **14,** 510 (1959); ibid., **15,** 489 (1960).

41. W. Rigole and P. Van der Wee, *J. Colloid Sci.*, **20,** 145 (1965).

42. T. Gillespie and R. M. Wiley, *J. Phys. Chem.*, **66,** 1077 (1962).

43. Y. H. Kim, D. T. Wasan, and P. J. Breen, *Colloids & Surfaces*, **A95,** 235 (1995).

44. Y. H. Kim, A. D. Nikolov, D. T. Wasan, H. Diaz-Arauzo, and C. S. Shetty, *J. Dispers. Sci. Technol.*, **17,** 33 (1996).

45. K. G. N. B, J. Sjöblom, and P. Stenius, *Colloids & Surfaces*, **63,** 241 (1992).

46. A. J. Scarlett, W. L. Morgan, and J. H. Hildebrand, *J. Phys. Chem.*, **31,** 1566 (1927).

47. E. H. Lucassen-Reynders and M. van den Tempel, *J. Phys. Chem.*, **67,** 731 (1963).

48. See W. D. Harkins and N. Beeman, *J. Am. Chem. Soc.*, **51,** 1674 (1929).

49. J. H. Schulman and J. Leja, *Trans. Faraday Soc.*, **50,** 598 (1954).

50. M. V. Ostrovsky and R. J. Good, *J. Disp. Sci. Tech.*, **7,** 95 (1986).

51. I. B. Ivanov, R. K. Jain, and P. Somasundaran, in *Solutions Chemistry of Surfactants*, K. L. Mittal, ed., Plenum, New York, 1979.

52. M. von Smoluchowski, *Phys. Z.*, **17,** 557, 585 (1916); *Z. Phys. Chem.*, **92,** 129 (1917).

53. H. R. Kruyt, *Colloid Science*, Elsevier, Amsterdam, 1952, p. 378.

54. L. A. Spielman, *J. Colloid Interface Sci.*, **33,** 562 (1970).

55. E. P. Honig, P. H. Wiersma, and G. J. Roeberson, *J. Colloid Interface Sci.*, **36,** 97 (1971).

56. D. F. Bernstein, W. I. Higuchi, and N. F. H. Ho, *J. Colloid Interface Sci.*, **39,** 439 (1972).

57. V. G. Babak and V. A. Chlenov, *J. Dispers. Sci. Technol.*, **4,** 221 (1983).

58. M. Vold, *Langmuir*, **1,** 74 (1985).

59. H. M. Princen, *J. Colloid Sci.*, **18,** 178 (1963).

60. S. P. Frankel and K. J. Mysels, *J. Phys. Chem.*, **66,** 190 (1962).

61. S. Hartland and D. K. Vohra, *J. Colloid Interface Sci.*, **77,** 295 (1980).

62. F. Sebba, *J. Colloid Interface Sci.*, **40,** 468 (1972).

63. F. Sebba, *J. Theor. Biol.*, **78,** 375 (1979).

64. F. Sebba, *Foams and Biliquid Foams—Aphrons*, Wiley, New York, 1987.

65. T. Green, R. P. Harker, and F. O. Howitt, *Nature*, **174,** 659 (1954).

66. H. F. E. Nodland, J. Sjöblom, and O. M. Kvalheim, *J. Colloid Interface Sci.*, **173,** 396 (1995).

67. W. C. Griffin, *J. Soc. Cosmet. Chem.*, **1,** 311 (1949); ibid., **5,** 249 (1954).

68. P. Becher, in *Encyclopedia of Emulsion Technology*, Vol. 2, P. Becher, ed., Marcel Dekker, New York, 1985.

69. J. T. Davies, *Proc. 2nd Int. Congr. Surf. Act., London*, Vol. 1, p. 426.

70. P. Becher, in *Surfactants in Solution*, Vol. 3, K. L. Mittal, ed., Plenum, New York, 1984; *J. Dispers. Sci. Technol.*, **5,** 81 (1984).

71. K. Shinoda and H. Kunieda, "Phase Properties of Emulsions: PIT and HLB," in *Encyclopedia of Emulsion Technology*, Vol. 1, P. Becher, ed., Marcel Dekker, New York, 1983.

72. C. Parkinson and P. Sherman, *J. Colloid Interface Sci.*, **41,** 328 (1972).

73. K. Shinoda and H. Sagitani, *J. Colloid Interface Sci.*, **64,** 68 (1978).

74. J. Boyd, C. Parkinson, and P. Sherman, *J. Colloid Interface Sci.*, **41,** 359 (1972).

75. R. G. Laughlin, *J. Soc. Cosmet. Chem.*, **32,** 371 (1981).

76. P. A. Winsor, *Trans. Faraday Soc.*, **44,** 376 (1948).

77. J. E. Bowcott and J. H. Schulman, *Z. Elektrochem.*, **59,** 283 (1955).

78. J. H. Schulman, W. Stoeckenius, and L. M. Prince, *J. Phys. Chem.*, **63,** 1677 (1959).

78a. D. H. Smith, *J. Colloid Interface Sci.*, **108,** 471 (1985); *Microemulsion Systems*, H. L. Rosano and M. Clausse, eds., Marcel Dekker, New York, 1987.

78b. D. H. Smith and Y.-H. C. Wang, *J. Phys. Chem.*, **98,** 7214 (1994).

79. S. E. Friberg and R. L. Venable, *Encyclopedia of Emulsion Technology*, Vol. 1, Marcel Dekker, New York, 1983.

80. D. Langevin, D. Guest, and J. Meunier, *Colloids & Surfaces*, **19,** 159 (1986); O. Abillon, B. P. Binks, C. Otero, and D. Langevin, *J. Phys. Chem.*, **92,** 4411 (1988).

81. D. F. Evans, D. J. Mitchell, and B. W. Ninham, *J. Phys. Chem.*, **90,** 2817 (1986).

82. J. Israelachvili, in *Surfactants in Solution*, K. L. Mittal and P. Bothorel, eds., Plenum, New York, 1987.

83. J. Sjöblom, Per Stenius, and I. Danielsson, in *Nonionic Surfactants*, 2nd ed., M. Schick, ed., Marcel Dekker, New York, 1987.

84. P. K. Kilpatrick, H. T. Davis, L. E. Scriven, and W. G. Miller, *J. Colloid Interface Sci.*, **118,** 270 (1987).

85. J. T. G. Overbeek, P. L. de Bruyn, and F. Verhoeck, *Surfactants*, Academic, New York, 1984; J. T. G. Overbeek, *Proc. Koninklijke Nederlandse Akademie Wetenschappen, Ser. B*, **89,** 61 (1986).

86. H. F. Eicke, in *Microemulsions*, I. D. Robb, ed., Plenum, New York, 1982.

87. E. Ruckenstein, *Fluid Phase Equilibria*, **20,** 189 (1985).

88. J. C. Lang, *Physics of Amphiphiles: Micelles, Vesicles and Microemulsions*, Soc. Italiana di Fisica, XC Corso, Bologna, 1985.

89. D. Andelman, M. E. Cates, D. Roux, and S. A. Safran, *J. Chem. Phys.*, **87,** 7229 (1987).

90. E. Dickinson, *J. Colloid Interface Sci.*, **84,** 284 (1981).

91. M. Kahlweit, R. Strey, P. Firman, and D. Haase, *Langmuir*, **1,** 281 (1985); M. Kahlweit and R. Strey, *Angew. Chem. Int. Ed. Engl.*, **24,** 654 (1985).

92. O. Ghosh and C. A. Miller, *J. Phys. Chem.*, **91,** 4528 (1987); P. K. Kilpatrick, C. A. Gorman, H. T. Davis, L. E. Scriven, and W. G. Miller, ibid., **90,** 5292 (1986).

93. J. C. Lang, Jr. and B. Widom, *Physica*, **81A,** 190 (1975).

94. K. R. Wormuth and E. W. Kaler, *J. Phys. Chem.*, **91,** 611 (1987).

95. V. Chen, D. F. Evans, and B. W. Ninham, *J. Phys. Chem.*, **91,** 1823 (1987).

96. S. I. Chou and D. O. Shah, *J. Colloid Interface Sci.*, **80,** 49 (1981).

97. R. Leung and D. O. Shah, *J. Colloid Interface Sci.*, **120,** 320 (1987); ibid., p. 330.

98. R. Skurtveit, J. Sjöblom, J. Bouwstra, G. Gooris, and M. H. Selle, *J. Colloid Interface Sci.*, **152,** 205 (1992).

99. K. Shinoda, M. Araki, A. Sadaghiani, A. Khan, and B. Lindman, *J. Phys. Chem.*, **95,** 989 (1991).

100. M. J. Hou, M. Kim, and D. O. Shah, *J. Colloid Interface Sci.*, **123,** 398 (1988).

101. M. Almgren and R. Jóhannsson, *J. de Phys. IV, C1*, **3**, 81 (1993).

102. D. H. Smith, S. N. Nwosu, G. K. Johnson, and K.-H. Lim, *Langmuir*, **8**, 1076 (1992).

103. L. M. Prince, *J. Colloid Interface Sci.*, **52**, 182 (1975).

104. A. W. Adamson, *J. Colloid Interface Sci.*, **29**, 261 (1969).

105. W. C. Tosch, S. C. Jones, and A. W. Adamson, *J. Colloid Interface Sci.*, **31**, 297 (1969).

106. L. E. Scriven, *Nature*, **263**, 123 (1976).

107. D. Guest, L. Auvray, and D. Langevin, *J. Phys. Lett.*, **46**, L-1055 (1985).

108. A. Cazabat, D. Chatenay, D. Langevin, and J. Meunier, *Faraday Disc. Chem. Soc.*, **76**, 291 (1982).

109. S. J. Chen, D. F. Evans, B. W. Ninham, D. J. Mitchell, F. D. Blum, and S. Pickup, *J. Phys. Chem.*, **90**, 842 (1986).

110. P. K. Vinson, J. G. Sheehan, W. G. Miller, L. E. Scriven, and H. T. Davis, *J. Phys. Chem.*, **95**, 2546 (1991).

111. W. Jahn and R. Strey, *J. Phys. Chem.*, **92**, 2294 (1988).

112. Y. Talmon and S. Prager, *J. Chem. Phys.*, **69**, 2984 (1978); **76**, 1535 (1982).

113. S. T. Hyde, *J. Phys. Chem.*, **93**, 1458 (1989).

114. D. S. Bohlen, H. T. Davis, and L. E. Scriven, *Langmuir*, **8**, 982 (1992).

115. D. Roux, *Physica A*, **213**, 168 (1995).

116. D. Roux, C. Coulon, and M. E. Cates, *J. Phys. Chem.*, **96**, 4174 (1992).

117. P. Pieruschka and S. A. Safran, *J. Phys. Condens. Matter*, **6**, A357 (1994).

118. P. Pieruschka and S. A. Safran, *Europhys. Lett.*, **31**, 207 (1995).

119. W. K. Kegel, I. Bodnàr, and H. N. W. Lekkerkerker, *J. Phys. Chem.*, **99**, 3272 (1995).

120. S. A. Safran, "Theory of Structure and Phase Transitions in Globular Microemulsions," in *Micellar Solutions and Microemulsions*, S. H. Chen and R. Rajagopalan, eds., Springer-Verlag, New York, 1990, Chapter 9.

121. S. A. Safran, *Phys. Rev. A*, **43**, 2903 (1991).

122. R. Menes, S. A. Safran, and R. Strey, *Phys. Rev. Lett.*, **74**, 3399 (1995).

123. E. Caponetti, L. J. Magid, J. B. Hayter, and J. S. Johnson, Jr., *Langmuir*, **2**, 722 (1986).

124. T. N. Zemb, S. T. Hyde, P. Derian, I. S. Barnes, and B. W. Ninham, *J. Phys. Chem.*, **91**, 3814 (1987).

125. P. Baglioni, L. Dei, and C. M. C. Gambi, *J. Phys. Chem.*, **99**, 5035 (1995).

126. C. Vinches, C. Coulon, and D. Roux, *J. Phys. II France*, **4**, 1165 (1994).

127. E. Y. Sheu, S. H. Chen, J. S. Huang, and J. C. Sung, *Phys. Rev. A*, **39**, 5867 (1989).

128. J. S. Huang, *J. Chem. Phys.*, **82**, 480 (1985).

129. H. M. Cheung, S. Qutubuddin, R. V. Edwards, and J. A. Moran, Jr., *Langmuir*, **3**, 744 (1987).

130. K. P. Das, A. Ceglie, B. Lindman, and S. E. Friberg, *J. Colloid Interface Sci.*, **116**, 390 (1987).

131. F. D. Blum, S. Pickup, B. Ninham, S. J. Chen, and D. F. Evans, *J. Phys. Chem.*, **89,** 711 (1985).

132. P. Guering and B. Lindman, *Langmuir*, **1,** 464 (1985).

133. E. Cheever, F. D. Blum, K. R. Foster, and R. A. Mackay, *J. Colloid Interface Sci.*, **104,** 121 (1985).

134. J. F. Bodet, J. R. Bellare, H. T. Davis, L. E. Scriven, and W. G. Miller, *J. Phys. Chem.*, **92,** 1898 (1988).

135. B. Lindman, K. Shinoda, U. Olsson, D. Anderson, G. Karlström, and H. Wennerström, *Colloids & Surfaces*, **38,** 205 (1989).

136. S. I. Chou and D. O. Shah, *J. Phys. Chem.*, **85,** 1480 (1981).

137. J. Sjöblom, R. Skurtveit, J. O. Saeten, and B. Gestblom, *J. Colloid Interface Sci.*, **141,** 329 (1991).

138. J. O. Saeten, H. Fördedal, T. Skodvin, J. Sjöblom, A. Amran, and S. E. Friberg, *J. Colloid Interface Sci.*, **154,** 167 (1992).

139. R. Jóhannsson and M. Almgren, *Langmuir*, **9,** 2879 (1993).

140. B. S. Murray, C. J. Drummond, F. Grieser, and L. R. White, *J. Phys. Chem.*, **94,** 6804 (1990).

141. M. Almgren, F. Grieser, and J. K. Thomas, *J. Am. Chem. Soc.*, **102,** 3188 (1980).

142. A. Barelli and H. Eicke, *Langmuir*, **2,** 780 (1986).

143. C. Ramachandran, S. Vijayan, and D. O. Shah, *J. Phys. Chem.*, **84,** 1561 (1980).

144. P. Baglioni, C. M. C. Gambi, and D. Goldfarb, *J. Phys. Chem.*, **95,** 2577 (1991).

145. P. Baglioni, C. M. C. Gambi, and D. Goldfarb, *Prog. Colloid Polym. Sci.*, **84,** 133 (1991).

146. D. H. Smith, G. K. Johnson, Y. H. C. Wang, and K.-H. Lim, *Langmuir*, **10,** 2516 (1994).

147. G. K. Johnson, D. B. Dadyburjor, and D. H. Smith, *Langmuir*, **10,** 2523 (1994).

148. J. W. Falco, R. D. W., Jr., and D. O. Shah, *AIChE J.*, **20,** 510 (1974).

149. H. T. Davis, J. F. Bodet, L. E. Scriven, and W. G. Miller, *Physica A*, **157,** 470 (1989).

150. C. Vinches, C. Coulon, and D. Roux, *J. Phys. II France*, **2,** 453 (1992).

151. D. H. Smith and G. L. Covatch, *J. Colloid Interface Sci.*, **162,** 372 (1994).

152. D. H. Smith and G. L. Covatch, *J. Chem. Phys.*, **93,** 6870 (1990).

153. J. L. Salager, J. C. Morgan, R. S. Schechter, and W. H. Wade, *Soc. Petr. Eng. J.*, April 1979, p. 107.

154. M. Allen, D. F. Evans, D. J. Mitchell, and B. W. Ninham, *J. Phys. Chem.*, **90,** 4581 (1986).

155. G. Sundar and B. Widom, *J. Phys. Chem.*, **91,** 4802 (1987).

156. D. Langevin, *Physics of Amphiphiles: Micelles, Vesicles, and Microemulsions*, Soc. Italiana di Fisica, XC Corso, Bologna, 1985.

157. E. Ruckenstein, *Soc. Petr. Eng. J.*, Oct. 1981, p. 593.

158. B. Widom, *Langmuir*, **3,** 12 (1987).

159. M.-J. Hou and D. O. Shah, *Langmuir*, **3,** 1086 (1987).

160. K. Shinoda, H. Kunieda, T. Arai, and H. Saijo, *J. Phys. Chem.*, **88,** 5126 (1984).

161. D. H. Smith and S. A. Templeton, *J. Colloid Interface Sci.*, **68**, 59 (1979).

162. R. Leung, D. O. Shah, and J. P. O'Connell, *J. Colloid Interface Sci.*, **111**, 286 (1986).

163. L. J. Magid, K. Kon-no, and C. A. Martin, *J. Phys. Chem.*, **85**, 1434 (1981).

164. T. Skodvin, J. Sjöblom, J. O. Saeten, and B. Gestblom, *J. Colloid Interface Sci.*, **155**, 392 (1993).

165. A. Jayakrishnan, K. Kalaiarasi, and D. O. Shah, *J. Soc. Cosmet. Chem.*, **34**, 335 (1983).

166. D. Mandler, Y. Degani, and I. Willner, *J. Phys. Chem.*, **88**, 4366 (1984).

167. J. Kiwi and M. Grätzel, *J. Phys. Chem.*, **84**, 1503 (1980).

168. P. Lianos and S. Modes, *J. Phys. Chem.*, **91**, 6088 (1987).

169. W. B. Gogarty and H. Surkalo, *J. Petr. Technol.*, **24**, 1161 (1972) and references therein.

170. D. O. Shah and W. C. Hsieh, "Microemulsions, Liquid Crystals and Enhanced Oil Recovery," in *Theory, Practice, and Process Principles for Physical Separations*, Engineering Foundation, New York, 1977.

171. Y. K. Pithapurwala and D. O. Shah, *J. Am. Oil Chem. Soc.*, **61**, 1399 (1984).

172. V. Pillai, P. Kumar, M. J. Hou, P. Ayyub, and D. O. Shah, *Adv. Colloid Interface Sci.*, **55**, 241 (1995).

173. V. Pillai, P. Kumar, M. S. Multani, and D. O. Shah, *Colloids & Surfaces*, **A80**, 69 (1993).

174. P. Kumar, V. Pillai, and D. O. Shah, *Appl. Phys. Lett.*, **62**, 765 (1993).

175. E. Manegold, *Schaum, Strassenbau, Chemie und Technik*, Heidelberg, 1953, p. 83.

176. F. Sebba, *J. Colloid Interface Sci.*, **35**, 643 (1971); ACS Symposium Series No. 9, 1975.

177. J. Plateau, *Mem. Acad. Roy. Soc. Belg.*, **33**, (1861), sixth series and preceding papers.

178. J. W. Gibbs, *Collected Works*, Vol. 1, Longmans, Green, New York, 1931, pp. 287, 301, 307.

179. E. B. Matzke, *Am. J. Bot.*, **33**, 58 (1946).

180. C. H. Desch, *Rec. Trav. Chim.*, **42**, 822 (1923).

181. See A. F. Wells, *The Third Dimension in Chemistry*, Clarendon Press, Oxford, 1956, p. 57; and D. W. Thompson, *Growth and Crystal Form*, Cambridge University Press, Cambridge, England, 1943, p. 551.

182. J. W. S. Rayleigh, *Scientific Papers*, Vol. 2, Dover, New York, 1964.

183. A. W. Reinhold and A. W. Rucker, *Trans. Roy. Soc.* (London), **172**, 447 (1881).

184. J. J. Bikerman, *Foams*, Reinhold, New York, 1953, p. 137.

185. H. M. Princen and S. G. Mason, *J. Colloid Sci.*, **20**, 453 (1965).

186. A. Waltermo, E. Manev, R. Pugh, and P. Claesson, *J. Dispers. Sci. Technol.*, **15**, 273 (1994).

187. E. D. Manev and R. J. Pugh, *Langmuir*, **8**, 2253 (1992).

188. R. Cohen, R. Koynova, B. Tenchov, and D. Exerowa, *Eur. Biophys. J.*, **20**, 203 (1991).

189. J. Th. G. Overbeek, *J. Phys. Chem.*, **64**, 1178 (1960).

190. K. J. Mysels, *J. Phys. Chem.*, **68**, 3441 (1964).

191. D. Exerowa, T. Kolarov, and Khr. Khristov, *Colloids & Surfaces*, **22**, 171 (1987).

192. T. Kolarov, R. Cohen, and D. Exerowa, *Colloids & Surfaces*, **A42**, 49 (1989).

193. J. Perrin, *Ann. Phys.*, **10**(9), 160 (1918).

194. M. N. Jones, K. J. Mysels, and P. C. Scholten, *Trans. Faraday. Soc.*, **62**, 1336 (1966).

195. K. Khristov, D. Exerowa, and P. M. Kruglyakov, *Colloids & Surfaces*, **A78**, 221 (1993).

196. D. Platikanov, H. A. Graf, and A. Weiss, *Colloid Polym. Sci.*, **268**, 760 (1990).

197. A. D. Nikolov and D. T. Wasan, *J. Colloid Interface Sci.*, **133**, 1 (1989).

198. X. L. Chu, A. D. Nikolov, and D. T. Wasan, *J. Chem. Phys.*, **103**, 6653 (1995).

199. P. A. Kralchevsky, A. D. Nikolov, D. T. Wasan, and I. B. Ivanov, *Langmuir*, **6**, 1180 (1990).

200. D. Platikanov, H. A. Graf, A. Weiss, and D. Clemens, *Colloid Polym. Sci.*, **271**, 106 (1993).

201. D. Exerowa, A. Nikolov, and M. Zacharieva, *J. Colloid Interface Sci.*, **81**, 419 (1981).

202. P. G. de Gennes, *CR Acad. Sc. Paris*, **303**, 1275 (1986).

203. M. Nedyalkov, R. Krustev, A. Stankova, and D. Platikanov, *Langmuir*, **8**, 3142 (1992).

204. R. Krustev, D. Platikanov, and M. Nedyalkov, *Colloids & Surfaces*, **A79**, 129 (1993).

205. D. Exerowa, D. Kashchiev, and D. Platikanov, *Adv. Colloid Interface Sci.*, **40**, 201 (1992).

206. D. Exerowa, Z. Lalchev, B. Marinov, and K. Ognyanov, *Langmuir*, **2**, 664 (1986); see also ibid., p. 668.

207. See K. J. Mysels, K. Shinoda, and S. Frankel, *Soap Films, Studies of Their Thinning and a Bibliography*, Pergamon, New York, 1959.

208. J. M. di Meglio and P. Baglioni, *J. Phys. Condens. Matter*, **6**, A375 (1994).

209. K. Khristov and D. Exerowa, *Colloids & Surfaces*, **A94**, 303 (1995).

210. P. M. Kruglyakov, D. Exerowa, and K. Khristov, *Adv. Colloid Interface Sci.*, **40**, 257 (1992).

211. K. J. Mysels, *J. Colloid Interface Sci.*, **35**, 159 (1971).

212. G. Frens, K. J. Mysels, and B. R. Vdayendran, *Spec. Disc. Faraday Soc.*, **1**, 12 (1970).

213. I. Penev, D. Exerowa, and D. Kashchiev, *Colloids & Surfaces*, **25**, 67 (1987).

214. A. T. Florence and G. Frens, *J. Phys. Chem.*, **76**, 3024 (1972).

215. G. Frens, *J. Phys. Chem.*, **78**, 1949 (1974).

216. J. F. Joanny and P. G. de Gennes, *Physica*, **147A**, 238 (1987).

217. A. V. Prokhorov and B. V. Derjaguin, *J. Colloid Interface Sci.*, **125**, 111 (1988).

218. D. Exerowa and D. Kashchiev, *Contemp. Phys.*, **27**, 429 (1986).

219. A. Sharma and E. Ruckenstein, *Langmuir*, **2**, 480 (1986).

220. A. Prins and M. van den Tempel, *J. Phys. Chem.*, **73,** 2828 (1969).

221. R. L. Ternes and J. C. Berg, *J. Colloid Interface Sci.*, **98,** 471 (1984).

222. K. Wantke, K. Malysa, and K. Lunkenheimer, *Colloids & Surfaces*, **A82,** 183 (1994).

223. K. Malysa, K. Khristov, and D. Exerowa, *Colloid Polym. Sci.*, **269,** 1045 (1991), ibid., p. 1055.

224. A. G. Brown, W. C. Thuman, and J. W. McBain, *J. Colloid Sci.*, **8,** 491 (1953).

225. A. Monsalve and R. S. Schechter, *J. Colloid Interface Sci.*, **97,** 327 (1984).

226. G. Narshimhan and E. Ruckenstein, *Langmuir*, **2,** 230, 294 (1986).

227. D. J. Durian, D. A. Weitz, and D. J. Pine, *Science*, **252,** 686 (1991).

228. S. E. Friberg, I. Blute, and H. Kunieda, *Langmuir*, **2,** 659 (1986).

229. A. M. Kraynik, *Ann. Rev. Fluid Mech.*, **20,** 325 (1988).

230. D. J. Durian, *Phys. Rev. Lett.*, **75,** 4780 (1995).

231. A. D. Gopal and D. J. Durian, *Phys. Rev. Lett.*, **75,** 2610 (1995).

232. S. Ross, A. F. Hughes, M. L. Kennedy, and A. R. Mardoian, *J. Phys. Chem.*, **57,** 684 (1953).

233. J. J. Bikerman, *Foams*, Springer-Verlag, New York, 1973.

234. K. Koczo, L. A. Lobo, and D. T. Wasan, *J. Colloid Interface Sci.*, **150,** 492 (1992).

235. K. Koczo, J. K. Koczone, and D. T. Wasan, *J. Colloid Interface Sci.*, **166,** 225 (1994).

236. I. Blute, M. Jansson, S. G. Oh, and D. O. Shah, *J. Am. Oil Chem. Soc.*, **71,** 41 (1994).

237. M. K. Sharma, D. O. Shah, and W. E. Brigham, *Ind. Eng. Chem. Fundam.*, **23,** 213 (1984).

238. M. K. Sharma, D. O. Shah, and W. E. Brigham, *AIChE J.*, **31,** 222 (1985).

239. S. K. Friedlander, *Ann. N.Y. Acad. Sci.*, **404,** 354 (1983).

240. G. O. Rubel and J. W. Gentry, *J. Aerosol. Sci.*, **18,** 23 (1987).

241. G. Narsimhan and E. Ruckenstein, *J. Colloid Interface Sci.*, **116,** 278, 288 (1987).

242. P. E. Wagner and M. Kerker, *J. Chem. Phys.*, **66,** 638 (1977).

243. S. W. Davison and J. W. Gentry, *Aerosol. Sci. Tech.*, **4,** 157 (1985).

244. M. P. Sinha and S. K. Friedlander, *J. Colloid Interface Sci.*, **112,** 573 (1986).

245. A. T. Shen and T. A. Ring, *Aerosol Sci. Tech.*, **5,** 477 (1986).

246. J. W. Gentry and R. V. Calabrese, *Part. Charact.* **3,** 104 (1986).

247. O. O. Olawoyin, T. M. Raunemaa, and P. K. Hopke, *Aerosol. Sci. Tech.*, **23,** 121 (1995).

248. W. Li, N. Montassier, and P. K. Hopke, *Aerosol. Sci. Tech.*, **17,** 25 (1992).

249. E. J. Davis, in *Surface and Colloid Science*, Vol. 14, E. Matijevic, ed., Plenum, New York, 1987.

250. S. Arnold and L. M. Folan, *Rev. Sci., Instrum.*, **58,** 1732 (1987).

251. L. K. Sun and H. Reiss, *J. Colloid Interface Sci.*, **99,** 515 (1984).

252. A. B. Pluchino and S. Arnold, *Opt. Lett.*, **10,** 261 (1985).

253. S. Arnold, M. Neuman, and A. B. Pluchino, *Opt. Lett.*, **9,** 4 (1984).

CHAPTER XV

Macromolecular Surface Films, Charged Films, and Langmuir–Blodgett Layers

1. Introduction

This chapter concludes our discussion of applications of surface chemistry with the possible exception of some of the materials on heterogeneous catalysis in Chapter XVIII. The subjects touched on here are a continuation of Chapter IV on surface films on liquid substrates. There has been an explosion of research in this subject area, and, again, we are limited to providing just an overview of the more fundamental topics.

Many complex systems have been spread on liquid interfaces for a variety of reasons. We begin this chapter with a discussion of the behavior of synthetic polymers at the liquid–air interface. Most of these systems are linear macromolecules; however, rigid-rod polymers and more complex structures are of interest for potential optoelectronic applications. Biological macromolecules are spread at the liquid–vapor interface to fabricate sensors and other biomedical devices. In addition, the study of proteins at the air–water interface yields important information on enzymatic recognition, and membrane protein behavior. We touch on other biological systems, namely, phospholipids and cholesterol monolayers. These systems are so widely and routinely studied these days that they were also mentioned in some detail in Chapter IV. The closely related matter of bilayers and vesicles is also briefly addressed.

Films spread at liquid–liquid interfaces or on liquids other than water are discussed followed by the important effects of charged monolayers on water. Finally, the most technologically important application of Langmuir films, the Langmuir–Blodgett film deposited on a solid substrate, is reviewed.

2. Langmuir Films of Polymers

A. Adsorption and Phase Behavior

Pressure–area isotherms for many polymer films lack the well-defined phase regions shown in Fig. IV-16; such films give the appearance of being rather amorphous and plastic in nature. At low pressures, non-ideal-gas behavior is approached as seen in Fig. XV-1 for poly(methyl acrylate) (PMA). The limiting slope is given by a virial equation

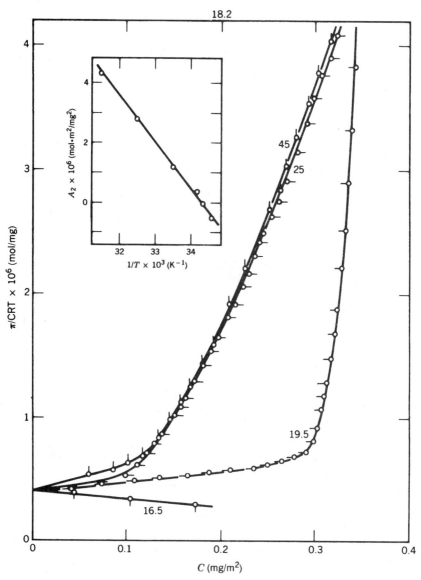

Fig. XV-1. Plots of π/CRT vs. C for a fractionated poly(methyl acrylate) polymer at the indicated temperatures in degrees Celsius. [From A. Takahashi, A. Yoshida, and M. Kawaguchi, *Macromolecules*, **15,** 1196 (1982) (Ref. 1). Copyright 1982, American Chemical Society.]

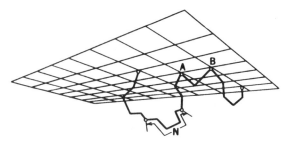

Fig. XV-2. Adsorption loops in a cubic lattice. The diagrams depict the surface sites as seen from below. (From Ref. 4.)

$$\frac{\pi}{C} = RT \left(\frac{1}{M} + A_2C + \cdots \right) \tag{XV-1}$$

where C is in mg/cm^2, M is the molecular weight, and A_2 is the second virial coefficient. In the case of the PMA film, extrapolation to $\pi = 0$ gave $M = 2400$ g/mol and a Θ (theta) temperature (where $A_2 = 0$) of 18.2°C [1]. The low concentration domain was treated long ago by Singer [2] in an extension of the Flory–Huggins treatment of polymer solutions. Frisch and Simha [3] considered a more general treatment that allows portions of the polymer chain to reside in the solution phase. Silberberg [4] considered the impact of various structural restrictions on adsorption equilibrium; an example of a configuration having loops extending into the solution is shown in Fig. XV-2. Raphaël and co-workers developed a mean-field model for polymer adsorption at liquid–air interfaces for good, ideal (theta), and bad solvents [5]. Their isotherms document the non-ideal-gas behavior but cannot be extended to high compressions.

Some polymers, such as functionally terminated poly(dimethyl siloxane) (PDMS), exhibit more complex isotherms such as those due to Koberstein and co-workers [6] shown in Fig. XV-3. Up to six inflection points or transitions, labeled A–F, were observed in these systems. Generally, at large areas the polymer lays flat on the water and between A and B makes a transition to a zigzag structure with every other oxygen or silicon at the surface. Between B and C PDMS coils into helices, and from C to D these helices begin to slide past one another and the monolayer collapses. Shorter chains make a transition to upright configurations at E and finally collapse at F. Thus, depending on the functional groups and the subphase, end-functionalized PDMS shows transitions of the macromolecule on the surface as well as transitions involving vertical orientations of whole chains.

A great many polymers appear to form films having a flat molecular configuration. Thus various polyesters [7] gave extrapolated areas of about 2.5 m^2/mg corresponding to about the calculated 60–70 Å^2 area per segment, or monolayer thickness of 3–5 Å. A similar behavior was noted for poly(vinyl acetate)

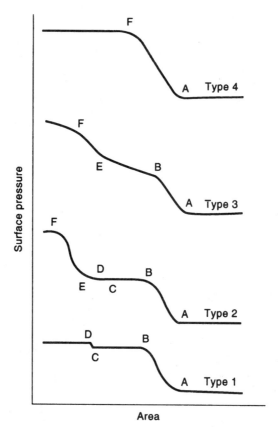

Fig. XV-3. Four types of pressure–area isotherms observed in end-functionalized PDMS monolayers. Up to six transitions (*A–F*) are observed and described in the text. (From Ref. 6.)

monolayers, yet by contrast, it was found that the behavior of poly(vinyl benzoate) was quite different [8]. The poly(vinyl benzoate) gave a very compact monolayer of extrapolated area 9 $Å^2$ per monomer unit, corresponding to a film thickness of about 20 Å; its compressibility was more like that of stearic acid, that is, about 0.006 cm/dyn, rather than the usual polymer film value of about 0.02–0.1 cm/dyn. Apparently, in this case, close packing of the benzene rings occurred. Ries and co-workers [9] have also studied stereoregular poly(methyl methacrylates) and found quite different π–σ plots for the isotactic and the syndiotactic forms.

An important issue in the measurement of macromolecular film isotherms has been raised by Peng and Barnes [10, 11]. They found that during compression, monolayers of various polymers sustained long-lasting pressure gradients across the film. These gradients produce inhomogeneous films that are difficult to characterize with pressure–area isotherms. The magnitude of the surface pressure gradient is reduced somewhat by mixing the polymer with a surfactant [11] or its monomer [12].

Gaines [13] has reported on dimethylsiloxane-containing block copolymers. Interestingly, if the organic block would not in itself spread, the area of the block polymer was simply proportional to the siloxane content, indicating that the organic blocks did not occupy any surface area. If the organic block was separately spreadable, then it contributed, but nonadditively, to the surface area of the block copolymer.

The "hairy rod" polymers of poly(γ-methyl-L-glutamate-co-γ alkyl-L-glutamate) PMLG-co-PALG, have generated some interest as spread films. The polymers earn their names by having a fairly rigid helical backbone with alkyl side-chain projections. They undergo interesting phase transitions such as that occurring when the alkyl "hairs" begin to interdigitate. Their phase transitions have been studied by simultaneous ellipsometry and $\pi - \sigma$ measurement [14]. The detailed conformations of PMLG in monolayers has been studied by Baglioni and co-workers [15] as well as the effect of conformation on the collapse process [16].

Mixtures of polymers at surfaces provide the interesting possibility of exploring polymer miscibility in two dimensions. Baglioni and co-workers [17] have shown that polymers having the same "orientation" at the interface are compatible while those having different orientations are not. Some polymers have their hydrophobic portions parallel to the surface, while others have a perpendicular disposition. The surface orientation effect is also present in mixtures of poly(methyl methacrylate), PMMA, and fatty acids.

B. Dynamics and Rheology

The dynamics and surface rheology of polymeric monolayers is of both practical and fundamental interest. On the practical side, the process of Langmuir–Blodgett deposition (Section XV-8) is an important means of forming polymeric thin films and the nature of the film transfer depends on flow processes in the monolayer. One way to monitor the polymer orientation in a monolayer undergoing extensional or stretching flow has been introduced by Fuller and co-workers [18]. They measure the optical anisotropy in the monolayer through the dichroism, or anisotropy in the imaginary (absorptive) part of the refractive index. With this in situ technique, they can monitor orientation of rodlike polymers along the extension axis during flow and their relaxation on cessation of flow. More conventional stress relaxation measurements have been performed on poly(vinyl stearate), PVS [19]. Dynamic measurements on PVS and PMMA with a torsion pendulum (see Section IV-3) showed shear moduli 100–500 times that of simple amphiphile monolayers and viscosities 3000–20,000 times greater [20]. These viscoelastic properties are correlated with the development of surface pressure gradients on compression described above.

The dynamics of polymers at surfaces can be studied via dynamic light scattering (DLS), as described in Section IV-3C. A modification of surface DLS using an evanescent wave to probe the solution in a region near the interface has been developed by Rice and co-workers [21]. In a study of spread polystyrene-PMMA, PS-PMMA diblock copolymers, they found a micellar structure they characterized as a "furry disk" [21]. They use these disk-like structures to probe the fundamental aspects of two dimensional diffusion; they find that, as anticipated theoretically, the self-diffusion in two dimensions diverges, that is, becomes exceedingly large, as time goes to infinity [22].

Other interesting Langmuir monolayer systems include spread thermotropic liquid crystals where a foam structure forms on expansion from a collapsed state [23]. Spread monolayers of clay dispersions form a layer of overlapping clay platelets that can be subsequently deposited onto solid substrates [24].

3. Langmuir Films of Proteins

Proteins, like other macromolecules, can be made into monolayers at the air–water interface either by spreading, adsorption, or specific binding. Proteins, while complex polymers, are interesting because of their inherent surface activity and amphiphilicity. There is an increasing body of literature on proteins at liquid interfaces, and here we only briefly discuss a few highlights.

The adsorption of proteins from a bulk aqueous solution onto the air–solution interface is important in many biotechnological and food applications. The techniques described in Section II-7 for the measurement of interfacial tension can be applied to the study of protein adsorption at the air–water interface. In particular, Neumann and co-workers have applied the axisymmetric drop shape analysis (ASDA) technique to the study of the adsorption kinetics of human albumin (HA) at the free surface [25]. They were able to probe a regime of adsorption where the adjustment of the HA concentration at the surface is not diffusion-controlled, providing some insight into the rate-limiting conformational changes. In a similar study on insulin, ASDA provides a means to see the difference between insulin analogs having single amino acid substitutions [26]. The manipulation of the insulin surface tension through these substitutions is an important attempt to reduce adsorption onto infusion equipment and the concomitant loss of insulin from the infusion fluids. Care must be taken when interpreting protein adsorption data as sometimes commercial proteins, such as β-lactoglobulin, are contaminated with surfactants [27] and must be treated with an extraction step to determine the true nature of the protein.

A major breakthrough in protein crystallization was the development over a decade ago of a technique for crystallizing proteins in two dimensions [28]. First developed to crystallize antibodies, this approach initially relied on the synthesis of lipids having ligands for binding the protein of interest. The lipid is mixed with lipids having no ligands, and they are spread on a water surface and compressed to the desired pressure. The proteins are introduced into the aqueous subphase and, on binding to the lipid monolayer, form a dense monolayer that subsequently orders. The concentration and orientation afforded by binding to the lipid monolayer causes some proteins to crystalize more readily in two dimensions than in three. Proteins adsorbed to the lipid layer by electrostatic binding crystallize as well, eliminating the need for a ligand in some cases [29].

While a number of proteins have been crystallized in this manner, the majority of studies have focused on a robust system comprising the tetrameric protein streptavidin and the vitamin biotin. The choice of this system is primarily motivated by the strong bond between biotin and streptavidin (having an association equilibrium constant, $K_a \sim 10^{15} M^{-1}$). The binding properties were recently characterized by Lösche and co-workers with neutron reflectivity and fluorescence microscopy [30] and a schematic diagram of four streptavidin molecules in a monolayer above a subphase of heavy water is presented in Fig XV-4. Among the studies on streptavidin crystallization are investigations of the growth dynamics through the discovery of a two dimensional crystal growth instability analogous to snowflake formation [31]. The addition of an uncrystallizable impurity, avidin, in the monolayer with streptavidin led to dendritic growth to form large X-shaped crystals with secondary dendrites as illustrated in Fig. XV-5. The anisotropic shape of these dendrites occurs because of the slight difference in interaction between the subunits bound to biotin and those facing away from the lipid that are free from biotin (see Fig. XV-4) [32]. These two-dimensional (2D) crystals can be used as templates for the growth of three-dimensional (3D) crystals suitable for x-ray-diffraction studies. Edwards and co-workers [33] have found that 2D arrays of streptavidin are very effective nucleating agents for 3D crystal growth and that growth appears to occur by epitaxy.

Membrane proteins comprise another important class of protein crystallized in 2D. These proteins perform important functions as membrane channels and recognition sites for cells. Unlike the streptavidin crystals, membrane proteins

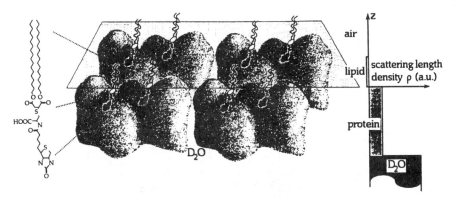

Fig. XV-4. Schematic drawing of four streptavidin molecules bound to biotinylated lipid in a monolayer above heavy water. The scattering length density for neutron reflectivity is shown at the side. (From Ref. 30.)

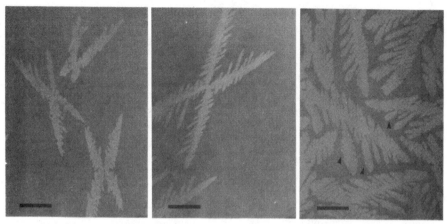

Fig. XV-5. Fluorescence micrographs illustrating morphologies of two-dimensional (2D) streptavidin crystals at three streptavidin/avidin ratios 15/85, 25/75, 40/60 from left to right. Scale bar is 100 μm (From Ref. 31.)

are inserted directly into the lipid bilayer where they interact through the surrounding sea of lipids [34].

4. Films of Other Biological Substances

There is quite a large body of literature on films of biological substances and related model compounds, much of it made possible by the sophisticated microscopic techniques discussed in Section IV-3E. There is considerable interest in biomembranes and how they can be modeled by lipid monolayers [35]. In this section we briefly discuss lipid monolayers, lipolytic enzyme reactions, and model systems for studies of biological recognition. The related subjects of membranes and vesicles are covered in the following section.

An essential component of cell membranes are the lipids, lecithins, or phosphatidylcholines (PC). The typical π–σ behavior shown in Fig. XV-6 is similar to that for the simple fatty-acid monolayers (see Fig. IV-16) and has been modeled theoretically [36]. Branched hydrocarbons tails tend to expand the monolayer [38], but generally the phase behavior is described by a fluid-gel transition at the plateau [39] and a semicrystalline phase at low σ. As illustrated in Fig. XV-7, the areas of the dense phase may initially be highly branched, but they anneal to a circular shape on recompression [40]. The theoretical evaluation of these shape transitions is discussed in Section IV-4F.

A study by Barnes and co-workers of the equilibrium spreading behavior of dimyristol phosphatidylcholine (DMPC) reconciles the differences between spreading of bulk solids and dispersions of liposomes [41]. This study shows the formation of multibilayers below the monolayer at the air–water interface. An incipient phase separation, undetectable by microscopy, in DMPC–cholesterol

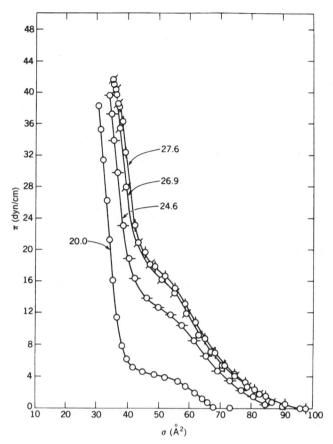

Fig. XV-6. Pressure–area isotherms for a synthetic lecithin at the indicated temperatures in degrees Celsius. [From H. E. Ries, Jr., M. Matsumoto, N. Uyeda, and E. Suito, *Adv. Chem. Ser.*, No. 144, ACS, 1975, p. 286 (Ref. 37). Copyright 1975, American Chemical Society.]

monolayers has been elucidated by Merkel and Sackmann via eximer forma-tion dynamics using a pyrene labeled lecithin [42]. Fluorescence photobleaching studies allow surface diffusion measurements [43] and the viscoelastic proper-ties of phospholipid monolayers have also been studied [44]. Mixed monolayers of phospholipids with helical peptides have also been studied via rheological and electrical measurements with the goal of understanding the lipid–protein interactions [45].

Remarkable chiral patterns, such as those in Figs. IV-15 and XV-8, are found in mixtures of cholesterol and *S*-dipalmitoyl PC (DPPC) on compression to the plateau region (as in Fig. XV-6). As discussed in Section IV-4F, this behavior has been modeled in terms of an anisotropic line tension arising from molecular symmetry [46–49].

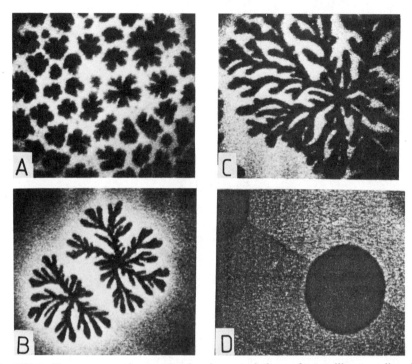

Fig. XV-7. Fluorescence micrographs showing morphology of crystalline L-α-dimyris-tolphosphatidylethanolamine domains following a π jump to the plateau region of the π–σ plot: (*a*) after 2 sec; (*b*) after 1 min; (*c*) after 20 min; (*d*) following a second pressure jump after condition (*c*). (From Ref. 40.)

Other lipid-containing systems that have been studied include mixed films of di-oleylphosphatidylcholine with retinal [50] and with cytochrome *c* (including surface potential and ellipsometric measurements) [51] and of lipid films with porphyrins (fluorescence studies) [52]. Monolayer studies have also been reported on other membrane constituents such as valine-gramiciden A and Valinomycin [53]. Finally, a bridged porphyrin, mesoporphyrin II, a somewhat spherical molecule, showed monolayer π–σ behavior surprisingly like that of the long-chain fatty acids [53a] An important mixture of lipids and proteins makes up the lung surfactant lining the pulmonary air spaces of mammals. This mixture includes DPPC, unsaturated PCs, unsaturated phosphatidyl-glycerols (PGs), palmitic acid (PA), and proteins. The lack of this surfactant causes respiratory distress in premature infants and the search for suitable replacements motivates much research. Zasadzinski and co-workers carried out a systematic study of the effect of the lung surfactant protein, SP-B, on pressure-area isotherms of DPPC, PG, and PA monolayers [54]. Through mutant studies they suggest that the stabilizing effect of the protein occurs through a specific charge interaction between the protein and lipid. The appearance of the isotherms is remarkably similar to that of simple poly-cations suggesting a potentially useful synthetic replacement. The protein appears to inhibit phase separation in PA monolayers by reducing the line tension between the fluid and condensed phases [55]. This mechanism is supported by the formation of a

Fig. XV-8. Fluorescence micrographs of crystalline domains of an *S*-DPPC monolayer containing 2% cholesterol and compressed to the plateau region. [From H. McConnell, D. Keller, and H. Gaub, *J. Phys. Chem.*, **90,** 1717 (1986) (Ref. 49). Copyright 1986, American Chemical Society.]

"stripe phase" (see Section IV-4F and Ref. 48) in the presence of SP-B as illustrated in Fig. XV-9.

The action of lipolytic enzymes on lipid monolayers has been studied. Phospholipase A$_2$ (PLA$_2$) is an interfacially activated enzyme existing in the pancreas of mammals as well as in cobra and bee venom. It catalyzes hydrolysis of the ester bond in glycerolphospholipids. Striking fluorescence micrographs by Grainger and co-workers [56] show that PLA$_2$ targets the solid-phase lipid domains. The enzyme binds electrostatically to the hydrolized lipid monolayer and forms 2D domains in regions enriched in fatty-acid reaction products [57].

Protein binding to monolayers is our final area of discussion in this section. Surface recognition and antibody binding are essential processes in immune response and targeted drug delivery (for a review, see Ref. 58). An interesting model system involves the fluorescent tag fluorescein and the monoclonal antibody antifluorescein. In a fluorescence quenching study, Ringsdorf and co-workers studied the penetration of fluorescein into the antibody binding site [59]. The behavior of various fluorescein-tagged lipids showed the importance of a long flexible hydrophilic spacer between the chromophore and lipid for effective antibody recognition. Molecular recognition is also studied by

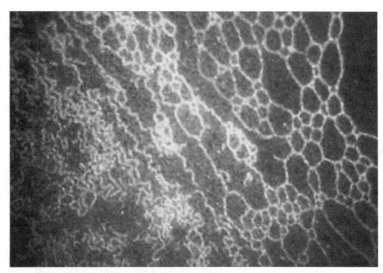

Fig. XV-9. Fluorescence micrograph of the stripe patterns observed in a monolayer from a mixture of PA and SP–B$_{1-25}$ (20% by weight peptide) on a buffered saline subphase at 16°C and zero surface pressure. (From Ref. 55.)

small molecular analogs such as Kemp's triacid (see Section XIV-3B) [60], a guanidinium-functionalized monolayer [61] and resorcinol–dodecanal cyclo-tetramer [62]. Some host molecules can specifically bind nucleic acids via *hydrogen bonding* [63].

5. Membranes, Bilayers, and Vesicles

There has been a surge of research activity in the physical chemistry of membranes, bilayers, and vesicles. In addition to the fundamental interest in cell membranes and phospholipid bilayers, there is tremendous motivation for the design of supported membrane biosensors for medical and pharmaceutical applications (see the recent review by Sackmann [64]). This subject, in particular its biochemical aspects, is too vast for full development here; we will only briefly discuss some of the more physical aspects of these systems. The reader is referred to the general references and some additional reviews [65–69].

Phospholipid molecules form bilayer films or membranes about 5 nm in thickness as illustrated in Fig. XV-10. Vesicles or liposomes are closed bilayer shells in the 100–1000-nm size range formed on sonication of bilayer forming amphiphiles. Vesicles find use as controlled release and delivery vehicles in cosmetic lotions, agrochemicals, and, potentially, drugs. The advances in cryoelectron microscopy (see Section VIII-2A) in recent years have aided their characterization [70–72]. Additional light and x-ray scattering measurements reveal bilayer thickness and phase transitions [70, 71]. Differential thermal analysis

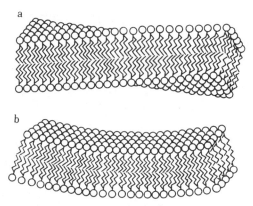

Fig. XV-10. Illustration of a bilayer membrane and two of its deformation modes: (a) twist; (b) splay. (From Ref. 75.)

and electron spin resonance (ESR) have been applied to the characterization of vesicle phase transitions [73].

While most vesicles are formed from double-tail amphiphiles such as lipids, they can also be made from some single chain fatty acids [73], surfactant–cosurfactant mixtures [71], and bola (two-headed) amphiphiles [74]. In addition to the more common spherical shells, tubular vesicles have been observed in DMPC–alcohol mixtures [70]. Polymerizable lipids allow photo- or chemical polymerization that can sometimes stabilize the vesicle [65]; however, the structural change in the bilayer on polymerization can cause giant vesicles to bud into smaller shells [76]. Multivesicular liposomes are collections of hundreds of bilayer enclosed water-filled compartments that are suitable for localized drug delivery [77]. The structures of these water-in-water vesicles resemble those of foams (see Section XIV-7) with the polyhedral structure persisting down to molecular dimensions as shown in Fig. XV-11.

The interest in vesicles as models for cell biomembranes has led to much work on the interactions within and between lipid layers. The primary contributions to vesicle stability and curvature include those familiar to us already, the electrostatic interactions between charged head groups (Chapter V) and the van der Waals interaction between layers (Chapter VI). An additional force due to thermal fluctuations in membranes produces a steric repulsion between membranes known as the *Helfrich* or *undulation* interaction. This force has been quantified by Sackmann and co-workers using reflection interference contrast microscopy to monitor vesicles weakly adhering to a solid substrate [78]. Membrane fluctuation forces may influence the interactions between proteins embedded in them [79]. Finally, in balance with these forces, bending elasticity helps determine shape transitions [80], interactions between inclusions [81], aggregation of membrane junctions [82], and unbinding of pinched membranes [83]. Specific interactions between membrane embedded receptors add an additional complication to biomembrane behavior. These have been stud-

Fig. XV-11. Electron micrograph of a freeze fracture replica of a region inside a multivesicular liposome. Note the tetrahedral coordination; nearly every vertex has three
edges, and each face is connected to three others. The average number of edges per
face is 5.1. (From Ref. 77.)

ied with the model streptavidin-biotin complex (see Section XV-3) via surface force measurements (see Section VI-3C) [84] and by cryoelectron microscopy [72].

Lipid bilayer dynamics have been characterized with NMR [75, 85], photon correlation spectroscopy [86], and fluorescence recovery after pattern photobleaching [87]. Typical membrane proteins limited by the membrane lipid viscosity diffuse with a diffusion coefficent $\mathcal{D} \approx 10^{-9} \mathrm{cm^2/sec}$, while those undergoing transient binding in the membrane slow to $\mathcal{D} \approx 3 \times 10^{-10} \mathrm{cm^2/sec}$. The observation of the motion of nanometer-size gold colloids attached to lipids by a technique known as *nanovid microscopy* (nanoparticle video-enhanced microscopy) provides a means to track the actual Brownian trajectory in a membrane [87–89]. This technique, where 30–40-nm gold particles are anchored to fluorescein labeled lipids via antifluorescein, aids in the determination of restricted motion. Lateral and rotational diffusion as well as out-of-plane fluctuations can be revealed by quasi-elastic neutron scattering [90].

6. Films at Liquid–Liquid Interfaces and on Nonaqueous Liquid Surfaces

The behavior of insoluble monolayers at the hydrocarbon–water interface has been studied to some extent. In general, σ values for straight-chain acids and alcohols are greater at a given film pressure than if spread at the water–air interface. This is perhaps to be expected since the nonpolar phase should tend to reduce the cohesion between the hydrocarbon tails. See Ref. 91 for early reviews. Takenaka [92] has reported polarized resonance Raman spectra for an azo dye monolayer at the CCl_4–water interface; some conclusions as to orientation were possible. A mean-field theory based on Lennard–Jones potentials has been used to model an amphiphile at an oil–water interface; one conclusion was that the depth of the interfacial region can be relatively large [93].

As has been noted, much of the interest in films of proteins, steroids, lipids, and so on, has a biological background. While studies at the air–water interface have been instructive, the natural systems approximate more closely to a water–oil interface. A fair amount of work has therefore been reported for such interfaces in spite of the greater experimental difficulties.

Protein monolayers tend to be expanded relative to those at the air–water interface [94]. Davies [95] studied hemoglobin, serum albumin, gliadin, and synthetic polypeptide polymers at the water–petroleum ether interface with the view of determining the behavior of the biologically important ϵ-NH_2 groups and concluded that on compression they were forced into the oil phase. Neumann and co-workers found that interfacial pressures were higher for human albumin (HA) at the HA solution–decane interface than at the solution–air interface [96]. Hermel and Miller investigated the effect of the secondary structure of poly-L-lysine on its adsorption at the solution–dodecane interface [97]. They found that the α-helix occupies a smaller area per residue and yields a greater interfacial tension depression than the β-sheet conformation, perhaps because of its more compact structure.

Stigter and Dill [98] studied phospholipid monolayers at the n-heptane–water interface and were able to treat the second and third virial coefficients (see Eq. XV-1) in terms of electrostatic, including dipole, interactions. At higher film pressures, Pethica and co-workers [99] observed quasi-first-order phase transitions, that is, a much flatter plateau region than shown in Fig. XV-6.

Glycerol has been used as a polar substrate. Glycerol is a very poor electrical conductor, and this made it possible to measure the conductivity of spread monolayers; films of some charge-transfer complexes were found to be highly conducting [100]. Ellison and Zisman [101] reported studies of monolayers of polymethylsiloxane polymer and of the protein zein on such substrates as white mineral oil, n-hexadecane, and tricresyl phosphate. Jarvis and Zisman [102] report qualitative spreading studies with a number of fluorinated organic compounds at a variety of organic liquid-air interfaces and π–σ data for some of the Gibbs monolayer systems.

Brooks and Pethica [103] used a film balance type of trough such that the oil-water interface could be swept and interfacial films could be directly compressed. They used a hydrophobic Wilhelmy slide for measuring γ and hence π values (as did Takenaka [92]). The procedure was claimed to be superior to the fixed area interface one, where film pressure is built up by successive addition of film-forming material, since spreading against an existing high surface pressure may not always be complete. Interfacial potentials have been measured by means of the vibrating electrode [104]; with polar oils direct measurements with a high-impedance voltmeter are possible [105]. For film viscosity, a torsion pendulum may be used [106].

There is a fair amount of work reported with films at the mercury–air interface. Rice and co-workers [107] used grazing incidence x-ray diffraction to determine that a crystalline stearic acid monolayer induces order in the Hg substrate. Quinone derivatives spread at the mercury-n-hexane interface form crystalline structures governed primarily by hydrogen bonding interactions [108].

As a point of interest, it is possible to form very thin films or membranes in water, that is, to have the water-film-water system. Thus a solution of lipid can be stretched on an underwater wire frame and, on thinning, the film goes through a succession of interference colors and may end up as a black film of 60–90 Å thickness [109]. The situation is reminiscent of soap films in air (see Section XIV-9); it also represents a potentially important modeling of biological membranes. A theoretical model has been discussed by Good [110].

Surfaces can be active in inducing blood clotting, and there is much current searching for thromboresistant synthetic materials for use in surgical repair of blood vessels (see Ref. 111). It may be important that a protective protein film be strongly adsorbed [112]. The role of water structure in cell-wall interactions may be quite important as well [113].

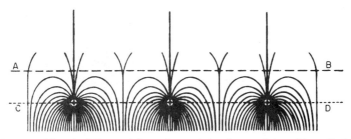

Fig. XV-12. Electrical lines of force for a charged monolayer. (From Ref. 114)

7. Charged Films

Most of the Langmuir films we have discussed are made up of charged amphiphiles such as the fatty acids in Chapter IV and the lipids in Sections XV-4 and 5. Depending on the pH and ionic strength of the subphase, electrostatic effects can become quite important. Here we develop the theoretical foundation for charged films with the Donnan relationship. Then we mention the influence of subphase pH on film behavior.

A. Equation of State for Charged Films

A picture of the electrical lines of force is given in Fig. XV-12 [114]; in the plane CD of the ionic groups, it will be a periodic field, whereas a little further into the solution the effect will be more that of a uniformly charged surface. The Donnan treatment is probably best justified if it is supposed that ions from solution penetrate into the region of CD itself and might in fact, lie between CD and AB.

The Donnan effect acts to exclude like-charged substrate ions from a charged surface region, and this exclusion, as well as the concentration of oppositely charged ions, can be expressed in terms of a Donnan potential ψ_D. Thus for a film of positively charged surfactant ions S^+ one can write

$$\frac{c_s^+}{c^+} = e^{-e\psi_D/kT} \tag{XV-2}$$

and

$$\frac{c_s^-}{c^-} = e^{e\psi_D/kT} \tag{XV-3}$$

where c^+ and c^- are the concentrations of nonsurfactant ions and the subscript s denotes that the concentration is for an interfacial region of thickness τ. On multiplying the two equations, one obtains the required condition that, neglecting activity coefficients, $(c^+)(c^-) = (c_s^+)(c_s^-)$ or, in the case of an insoluble surfactant ion, since $c^+ = c^- = c$, the bulk electrolyte concentration, the condition becomes

$$c^2 = (c_s^+)(c_s^-) \tag{XV-4}$$

In the interfacial region, electroneutrality requires that $c_s^- = (S^+) + c_s^+$, so that Eq. XV-4 becomes

$$(S^+) = \frac{c^2}{c_s^+} - c_s^+ \tag{XV-5}$$

or, using Eq. XV-2

$$(S^+) = (e^{e\psi_D/kT} - e^{-e\psi_D/kT})c$$

$$= 2c \sinh \left(\frac{e\psi_D}{kT} \right)^{\dagger} \tag{XV-6}_{\text{r}}$$

Now, $(S^+) = 1000\Gamma/\tau$, where Γ is the surface excess in moles per square centimeter, or $(S^+) = 1000 \times 10^{16}/N\tau\sigma$, where σ is in Å^2 per molecule. With these substitutions, Eq. XV-6 may be solved for ψ_D to give

$$\psi_D = \frac{kT}{e} \sinh^{-1} \left(\frac{1000 \times 10^{16}}{2N\tau\sigma c} \right) \tag{XV-7}$$

At this point an interesting simplification can be made if it is assumed that τ, as representing the depth in which the ion discrimination occurs, is taken to be just equal to $1/\kappa$, the ion atmosphere thickness given by Debye-Hückel theory (see Section V-2). In the present case of a 1 : 1 electrolyte, $\kappa = (8\pi e^2 N/1000\epsilon kT)^{1/2}c^{1/2}$, and on making the substitution into Eq. XV-7 and inserting numbers (for the case of water at 20°C), one obtains, for ψ_D in millivolts:

$$\psi_D = 25.2 \sinh^{-1} \frac{2 \times 134}{\sigma c^{1/2}} \tag{XV-8}$$

We can now calculate the Donnan contribution to film pressure through the use of Eq. III-113 in the approximate form:

$$\pi_{Os} V_1 = RTN_2^s = \frac{RTn_2^s}{n_1^0} \tag{XV-9}$$

or

---

†The reader is reminded that $\sinh(x) = \frac{1}{2}(e^x - e^{-x})$ and that $\cosh x = \frac{1}{2}(e^x + e^{-x})$.

$$\pi_{Os} = RT \frac{n_2^s}{\mathcal{A}} \tag{XV-10}$$

The total moles n_2 in the surface region is given by $C\mathcal{A}\tau/1000$, where C is the sum of the concentrations of the ionic species present:

$$C = (S^+) + c_s^+ + c_s^- = (S^+) + 2c \cosh\left(\frac{e\psi_D}{kT}\right) \tag{XV-11}$$

Actually, it is the *net* concentration that is needed:

$$C_{net} = (S^+) + 2c\left(\cosh\left(\frac{e\psi_D}{kT}\right) - 1\right) \tag{XV-12}$$

On combining these results with Eq. XV-10

$$\pi = RT\Gamma_s^+ + \frac{2c\tau RT}{1000}\left(\cosh\left(\frac{e\psi_D}{kT}\right) - 1\right) \tag{XV-13}$$

The contribution of the Donnan effect is that of the second term in Eq. XV-13, that is,

$$\pi_D = \frac{2c\tau RT}{1000}\left(\cosh\left(\frac{e\psi_D}{kT}\right) - 1\right) \tag{XV-14}$$

and in combination with Eq. XV-8 for water at 20°C,

$$\pi_D = 1.52c^{1/2}\left(\cosh\sinh^{-1}\left(\frac{2 \times 134}{\sigma c^{1/2}}\right) - 1\right) \tag{XV-15}$$

again taking τ to be $1/\kappa$.

If $2 \times 134/\sigma c^{1/2}$ is large enough (about 4), Eq. XV-15 reduces to

$$\pi_D = \frac{kT}{\sigma} - 1.52c^{1/2}$$

or, including the contribution from (S^+),

$$\pi_{ions} = \frac{2kT}{\sigma} - 1.52c^{1/2} \tag{XV-16}$$

This treatment assumes the surface charge to be diffused over a thickness τ

$= 1/\kappa$. As an alternative, the charge may be taken to be spread uniformly on the plane CD, in which case the counterions are treated as a diffuse double layer (see Section V-2), and the potential, now called the *Gouy potential*, is

$$\psi_G = 50.4 \sinh^{-1} \left(\frac{134}{\sigma c^{1/2}} \right) \qquad \text{(XV-17)}$$

again, for water at 20°C and assuming a 1:1 electrolyte and expressing ψ_G in millivolts. This is not very different from ψ_D as given by Eq. XV-8. Thus, if $134/\sigma c^{1/2}$ is 2 (e.g., $\sigma = 67$ Å^2 per molecule and substrate concentration 1M), ψ_G is 73 mV and ψ_D is 53 mV. The potential ψ_D should be smaller because the interfacial charge is diffused in a thickness τ rather than concentrated on a plane.

Referring to Section V-2, the double-layer system associated with a surface whose potential is some value ψ_0 requires for its formation a free energy per unit area or a π of

$$\pi_e = 6.10 c^{1/2} \left(\cosh \left(\frac{e\psi_0}{2kT} \right) - 1 \right) \qquad \text{(XV-18)}$$

(again water at 20°C and a 1:1 electrolyte). Equations XV-18 and XV-17 may be combined to give

$$\pi_e = 6.10 c^{1/2} \left(\cosh \sinh^{-1} \left(\frac{134}{\sigma c^{1/2}} \right) - 1 \right) \qquad \text{(XV-19)}$$

Equation XV-19 was given by Davies [114], assuming that ψ_G and ψ_0 were the same.

The various treatments were discussed by Stigter and Dill [115]. They tested the data of Mingins et al. for sodium octadecyl sulfate at the *n*-heptane–water interface [116] using both a virial equation and the Davies approach, modifying the latter to allow for the discrete size of the polar head group. The corrected Davies equation gave a much improved fit to the data. Alternatively, Gaines [117] applied the surface phase approach to ionized films, obtaining fairly good agreement with data.

While the π–σ plots for ionized monolayers often show no distinguishing features, it is entirely possible for such to be present and, in fact, for actual phase transitions to be observed. This was the case for films of poly(4-vinylpyridinium) bromide at the air–aqueous electrolyte interface [118]. In addition, electrostatic interactions play a large role in the stabilization of solid-supported lipid monolayers [119] as well as in the interactions between bilayers [120].

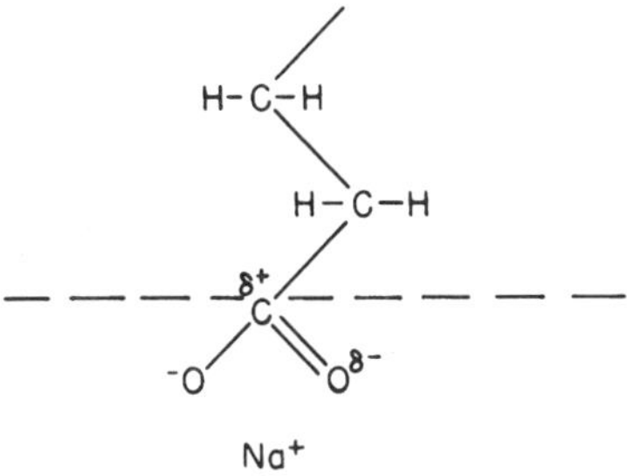

Fig. XV-13.

B. Influence of Subphase pH on the State of Monomolecular Films

Because of the charged nature of many Langmuir films, fairly marked effects of changing the pH of the substrate phase are often observed. An obvious case is that of the fatty-acid monolayers; these will be ionized on alkaline substrates, and as a result of the repulsion between the charged polar groups, the film reverts to a gaseous or liquid expanded state at a much lower temperature than does the acid form [121]. Also, the surface potential drops since, as illustrated in Fig. XV-13, the presence of nearby counterions introduces a dipole opposite in orientation to that previously present. A similar situation is found with long-chain amines on acid substrates [122].

The effect is more than just a matter of pH. As shown in Fig. XV-14, phospholipid monolayers can be expanded at low pH values by the presence of phosphotungstate ions [123], which disrupt the structural order in the lipid film [124]. Uranyl ions, by contrast, contract the low-pH expanded phase presumably because of a type of counterion condensation [123]. These effects caution against using these ions as "stains" in electron microscopy. Clearly the nature of the counterion is very important. It is dramatically so with fatty acids that form an insoluble salt with the ion; here quite low concentrations ($10^{-4}M$) of divalent ions lead to the formation of the metal salt unless the pH is quite low. Such films are much more condensed than the fatty-acid monolayers themselves [125–127].

Mixing fatty acids with fatty bases can dissolve films as the resulting complexes become water-soluble; however, in some cases the mixed Langmuir film is stabilized [128]. The application of an electric field to a mixed lipid monolayer can drive phase separation [129].

8. Langmuir–Blodgett Films

The transfer of Langmuir films from the air–liquid surface to a solid substrate has come to be known as *Langmuir–Blodgett* deposition, after its developers [130, 131]. The solid substrates are usually hydrophilic surfaces such as

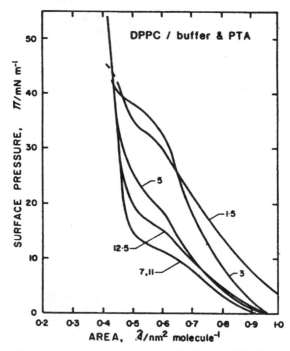

Fig. XV-14. Surface pressure–area isotherms at 298 K for a DPPC monolayer on phosphotungstic acid ($10^{-3}M$) at the pH values shown with $10^{-1}M$ NaCl added. (From Ref. 123.)

glass, chromium, quartz, or silicon. The film is transferred to the solid as it is passed vertically through the liquid interface as illustrated schematically in Fig. XV-15. One may then dip the—now hydrophobic—coated surface back through the film-covered interface and deposit a second layer "back to back" with the first. The successive layers that can be built up by this process were termed Y films by Katherine Blodgett [131]. Such films have either hydrophobic or hydrophilic surfaces depending on which direction the plate was last moved through the surface. Similarly, monolayers of like orientation can be formed; these are termed X and Z layers for those deposited on the downstroke and upstroke, respectively. These *built-up films* can be formed from hundreds of layers. Langmuir–Blodgett (LB) films are the subject of much research activity in recent years and the interested reader is referred to the general references in this area and a recent review [132].

A. Structure and Characterization of LB Films

Deposited Langmuir–Blodgett films take on many of the same structures as the Langmuir monolayers discussed in Section IV-4C, and they are often compared to the self-assembling monolayers described in Section XI-1B. The area

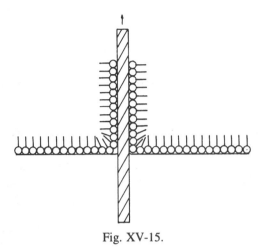

Fig. XV-15.

per molecule is generally about the same as it was on the water surface, that is, the *transfer ratio* is near unity [133–136]. The development of scanning probe microscopy (Section VIII-2B), in particular, atomic force microscopy (AFM), has provided atomic-scale resolution of LB film structure. The probes have elucidated the hexatic [137, 138], triclinic, and herringbone structures [132] and the structure near the dipping line [139] and allowed comparison of transfer procedures [140].

The characterization of LB films by spectroscopic techniques has been quite well developed. Disorder in *Y*-type bilayer assemblies of arachidic acid [141] has been investigated by electron diffraction [142], infrared [143, 144], Raman [145], and infrared optoacoustic [146] spectroscopies. Electron diffraction [147, 148], EELs [149], and infrared spectroscopy [135, 150–152] have all contributed to a molecular picture of LB film structure. Knobloch and Knoll's use of a surface plasmon propagating in a metallic substrate to excite Raman scattering allows some spatial resolution [153, 154] while surface plasmon microscopy [155, 156], optical waveguide microscopy [157] and Brewster angle microscopy [139] all provide 5–10-μm resolution. As with Langmuir films, X-ray diffraction reveals the tilt behavior in LB films [158] while X-ray photoelectron spectroscopy can be used to determine their chemical profiles [159]. Time-of-flight secondary-ion mass spectroscopy (TOFSIMS) has been used to get the molecular weight distribution of LB films of oligomers [160]. Streaming potential measurements (see Section V-6C) in LB-coated rectangular capillaries shows the dependence of the pH of neutralization on counterion binding [161], and combination with a spectroscopic probe indicates hysteretic neutralization and reionization [162].

Deposited monolayers of such RX-type compounds as fatty acids and amines can be extremely tenaciously held, as evident for example, in frictional wear experiments (see Section XII-7) and in their stability against evaporation under high vacuum [163]. The nature of the substrate may be important; fatty acid films deposited on a NiO surface (with which reaction undoubtedly occurs) showed an alternation in surface potential between odd- and even-chain lengths, but no such effect was seen if an inert substrate such as Pt was used [164]. Even hydrocarbons, such as hexane, form tenacious monolayers on metal surfaces illustrating the hazard in assuming that films deposited from a solution will be free of solvent contamination. Somewhat by contrast, however, it has been shown that deposited monolayers on mica and on various metal plates can transfer from the original solid substrate to a second one [165]. The suggested mechanism was one of surface diffusion across bridges formed by points of contact between the two solids. In the case of fatty acids on silica gel, there was evidence that the mechanism was one of vapor transport [166] (see also Ref. 167).

Chemical properties of deposited monolayers have been studied in various ways. The degree of ionization of a substituted coumarin film deposited on quartz was determined as a function of the pH of a solution in contact with the film, from which comparison with Gouy–Chapman theory (see Section V-2) could be made [151]. Several studies have been made of the UV-induced polymerization of monolayers (as well as of multilayers) of diacetylene amphiphiles (see Refs. 168, 169). Excitation energy transfer has been observed in a mixed monolayer of donor and acceptor molecules in stearic acid [170]. Electrical properties have been of interest, particularly the possibility that a suitably asymmetric film might be a unidirectional conductor, that is, a rectifier (see Refs. 171, 172). Optical properties of interest include the ability to make planar optical waveguides of thick LB films [173, 174].

B. Mixed LB Films and Films of Polymers and Colloids

Most LB-forming amphiphiles have hydrophobic tails, leaving a very hydrophobic surface. In order to introduce polarity to the final surface, one needs to incorporate bipolar components that would not normally form LB films on their own. Berg and co-workers have partly surmounted this problem with two- and three-component mixtures of fatty acids, amines, and bipolar alcohols [175, 176]. Interestingly, the type of deposition depends on the contact angle of the substrate, and, thus, when relatively polar monolayers are formed, they are deposited as Z-type multilayers. Phase-separated LB films of hydrocarbon–fluorocarbon mixtures provide selective adsorption sites for macromolecules, due to the formation of a step site at the domain boundary [177].

Film stability is a primary concern for applications. LB films of photopolymerizable polymeric amphiphiles can be made to crosslink under UV radiation to greatly enhance their thermal stability while retaining the ordered layered structure [178]. Low-molecular-weight perfluoropolyethers are important industrial lubricants for computer disk heads. These small polymers attached to a polar head form continuous films of uniform thickness on LB deposi-

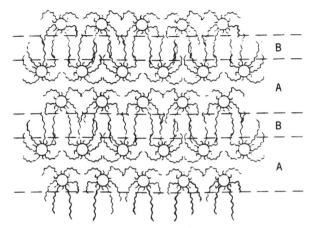

Fig. XV-16. A schematic drawing of the arrangement of polyglutamates having alkyl side chains in an LB film. The circles represent the rodlike polyglutamate backbone oriented perpendicular to the page; the wiggly lines are the alkyl sidechains. (From Ref. 182.)

tion. Goedel and co-workers [179] have studied this system and estimated the contribution to the surface pressure from the stretching of the polymer tails tethered to the interface. They found that for these short polymers the tail entropy was negligible compared to the head-group interactions [179]. Another study of these polymer "brushes" was made by Kurihara [180] using LB deposited poly(methacrylic acid) to investigate polyelectrolyte brushes in the surface force apparatus (SFA). Interesting brushes of LB deposited, end-anchored poly(glutamic acid) show the influence of secondary structure (α helix and β sheet) on the forces and elastic properties between layers measured in the SFA [181].

A unique but widely studied polymeric LB system are the polyglutamates or "hairy rod" polymers. These polymers have a hydrophilic rod of helical polyglutamate with hydrophobic alkyl side chains. Their rigidity and amphiphilicity imparts order (lyotropic and thermotropic) in LB films and they take on a Y-type structure such as that illustrated in Fig. XV-16 [182]. These LB films are useful for waveguides, photoresists, and chemical sensors. LB films of these polymers are very thermally stable, as was indicated by the lack of interdiffusion up to 414 K shown by neutron reflectivity of alternating hydrogenated and deuterated layers [183]. AFM measurements have shown that these films take on different structures if directly deposited onto silicon or onto LB films of cadmium arachidate [184].

Among the many applications of LB films, the creation or arrangement of colloidal particles in these films is a unique one. On one hand, colloidal particles such as 10-nm silver sols stabilized by oleic acid can be spread at the air–water interface and LB deposited to create unique optical and electrooptical properties for devices [185].

Another approach is to use the LB film as a template to limit the size of growing colloids such as the Q-state semiconductors that have applications in nonlinear optical devices. Furlong and co-workers have successfully synthesized CdSe [186] and CdS [187] nanoparticles (<5 nm in radius) in Cd arachidate LB films. Finally, as a low-temperature ceramic process, LB films can be converted to oxide layers by UV and ozone treatment; examples are polydimethylsiloxane films to make SiO_x [188] and Cd arachidate to make CdO_x [189].

C. Studies of the LB Deposition Process

An important issue in the formation of LB films is their stability and deformation during the transfer process. This has motivated research in the areas of process feedback control, flow in Langmuir monolayers, and the structure of transferred films. In addition to the LB deposition technique illustrated in Fig. XV-15, freely suspended films (in effect, soap films) can be transferred from an aperture to a surface. The transfer of freely suspended films creates layers having fewer defects since they can equilibrate prior to deposition [140]. The control and optimization of the LB film transfer has been studied by Koberstein and co-workers [190, 191]. They found the optimum control parameters to minimize pressure excursions on compression [190], and they investigated the effect of the monolayer properties on those parameters. The behavior of the monolayer during transfer can be related to their surface rheology (see Sections XV-2B and IV-3C). A new apparatus for measuring shear modulus has been developed by Ketterson and co-workers [192]; this versatile torsion device has a large dynamic range and high sensitivity (capable of measuring moduli from 0.005 to 1000 dyn/cm!). Fuller and co-workers have used Brewster angle microscopy to directly visualize the anisotropy induced in a fatty-acid monolayer by extensional flow [193]. This type of flow is important in the deposition process and can alter both domain and molecular orientation.

9. Problems

1. A monolayer of Streptavidin containing 1.75 mg of protein/m^2 gives a film pressure of 0.070 erg/m^2 at 15°C. Calculate the molecular weight of the protein, assuming ideal-gas behavior.

2. Estimate M for the PMA film of Fig. XV-1 and also the value of A in Eq. XV-1 at 25°C.

3. Describe situations that might result in the four types of isotherms shown in Fig. XV-3.

4. Looking at the multivesicular liposome in Fig. XV-11, estimate the average number of edges per face, q, and use your value of q to determine the average number of faces per polyhedra, F, using the Euler relation $V - E + F = 2$, where V is the number of vertices and E is the number of edges. Tetrahedral coordination restricts $3V = 2E = qF$. Compare your result with the experimental finding that there are 5.1 faces per pair of edges.

5. Show under what limiting condition Eqs. XV-8 and XV-17 become the same.

6. At what molecular area should a fatty-acid film spread on $0.05M$ NaOH have a film pressure of 4 dyn/cm, according to the Donnan treatment and to the Gouy treatment? Assume the hydrocarbon part of the film to behave as an ideal gas. Assume 20°C.

7. Davies [114] found that the rates of desorption of sodium laurate and of lauric acid films were in the ratio $6.70:1$ at 21.5°C at molecular areas of 90 and 60 Å^2 per molecule, respectively. Calculate ψ_0, the potential at the plane CD in Fig. XV-12.

8. Investigate the differences between LB films and self-assembled monolayer SAMs (Chapter XI). Which are finding more practical use, and what are the potential applications of each?

9. From what you know about wetting, contact angles, and spread monolayers, explain why a Langmuir–Blodgett layer will deposit as a Y type if there are nonpolar fatty acids, yet will switch to a Z-type deposition if bipolar alcohols or amines are added (see Refs. 175, 176). What defines the critical contact angle for the deposition change?

10. Discuss the role of the various components of lung surfactants in making a resilient film.

General References

G. L. Gaines, Jr. *Insoluble Monolayers at Gas-Liquid Interfaces*, Interscience, New York, 1966.

W. M. Gelbart, A. Ven-Shaul, and D. Roux, eds., *Micelles, Membranes, Microemulsions and Monolayers*, Springer-Verlag, Berlin, 1994.

D. D. Lasic, *Liposomes: From Physics to Applications*, Elsevier, Amsterdam, 1993.

G. G. Roberts, *Langmuir-Blodgett Films*, Plenum, New York, 1990.

S. A. Safran, *Statistical Thermodynamics of Surfaces, Interfaces and Membranes*, Addison-Wesley, Reading, MA, 1994.

A. Ulman, *An Introduction to Ultrathin Organic Films*, Academic, Boston, 1991.

Textual References

1. A. Takahashi, A. Yoshida, and M. Kawaguchi, *Macromolecules*, **15,** 1196 (1982).

2. S. J. Singer, *J. Chem. Phys.*, **16,** 872 (1948).

3. H. L. Frisch and R. Simha, *J. Chem. Phys.*, **27,** 702 (1957).

4. A. Silberberg, *J. Phys. Chem.*, **66,** 1872 (1962).

5. V. Aharonson, D. Andelman, A. Zilman, P. A. Pincus, and E. Raphaël, *Physica A*, **204,** 1 (1994).

6. T. J. Lenk, D. H. T. Lee, and J. T. Koberstein, *Langmuir*, **10,** 1857 (1994).

7. W. M. Lee, R. R. Stromberg, and J. L. Shereshefsky, *J. Res. Natl. Bur. Stand.* **66A,** 439 (1962).

8. H. E. Ries, Jr., N. Beredjick, and J. Gabor, *Nature*, **186,** 883 (1960).

9. N. Beredjick, R. A. Ahlbeck, T. K. Kwei, and H. E. Ries, Jr., *J. Polymer Sci.*, **46,** 268 (1960).

10. J. B. Peng and G. T. Barnes, *Langmuir*, **6,** 578 (1990).

11. J. B. Peng and G. T. Barnes, *Langmuir*, **7,** 1749 (1991), 3090 (1991).

12. J. B. Peng and G. T. Barnes, *Colloids & Surfaces*, **A102,** 75 (1995).

13. G. L. Gaines, Jr., *Adv. Chem.*, **144,** 338 (1975).

14. H. Motschmann, R. Reiter, R. Lawall, G. Duda, M. Stamm, G. Wegner, and W. Knoll, *Langmuir*, **7,** 2743 (1991).

15. P. Baglioni, E. Gallori, G. Gabrielli, and E. Ferroni, *J. Colloid Interface Sci.*, **88,** 221 (1982).

16. P. Baglioni, L. Dei, and G. Gabrielli, *J. Colloid Interface Sci.*, **93,** 402 (1983).

17. G. Gabrielli, M. Puggelli, and P. Baglioni, *J. Colloid Interface Sci.*, **86,** 485 (1982).

18. M. C. Friedenberg, G. G. Fuller, C. W. Frank, and C. R. Robertson, *Macromolecules*, **29,** 705 (1996).

19. K. C. O'Brien and J. B. Lando, *Langmuir*, **1,** 453 (1985).

20. J. B. Peng, G. T. Barnes, and B. M. Abraham, *Langmuir*, **9,** 3547 (1993).

21. B. Lin, S. A. Rice, and D. A. Weitz, *J. Chem. Phys.*, **99,** 8308 (1993).

22. B. Lin, S. A. Rice, and D. A. Weitz, *Phys. Rev. E*, **51,** 423 (1995).

23. M. C. Friedenberg, G. G. Fuller, C. W. Frank, and C. R. Robertson, *Langmuir*, **10,** 1251 (1994).

24. N. A. Kotov, F. C. Meldrum, J. H. Fendler, E. Tombacz, and I. Dekany, *Langmuir*, **10,** 3797 (1994).

25. R. Miller, Z. Policova, R. Sedev, and A. W. Neumann, *Colloids & Surfaces*, **A76,** 179 (1993).

26. F. Y. H. Lin, D. Y. Kwok, Z. Policova, W. Zingg, and A. W. Neumann, *Colloids & Surfaces*, **B3,** 281 (1995).

27. D. C. Clark, F. Husband, P. J. Wilde, M. Cornec, R. Miller, J. Krägel, and R. Üstneck, *J. Chem. Soc., Faraday Trans.*, **91,** 1991 (1995).

28. E. E. Uzgiris and R. D. Kornberg, *Nature*, **301,** 125 (1983).

29. S. A. Darst, H. O. Ribi, D. W. Pierce, and R. D. Kornberg, *J. Mol. Biol.*, **203,** 269 (1988).

30. D. Vaknin, J. Als-Nielsen, M. Piepenstock, and M. Lösche, *Biophys. J.*, **60,** 1545 (1991).

31. A. C. Ku, S. A. Darst, R. D. Kornberg, C. R. Robertson, and A. P. Gast, *Langmuir*, **8,** 2357 (1992).

32. A. C. Ku, S. A. Darst, C. R. Robertson, A. P. Gast, and R. D. Kornberg, *J. Phys. Chem.*, **97,** 3013 (1993).

33. S. A. Hemming, A. Bockharev, S. A. Darst, R. D. Kornberg, P. Ala, D. S. C. Yang, and A. M. Edwards, *J. Mol. Biol.*, **246,** 308 (1995).

34. A. Engel, A. Hoenger, A. Hefti, C. Henn, R. C. Ford, J. Kistler, and M. Zulauf, *J. Struct. Biol.*, **109,** 219 (1992).

35. D. A. Cadenhead, in *Structures and Properties of Cell Membranes*, Vol. III, G. Benga, ed., CRC Press, Boca Raton, FL, 1985, p. 22.

36. R. S. Canto and K. A. Dill, *Langmuir*, **2,** 331 (1986).

37. H. E. Ries, Jr., M. Matsumoto, N. Uyeda, and E. Suito, *Adv. Chem. Ser.*, No. 144, American Chemical Society, 1975, p. 286.

38. D. M. Balthasar, D. A. Cadenhead, R. N. A. H. Lewis, and R. N. McElhaney, *Langmuir*, **4**, 180 (1988).

39. K. Kjaer, J. Als-Nielsen, C. A. Helm, L. A. Laxhuber, and H. Möhwald, *Phys. Rev. Lett.*, **58**, 2224 (1987).

40. A. Miller and H. Möhwald, *J. Chem. Phys.*, **86**, 4258 (1987).

41. G. T. Barnes, G. A. Lawrie, B. J. Battersby, S. M. Sarge, H. K. Cammenga, and P. B. Schneider, *Thin Solid Films*, **242**, 201 (1994).

42. R. Merkel and E. Sackmann, *J. Phys. Chem.*, **98**, 4428 (1994).

43. R. Peters and K. Beck, *Proc. Natl. Acad. Sci. USA*, **80**, 7183 (1983).

44. B. M. Abraham and J. B. Ketterson, *Langmuir*, **2**, 801 (1986).

45. P. Baglioni, L. Dei, E. Ferroni, and G. Gabrielli, *Colloids & Surfaces*, **60**, 399 (1991).

46. R. E. Goldstein and D. P. Jackson, *J. Phys. Chem.*, **98**, 9626 (1994).

47. H. Möhwald, *Annu. Rev. Phys. Chem.*, **41**, 441 (1990).

48. H. M. McConnell, *Annu. Rev. Phys. Chem.*, **42**, 171 (1991).

49. H. M. McConnell, D. Keller, and H. Gaub, *J. Phys. Chem.*, **90**, 1717 (1986).

50. P. Tancrède, Luc Parent, and R. M. Leblanc, *J. Colloid Interface Sci.*, **89**, 117 (1982).

51. F. Lamarche, F. Techy, J. Aghion, and R. M. Leblanc, *Colloids & Surfaces*, **30**, 209 (1988).

52. H. Möhwald, A. Miller, W. Stich, W. Knoll, A. Ruaudel-Teixier, T. Lehmann, and J. H. Fuhrhop, *Thin Solid Films*, **4**, 261 (1986).

53. H. E. Ries, Jr. and H. Swift, *J. Colloid Interface Sci.*, **117**, 584 (1987).

53a. H. E. Ries, Jr., *J. Colloid Interface Sci.*, **106**, 273 (1985).

54. M. L. Longo, A. M. Bisagno, J. A. Zasadzinski, R. Bruni, and A. J. Waring, *Science*, **261**, 453 (1993).

55. M. M. Lipp, K. Y. C. Lee, J. A. Zasadzinski, and A. J. Waring, *Science*, **273**, 1196 (1996).

56. D. W. Grainger, A. Reichert, H. Ringsdorf, and C. Salesse, *Fed. Eur. Biochem. Sci. Lett.*, **252**, 73 (1989).

57. K. M. Maloney, M. Grandbois, D. W. Grainger, C. Salesse, K. A. Lewis, and M. F. Roberts, *Biochim. Biophys. Acta*, **1235**, 395 (1995).

58. D. W. Grainger, M. Ahlers, R. Blankenburg, P. Meller, A. Reichert, H. Ringsdorf, and C. Salesse, "Modeling of Cell Membrane Targeting: Specific Recognition, Binding and Protein Domain Formation in Ligand-Containing Model Biomembranes," in *Targeting of Drugs*, by G. G. Gregoriadis et al., Plenum, New York, 1990.

59. M. Ahlers, D. W. Grainger, J. N. Herron, K. Lim, H. Ringsdorf, and C. Salesse, *Biophys. J.*, **63**, 823 (1992).

60. Y. Ikeura, K. Kurihara, and T. Kunitake, *J. Am. Chem. Soc.*, **113**, 7342 (1991).

61. D. Y. Sasaki, K. Kurihara, and T. Kunitake, *J. Am. Chem. Soc.*, **113**, 9685 (1991).

62. K. Kurihara, K. Ohto, Y. Tanaka, Y. Aoyama, and T. Kunitake, *J. Am. Chem. Soc.*, **113,** 444 (1991).

63. K. Kurihara, K. Ohto, Y. Honda, and T. Kunitake, *J. Am. Chem. Soc.*, **113,** 5077 (1991).

64. E. Sackmann, *Science*, **271,** 43 (1996).

65. J. H. Fendler, *Chem. Rev.*, **87,** 877 (1987).

66. K. Tajima and N. L. Gershfeld, *Biophys. J.*, **47,** 203 (1985).

67. Y. Okahata, *Accts. Chem. Res.*, **19,** 57 (1986).

68. R. P. Rand, *Ann. Rev. Biophys. Bioeng.*, **10,** 277 (1981).

69. J. Fuhrhop and D. Fritsch, *Accts. Chem. Res.*, **19,** 130 (1986).

70. S. Chiruvolu, H. E. Warriner, E. Naranjo, S. H. J. Idziak, J. O. Rädler, R. J. Plano, J. A. Zasadzinski, and C. R. Safinya, *Science*, **266,** 1222 (1994).

71. T. Imae, T. Iwamoto, G. Platz, and C. Thunig, *Colloid Polym. Sci.*, **272,** 604 (1994).

72. S. Chiruvolu, S. Walker, J. Israelachvili, F.-J. Schmitt, D. Leckband, and J. A. Zasadzinski, *Science*, **264,** 1753 (1994).

73. K. Nakazawa and T. Imae, *J. Colloid Interface Sci.*, **151,** 571 (1992).

74. P. L. Nostro and G. Gabrielli, *Colloids & Surfaces*, **44,** 119 (1990).

75. M. F. Brown, A. A. Ribeiro, and G. D. Williams, *Proc. Natl. Acad. Sci.* (USA), **80,** 4325 (1983).

76. H. Meier, I. Sprenger, M. Bärmann, and E. Sackmann, *Macromolecules*, **27,** 7581 (1994).

77. M. S. Spector, J. A. Zasadzinski, and M. B. Sankaram, *Langmuir*, **12,** 4704 (1996).

78. J. O. Rädler, T. J. Feder, H. H. Strey, and E. Sackmann, *Phys. Rev. E*, **51,** 4526 (1995).

79. K. M. Palmer, M. Goulian, and P. Pincus, *J. Phys. II France*, **4,** 805 (1994).

80. E. Sackmann, *Fed. Eur. Biochem. Soc. Lett.*, **346,** 3 (1994).

81. N. Dan, P. Pincus, and S. A. Safran, *Langmuir*, **9,** 2768 (1993).

82. R. Bruinsma, M. Goulian, and P. Pincus, *Biophys. J.*, **67,** 746 (1994).

83. R. Bar-Ziv, R. Menes, E. Moses, and S. A. Safran, *Phys. Rev. Lett.*, **75,** 3356 (1995).

84. D. Leckband, W. Müller, F.-J. Schmitt, and H. Ringsdorf, *Biophys. J.*, **69,** 1162 (1995).

85. A. Salmon, S. W. Dodd, G. D. Williams, J. M. Beach, and M. F. Brown, *J. Am. Chem. Soc.*, **109,** 2600 (1987).

86. G. E. Crawford and J. C. Earnshaw, *Biophys. J.*, **52,** 87 (1987).

87. F. Zhang, G. M. Lee, and K. A. Jacobson, *BioEssays*, **15,** 579 (1993).

88. G. M. Lee, A. Ishihara, and K. A. Jacobson, *Proc. Natl. Acad. Sci.* (USA), **88,** 6274 (1991).

89. G. M. Lee, F. Zhang, A. Ishihara, C. L. McNeil, and K. A. Jacobson, *J. Cell Biol.*, **120,** 25 (1993).

90. W. Pfeiffer, T. Henkel, E. Sackmann, W. Knoll, and D. Richter, *Europhys. Lett.*, **8,** 201 (1989).

91. E. Hutchinson, *J. Colloid Sci.*, **3,** 219 (1948); H. Sobotka, ed., *Monomolecular*

Layers, American Association for the Advancement of Science, Washington, DC, 1954, p. 161.

92. T. Takenaka, *Chem. Physics Lett.*, **55,** 515 (1978).

93. M. M. Telo da Gama and K. E. Gubbins, *Mol. Phys.*, **59,** 227 (1986).

94. C. H. Bamford, A. Elliot, and W. E. Hanby, *Synthetic Polypeptides*, Academic, New York, 1956.

95. J. T. Davies, *Biochem. J.*, **56,** 509 (1954).

96. M. A. Cabrerizo–Vilchez, Z. Policova, D. Y. Kwok, P. Chen, and A. W. Neumann, *Colloids & Surfaces*, **B5,** 1 (1995).

97. H. Hermel and R. Miller, *Colloid Polym. Sci.*, **273,** 387 (1995).

98. D. Stigter and K. A. Dill, *Langmuir*, **4,** 200 (1988).

99. B. Y. Yue, C. M. Jackson, J. A. G. Taylor, J. Mingins, and B. A. Pethica, *J. Chem. Soc., Faraday Trans. I*, **72,** 2685 (1976).

100. A. Barraud, J. Leloup, and P. Lesieur, *Thin Solid Films*, **133,** 113 (1985).

101. A. H. Ellison and W. A. Zisman, *J. Phys. Chem.*, **60,** 416 (1956).

102. N. L. Jarvis and W. A. Zisman, *J. Phys. Chem.*, **63,** 727 (1959).

103. J. H. Brooks and B. A. Pethica, *Trans. Faraday Soc.*, **60,** 208 (1964).

104. J. T. Davies, *Trans. Faraday Soc.*, **48,** 1052 (1952).

105. J. Mingins, F. G. R. Zobel, B. A. Pethica, and C. Smart, *Proc. Roy. Soc.* (London), **A324,** 99 (1971).

106. J. T. Davies and G. R. A. Mayers, *Trans. Faraday Soc.*, **56,** 691 (1960).

107. S. W. Barton, B. N. Thomas, E. Flom, F. Novak, and S. A. Rice, *Langmuir*, **4,** 233 (1988).

108. P. Baglioni, G. Cardini, G. Gabrielli, and G. Taddel, *J. Phys. Chem.*, **86,** 4684 (1982).

109. P. Mueller, D. O. Rudin, H. T. Tien, and W. C. Wescott, *J. Phys. Chem.*, **67,** 535 (1963).

110. R. J. Good, *J. Colloid Interface Sci.*, **31,** 540 (1969).

111. R. E. Baier, *Surface Chemistry of Biological Systems*, Plenum, New York, p. 235, 1970; see also *J. Biomed. Res.*, **9,** 327 (1975).

112. L. Vroman and A. L. Adams, *J. Biomed. Mater. Res.*, **3,** 43 (1969).

113. W. Drost-Hansen, *Fed. Proc.*, **30,** 1539 (1971).

114. J. T. Davies, *Proc. Roy. Soc.* (London), **A208,** 224 (1951).

115. D. Stigter and K. A. Dill, *Langmuir*, **2,** 791 (1986).

116. J. Mingins, J. A. G. Taylor, N. F. Owens, and J. H. Brooks, *Adv. Chem.*, **144,** 28 (1975).

117. G. L. Gaines, Jr., *J. Chem. Phys.*, **69,** 2627 (1978).

118. M. Kawaguchi, S. Itoh, and A. Takahashi, *Macromolecules*, **20,** 1052, 1056 (1987).

119. W. Frey and E. Sackmann, *J. Colloid Interface Sci.*, **174,** 378 (1995).

120. W. K. Kegel and H. N. W. Lekkerkerker, *Europhys. Lett.*, **26,** 425 (1994).

121. N. K. Adam and J. G. F. Miller, *Proc. Roy. Soc.* (London), **142,** 401 (1933).

122. J. J. Betts and B. A. Pethica, *Trans. Faraday Soc.*, **52,** 1581 (1956).

123. D. Gorwyn and G. T. Barnes, *Langmuir*, **6,** 222 (1990).

124. G. T. Barnes, I. R. Gentle, C. H. L. Kennard, J. B. Peng, and I. M. Jamie, *Langmuir*, **11**, 281 (1995).

125. E. D. Goddard and J. A. Ackilli, *J. Colloid Chem.*, **18**, 585 (1963).

126. J. A. Spink, *J. Colloid Sci.*, **18**, 512 (1963).

127. N. W. Rice and F. Sebba, *J. Appl. Chem.* (London), **15**, 105 (1965); F. Sebba, *Nature*, **184**, 1062 (1959).

128. H. Ebeltoft, J. Sjöblom, J. O. Saeten, and G. Olofsson, *Langmuir*, **10**, 2262 (1994).

129. K. Y. C. Lee, J. Klinger, and H. M. McConnell, *Science*, **263**, 655 (1994).

130. I. Langmuir, *Trans. Far. Soc.*, **15**, 62 (1920).

131. K. B. Blodgett, *J. Am. Chem. Soc.*, **57**, 1007 (1935).

132. J. A. Zasadzinski, R. Viswanathan, L. Madsen, J. Garnaes, and D. K. Schwartz, *Science*, **263**, 1726 (1994).

133. R. D. Neuman and J. W. Swanson, *J. Colloid Interface Sci.*, **74**, 244 (1980).

134. M. A. Richard, J. Deutch, and G. M. Whitesides, *J. Am. Chem. Soc.*, **100**, 6613 (1978).

135. E. Okamura, J. Umemura, and T. Takenaka, *Biochimica Biophysica Acta*, **812**, 139 (1985).

136. J. B. Peng, B. M. Abraham, P. Dutta, and J. B. Ketterson, *Thin Solid Films*, **134**, 187 (1985); E. P. Honig, *Langmuir*, **5**, 882 (1989).

137. J. B. Peng and G. T. Barnes, *Thin Solid Films*, **252**, 44 (1994).

138. R. Viswanathan, L. Madsen, J. A. Zasadzinski, and D. K. Schwartz, *Science*, **269**, 51 (1995).

139. L. Wolthaus, A. Schaper, and D. Möbius, *J. Phys. Chem.*, **98**, 10809 (1994).

140. R. Overney, E. Meyer, J. Frommer, H. J. Güntherodt, G. Decher, J. Reibel, and U. Sohling, *Langmuir*, **9**, 341 (1993).

141. R. F. Fischetti, V. Skita, A. F. Garito, and J. K. Blasie, *Phys. Rev. B*, **37**, 4788 (1988); V. Skita, M. Filipkowski, A. F. Garito, and J. K. Blasie, ibid., **34**, 5826 (1986).

142. I. R. Peterson, *Br. Polym. J.*, **19**, 391 (1987).

143. N. Nakashima, N. Yamada, T. Kunitake, J. Umemura, and T. Takenaka, *J. Phys. Chem.*, **90**, 3374 (1986).

144. D. D. Saperstein, *J. Phys. Chem.*, **90**, 1408 (1986).

145. J. P. Rabe, J. D. Swalen, and J. F. Rabolt, *J. Chem. Phys.*, **86**, 1601 (1987).

146. L. Rothberg, G. S. Higashi, D. L. Allara, and S. Garoff, *Chem. Phys. Lett.*, **133**, 67 (1987).

147. S. Garoff, H. W. Deckman, J. H. Dunsmuir, and M. S. Alvarez, *J. Physique*, **47**, 701 (1986).

148. V. Vogel and C. Wöll, *J. Chem. Phys.*, **84**, 5200 (1986).

149. J. H. Wandass and J. A. Gardella, Jr., *Surf. Sci.*, **150**, L107 (1985).

150. J. F. Rabolt, F. C. Burns, N. E. Schlotter, and J. D. Swalen, *J. Chem. Phys.*, **78**, 946 (1983).

151. B. Lovelock, F. Grieser, and T. W. Healy, *Langmuir*, **2**, 443 (1986).

152. S. J. Mumby, J. D. Swalen, and J. F. Rabolt, *Macromolecules*, **19**, 1054 (1986).

153. H. Knobloch and W. Knoll, *Makromol. Chem. Macromol. Symp.*, **46,** 389 (1991).

154. A. Nemetz, H. Knobloch, and W. Knoll, *Thin Solid Films*, **226,** 48 (1993).

155. B. Rothenhäusler and W. Knoll, *Nature*, **332,** 615 (1988).

156. W. Hickel and W. Knoll, *J. Appl. Phys.*, **67,** 3572 (1990).

157. W. Hickel and W. Knoll, *Appl. Phys. Lett.*, **57,** 1286 (1990).

158. G. J. Foran, J. B. Peng, R. Steitz, G. T. Barnes, and I. R. Gentle, *Langmuir*, **12,** 774 (1996).

159. K. Kurihara, T. Kawahara, D. Y. Sasaki, K. Ohto, and T. Kunitake, *Langmuir*, **11,** 1408 (1995).

160. J. F. Elman, D. H. T. Lee, and J. T. Koberstein, *Langmuir*, **11,** 2761 (1995).

161. P. G. Scales, F. Grieser, and T. W. Healy, *Thin Solid Films*, **215,** 223 (1992).

162. B. S. Murray, F. Grieser, T. W. Healy, and P. G. Scales, *Langmuir*, **8,** 217 (1992).

163. G. L. Gaines, Jr., and R. W. Roberts, *Nature*, **197,** 787 (1963).

164. C. O. Timmons and W. A. Zisman, *J. Phys. Chem.*, **69,** 984 (1965).

165. E. K. Rideal and J. Tadayon, *Proc. Roy. Soc.* (London), **A225,** 346, 357 (1954).

166. A. W. Adamson and V. Slawson, *J. Phys. Chem.*, **85,** 116 (1981).

167. Young, J. E., *Aust. J. Chem.*, **8,** 173 (1955).

168. D. Day and H. Ringsdorf, *J. Polym. Sci. Polym. Lett. Ed.*, **16,** 205 (1978).

169. B. Tieke and K. Weiss, *J. Colloid Interface Sci.*, **101,** 129 (1984).

170. N. Tamai, T. Yamazaki, and I. Yamazaki, *J. Phys. Chem.*, **91,** 841 (1987).

171. R. M. Metzger, R. R. Schumaker, M. P. Cava, R. K. Laidlaw, C. A. Panetta, and E. Torres, *Langmuir*, **4,** 298 (1988).

172. R. M. Metzger, C. A. Panetta, N. E. Heimer, A. M. Bhatti, E. Torres, G. F. Blackburn, S. K. Tripathy, and L. A. Samuelson, *J. Molecular Electronics*, **2,** 119 (1986).

173. J. D. Swalen, *J. Mol. Electronics*, **2,** 155 (1986).

174. W. Hickel, G. Duda, M. Jurich, T. Kröhl, K. Rochford, G. I. Stegeman, J. D. Swalen, G. Wegner, and W. Knoll, *Langmuir*, **6,** 1403 (1990).

175. J. M. Berg and L. G. Tomas Eriksson, *Langmuir*, **10,** 1213 (1994).

176. J. M. Berg, L. G. Tomas Eriksson, P. M. Claesson, and K. G. Nordli B, *Langmuir*, **10,** 1225 (1994).

177. J. Frommer, R. Lüthi, E. Meyer, D. Anselmetti, M. Dreier, R. Overney, H. J. Güntherodt, and M. Fujihara, *Nature*, **364,** 198 (1993).

178. T. Hayashi, M. Mabuchi, M. Mitsuishi, S. Ito, M. Yamamoto, and W. Knoll, *Macromolecules*, **28,** 2537 (1995).

179. W. A. Goedel, H. Wu, M. C. Friedenberg, G. G. Fuller, M. Foster, and C. W. Frank, *Langmuir*, **10,** 4209 (1994).

180. K. Kurihara, "Direct Measurement of Surface Forces of Supramolecular Systems: Structures and Interactions," in *Microchemistry*, H. Masuhara et al., Elsevier Science, 1994.

181. T. Abe, K. Kurihara, N. Higashi, and M. Niwa, *J. Phys. Chem.*, **99,** 1820 (1995).

182. S. Lee, J. R. Dutcher, G. I. Stegeman, G. Duda, G. Wegner, and W. Knoll, *Phys. Rev. Lett.*, **70,** 2427 (1993).

183. A. Schmidt, K. Mathauer, G. Reiter, M. D. Foster, M. Stamm, G. Wegner, and W. Knoll, *Langmuir*, **10,** 3820 (1994).

184. V. V. Tsukruk, M. D. Foster, D. H. Reneker, A. Schmidt, and W. Knoll, *Langmuir*, **9,** 3538 (1993).

185. F. C. Meldrum, N. A. Kotov, and J. H. Fendler, *Langmuir*, **10,** 2035 (1994).

186. R. S. Urquhart, D. N. Furlong, T. Gegenbach, N. J. Geddes, and F. Grieser, *Langmuir*, **11,** 1127 (1995).

187. R. S. Urquhart, D. N. Furlong, H. Mansur, F. Grieser, K. Tanaka, and Y. Okahata, *Langmuir*, **10,** 899 (1994).

188. C. L. Mirley and J. T. Koberstein, *Langmuir*, **11,** 1049 (1995).

189. C. L. Mirley and J. T. Koberstein, *Langmuir*, **11,** 2837 (1995).

190. C. L. Mirley, M. G. Lewis, J. T. Koberstein, and D. H. T. Lee, *Langmuir*, **10,** 2370 (1994).

191. C. L. Mirley, M. G. Lewis, J. T. Koberstein, and D. H. T. Lee, *Langmuir*, **11,** 2755 (1995).

192. R. S. Ghaskadvi, T. M. Bohanon, P. Dutta, and J. B. Ketterson, *Phys. Rev. E*, **54,** 1770 (1996).

193. M. C. Friedenberg, G. G. Fuller, C. W. Frank, and C. R. Robertson, *Langmuir*, **12,** 1594 (1996).

CHAPTER XVI

The Solid–Gas Interface—General Considerations

1. Introduction

These concluding chapters deal with various aspects of a very important type of situation, namely, that in which some adsorbate species is distributed between a solid phase and a gaseous one. From the phenomenological point of view, one observes, on mechanically separating the solid and gas phases, that there is a certain distribution of the adsorbate between them. This may be expressed, for example, as n_a, the moles adsorbed per gram of solid versus the pressure P. The distribution, in general, is temperature dependent, so the complete empirical description would be in terms of an adsorption function $n_a = f(P, T)$.

While a *thermodynamic* treatment can be developed entirely in terms of $f(P, T)$, to apply adsorption *models*, it is highly desirable to know n_a on a per square centimeter basis rather than a per gram basis or, alternatively, to know θ, the fraction of surface covered. In both the physical chemistry and the applied chemistry of the solid–gas interface, the specific surface area Σ is thus of extreme importance.

All gases below their critical temperature tend to adsorb as a result of general van der Waals interactions with the solid surface. In this case of *physical adsorption*, as it is called, interest centers on the size and nature of adsorbent–adsorbate interactions and on those between adsorbate molecules. There is concern about the degree of heterogeneity of the surface and with the extent to which adsorbed molecules possess translational and internal degrees of freedom.

If the adsorption energy is large enough to be comparable to chemical bond energies, we now speak of *chemisorption*. The adsorbate tends to be localized at particular sites (although some surface diffusion or mobility may still be present), and the equilibrium gas pressure may be so low that the adsorbent-adsorbate system can be studied under high-vacuum conditions. This allows the many diffraction and spectroscopic techniques described in Chapter VIII to be used to determine what actual species are present on the surface and their packing and chemical state. This is also true for physisorption systems if the surface is well defined and the temperature low enough that the equilibrium pressure is very low, see Fig. XVII-17 for example.

It has become increasingly appreciated in recent years that the surface structure of the adsorbent may be altered in the adsorption process. Qualitatively,

TABLE XVI-1
Types of Adsorption Systems

Type of Adsorbate-Adsorbent Interaction	Type of Adsorbent	
	Molecular	Refractory
Van der Waals (physical adsorption)	Surface restructuring on adsorption	No change in surface structure on adsorption
Chemical (chemisorption)	Chemical reaction modifying the adsorbent	Surface restructuring on adsorption

such structural perturbation is apt to occur if the adsorption energy is comparable to the surface energy of the adsorbent (on a per molecular unit basis). As summarized in Table XVI-1, physical adsorption (sometimes called *physisorption*) will likely alter the surface structure of a molecular solid adsorbent (such as ice, paraffin, and polymers) but not that of high-surface-energy, refractory solids (such as the usual metals and metal oxides and carbon black). Chemisorption may alter the surface structure of refractory solids.

The function of this chapter is to summarize some of the general approaches to the determination of the physical and chemical state in both of the types of adsorption systems described.

2. The Surface Area of Solids

The specific surface area of a solid is one of the first things that must be determined if any detailed physical chemical interpretation of its behavior as an adsorbent is to be possible. Such a determination can be made through adsorption studies themselves, and this aspect is taken up in the next chapter; there is a number of other methods, however, that are summarized in the following material. Space does not permit a full discussion, and, in particular, the methods that really amount to a *particle* or *pore size* determination, such as optical and electron microscopy, x-ray or neutron diffraction, and permeability studies are largely omitted.

A. The Meaning of Surface Area

It will be seen that each method for surface area determination involves the measurement of some property that is observed qualitatively to depend on the extent of surface development and that can be related by means of theory to the actual surface area. It is important to realize that the results obtained by different methods differ, and that one should in general *expect* them to differ. The problem is that the concept of surface area turns out to be a rather elusive one as soon as it is examined in detail.

The difficulty in stating just what is meant by the "surface area" of a solid, in

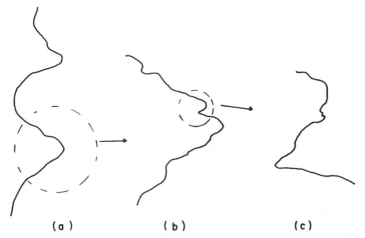

Fig. XVI-1. Successive enlargements of a hypothetical coastline. (From Ref. 1.)

one aspect, can be illustrated by considering the somewhat analogous question of what is meant by the "length" of a particular section of coastline. Superficially, it seems easy to say that the length is simply the distance along the shore between two points. Geography books, however, are prone to make such statements as "because of its many indentations, the coast of Maine is actually some 2000 miles long." It seems reasonable that the coast of Maine should be considered as being greater than that of a straight line drawn from border to border and that the added length due to harbors, bays, points, and so on should be included. However, once one begins to consider irregularities, the problem is when to stop. Figure XVI-1 shows how successive magnifications of a hypothetical stretch of coastline might look. Quite obviously there are indentations on indentations, and with each successive magnification one increases the computed coastline by a more or less constant factor.

There seems to be no logical stopping point to the sequence of successive enlargements if one is after some "absolute" value for the length of the shoreline. One thus arrives at the point of considering irregularities due to individual stones on the beach, due to the individual grains of sand, and due to the surface roughness of each grain and, finally, of considering the surface area of individual atoms and molecules. In our particular example, the tides and waves would prevent the exact carrying out of the sequence, but apart from this, the Heisenberg principle comes in eventually to inform us that there is considerable uncertainty in the location of electrons and that the "surface area" of an atom is a somewhat philosophical concept.

Quite obviously, the geographer solves this logical difficulty by arbitrarily deciding that he will not consider irregularities smaller than some specified size. A similar and equally arbitrary choice is frequently made in the estimation of surface areas. The minimum size of irregularities to be considered may be defined in terms of the resolving power of a microscope or the size of adsorbate molecules.†

Another approach is the iterative one. In the case of the Maine coastline, one might count the number of coastal villages or landing jetties and take this number as a measure of the length of a stretch of coast. In the case of a solid surface, one might apply a method that attempts to count the number of atoms or molecules in the surface. The result obtained is undoubtedly useful, but again, it does not have the quality of being a unique measure of absolute surface area. Perhaps the closest approach to such a measure of surface area is the geometric area of a plane liquid surface or of a perfect crystalline plane.

Finally, in the case of solids, there is the difficulty that surface atoms and molecules differ in their properties from one location to another. The discussion in Section VII-4 made clear the variety of surface heterogeneities possible in the case of a solid. Those measurements that depend on the state of surface atoms or molecules will generally be influenced differently by such heterogeneities. Different methods of measuring surface area will thus often not only give different absolute values, but may also give different *relative* values for a series of solids.

The preceding discussion serves to emphasize the point that, although the various topics to be discussed in this and succeeding chapters relate to the common property, surface area, the actual meaning of the answer arrived at is generally characteristic of the method and associated theory. Fortunately, one is very commonly more interested in relative surface areas than in accurate "absolute" values. Any given method will usually be valuable in determining ratios of areas between various materials (subject to the reservation made in the preceding paragraph). The comparisons, moreover, will be particularly valid when they are to be applied to the prediction or correlation of properties closely related to the one used in the measurement. Thus surface areas estimated by adsorption studies are more likely to lead to valid predictions of how additional, similar adsorbates will behave than to accurate predictions of, say, how the relative heats of immersion or the relative microscopic areas should vary.

B. Surfaces as Having a Fractal Geometry

Figure XVI-1 and the related discussion first appeared in 1960 [1], and since then a very useful mathematical approach to irregular surfaces has been applied to the matter of surface area measurement. Figure XVI-1 suggests that a coastline might appear similar under successive magnifications, and one now proceeds to assume that this similarity is *exact*. The result, as discussed in Section VII-4C and illustrated in Fig. VII-6, is a self-similar line, or in the present case, a *self-similar* surface. Equation VII-21 now applies and may be written in the form

---

†Reference 9 shows a needlepoint embroidery (by V. L. Adamson) illustrating the point that a large adsorbate molecule would "see" less of a rough surface than would a small one.

$$\Sigma = N(\sigma)\sigma = C'\sigma^{(1-D/2)} \tag{XVI-1}$$

where $N(\sigma)$ is the number of adsorbate molecules of area σ needed to "cover" one gram of the surface. For a perfect plane surface, $D = 2$ and $\Sigma = C'$, that is, it does not depend on the area of the adsorbate molecules. In the case of a self-similar surface such that $2 < D < 3$, Σ will increase with decreasing σ and D may be obtained from a plot of $\ln \Sigma$ versus $\ln \sigma$. To the extent that Eq. XVI-1 is obeyed, C' rather than Σ now characterizes the adsorbent. The specific surface area Σ will also vary with the particle size, and a more general equation is

$$\Sigma = C''\sigma^{(1-D/2)}R^{(D-3)} \tag{XVI-2}$$

where R is the particle radius. If the particle surface is smooth, $D = 2$ and Σ varies inversely as R, as expected. In the limit of very porous particles, $D \rightarrow 3$, however, and Σ becomes independent of the particle size. A plot of $\ln \Sigma$ versus $\ln R$ thus provides an alternative method for determining D. The question of what σ to use is discussed in Ref. 2.

The above methods for obtaining D, as well as other ones, are reviewed in Refs. 3–12, and Refs. 7–9 give tables of D values for various adsorbents. For example, D is close to 3 for the highly porous silica gels and close to 2 for nonporous fumed silica and for graphitized carbon black; coconut charcoal and alumina were found to have D values of 2.67 and 2.79, respectively [7].

The chemical reactivity of a self-similar surface should vary with its fractional dimension. Consider a reactive molecule that is approaching a surface to make a "hit." Taking Fig. VII-6d as an illustration, it is evident that such a molecule can "see" only a fraction of the surface. The rate of dissolving of quartz in HF, for example, is proportional to $R^{(D_r - 3)}$, where D_r, the reactive dimension, is about the same as D determined from adsorption studies (see Refs. 7–12). Values of D have also been obtained from bimolecular energy transfer reactions where acceptor and donor molecules must make an encounter by diffusing along the surface and also from the variation of x-ray scattering with scattering angle (see Refs. 4–15). The common factor in all of these approaches is that an analysis of the change in some measurable quantity as the yardstick size (adsorbent size, adsorbate size, diffusion distance, scattering distance) is varied.

It might be thought that self-similarity is too demanding a mathematical property to be likely to occur in nature, and some reservations are, indeed, appropriate. If the range is small, data may only appear to fit the linear log–log plot used to obtain a D value. Other explanations, including surface chemical heterogeneity, may be important. Yet it does appear that approximate self-similarity over a fair range of yardstick size is not uncommon. Nor is it actually unexpected. Computer simulations of particle growth by nonequilibrium accretion or aggregation processes do lead to self-similar surfaces, for example (see Refs. 11–18).

C. Methods Requiring Knowledge of the Surface Free Energy or Total Energy

A number of methods have been described in earlier sections whereby the surface free energy or total energy could be estimated. Generally, it was necessary to assume that the surface area was known by some other means; conversely, if some estimate of the specific thermodynamic quantity is available, the application may be reversed to give a surface area determination. This is true if the heat of solution of a powder (Section VII-5B), its heat of immersion (Section X-3A), or its solubility increase (Section X-2) are known.

Two approaches of this type, purporting to give absolute surface areas, might be mentioned. Bartell and Flu [19] proposed that the heat of immersion of a powder in a given liquid

$$q_{imm} = \Sigma(E_{SV} - E_{SL}) \tag{XVI-3}$$

can be combined with data on the temperature dependence of the contact angle for that liquid to obtain the area Σ of the sample. Thus from Eq. X-39,

$$q_{imm} = \Sigma\left(E_L \cos\theta - T\gamma_{LV} \frac{d\cos\theta}{dT} \right) \tag{XVI-4}$$

where q_{imm} is on a per gram basis so that Σ is area per gram. The actual use of Eq. XVI-4 to obtain Σ values is far from easy (see Ref. 20).

The general type of approach, that is, the comparison of an experimental heat of immersion with the expected value per square centimeter, has been discussed and implemented by numerous authors [21,22]. It is possible, for example, to estimate $E_{SV} - E_{SL}$ from adsorption data or from the so-called isosteric heat of adsorption (see Section XVII-12B). In many cases where approximate relative areas only are desired, as with coals or other natural products, the heat of immersion method has much to recommend it. In the case of microporous adsorbents surface areas from heats of immersion can be larger than those from adsorption studies [23], but the former are the more correct [24].

A second method that has been suggested as giving absolute surface areas involves the following procedure. If the solid is first equilibrated with saturated vapor, the molar free energy of the adsorbed material must be equal to that of the pure liquid, and it is reasonable (but nonthermodynamic) argument that therefore the interfacial energy E_L for the adsorbed film-vapor interface will be the same as for the liquid-vapor interface. Note that the adsorbed film is thought of as being duplex in nature in that the film-vapor interface is regarded as distinct from the film-solid interface. If the solid is now immersed in pure liquid adsorbate, the film-vapor interface is destroyed, and the heat liberated should correspond to $\mathcal{A} E_L$, thus giving the area $\mathcal{A}$.

This assumption was invoked by Harkins and Jura [21] in applying the method, with some success, to a nonporous powder. A concern, however, is that if the adsorbed film is thick, interparticle and capillary condensation will occur to give too low an apparent area. Roquerol and co-workers, however, have shown that in the case of a ground quartz–water system, the heat of immersion became constant after only about 1.5 statistical monolayers of adsorbed water [25]; see also Ref. 24.

D. Rate of Dissolving

A rather different method from the preceding is that based on the rate of dissolving of a soluble material. At any given temperature, one expects the initial dissolving rate to be proportional to the surface area, and an experimental verification of this expectation has been made in the case of rock salt (see Refs. 26, 27). Here, both forward and reverse rates are important, and the rate expressions are

$$R_f = -\frac{d(\text{NaCl})}{dt} = k_1 \mathcal{A} \qquad\qquad \text{(XVI-5)}$$

$$R_b = \frac{d(\text{NaCl})}{dt} = k_2 \mathcal{A} a_{\text{NaCl}} \qquad\qquad \text{(XVI-6)}$$

where a_{NaCl} denotes the activity of the dissolved salt. By employing a single cubic piece of rock salt, it was possible to confirm the above rate expressions.

In such cases, the rate constants k_1 and k_2 are found to depend on the degree of stirring, although their ratio at equilibrium, of course, does not. The rate-controlling step is therefore probably one of diffusion through a boundary liquid film, since the liquid layers next to the solid surface will possess very little velocity normal to the surface. In dissolving, a concentration C_0, equal to that of the saturated solution, is built up in the layer immediately adjacent to the solid, whereas at some distance τ away the concentration is equal to that of the bulk solution C. The dissolving rate is determined by the rate of diffusion of solute across this concentration gradient.

Although the rate of dissolving measurements do thus give a quantity identified as the total surface area, this area must include that of a film whose thickness is on the order of a few micrometers but basically is rather indeterminate. Areas determined by this procedure thus will not include microscopic roughness (or fractal nature).

The slow step in dissolving, alternatively, may be that of the chemical process of dissolving itself. This apparently was the case in a study [28] of the rate of dissolving of silica by hydrofluoric acid. The method was calibrated by determining the rate of dissolving of silica rods of geometrically determined area and was then applied to samples of powdered material. The rate was not affected by stirring rate and, moreover, the specific surface area of the silica powder so obtained, about 0.5 m^2/g, was found to agree well with that estimated from adsorption isotherms. As noted in Section XVI-2B, however, chemical dissolving rate constants may vary with particle size.

E. The Mercury Porosimeter

The method to be described determines the pore size distribution in a porous material or compacted powder; surface areas may be inferred from the results, however. The method is widely used and, moreover, does make use of one of the fundamental equations of surface chemistry, the Laplace equation (Eq. II-7). It will be recalled that Bartell and co-workers [29,30] determined an average pore radius from the entry pressure of a liquid into a porous plug (Section X-6A).

A procedure that is more suitable for obtaining the actual distribution of pore sizes involves the use of a nonwetting liquid such as mercury—the contact angle on glass being about 140° (Table X-2) (but note Ref. 31). If all pores are equally accessible, only those will be filled for which

$$r > \frac{2\gamma |\cos \theta|}{P} \qquad\qquad \text{(XVI-7)}$$

so that each increment of applied pressure causes the next smaller group of pores to be filled, with a concomitant increase in the total volume of mercury penetrated into the solid. See Refs. 32–35 for experimental details; commercial equipment is available (see Refs. 36, 37).

The analysis of the direct data, namely, volume penetrated versus pressure, is as follows. Let dV be the volume of pores of radii between r and $r - dr$; dV will be related to r by some distribution function $D(r)$:

$$dV = -D(r)\, dr \qquad\qquad \text{(XVI-8)}$$

For constant θ and γ (the contact angle was found not to be very dependent on pressure), one obtains from the Laplace equation,

$$\mathbf{P}\, dr + r\, d\mathbf{P} = 0 \qquad\qquad \text{(XVI-9)}$$

$$\Rightarrow dr = -\frac{dP}{P}\, r$$

which, in combination with Eq. XVI-8, gives

$$dV = D(r)\, \frac{r}{\mathbf{P}}\, d\mathbf{P} = D(r)\, \frac{2\gamma \cos \theta}{\mathbf{P}^2}\, d\mathbf{P} \qquad\qquad \text{(XVI-10)}$$

$$r = \frac{2\gamma \cos\theta}{P}$$

or

$$D(r) = \frac{(\mathbf{P}/r)\, dV}{d\mathbf{P}} \qquad\qquad \text{(XVI-11)}$$

Thus $D(r)$ is given by the slope of the V versus $\mathbf{P}$ plot. The same distribution function can be calculated from an analysis of vapor adsorption data showing hysteresis due to capillary condensation (see Section XVII-16). Joyner and co-workers [38] found that the two methods gave very similar results in the case of charcoal, as illustrated in Fig. XVI-2. See Refs. 36 and 39 for more recent such comparisons. There can be some question as to what the local contact angle is [31,40]; an error here would shift the distribution curve.

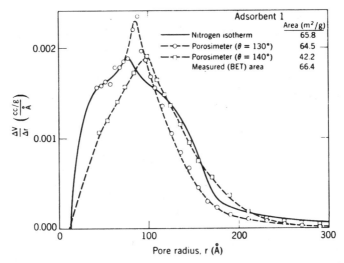

	Adsorbent 1
	Area (m^2/g)
—— Nitrogen isotherm	65.8
—O— Porosimeter ($\theta = 130°$)	64.5
—□— Porosimeter ($\theta = 140°$)	42.2
Measured (BET) area	66.4

Fig. XVI-2. Comparison of the pore volume distribution curves obtained from porosimeter data assuming contact angles of 140° and 130° with the distribution curve obtained by the isotherm method for a charcoal. (From Ref. 38.)

The foregoing analysis regards the porous solid as equivalent to a bundle of capillaries of various size, but apparently not very much error is introduced by the fact that such solids in reality consist of interconnected channels, provided that all pores are equally accessible (see Refs. 41, 42); by this it is meant that access to a given-sized pore must always be possible through pores that are as large or larger. An extreme example of the *reverse* situation occurs in an "ink bottle" pore, as illustrated in Fig. XVI-3. These are pores that are wider in the interior than at the exit, so that mercury cannot enter until the pressure has risen to the value corresponding to the radius of the entrance capillary. Once this pressure is realized, however, the entire space fills, thus giving an erroneously high apparent pore volume for capillaries of that size. Such a situation should also lead to a hysteresis effect, that is, on reducing the pressure, mercury would leave the entrance capillary at the appropriate pressure, but the mercury in the ink-bottle part would be trapped. Indeed, there is often some hysteresis in that

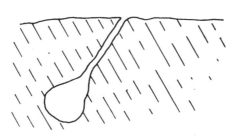

Fig. XVI-3. An ink bottle pore.

the plot of V versus P is not retraced on depressurization (see Ref. 43). In some cases, hysteresis can be explained as being due to a difference between the advancing and receding contact angle [44].

Interestingly, a general thermodynamic relationship allows the surface area of a porous system (without ink bottles) to be calculated from porosimetry data, note Section XVII-16B. The equation is [45]

$$\mathcal{A} = -\frac{1}{\gamma \cos \theta} \int_0^{V_m} P \, dV \qquad \text{(XVI-12)}$$

where V_m is the total volume penetrated. There is good agreement between areas obtained this way and from nitrogen adsorption [37].

F. Other Methods of Surface Area Estimation

There is a number of other ways of obtaining an estimate of surface area, including such obvious ones as direct microscopic or electron-microscopic examination. The rate of charging of a polarized electrode surface can give relative areas. Bowden and Rideal [46] found, by this method, that the area of a platinized platinum electrode was some 1800 times the geometric or apparent area. Joncich and Hackerman [47] obtained areas for platinized platinum very close to those given by the BET gas adsorption method (see Section XVII-5). The diffuseness of x-ray diffraction patterns can be used to estimate the degree of crystallinity and hence particle size [48,49]. One important general approach, useful for porous media, is that of permeability determination; although somewhat beyond the scope of this book, it deserves at least a brief mention.

A simple law, known as Darcy's law (1956), states that the volume flow rate per unit area is proportional to the pressure gradient if applied to the case of viscous flow through a porous medium treated as a bundle of capillaries,

$$Q = \frac{r^2 A}{8\eta} \frac{\Delta P}{l} = K \frac{A \Delta P}{\eta l} \qquad \text{(XVI-13)}$$

where Q is the volume flow rate, r is the radius of the capillaries, A is the total cross-sectional area of the porous medium, l is its length, ΔP is the pressure drop, η is the viscosity of the fluid, and K is called the permeability of the medium.

In applying Eq. XVI-13 to an actual porous bed, r is taken to be proportional to the volume of void space $A l \epsilon$, where ϵ is the porosity, divided by the amount of surface; alternatively, then,

$$r = \frac{k \epsilon}{A_0 (1 - \epsilon)} \qquad \text{(XVI-14)}$$

where A_0 is the specific surface area of the particles that make up the porous bed, that is, the area divided by the volume of the particles $A l (1 - \epsilon)$. With the further assumption that the effective or open area is ϵA, K in Eq. XVI-13 becomes

$$K = \frac{1}{8} \left(\frac{k}{A_0} \right)^2 \frac{\epsilon^3}{(1 - \epsilon)^2} \qquad (\text{XVI-15})$$

The detailed consideration of these equations is due largely to Kozeny [50]; the reader is also referred to Collins [51]. However, it is apparent that, subject to assumptions concerning the topology of the porous system, the determination of K provides an estimate of A_0. It should be remembered that A_0 will be the external area of the particles and will not include internal area due to pores (note Ref. 52). Somewhat similar equations apply in the case of gas flow; the reader is referred to Barrer [53] and Kraus and co-workers [54].

3. The Structural and Chemical Nature of Solid Surfaces

We have considered briefly the important macroscopic description of a solid adsorbent, namely, its specific surface area, its possible fractal nature, and if porous, its pore size distribution. In addition, it is important to know as much as possible about the microscopic structure of the surface, and contemporary surface spectroscopic and diffraction techniques, discussed in Chapter VIII, provide a good deal of such information (see also Refs. 55 and 56 for short general reviews, and the monograph by Somorjai [57]). Scanning tunneling microscopy (STM) and atomic force microscopy (AFT) are now widely used to obtain the structure of surfaces and of adsorbed layers on a molecular scale (see Chapter VIII, Section XVIII-2B, and Ref. 58). On a less informative and more statistical basis are site energy distributions (Section XVII-14); there is also the somewhat large-scale type of structure due to surface imperfections and dislocations (Section VII-4D and Fig. XVIII-14).

The composition and chemical state of the surface atoms or molecules are very important, especially in the field of heterogeneous catalysis, where mixed-surface compositions are common. This aspect is discussed in more detail in Chapter XVIII (but again see Refs. 55, 56). Since transition metals are widely used in catalysis, the determination of the valence state of surface atoms is important, such as by ESCA, EXAFS, or XPS (see Chapter VIII and note Refs. 59, 60).

Many solids have foreign atoms or molecular groupings on their surfaces that are so tightly held that they do not really enter into adsorption-desorption equilibrium and so can be regarded as part of the surface structure. The partial surface oxidation of carbon blacks has been mentioned as having an important influence on their adsorptive behavior (Section X-3A); depending on conditions, the oxidized surface may be acidic or basic (see Ref. 61), and the surface pattern of the carbon rings may be affected [62]. As one other example, the chemical nature of the acidic sites of silica-alumina catalysts has been a subject of much discussion. The main question has been whether the sites represented Brønsted (proton donor) or Lewis (electron–acceptor) acids. Hall et al. [63] have addressed this type of question in the case of zeolites; see also Refs. 64 and 65.

4. The Nature of the Solid–Adsorbate Complex

Before entering the detailed discussion of physical and chemical adsorption in the next two chapters, it is worthwhile to consider briefly and in relatively general terms what type of information can be obtained about the chemical and structural state of the solid–adsorbate complex. The term *complex* is used to avoid the common practice of discussing adsorption as though it occurred on an inert surface. Three types of effects are actually involved: (1) the effect of the adsorbent on the molecular structure of the adsorbate, (2) the effect of the adsorbate on the structure of the adsorbent, and (3) the character of the direct bond or local interaction between an adsorption site and the adsorbate.

A. Effect of Adsorption on Adsorbate† Properties

Statistical Thermodynamics of Adsorbates. First, from a thermodynamic or statistical mechanical point of view, the internal energy and entropy of a molecule should be different in the adsorbed state from that in the gaseous state. This is quite apart from the energy of the adsorption bond itself or the entropy associated with confining a molecule to the interfacial region. It is clear, for example, that the adsorbed molecule may lose part or all of its freedom to rotate.

It is of interest in the present context (and is useful later) to outline the statistical mechanical basis for calculating the energy and entropy that are associated with rotation [66]. According to the Boltzmann principle, the time average energy of a molecule is given by

$$\bar{\epsilon} = \frac{\sum N_j \epsilon_j}{\sum N_j} = \frac{\sum g_j \epsilon_j e^{-\epsilon_j/kT}}{\sum g_j e^{-\epsilon_j/kT}} \qquad \text{(XVI-16)}$$

where the summations are over the energy states ϵ_j, and g_j is the statistical weight of the jth state. It is the assumption of statistical thermodynamics that this time–average energy for a molecule is the same as the instantaneous average energy for a system of such molecules. The average molar energy for the system, E, is taken to be $N_0 \bar{\epsilon}$. Equation XVI-16 may be written in the form

$$E = RT^2 \frac{\partial \ln Q}{\partial T} \qquad \text{(XVI-17)}$$

---

†An alternative term for *adsorbate* is *adsorptive*; it is more commonly used in European literature than in the U.S. literature.

where $\mathbf{Q}$ is the partition function and is defined as the term in the denominator of Eq. XVI-16. The heat capacity is then

$$C_v = \left(\frac{\partial E}{\partial T}\right)_V = \frac{R}{T^2}\frac{\partial^2 \ln \mathbf{Q}}{\partial (1/T)^2} \qquad \text{(XVI-18)}$$

from which the entropy is found to be

$$S = \int_0^T C_v \, d\ln T = \frac{E}{T} + R\ln \mathbf{Q} \qquad \text{(XVI-19)}$$

if the entropy at 0 K is taken to be zero. The Helmholtz free energy A is just $E - TS$, so

$$A = -RT\ln \mathbf{Q} \qquad \text{(XVI-20)}$$

The rotational energy of a rigid molecule is given by $J(J + 1)h^2/8\pi^2 IkT$, where J is the quantum number and I is the moment of inertia, but if the energy level spacing is small compared to kT, integration can replace summation in the evaluation of $\mathbf{Q}_{rot}$, which becomes

$$\mathbf{Q}_{rot} = \frac{1}{\pi\sigma}\left(\frac{8\pi^3 IkT}{h^2}\right)^{n/2} \qquad \text{(XVI-21)}$$

Equation XVI-21 provides for the general case of a molecule having n independent ways of rotation and a moment of inertia I that, for an asymmetric molecule, is the (geometric) mean of the principal moments. The quantity σ is the symmetry number, or the number of indistinguishable positions into which the molecule can be turned by rotations. The rotational energy and entropy are [66,67]

$$E_{rot} = \frac{nRT}{2} \qquad \text{(XVI-21a)}$$

$$S_{rot} = R\left\{\ln\left[\frac{1}{\pi\sigma}\left(\frac{8\pi^3 IkT}{h^2}\right)^{n/2}\right] + \frac{n}{2}\right\} \qquad \text{(XVI-22)}$$

Molecular moments of inertia are about 10^{-39} g/cm^2; thus I values for benzene, N_2, and NH_3 are 18, 1.4, and 0.28, respectively, in those units. For the case of benzene gas, $\sigma = 6$ and $n = 3$, and S_{rot} is about 21 cal K^{-1} mol^{-1} at 25°C. On adsorption, all of this entropy would be lost if the benzene were unable to rotate, and part of it if, say, rotation about only one axis were possible (as might be the situation if the benzene was subject only to the constraint of lying flat on the surface). Similarly, all or part of the rotational energy of 0.88 kcal/mol would be liberated on adsorption or perhaps converted to vibrational energy.

Vibrational energy states are too well separated to contribute much to the entropy or the energy of small molecules at ordinary temperatures, but for higher temperatures this may not be so, and both internal entropy and energy changes may occur due to changes in vibrational levels on adsorption. From a somewhat different point of view, it is clear that even in physical adsorption, adsorbate molecules should be polarized on the surface (see Section VI-8), and in chemisorption more drastic perturbations should occur. Thus internal bond energies of adsorbed molecules may be affected.

Vibrational and Related Spectroscopies. The determination of the characteristic vibrational frequencies of adsorbed molecules is useful in at least two general ways. The spectrum can serve to identify what molecular species are present, which is important when a chemical reaction may have occurred; changes in molecular symmetry and bond strengths may manifest themselves as well as the new adsorbent–adsorbate bond that has formed. A common experimental approach is simply that of infrared absorption spectroscopy—see Refs. 68 and 69 and, more recently, Refs. 57, 70, and 71. As an example, the proportion of Brønsted and Lewis acid sites on a promoted ZrO_2 catalyst was determined using pyridine as an indicator [71a]. While transmission infrared spectroscopy can often be used for high-surface-area oxide-type adsorbents, in other cases (such as with metal adsorbents) RAIRS (reflection–absorption infrared spectroscopy) or EELS (electron-energy-loss spectroscopy) or VEELS (vibrational EELS) is necessary. Figure XVI-4 shows the RAIRS spectrum for propene adsorbed on a Pt(111) surface along with transmission IR spectra for propene and propane adsorbed on SiO_2-supported Pt. Of interest was the spectral evidence for the presence of surface propylidyne. Other examples are found in Section XVIII-2C, and see Ref. 72 for a review on EELS.

An interesting point is that infrared absorptions that are symmetry-forbidden and hence that do not appear in the spectrum of the gaseous molecule may appear when that molecule is adsorbed. Thus Sheppard and Yates [74] found that normally forbidden bands could be detected in the case of methane and hydrogen adsorbed on glass; this meant that there was a decrease in molecular symmetry. In the case of the methane, it appeared from the band shapes that some reduction in rotational degrees of freedom had occurred. Figure XVII-16 shows the IR spectrum for a physisorbed H_2 system, and Refs. 69 and 75 give the IR spectra for adsorbed N_2 (on Ni) and O_2 (in a zeolite), respectively.

Vibrational spectra of adsorbed species may also be obtained by Raman spectroscopy (see Refs. 75, 76, and 78). Here, irradiation in the wavelength region of an adsorbate electronic absorption band can show features representing concomitant changes in vibrational states, or a resonance Raman effect. Figure XVI-5 shows spectra obtained with *p*-nitrosodimethylaniline, NDMA, as the absorbing or probe molecule, which was adsorbed on ZnO [79]. The intensity changes of the 1600, 1445, and 1398 cm^{-1} peaks indicated that on adsorption of NH_3, the NDMA was displaced from the acidic sites that it originally occupied.

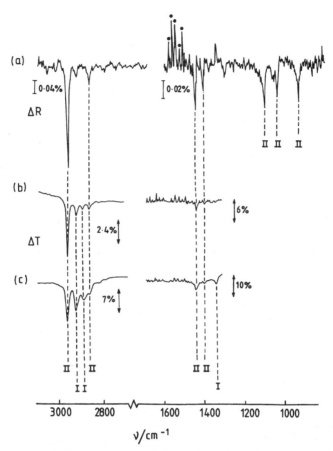

Fig. XVI-4. (*a*) RAIRS for propene on Pt(III) at 340 K. (*b*) Transmission IR of propane on Pt/SiO$_2$ at ~300 K; (*c*) same as (*b*) but with propene adsorbed. II denotes propylidyne assignments. (From Ref. 73.) (Reproduced by permission of Royal Society of Chemistry.)

See Refs. 80 and 81 for other examples. Surface-enhanced Raman spectroscopy is discussed in Section XVI-4C.

Information about molecular perturbations may also be found from UV-visible spectroscopy of adsorbed species. Kiselev and co-workers [82] observed shifts of 200–300 cm^{-1} (about 700 cal/mol) to longer wavelengths in the ultraviolet absorption spectra of benzene and other aromatic species on adsorption on Aerosil (a silica). Leermakers and co-workers [83] made extensive studies of the spectra and photochemistry of adsorbed organic species. The absorption spectrum of Ru(bipyridine)$_3^{2+}$ is shifted slightly on adsorption on silica gel [84]. Absorption peak positions give only the difference in energy between the ground and excited states, so the shifts, while relatively small, could reflect larger but parallel changes in the absolute energies of the two states. In other

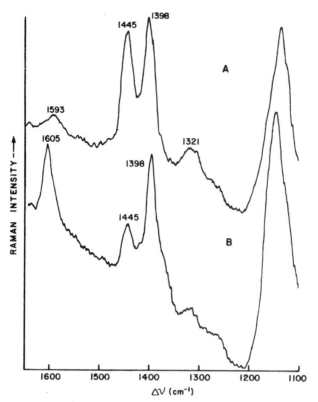

Fig. XVI-5. Resonance Raman spectra of NDMA adsorbed on ZnO; (*a*) in the presence of 100 torr of NH_3; (*b*) after evacuation of the NH_3 from the cell. [Reprinted with permission from J. F. Brazdil and E. B. Yeager, *J. Phys. Chem.*, **85**, 1005 (1981) (Ref. 79). Copyright 1981, American Chemical Society.]

words, the shifts suggest that appreciable changes in the ground-state energies occurred on adsorption. As with the lost rotational energy, these would appear as contributions to the heat of adsorption.

Electron Spin Resonance Spectroscopy. Several ESR studies have been reported for adsorption systems [85–90]. ESR signals are strong enough to allow the detection of quite small amounts of unpaired electrons, and the shape of the signal can, in the case of adsorbed transition metal ions, give an indication of the geometry of the adsorption site. Ref. 91 provides a contemporary example of the use of ESR and of electron spin echo modulation (ESEM) to locate the environment of Cu(II) relative to ^{31}P in a microporous aluminophosphate molecular sieve.

Nuclear Magnetic Resonance. NMR studies are limited to atoms having a nonzero nuclear magnetic moment, common ones being ^1H, ^2H, ^{13}C, ^{15}N. ^{17}O, ^{19}F, ^{29}Si, ^{31}P, and more esoteric ones such as ^{95}Mo and ^{129}Xe. There are two types of problems. One is that the intensity of the NMR signal may not be strong enough unless high surface area adsorbents are used, such as a silica gel or a zeolite, and the other is that line widths tend to be such as to obscure chemical shifts (but see further below). The early work concentrated on proton NMR (e.g., Ref. 92). In the case of n-heptane adsorbed on Spheron or diamond, the line width for the first adsorbed layer was very large, but with three or more adsorbed layers, a line structure developed [93]. Again, a narrow line is associated with sufficient mobility that local magnetic fields are averaged, and the interpretation was that adsorbed layers beyond the second were mobile. The longitudinal and transverse relaxation times T_1 and T_2 can be interpreted in terms of a correlation time t, which is about the time required for a molecule to turn through a radian or to move a distance comparable with its dimensions. Zimmerman and Lasater [94], studying water adsorbed on a silica gel, found evidence for two kinds of water for surface coverages greater than 0.5 and, as illustrated in Fig. XVI-6, a variation of calculated correlation time with coverage, which suggested that surface mobility was fully developed at about the expected monolayer point. A tentative phase identification can be made on the basis that t for water is about 10^{-11} sec at 25°C, while for ice (or "rigid" water) it is about 10^{-5} sec. That is, inspection of Fig. XVI-6 would suggest that below $\theta = 0.5$, the adsorbed water was solidlike, while above $\theta = 1$ it was liquidlike. As an example of the sophistication of more recent work, Dybowski et al. [95] studied the

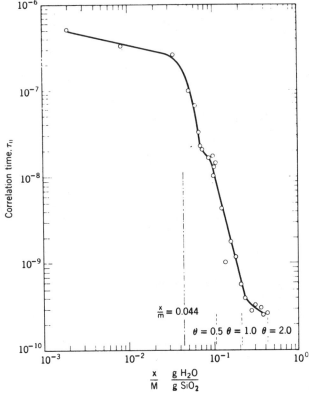

Fig. XVI-6. Nuclear correlation times for water adsorbed on silica gel. (From Ref. 94.)

heteronuclear coupling of ^{1}H and ^{195}Pt for benzene adsorbed on a Pt/Al$_2$O$_3$ catalyst to draw conclusions as to the nature of the adsorption site and the Pt–benzene distance. Some useful general references are those by Hunter, Dybowski, and co-workers [96–98].

There has been much interest in the mobility of species adsorbed on or in layer and cage minerals such as clays and zeolites (see Fig. XVII-26 for an illustrative structure of the latter). As one example, in the case of benzene adsorbed on a zeolite, ^{13}C NMR showed a line broadening (and hence correlation time) which increased with n, the number of molecules per cage, up to $n = 4$. Beyond this point there was a rapid line narrowing, indicating that a new, highly mobile phase was present [99]. In a study of chemisorbed methanol on MgO, ^{13}C NMR indicated molecules to be rigidly bound below a half monolayer coverage but that higher coverages produced isotropically rotating molecules [100]. As another illustration, zeolite rho is a relatively flexible zeolite with useful catalytic properties. Xenon-129 NMR spectroscopy indicates this flexibility and also that there can be rapid exchange between two kinds of adsorption sites [101–103]. Finally, it has been possible to determine the long-range diffusion coefficient of N$_2$ in commercial zeolite crystals, using pulsed field gradient NMR. Values of $\mathcal{D}$ ranging from 10^{-2} to 10^{-4} cm^2/sec were found [103a].

An important development has been that of *"magic angle" spinning* (MAS) [97]. A major source of line broadening is that of *dipole–dipole relaxation*, due to fluctuating local magnetic fields imposed on the externally applied field; the effect may be due to protons in the same molecule, for example. The rapid tumbling of molecules in fluid solution tends to average out the local field, giving sharp NMR lines. Molecules adsorbed on a solid, however, may not be able to tumble (or move around) on the NMR time scale—hence the typical line broadening that is observed. The local field B_{loc} at a nucleus A generated by a neighboring nucleus B is given by

$$B_{loc} = \pm\mu_B r_{AB}^{-3}(3\cos^2\theta - 1) \qquad \text{(XVI-23)}$$

where μ_B is the magnetic moment of B, r_{AB} the internuclear distance, and θ the angle between the internuclear vector and the applied field. The $(3\cos^2\theta - 1)$ term is zero for $\theta = 54°44'$. In solution, θ drifts rapidly over all values, averaging out the effect, but essentially the same result can be obtained with a solid sample by spinning it rapidly at the magic angle (about 3 Hz is needed). In many cases, relatively sharp lines can now be obtained. One may now see chemical shifts due to sites of differing acidity, for example [104]. Other examples are found in Refs. 96 and 105–109.

Dielectric Behavior of Adsorbed Water. Determination of the dielectric absorption of adsorbed water can yield conclusions similar to those from proton NMR studies and there is a considerable, although older literature on the subject. Figure XVI-7 illustrates how the dielectric constant for adsorbed water varies with the frequency used as well as with the degree of surface coverage. A characteristic relaxation time τ can be estimated

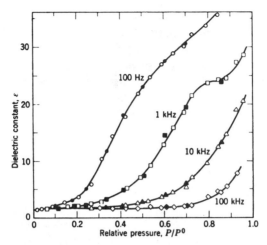

Fig. XVI-7. Dielectric isotherms of water vapor at 15°C adsorbed on α-Fe$_2$O$_3$ (solid points indicate desorption). A complete monolayer was present at $P/P^0 = 0.1$, and by $P/P^0 = 0.8$ several layers of adsorbed water were present. (From Ref. 110.)

from the frequency dependence of the dielectric constant, especially in combination with that of the dielectric loss. This relaxation time is essentially that for the molecule to move or to reorient so as to follow a changing electric field. For water on α-Fe$_2$O$_3$, τ varied from about 1 sec at monolayer coverage to 10^{-4} sec when several adsorbed layers were present [110]. Since the characteristic time was far larger than for liquid bulk water, 10^{-10} sec, the conclusion was that the adsorbed water was present in an icelike, hydrogen-bonded structure. A more recent study indicated adsorption steps for water on FeOOH-type oxides [111].

Similar, very detailed studies were made by Ebert [112] on water adsorbed on alumina with similar conclusions. Water adsorbed on zeolites showed a dielectric constant of only 14–21, indicating greatly reduced mobility of the water dipoles [113]. Similar results were found for ammonia adsorbed in Vycor glass [114]. Klier and Zettlemoyer [114a] have reviewed a number of aspects of the molecular structure and dynamics of water at the surface of an inorganic material.

The state of an adsorbate is often described as "mobile" or "localized," usually in connection with adsorption models and analyses of adsorption entropies (see Section XVII-3C). A more direct criterion is, in analogy to that of the fluidity of a bulk phase, the degree of mobility as reflected by the surface diffusion coefficient. This may be estimated from the dielectric relaxation time; Resing [115] gives values of the diffusion coefficient for adsorbed water ranging from near bulk liquids values (10^{-5} cm^2/sec) to as low as 10^{-9} cm^2/sec.

B. Effect of the Adsorbate on the Adsorbent

It is evident from the preceding material that a great deal of interest has centered on the chemical and physical state of the adsorbate. There is no reason not to expect the adsorbate to affect properties of the adsorbent. For example, as discussed in Section XVII-12Aii, adsorbent thermodynamic properties *must* change as adsorption occurs. Of more interest is the matter of adsorbent surface structural changes accompanying adsorption.

First, it is entirely possible that surface heterogeneities and imperfections undergo some reversible redistribution during adsorption. As noted in Section VII-4B, Dunning [116] has considered that since the presence of an adsorbed molecule should alter the energy of special sites (such as illustrated in Fig. VII-5), above some critical temperature for surface mobility, their distribution should depend on the extent of adsorption. Also, the first few layers of the crystalline surface of a solid are distorted (see Section VII-3B), and this distortion should certainly be altered if an adsorbed layer is present; here no more than motions perpendicular to the surface may be involved. There is a scattering of early literature indication that changes in surface structure, specifically, surface *reconstruction*, can occur on adsorption. In the 1960's Lander and Morrison [117] concluded from a LEED study that considerable surface rearrangement of germanium surfaces took place on adsorption of iodine. Similarly, field emission work with tungston led Ehrlich and Hudda [118] to conclude that reconstruction could accompany the adsorption or desorption of nitrogen. Hydrogen atom adsorption on metals can lead to complex surface reconstructions, as for H on Ni(110) [56] and on W(100) [119]. Somorjai [120] gives a somewhat more detailed account of such effects.

Some fascinating effects occur in the case of CO on Pt(100). As illustrated in Fig. XVI-8, the clean surface is reconstructed naturally into a quasi-hexagonal pattern, but on adsorption of CO, this reconstruction is lifted to give the bulk termination structure of (100) planes [56]. As discussed in Section XVIII-9E very complicated changes in surface structure occur on the oxidation of CO

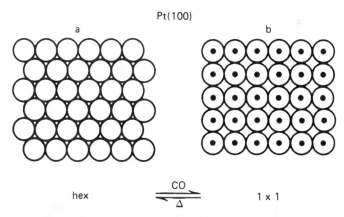

Fig. XVI-8. (*a*) The quasi-hexagonal surface structure of clean Pt(100) surface. (*b*) Adsorption of CO lifts this reconstruction to give the structure corresponding to the termination of (100) planes (from LEED studies). [Reprinted with permission from G. Ertl, *Langmuir*, **3**, 4 (1987) (Ref. 56). Copyright 1987, American Chemical Society.]

on various metal surfaces. The picture that emerges is that the adsorbent is not just a chemically active platform for catalyzed reaction but may execute a complicated pas à deux with the adsorbate.

As discussed in connection with Table XVI-1, one can expect important surface structural changes to accompany adsorption whenever the adsorbent-adsorbate bond is comparable in energy to the heat of sublimation of the adsorbent, that is, to the adsorbent-adsorbent bond energy. In the case of refractory solids, such as metals and ionic crystals and oxides, sublimation energies are in the range of 20–100 kcal/mol, which is also that for chemisorption. One thus expects surface reconstruction to be likely under chemisorption conditions. The energy of physical adsorption, which is due to van der Waals forces, is typically around 5–10 kcal/mol, about the same as the cohesion energy of molecular solids and of chain-chain interactions in polymers. One concludes that physical adsorption should not perturb the structure of a refractory adsorbent but should be able to do so in the case of a molecular solid. As an example, surface restructuring is indicated when n-hexane is adsorbed on ice above $-35°C$ [121,122]. The surface structure of polymers has been estimated from LEIS (see p. 308) measurements; here, one can expect the surface packing of polymer chains to be affected when physical adsorption of a vapor occurs.

C. The Adsorbate–Adsorbent Bond

The immediate site of the adsorbent-adsorbate interaction is presumably that between adjacent atoms of the respective species. This is certainly true in chemisorption, where actual chemical bond formation is the rule, and is largely true in the case of physical adsorption, with the possible exception of multilayer formation, which can be viewed as a consequence of weak, long-range force fields. Another possible exception would be the case of molecules where some electron delocalization is present, as with aromatic ring systems.

It has been customary to assign all of the observed heat of adsorption to the adsorption bond, although the preceding discussion has made it clear that other, less localized contributions can be important. Furthermore, the matter is complicated by the surface reconstruction that can occur. The various spectroscopies discussed in Section XVIII-2 give some information about chemisorption bond energies and bond lengths, and the chemisorption bond itself is covered in Section XVIII-6. There are no great surprises but there are complications if the adsorbate has more than one possible bonding atom and if bridging bonds must be considered, as in Fig. XVIII-5.

The physisorption bond, being relatively weak, is even more difficult to characterize than the chemisorption one. Some aspects of this are covered in Section XVII-10.

SERS. A phenomenon that certainly involves the adsorbent-adsorbate interaction is that of surface-enhanced resonance Raman spectroscopy, or SERS. The basic observation is that for pyridine adsorbed on surface-roughened silver, there is an amazing enhancement of the resonance Raman intensity (see Refs. 124–128). More recent work has involved other adsorbates and colloidal silver particles [129–132]. The intensity enhancement, around a factor of 10^5, has been attributed to a large electric fields generated by oscillations of the electrons in small metal structures [129,133,133a]. There is also the possibility that much of the effect is due to the actual surface area being larger than thought (see Ref. 134).

5. Problems

1. The following data have been obtained for powdered periclase (see Ref. 2) (mean particle diameter, μm; surface area, m^2/g): (3.39; 2.280), (4.77; 1.541), (8.03; 0.854), (11.8; 0.543), (18.0; 0.421), (22.6; 0.295). Make the appropriate plot and determine D.

2. The following data have been obtained for a sample of coal mine dust (see Ref. 2) (mean particle diameter, μm; surface area, m^2/g): (2.40; 8.61), (2.82; 6.93), (5.95; 5.28), (8.00; 4.51). Make the appropriate plot and obtain D. Comment on the result.

3. The following data have been obtained for a succession of adsorbates on a silica gel (see Ref. 4) (adsorbate; n, moles adsorbed at monolayer point; σ, molecular area of adsorbate, $Å^2$): (N_2; 5.29; 16.2), (CH_3OH; 4.09; 17.8), (CH_3CH_2OH; 2.82; 22.9), [$(CH_3)_3COH$; 1.80; 31.6], [$(CH_3)_2(CH_3CH_2)COH$; 1.50; 37.5]), [$(CH_3CH_2)_3COH$; 1.26; 42.7]. Make the appropriate plot and obtain D. Comment on the result.

4. Derive Eq. XVI-12.

5. In a mercury porosimetry experiment γ_{Hg} is taken to be 480 dyn/cm, θ is 130°. If pressure is measured in psi and volume intruded in cm^3/g, show that Eq. XVI-12 becomes $\Sigma = 0.0223 \int_0^v P\,dV$, where Σ is in m^2/g. The following data were obtained for a Sterling FT carbon (volume intruded, cm^3/g; applied pressure, psi): (0; 2), (0.2; 2.05), (0.35; 2.1), (0.36; 5), (0.37; 10), (0.39; 35), (0.52; 100), (0.85; 500), (1.10; 1000). Estimate the specific surface area of this sample.

6. The contact angle for water on single-crystal naphthalene is 87.7° at 35°C, and $d\theta/dT$ is −0.13 deg/K. Using data from Table III-1 as necessary, calculate the heat of immersion of naphthalene in water in cal/g if a sample of powdered naphthalene of 10 m^2/g is used for the immersion study. (Note Ref. 135.)

7. Bartell and Fu [19] were able to determine the adhesion tension, that is, $\gamma_{SV}-\gamma_{SL}$, for the water-silica interface to be 82.8 $ergs/cm^2$ at 20°C and its temperature change to be −0.173 erg cm^{-2} K^{-1}. The heat of immersion of the silica sample in water was 15.9 cal/g. Calculate the surface area of the sample in square centimeters per gram.

8. Harkins and Jura [21] found that a sample of TiO_2 having a thick adsorbed layer of water on it gave a heat of immersion in water of 0.600 cal/g. Calculate the specific surface area of the TiO_2 in square centimeters per gram.

9. The rate of dissolving of a solid is determined by the rate of diffusion through a boundary layer of solution. Derive the equation for the net rate of dissolving. Take C_0 to be the saturation concentration and d to be the effective thickness of the diffusion layer; denote diffusion coefficient by $\mathcal{D}$.

10. Using the curve given by the square points in Fig. XVI-2, make the qualitative reconstruction of the original data plot of volume of mercury penetrated per gram versus applied pressure.

11. Optical microscopic examination of a finely divided silver powder shows what

11. Optical microscopic examination of a finely divided silver powder shows what appears to be spheres 2 μm in diameter. The density of the material is 4 g/cm^3. The surface is actually rather rough on a submicroscopic scale, and suppose that a cross section through the surface actually has the appearance of the coastline shown in Fig. XVI-1, where the left arrow represents a distance of 0.1 μm and the successive enlargements are each by 10-fold.

Make a numerical estimate, with an explanation of the assumptions involved, of the specific surface area that would be found by (a) a rate of dissolving study, (b) Harkins and Jura, who find that at P^0 the adsorption of water vapor is 6.5 cm^3 STP/g (and then proceed with a heat of immersion measurement), and (c) a measurement of the permeability to liquid flow through a compacted plug of the powder.

12. Calculate the rotational contribution to the entropy of adsorption of benzene on carbon at 35°C, assuming that the adsorbed benzene has one degree of rotational freedom.

13. Calculate the rotational contribution to the entropy of adsorption of ammonia on silica at -30°C, assuming (a) that the adsorbed ammonia retains one degree of rotational freedom and (b) that it retains none. In case (a) assume that the nitrogen is bonded to the surface.

14. Explain why B_{loc} in Eq. XVI-23 should average to zero for a freely tumbling molecule.

15. According to Eq. XVI-2 the specific surface area of an adsorbate should vary with particle size R as R^{D-3}, using the same adsorbate for each sample. Justify this relationship by means of a qualitative argument (an exact derivation is not required).

General References

D. Avnir, ed., *The Fractal Approach to Heterogeneous Chemistry*, Wiley, New York, 1989.

Advances in Catalysis, Vol. 10, Academic, New York, 1958; Vol. 12, 1960; Vol. 16, 1966.

R. K. Chang and T. E. Furtek, eds., *Surface Enhanced Raman Scattering*, Plenum, New York, 1982.

R. J. Clark and R. E. Hester, eds., *Advances in Infrared and Raman Spectroscopy*, Vol. 9, Heyden, London, 1982.

R. E. Collins, *Flow of Fluids through Porous Materials*, Reinhold, New York, 1961.

C. Dyhowski and R. L. Lichter, eds., *NMR Spectroscopy Techniques*, Marcel Dekker, New York, 1987.

A. E. Flood, ed., *The Solid-Gas Interface*, Vol. 1, Marcel Dekker, New York, 1967.

S. J. Gregg and K. S. W. Sing, *Adsorption, Surface Area, and Porosity*, Academic, New York, 1982.

M. L. Hair, *Infrared Spectroscopy in Surface Chemistry*, Marcel Dekker, New York, 1967.

T. L. Hill, *Introduction to Statistical Thermodynamics*, Addison-Wesley, Reading, MA, 1962.

R. R. Irani and C. F. Callis, *Particle Size*, Wiley, New York, 1963.

S. Ross and J. P. Olivier, *On Physical Adsorption*, Interscience, New York, 1964.

J. K. M. Sanders and B. K. Hunter, *Modern NMR Spectroscopy*, 2nd ed., Oxford University Press, New York, 1993.

G. Somorjai, *Introduction to Surface Chemistry and Catalysis*, Wiley-Interscience, New York, 1994.

J. M. Thomas and R. M. Lambert, eds., *Characterization of Catalysts*, Wiley, New York, 1980.

A. Zangwill, *Physics at Surfaces*, Cambridge University Press, New York, 1988.

Textual References

1. A. W. Adamson, *The Physical Chemistry of Surfaces*, Interscience, New York, 1960, p. 427.

2. A. Y. Meyer, D. Farin, and D. Avnir, *J. Am. Chem. Soc.*, **108,** 7987 (1986).

3. D. Avnir and D. Farin, *J. Colloid Interface Sci.*, **103,** 112 (1985).

4. P. Pfeifer, in *Preparative Chemistry Using Supported Reagents*, P. Laszlo, ed., Academic, New York, 1987.

5. H. Van Damme, P. Levitz, F. Bergaya, J. F. Alcover, L. Gatineau, and J. J. Fripiat, *J. Chem. Phys.*, **85,** 616 (1986).

6. P. Meakin, *Multiple Scattering of Waves in Random Media and Random Rough Surfaces*, The Pennsylvania State University Press, State College, PA, 1985.

7. D. Avnir, D. Farin, and P. Pfeifer, *Nature*, **308,** 261 (1984).

8. D. Avnir, in *Better Ceramics Through Chemistry*, C. J. Brinker, ed., Material Research Society, Pittsburgh, PA, 1986.

9. D. Avnir, D. Farin, and P. Pfeifer, *New J. Chem.*, **16,** 439 (1992).

10. D. Farin and D. Avnir, *J. Phys. Chem.*, **91,** 5517 (1987).

11. D. Avnir, ed., *The Fractal Approach to Heterogeneous Chemistry*, Wiley, New York, 1989.

12. D. Farin and D. Avnir, *J. Chromatog.*, **406,** 317 (1987).

13. D. Rojanski, D. Huppert, H. D. Bale, X. Dacai, P. W. Schmidt, D. Farin, A. Seri-Levy, and D. Avnir, *Phys. Rev. Lett.*, **56,** 2505 (1986).

14. D. W. Schaefer, B. C. Bunker, and J. P. Wilcoxon, *Phys. Rev. Lett.*, **58,** 284 (1987).

15. P. Meakin, *CRC Critical Reviews in Solid State and Materials Science*, Vol. 13, Grand Rapids, MI, 1986, p. 143.

16. P. Meakin, H. E. Stanley, A. Coniglio, and T. A. Witten, *Phys. Rev. A*, **32,** 2364 (1985).

17. R. Kopelman, *Science*, **241,** 1620 (1988).

18. P. Meakin and J. M. Deutsch, *J. Chem. Phys.*, **85,** 2320 (1986).

19. F. E. Bartell and Y. Fu, in *The Specific Surface Area of Activated Carbon and Silica*, in *Colloid Symposium Annual*, H. B. Weiser, ed., Vol. 7, Wiley, New York, 1930.

20. J. W. Whalen and W. H. Wade, *J. Colloid Interface Sci.*, **24,** 372 (1967).

21. See G. Jura, in *The Determination of the Area of the Surfaces of Solids* in *Physical*

Methods in Chemical Analysis, W. G. Berl, ed., Vol. 2, Academic, New York, 1951.

22. S. Ergun, *Anal. Chem.*, **24**, 388 (1952).

23. A. J. Tyler, J. A. G. Taylor, B. A. Pethica, and J. A. Hockey, *Trans. Faraday Soc.*, **67**, 483 (1971).

24. R. Denoyel, J. Fernandez-Colinas, Y. Grillet, and J. Rouquerol, *Langmuir*, **9**, 515 (1993).

25. S. Partyea, F. Rouquerol, and J. Rouquerol, *J. Colloid Interface Sci.*, **68**, 21 (1979).

26. A. B. Zdanovskii, *Zh. Fiz. Khim.*, **25**, 170 (1951); *Chem. Abstr.*, **48**, 4291c (1964).

27. A. R. Copper, Jr., *Trans. Faraday Soc.*, **58**, 2468 (1962).

28. W. G. Palmer and R. E. D. Clark, *Proc. Roy. Soc.* (London), **A149**, 360 (1935); see also E. Darmois, *Chim. Ind.* (Paris), **59**, 466 (1948) and I. Bergman, J. Cartwright, and R. A. Bentley, *Nature*, **196**, 248 (1962).

29. F. E. Bartell and C. W. Walton, Jr., *J. Phys. Chem.*, **38**, 503 (1934).

30. See F. E. Bartell and L. S. Bartell, *J. Am. Chem. Soc.*, **56**, 2202 (1934).

31. J. Kloubek, *Powder Tech.*, **29**, 89 (1981).

32. H. L. Ritter and L. C. Drake, *Ind. Eng. Chem., Anal. Ed.*, **17**, 782 (1945).

33. L. C. Drake, *Ind. Eng. Chem.*, **41**, 780 (1949).

34. N. M. Winslow and J. J. Shapiro, *ASTM Bull.*, No. 236, p. 39 (1959).

35. L. A. de Wit and J. J. F. Scholten, *J. Catal.*, **36**, 36 (1975).

36. J. Kloubek and J. Medek, *Carbon*, **24**, 501 (1986).

37. D. R. Milburn, B. D. Adkins, and B. H. Davis, in *Fundamentals of Adsorption*, A. I. Liapis, ed., Engineering Foundation, New York, 1987.

38. L. G. Joyner, E. P. Barrett, and R. Skold, *J. Am. Chem. Soc.*, **73**, 3155 (1951).

39. B. D. Adkins and B. H. Davis, *Adsorp. Sci. Technol.*, **5**, 76, 168 (1988).

40. R. J. Good and R. Sh. Mikhail, *Powder Tech.*, **29**, 53 (1981).

41. D. H. Everett, in *The Solid-Gas Interface*, E. A. Flood, ed., Marcel Dekker, New York, 1967.

42. D. N. Winslow, *J. Colloid Interface Sci.*, **67**, 42 (1978).

43. H. M. Rootare and C. F. Prenzlow, *J. Phys. Chem.*, **71**, 2733 (1967); M. P. Astier and K. S. M. Sing, in *Adsorption at the Gas-Solid and Liquid-Solid Interface*, J. Rouquerol and K. S. W. Sing, eds., Elsevier, Amsterdam, 1982.

44. S. Lowell and J. E. Shields, *J. Colloid Interface Sci.*, **80**, 192 (1981).

45. L. C. Drake and H. L. Ritter, *Ind. Eng. Chem., Anal. Ed.*, **17**, 787 (1945).

46. F. P. Bowden and E. K. Rideal, *Proc. Roy. Soc.* (London), **A120**, 59 (1928).

47. M. J. Joncich and N. Hackerman, *J. Electrochem. Soc.*, **111**, 1286 (1964).

48. D. J. C. Yates, *Can. J. Chem.*, **46**, 1695 (1968).

49. R. J. Matyi, L. H. Schwartz, and J. B. Butt, *Catal. Rev. Sci. Eng.*, **29**, 41 (1987).

50. J. Kozeny, *Sitzber. Akad. Wiss. Wien, Wasserwirtsch. Math. Nat., Kl.* **IIa,** 136, 271 (1927); *Wasserwirtschaft*, **22**, 67, 86 (1927).

51. R. E. Collins, *Flow of Fluids Through Porous Materials*, Reinhold, New York, 1961.

52. B. V. Derjaguin, D. V. Fedoseev, and S. P. Vnukov, *Powder Tech.*, **14**, 169 (1976).

53. R. M. Barrer, *Discuss. Faraday Soc.*, **3,** 61 (1948); see also R. M. Barrer and D. M. Grove, *Trans. Faraday Soc.*, **47,** 826 (1951).

54. G. Kraus, J. W. Ross, and L. A. Girifalco, *J. Phys. Chem.*, **57,** 330 (1953); see also G. Kraus and J. W. Ross, *J. Phys. Chem.*, **57,** 334 (1953).

55. G. A. Somorjai and S. M. Davis, *Chemtech*, **13,** 502 (1983); S. R. Bare and G. A. Somorjai, *Encyclopedia of Physical Science and Technology*, Vol. 13, Academic, New York, 1987.

56. G. Ertl, *Langmuir*, **3,** 4 (1987).

57. G. A. Somorjai, *Introduction to Surface Chemistry and Catalysis*, Wiley-Interscience, New York, 1994.

58. S. N. Magonov and M.-H. Whangbo, *Surface Analysis with STM and AFM*, VCH, New York, 1996.

59. J. N. Fiedor, A. Proctor, M. Houalla, P. M. A. Sherwood, F. M. Mulcahy, and D. M. Hercules, *J. Phys. Chem.*, **96,** 10967 (1992).

60. A. Yu. Stakheev and W. M. H. Sachtler, *J. Chem. Soc., Faraday Trans.*, **87,** 3703 (1991).

61. H. P. Boehm, *Adv. Catal.*, **16,** 179 (1966).

62. G. R. Henning, *Z. Elektrochem.*, **66,** 629 (1962).

63. B. Umansky, J. Engelhardt, and W. K. Hall, *J. Catal.*, **127,** 128 (1991).

64. F. Boccuzzi, S. Coluccia, G. Ghiotti, C. Morterra, and A. Zecchina, *J. Phys. Chem.*, **82,** 1298 (1978).

65. B. A. Morrow, I. A. Cody, and L. S. M. Lee, *J. Phys. Chem.*, **80,** 2761 (1976).

66. A. W. Adamson, *Textbook of Physical Chemistry*, 3rd ed., Academic Press, New York, 1986, p. 137.

67. C. Kemball, *Proc. Roy. Soc.* (London), **A187,** 73 (1946).

68. L. H. Little, *Infrared Spectra of Adsorbed Molecules*, Academic, New York, 1966.

68a. M. L. Hair, *Infrared Spectroscopy in Surface Chemistry*, Marcel Dekker, New York, 1967.

69. R. P. Eischens, *Acc. Chem. Res.*, **5,** 74 (1972).

70. M. K. Weldon and C. M. Friend, *Chem. Rev.*, **96,** 1391 (1996).

71. N. Sheppard, *Annu. Rev. Phys. Chem.*, **39,** 589 (1988).

71a. K. Ebitani, H. Hattori, and K. Tanabe, *Langmuir*, **6,** 1743 (1990).

72. M. A. Chesters and N. Sheppard, in *Spectroscopy of Surfaces*, R. J. H. Clark and R. E. Hester, eds., Wiley, New York, 1988, p. 377.

73. M. A. Chesters, C. De La Cruz, P. Gardner, E. M. McCash, P. Pudney, G. Shahid, and N. Sheppard, *J. Chem. Soc. Faraday Trans.*, **86,** 2757 (1990).

74. N. Sheppard and D. J. C. Yates, *Proc. Roy. Soc.* (London), **A238,** 69 (1956).

75. F. Jousse and E. C. De Lara, *J. Phys. Chem.*, **100,** 233 (1996).

76. P. J. Hendra, I. D. M. Turner, E. J. Loader, and M. Stacey, *J. Phys. Chem.*, **78,** 300 (1974).

77. J. F. Rabolt, R. Santo, and J. D. Swalen, *Appl. Spectrosc.*, **13,** 549 (1979).

78. R. B. Quincy, M. Houalla, and D. M. Hercules, *Fresenius J. Anal. Chem.*, **346,** 676 (1993).

79. J. F. Brazdil and E. B. Yeager, *J. Phys. Chem.*, **85**, 1005 (1981).

80. T. Takenaka and K. Yamasaki, *J. Colloid Interface Sci.*, **78**, 37 (1980).

81. J. F. Brazdil and E. B. Yeager, *J. Phys. Chem.*, **85**, 995, 1005 (1981).

82. V. N. Abramov, A. V. Kiselev, and V. I. Lygin, *Russ. J. Phys. Chem.*, **37**, 1507 (1963).

83. See L. D. Weis, T. R. Evans, and P. A. Leermakers, *J. Am. Chem. Soc.*, **90**, 6109 (1968) and references therein.

84. J. Namnath and A. W. Adamson, unpublished work.

85. G. P. Lozos and B. M. Hoffman, *J. Phys. Chem.*, **78**, 200 (1974).

86. M. B. McBride, T. J. Pinnavia, and M. M. Mortland, *J. Phys. Chem.*, **79**, 2430 (1975).

87. S. Abdo, R. B. Clarkson, and W. K. Hall, *J. Phys. Chem.*, **80**, 2431 (1976).

88. M. F. Ottaviani and G. Martini, *J. Phys. Chem.*, **84**, 2310 (1980).

89. L. Kevan and S. Schlick, *J. Phys. Chem.*, **90**, 1998 (1986).

90. M. Che and A. J. Tench, *Advances in Catalysis*, Academic, New York, 1983, Vol. 32.

91. T. Wasowica, S. J. Kim, S. B. Hong, and L. Kevan, *J. Phys. Chem.*, **100**, 15954 (1996).

92. W. S. Brey, Jr. and K. D. Lawson, *J. Phys. Chem.*, **68**, 1474 (1964).

93. D. Graham and W. D. Phillips, *Proc. 2nd Int. Congr. Surf. Act.*, London, Vol. II, p. 22.

94. J. R. Zimmerman and J. A. Lasater, *J. Phys. Chem.*, **62**, 1157 (1958).

95. C. F. Tirendi, G. A. Mills, and C. Dyhowski, *J. Phys. Chem.*, **96**, 5045 (1992).

96. J. K. M. Sanders and B. K. Hunter, *Modern NMR Spectroscopy*, Oxford University Press, Oxford, 1987.

97. C. Dybowski and R. L. Lichter, eds., *NMR Spectroscopy Techniques*, Marcel Dekker, New York, 1987.

98. J. K. M. Sanders and B. K. Hunter, *Modern NMR Spectroscopy*, Oxford University Press, New York, 1993.

99. V. Yu. Borovkov, W. K. Hall, and V. B. Kazanski, *J. Catal.*, **51**, 437 (1978).

100. I. D. Gay, *J. Phys. Chem.*, **84**, 3230 (1980).

101. C. Tsiao, J. S. Kauffman, D. R. Corbin, L. Abrams, E. E. Carroll, Jr., and C. Dybowski, *J. Phys. Chem.*, **95**, 5586 (1991).

102. M. L. Smith, D. R. Corgin, and C. Dybowski, *J. Phys. Chem.*, **97**, 9045 (1993).

103. M. L. Smith, D. R. Corbin, L. Abrams, and C. Dybowski, *J. Phys. Chem.*, **97**, 7793 (1993).

103a. P. L. McDaniel, C. G. Coe, J. Kärger, and J. D. Moyer, *J. Phys. Chem.*, **100**, 16263 (1996).

104. W. L. Earl, P. O. Fritz, A. A. V. Gibson, and J. H. Lunsford, *J. Phys. Chem.*, **91**, 2091 (1987).

105. W. P. Rothwell, W. Shen, and J. H. Lunsford, *J. Am. Chem. Soc.*, **106**, 2452 (1984).

106. D. Slotfeldt-Ellingsen and H. A. Resing, *J. Phys. Chem.*, **84**, 2204 (1980); J. B. Nagy, E. G. Derouane, H. A. Resing, and G. R. Miller, ibid., **87**, 833 (1983).

107. E. C. Kelusky and C. A. Fyfe, *J. Am. Chem. Soc.*, **108,** 1746 (1986).

108. D. Slotfeldt-Ellingsen, H. A. Resing, K. Unger, and J. Frye, *Langmuir*, **5,** 1324 (1989).

109. J. C. Edwards and P. D. Ellis, *Langmuir*, **7,** 2117 (1991).

110. E. McCafferty and A. C. Zettlemoyer, *Disc. Faraday Soc.*, 239 (1971).

111. K. Kaneko, M. Serizawa, T. Ishikawa, and K. Inouye, *Bull. Chem. Soc. Jpn.*, **48,** 1764 (1975).

112. G. Ebert, *Kolloid-Z.*, **174,** 5 (1961).

113. K. R. Foster and H. A. Resing, *J. Phys. Chem.*, **80,** 1390 (1976).

114. I. Lubezky, U. Feldman, and M. Folman, *Trans. Faraday Soc.*, **61,** 1 (1965).

114a. K. Klier and A. C. Zettlemoyer, *J. Colloid Interface Sci.*, **58,** 216 (1977).

115. H. A. Resing, *Adv. Mol. Relaxation Processes*, **1,** 109 (1967–68); ibid., **3,** 199 (1972).

116. W. J. Dunning, *J. Phys. Chem.*, **67,** 2023 (1963).

117. J. J. Lander and J. Morrison, *J. Appl. Phys.*, **34,** 1411 (1963) and preceding papers.

118. G. Ehrlich and F. G. Hudda, *J. Chem. Phys.*, **35,** 1421 (1961).

119. J. J. Arrecis, Y. J. Chabal, and S. B. Christman, *Phys. Rev. B*, **33,** 7906 (1986).

120. G. A. Somorjai, *Introduction to Surface Chemistry and Catalysis*, Wiley-Interscience, New York, 1994, pp. 55ff, 412 ff.

121. A. W. Adamson and M. W. Orem, in *Progress in Surface and Membrane Science*, Vol. 8. D. A. Cadenhead, J. F. Danielli, and M. D. Rosenberg, eds., Academic, New York, 1974.

122. A. W. Adamson, L. M. Dormant, and M. W. Orem, *J. Colloid Interface Sci.*, **25,** 206 (1966); M. W. Orem and A. W. Adamson, ibid., **31,** 278 (1969).

123. K. J. Hook, J. A. Gardella, Jr., and L. Salvati, Jr., *Macromolecules*, **20,** 2112 (1987).

124. J. A. Creighton, C. G. Blatchford, and M. G. Albrecht, *J. Chem. Soc., Faraday Trans. II*, **75,** 790 (1979).

125. C. S. Allen and R. P. van Duyne, *Chem. Phys. Lett.*, **63,** 455 (1979).

126. A. Otto, *Surf. Sci.*, **75,** L392 (1978).

127. I. Pockrand, in *Springer Tracts in Modern Physics*, G. Höhler. ed., Vol. 104, Springer-Verlag, 1984.

128. M. Moskovits, *Rev. Mod. Phys.*, **57,** 783 (1985).

129. O. Siiman, R. Smith, C. Blatchford, and M. Kerker, *Langmuir*, **1,** 90 (1985).

130. M. Kerker, *Acc. Chem. Res.*, **17,** 271 (1984).

131. A. Lepp and O. Siiman, *J. Phys. Chem.*, **89,** 3493 (1985).

132. N. Neto, M. Muniz-Miranda, and G. Sbrana, *J. Phys. Chem.*, **100,** 9911 (1996).

133. D. A. Weitz and S. Garoff, *J. Chem. Phys.*, **78,** 5324 (1983).

133a. T. Xiao, Qi Ye, and Li Sun, *J. Phys. Chem. B*, **101,** 632 (1997).

134. M. R. Mahoney, M. W. Howard, and R. P. Cooney, *Chem. Phys. Lett.*, **71,** 59 (1980); M. W. Howard, R. P. Cooney, and A. J. McQuillan, *J. Raman Spectrosc.*, **9,** 273 (1980).

135. T. C. Wong and T. T. Ang, *J. Phys. Chem.*, **89,** 4047 (1985).

CHAPTER XVII

Adsorption of Gases and Vapors on Solids

1. Introduction

The subject of gas adsorption is, indeed, a very broad one, and no attempt is made to give complete coverage to the voluminous literature on it. Instead, as in past chapters, the principal models or theories are taken up partly for their own sake and partly as a means of introducing characteristic data.

As stated in the introduction to the previous chapter, adsorption is described phenomenologically in terms of an empirical adsorption function $n = f(P, T)$ where n is the amount adsorbed. As a matter of experimental convenience, one usually determines the adsorption *isotherm* $n = f_T(P)$; in a detailed study, this is done for several temperatures. Figure XVII-1 displays some of the extensive data of Drain and Morrison [1]. It is fairly common in physical adsorption systems for the low-pressure data to suggest that a limiting adsorption is being reached, as in Fig. XVII-1a, but for continued further adsorption to occur at pressures approaching the saturation or condensation pressure P^0 (which would be close to 1 atm for N_2 at 75 K), as in Fig. XVII-1b.

Alternatively, data may be plotted as n versus T at constant pressure or as P versus T at constant n. One thus has adsorption *isobars* and *isosteres* (note Problem XVII-2).

As also noted in the preceding chapter, it is customary to divide adsorption into two broad classes, namely, *physical adsorption* and *chemisorption.* Physical adsorption equilibrium is very rapid in attainment (except when limited by mass transport rates in the gas phase or within a porous adsorbent) and is reversible, the adsorbate being removable without change by lowering the pressure (there may be hysteresis in the case of a porous solid). It is supposed that this type of adsorption occurs as a result of the same type of relatively nonspecific intermolecular forces that are responsible for the condensation of a vapor to a liquid, and in physical adsorption the heat of adsorption should be in the range of heats of condensation. Physical adsorption is usually important only for gases below their critical temperature, that is, for vapors.

Chemisorption may be rapid or slow and may occur above or below the critical temperature of the adsorbate. It is distinguishable, qualitatively, from physical adsorption in that chemical specificity is higher and that the energy of adsorption is large enough to suggest that full chemical bonding has occurred. Gas that is chemisorbed may be difficult to remove, and desorption may be

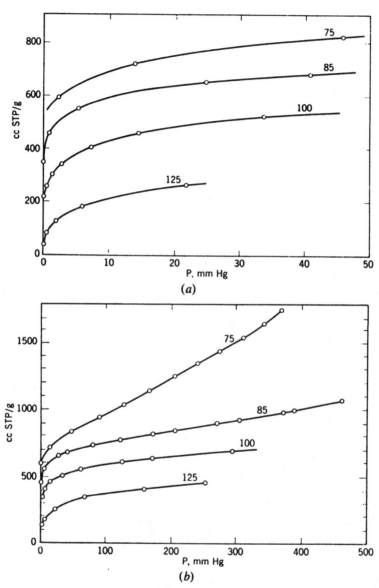

Fig. XVII-1. Adsorption of N_2 on rutile; temperatures indicated are in degrees Kelvin. (a) Low-pressure region; (b) high-pressure region. (From Ref. 1.).

accompanied by chemical changes. For example, oxygen adsorbed on carbon is held very strongly; on heating it comes off as a mixture of CO and CO_2 [2]. Because of its nature, chemisorption is expected to be limited to a monolayer; as suggested in connection with Fig. XVII-1, physical adsorption is not so limited and, in fact, may occur on top of a chemisorbed layer (note Section

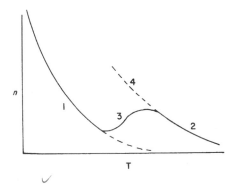

Fig. XVII-2. Transition between physical and chemical adsorption.

XI-1C) as well as alongside it. The infrared spectrum of CO_2 adsorbed on γ-alumina suggests the presence of both physically and chemically adsorbed molecules [3].

Chemisorption may be slow and the rate behavior indicative of the presence of an activation energy; it may, in fact, be possible for a gas to be physically adsorbed at first, and then, more slowly, to enter into some chemical reaction with the surface of the solid. At low temperatures, chemisorption may be so slow that for practical purposes only physical adsorption is observed, whereas, at high temperatures, physical adsorption is small (because of the low adsorption energy), and only chemisorption occurs. An example is that of hydrogen on nickel, for which a schematic isobar is shown in Fig. XVII-2. Curve 1 shows the normal decrease in physical adsorption with temperature, and curve 2, that for chemisorption. In the transition region, curve 3, the rate of chemisorption is slow, but not negligible, so the location of the points depends on the equilibration time allowed; curve 3 is therefore not an equilibrium one and is not retraced on cooling, but rather some curve between 3 and 4 is followed, depending on the rate.

As is made evident in the next section, there is no sharp dividing line between these two types of adsorption, although the extremes are easily distinguishable. It is true that most of the experimental work has tended to cluster at these extremes, but this is more a reflection of practical interests and of human nature than of anything else. At any rate, although this chapter is ostensibly devoted to physical adsorption, much of the material can be applied to chemisorption as well. For the moment, we do assume that the adsorption process is reversible in the sense that equilibrium is reached and that on desorption the adsorbate is recovered unchanged.

2. The Adsorption Time

A useful approach to the phenomenon of adsorption is from the point of view of the adsorption time, as discussed by de Boer [4]. Consider a molecule in the gas phase that is approaching the surface of the solid. If there were no attractive forces at all between the molecule and the solid, then the time of stay of the molecule in the vicinity of the surface would be of the order of a molecular vibration time, or about 10^{-13} sec, and its accommodation coefficient would be zero. This means the molecule retains its original energy. A "hot" molecule striking a cold surface should then rebound with its original energy, and its reflection from the surface would be specular. The definition of accommodation coefficient, α, as given by Knudsen [5] (see Ref. 6), is

$$\alpha = \frac{T_3 - T_1}{T_2 - T_1} \tag{XVII-1}$$

where T_1, T_2, and T_3 are the temperatures of the gas molecules before they strike the surface, of the surface, and of the molecules that leave the surface, respectively. Yasumoto [6] cites experimental α values of 0.48 and 0.18 for He on carbon and Pt, respectively, as compared to 0.90 and 0.71 for Ar, again respectively.

In the case of polyatomic molecules, one may consider separately the accommodation coefficients for translational and for vibrational energy. Values of the latter, α_v, are discussed by Nilsson and Rabinovitch [7].

If attractive forces are present, then according to an equation by Frenkel (see Ref. 2), the average time of stay τ of the molecule on the surface will be

$$\tau = \tau_0 e^{Q/RT} \tag{XVII-2}$$

where τ_0 is 10^{-12} to 10^{-13} sec and Q is the interaction energy, that is, the energy of adsorption. If τ is as large as several vibration periods, it becomes reasonable to consider that adsorption has occurred; temperature equilibration between the molecule and the surface is approached and, on desorption the molecule leaves the surface in a direction that is independent of that of its arrival; the accommodation coefficient is now said to be unity.

In addition to Q and τ, a quantity of interest is the surface concentration Γ, where

$$\Gamma = Z\tau \tag{XVII-3}$$

Here, if Z is expressed in moles of collisions per square centimeter per second, Γ is in moles per square centimeter. We assume the condensation coefficient to be unity, that is, that all molecules that hit the surface stick to it. At very low Q values, Γ as given by Eq. XVII-3 is of the order expected just on the basis that the gas phase continues uniformly up to the surface so that the net surface concentration (e.g., Γ_2^s in Eq. XI-24) is essentially zero. This is the situation

TABLE XVII-1
The Adsorption Spectrum

Q (kcal/mol)	τ, 25°C (sec)	Γ, net (mol/cm^2)	Comments
0.1	10^{-13}	0	Adsorption nil; specular reflection; accommodation coefficient zero
1.5	10^{-12}	0	Region of physical adsorption;
3.5	4×10^{-11}	10^{-12}	accommodation coefficient
9.0	4×10^{-7}	10^{-8}	unity
20.0	100		Region of chemisorption
40.0	10^{17}		

prevailing in the first two rows of Table XVII-1. The table, which summarizes the spectrum of adsorption behavior, shows that with intermediate Q values of the order of a few kilocalories, Γ rises to a level comparable to that for a complete monolayer. This intermediate region corresponds to one of physical adsorption.

The third region is one for which the Q values are of the order of chemical bond energies; the τ values become quite large, indicating that desorption may be slow, and Γ as computed by Eq. XVII-3 becomes preposterously large. Such values are evidently meaningless, and the difficulty lies in the assumption embodied in Eq. XVII-3 that the collision frequency gives the number of molecules hitting and *sticking* to the surface. As monolayer coverage is approached, it is to be expected that more and more impinging molecules will hit occupied areas and rebound without experiencing the full Q value. One way of correcting for this effect is taken up in the next section, which deals with the Langmuir adsorption equation.

3. The Langmuir Adsorption Isotherm

The following several sections deal with various theories or models for adsorption. It turns out that not only is the adsorption isotherm the most convenient form in which to obtain and plot experimental data, but it is also the form in which theoretical treatments are most easily developed. One of the first demands of a theory for adsorption then, is that it give an experimentally correct adsorption isotherm. Later, it is shown that this test is insufficient and that a more sensitive test of the various models requires a consideration of how the energy and entropy of adsorption vary with the amount adsorbed. Nowadays, a further expectation is that the model not violate the molecular picture revealed by surface diffraction, microscopy, and spectroscopy data, see Chapter VIII and Section XVIII-2; Steele [8] discusses this picture with particular reference to physical adsorption.

A. Kinetic Derivation

The derivation that follows is essentially that given by Langmuir [9] in 1918, in which one writes separately the rates of evaporation and of condensation. The surface is assumed to consist of a certain number of sites S of which S_1 are occupied and $S_0 = S - S_1$ are free. The rate of evaporation is taken to be proportional to S_1, or equal to $k_1 S_1$, and the rate of condensation proportional to the *bare surface* S_0 and to the gas pressure, or equal to $k_2 P S_0$. At equilibrium,

$$k_1 S_1 = k_2 P S_0 = k_2 P(S - S_1) \tag{XVII-4}$$

Since S_1/S equals θ, the fraction of surface covered, Eq. XVII-4 can be written in the form

$$\theta = \frac{bP}{1 + bP} \tag{XVII-5}$$

where

$k_1 S_1 = k_2 P (S - S_1)$

$k_1 \theta = k_2 P (1 - \theta)$

$\therefore \dfrac{k_2 P}{(k_1 + k_2 P)} = \theta$

$\Rightarrow \theta = \dfrac{bP}{1 + bP}$ $w.b = \dfrac{k_2}{k_1}$

$$b = \frac{k_2}{k_1} \tag{XVII-6}$$

Alternatively, θ can be replaced by n/n_m, where n_m denotes the moles per gram adsorbed at the monolayer point. Thus   *n. actual adsorbed amount*

$$n = \frac{n_m bP}{1 + bP} \tag{XVII-7}$$

It is of interest to examine the algebraic behavior of Eq. XVII-7. At low pressure, the amount adsorbed becomes proportional to the pressure

$$n = n_m bP \tag{XVII-8}$$

whereas at high pressure, n approaches the limiting value n_m. Some typical shapes are illustrated in Fig. XVII-3. For convenience in testing data, Eq. XVII-7 may be put in the linear form

$n + nbP = n_m bP$ $\dfrac{}{bn_m n}$

$$\frac{P}{n} = \frac{1}{bn_m} + \frac{P}{n_m} \tag{XVII-9}$$

A plot of P/n versus P should give a straight line, and the two constants n_m and b may be evaluated from the slope and intercept. In turn, n_m may be related to the area of the solid:

Surface

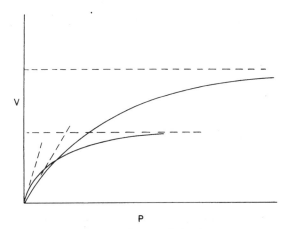

Fig. XVII-3. Langmuir isotherms.

$$\Sigma = N_0 \sigma^0 A_m \theta n$$

$$n_m = \frac{\Sigma \quad 0.1 \, n_m}{N_0 \sigma^0} \tag{XVII-10}$$

where Σ denotes the specific surface area of the solid, and σ^0 is the area of a site. If σ^0 can be estimated, Σ can be calculated from the experimental n_m value.

If there are several competing adsorbates,† a derivation analogous to the foregoing gives

$$n_i = \frac{n_{mi} b_i P_i}{1 + \sum_j b_j P_j} \qquad \checkmark \text{(XVII-11)}$$

The rate constants k_1 and k_2 may be related to the concepts of the preceding section as follows. First, k_1 is simply the reciprocal of the adsorption time, that is,

$$k_1 = \left(\frac{1}{\tau_0}\right) e^{-Q/RT} \tag{XVII-12}$$

In evaluating k_2, if a site can be regarded as a two-dimensional potential box, then the rate of adsorption will be given by the rate of molecules impinging on the site area σ_0. From gas kinetic theory,

---

†The term *adsorbate* is used here to denote the species adsorbed; an alternative term, more popular in Europe than in the United States, is *adsorptive*. The distinction may be made that *adsorbate* refers to the species actually adsorbed, while *adsorptive* refers to a gas-phase species that is capable of being adsorbed. [See Gregg and Sing (1983) in General References.]

$$k_2 = \frac{N_0 \sigma^0}{(2\pi MRT)^{1/2}} \qquad \text{(XVII-13)}$$

and the Langmuir constant b becomes

$$\frac{k_2}{k_1} = b = \frac{N_0 \sigma^0 \tau_0 e^{Q/RT}}{(2\pi MRT)^{1/2}} = k_2 \tau \qquad \text{(XVII-14)}$$

It will be convenient to write b as

$$b = b_0 e^{Q/RT} \qquad b_0 = \frac{N_0 \sigma^0 \tau_0}{(2\pi MRT)^{1/2}} \qquad \text{(XVII-15)}$$

M?

where b_0 is of the nature of a frequency factor. Thus for nitrogen at its normal boiling point of 77 K, b_0 is 9.2×10^{-4}, with pressure in atmospheres, σ^0 taken to be 16.2 Å^2 per site (actually, this is the estimated molecular area of nitrogen), and τ_0, as 10^{-12} sec.   Å : astrongm ??

B. Statistical Thermodynamic Derivation

The preceding derivation, being based on a definite mechanical picture, is easy to follow intuitively; kinetic derivations of an *equilibrium* relationship suffer from a common disadvantage, namely, that they usually assume more than is necessary. It is quite possible to obtain the Langmuir equation (as well as other adsorption isotherm equations) from examination of the statistical thermodynamics of the two states involved.

The following derivation is modified from that of Fowler and Guggenheim [10,11]. The adsorbed molecules are considered to differ from gaseous ones in that their potential energy and local partition function (see Section XVI-4A) have been modified and that, instead of possessing normal translational motion, they are confined to *localized* sites without any interactions between adjacent molecules but with an adsorption energy Q.

Since translational and internal energy (of rotation and vibration) are independent, the partition function for the gas can be written

$$Q^g = Q^g_{\text{trans}} Q^g_{\text{int}} \qquad \text{(XVII-16)}$$

We write for the adsorbed or surface state

$$Q^s = Q^s_{\text{site}} Q^s_{\text{int}} e^{Q/RT} \qquad \text{(XVII-17)}$$

where the significance of the site partition function Q^s_{site} is explained later and the inclusion of the term $e^{Q/RT}$ means that $\mathbf{Q}^s$ is referred to the gaseous state. Furthermore, $\mathbf{Q}^s$ is a function of temperature only and not of the degree of occupancy of the sites. The complete partition function is obtained by multiplying $\mathbf{Q}^s$ by the number of distinguishable ways of placing N molecules on S sites. This number is obtained as follows: there are S ways of placing the first molecule, $(S - 1)$ for the second, and so on; and for N molecules, the number of ways is

$$S(S - 1)\cdots[S - (N - 1)] \quad \text{or} \quad S!/(S - N)!$$

Of these, $N!$ are indistinguishable since the molecules are not labeled, and the complete partition function for N molecules becomes

$$\mathbf{Q}^s_{\text{tot}} = \frac{S!}{(S - N)!N!} (\mathbf{Q}^s)^N \tag{XVII-18}$$

The Helmholtz free energy of the adsorbed layer is given by $-kT\ln \mathbf{Q}^s_{\text{tot}}$ (Eq. XVI-20), and with the use of Stirling's approximation for factorials $x! = (x/e)^x$ one obtains

$$A^s = kT[-S \ln S + N \ln N + (S - N) \ln(S - N) - N \ln \mathbf{Q}^s] \tag{XVII-19}$$

The chemical potential μ^s is given by $(\partial A^s/\partial N)_T$, so that

$$\mu^s = kT\ln \frac{N}{S - N} - kT\ln \mathbf{Q}^s \tag{XVII-20}$$

For the gas phase,

$$\mu^g = -kT\ln \mathbf{Q}^g \tag{XVII-21}$$

and on equating the two chemical potentials (remembering that $\theta = N/S$) one obtains

$$\frac{\theta}{1 - \theta} = \frac{\mathbf{Q}^s}{\mathbf{Q}^g} \tag{XVII-22}$$

It is now necessary to examine the partition function in more detail. The energy states for translation are assumed to be given by the quantum-mechanical picture of a particle in a box. For a one-dimensional box of length a,

$$\epsilon_n = \frac{n^2 h^2}{8a^2 m} \tag{XVII-23}$$

so that the one-dimensional translational partition function is

$$Q_{\substack{trans \\ 1\,dim}} = \sum_n \exp\left(-\frac{\epsilon_n}{kT}\right) \approx \int_0^\infty \exp\left(-\frac{n^2 h^2}{8a^2 mkT}\right) dn$$

$$= \left(\frac{2\pi mkT}{h^2}\right)^{1/2} a \qquad\qquad\qquad (XVII\text{-}24)$$

where ordinarily the states are so close together that the sum over the quantum numbers n may be replaced by the integral (perhaps contrary to intuition, this is true even if a is of molecular magnitude). For a two-dimensional box,

$$Q_{\substack{trans \\ 2\,dim}} = \left(\frac{2\pi mkT}{h^2}\right) a^2 \qquad\qquad\qquad (XVI\text{-}25)$$

and in three dimensions,

$$Q_{\substack{trans \\ 3\,dim}} = \left(\frac{2\pi mkT}{h^2}\right)^{3/2} \frac{kT}{P} \qquad\qquad (XVII\text{-}26)$$

where a^3 is now replaced by the volume and, in turn, by kT/P. Substitution of Eqs. XVII-26 and XVII-17 into Eq. XVII-22 gives

$$\frac{\theta}{1-\theta} = \frac{(h^2/2\pi mkT)^{3/2}}{kT} \frac{Q_{site}^s Q_{int}^s}{Q_{int}^g} e^{Q/RT} P \qquad (XVII\text{-}27)$$

or

$$\frac{\theta}{(1-\theta)} = bP \qquad\qquad\qquad (XVII\text{-}28)$$

which is the same as the Langmuir equation, Eq. XVII-5, but with b_0 of Eq. XVII-15 given by

$$b_0' = \frac{(h^2/2\pi mkT)^{3/2}}{kT} \frac{Q_{site}^s Q_{int}^s}{Q_{int}^g} \qquad (XVII\text{-}29)$$

The two expressions for b_0 may be brought into formal identity as follows. On adsorption, the three degrees of translational freedom can be supposed to appear as two degrees of translational motion within the confines of a two-dimensional box of area $a^2 = \sigma^0$, plus one degree of vibration in the adsorption bond, normal to the surface. The partition function for the first is given by Eq. XVII-25. For one degree of vibrational freedom the energy states are

$$\epsilon_n = (n + \tfrac{1}{2})h\upsilon^0 \tag{XVII-30}$$

in the case of a harmonic oscillator. The partition function is

$$\mathbf{Q}_{\text{vib}} = \sum_n \exp\frac{-\epsilon_n}{kT} = \frac{e^{h\upsilon^0/2kT}}{e^{h\upsilon^0/kT} - 1} \tag{XVII-31}$$

If the adsorption bond is weak so that $h\upsilon^0/kT \ll 1$, expansion of Eq. XVII-31 gives $\mathbf{Q}_{\text{vib}} \simeq kT/h\upsilon^0$. If we now make the following identification:

$$\mathbf{Q}^s_{\text{site}} = \underset{\sigma^0 \text{ box}}{\mathbf{Q}^s_{\text{trans}}} \times \underset{\text{ads bond}}{\mathbf{Q}^s_{\text{vib}}} \tag{XVII-32}$$

then

$$\mathbf{Q}^s_{\text{site}} = \frac{2\pi mkT}{h^2}\,\sigma^0\,\frac{kT}{h\upsilon^0} \tag{XVII-33}$$

Since υ^0 corresponds to $1/\tau_0$, we have

$$b'_0 = \frac{N\sigma^0\tau_0}{(2\pi MRT)^{1/2}}\left(\frac{\mathbf{Q}^s_{\text{int}}}{\mathbf{Q}^g_{\text{int}}}\right) = b_0\frac{\mathbf{Q}^s_{\text{int}}}{\mathbf{Q}^g_{\text{int}}} \tag{XVII-34}$$

Thus the kinetic and statistical mechanical derivations may be brought into identity by means of a specific series of assumptions, including the assumption that the internal partition functions are the same for the two states (see Ref. 12). As discussed in Section XVI-4A, this last is almost certainly not the case because as a minimum effect some loss of rotational degrees of freedom should occur on adsorption.

C. Adsorption Entropies

i. Configurational Entropies.   The factorial expression in Eq. XVII-18 may be called the *configurational partition function*; it is that part of the partition function that has to do with the ways of arranging a given state. Thus

$$\mathbf{Q}^s_{\text{config}} = \frac{S!}{(S-N)!N!} \tag{XVII-35}$$

Since Q^s_{config} has no temperature dependence, we have from Eq. XVII-20 that $S^s_{config} = k \ln Q^s_{config}$. On applying Stirling's approximation and dividing through by N, to obtain S^s_{config} on a per molecule basis, the result is

$$S^s_{config} = -k \left[\frac{1 - \theta}{\theta} \ln(1 - \theta) + \ln \theta \right] \qquad \text{(XVII-36)}$$

This is an *integral* entropy; the differential entropy is obtained by the operation $\overline{S} = \partial(NS)/\partial N = S + N(\partial S/\partial N)$, which yields

$$\overline{S}^s_{config} = -k \ln \frac{\theta}{1 - \theta} \qquad \text{(XVII-37)}$$

(The same result can be obtained from Eq. XVII-20 since $\overline{S} = -\partial\mu/\partial T$, considering only the configurational term in that equation.)

Thus the thermodynamic description of the Langmuir model is that the energy of adsorption Q is constant and that the entropy of adsorption varies with θ according to Eq. XVII-37.

To further emphasize the special nature of the configurational entropy assumption embodied in the Langmuir model, let us repeat the derivation assuming instead that the surface molecules are mobile. In terms of the kinetic derivation, this amounts to setting the rate of condensation equal to $k_2 PS$ rather than to $k_2 PS_0$, with the result that $\theta = bP$. The effect on the statistical thermodynamic derivation is first that the factorial grouping in Eq. XVII-18 becomes just $1/N!$ multiplied by S^N (since there are now S ways of placing each of the N molecules), with the result that Eq. XVII-22 becomes

$$\theta = \frac{Q^s}{Q^g} \qquad \text{(XVII-38)}$$

and second that in obtaining Q^s_{site}, a^2 in Eq. XVII-25 is replaced by the total area $\mathcal{A} = S\sigma^0$ rather than just by the site area σ^0. On making these substitutions, the result is

$$\theta = bP \qquad b = b'_0 e^{Q/RT} \qquad \text{(XVII-39)}$$

The configurational entropies are now

$$S^s_{\substack{config \\ mobile}} = -k \ln \theta + k \qquad \text{(XVII-40)}$$

$$\overline{S}^s_{\substack{config \\ mobile}} = -k \ln \theta \qquad \text{(XVII-41)}$$

All of these entropies may be put on a per mole basis by replacing k by R.

The case of a vapor adsorbing on its own liquid surface should certainly correspond to mobile adsorption. Here, θ is unity and $P = P^0$, the vapor pressure. The energy of adsorption is now that of condensation Q_v, and it will be convenient to define the Langmuir constant for this case as b^0; thus, from Eq. XVII-39,

$$1 = b^0 P^0 = b_0 P^0 \exp \frac{Q_v}{RT} \qquad \text{(XVII-42)}$$

If, furthermore, we write $c = b/b^0$ and $x = P/P^0$, the Langmuir equation can be put in the form

$$\theta = \frac{cx}{1 + cx} \qquad \text{(XVII-43)}$$

ii. Entropies of Adsorption.   The Langmuir model is not usually considered to imply any particular value for the *total* entropy change on adsorption, but the statistical thermodynamic approach makes it easy to postulate various possible values. In considering, for example, the differential entropy change that should occur when 1 mole of gas at 1 atm pressure adsorbs at surface coverage θ, the matter becomes one of assembling the various possible contributions.

First, the total translational partition function for 1 mol of gas is

$$Q_{tot}^{g} = \frac{1}{N!} \left[\left(\frac{2\pi m k T}{h^2} \right)^{3/2} V \right]^N \qquad \text{(XVII-44)}$$

On applying Stirling's approximation, replacing V/N by kT/P, and using Eq. XVII-20 and setting $P = 1$ atm, the final result, known as the Sackur-Tetrode equation, is

$$\overline{S}_{trans}^{0, g} = R \ln(T^{5/2} M^{3/2}) - 2.30 \qquad \text{(XVII-45)}$$

For nitrogen at 77 K, $\overline{S}_{trans}^{0, g}$ is 29.2 EU(cal K^{-1} mol^{-1}).

For localized adsorption, we need the contribution from the adsorption bond. Per degree of vibrational freedom, we obtain on applying Eq. XVII-20 to Eq. XVII-31,

$$S_{vib}^{s} = R \left[\frac{h v^0/kT}{e^{h v^0/kT} - 1} - \ln(1 - e^{-h v^0/kT}) \right] \qquad \text{(XVII-46)}$$

which approximates to $R[1 - \ln(h v^0/kT)]$. If T is 77 K and v^0 is taken to be 10^{12}

sec^{-1} (see Ref. 13), then S_{vib}^s is 2.9 EU per degree of freedom. Second, if the adsorption site is regarded as a two-dimensional potential box, as implied by the kinetic derivation given above, the corresponding translational entropy must be evaluated. The total partition function for N molecules is just $[(2\pi mkT/h^2)\sigma^0]^N$, and the usual operation yields

$$\overline{S}_{\substack{trans \\ \sigma^0 \, box}}^s = R\ln(MT\sigma^0) + 63.8 \qquad \text{(XVII-47)}$$

For nitrogen at 77 K, with $\sigma^0 = 16.2$ Å^2, this entropy becomes 11.4 EU.

If the adsorbed gas is mobile, the total partition function is $(S^N/N!)$ $[(2\pi mkT/h^2)\mathcal{A}]^N$, which gives (see Ref. 11)

$$S_{\substack{trans \\ 2\,dim}}^s = -R\ln\theta + R\ln(MT\sigma^0) + 65.8 \qquad \text{(XVII-48)}$$

remembering that $\mathcal{A}/N = \sigma^0/\theta$. Alternatively,

$$\overline{S}_{\substack{trans \\ 2\,dim}}^s = -R\ln\theta + R\ln(MT\sigma^0) + 63.8 \qquad \text{(XVII-49)}$$

Notice that the nonconfigurational part of Eq. XVII-49 is just the entropy given by Eq. XVII-47.

We can now proceed with various estimates of the entropy of adsorption $\Delta\overline{S}_{ads}^0$. Two extreme positions are sometimes taken (see Ref. 14). First, one assumes that for localized adsorption the only contribution is the configurational entropy. Thus

$$\Delta\overline{S}_{\substack{ads \\ local}}^0 = -R\ln\left(\frac{\theta}{1-\theta}\right) - \overline{S}_{trans}^{0,g}$$

$$= 0 - 29.2$$

$$= -29.2 \text{ cal K}^{-1} \text{ mol}^{-1} \qquad \text{(XVII-50)}$$

where the numbers are for nitrogen at 77 K and $\theta = 0.5$. This contrasts with the value for a mobile film:

$$\Delta\overline{S}_{\substack{ads \\ mobile}}^0 = -R\ln\theta + S_{\substack{trans \\ \sigma^0 \, box}}^s - \overline{S}_{trans}^{0,g}$$

$$= 1.4 + 11.4 - 29.2$$

$$= -16.4 \text{ cal K}^{-1} \text{ mol}^{-1} \qquad \text{(XVII-51)}$$

This difference looks large enough to be diagnostic of the state of the adsorbed film. However, to be consistent with the kinetic derivation of the Langmuir equation, it was necessary to suppose that the site acted as a potential box and, furthermore, that a weak adsorption bond of v^0 corresponding to $1/\tau_0$ was present. With these provisions we obtain

$$\Delta \overline{S}^0_{\substack{\text{ads} \\ \text{local}}} = -R \ln \left(\frac{\theta}{1 - \theta} \right) + S^s_{\substack{\text{trans} \\ \sigma^0 \text{ box}}} + S^s_{\substack{\text{vib} \\ \text{ads bond}}} - \overline{S}^{0,\text{g}}_{\text{trans}}$$

$$= 0 + 11.4 + 2.9 - 29.2$$

$$= -14.9 \text{ cal K}^{-1} \text{ mol}^{-1} \qquad \text{(XVII-52)}$$

Thus the entropy of "localized" adsorption can range widely, depending on whether the site is viewed as equivalent to a strong adsorption bond of negligible entropy or as a potential box plus a weak bond (see Ref. 12). In addition, estimates of $\Delta \overline{S}^0_{\text{ads}}$ should include possible surface vibrational contributions in the case of mobile adsorption, and all calculations are faced with possible contributions from a loss in rotational entropy on adsorption as well as from change in the adsorbent structure following adsorption (see Section XVI-4B). These uncertainties make it virtually impossible to affirm what the state of an adsorbed film is from entropy measurements alone; for this, additional independent information about surface mobility and vibrational surface states is needed. (However, see Ref. 15 for a somewhat more optimistic conclusion.)

D. Lateral Interaction

It is assumed in the Langmuir model that while the adsorbed molecules occupy sites of energy Q they do not interact with each other. An approach due to Fowler and Guggenheim [10] allows provision for such interaction. The probability of a given site being occupied is N/S, and if each site has z neighbors, the probability of a neighbor site being occupied is zN/S, so the fraction of adsorbed molecules involved is $z\theta/2$, the factor one-half correcting for double counting. If the lateral interaction energy is ω, the added energy of adsorption is $z\omega\theta/2$, and the added differential energy of adsorption is just $z\omega\theta$.

The modified Langmuir equation becomes

$$\frac{\theta}{1 - \theta} = b'P$$

$$b' = b_0 \exp \left(\frac{Q + z\omega\theta}{RT} \right) = b \exp \frac{z\omega\theta}{RT} \qquad \text{(XVII-53)}$$

It is convenient to illustrate the lateral interaction effect by plotting θ versus

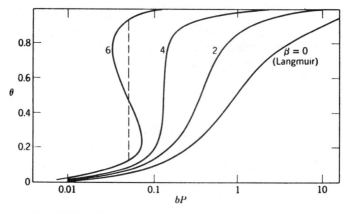

Fig. XVII-4. Langmuir plus lateral interaction isotherms.

the Langmuir variable bP for various values of $\beta = z\omega/RT$, as shown in Fig. XVII-4. For $\beta < 4$, the isotherms are merely steepened versions of the Langmuir equation, but if $\beta > 4$, a maximum and a minimum in bP appear so that a two-phase equilibrium is implied, with phases of θ values lying at the ends of the dotted line.

The quantity $z\omega$ will depend very much on whether adsorption sites are close enough for neighboring adsorbate molecules to develop their normal van der Waals attraction; if, for example, $z\omega$ is taken to be about one-fourth of the energy of vaporization [16], β would be 2.5 for a liquid obeying Trouton's rule and at its normal boiling point. The critical pressure P_c, that is, the pressure corresponding to $\theta = 0.5$ with $\beta = 4$, will depend on both Q and T. A way of expressing this follows, with the use of the definitions of Eqs. XVII-42 and XVII-43 [17]:

$$\frac{P_c}{P^0} = \frac{1}{e^2 c} \qquad\qquad \text{(XVII-54)}$$

This is useful since c can be estimated by means of the BET equation (see Section XVII-5). A number of more or less elaborate variants of the preceding treatment of lateral interaction have been proposed. Thus, Kiselev and co-workers, in their very extensive studies of physical adsorption, have proposed an equation of the form

$$K_1 x = \frac{\theta}{1 - \theta} - K_1 K_n \theta x \qquad\qquad \text{(XVII-55)}$$

where K_n reflects the degree of adsorbate-adsorbate interaction and $c = K_1 + K_1 K_n$ so that Eq. XVII-55 reduces to Eq. XVII-43 if K_n is zero [18]. Misra [19] has summarized other (and semiempirical) Langmuir-like adsorption isotherms.

A fundamental approach by Steele [8] treats monolayer adsorption in terms of interatomic potential functions, and includes pair and higher order interactions. Young and Crowell [11] and Honig [20] give additional details on the general subject; a recent treatment is by Rybolt [21].

E. Experimental Applications of the Langmuir Equation

A variety of experimental data has been found to fit the Langmuir equation reasonably well. Data are generally plotted according to the linear form, Eq. XVII-9, to obtain the constants b and n_m from the best fitting straight line. The specific surface area, Σ, can then be obtained from Eq. XVII-10. A widely used practice is to take σ^0 to be the molecular area of the adsorbate, estimated from liquid or solid adsorbate densities. On the other hand, the Langmuir model is cast around the concept of adsorption sites, whose spacing one would suppose to be characteristic of the adsorbent. See Section XVII-5B for an additional discussion of the problem.

A true fit to the Langmuir equation implies that n_m and Q are independent of temperature, and systems obeying the form of Eq. XVII-9 often fail this more severe test. In some cases, multilayer formation, discussed in Section XVII-5, is a source of trouble, but in general the Langmuir model is too simple for really detailed agreement to be expected with experimental systems. However, more sophisticated models also introduce more semiempirical parameters, so the net gain, apart from data fitting, has not been great. The simple Langmuir model has therefore retained great general utility as well as providing the point of departure from many of the proposed refinements (as, for example, Eq. XVII-53).

4. Experimental Procedures

The remainder of the chapter is concerned with increasingly specialized developments in the study of gas adsorption, and before proceeding to this material, it seems desirable to consider briefly some of the experimental techniques that are important in obtaining gas adsorption data. See Ref. 22 for a review of traditional methods, and Ref. 23 for IUPAC (International Union of Pure and Applied Chemistry) recommendations for symbols and definitions.

Adsorption isotherms conventionally have been determined by means of a vacuum line system whereby pressure–volume measurements are made before and after admitting the adsorbate gas to the sample. For some recent experimental papers, see Refs. 24 and 25.

If the total surface area is small (say, a few hundred square centimeters), the amount adsorbed becomes so little that measurements are difficult by normal procedures. Thus the change in pressure-volume product on admitting gas to the adsorbent becomes so small that precision is impaired.

One way of avoiding this problem is to set T_1 so that the vapor pressure of the liquid adsorbate is very low (e.g., krypton at $-195°C$ [26, 27]. Monolayer

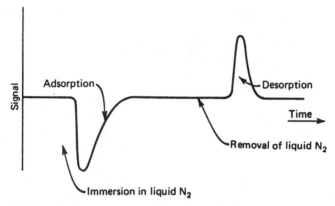

Fig. XVII-5. Schematic detector response in a determination of nitrogen adsorption and desorption. A flow of He and N_2 is passed through the sample until the detector reading is constant; the sample is then cooled in a liquid nitrogen bath. For desorption, the bath is removed. (From Ref. 28. Reprinted with permission from John Wiley & Sons, copyright 1995.)

formation usually occurs when a few tenths of saturation pressure is reached, regardless of its *absolute* value; and if P^0 is very small, even small amounts of adsorption will cause relatively large changes in the pressure-volume product of the gas in the vacuum line.

Alternatively, gas chromatography may be used; Fig. XVII-5 shows a schematic readout of the thermal conductivity detector, the areas under the peaks giving the amount adsorbed or desorbed.

A second general type of procedure, due to McBain [29], is to determine n by a direct weighing of the amount of adsorption. McBain used a delicte quartz spiral spring, but modern equipment generally makes use of a microbalance or a transducer. An illustrative schematic is shown in Fig. XVII-6.

Ultrahigh vacuum techniques have become common, especially in connection with surface spectroscopic and diffraction studies, but also in adsorption on very clean surfaces. The techniques have become rather specialized and the reader is referred to Ref. 8 and citations therein.

The heat of adsorption is an important experimental quantity. The heat evolution with each of successive admissions of adsorbate vapor may be measured directly by means of a calorimeter described by Beebe and co-workers [31]. Alternatively, the heat of immersion in liquid adsorbate of adsorbent having various amounts preadsorbed on it may be determined. The difference between any two values is related to the integral heat of adsorption (see Section X-3A) between the two degrees of coverage. See Refs. 32 and 33 for experimental papers in this area.

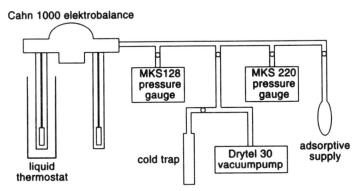

Fig. XVII-6. Schematic of gravimetric apparatus for adsorption measurements. (From Ref. 30. Reprinted with permission from American Chemical Society, copyright 1995.)

5. The BET and Related Isotherms

Adsorption isotherms are by no means all of the Langmuir type as to shape, and Brunauer [34] considered that there are five principal forms, as illustrated in Fig. XVII-7. Type I is the Langmuir type, roughly characterized by a monotonic approach to a limiting adsorption that presumably corresponds to a complete monolayer. Type II is very common in the case of physical adsorption

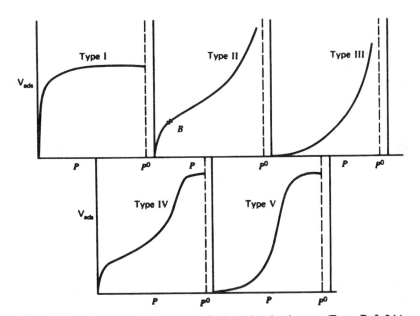

Fig. XVII-7. Brunauer's five types of adsorption isotherms. (From Ref. 34.)

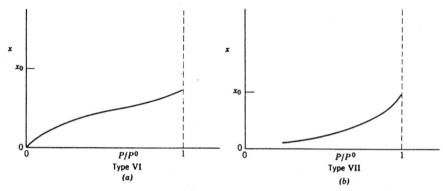

Fig. XVII-8. Two additional types of adsorption isotherms expected for nonwetting adsorbate–adsorbent systems. (From Ref. 37.)

and undoubtedly corresponds to multilayer formation. For many years it was the practice to take point B, at the knee of the curve, as the point of completion of a monolayer, and surface areas obtained by this method are fairly consistent with those found using adsorbates that give type I isotherms. Type III is relatively rare—an example is that of the adsorption of nitrogen on ice [35]—and seems to be characterized by a heat of adsorption equal to or less than the heat of liquefaction of the adsorbate. Types IV and V are considered to reflect capillary condensation phenomena in that they level off before the saturation pressure is reached and may show hysteresis effects.

This description is traditional, and some further comment is in order. The flat region of the type I isotherm has never been observed up to pressures approaching P^0; this type typically is observed in chemisorption, at pressures far below P^0. Types II and III approach the P^0 line asymptotically; experimentally, such behavior is observed for adsorption on powdered samples, and the approach toward infinite film thickness is actually due to interparticle condensation [36] (see Section X-6B), although such behavior is expected even for adsorption on a flat surface if bulk liquid adsorbate wets the adsorbent. Types IV and V specifically refer to porous solids. There is a need to recognize at least the two additional isotherm types shown in Fig. XVII-8. These are two simple types possible for adsorption on a flat surface for the case where bulk liquid adsorbate rests on the adsorbent with a finite contact angle [37, 38].

A. Derivation of the BET Equation

Because of their prevalence in physical adsorption studies on high-energy, powdered solids, type II isotherms are of considerable practical importance. Brunauer, Emmett, and Teller (BET) [39] showed how to extend Langmuir's approach to multilayer adsorption, and their equation has come to be known as the BET equation. The derivation that follows is the traditional one, based on a detailed balancing of forward and reverse rates.

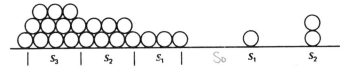

Fig. XVII-9. The BET model.

The basic assumption is that the Langmuir equation applies to each layer, with the added postulate that for the first layer the heat of adsorption Q may have some special value, whereas for all succeeding layers, it is equal to Q_v, the heat of condensation of the liquid adsorbate. A further assumption is that evaporation and condensation can occur only from or on exposed surfaces. As illustrated in Fig. XVII-9, the picture is one of portions of uncovered surface S_0, of surface covered by a single layer S_1, by a double-layer S_2, and so on.†　The condition for equilibrium is taken to be that the amount of each type of surface reaches a steady-state value with respect to the next-deeper one. Thus for S_0

$$a_1 P S_0 = b_1 S_1 e^{-Q_1/RT} \qquad \text{Langmuir's} \qquad \text{(XVII-56)}$$

and for all succeeding surfaces,

$$a_i P S_{i-1} = b_i S_i e^{-Q_v/RT} \qquad i = 2, 3 \cdots \quad \text{(XVII-57)}$$

$$P = x_A P_A^\circ$$

It then follows that

$$S_1 = y S_0 \qquad S_2 = x S_1$$

and

$$S_i = x^{i-1} S_1 = y x^{i-1} S_0 = c x^i S_0$$

where

$$C = \frac{y}{x}$$

$$y = \frac{a_1}{b_1} P e^{Q_1/RT} \qquad x = \frac{a_i}{b_i} P e^{Q_v/RT} \qquad \text{(XVII-58)}$$

and

†On the right of Fig. XVII-9 is shown an isolated stack of two molecules. While physically unrealistic, such stacks are implicit in the BET model and, in fact, have been reproduced in a Monte Carlo simulation [40].

$$c = \frac{y}{x} = \left(\frac{a_1 b_i}{b_1 a_i}\right) e^{(Q_1 - Q_v)/RT} \simeq e^{(Q_1 - Q_v)/RT} = C \qquad \text{(XVII-59)}$$

Then

$$\frac{n}{n_m} = \frac{\sum_{i=1}^{\infty} i S_i}{\sum_{i=0}^{\infty} S_i} = c S_0 \frac{\sum_{i=1}^{\infty} i x^i}{S_0 + S_0 c \sum_{i=1}^{\infty} x^i} \qquad \text{(XVII-60)}$$

Insertion of the algebraic equivalents to the sums yields

$$\frac{n}{n_m} = \frac{cx/(1-x)^2}{1 + cx/(1-x)} \qquad \text{(XVII-61)}$$

which rearranges to

$$\theta = \frac{v}{v_m} = \frac{n}{n_m} = \frac{cx}{(1-x)[1 + (c-1)x]} \qquad x = \frac{P}{P^0} \qquad \text{(XVII-62)}$$

where v denotes cm^3 STP (standard temperature and pressure) of gas adsorbed. The essential next step is to take the ratio of frequency terms a_i/b_i to be the same as for liquid adsorbate-vapor equilibrium so that combination of Eqs. XVII-58 and XVII-42 (where $b_0^0 = a_i/b_i$) identifies x as equal to P/P^0. Usually, also, it is assumed that $a_1/b_1 = a_i/b_i$ in that the approximate form of Eq. XVII-59 is used in interpreting the constant c.

Although the preceding derivation is the easier to follow, the BET equation also may be derived from statistical mechanics by a procedure similar to that described in the case of the Langmuir equation [41,42].

B. Properties of the BET Equation

The BET equation filled an annoying gap in the interpretation of adsorption isotherms, and at the time of its appearance in 1938 it was also hailed as a general method for obtaining surface areas from adsorption data. The equation can be put in the form

$$\frac{x}{n(1-x)} = \frac{1}{cn_m} + \frac{(c-1)x}{cn_m} \qquad \text{(XVII-63)}$$

so that n_m and c can be obtained from the slope and intercept of the straight line best fitting the plot of $x/n(1-x)$ versus x. The specific surface area can then be obtained through Eq. XVII-10 if σ^0 is known. In the case of multilayer adsorption it seems reasonable to take σ^0 as an adsorbate (as opposed to an adsorbent site) area, based on either the solid or the liquid density, depending on the temperature. Values that give reasonably self-consistent areas are (in square angstroms per molecule): N_2, 16.2; O_2, 14.1; Ar, 13.8; Kr, 19.5; n-C_4H_{10}, 18.1. These and values for other adsorbates are reviewed critically by McClellan and Harnsberger [43]. The preceding values are close to the calculated ones from the liquid densities at the boiling points and are, in this respect, reasonable for a multilayer adsorption situation. There may be special cases. Pierce and Ewing [44] suggest that for graphite surfaces it is the lattice spacing of the adsorbent that controls and that the effective molecular area for N_2 becomes 20 $\mathring{A}^2$ rather than the usual 16.2 $\mathring{A}^2$. Amati and Kováts [45] found the effective σ_0 for N_2 to vary from 16.2 to 21 $\mathring{A}^2$ as γ_s was decreased by surface modification of a silica. Rouquerol et al. [46] conclude that, depending on the adsorbent, the effective σ^0 for N_2 may vary from 11.2 to 16.2 $\mathring{A}^2$ since the N_2 may in some case be adsorbed end-on.

From the experimental point of view, the BET equation is easy to apply, and the surface areas so obtained are reasonably consistent (see Section XVII-8 for further discussion). The equation in fact has become the standard one for practical surface area determinations, usually with nitrogen at 77 K as the adsorbate, but in general with any system giving type II isotherms. On the other hand, the region of fit usually is not very great—the linear region of a plot according to Eq. XVII-63 typically lies between a P/P^0 of 0.05 and 0.3, as illustrated in Fig. XVII-10. The typical deviation is such that the best-fitting BET equation predicts too little adsorption at low pressures and too much at high pressures.

The BET equation also seems to cover three of the five isotherm types described in Fig. XVII-7. Thus for c large, that is, $Q_1 \gg Q_v$, it reduces to the Langmuir equation, Eq. XVII-43, and for small c values, type III isotherms result, as illustrated in Fig. XVII-10. However, the adsorption of relatively inert gases such as nitrogen and argon on polar surfaces generally gives c values around 100, corresponding to type II isotherms. For such systems the approximate form of Eq. XVII-62,

$$\frac{n}{n_m} = \frac{1}{1-x} \tag{XVII-64}$$

works quite well in the usual region of fit of the BET equation, and a "one point" method of surface area estimation thus follows (see Ref. 47). The method has been incorporated in commercial rapid surface area determination equipment.

C. Modifications of the BET Equation

The very considerable success of the BET equation stimulated various investigators to consider modifications of it that would correct certain approximations and give a better fit to type II isotherms. Thus if it is assumed that multilayer formation is limited to n layers, perhaps because of the opposing walls of a capillary being involved, one

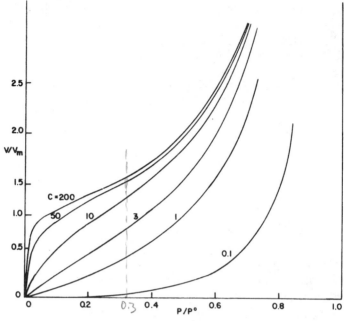

Fig. XVII-10. BET isotherms.

obtains (1)

$$v = \frac{[v_m c x/(1 - x)][1 - (n + 1)x^n + nx^{n+1}]}{1 + (c - 1)x - cx^{n+1}}$$ (XVII-65)

By choosing the appropriate n value, the amount of adsorption predicted at large P/P^0 values is reduced, and a better fit to data can usually be obtained. Kiselev and co-workers have proposed an equation paralleling Eq. XVII-55, but that reduces to the BET equation if K_n, the lateral interaction parameter, is zero [18]. There are other modifications that also give better fits to data but also introduce one or more additional parameters. Young and Crowell [47a] and Gregg and Sing [48] may be referred to for a more detailed summary of such modifications; see Sircar [49] for a recent one. One interesting example is the following. If the adsorbed film is not quite liquidlike so that $a_i/b_i \neq b_0^0$, its effective vapor pressure will be $P^{0\prime}$ rather than P^0 (note Section X-6C), and the effect is to multiply x by a factor k [50]. Redhead [51] discusses a composite model that approaches the BET equation at one limit and the Frenkel–Halsey–Hill one at another (see Section XVII-7B).

6. Isotherms Based on the Equation of State of the Adsorbed Film

It was shown in Section X-3B that a formal calculation can be made of the film pressure, or reduction in the surface free energy of the solid, by evaluating the integral of $nd \ln P$. Conversely, if it is assumed that the adsorbed material can be treated as a two-dimensional film of one of the types found with monolayers on liquid substrates, one has a large new source of isotherm equations.

A. Film Pressure–Area Diagrams from Adsorption Isotherms

According to Eq. X-12, straightforward application of the Gibbs equation gives

$$\pi_{\text{at } P_1} = \frac{RT}{\Sigma} \int_0^{P_1} nd \ln P \qquad\qquad \text{(XVII-66)}$$

If the isotherm is linear, Eq. XVII-66 reduces to

$$\pi = \frac{RTn_1}{\Sigma} = \Gamma RT \quad \text{or} \quad \pi\sigma = RT \qquad\qquad \text{(XVII-67)}$$

Gregg [52] (see also Ref. 48) has surveyed the types of force–area equations obtainable from adsorption isotherms.

B. Adsorption Isotherms from Two-Dimensional Equations of State

Numerous types of equations of state apply to solids, liquids, and gases; a few of these are considered here.

Ideal-Gas Law. Here Eq. XVII-67 applies, and on reversing the procedure that led to it, one finds

$$kP = \theta \qquad\qquad \text{(XVII-68)}$$

The ideal gas law equation of state thus leads to a linear or Henry's law isotherm. A natural modification adds a co-area term:

$$\pi(\sigma - \sigma^0) = RT \qquad\qquad \text{(XVII-69)}$$

from which one obtains (remembering that $\theta = \sigma^0/\sigma$)

$$kP = \frac{\theta}{1 - \theta} e^{\theta/(1 - \theta)} \qquad\qquad \text{(XVII-70)}$$

Equation XVII-70 bears a strong resemblance to the Langmuir equation (see Ref. 4)—to the point that it is doubtful whether the two could always be distinguished experimentally. An equivalent form obtained by Volmer [53] worked well for data on the adsorption of various organic vapors on mercury [54] (see Problem XVII-40).

Van der Waals Equations of State. A logical step to take next is to consider equations of state that contain both a covolume term and an attractive force term, such as the van der Waals equation. De Boer [4] and Ross and Olivier [55] have given this type of equation much emphasis.

TABLE XVII-2

Two-Dimensional Equations of State and Corresponding Isotherms

Equation of State	Corresponding Isotherm
Ideal-gas type	
$\pi\sigma = RT$	$\ln kP = \ln\theta$
$\pi(\sigma - \sigma^0) = RT$	$\ln kP = \theta/(1-\theta) + \ln[\theta/(1-\theta)]$
Van der Waals type	
$(\pi + a/\sigma^2)(\sigma - \sigma^0) = RT$	$\ln kP = \theta/(1-\theta) + \ln[\theta/(1-\theta)] - c\theta$
$(\pi + a/\sigma^3)(\sigma - \sigma^0) = RT$	$\ln kP = \theta/(1-\theta) + \ln[\theta(1-\theta)] - c\theta^2$
$(\pi + a/\sigma^3)(\sigma - \sigma^0/\sigma) = RT$	$\ln kP = 1/(1-\theta) + \frac{1}{2}\ln[\theta/(1-\theta)] - c\theta$
	$(c = 2a/\sigma^0 RT)$
Virial type	
$\pi\sigma = RT + \alpha\pi - \beta\pi^2$	$\ln kP = (\phi^2/2\omega) + (\frac{1}{2}\omega)(\phi + 1)$
	$[(\phi - 1)^2 + 2\omega)]^{1/2}$
	$\quad - \ln\{(\phi - 1) + [(\phi - 1)^2 + 2\omega]^{1/2}\}$
	$(\phi = 1/\theta, \omega = 2\beta RT/\alpha^2)$
Condensed film	(see Ref. 56)
$\pi = b - a\sigma$	$\ln(P/P^0) = B - (A/n^2)$

It must be remembered that, in general, the constants a and b of the van der Waals equation depend on volume and on temperature. Thus a number of variants are possible, and some of these and the corresponding adsorption isotherms are given in Table XVII-2. All of them lead to rather complex adsorption equations, but the general appearance of the family of isotherms from any one of them is as illustrated in Fig. XVII-11. The dotted line in the figure represents the presumed actual course of that particular isotherm and corresponds to a two-dimensional condensation from gas to liquid. Notice the general similarity to the plots of the Langmuir plus the lateral interaction equation shown in Fig. XVII-4.

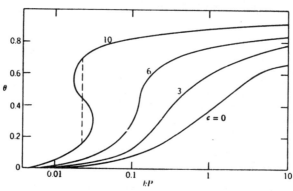

Fig. XVII-11. The van der Waals equation of state isotherm.

Equations such as in Table XVII-2 may well apply both experimentally and in concept to selected cases of adsorption in the submonolayer region (see Ref. 57 for example). The difficulty is that, as with the various modified BET equations, there are enough possibilities for algebraic variations and enough parameters that it is difficult to know how much significance should be attached to ability to fit data. The matter of possible phase changes is discussed in Section XVII-11 and that of the complicating effect of surface heterogeneity and of surfaces having a fractal dimension in Section XVII-15.

7. The Potential Theory

A. The Polanyi Treatment

A still different approach to multilayer adsorption considers that there is a potential field at the surface of a solid into which adsorbate molecules "fall." The adsorbed layer thus resembles the atmosphere of a planet—it is most compressed at the surface of the solid and decreases in density outward. The general idea is quite old, but was first formalized by Polanyi in about 1914—see Brunauer [34]. As illustrated in Fig. XVII-12, one can draw surfaces of equipotential that appear as lines in a cross-sectional view of the surface region. The space between each set of equipotential surfaces corresponds to a definite volume, and there will thus be a relationship between potential U and volume ϕ.

If we consider the case of a gas in adsorption equilibrium with a surface, there must be no net free energy change on transporting a small amount from one region to the other. Therefore, since the potential U_x represents the work done by the adsorption forces when adsorbate is brought up to a distance x from the surface, there must be a compensating compressional increase in the free energy of the adsorbate. Thus

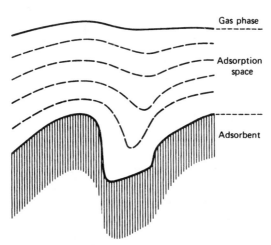

Fig. XVII-12. Isopotential contours. (From Ref. 34.)

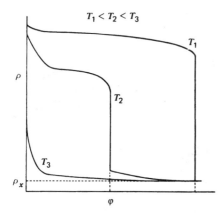

Fig. XVII-13. Variation of the density with the adsorbed phase according to the potential theory.

$$U_x = \int_{\mathbf{P}g}^{\mathbf{P}x} V \, d\mathbf{P} \tag{XVII-71}$$

The mass of adsorbate present is given by

$$w = \Sigma \int_0^\infty (\rho_x - \rho_g) \, dx \tag{XVII-72}$$

and the equation of state of the adsorbate gives the necessary relationship between the density ρ and the pressure. These equations allow any adsorption isotherm to be converted to a U–ϕ plot, called the *characteristic curve* or, alternatively, to a ρ versus ϕ plot. A schematic example of the latter is given in Figure XVII-13 to illustrate that at high temperatures the adsorbate remains gaseous and is merely compressed in the potential field, but at some lower temperature the pressure of the adsorbate will reach P^0 and condensation to liquid will occur.

The treatment at this point is essentially thermodynamic—it does not imply any particular function for $U(\phi)$, neither does it provide a determination of Σ. However, the function $U(\phi)$ should be temperature independent, that is, for a given system all isotherms should give the same characteristic function. In many cases this seems to be true. Thus McGavack and Patrick [57a] found that for sulfur dioxide on silica gel the same characteristic curve was obtained for eight temperatures between 373 and 193 K. The treatment can be applied to adsorption in the submonolayer region, in which case U is really a function of θ rather than of ϕ, that is, one is dealing with a heterogeneous surface. It turns out that the calculation of a site energy distribution for a heterogeneous surface amounts to obtaining a characteristic curve for the surface (see Section XVII-14).

In the case of multilayer adsorption it seems reasonable to suppose that condensation to a liquid film occurs (as in curves T_1 or T_2 of Fig. XVII-13). If one now assumes that the amount adsorbed can be attributed entirely to such a film, and that the liquid is negligibly compressible, the thickness x of the film is related to n by

$$n = \frac{\Sigma x}{V_l} = \frac{\phi}{V_l} \qquad \text{(XVII-73)}$$

where V_l is the molar volume of the liquid, and the potential at x, being just sufficient to cause condensation, is given by

$$U_x = RT \ln \frac{P^0}{P_g} \qquad \text{(XVII-74)}$$

Where P_g is the pressure of vapor in equilibrium with the adsorbed film. The characteristic curve is now just $RT \ln P_0/P$ versus x (or against ϕ, if Σ is not known). Dubinin and co-workers (see Ref. 58) have made much use of a semiempirical relation between ϕ and U_x:

$$\phi = \phi_0 \exp(-bU^2) \qquad \text{(XVII-75)}$$

where, for porous solids, ϕ_0 is taken to be the pore volume, and ϕ, the volume adsorbed at a given P^0/P_g value. See Ref. 59 for a variant of Eq. XVII-75 known as the *Dubinin–Radushkevich* equation. In another variant, it is proposed that n be taken as proportional to $\ln(U/RT)$ [60].

B. Isotherms Based on an Assumed Variation of Potential with Distance

A monolayer can be regarded as a special case in which the potential is a square well; however, the potential well may take other forms. Of particular interest now is the case of multilayer adsorption, and a reasonable assumption is that the principal interaction between the solid and the adsorbate is of the dispersion type, so that for a plane solid surface the potential should decrease with the inverse cube of the distance (see Section VI-3A). To avoid having an infinite potential at the surface, the potential function may be written

$$U(x) = \frac{U_0}{(a + x)^3} \qquad \text{(XVII-76)}$$

where a is a distance of the order of a molecular radius.†

---

†Equation XVII-76 is based on the assumption that the solid is continuous and actually is a rather poor approximation for the case of a molecule in the first few layers of a multilayer on a crystalline surface.

On combining Eqs. XVII-74 and XVII-76, one obtains

$$RT\ln\frac{P^0}{P} = \frac{U_0}{(a+x)^3} \qquad\qquad \text{(XVII-77)}$$

where P is the gas pressure, the subscript g no longer being needed for clarity. On solving for x and substituting into Eq. XVII-74,

$$n = -\alpha + \beta\omega^{-1/3} \qquad\qquad \text{(XVII-78)}$$

where

$$\alpha = \frac{a\Sigma}{V_l} \qquad \beta = \frac{\Sigma}{V_l}\left(\frac{U_0}{RT}\right)^{1/3} \qquad \omega = \ln\frac{P^0}{P}$$

The physical model is thus that of a liquid film, condensed in the inverse cube potential field, whose thickness increases to infinity as P approaches P^0.

Equation XVII-78 turns out to fit type II adsorption isotherms quite well—generally better than does the BET equation. Furthermore, the exact form of the potential function is not very critical; if an inverse square dependence is used, the fit tends to be about as good as with the inverse-cube law, and the equation now resembles that for a condensed film in Table XVII-2. Here again, quite similar equations have resulted from deductions based on rather different models.

The general approach goes back to Frenkel [63] and has been elaborated on by Halsey [64], Hill [65], and McMillan and Teller [66]. A form of Eq. XVII-78, with $a = 0$,

$$\theta = \left(\frac{n}{n_m}\right)^n = \frac{A}{\ln(P^0/P)} \qquad A = \frac{U_0}{x_m^n RT} \qquad \checkmark \text{(XVII-79)}$$

where x_m is the film thickness at the monolayer point, is frequently referred to as the *Frenkel–Halsey–Hill* equation. In this form, the power n in $U = U_0/r^n$ may be left as an empirical parameter, and in Halsey's tabulation of n values for various systems [64], values between 2 and 3 are fairly common. Pierce [67] finds $n = 2.75$ for nitrogen essentially independent of the solid; additional n values are given in Refs. 68–70.

As with the BET equation, a number of modifications of Eqs. XVII-77 or XVII-79 have been proposed, for example Ref. 71. FHH-type equations go to infinite film thickness (i.e., bulk liquid), as $P \rightarrow P^0$ and this cannot be the case if the liquid does not wet the solid, and Adamson [72] proposed

$$kT\ln\frac{P^0}{P} = \frac{g}{x^3} + U_0 e^{-ax} - \beta e^{-\alpha x} \qquad \text{(XVII-80)}$$

The first term on the right is the common inverse cube law, the second is taken to be the empirically more important form for moderate film thickness (and also conforms to the polarization model, Section XVII-7C), and the last term allows for structural perturbation in the adsorbed film relative to bulk liquid adsorbate. In effect, the vapor pressure of a thin multilayer film is taken to be $P^{0\prime}$ and to relax toward P^0 as the film thickens. The equation has been useful in relating adsorption isotherms to contact angle behavior (see Section X-7). Roy and Halsey [73] have used a similar equation; earlier, Halsey [74] allowed for surface heterogeneity by assuming a distribution of U_0 values in Eq. XVII-79. Dubinin's equation (Eq. XVII-75) has been mentioned; another variant has been used by Bonnetain and co-workers [75].

The potential model has been applied to the adsorption of mixtures of gases. In the *ideal adsorbed solution* model, the adsorbed layer is treated as a simple solution, but with potential parameters assigned to each component (see Refs. 76–79).

C. The Polarization Model

An interesting alternative method for formulating $U(x)$ was proposed in 1929 by de Boer and Zwikker [80], who suggested that the adsorption of nonpolar molecules be explained by assuming that the polar adsorbent surface induces dipoles in the first adsorbed layer and that these in turn induce dipoles in the next layer, and so on. As shown in Section VI-8, this approach leads to

$$U(x) = U_0 e^{-ax} \qquad \text{(XVII-81)}$$

which, in combination with Eq. XVII-74 gives

$$RT\ln\frac{P^0}{P} = U_0 e^{-ax} \qquad \text{(XVII-82)}$$

On using Eq. VI-44 and taking $V_l/d_0\Sigma$ to be n/n_m, one obtains

$$\ln\ln\frac{P^0}{P} = \ln\frac{U_0}{RT} - \left[\ln\left(\frac{d^3}{\alpha}\right)^2\right]\left(\frac{n}{n_m}\right) \qquad \text{(XVII-83)}$$

Thus a plot of log log (P^0/P) versus n should give a straight line (see Fig. XVII-14). If Eq. XVII-83 is applied to typical type II isotherms, such as CO or N_2 on silica "C" [81] or N_2 on KCl [82], α/d^3 comes out to be about 0.4, from which d is about an atomic radius.

The polarization model suggests strongly that orientational effects should be present in multilayers. As seen in Section X-6, such perturbations are essential to the explanation of contact angle phenomena.

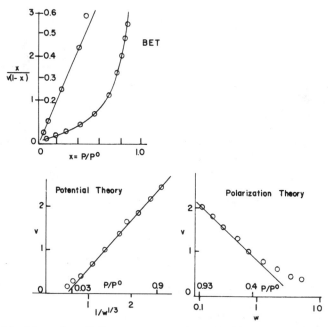

Fig. XVII-14. Adsorption of nitrogen on potassium chloride at 79 K, plotted according to various equations. (Data from Ref. 82.)

8. Comparison of the Surface Areas from the Various Multilayer Models

It has been emphasized repeatedly that the fact that an isotherm equation fits data is an insufficient test of its validity (although not of its practical usefulness). To further stress this point, the data for a particular isotherm for the adsorption of nitrogen on powdered potassium chloride at 78 K are plotted in Fig. XVII-14 according to four equations; in each case there is a satisfactory fit. For simple gases such as nitrogen, oxygen, and argon, the BET equation generally fits over the range of P/P^0 values of 0.05 to 0.3; the potential theory (Eq. XVII-78), over the range 0.1 to 0.8; and similarly for the polarization equation, Eq. XVII-83. There is thus little to choose from between the various models, but partly because of tradition and familiarity, and partly because n_m enters in it so explicitly, the BET equation is in fact almost exclusively used, even though BET Σ values can differ significantly from those obtained by other methods [83].

A seemingly more stringent test would be to determine whether the ratios of areas for various solids as obtained by means of a given isotherm equation are independent of the nature of the adsorbate. The data summarized in Table XVII-3 were selected from the literature mainly because each author had obtained areas for two or more solids using two or more adsorbates. It is seen

TABLE XVII-3

Gas/K	BET	Potential Theory[b]	Polarization Theory[c]	Characteristic Isotherm
			Ratio of Values[a]	
Σ of egg albumin[d] /Σ of KCl 2				
O_2/90	4.0	5.5	5.3	5.2
Ar/90	4.0	5.7	5.6	6.7
N_2/78	5.3	5.7	4.4	5.4
	4.4	5.6	5.1	5.8
	±15%	±2%	±6%	±10%
Σ of TiO_2/Σ of KCl 2				
O_2/78	5.2	8.1	6.7	8.1
N_2/78	6.4	7.9	6.7	7.1
	5.8	8.0	6.7	7.6
	±10%	±1%	±0%	±7%
Σ of Si(C)/Σ Sterling S450				
CO/78	3.0	2.9	2.7	3.0
N_2/78	2.9	2.6	2.8	3.0
C_2H_5Cl/195	2.7	3.2	2.5	2.4
	2.9	2.9	2.7	2.8
	±4%	±7%	±4%	±9%

[a]Data are from the following sources: egg albumin, Ref. 84; KCL, Ref. 82; TiO$_2$, Ref. 84; Si(C) and Sterling S450, Ref. 85.
[b]Eq. XVII-77.
[c]Eq. XVI-83.
[d]This was a lyophilized, dry protein.

that the BET equation gives ratios of areas that may vary by 10–15% from one adsorbate to another. The performance of the potential theory in its two forms is somewhat better (the characteristic isotherm, discussed in the next section, is a form of potential theory). Note that the BET analysis gives ratios noticeably different from those given by the other four; this is probably a reflection of the fact that the region of fit of the BET equation is over a lower P/P^0 range than that for other models—this is to be expected if the surfaces are fractal in nature (see Section XVI-2B). The relative merits of various models are discussed in more detail in Section XVII-14.

9. The Characteristic Isotherm and Related Concepts

The relatively uniform success of these various plots suggests that, except as modified by changes in Σ, the shape of the isotherm in the multilayer region tends to be characteristic of the adsorbate and independent of the nature of the

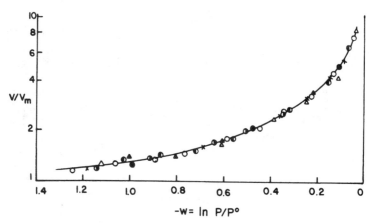

Fig. XVII-15. Characteristic isotherm for nitrogen at 78 K on various solids: ○, RCl-1 (Ref. 82); △, egg albumin 61 (Ref. 82); ◉, bovine albumin 68 (Ref. 84); ×, titanium dioxide (Ref. 88); ◎, Graphon (Ref. 89); ▲, egg albumin, 59 (Ref. 84); ●, polyethylene. [From A. C. Zettlemoyer, A. Chand, and E. Gamble, *J. Am. Chem. Soc*, **72,** 2752 (1950).]

solid. The test of this is that on plotting log n versus some arbitrary function of P/P^0, the isotherms for a given adsorbate and various solids should be identical except for a vertical displacement. As noted in the first edition (1960; see Ref. 86), this prediction turns out amazingly well. Figure XVII-15 shows the superposition of a number of nitrogen adsorption isotherms. Each adsorbent gives the same isotherm shape, to within experimental error, and if each is adjusted vertically by choosing the best value of n_m, all of the data fall on a single curve over the range of 0.3–0.95 in P/P^0. Below P/P^0 about 0.3 there are increasing differences between the curves, presumably as the individual nature of the various solids make themselves felt. Somewhat similar early observations have been made by Pierce [67] and by Halász and Schay [87], and by Dubinin [58].

The curve of Fig. XVII-15 is essentially a characteristic curve of the Polanyi theory, but in the form plotted in might better be called a *characteristic isotherm*. Furthermore, as would be expected from the Polanyi theory, if the data for a given adsorbate are plotted with $RT \ln(P/P^0)$ as the abscissa instead of just $\ln(P/P^0)$, then a nearly invariant shape is obtained for different temperatures. The plot might then be called the *characteristic adsorption curve*.

The existence of this situation (for nonporous solids) explains why the ratio test discussed above and exemplified by the data in Table XVII-3 works so well. Essentially, *any* isotherm fitting data in the multilayer region *must* contain a parameter that will be found to be proportional to surface area. In fact, this observation explains the success of the point B method (as in Fig. XVII-7) and other single-point methods, since for *any* P/P^0 value in the characteristic isotherm region, the measured n is related to the surface area of the solid by a proportionality constant that is independent of the nature of the solid.

TABLE XVII-4
The t Plot for N_2 at 78 K^a

P/P^0	t (Å)	P/P^0	t (Å)	P/P^0	t (Å)	P/P^0	t (Å)
0.08	3.51	0.32	5.14	0.56	6.99	0.80	10.57
0.10	3.68	0.34	5.27	0.57	7.17	0.82	11.17
0.12	3.83	0.36	5.41	0.60	7.36	0.84	11.89
0.14	3.97	0.38	5.56	0.62	7.56	0.86	12.75
0.16	4.10	0.40	5.71	0.64	7.77	0.88	13.82
0.18	4.23	0.42	5.86	0.66	8.02	0.90	14.94
0.20	4.36	0.44	6.02	0.68	8.26	0.92	16.0^b
0.22	4.49	0.46	6.18	0.70	8.57	0.94	17.5^b
0.24	4.62	0.48	6.34	0.72	8.91	0.96	19.8^b
0.26	4.75	0.50	6.50	0.74	9.27	0.98	22.9^b
0.28	4.88	0.52	6.66	0.76	9.65		
0.30	5.01	0.54	6.82	0.78	10.07		

^{a}From Ref. 90.
^{b}These are extrapolated values and undoubtedly contain an important contribution from interparticle condensation.

The characteristic isotherm concept was elaborated by de Boer and co-workers [90]. By accepting a reference n_m from a BET fit to a standard system and assuming a density for the adsorbed film, one may convert n/n_m to film thickness t. The characteristic isotherm for a given adsorbate may then be plotted as t versus P/P^0. For any new system, one reads t from the standard t-curve and n from the new isotherm, for various P/P^0 values. De Boer and co-workers' t values are given in Table XVII-4. A plot of t versus n should be linear if the experimental isotherm has the same shape as the reference characteristic isotherm, and the slope gives Σ:

$$\Sigma = \frac{15.47v}{t} = \frac{3.26 \times 10^5 n}{t} \qquad \text{(XVII-84)}$$

where Σ is in square meters per gram, v in cubic centimeters STP per gram, and t, in angstrom units. Sing [91] has reviewed much of the literature on such uses of the characteristic isotherm. To avoid referencing to the BET equation, he proposed the use of a quantity $\alpha_s = n/n_x$ instead of t, where n_x is the amount adsorbed at $P/P^0 = 0.4$. For nonporous carbons, Eq. XVII-84 becomes [92, 93]

$$\Sigma = \frac{2.86v}{\alpha_s} \qquad \text{(XVII-85)}$$

(See Refs. 92–94 for additional tables of the type of Table XVII-4.)

The existence of a characteristic isotherm for each adsorbate is encouraging in the sense that it appears that investigators interested primarily in relative surface area values can choose their isotherm equation on the basis of convenience and with little concern as to whether it is a fundamentally correct one. On the other hand, one would like to know if there are any points of experimental distinction that are diagnostic of the correctness of an isotherm model. Such an evaluation is only possible to some extent and is discussed in Section XVII-13.

10. Chemical Physics of Submonolayer Adsorption

The treatment of physical adsorption has so far been based on more or less plausible physical models leading to expressions for an adsorption isotherm. Historically, this has been the productive approach, focused on surface area determination. The multilayer region was the one of interest, with submonolayer adsorption viewed mainly as a means of exploring adsorbent heterogeneity (see Section XVII-14). We return to this phenomenological approach in following sections, but recognize here the important developments in recent years in which the methods of chemisorption (note Section XVIII-2) have been applied to physisorption systems. Clean, well-defined surfaces are used, the adsorption is studied at sufficiently low temperatures that the ambient vapor pressure is low enough to permit the use of the various diffraction and spectroscopic techniques as well as of the various microscopies such as scanning tunneling (STM) and atomic force (AFM).

Physical adsorption may now, in fact, be seen as a preamble to chemisorption in heterogeneous catalysis, that is, as a *precursor* state (see Section XVIII-4). Physical adsorption is not considered to involve chemical bond formation, however, and the current theoretical approaches deal mainly with the relatively long range electrostatic and van der Waals types of forces. The quantum mechanics of chemical bonding is thus largely missing although aspects of it appear in the treatment of physisorption on metal surfaces. Steele [8] provides an extensive review of molecular interactions in physical adsorption generally, and for the case of molecules adsorbed on the graphite basal plane in particular [95].

Infrared Spectroscopy. The infrared spectroscopy of adsorbates has been studied for many years, especially for chemisorbed species (see Section XVIII-2C). In the case of physisorption, where the molecule remains intact, one is interested in how the molecular symmetry is altered on adsorption. Perhaps the conceptually simplest case is that of H_2 on NaCl(100). Being homopolar, H_2 by itself has no allowed vibrational absorption (except for some weak collision-induced transitions) but when adsorbed, the reduced symmetry allows a vibrational spectrum to be observed. Fig. XVII-16 shows the infrared spectrum at 30 K for various degrees of monolayer coverage [96] (the adsorption is Langmuirian with half-coverage at about 10^{-4} atm). The bands labeled sf are for transitions of H_2 on a smooth face and are from the $J = 0$ and $J = 1$ rotational states; $Q_{sf}(R)$ is assigned as a combination band. The bands labeled

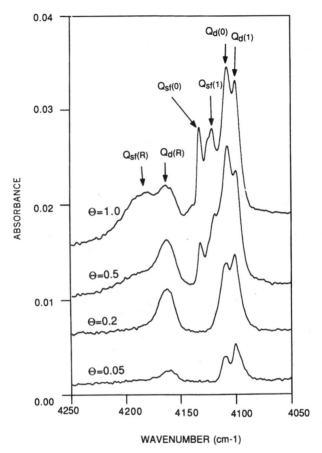

Fig. XVII-16. Infrared absorption spectra of H_2 physisorbed on NaCl(100) at 30 K. See text for explanations. (From Ref. 96. Reprinted with permission from American Institute of Physics, copyright 1993.)

d are for H_2 on a defect site. The hydrogen molecule is believed to be adsorbed over a Na^+ ion and with the molecular axis parallel to the surface. The vibrational transitions have become allowed because of the surface electric field; the mechanism is much the same as for the gas in a strong external field.

Two other examples will suffice. Methane physisorbs on NaCl(100) and an early study showed that the symmetrical, IR-inactive ν_1 mode could now be observed [97]. In more recent work, polarized FTIR reflection spectroscopy was used to determine that on being adsorbed, the three-fold degeneracies of the ν_3 and ν_4 modes were partially removed [98]. This finding allowed consideration of possible adsorbate–adsorbent geometries; one was that of a tripod with three of the methane hydrogens on the surface. The systems were at between 4 and 40 K so that the equilibrium pressure was very low, about 10^{-10} atm.

As a final example, similar spectroscopy was carried out for CO_2 physisorbed on MgO(100) [99]. Temperatures were around 80 K and equilibrium pressures, as low as 10^{-13} atm (at higher temperatures, CO_2 chemsorbs to give surface carbonate). Here, the variation of the absorbance of the infrared bands with the polarization of the probe beam indicated that the surface CO_2 phase was highly oriented.

Submonolayer Phases and Phase Transitions. There is now a considerable literature on two-dimensional (2D) phases and phase transformations for a variety of species physically adsorbed on clean, smooth surfaces such as those of alkali halide crystals, exfoliated graphite, and cleaved MgO. As an example, the phase diagram for O_2 on graphite is shown in Fig. XVII-17. Graphite is a popular surface because of the freedom of the basal plane surface from defects. Note that most of the diagram involves temperatures below the three-dimensional (3D) or bulk melting point of 55 K. There is a low-density solid phase, δ, and several higher-density solid phases ($\epsilon, \zeta_1, \zeta_2$) having different unit cell orientations but the same density. The O_2 molecules lie flat on the surface in the δ phase, but in the higher-density phases they are perpendicular to it and their lattice is not *commensurate* (i.e., is not in register with the graphite lattice). There is a 2D liquid phase and a critical temperature of 64 K. Thomy and Duval, in a useful review [100], note that the 2D critical temperature is usually about 0.4 of the 3D T_c.

The O_2–graphite system has been studied by means of quite a few techniques: LEED, RHEED, EELS, and NEXAFS (see Ref. 101). It was concluded,

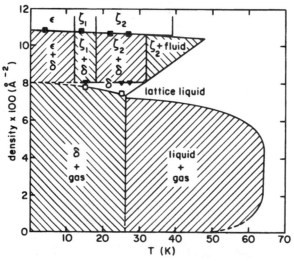

Fig. XVII-17. Schematic phase diagram for O_2 on graphite (see text). (From Ref. 95. Reprinted with permission from American Chemical Society, copyright 1996.)

from results using the latter technique, that in the δ phase the mean molecular tilt of the O_2 was 20° from the surface, while in the ζ phases, it was 37° from the surface normal [101]. The results for a variety of other systems can be found in Refs. 8 and 95–104. The field now resembles that of chemisorption in many respects. One exception is that of commensurate–incommensurate transitions without change in the adsorbent surface structure—by contrast, in chemisorption a change in film structure is likely to be accompanied by a reconstruction of the solid surface (note Section XVIII-9E). Another difference is in the theoretical approach to the adsorption bond, discussed below.

Potential Theory Approach. Potential theory may be used to treat submonolayer adsorbtion; the formal statement is

$$n_a = \mathcal{A} \int_a^\infty (C - C_g) \, dz \qquad \text{(XVII-86)}$$

where

$$C = C_g e^{-U/kT}$$

The concentration in the gas phase C_g is given by P/RT, and we have

$$K\mathcal{A} = \frac{n_a}{P} = \mathcal{A} \, \frac{1}{RT} \left[\int_a^\infty (e^{-U/kT} - 1) \, dz \right] \qquad \text{(XVII-87)}$$

where K is the Henry's law constant, that is, the limiting slope of the adsorption isotherm. It is assumed that adsorption is occurring at a sufficiently low θ that θ is proportional to pressure, and, of course, that the surface is uniform. The potential U is some function of distance z normal to the surface, and a is taken to be that at which $U(z)$ is zero; that is, van der Waals attraction and the short-range repulsion exactly balance.

The Lennard-Jones potential (Eq. III-46) is commonly used:

$$U(z) = -4U_0 \left[\left(\frac{\sigma}{z} \right)^6 - \left(\frac{\sigma}{z} \right)^{12} \right] \qquad \text{(XVII-88)}$$

where U_0 is the depth of the potential well (see Fig. III-6) and σ is an effective molecular diameter. More accurately, an exponential repulsion term may be used:

$$U(z) = -\frac{A}{z^6} + Be^{-Cz} \qquad \text{(XVII-89)}$$

where A, B, and C are constants. The attractive term represents the dispersion interaction and, being long range, must be summed pairwise between the atoms in the adsorbent and in the adsorbate, so that Eq. XVII-89 takes the form (note Section VI-3A regarding such summations or integrations)

$$U(z) = \frac{A'}{z^3} + Be^{-Cz} \qquad\qquad \text{(XVII-90)}$$

The constants in Eqs. XVII-88–XVII-90 may be calculated from theory to give the Henry's law constant K; from Eq. XVII-87, the experimental n_a/P then gives the surface area. Alternatively, the constants may be arrived at from an experimental K (assuming that $\mathcal{A}$ is known) and either the isosteric heat of adsorption

$$q_{st} = -RT^2 \left(\frac{\partial \ln K}{\partial T} \right)_{n_a} \qquad\qquad \text{(XVII-91)}$$

or the desorption energy from thermal desorption measurements. The most accurate rendering of $U(z)$, however, has come from molecular beam experiments where H, H_2, or HD is scattered inelastically by the surface. Vidali et al. [104] discuss this in detail and give an extensive table of parameters for $U(z)$. Contemporary computing power is such that many detailed contour maps have been obtained for the variation in adsorption energy as an adsorbate is moved around on the surface of a crystalline surface. Fig. XVII-18 shows the case of Kr on graphite.

As usual, there are complications in the details. The inverse 6th power term in Eq. XVII-88 for the dispersion attraction is not accurate at small distances since higher-order terms become significant (note Eq. VI-19). The *corrugation* of the surface, either topological or energetic, makes the interpretation of scattering experiments complicated. In the case of polyatomic adsorbates, $U(z)$ also depends on the molecular orientation; and in calculating the coefficient for the dispersion term, bond as well as atomic polarizabilities must be known. The case of molecular solids as adsorbents is interesting. Here, there are no broken bonds at the surface, nor ions. The results for N_2 adsorbed on NH_3, CO_2, CH_3OH and I_2 did not agree with dispersion theory [105,106]. There are surface bond dipoles, however, and perhaps dipole-induced dipole interactions were important. Finally, physisorption on metals has been a difficult subject to treat. While experimental potentials have been obtained, the theory is made complicated by the matter of image forces, and the approach is now rather quantum-mechanical, such as using density functional theory (see Section XVIII-6A). References 8, 104, and 107 discuss the topic.

We conclude with the matter of adsorbate–adsorbate interactions; these give rise to deviations from Henry's law behavior. These may be expressed in the form of a virial equation, much as is done for imperfect gases. Following Steele [8], one can write

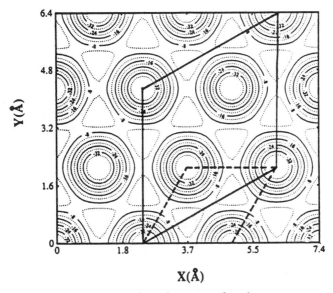

Fig. XVII-18. Contours of constant adsorption energy for a krypton atom over the basal plane of graphite. The carbon atoms are at the centers of the dotted triangular regions. The rhombuses show the unit cells for the graphite lattice and for the commensurate adatom lattice. (From Ref. 8. Reprinted with permission from American Chemical Society, copyright 1993.)

$$P = \frac{n_a}{K} \exp\left[2B_{2D} \frac{n_a}{\mathcal{A}} + \frac{3}{2} C_{2D} \left(\frac{n_a}{\mathcal{A}}\right)^2 + \cdots \right] \qquad (XVII\text{-}92)$$

where B_{2D}, and C_{2D}, and so on are the virial coefficients for adsorbate pairs, triplets, and so forth. The coefficient B_{2D} may be calculated from one or another assumed potential function (including angular dependencies in the case of diatomic or polyatomic adsorbates [108]). There is the matter of choice of a molecular size, now in the form of a two-dimensional collision cross section; this may either be taken from data on the deviation from ideality of the gaseous adsorbate, or by use of the preceding analysis of K itself. An added complication is that the adsorbed molecules do not really move in a plane but have vibrational motions perpendicular to the surface, so that $n_a/\mathcal{A}$ does not give the true average distance apart. The procedure has now become very complex, with several cycles of approximations needed to find that set of parameters giving the best fit to K, to the deviation from K, and to their temperature dependencies.

11. Phase Transformations in the Multilayer Region

The multilayer isotherms illustrated thus far have all been of a continuous appearance—it was such isotherms that the BET, FHH, and other equations treated. About 30 years ago, however, multilayer adsorption on smooth surfaces such as those of graphite and boron nitride began to be reported to give stepped isotherms as in Fig. XVII-19, the steps in some cases varying with temperature, indicating that phase transformations occurred. In fact, one of the first stepped isotherms was reported in 1952 by Bonnetain; this history is reviewed in Ref. 100. Halsey, in his 1965 Kendall Award paper, assembled a hypothetical isotherm showing the then known types of phase changes in the multilayer region [109]. Figure XVII-19 shows a more recent one, for CH$_4$ on MgO at 77 K. Such steps indicate a first-order phase transition; often the temperature is below the bulk melting point of the adsorbate and the surface phase changes could be from solid-like to liquid-like, or between two liquid-like phases. A change from a commensurate to an incommensurate phase may also be involved. Typically, the steps become smaller as the temperature is raised, and disappear at a critical temperature.

The temperature may be low enough to permit spectroscopic examination of

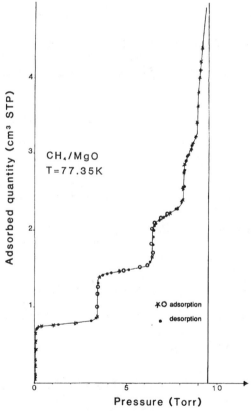

Fig. XVII-19. Adsorption of CH$_4$ on MgO(100) at 77.35 K. The vertical line locates P^0; each vertical step corresponds to the condensation of a monolayer. There was no hysteresis. Desorption points are shown as ●. (From Ref. 110.)

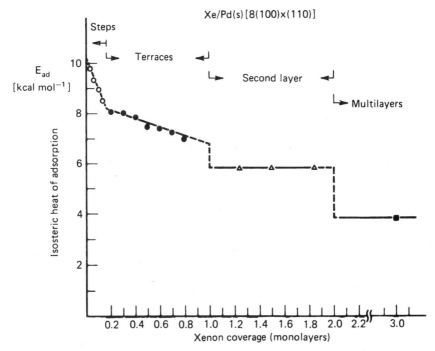

Fig. XVII-20. Isosteric heat of adsorption of Xe on a stepped Pd surface [8(100) × (110)]. (From Ref. 111.)

the film. Figure XVII-20 shows stepped isosteric heat changes for Xe on a Pd surface, the phases and to some extent their structure having been studied by means of Auger spectroscopy (AES). As another example, combined adsorption and calorimetric data were obtained for Ar on BN at 77.6 K; phase changes were found around the completion of the first layer and between the second and third layers [112].

Some general points are the following. One precondition for a vertical step in an isotherm is presumably that the surface be sufficiently uniform that the transition does not occur at different pressures on different portions, with a resulting smearing out of the step feature. It is partly on this basis that graphitized carbon, BN, MgO, and certain other adsorbents have been considered to have rather uniform surfaces. Sharp LEED patterns are another indication.

12. Thermodynamics of Adsorption

A. Theoretical Considerations

We take up here some aspects of the thermodynamics of adsorption that are of special relevance to gas adsorption. Two types of processes are of interest:

the first may be called integral adsorption and may be written, for an adsorbent X_1 and an adsorbate X_2,

$$n_1^s X_1 \text{ (adsorbent at } T) + n_2^g X_2 \text{ (gaseous adsorbate at } P, T)$$
$$= n_2^s X_2 \text{ adsorbed on } n_1^s X_1 \qquad (XVII\text{-}93)$$

If the process is carried out at constant volume, the heat evolved Q_i will be equal to an energy change ΔE_2 or, per mole of adsorbate, $q_i = \Delta E_2$ (small capital letters will be used to denote mean molar quantities). Alternatively, the process may be

$$n_2^g X_2 (\text{gas}, P, T) = n_2^s (\text{adsorbed at composition } n_2^{\prime s}/n_1^{\prime s} \text{ constant}) \qquad (XVII\text{-}94)$$

The heat evolved will now be a *differential heat* of adsorption, equal at constant volume to Q_d or per mole, to $q_d = \Delta \bar{E}_2$, where $\Delta \bar{E}_2$ is the change in partial molar energy. It follows that

$$q_d = \left(\frac{\partial Q_i}{\partial n_2^s} \right)_{T,V} \qquad (XVII\text{-}95)$$

i. Adsorption Heats and Entropies. It is not necessary, phenomenologically, to state whether the process is adsorption, absorption, or solution, and for the adsorbent-adsorbate complex formal equations can be written, such as

$$dG^s = \bar{G}_1^s \, dn_1^s + \bar{G}_2^s \, dn_2^s \qquad (XVII\text{-}96)$$

and

$$G^s = n_1^s \bar{G}_1^s + n_2^s \bar{G}_2^s \qquad (XVII\text{-}97)$$

However, a body of thermodynamic treatment has been developed on the basis that the adsorbent is inert and with attention focused entirely on the adsorbate. The abbreviated presentation given here is based on that of Hill (see Refs. 65 and 113) and of Everett [114]. First, we have the defining relationships:

$$dE^s = T \, dS^s - \pi \, d\mathcal{A} + \mu_2 \, dn_2^s \qquad E^s = TS^s - \pi\mathcal{A} + \mu_2 n_2^s \qquad (XVII\text{-}98)$$
$$dH^s = T \, dS^s - \pi \, d\mathcal{A} + \mu_2 \, dn_2^s \qquad H^s = TS^s - \pi\mathcal{A} + \mu_2 n_2^s \qquad (XVII\text{-}99)$$
$$dA^s = -S^s \, dT - \pi \, d\mathcal{A} + \mu_2 \, dn_2^s \qquad A^s = -\pi\mathcal{A} + \mu_2 n_2^s \qquad (XVII\text{-}100)$$
$$dG^s = -S^s \, dT - \pi \, d\mathcal{A} + \mu_2 \, dn_2^s \qquad G^s = -\pi\mathcal{A} + \mu_2 n_2^s \qquad (XVII\text{-}101)$$

The second set of equations is obtained from the first set by the Gibbs integration at constant intensive variables, as was done in obtaining Eq. III-77. It is convenient, in dealing with a surface species, to introduce some special definitions, two of which are

$$d\mathcal{H}^s = T\,dS^s + \mathcal{A}\,d\pi + \mu_2\,dn_2^s \qquad \mathcal{H}^s = TS^s + \mu_2 n_2^s \qquad \text{(XVII-102)}$$

$$dG^s = -S^s\,dT + \mathcal{A}\,d\pi + \mu_2\,dn_2^s \qquad G^s = \mu_2 n_2^s \qquad \text{(XVII-103)}$$

The expressions for $\mathcal{H}^s$ and G^s now have the same appearance as those for H and G in a bulk system.

Now, considering process XVII-93, where P is the equilibrium pressure of the adsorbate, we have

$$\Delta\bar{G}_2 = \bar{G}_2^s - \bar{G}_2^{0,\,g} = RT\ln P \qquad \text{(XVII-104)}$$

where the quantity $\bar{G}_2^s$ is defined as $(\partial G_2^s/\partial n_2^s)_{T,A} = \mu_2$. A standard thermodynamic relation is that

$$\left[\frac{\partial(\bar{G}_2^{0,\,g}/T)}{\partial T}\right]_{P,\,n_2^s} = -\frac{\bar{H}_2^{0,\,g}}{T^2} \qquad \text{(XVII-105)}$$

where the superscript zero means that a standard state (gas at 1 atm) is referred to. It follows from the preceding definitions that

$$\left[\frac{\partial(\bar{G}_2^s/T)}{\partial T}\right]_{A,\,n_2^s} = -\frac{\bar{H}_2^s}{T^2} \qquad \text{(XVII-106)}$$

In combination with Eq. XVII-104 we obtain

$$\left(\frac{\partial\ln P}{\partial T}\right)_\Gamma = -\frac{\bar{H}_2^s - \bar{H}_2^g}{RT^2} = -\frac{\Delta\bar{H}_2}{RT^2} = \frac{q_{st}}{RT^2} \qquad \text{(XVII-107)}$$

dropping unnecessary superscripts for the gas phase, assumed to be ideal, and remembering that $n_2^s/\mathcal{A} = \Gamma$; q_{st} is called the *isosteric heat of adsorption*.

Since $\bar{G}_2^s = (\partial G_2^s/\partial n_2^s)_\pi = \mu_2$ and $G_2^s = \bar{G}_2^s$, an alternative statement to Eq. XVII-104 is

$$\Delta G_2 = G_2^s - G_2^{0,\,g} = RT\ln P \qquad \text{(XVII-108)}$$

and

$$\left[\frac{\partial(G_2^s/T)}{\partial T}\right]_{\pi} = -\frac{\mathcal{H}_2^s}{T^2} \qquad \text{(XVII-109)}$$

it follows that

$$\left(\frac{\partial \ln P}{\partial T}\right)_{\pi} = -\frac{\mathcal{H}_2^s - H_2^g}{RT^2} = -\frac{\Delta\mathcal{H}_2}{RT^2} = \frac{q_{\pi}}{RT^2} \qquad \text{(XVII-110)}$$

The quantity $\Delta\mathcal{H}_2$ has been called (by Hill) the *equilibrium heat of adsorption*. It follows from the foregoing definitions that

$$\mathcal{H}_2^s = H_2^s + \frac{\pi}{\Gamma} \qquad \text{(XVII-111)}$$

also, it can be shown [115] that

$$\Delta\mathcal{H}_2 = \Delta\bar{H}_2 + \frac{T}{\Gamma}\left(\frac{\partial \pi}{\partial T}\right)_{P,\Gamma} \qquad \text{(XVII-112)}$$

To summarize, the four common heat quantities are

1. Integral calorimetric heat

$$q_i = \left(\frac{Q_i}{n_2^s}\right)_V \qquad \text{(XVII-113)}$$

2. Differential calorimetric heat

$$q_d = \left(\frac{\partial Q_i}{\partial n_2^s}\right)_{T,V} \qquad \text{(XVII-95)}$$

3. Isosteric or differential thermodynamic heat

$$q_{st} = -\Delta\bar{H}_2 = q_d + RT \qquad \text{(XVII-114)}$$

4. Integral thermodynamic heat

$$q_\pi = -\Delta\mathcal{H}_2 = q_i + RT - \frac{\pi}{\Gamma} \qquad \text{(XVII-115)}$$

It follows from the defining relationships that

$$\overline{H}_2^s = T\overline{s}_2^s + \overline{G}_2^s \qquad \text{(XVII-116)}$$

and

$$\mathcal{H}_2^s = Ts_2^s + G_2^s \qquad \text{(XVII-117)}$$

so that

$$\Delta\overline{s}_2 = \overline{s}_2^s - s_2^{0,\,g} = \frac{\Delta H_2 - \Delta\overline{G}_2}{T} \qquad \text{(XVII-118)}$$

and

$$\Delta s_2 = s_2^s - \overline{s}_2^{0,\,g} = \frac{\Delta\mathcal{H}_2 - \Delta G_2}{T} \qquad \text{(XVII-119)}$$

where

$$\Delta\overline{G}_2 = \Delta G_2 = RT\ln P \qquad \text{(XVII-120)}$$

Thus from an adsorption isotherm and its temperature variation, one can calculate either the differential or the integral entropy of adsorption as a function of surface coverage. The former probably has the greater direct physical meaning, but the latter is the quantity usually first obtained in a statistical thermodynamic adsorption model.

The adsorbed state often seems to resemble liquid adsorbate, as in the approach of the heat of adsorption to the heat of condensation in the multilayer region. For this reason, a common choice for the standard state of free adsorbate is the pure liquid. We now have

$$\Delta\overline{G}_{2(l)} = \Delta G_{2(l)} = RT\ln x \qquad \text{(XVII-121)}$$

and

$$\Delta \overline{H}_{2(l)} = \overline{H}_2^s - \overline{H}_2^{0,\,l} \qquad \Delta \mathcal{H}_{2(l)}^0 = \mathcal{H}_2^s - \overline{H}_2^{0,\,l} \qquad \text{(XVII-122)}$$

$$\Delta \overline{s}_{2(l)} = \overline{s}_2^s - \overline{s}_2^{0,\,l} \qquad \Delta s_{2(l)} = \overline{s}_2^s - s_2^{0,\,l} \qquad \text{(XVII-123)}$$

Also

$$q_{st(l)} = RT^2 \left(\frac{\partial \ln x}{\partial T} \right)_{\Gamma} = q_{st} - \Delta H_v \qquad \text{(XVII-124)}$$

and

$$q_{\pi(l)} = RT^2 \left(\frac{\partial \ln x}{\partial T} \right)_{\pi} = q_{\pi} - \Delta H_v \qquad \text{(XVII-125)}$$

Thus the new thermodynamic heats and entropies of adsorption differ from the preceding ones by the heats and entropies of vaporization of liquid adsorbate.

There are alternative ways of defining the various thermodynamic quantities. One may, for example, treat the adsorbed film as a phase having volume, so that P, V terms enter into the definitions. A systematic treatment of this type has been given by Honig [116], who also points out some additional types of heat of adsorption.

Finally, it is perfectly possible to choose a standard state for the surface phase. De Boer [14] makes a plea for taking that value of π^0 such that the average distance apart of the molecules is the same as in the gas phase at STP. This is a hypothetical standard state in that π^0 for an ideal two-dimensional gas with this molecular separation would be 0.338 dyn/cm at 0°C. The standard molecular area is then $4.08 \times 10^{-16}T$. The main advantage of this choice is that it simplifies the relationship between translational entropies of the two- and the three-dimensional standard states.

Thermodynamic treatments may, of course, be extended to multicomponent systems. See Ref. 117 for an example.

ii. Thermodynamic Qualities for the Adsorbent.   It is also possible to calculate the change in thermodynamic quantities for the *adsorbent*, in the adsorption process, and this has been discussed by Copeland and co-workers [118, 119]. One problem, however, is how to define quantities such as $\overline{G}_1^s$ (in Eq. XVII-96) when the system consists of adsorbent particles so that there is no way of making a minute change dn_1^s without changing the specific surface area. This is handled by taking the change Δn_1^s, corresponding to the addition of one particle, but treating the thermodynamic quantities as continuous functions by basing them on the locus through the points representing successive increments of Δn_1^s. On this basis, one can proceed to apply ordinary two-component thermodynamics. If the analysis is applied to the data of Drain and Morrison from Ref. 1, $-\Delta \overline{H}_2$ (for the adsorbate, N_2) is found to decrease monotonically with coverage and is about 4 kcal at the monolayer point. For the adsorbent, rutile, $-\Delta\overline{H}_1$ increases steadily from zero and is small, as it is computed per *mole* of adsorbent; however, if normalized to be per mole of adsorbate, it reaches about 0.8 kcal, again at the monolayer point. The *local* $\Delta\overline{H}_1$, that is, around each adsorbate, is thus not negligible [120]. A possible way to see the effect experimentally would be to look for vapor pressure changes for a slightly volatile *adsorbent* as adsorption occurs, such as in the case of adsorption on low temperature powdered ice or benzene.

B. Experimental Heats and Entropies of Adsorption

Before taking up the results of measurements of heats and entropies of adsorption, it is perhaps worthwhile to review briefly the various alternative procedures for obtaining these quantities.

The *integral heat* of adsorption Q_i may be measured calorimetrically by determining directly the heat evolution when the desired amount of adsorbate is admitted to the clean solid surface. Alternatively, it may be more convenient to measure the heat of immersion of the solid in pure liquid adsorbate. Immersion of clean solid gives the integral heat of adsorption at $P = P_0$, that is, $Q_i(P_0)$ or $q_i(P_0)$, whereas immersion of solid previously equilibrated with adsorbate at pressure P gives the difference $[q_i(P_0) - q_i(P)]$, from which $q_i(P)$ can be found [121, 122]. The *differential heat* of adsorption q_d may be obtained from the slope of the Q_i–n_2 plot, or by measuring the heat evolved as small increments of adsorbate are added [123].

Alternatively, q_{st} may be obtained from the application of Eq. XVII-107 to adsorption data at two or more temperatures (see Ref. 89). Similarly, q_π is obtainable from isotherm data by means of Eq. XVII-115, but now only provided that isotherms down to low pressures are available so that Gibbs integrations to obtain π values are possible.

The *partial molar* entropy of adsorption $\Delta\bar{s}_2$ may be determined from q_d or q_{st} through Eq. XVII-118, and hence is obtainable either from calorimetric heats plus an adsorption isotherm or from adsorption isotherms at more than one temperature. The *integral entropy* of adsorption can be obtained from isotherm data at more than one temperature, through Eqs. XVII-110 and XVII-119, in which case complete isotherms are needed. Alternatively, ΔS_2 can be obtained from the calorimetric q_i plus a single complete adsorption isotherm, using Eq. XVII-115. This last approach has been recommended by Jura and Hill [121] as giving more accurate integral entropy values (see also Ref. 124).

Turning now to the results of such measurements, perhaps the first point of interest is whether the calorimetric and thermodynamic heats of adsorption do, in fact agree according to Eqs. XVII-113 and XVII-115. This appears to be the case. Brunauer [34] gives several examples involving vapors such as carbon dioxide, methanol, nitrogen, and water, adsorbed on charcoal, in which q_d and q_{st} agree within experimental error. Greyson and Aston [127] found that the detailed and rather complex variation of q_d with amount adsorbed, in the case of neon on graphitized carbon, agreed very closely with their q_{st} values. Note in Fig. XVII-21a that both isosteric and calorimetric values are shown.

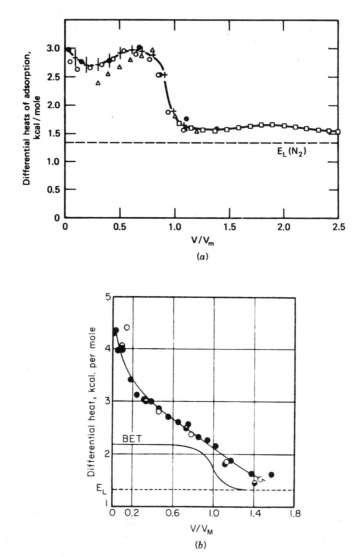

Fig. XVII-21. (*a*) Differential heat of adsorption of N_2 on Graphon, except for ○ and ●, which were determined calorimetrically. (From Ref. 89.) (*b*) Differential heat of adsorption of N_2 on carbon black (Spheron 6) at 78.5 K (From Ref. 124).

The question is not trivial; such agreement is not assured in the case of systems showing hysteresis (see Section XVII-16), and it has been difficult to affirm it on rigorous thermodynamic grounds in the case of a heterogeneous surface.

Differential heats of adsorption generally decrease steadily with increasing amount adsorbed and, in the case of physical adsorption tend to approach the heat of liquefaction of the adsorbate as P approaches P^0. Some illustrative data

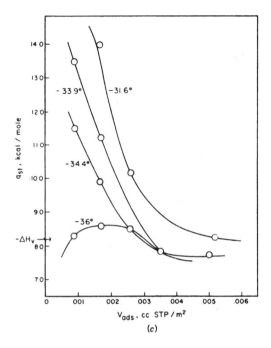

(c)

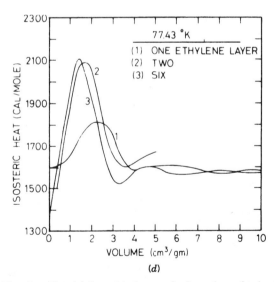

(d)

Fig. XVII-21. (*Continued*)   (c) Isosteric heats of adsorption of *n*-hexane on ice powder; $v_m = 0.073$ cm^3 STP. (From Ref. 125). (d) Isosteric heats of adsorption of Ar on graphitized carbon black having the indicated number of preadsorbed layers of ethylene. (From Ref. 126.)

are shown in Fig. XVII-21. The presumed monolayer point may be marked by a sharp decrease in q_{st}, as in Fig. XVI-21a; a more steady decrease, as in Figs. XVII-21b, probably indicates surface heterogeneity. At very low coverages, q_{st} may rise further, as with O_2 on nongraphitized carbon [128]. On very uniform surfaces, q_{st} may fall *below* the heat of liquefaction [129]. The dramatic change in behavior around $-35°C$ for n-hexane on ice (Fig. XVII-21c) is attributed to surface clathrate formation at the higher temperatures. Figure XVII-21d shows that only two monolayers of ethylene were sufficient to shield Ar from graphitized carbon black since no further change in the q_{st} behavior occurred. A few angstroms-thick coating of silica on rutile is enough to establish a new surface behavior [130]. Gregg and Sing [48] give other examples.

It is noted in Sections XVII-10 and 11 that phase transformations may occur, especially in the case of simple gases on uniform surfaces. Such transformations show up in q_{st} plots, as illustrated in Fig. XVII-22 for Kr adsorbed on a graphitized carbon black. The two plots are obtained from data just below and just above the limit of stability of a "solid" phase that is in registry with the graphite lattice [131].

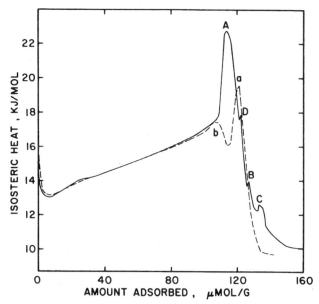

Fig. XVII-22. Isosteric heats of adsorption for Kr on graphitized carbon black. Solid line: calculated from isotherms at 110.14, 114.14, and 117.14 K; dashed line: calculated from isotherms at 122.02, 125.05, and 129.00 K. Point *A* reflects the transition from a fluid to an in-registry solid phase; points *B* and *C* relate to the transition from the in-registry to and out-of-registry solid phase. The normal monolayer point is about 124 mol/g. [Reprinted with permission from T. P. Vo and T. Fort, Jr., *J. Phys. Chem.*, **91**, 6638 (1987) (Ref. 131). Copyright 1987, American Chemical Society.]

Some representative plots of entropies of adsorption are shown in Fig. XVII-23, in general, $T \, \Delta \bar{s}_2$ is comparable to ΔH_2, so that the entropy contribution to the free energy of adsorption is important. Notice in Figs. XVII-23a and b how nearly the entropy plot is a mirror image of the enthalpy plot. As a consequence, the maxima and minima in the separate plots tend to cancel to give a smoothly varying free energy plot, that is, adsorption isotherm.

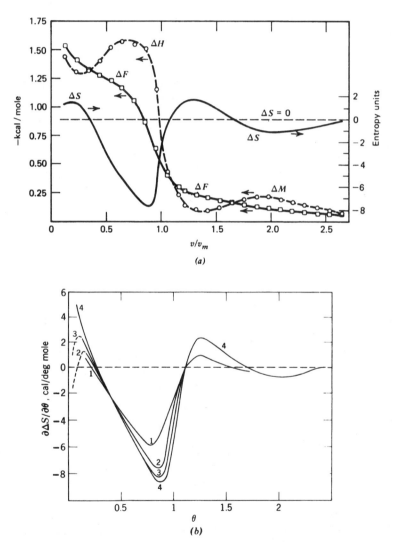

Fig. XVII-23. (a) Entropy; enthalpy, and free energy of adsorption relative to the liquid state of N_2 on Graphon at 78.3 K (From Ref. 89.) (b) Differential entropies of adsorption of n-hexane on (1) 1700°C heat-treated Spheron 6, (2) 2800°C heat-treated, (3) 3000°C heat-treated, and (4) Sterling MT-1, 3100°C heat-treated. (From Ref. 18.)

As with enthalpies of adsorption, the entropies tend to approach the entropy of condensation as P approaches P^0, in further support of the conclusion that the nature of the adsorbate is approaching that of the liquid state.

Finally, one frequently observes that $\Delta \bar{s}_2$ goes through a minimum at or near $n = n_m$. An interesting example of this effect is provided by some data of Isirikyan and Kiselev [18] for the adsorption of n-hexane on various carbons, illustrated in Fig. XVII-23b. The minimum deepens with degree of graphitization and hence with presumed increasing degree of surface uniformity. Hill et al. [132] report a very clear-cut minimum in $\Delta \bar{s}_2$ in the case of nitrogen on Graphon—another carbon that appears to have a very uniform surface.

The accepted explanation for the minimum is that it represents the point of complete coverage of the surface by a monolayer; according to Eq. XVII-37, $\bar{s}_{config}$ should go to minus infinity at this point, but in real systems an onset of multilayer adsorption occurs, and this provides a countering positive contribution. Some further discussion of the behavior of adsorption entropies in the case of heterogeneous adsorbents is given in Section XVII-14.

13. Critical Comparison of the Various Models for Adsorption

As pointed out in Section XVII-8, agreement of a theoretical isotherm equation with data at one temperature is a necessary but quite insufficient test of the validity of the premises on which it was derived. Quite differently based models may yield equations that are experimentally indistinguishable and even algebraically identical. In the multilayer region, it turns out that in a number of cases the isotherm shape is relatively independent of the nature of the solid and that any equation fitting it can be used to obtain essentially the same relative surface areas for different solids, so that consistency of surface area determination does not provide a sensitive criterion either.

The data on heats and entropies of adsorption do allow a more discriminating test of an adsorption model, although even so only some rather qualitative conclusions can be reached. The discussion of these follows.

A. The Langmuir–BET Model

This model calls for localized adsorption in all layers, with $\Delta \bar{H}_2$ a constant for the first layer and equal to the heat of condensation for all succeeding ones. The Langmuir model is perfectly acceptable for adsorption at low P/P^0 values, but as was seen in Section XVII-3C, it is not easily distinguished from mobile adsorption in terms of $\Delta \bar{s}_2$ and, in Section XVII-6B, it was noted that the distinction is also difficult in terms of algebraic form. It appears that independent data on the surface diffusion coefficient or on the energy states of the adsorbent–adsorbate complex are essential to any firm diagnosis. Theoretical calculations can be made on the structure of submonolayer films on perfect lattices, including commensurate–incommensurate phase changes [133] but this does not help much with the present matter. The complicating effect of surface heterogeneity is discussed in Section XVII-14.

Turning to the multilayer region, the actual assumptions of the BET model are not realistic. As illustrated in Fig. XVII-21b, it does not give the correct variation of q_d with θ (but partly because of the surface heterogeneity) or the correct value of $\Delta \bar{s}_2$. Basically, the available evidence suggests that the adsorbed film approaches bulk liquid in properties as P approaches P^0, and while the BET assumption as to the adsorption energy correctly reflects this behavior, the assumption of localized multilayers is not consistent with it and gives an erroneous configurational entropy. Related to this is a catastrophe that Cassel [134] has pointed out, namely that the integral I of Eq. X-40 is infinity in the BET model. Some further evidence of the liquidlike state of multilayer films was provided by Arnold [135], who found that his data on the desorption of oxygen–nitrogen mixtures on titanium dioxide could be accounted for by assuming the second and third layers to possess the molar entropy of normal liquid and that Raoult's law applied, whereas the BET treatment gave much poorer agreement.

Brunauer (see Refs. 136–138) defended these defects as deliberate approximations needed to obtain a practical two-constant equation. The assumption of a constant heat of adsorption in the first layer represents a balance between the effects of surface heterogeneity and of lateral interaction, and the assumption of a constant instead of a decreasing heat of adsorption for the succeeding layers balances the overestimate of the entropy of adsorption. These comments do help to explain why the model works as well as it does. However, since these approximations are inherent in the treatment, one can see why the BET model does not lend itself readily to any detailed insight into the real physical nature of multilayers. In summary, the BET equation will undoubtedly maintain its usefulness in surface area determinations, and it does provide some physical information about the nature of the adsorbed film, but only at the level of approximation inherent in the model. Mainly, the c value provides an estimate of the first layer heat of adsorption, averaged over the region of fit.

B. Two-Dimensional Equation of State Treatments

There is little doubt that, at least with type II isotherms, we can tell the approximate point at which multilayer adsorption sets in. The concept of a two-dimensional phase seems relatively sterile as applied to multilayer adsorption, except insofar as such isotherm equations may be used as empirically convenient, since the thickness of the adsorbed film is not easily allowed to become variable.

On the other hand, as applied to the submonolayer region, the same comment can be made as for the localized model. That is, the two-dimensional non-ideal-gas equation of state is a perfectly acceptable concept, but one that, in practice, is remarkably difficult to distinguish from the localized adsorption picture. If there can be even a small amount of surface heterogeneity the distinction becomes virtually impossible (see Section XVII-14). Even the cases of phase change are susceptible to explanation on either basis.

Ross and Olivier [55], in their extensive development of the van der Waals equation of state model have, however, provided a needed balance to the Langmuir picture.

C. The Potential Model

Returning to multilayer adsorption, the potential model appears to be fundamentally correct. It accounts for the empirical fact that systems at the same value of $RT \ln (P/P^0)$ are in essentially corresponding states, and that the multilayer approaches bulk liquid in properties as P approaches P^0. However, the specific treatments must be regarded as still somewhat primitive. The various proposed functions for $U(r)$ can only be rather approximate. Even the general-appearing Eq. XVII-79 cannot be correct, since it does not allow for structural perturbations that make the film different from bulk liquid. Such perturbations should in general be present and *must* be present in the case of liquids that do not spread on the adsorbent (Section X-7). The last term of Eq. XVII-80, while reasonable, represents at best a semiempirical attempt to take structural perturbation into account.

The existence of a characteristic isotherm (or of a t-plot) gives a very important piece of information about the adsorption potential, at least for polar solids for which the observation holds. The direct implication is that film thickness t, or alternatively n/n_m is determined by P/P^0 *independent of the nature of the adsorbent*. We can thus write

$$\frac{n}{n_m} = f\left(\frac{P}{P^0}\right) \qquad\qquad \text{(XVII-126)}$$

where the function f (and any associated constants) is independent of the nature of the solid (but may vary with that of the adsorbate).

Unfortunately none of the various proposed forms of the potential theory satisfy this criterion! Equation XVII-78 clearly does not; Eq. XVII-79 would, except that f includes the constant A, which contains the dispersion energy U_0, which, in turn, depends on the nature of the adsorbent. Equation XVII-82 fares no better if, according to its derivation, U_0 reflects the surface polarity of the adsorbent (note Eq. VI-40). It would seem that after one or at most two layers of coverage, the adsorbate film is effectively insulated from the adsorbent.

This insulation does, in fact, seem to occur. It was remarked on in connection with Fig. XVII-21 and was very explicitly shown in some results of Dubinin and co-workers [139]. Nitrogen adsorption isotherms at $-185°C$ were determined for a carbon black having various amounts of preadsorbed benzene. After only about 1.5 statistical monolayers of benzene, no further change took place in the nitrogen isotherm. A similar behavior was reported by Halsey and co-workers [140] for Ar and N_2 adsorption on TiO_2 having increasing amounts of preadsorbed water. In this case, a limiting isotherm was reached with about four statistical layers of water, the approach to the limiting form being approximately exponential. Interestingly, the final isotherm, in the case of

N_2 as adsorbate, was quite similar to that for N_2 on a directly prepared and probably amorphous ice powder [35, 141]. On the other hand, N_2 adsorption on carbon with increasing thickness of preadsorbed methanol decreased steadily—no limiting isotherm was reached [139].

Clearly, it is more desirable somehow to obtain detailed structural information on multilayer films so as perhaps to settle the problem of how properly to construct the potential function. Some attempts have been made to develop statistical mechanical other theoretical treatments of condensed layers in a potential field; success has been reasonable (see Refs. 142, 143).

14. Physical Adsorption on Heterogeneous Surfaces

The discussion on adsorption models in the preceding section is rather restrained because it turns out that for nearly all systems that are of practical interest an overriding effect makes it virtually impossible to make an experimental verification of the validity of a model or to set up any but remotely austere fundamental theoretical treatments. The effect is that of surface heterogeneity. It has been made very clear that solid surfaces are not in general uniform (e.g., Section VII-4) and the data of Figs. XVII-21 and XVII-23 provide a direct indication of nonuniformity of heats and entropies of adsorption. Surfaces may also be *geometrically* heterogeneous, as in the case of fractal behavior. This aspect is discussed briefly in Section XVII-14D.

A. Distribution of Site Energy Patches

To keep the situation manageable, we confine ourselves here to adsorption in the submonolayer region and assume that there are patches having various adsorption energies, each patch, however, being itself uniform. This would be the case, for example, if the adsorbent consisted of crystallites exhibiting various facets or if it were a mixture, even a microscopic one, of different kinds of adsorbents. Adsorption can then be formulated in a general way by noting that, regardless of model, the fraction of surface covered should be some function θ of Q, P, and T, where Q is an adsorption energy. If the surface is heterogeneous, the probability of there being an adsorption energy between Q and $Q + dQ$ can be described by a distribution function $f(Q)\, dQ$. The experimentally observed adsorption will be the sum of all the adsorptions on the different kinds of surface, and so will be a function Θ of P and T. Thus

$$\Theta(P, T) = \int_0^\infty \theta(Q, P, T) f(Q)\, dQ \qquad \text{(XVII-127)}$$

Alternatively, an integral distribution function F may be defined as giving the fraction of surface for which the adsorption energy is greater than or equal to a given Q,

$$f(Q) = \frac{dF}{dQ} \tag{XVII-128}$$

whence

$$\Theta(P,T) = \int_0^1 \theta(Q,P,T) \, dF \tag{XVII-129}$$

The variation in q_d with Θ will not in general be the same as $F(\Theta)$, since some adsorption will be occurring on all portions of the surface so that the heat liberated on adsorption of dn_2^s moles will be a weighted average. There is one exception, however, namely adsorption at 0 K; the adsorption will occur sequentially on portions of increasing Q value so that $q_d(\Theta)$ now gives $F(\Theta)$. This circumstance was used by Drain and Morrison [1], who determined $q_d(\Theta)$ for argon, nitrogen, and oxygen on titanium dioxide at a series of temperatures and extrapolated to 0 K. The procedure is a difficult one and not without some approximations in the extrapolation. Clearly, it would be very desirable to find a way of solving the integral equation so that site or adsorption energy distributions could be obtained from data at customary temperatures.

Equation XVII-127 connects the functions $\Theta(P, T)$, $\theta(Q, P, T)$ and $f(Q)$ and, in principle, if any *two* are known or can be assumed, the remaining one can be calculated. As may be imagined, many choices of such pairs of functions have been examined, often designed so that Eq. XVII-127 can be handled analytically; alternatively, various schemes of successive approximations may be used. The field has become somewhat of a happy hunting ground for physical chemists and there are numerous reviews of the now-extensive literature (see Refs. 144–147; the last is a personalized account). For this reason only some generic approaches will be discussed here.

One may choose $\theta(Q, P, T)$ and $f(Q)$. Thus if $f(Q) = \alpha e^{-Q/nRT}$ and $\theta(Q, P, T)$ is the Langmuir equation, then the Freundlich equation (Eq. XI-12) results. Ross and Olivier [55] took $f(Q)$ to be Gaussian and $\theta(Q, P, T)$ to be the two-dimensional van der Waals equation and provided extensive tabulations of the solutions to Eqs. XVII-129 for various choices of the parameters. Sircar [148] used a gamma function for $f(Q)$ and the Langmuir equation. One can, alternatively, bypass the integral equation by taking Q to be some defined function of Θ. Temkin [149] assumed $Q = Q_0(1 - \alpha\Theta)$ and substituted directly into the expression for b in the Langmuir equation.

One may choose $\theta(Q, P, T)$ such that the integral equation can be inverted to give $f(Q)$ from the observed isotherm. Hobson [150] chose a local isotherm function that was essentially a stylized van der Waals form with a linear low-pressure region followed by a vertical step to $\theta = 1$. Sips [151] showed that Eq. XVII-127 could be converted to a standard transform if the Langmuir adsorption model was used. One writes

$$\Theta(P,T) = \int_0^\infty \frac{f(Q)bP}{1+bP} \, dQ = \int_0^\infty \frac{e^{Q/RT}f(Q)}{e^{Q/RT} + 1/b_0 P} \, dQ \qquad \text{(XVII-130)}$$

or

$$\frac{1}{RT} \Theta \frac{1}{b_0 y} = \int_1^\infty \frac{f(RT \ln x)}{x+y} \, dx \qquad \text{(XVII-131)}$$

where $y = 1/b_0 P$ and $x = e^{Q/RT}$. The equation is now in the form

$$\chi(y) = \int_1^\infty \frac{\phi(x) \, dx}{(x+y)} \qquad \text{(XVII-132)}$$

for which the solution is

$$\phi(x) = \frac{\chi(xe^{-\pi i}) - \chi(xe^{\pi i})}{2\pi i} \qquad \text{(XVII-133)}$$

It is necessary, for this procedure, to express Θ as an analytical function of P. If the Freundlich equation is used, $\Theta = P^c$, then Eq. XI-11 results. If one chooses $\Theta = AP^c/(1+AP^c)$ so as to have a form that gives a limiting Θ, the resulting $f(Q)$ is Gaussianlike. Honig and Reyerson [152] used $\Theta = [P/(A+P)]^c$ and found an $f(Q)$ that steadily decreased with increasing Q, in the case of nitrogen adsorbed on TiO_2. Toth [153] has made much use of the form $\Theta = kP/[1 + (kP)^m]^{1/m}$, which gives an $f(Q)$ that is Gaussianlike but truncated at the high energy side; another variant is due to Misra [154].

Mostly, investigators have used an assumed form for $\theta(Q,P,T)$ and a smoothed $\Theta(P,T)$ to obtain $f(Q)$. Smoothing is necessary since scatter in the experimental points will lead to artifacts in $f(Q)$. An early graphical method of successive approximations is due to Adamson and Ling [155], later formalized in a computer program known as HILDA (Heterogeneity Investigated at Loughborough by a Distribution Analysis) [156]. Another program, CAEDMON (Computed Adsorptive-Energy Distribution in the Monolayer), is based on taking $\theta(Q,P,T)$ to be a two-dimensional virial equation [157]. An alternative to taking $f(Q)$ to be continuous is to assume a small number of patches, 5–10, and write $\Theta(P,T)$ as a sum of the adsorption on each. One then takes a series of data points to obtain a set of simultaneous equations, which is then solved (computerwise) for the optimum set of Q values and extent of each corresponding patch. Koopal and Vos [144] describe several programs (CAEDMON-W, REMEDI, CAESAR, etc.) and discuss their performance, as do Jaroniec and Maday [145a].

There is a limit to the usefulness in elaborating on procedures for solving Eq. XVII-127 since in general there is *no* exactly correct solution. Experimental data have error, only arbitrarily improved by smoothing; the assumed $\Theta(Q, P, T)$ will not be exactly correct. There is one redeeming situation. In the case of a very heterogeneous surface, the site energy distribution obtained is relatively *independent* of what adsorption model is used—one may even use just a step function (sometimes known as the *condensation approximation*—see Refs. 158 and 159). A gross error in the assumed entropy of adsorption (e.g., b_0) can be detected in that the distribution will fail to be temperature independent, and if a model providing for lateral interactions is used, the distribution will be shifted along the Q scale by an amount corresponding to the average lateral interaction energy.

The analysis is thus relatively exact for heterogeneous surfaces and is especially valuable for analyzing changes in an adsorbent following one or another treatment. An example is shown in Fig. XVII-24 [160]. This type of application has also been made to carbon blacks and silica–alumina catalysts [106a]. House and Jaycock [161] compared the Ross–Olivier [55] and Adamson–Ling

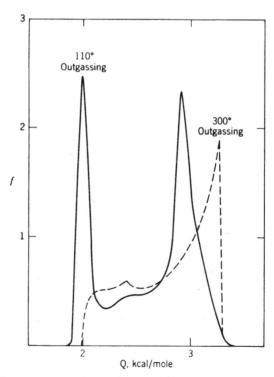

Fig. XVII-24. Site energy distribution for nitrogen adsorbed on Silica SB. (From Ref. 160.) (Reprinted with permission from *J. Phys. Chem.* Copyright by the American Chemical Society.)

[155] procedures as applied to adsorption on TiO$_2$ (anatase); each could represent the data, but the obtained site energy distribution did depend somewhat on the method and choice of local isotherm function.

Turning to the case of relatively homogeneous surfaces, the use of the site energy distribution analysis illuminates an awkward situation. As before, various adsorption models may be used to obtain an $f(Q)$, but since $f(Q)$ now represents a fairly narrow site energy distribution, different adsorption models give noticeably different distributions. This is illustrated in Fig. XVII-25 for the case of nitrogen on BN [162] (data from Ref. 163); the three distributions and associated choices for $\theta(Q, P, T)$ gave entirely comparable agreements with the experimental results. The somewhat paradoxical situation is thus that in the case of a nearly homogeneous surface, *neither* the adsorption model *nor* the site energy distribution can be affirmed with any great assurance.

B. Thermodynamics of Adsorption on Heterogeneous Surfaces

It is generally assumed that isosteric thermodynamic heats obtained for a heterogeneous surface retain their simple relationship to calorimetric heats (Eq. XVII-114), although it may be necessary in a thermodynamic proof of this to assume that the chemical potential of the adsorbate does not show discontinu-

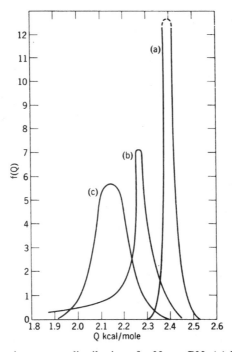

Fig. XVII-25. Interaction energy distributions for N$_2$ on BN: (*a*) Langmuir; (*b*) Langmuir plus lateral interaction; (*c*) van der Waals. (From Ref. 162.)

ities as Θ is varied [164]. An analytical proof for the special case in which $\theta(Q, P, T)$ is the Langmuir equation has been given [165].

There is a dramatic effect of surface heterogeneity on the adsorption entropy, as noted by Everett [13]. The main concern here is with how the configurational entropy varies with Θ rather than with absolute adsorption entropies, and the point is that $\bar{S}_{config}$ is so altered on a heterogeneous surface that a distinction between models is not possible. The physical explanation is that in the case of a heterogeneous surface, each added increment of adsorbate is confined to the small portion of the surface that is being filled. For an extremely heterogeneous surface, $\bar{S}_{config}$ would be zero.

There is a positive aspect to the situation. Since for a heterogeneous surface neither the site energy distribution nor the configurational entropy are very model sensitive, the former gives relatively unambiguous information about the energetics of the system, and the entropy of adsorption is now assignable to the entropy change in forming the adsorbent-adsorbate complex without it being necessary to estimate configurational entropy corrections.

C. Point versus Patch Site Energy Distributions

The preceding discussion on site energy distributions has made the implicit assumption that each kind of site occurred in patches at least large enough that boundary effects could be neglected. In effect, each patch is treated as a small adsorption system within which lateral interaction between adsorbate molecules would occur. At the other extreme, sites of varying energy are randomly distributed on the surface, and only as lower energy sites became occupied would there be a developing probability of adjacent sites being occupied. A complete description of surface heterogeneity would have to include the distribution of site energies adjacent to a site of given energy. At this point there are probably too many variables to be extracted from adsorption data alone, although the comparison of isotherms for adsorbates of varying size may help (e.g., nitrogen versus butane—see Ref. 162). A formal statistical mechanical approach to the problem was made by Steel [166].

D. Geometric Heterogeneity

The preceding material has been couched in terms of site energy distributions—the implication being that an adsorbent may have chemically different kinds of sites. This is not necessarily the case—if micropores are present (see Section XVII-16) adsorption in such may show an increased Q because the adsorbate experiences interaction with surrounding walls of adsorbent. To a lesser extent this can also be true for a nonporous but very rough surface.

The currently useful model for dealing with rough surfaces is that of the self-similar or *fractal* surface (see Sections VII-4C and XVI-2B). This approach has been very useful in dealing with the variation of apparent surface area with the size of adsorbate molecules used and with adsorbent particle size. All adsorbate molecules have access to a plane surface, that is, one of fractal dimension 2. For surfaces of $D > 2$, however, there will be regions accessible to small molecules

TABLE XVII-5
Fractal Dimensions for Selected Adsorbents[a]

Adsorbent	D	Ref.	Adsorbent	D	Ref.
Active carbon			Silica 40	3.0	173
(coconut shell)	2.7	169	Silica gel	3.0	167
Coal	2.4	170	Vitreous silica	2.0	174
Graphitized			α Fe$_2$O$_3$	2.4–2.7	167
carbon black	2.0–2.1	167	Kaolinite	2.1, 2.4	175
Al-pillared			Montmorillonite	2.1, 2.3	175
montmorillonite	1.9	171			
Synthetic					
faujasite	2.0	172			

[a]Taken from Ref. 167; D values obtained by varying adsorbate size, Eq. XVI-1.

but not to larger ones; the point is illustrated in Ref. 167,† which also tabulates D values for a number of adsorbents, determined by various means. Table XVII-5 gives a few of the cases where D was determined by varying the adsorbate size.

In the case of mixtures of gases of different molecular size, an adsorbent of $D > 2$ will effect some segregation by size. This segregation will also affect the probability of bimolecular reactions between molecules of different sizes [168].

15. Rate of Adsorption

The rate of physical adsorption may be determined by the gas kinetic surface collision frequency as modified by the variation of sticking probability with surface coverage—as in the kinetic derivation of the Langmuir equation (Section XVII-3A)—and should then be very large unless the gas pressure is small. Alternatively, the rate may be governed by boundary layer diffusion, a slower process in general. Such aspects are mentioned in Ref. 146.

Rate effects may not be chemical kinetic ones. Benson and co-worker [84], in a study of the rate of adsorption of water on lyophilized proteins, comment that the empirical rates of adsorption were very markedly complicated by the fact that the samples were appreciably heated by the heat evolved on adsorption. In fact, it appeared that the actual adsorption rates were very fast and that the time dependence of the adsorbate pressure above the adsorbent was simply due to the time variation of the temperature of the sample as it cooled after the initial heating when adsorbate was first introduced.

Deitz and Carpenter [176] found that argon and nitrogen adsorbed only

---

†One of us is pleased that this paper shows an illustrative needlepoint embroidery created by his wife.

slowly on diamond under conditions such that the final θ was only a few percent. They were able to rule out heat transfer at rate controlling and concluded that while the adsorption process per se was very rapid, the nature of the diamond surface changed slowly to a more active one on the cooling of the sample prior to an adsorption run. The change appeared to be reversible and may constitute a type of illustration of the Dunning effect discussed in Sections VII-4B and XVI-4B. A related explanation was invoked by Good and co-workers [177]. In studies of the heat of immersion of Al_2O_3 and SiO_2 in water, they noticed a slow residual heat evolution, which they attributed to slow surface hydration.

In conclusion, any observation of slowness in attainment of physical adsorption equilibrium should be analyzed with caution and in detail. When this has been done, the phenomenon has either been found to be due to trivial causes or else some unsuspected and interesting other effects were operative.

16. Adsorption on Porous Solids—Hysteresis

As a general rule, adsorbates above their critical temperatures do not give multilayer type isotherms. In such a situation, a porous absorbent behaves like any other, unless the pores are of molecular size, and at this point the distinction between adsorption and absorption dims. Below the critical temperature, multilayer formation is possible and capillary condensation can occur. These two aspects of the behavior of porous solids are discussed briefly in this section. Some IUPAC (International Union of Pure and Applied Chemistry) recommendations for the characterization of porous solids are given in Ref. 178.

A. Molecular Sieves

Some classes of adsorbents have internal surface accessible by pores small enough to act as molecular sieves, so that different apparent surface areas are obtained according to the size of the adsorbate molecule. This is a screening effect and not one of surface roughness as might be described in terms of fractal geometry (Section XVI-2B). Zeolites have been of much interest in this connection because the open way in which the $(Al, Si)O_4$ tetrahedra join gives rise to large cavities and large windows into the cavities. This is illustrated in Fig. XVII-26 [179]. As a specific example, chabasite ($CaAl_2Si_4O_{12}$) has cages about 10 Å in diameter, with six openings into each or windows of about 4-Å diameter. Monatomic and diatomic gases, water, and n-alkanes can enter into such cavities, but larger molecules do not. Thus isobutane can be separated from n-alkanes and, on the basis of rates, even propane from ethane [180]. The replacement of the calcium by other ions (zeolites have ion exchange properties—note Section XI-6C) considerably affects the relative adsorption behavior. Various synthetic zeolites having various window diameters in the range of 4 to 10 Å are available under the name of Linde Molecular Sieves (see Ref. 179). Listings of types of zeolites and their geometric properties may be found in Refs. 181 and 182 and the related properties of expanded clay

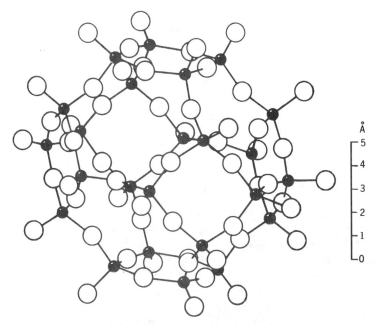

Fig. XVII-26. The arrangement of (Al, Si)O$_4$ tetrahedra that gives the cubooctahedral cavity found in some felspathoids and zeolites. (From Ref. 179.)

minerals in Ref. 183. A more recent addition to commercial porous materials is called M41S (MCM-41). These are silicious but formed on surfactant liquid crystal templates (184,185). The pores are parallel cylinderlike, a cross-section looking much like a honey comb. It can be synthesized with pore diameters in the *mesoporous* range, varying from 15 to 100 Å, each type of preparation having a narrow size distribution. FTIR and other studies of water adsorption indicated a relatively hydrophobic character to the pores [185a].

The adsorption isotherms are often Langmuirian in type (under conditions such that multilayer formation is not likely), and in the case of zeolites, both n_m and b vary with the cation present. At higher pressures, capillary condensation typically occurs, as discussed in the next section. Some N_2 isotherms for M41S materials are shown in Fig. XVII-27; they are Langmuirian below P/P^0 of about 0.2. In the case of a microporous carbon (prepared by carbonizing olive pits), the isotherms for He at 4.2 K and for N_2 at 77 K were similar and Langmuirlike up to P/P^0 near unity, but were fit to a modified Dubninin–Radushkevich (DR) equation (see Eq. XVII-75) to estimate micropore sizes around 40 Å [186].

A special case of adsorption in cavities is that of clathrate compounds. Here, cages are present, but without access windows, so for "adsorption" to occur the solid usually must be crystallized in the presence of the "adsorbate." Thus quinol crystallizes in such a manner that holes several angstroms in diameter occur and, if crystallization takes place in the presence of solvent or gas

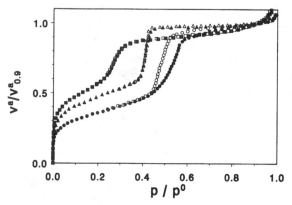

Fig. XVII-27. Nitrogen adsorption at 77 K for a series of M41S materials. Average pore diameters: squares, 25 Å; triangles, 40 Å; circles, 45 Å. Adsorption: solid symbols; desorption: open symbols. The isotherms are normalized to the volume adsorbed at P/P^0 = 0.9. (From Ref. 187. Reprinted with kind permission from Elsevier Science-NL, Sara Burgerhartstraat 25, 1055 KV Amsterdam, The Netherlands.)

molecules of small enough size, one or more such molecules may be incorporated into each hole. The union may thus be stoichiometric, but topological rather than specific chemical factors are involved; in the case of quinol, such diverse species as sulfur dioxide, methanol, formic acid, and nitrogen may form clathrate compounds [188].

A particularly interesting clathrating substance is ice; the cavities consist of six cages of 5.9 Å diameter and two of 5.2 Å per unit cell [189], and a variety of molecules are clathrated, ranging from rare gases to halogens and hydrocarbons. The ice system provides an illustration that it is not always necessary to carry out an in situ crystallization to obtain a clathrate. Barrer and Ruzicka [190] report a spontaneous clathrate formation between ice powder and xenon and krypton at −78°C, and in an attempted adsorption study of ethane on ice at −96°C it was found that, again, spontaneous ethane hydrate formation occurred [191]; similar behavior was found for the adsorption of CO_2 by ice [192]. These cases could be regarded as extreme examples of an adsorbate-induced surface rearrangement!

B. Capillary Condensation

Below the critical temperature of the adsorbate, adsorption is generally multilayer in type, and the presence of pores may have the effect not only of limiting the possible number of layers of adsorbate (see Eq. XVII-65) but also of introducing capillary condensation phenomena. A wide range of porous adsorbents is now involved and usually having a broad distribution of pore sizes and shapes, unlike the zeolites. The most general characteristic of such adsorption systems is that of hysteresis; as illustrated in Fig. XVII-27 and, more gener-

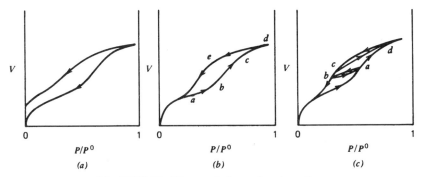

Fig. XVII-28. Hysteresis loops in adsorption.

ally, in Fig. XVII-28, the desorption branch lies to the left of the adsorption one. The isotherms may tend to flatten as P/P^0 approaches unity (type IV of Fig. XVII-7). We are concerned in this section mainly with loops of the type shown in Fig. XVII-28b and c. The open loop of Fig. XVI-28a has at least two types of explanation. One is that "ink-bottle" pores (see Fig. XVI-3) may be present, that can trap adsorbate. A perhaps more plausible explanation in the case of vapors (as contrasted to liquid mercury) is that irreversible change may occur in the pore structure on adsorption, so that the desorption situation is truly different from the adsorption one [193].

The adsorption branch of isotherms for porous solids has been variously modeled. Again, the DR equation (Eq. XVII-75) and related forms have been used [186,194]. With respect to desorption, the variety of shapes of loops of the closed variety that may be observed in practice is illustrated in Fig. XVII-29 (see also Refs. 195 and 197).

Explanations of hysteresis in capillary condensation probably begin in 1911 with Zsigmondy [198], who attributed the effect to contact angle hysteresis due to impurities; this might account for behavior of the type shown in Fig. XVII-28a, but not in general for the many systems having retraceable closed hysteresis loops. Most of the early analyses and many current ones are in terms of a model representing the adsorbent as a bundle of various-sized capillaries. Cohan [199] suggested that the adsorption branch—curve abc in Fig. XVII-28b—represented increasingly thick film formation whose radius of curvature would be that of the capillary r, so that at each stage the radius of capillaries just filling would be given by the corresponding form of the Kelvin equation (Eq. III-19):

$$x_a = e^{-\gamma V/rRT} \qquad\qquad (XVII\text{-}134)$$

At c all such capillaries would be filled and on desorption would empty by retreat of a meniscus of curvature $2/r$, so that at each stage of the desorption branch dea the radius of the capillaries emptying would be

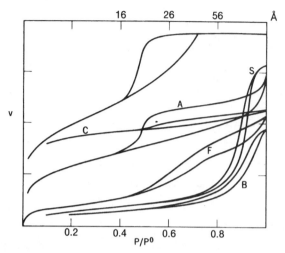

Fig. XVII-29. Nitrogen isotherms; the volume adsorbed is plotted on an arbitrary scale. The upper scale shows pore radii corresponding to various relative pressures. Samples: A, Oulton catalyst; B, bone char number 452; C, activated charcoal; F, Alumina catalyst F12; G, porous glass; S, silica aerogel. (From Ref. 196).

$$x_d = e^{-2\gamma V/rRT} \qquad\qquad (XVII\text{-}135)$$

The section cd can be regarded as due to relatively large cone-shaped pores that would fill and empty without hysteresis. At the end of section cd, then, all pores should be filled, and the adsorbent should hold the same volume of any adsorbate. See Ref. 200 for a discussion of this conclusion, sometimes known as the *Gurvitsch rule*.

The effect of the bundle of capillaries picture is to stress the use of the desorption branch to obtain a pore-size distribution. The basic procedure stems from those of Barrett et al. [201] and Pierce [196]; see also Ref. 58 wherein the effective meniscus curvature is regarded as given by the capillary radius minus the thickness of ordinary multilayer adsorption expected at that P/P^0. This last can be estimated from adsorption data on similar but nonporous material, and for this de Boer's t-curve (see Table XVII-4) is widely used. The general calculation is as follows. After each stepwise decrease in P_d (pressure on desorption), an effective capillary radius is calculated from Eq. XVII-135, and the true radius is obtained by adding the estimated multilayer thickness. The exposed pore volume and pore area can then be calculated from the volume desorbed in that step. For all steps after the first, the desorbed volume is first corrected for that from multilayer thinning on the sum of the areas of previously exposed pores. In this way a tabulation of cumulative pore volume of pores of radius greater than a given r is obtained and, from the slopes of the corresponding plot, a pore-size distribution results. Such a distribution was compared in Fig. XVI-2 to the one from mercury porosimetry, with excellent agreement. See Refs. 202 and 203 for additional examples. Also, Dubinin [204] has elaborated on the de Boer type of approach. It might be noted that nonporous adsorbents may show hysteresis due to interparticle capillary condensation (see Ref. 205). A general complication in calculating pore volumes is that the density of capillary-condensed liquid may be significantly diminished because of being under a high *negative* pressure [206].

Sing (see Ref. 207 and earlier papers) developed a modification of the de Boer *t*-plot idea. The latter rests on the observation of a characteristic isotherm (Section XVII-9), that is, on the conclusion that the adsorption isotherm is independent of the adsorbent in the multilayer region. Sing recognized that there were differences for different adsorbents, and used an appropriate "standard isotherm" for each system, the standard isotherm being for a nonporous adsorbent of composition similar to that of the porous one being studied. He then defined a quantity $\alpha_s = (n/n_x)_s$ where n_x is the amount adsorbed by the nonporous reference material at the selected P/P^0. The α_s values are used to correct pore radii for multilayer adsorption in much the same manner as with de Boer. Lecloux and Pirard [208] have discussed further the use of standard isotherms.

Everett [209] has pointed out that the bundle of capillaries model can be outrageously wrong for real systems so that the results of the preceding type of analysis, while internally consistent, may not give more than the roughest kind of information about the real pore structure. One problem is that of "ink bottle" pores (Fig. XVI-3), which empty at the capillary vapor pressure of the access channel but then discharge the contents of a larger cavity. Barrer et al. [210] have also discussed a variety of geometric situations in which filling and emptying paths are different. In a rather different perspective, Seri-Levy and Avnir [210a] have shown that hysteresis can occur with adsorption on a fractal surface due to "pores" that are highly irregular ink bottle in shape.

Brunauer and co-workers [211, 212] proposed a "modelless" method for obtaining pore size distributions; no specific capillary shape is assumed. Use is made of the general thermodynamic relationship due to Kiselev [213]

$$\gamma \, d\mathcal{A} = \Delta\mu \, dn \qquad (\text{XVII-136})$$

where $d\mathcal{A}$ is the surface that disappears when a pore is filled by capillary condensation, $\Delta\mu$ is the change in chemical potential, equal to $RT \ln P/P^0$, and dn is the number of moles of liquid taken up by the pore. It can be shown that the Kelvin equation is a special case of Eq. XVII-136 (Problem XVII-33). The integral form of Eq. XVII-136 is

$$\mathcal{A} = -\frac{RT}{\gamma} \int_{n_h}^{n_s} \ln \left(P/P^0 \right) dn \qquad (\text{XVII-137})$$

One now defines a hydraulic pore radius r_h as

$$r_h = \frac{V}{\mathcal{A}}$$

(XVII-138)

where V is the volume of a set of pores and $\mathcal{A}$, their surface area. The remaining procedure is similar to that described. For each of successive steps along the desorption branch, the volume of that group of pores that empty is given by the change in n, and their area by Eq. XVII-138. Their hydraulic radius follows from Eq. XVII-137. The total pore surface area can be estimated from application of the BET equation to the section of the isotherm before the hysteresis loop, and a check on the pore distribution analysis is that the total pore area calculated from the distribution agrees with the BET area. In one test of the method, the pore size distribution for a hardened Portland cement paste agreed well with that obtained by the traditional method [211].

More detailed information about the pore system can be obtained from "scanning curves," illustrated in Fig. XVII-28c. Thus if adsorption is carried only up to point a and then desorption is started, the lower curve ab will be traced; if at b absorption is resumed, the upper curve ab is followed, and so on. Any complete model should account in detail for such scanning curves and, conversely, through their complete mapping much more information can be obtained about the nature of the pores. Rao [214] and Emmett [215] have summarized a great deal of such behavior.

A potentially powerful approach is that of Everett [216], who treats the pore system as a set of domains, independently acting in a first approximation. Each domain consists of those elements of the adsorbent that fill at a particular $x_{a(j)}$ and empty at a particular other relative pressure $x_{d(j)}$, the associated volume being V_j. Each domain is thus characterized by these three variables, and a plot of the function $V(x_a, x_d)$ would produce a surface in three dimensions something like a relief map. On increasing the vapor pressure from x_a to $x_a + dx_a$, all domains of filling pressure in this interval should fill, but these domains can and in general would have a range of emptying pressure x_d ranging from $x_d = 0$ to $x_d = x_a$. Since the x_d of any domain cannot exceed its x_a (it cannot empty at a higher pressure than it fills!), the base of the topological map must be a $45°$ triangle. The detailed map contains in principle full information about the adsorption and desorption branches, as well as about all possible scanning loops. The problem of *deducing* such a map from data is a massive one, however, and seems not yet to have been done. Other approaches to the problem treat a porous adsorbent as a network of various size capillaries [217, 218]. Finally, capillary condensation where the vapor consists of two or more components—a rather complex subject—is discussed in several papers in Ref. 219.

A concluding comment might be made on the temperature dependence of adsorption in such systems. One can show by setting up a piston and cylinder experiment that mechanical work must be lost (i.e., converted to heat) on carrying a hysteresis system through a cycle. An irreversible process is thus involved, and the entropy change in a small step will not in general be equal to q/T. As was pointed out by LaMer [220], this means that second-law equations such as Eq. XVII-107 no longer have a simple meaning. In hysteresis systems, of course, two sets of q_{st} values can be obtained, from the adsorption and from the desorption branches. These usually are not equal and neither of them in general can be expected to equal the calorimetric heat. Another way of stating the problem is that the system is not locally reversible. The adsorption following an *increase* of x by δx is not retraced on *decreasing* the pressure by δx. This means that extreme caution should be exercised in treating q_{st} values as though they represented physical heat quantities, although it is certainly possible that in individual cases or in terms of particular models the discrepancy between q_{st} and a calorimetric heat may not be serious (see Ref. 221).

C. Micropore Analysis

Adsorbents such as some silica gels and types of carbons and zeolites have pores of the order of molecular dimensions, that is, from several up to 10–15 Å in diameter. Adsorption in such pores is not readily treated as a capillary condensation phenomenon—in fact, there is typically no hysteresis loop. What happens physically is that as multilayer adsorption develops, the pore becomes filled by a meeting of the adsorbed films from opposing walls. Pores showing this type of adsorption behavior have come to be called *micropores*—a conventional definition is that micropore diameters are of width not exceeding 20 Å (larger pores are called *mesopores*), see Ref. 221a.

Adsorption isotherms in the micropore region may start off looking like one of the high BET c-value curves of Fig. XVII-10, but will then level off much like a Langmuir isotherm (Fig. XVII-3) as the pores fill and the surface area available for further adsorption greatly diminishes. The BET-type equation for adsorption limited to n layers (Eq. XVII-65) will sometimes fit this type of behavior. Currently, however, more use is made of the Dubinin-Raduschkevich or DR equation. This is Eq. XVII-75, but now put in the form

$$\frac{V}{V_0} = \exp\left[-B\left(\frac{T}{\rho}\right)^2 \log^2\left(\frac{P^0}{P}\right)\right] \tag{XVII-139}$$

(see Refs. 222–224). Here, β is a "similarity" coefficient characteristic of the adsorbate and B is a constant characteristic of the adsorbent; V and V_0 are the amounts adsorbed expressed as a liquid volume, at a given P/P^0, and when micropore filling is complete. For a single set of slitlike pores, B appears to be proportional to the pore width [224, 225]. Some recent work [226] on microporous carbon having slitlike pores about three adsorbate molecules thick suggests that pore filling may be autocatalytic. As opposing walls acquire one adsorbed layer each, filling of the middle region then becomes easy. This was the explanation of S-shaped isotherms, that is, sections where addition of more N_2 adsorbate led to a *decrease* of the final pressure (note Problem XVII-30).

Most microporous adsorbents have a range of micropore size, as evidenced, for example, by a variation in q_{st} or in calorimetric heats of adsorption with amount adsorbed [227]. As may be expected, a considerable amount of effort has been spent in seeing how to extract a size distribution from adsorption data.

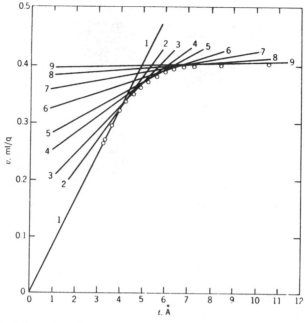

Fig. XVII-30. Adsorption of N_2 on a silica gel at 77.3 K, expressed as a v–t plot, illustrating a method for micropore analysis. (From Ref. 230.)

One approach has been to build on the DR equation by adding a Gaussian [228] or gamma function-type [224] distribution of B values. Another approach makes use of a modified Fowler–Guggenheim equation (see Eq. XVII-53) [229]. A rather different method for obtaining a micropore size distribution was proposed by Mikhail, Brunauer, and Bodor [230], often known as the MP method. The method is an extension of the t-curve procedure for obtaining surface areas (Section XVII-9); a plot of cubic centimeters STP adsorbed per gram v versus the value of t for the corresponding P/P^0 (as given, for example, by Table XVII-4) should, according to Eq. XVII-84, give a straight line of slope proportional to the specific surface area Σ. As illustrated in Fig. XVII-30, such plots may bend over. This is now interpreted not as a deviation from the characteristic isotherm principle but rather as an indication that progressive reduction in surface area is occurring as micropores fill. The proposal of Mikhail et al. was that the slope at each *point* gave a correct surface area for the P/P^0 and v value. The drop in surface area between successive points then gives the volume of micropores that filled at the average P/P^0 of the two points, and the average t value, the size of the pores that filled. In this way a pore size distribution can be obtained. Figure XVII-31a shows adsorption isotherms obtained for an adsorbent consisting of α-FeOOH dispersed on carbon fibers, and Fig. XVII-31b, the corresponding distribution of micropore diameters [231].

Finally, as might be expected, a fractal geometry approach has been made.

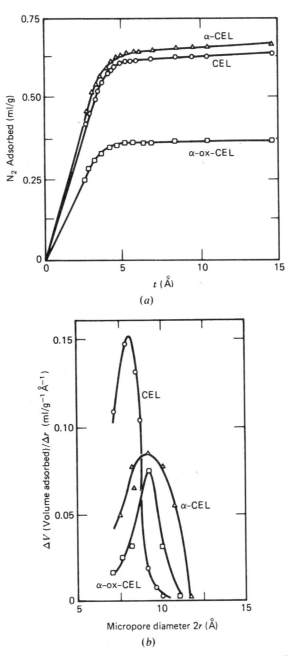

Fig. XVII-31. (*a*) Nitrogen adsorption isotherms expressed as *t*-plots for various samples of α-FeOOH dispersed on carbon fibers. (*b*) Micropore size distributions as obtained by the MP method. [Reprinted with permission from K. Kaneko, *Langmuir*, **3,** 357 (1987) (Ref. 231.) Copyright 1987, American Chemical Society.]

This may be based on Eq. XVI-2 [232] or on related equations with film thickness given by some version of the Frenkel–Halsey–Hill equation (Eq. XVII-79) [233,234].

There has been fierce debate (see Refs. 232, 235–237) over the usefulness of the preceding methods and the matter is far from resolved. On the one hand, the use of algebraic models such as modified DR equations imposes artificial constraints, while on the other hand, the assumption of the validity of the t-plot in the MP method is least tenable just in the relatively low P/P^0 region where micropore filling should occur.

17. Problems

1. The accommodation coefficient for Kr on a carbon filament is determined experimentally as follows. The electrically heated filament at temperature T_2 is stretched down the center of a cylindrical cell containing Kr gas at T_1. Gas molecules hitting the filament cool it, and to maintain its temperature a resistance heating of Q cal sec^{-1} cm^{-2} is needed. Derive from simple gas kinetic theory the expression

$$\alpha = AQ(MT_1)^{1/2} \left[\frac{1}{P(T_2 - T_1)} \right] \qquad \text{(XVII-140)}$$

Evaluate the constant A if M is molecular weight and P is in dyn/cm^2.

2. Read off points from Fig. XVII-1 and plot a set of corresponding isosteres and isobars.

3. Derive the general form of the Langmuir equation (Eq. XVII-11).

4. The separate adsorption isotherms for gases A and B on a certain solid obey the Langmuir equation, and it may be assumed that the mixed or competitive adsorption obeys the corresponding form of the equation.

Gas A, by itself, adsorbs to a θ of 0.02 at $P = 200$ mm Hg, and gas B, by itself, adsorbs to $\theta = 0.02$ at $P = 20$ mm Hg; T is 77 K in both cases. (a) Calculate the difference between Q_A and Q_B, the two heats of adsorption. Explain briefly any assumptions or approximations made. (b) Calculate the value for θ_A when the solid, at 77 K, is equilibrated with a mixture of A and B such that the final pressures are 200 mm Hg each. (c) Explain whether the answer in b would be raised, lowered, or affected in an unpredictable way if all of the preceding data were the same but the surface was known to be heterogeneous. The local isotherm function can still be assumed to be the Langmuir equation.

5. Calculate the value of the first three energy levels according to the wave mechanical picture of a particle in a one-dimensional box. Take the case of nitrogen in a 3.5-Å box. Calculate $Q_{\text{trans}}_{\text{(1 dim)}}$ at 78 K using the integration approximation and by directly evaluating the sum $\Sigma_n \exp(-\epsilon_n/kT)$.

6. Calculate $\Delta \bar{S}_2$ at $\theta = 0.1$ for argon at 77 K that forms a weak adsorption bond with the adsorbent, having three vibrational degrees of freedom.

7. Discuss the physical implications of the conditions under which b_0 as given by

the kinetic derivation of the Langmuir equation is the same as b_0', the constant as given by the statistical thermodynamic derivation.

8. The standard entropy of adsorption $\Delta \bar{S}_2$ of benzene on a certain surface was found to be -25.2 EU at 323.1 K; the standard states being the vapor at 1 atm and the film at an area of $22.5 \times T$ Å^2 per molecule. Discuss, with appropriate calculations, what the state of the adsorbed film might be, particularly as to whether it is mobile or localized. Take the molecular area of benzene to be 22 Å^2.

9. Derive Eq. XVII-54.

10. Show that the critical value of β in Eq. XVII-53 is 4, that is, the value of β above which a maximum and a minimum in bP appear. What is the critical value of θ?

11. Drain and Morrison (1) report the following data for the adsorption of N_2 on rutile at 75 K, where P is in millimeters of mercury and v in cubic centimeters STP per gram.

P	v	P	v	P	v	P	v
1.17	600.06	275.0	1441.14	455.2	2418.34	498.6	3499.13
14.00	719.54	310.2	1547.37	464.0	2561.64	501.8	3628.63
45.82	821.77	341.2	1654.15	471.2	2694.67		
87.53	934.68	368.2	1766.89	477.1	2825.39		
127.7	1045.75	393.3	1890.11	482.6	2962.94		
164.4	1146.39	413.0	2018.18	487.3	3107.06		
204.7	1254.14	429.4	2144.98	491.1	3241.28		
239.0	1343.74	443.5	2279.15	495.0	3370.38		

Plot the data according to the BET equation and calculate v_m and c, and the specific surface area in square meters per gram.

The saturation vapor pressure P^0 of N_2 is given by

$$\log P^0 = -\frac{339.8}{T} + 7.71057 - 0.0056286T$$

(Ref. 48), in mm Hg.

12. When plotted according to the linear form of the BET equation, data for the adsorption of N_2 on Graphon at 77 K give an intercept of 0.004 and a slope of 1.7 (both in cubic centimeters STP per gram). Calculate Σ assuming a molecular area of 16 Å^2 for N_2. Calculate also the heat of adsorption for the first layer (the heat of condensation of N_2 is 1.3 kcal/mol). Would your answer for v_m be much different if the intercept were taken to be zero (and the slope the same)? Comment briefly on the practical significance of your conclusion.

13. Consider the case of the BET equation with $c = 1$. Calculate for this case the heat of adsorption for the process:

A(liquid adsorbate at T) = A(adsorbed, in equilibrium with pressure P, at T)

for θ values of 0.1 and 1.5. Calculate also the entropies of adsorption for the same

θ values. Finally, derive the corresponding two-dimensional equation of state of the adsorbed film.

14. Make a set of plots analogous to those of Fig. XVII-10, but using Eq. XVII-65 with $c = 100$ and $n = 4, 7$, and infinity.

15. Make a plot of Eq. XVII-69 as θ versus P, and, for comparison, one of a Langmuir adsorption isotherm of same limiting or Henry's law slope. Comment on the comparison.

16. An equation very similar to the BET equation can be derived by assuming a multilayer structure, with the Langmuir equation applying to the first layer, and Raoult's law to succeeding layers, supposing that the escaping tendency of a molecule is unaffected by whether it is covered by others or not. Make this derivation, adding suitable assumptions as may be needed.

17. The equation of state for a solid film is often $\pi = b - a\sigma$ (note Section IV-4D). Derive the corresponding adsorption isotherm equation. Plot the data of Problem 11 according to your isotherm equation.

18. Show that $\bar{S}_{\text{config}} = 0$ for adsorption obeying the Dubinin equation (Eq. XVII-75).

19. Plot the data of Problem 11 according to Eq. XVII-76 and according to Eq. XVII-79. Comment.

20. Plot the data of Problem 11 according to Eq. XVII-83. Comment.

21. Construct a v versus t plot for the data of Problem 11 (assume that Table XVII-4 can be used) and calculate the specific surface area of the rutile.

22. An adsorption system follows Eq. XVII-79 in the form $\ln v = B - (1/n)\ln\ln(P^0/P)$ with $n = 2.75$ and $B = 3.2$. Assuming now that you are presented with data that fall on the curve defined by this equation, calculate the corresponding BET v_m and c values.

23. Plot the data of Table XVII-4 as v/v_m versus P/P^0, and plot according to (a) the BET equation, (b) Eq. XVII-79, and (c) Eq. XVII-83.

24. Use the data of Fig. XVII-1 to calculate q_{st} for a range of v values and, in conjunction with Problem 11, plot q_{st} versus n/n_m.

25. Show that Eq. XVII-85 is consistent with Eq. XVII-84.

26. Referring to Fig. XVII-17, use handbook data to calculate the vapor pressure of O_2 ordinary liquid at the melting point of the δ phase. Comment on the result. Locate the 2D S–L–V triple point.

27. Derive from Eq. XVII-132 the equation for $f(Q)$ if $\Theta = P^c$, where c is a constant.

28. Most calculations of $f(Q)$ for a heterogeneous surface, using an adsorption isotherm assume a patchwise distribution of sites. Explain for what kind of local isotherm functions, $f(Q, P, T)$ this assumption is not necessary, and for which it is necessary. Give examples.

29. The monolayer amount adsorbed on an aluminum oxide sample was determined using a small molecule adsorbate and then molecular-weight polystyrenes (much as shown in Ref. 169). The results are shown in the table. Calculate the fractal dimension of the oxide.

| Moles $\times 10^{-10} g^{-1}$ | 2.64×10^7 | 1200 | 167 | 6.92 |
| radius of gyration (Å) | 22.4 | 79.5 | 120 | 502 |

30. In Section XVII-16C there is mention of S-shaped isotherms being obtained. That is, as pressure increased, the amount adsorbed increased, then decreased, then increased again. If this is equilibrium behavior, explain whether a violation of the second law of thermodynamics is implied. A sketch of such an isotherm is shown in Ref. 226 for nitrogen adsorbed on a microporous carbon (see Ref. 226).

31. As a simple model of a heterogeneous surface, assume that 20% of it consists of sites of $Q = 2.5$ kcal/mol; 45% of sites $Q = 3.5$ kcal/mol; and the remainder, of sites of $Q = 4.5$ kcal/mol. Calculate $\Theta(P, T)$ for nitrogen at 77 K and at 90 K, assuming the adsorption to follow the Langmuir equation with b_0 given by Eq. XVII-15. Calculate q_{st} for several Θ values and compare the result with the assumed integral distribution function.

32. If $\Theta(P, T)$ is $\Theta = bP/(1 + bP)$, show what result for $\phi(x)$ follows from Eq. XVII-133.

33. Show what the hydraulic radius of a right circular cylinder is, relative to its diameter.

34. Convert the upper curve of Fig. XVII-31a to an adsorption isotherm, that is, to a plot of v versus P/P^0.

35. The projection of a domain plot onto its base makes a convenient two-dimensional graphical representation for describing adsorption–desorption operations. Here, the domain region that is filled can be indicated by shading the appropriate portion of the 45° base triangle. Indicate the appropriate shading for (a) adsorption up to $x_a = 0.8$; (b) such adsorption followed by desorption to $x_d = 0.5$; and (c) followed by readsorption from $x_d = 0.5$ to $x_a = 0.7$.

36. The nitrogen adsorption isotherm is determined for a finely divided, nonporous solid. It is found that at $\theta = 0.5$, P/P^0 is 0.05 at 77 K, and P/P^0 is 0.2 at 90 K. Calculate the isosteric heat of adsorption, and $\Delta \bar{S}^0$ and $\Delta \bar{G}^0$ for adsorption at 77 K. Write the statement of the process to which your calculated quantities correspond. Explain whether the state of the adsorbed N_2 appears to be more nearly gaslike or liquidlike. The normal boiling point of N_2 is 77 K, and its heat of vaporization is 1.35 kcal/mol.

37. Discuss physical situations in which it might be possible to observe a vertical step in the adsorption isotherm of a gas on a heterogeneous surface.

38. Derive Eq. XVII-136. Derive from it the Kelvin equation (Eq. III-18).

39. Use the data of Problem XVII-11 to obtain an MP plot, as in Fig. XVII-30. Comment on the result.

40. Show what two-dimensional equation of state corresponds to the isotherm:

$$\ln kP = \ln \pi + \frac{b\pi}{RT}$$

($Hint$: Refer to Section XVII-6.)

General References

D. Avnir, ed., *The Fractal Approach to Heterogeneous Chemistry*, Wiley, New York, 1989.

D. W. Breck, *Zeolite Molecular Sieves*, Wiley-Interscience, New York, 1974.

J. H. de Boer, *The Dynamical Character of Adsorption*, Clarendon Press, Oxford, 1953.

D. H. Everett and R. H. Ottewill, eds., *Surface Area Determination*, Butterworths, London, 1970.

W. H. Flank and T. Whyte, *Perspective in Molecular Sieve Science*, ACS Symposium Series, No. 135, American Chemical Society, Washington, DC, 1980.

E. A. Flood, ed., *The Solid–Gas Interface*, Vols. 1 and 2, Marcel Dekker, New York, 1967.

R. H. Fowler and E. A. Guggenheim, *Statistical Thermodynamics*, Cambridge University Press, Cambridge, England, 1952.

S. J. Gregg and K. S. W. Sing, *Adsorption, Surface Area, and Porosity*, 2nd ed., Academic, Orlando, FL, 1983.

M. Jaroniec and R. Maday, *Physical Adsorption on Porous Solids*, Elsevier, New York, 1988.

S. Ross and J. P. Olivier, *On Physical Adsorption*, Interscience, New York, 1964.

B. W. Rossiter and R. C. Baetzold, *Physical Methods of Chemistry*, 2nd ed., Vol. IXA, *Investigations of Surfaces and Interfaces*, Wiley-Interscience, New York, 1993.

G. A. Somorjai, *Introduction to Surface Chemistry and Catalysis*, Wiley-Interscience, New York, 1994.

W. A. Steele, *The Interaction of Gases with Solids Surfaces*, Pergamon, New York, 1974.

M. Suzuki, ed., *Fundamentals of Adsorption, Studies in Surface Science and Catalysis*, B. Delman and J. T. Yates, eds., Elsevier, New York, 1993.

D. M. Young and A. D. Crowell, *Physical Adsorption of Gases*, Butterworths, London, 1962.

Z. Zangwill, *Physics at Surfaces*, Cambridge University Press, Cambridge, UK, 1988.

Textual References

1. L. E. Drain and J. A. Morrison, *Trans. Faraday Soc.*, **49**, 654 (1953).

2. M. T. Coltharp and N. Hackerman, *J. Phys. Chem.*, **72**, 1171 (1968).

3. S. J. Gregg and J. D. F. Ramsay, *J. Phys. Chem.*, **73**, 1243 (1969).

4. J. H. de Boer, *The Dynamical Character of Adsorption*, Clarendon Press, Oxford, 1953.

5. M. Knudsen, *Ann. Phys.*, **34**, 593 (1911).

6. I. Yasumoto, *J. Phys. Chem.*, **91**, 4298 (1987).

7. W. B. Nilsson and B. S. Rabinovitch, *Langmuir*, **1**, 71 (1985).

8. W. Steele, *Chem. Rev.*, **93**, 2355 (1993).

9. I. Langmuir, *J. Am. Chem. Soc.*, **40**, 1361 (1918).

10. R. H. Fowler and E. A. Guggenheim, *Statistical Thermodynamics*, Cambridge University Press, Cambridge, England, 1952.

11. D. M. Young and A. D. Crowell, *Physical Adsorption of Gases*, Butterworths, London, 1962.

12. J. E. Lennard-Jones and A. F. Devonshire, *Proc. Roy. Soc.*, **A156**, 6, 29 (1936).

13. D. H. Everett, *Proc. Chem. Soc.* (London), **1957,** 38.

14. J. H. de Boer and S. Kruyer, *K. Ned. Akad. Wet. Proc.*, **55B,** 451 (1952).

15. D. Knox and D. B. Dadyburjor, *Chem. Eng. Commun.*, **11,** 99 (1981).

16. S. Ross and J. P. Olivier, *The Adsorption Isotherm*, Rensselaer Polytechnic Institute, Troy, NY, 1959, p. 39f.

17. T. L. Hill, *J. Chem. Phys.*, **15,** 767 (1947).

18. A. A. Isirikyan and A. V. Kiselev, *J. Phys. Chem.*, **66,** 205, 210 (1962).

19. D. N. Misra, *J. Colloid Interface Sci.*, **77,** 543 (1980).

20. J. M. Honig, in *The Solid–Gas Interface*, E. A. Flood, ed., Marcel Dekker, New York, 1967.

21. T. R. Rybolt, *J. Colloid Interface Sci.*, **107,** 547 (1985).

22. E. A. Flood, ed., *The Solid–Gas Interface*, Marcel Dekker, New York, 1967.

23. K. S. W. Sing, *Pure Appl. Chem.*, **57,** 603 (1985).

24. U. Müller, H. Reichert, E. Robens, K. K. Unger, Y. Grillet, F. Rouquerol, J. Rouquerol, D. Pan, and A. Mersmann, *Fresenius Z. Anal. Chem.*, **333,** 433 (1989).

25. P. L. Llewellyn, J. P. Coulomb, Y. Grillet, J. Patarin, G. Andre, and J. Rouquerol, *Langmuir*, **9,** 1852 (1993).

26. R. A. Beebe, J. B. Beckwith, and J. M. Honig, *J. Am. Chem.*, **67,** 1554 (1945).

27. A. J. Rosenberg, *J. Am. Chem. Soc.*, **78,** 2929 (1956).

28. M. A. Kaiser and C. Dybowski, *Modern Practice of Gas Chromatography*, 3rd ed., R. L. Grob, ed., Wiley, New York, 1995.

29. J. W. McBain and A. M. Bakr, *J. Am. Chem. Soc.*, **48,** 690 (1926).

30. L. J. M. Schlangen, L. K. Koopal, M. A. Cohen Stuart, J. Lyklema, M. Robin, and H. Toulboat, *Langmuir*, **11,** 1701 (1995).

31. G. H. Amberg, W. B. Spencer, and R. A. Beebe, *Can. J. Chem.*, **33,** 305 (1955).

32. S. Partyka, F. Rouquerol, and J. Rouquerol, *J. Colloid Interf. Sci.*, **68,** 21 (1979).

33. R. Denoyel, J. Fernandez-Colinas, Y. Grillet, and J. Rouquerol, *Langmuir*, **9,** 515 (1993).

34. S. Brunauer, *The Adsorption of Gases and Vapors*, Vol. 1, Princeton University Press, Princeton, NJ, 1945.

35. A. W. Adamson and L. Dormant, *J. Am. Chem. Soc.*, **88,** 2055 (1966).

36. W. H. Wade and J. W. Whalen, *J. Phys. Chem.*, **72,** 2898 (1968).

37. A. W. Adamson, *J. Colloid Interface Sci.*, **27,** 180 (1968).

38. M. E. Tadros, P. Hu, and A. W. Adamson, *J. Colloid Interface Sci.*, **49,** 184 (1974).

39. S. Brunauer, P. H. Emmett, and E. Teller, *J. Am. Chem. Soc.*, **60,** 309 (1938).

40. A. Seri-Levy and D. Avnir, *Langmuir*, **9,** 2523 (1993).

41. A. B. D. Cassie, *Trans. Faraday Soc.*, **41,** 450 (1945).

42. T. Hill, *J. Chem. Phys.*, **14,** 263 (1946), and succeeding papers.

43. A. L. McClellan and H. F. Harnsberger, *J. Colloid Interface Sci.*, **23**, 577 (1967).

44. C. Pierce and B. Ewing, *J. Phys. Chem.*, **68**, 2562 (1964).

45. D. Amati and E. Kováts, *Langmuir*, **3**, 687 (1987); **4**, 329 (1988).

46. J. Rouquerol, F. Rouquerol, Y. Grillet, and M. J. Torralvo, in *Fundamentals of Adsorption*, A. L. Myers and G. Belfort, eds., Engineering Foundation, New York, 1984, p. 501.

47. M. J. Katz, *Anal. Chem.*, **26**, 734 (1954).

47a. M. Young and A. D. Crowell, *Physical Adsorption of Gases*, Butterowrths, London, 1962.

48. S. J. Gregg and K. S. W. Sing, *Adsorption, Surface Area, and Porosity*, Academic, New York, 1967.

49. S. Sircar, *Adsorp. Sci. & Tech.*, **2**, 23 (1985).

50. R. B. Anderson, *J. Am. Chem. Soc.*, **68**, 686 (1946).

51. P. A. Redhead, *Langmuir*, **12**, 763 (1996).

52. S. J. Gregg, *J. Chem. Soc.*, **1942**, 696.

53. M. Volmer, *Z. Phys. Chem.*, **115**, 253 (1925).

54. C. Kemball and E. K. Rideal, *Proc. Roy. Soc.* (London), **A187**, 53 (1946).

55. S. Ross and J. P. Olivier, *On Physical Adsorption*, Interscience, New York, 1964.

56. W. D. Harkins and G. Jura, *J. Am. Chem. Soc.*, **66**, 1366 (1944).

57. S. Ross and W. Winkler, *J. Colloid Sci.*, **10**, 319 (1955); see also *J. Am. Chem. Soc.*, **76**, 2637 (1954).

57a. J. McGavack, Jr., and W. A. Patrick, *J. Am. Chem. Soc.*, **42**, 946 (1920).

58. M. M. Dubinin, *Russ. J. Phys. Chem.* (Engl. Transl. Engl.), **39**, 697 (1965); *Q. Rev.*, **9**, 101 (1955).

59. See M. J. Sparnaay, *Surf. Sci.*, **9**, 100 (1968).

60. E. L. Fuller, Jr., and K. A. Thompson, *Langmuir*, **3**, 713 (1987).

61. A. W. Adamson and I. Ling, *Advances in Chemistry Series*, No. 43, American Chemical Society, Washington, DC, 1964.

62. T. D. Blake, *J. Chem. Soc., Faraday Trans. I*, **71**, 192 (1975).

63. Y. L. Frenkel, *Kinetic Theory of Liquids*, Clarendon Press, Oxford, 1946 (reprinted by Dover Publications, 1955).

64. G. D. Halsey, Jr., *J. Chem. Phys.*, **16**, 931 (1948).

65. T. L. Hill, *Adv. Catal.*, **4**, 211 (1952).

66. W. G. McMillan and E. Teller, *J. Chem. Phys.*, **19**, 25 (1951).

67. C. Pierce, *J. Phys. Chem.*, **63**, 1076 (1959).

68. B. D. Adkins, P. J. Reucroft, and B. H. Davis, *Adsorp. Sci. Technol.*, **3**, 123 (1986).

69. D. N. Furlong, K. S. W. Sing, and G. D. Parfitt, *Adsorp. Sci. & Technol.*, **3**, 25 (1986).

70. P. J. M. Carrott, A. I. McLeod, and K. S. W. Sing, in *Adsorption at the Gas–Solid and Liquid–Solid Interface*, J. Rouquerol and K. S. W. Sing, eds., Elsevier, Amsterdam, 1982.

71. C. M. Greenlief and G. D. Halsey, *J. Phys. Chem.*, **74**, 677 (1970).

72. See A. W. Adamson, *J. Colloid Interface Sci.*, **44**, 273 (1973); A. W. Adamson and I. Ling in *Adv. Chem.*, **43**, 57 (1964).

73. N. N. Roy and G. D. Halsey, Jr., *J. Chem. Phys.*, **53**, 798 (1970).

74. G. D. Halsey, Jr., *J. Am. Chem. Soc.*, **73**, 2693 (1951).

75. J. Ginous and L. Bonnetain, *CR*, **272**, 879 (1971).

76. A. L. Myers and J. M. Prausnitz, *AIChE J.*, **11**, 121 (1965).

77. J. A. Ritter and R. T. Yang, *Ind. Eng. Chem.*, **26**, 1679 (1987).

78. M. Manes, in *Fundamentals of Adsorption*, A. L. Myers and G. Belfort, eds., Engineering Foundation, New York, 1984.

79. S. Sircar and A. L. Myers, *Chem. Eng. Sci.*, **28**, 489 (1973).

80. J. H. de Boer and C. Zwikker, *Z. Phys. Chem.*, **B3**, 407 (1929).

81. W. A. Steele, *J. Chem. Phys.*, **65**, 5256 (1976).

82. A. G. Keenan and J. M. Holmes, *J. Phys. Colloid Chem.*, **53**, 1309 (1949).

83. T. R. Rybolt, *J. Tennessee Academy Sci.*, **61**, 66 (1986).

84. S. W. Benson and D. A. Ellis, *J. Am. Chem. Soc.*, **72**, 2095 (1950).

85. C. Pierce and B. Ewing, *J. Phys. Chem.*, **68**, 2562 (1964).

86. A. W. Adamson, *The Physical Chemistry of Surfaces*, Interscience, New York, 1960, and unpublished work, 1954.

87. I. Halász and G. Schay, *Acta Chim. Acad. Sci. Hung.*, **14**, 315 (1956).

88. W. D. Harkins and G. Jura, *J. Am. Chem. Soc.*, **66**, 919 (1944).

89. L. G. Joyner and P. H. Emmett, *J. Am. Chem. Soc.*, **70**, 2353 (1948).

90. B. C. Lippens, B. G. Linsen, and J. H. de Boer, *J. Catal.*, **3**, 32 (1964); J. H. de Boer, B. C. Lippens, V. G. Linsen, J. C. P. Broekhoff, A. van den Heuvel, and Th. J. Osinga, *J. Colloid Interface Sci.*, **21**, 405 (1956). See also R. W. Cranston and F. A. Inkley, *Adv. Catal.*, **9**, 143 (1957).

91. M. R. Bhambhani, P. A. Cutting, K. S. W. Sing, and D. H. Turk, *J. Colloid Interface Sci.*, **38**, 109 (1972).

92. P. J. M. Carrott, R. A. Roberts, and K. S. W. Sing, *Carbon*, **25**, 769 (1987).

93. P. J. M. Carrott, R. A. Roberts and K. S. W. Sing, in *Particle Size Analysis*, P. J. Lloyd, ed., Wiley, New York, 1988.

94. P. J. M. Carrott, R. A. Roberts, and K. S. W. Sing, *Langmuir*, **4**, 740 (1988).

95. W. A. Steele, *Langmuir*, **12**, 145 (1996).

96. D.J. Dai and G. E. Ewing, *J. Chem. Phys.*, **98**, 5050 (1993).

97. L. M. Quattrocci and G. E. Ewing, *J. Chem. Phys.*, **96**, 4205 (1992).

98. S. Zehme, J. Heidberg, and H. Hartmann, *Forsh. Kolloid Polym.*, **55**, 65 (1971).

99. J. Heidberg and D. Meine, *Surf. Sci. Lettr.*, **279**, L175 (1992).

100. A. Thomy and X. Duval, *Surf. Sci.*, **299–300**, 415 (1994).

101. R. J. Guest, A. Nilsson, O. Björneholm, B. Hernnäs, A. Sandell, R. E. Palmer, and N. Mårtensson, *Surf. Sci.*, **269–270**, 432 (1992).

102. V. L. Eden and S. C. Fain, Jr., *J. Vac. Sci. Technol. A*, **10**, 2227 (1992).

103. J. Eckert, J. M. Nicol, J. Howard, and F. R. Trouw, *J. Phys. Chem.*, **100**, 10646 (1996).

104. G. Vidali, G. Ihm, H.-Y. Kim, and M. W. Cole, *Surf. Sci. Rep.*, **12**, 133 (1991).

105. A. W. Adamson and M. W. Orem, *Prog. Surf. Membrane Sci.*, **8**, 285 (1974).

106. L. M. Dormant and A. W. Adamson, *J. Colloid Interface Sci.*, **28**, 459 (1968).

107. A. Chizmeshya and E. Zaremba, *Surf. Sci.*, **268**, 432 (1992).

108. M. J. Bojan and W. A. Steele, *Langmuir*, **3**, 116 (1987).

109. G. D. Halsey, see *Twenty Years of Colloid and Surface Chemistry: The Kendall Award Adresses*, Amer. Chem. Soc., Washington, DC, 1973.

110. K. Madih, B. Croset, J. P. Coulomb, and H. I. Laute, *Europhys. Lett.*, **8**, 459 (1989).

111. R. Miranda, S. Daiser, K. Wandelt, and G. Ertl, *Surf. Sci.*, **131**, 61 (1983).

112. Y. Grillet and J. Rouquerol, *J. Colloid Interface Sci.*, **77**, 580 (1980).

113. T. L. Hill, *J. Chem. Phys.*, **17**, 520 (1949).

114. D. H. Everett, *Trans. Faraday Soc.*, **46**, 453 (1950).

115. R. N. Smith, *J. Am. Chem. Soc.*, **74**, 3477 (1952).

116. J. M. Honig, *J. Colloid Interface Sci.*, **70**, 83 (1979).

117. S. Sircar, *J. Chem. Soc., Faraday Trans. I*, **81**, 1527 (1985).

118. L. E. Copeland and T. F. Young, *Adv. Chem.*, **33**, 348 (1961).

119. Y. C. Wu and L. E. Copeland, *Adv. Chem.*, **33**, 357 (1961).

120. L. M. Dormant and A. W. Adamson, *J. Colloid Interface Sci.*, **75**, 23 (1980).

121. G. Jura and T. L. Hill, *J. Am. Chem. Soc.*, **74**, 1598 (1952).

122. A. C. Zettlemoyer, G. J. Young, J. J. Chessick, and F. H. Healey, *J. Phys. Chem.*, **57**, 649 (1953).

123. G. L. Kington, R. A. Beebe, M. H. Polley, and W. R. Smith, *J. Am. Chem. Soc.*, **72**, 1775 (1950).

124. E. Garrone, G. Ghiotti, E. Giamello, and B. Fubini, *J. Chem. Soc., Faraday Trans. I.*, **77**, 2613 (1981).

125. M. W. Orem and A. W. Adamson, *J. Colloid Interface Sci.*, **31**, 278 (1969).

126. C. Prenzlow, *J. Colloid Interface Sci.*, **37**, 849 (1971).

127. J. Greyson and J. G. Aston, *J. Phys. Chem.*, **61**, 610 (1957).

128. M. O'Neil, R. Lovrien, and J. Phillips, *Rev. Sci. Instr.*, **56**, 2312 (1985).

129. K. Miura and T. Morimoto, *Langmuir*, **2**, 824 (1986).

130. D. N. Furlong, K. S. W. Sing, and G. D. Parfitt, *Adsorp. Sci. Technol.*, **3**, 25 (1986); D. N. Furlong, F. Rouquerol, J. Rouquerol, and K. S. W. Sing. *J. Colloid Interface Sci.*, **75**, 68 (1980).

131. T. P. Vo and T. Fort, Jr., *J. Phys. Chem.*, **91**, 6638 (1987).

132. T. L. Hill, P. H. Emmett, and L. G. Joyner, *J. Am. Chem. Soc.*, **73**, 5102 (1951).

133. A. Zangwill, *Physics at Surfaces*, Cambridge University Press, New York, 1988, p. 266.

134. H. M. Cassel, *J. Chem. Phys.*, **12**, 115 (1944); *J. Phys. Chem.*, **48**, 195 (1944).

135. J. R. Arnold, *J. Am. Chem. Soc.*, **71**, 104 (1949).

136. S. Brunauer, L. E. Copeland, and D. L. Kantro, in *The Solid–Gas Interface*, E. A. Flood, ed., Marcel Dekker, New York, 1966.

137. S. Brunauer, in *Surface Area Determination, Proc. Int. Symp., Bristol*, 1969, Butterworths, London.

138. S. Brunauer, *Langmuir*, **3**, 3 (1987) (posthumous paper).

139. A. I. Sarakhov, M. M. Dubinin, and Yu. F. Bereskina, *Izv. Akad. Nauk SSSR, Ser. Khim.*, 1165 (July 1963).

140. F. E. Karasz, W. M. Champion, and G. D. Halsey, Jr., *J. Phys. Chem.*, **60**, 376 (1956).

141. A. W. Adamson, L. Dormant, and M. Orem, *J. Colloid Interface Sci.*, **25**, 206 (1967).

142. V. R. Bhethanabolta and W. A. Steele, *Langmuir*, **3**, 581 (1987).

143. V. R. Bhethanabotla and W. A. Steele, *Phys. Rev. B*, **41**, 9480 (1990).

144. L. K. Koopal and C. H. W. Vos, *Langmuir*, **9**, 2593 (1993).

145. G. F. Cerofolini and N. Re, *La Rivista del Nuovo Cimento della Società Italiana di Fisica*, **16**(7), 1 (1993). Note also G. F. Cerofolini, *Z. Phys. Chem.* (Leipzig), **258**, 937 (1977); G. F. Cerofolini, M. Jaroniec, and S. Sokolowski, *Colloid Polym. Sci.*, **256**, 471 (1978).

145a. M. Jaroniec and R. Maday, *Physical Adsorption on Heterogeneous Solids*, Elsevier, New York, p. 162, 1988.

146. M. Jaroniec and R. Madey, *Physical Adsorption on Heterogeneous Solids*, Elsevier, New York, 1988.

146. V. Sh. Mamleev and E. A. Bekturov, *Langmuir*, **12**, 3630 (1996).

147. A. W. Adamson, *Colloids & Surfaces*, **A118**, 193 (1996).

148. S. Sircar, *J. Chem. Soc., Faraday Trans. I*, **80**, 1101 (1984).

149. See B. M. W. Trapnell, *Chemisorption*, Academic, New York, 1955, p. 124.

150. J. P. Hobson, *Can. J. Phys.*, **43**, 1934 (1965).

151. R. Sips, *J. Chem. Phys.*, **16**, 490 (1948).

152. J. M. Honig and L. H. Reyerson, *J. Phys. Chem.*, **56**, 140 (1952); J. M. Honig and P. C. Rosenbloom, *J. Chem. Phys.*, **23**, 2179 (1955).

153. J. Toth, *Acta Chim. Hung.*, **32**, 31 (1962); ibid., **69**, 311 (1971). For variants, see J. Toth, *J. Colloid Interface Sci.*, **79**, 85 (1981).

154. D. N. Misra, *J. Chem. Phys.*, **52**, 5499 (1970).

155. A. W. Adamson and I. Ling, *Adv. Chem.*, **33**, 51 (1961).

156. W. A. House and M. J. Jaycock, *Colloid and Polymer Sci.*, **256**, 52 (1978).

157. S. Ross and I. D. Morrison, *Surf. Sci.*, **52**, 103 (1975).

158. L. B. Harris, *Surf. Sci.*, **10**, 129 (1968); ibid., **13**, 377 (1969); ibid., **15**, 182 (1969). M. Jaroniec, S. Sokolowski, and G. F. Cerofolini, *Thin Solid Films*, **31**, 321 (1976).

159. M. Jaroniec, S. Sokolowski, and G. F. Cerofolini, *Thin Solid Films*, **31**, 321 (1976).

160. J. W. Whalen, *J. Phys. Chem.*, **71**, 1557 (1967).

160a. P. Y. Hsieh, *J. Phys. Chem.*, **68**, 1068 (1964); *J. Catal.*, **2**, 211 (1963).

161. W. A. House and M. J. Jaycock, *J. Colloid Interface Sci.*, **47**, 50 (1974).

162. A. W. Adamson, I. Ling, L. Dormant, and M. Oren, *J. Colloid Interface Sci.*, **21**, 445 (1966).

163. S. Ross and W. W. Pultz, *J. Colloid Sci.*, **13**, 397 (1958).

164. D. H. Everett, private communication.

165. L. G. Helper, *J. Chem. Phys.*, **16**, 2110 (1955).

166. W. A. Steele, *J. Phys. Chem.*, **67**, 2016 (1963); see also J. M. Honig, *Adv. Chem.*, **33**, 239 (1961).

167. D. Avnir and D. Farin, *New J. Chem.*, **16**, 439 (1992).

168. D. Avnir, *J. Am. Chem. Soc.*, **109**, 2931 (1987).

169. D. Avnir, D. Farin, and P. Pfeifer, *Nature*, **308**, 261 (1984).

170. J. W. Larsen and P. Wernett, *Energy Fuels*, **2**, 719 (1988).

171. H. Van-Damme and J. J. Fripiat, *J. Chem. Phys.*, **82**, 2785 (1985).

172. D. Avnir, D. Farin, and P. Pfeifer, *J. Chem. Phys.*, **79**, 3566 (1983).

173. J. M. Drake, P. Levitz, and S. Sinha, in *Better Ceramics Through Chemistry*, C. J. Brinker, D. E. Clarck, and D. R. Ulrich, eds., *Mat. Res. Soc. Symp. Proc.*, **73**, 305 (1986).

174. D. Farin and D. Avnir, *J. Phys. Chem.*, **91**, 5517 (1987).

175. (a) Z. Sokolowska, *Gerderma*, **45**, 251 (1989); (b) Z. Sokolowska, J. Stawinski, A. Patrykiewic, and S. Sokolowski, *Int. Agrophys.*, **5**, 3 (1989).

176. V. R. Deitz and F. G. Carpenter, *Adv. Chem.*, **33**, 146 (1961).

177. C. A. Guderjahn, D. A. Paynter, P. E. Berghausen, and R. J. Good, *J. Phys. Chem.*, **63**, 2066 (1959).

178. *Recommendations for the Characterization of Porous Solids, Pure & Appl. Chem.*, **66**, 1739 (1994).

179. R. M. Barrer, *Proc. 10th Colston Symp.*, Butterworths, London, 1958, p. 6.

180. R. M. Barrer, *Discuss. Faraday Soc.*, **7**, 135 (1949); *Q. Rev.*, **3**, 293 (1949).

181. R. M. Barrer, *Zeolites*, **1**, 130 (1981).

182. R. M. Barrer, *J. Inclusion Phen.*, **1**, 105 (1983).

183. R. M. Barrer, *J. Inclusion Phen.*, **4**, 109 (1986).

184. C. T. Kresge, M. E. Leonowicz, W. J. Roth, J. C. Vartull, and J. S. Beck, *Nature*, **359**, 710 (1992).

185. J. S. Beck, J. C. Vartuli, W. J. Roth, M. E. Leonowicz, C. T. Kresge, K. D. Schmitt, C. T.-W. Chu, D. H. Olson, E. W. Sheppard, S. B. McCullen, J. B. Higgines, and J. L. Schlenker, *J. Am. Chem. Soc.*, **114**, 10834 (1992).

185a. P. L. Llewellyt, F. Schuth, V. Grillet, F. Rouquerol, J. Rouquerol, and K. K. Unger, *Langmuir*, **11**, 574 (1995).

186. N. Setoyama and K. Kaneko, *J. Phys. Chem.*, **100**, 10331 (1996).

187. P. L. Llewellyn, Y. Grillet, F. Schüth, H. Reichert, and K. K. Unger, *Microporous Mat.*, **3**, 345 (1994).

188. H. M. Powell, *J. Chem. Soc.*, **1954**, 2658.

189. R. M. Barrer and W. I. Stuart, *Proc. Roy. Soc.* (London), **A243**, 172 (1957).

190. R. M. Barrer and D. J. Ruzicka, *Trans. Faraday Soc.*, **58**, 2262 (1962).

191. A. W. Adamson, L. Dormant, and M. Orem, *J. Colloid Interface Sci.*, **25**, 206 (1967).

192. A. W. Adamson and B. R. Jones, *J. Colloid Interface Sci.*, **37**, 831 (1971).

193. A. Bailey, D. A. Cadenhead, D. H. Davies, D. H. Everett, and A. J. Miles, *Trans. Faraday Soc.*, **67**, 231 (1971).

194. J. J. Hacskaylo and M. D. LeVan, *Langmuir*, **1**, 97 (1985).

195. R. I. Razouk, Sh. Nashed, and F. N. Antonious, *Can. J. Chem.*, **44,** 877 (1966).

196. C. Pierce, *J. Phys. Chem.*, **57,** 149 (1953).

197. R. S. Schechter, W. H. Wade, and J. A. Wingrave, *J. Colloid Interface Sci.*, **59,** 7 (1977).

198. R. Zsigmondy, *Z. Anorg. Chem.*, **71,** 356 (1911).

199. L. H. Cohan, *J. Am. Chem. Soc.*, **60,** 433 (1938); see also ibid., **66,** 98 (1944).

200. L. Sliwinska and B. H. Davis, *Ambix*, **34,** 81 (1987).

201. E. P. Barrett, L. G. Joyner, and P. P. Halenda, *J. Am. Chem. Soc.*, **73,** 373 (1951).

202. H. Naono, T. Kadota, and T. Morimoto, *Bull. Chem. Soc. Jpn.*, **48,** 1123 (1975).

203. B. D. Adkins and B. H. Davis, *J. Phys. Chem.*, **90,** 4866 (1986).

204. M. M. Dubinin, *J. Colloid Interface Sci.*, **77,** 84 (1980).

205. W. D. Machin, *J. Chem. Soc., Faraday Trans. I*, **78,** 1591 (1982).

206. W. D. Machin and J. T. Stuckless, *J. Chem. Soc., Faraday Trans. I*, **81,** 597 (1985).

207. G. D. Parfitt, K. S. W. Sing, and D. Urwin, *J. Colloid Interface Sci.*, **53,** 187 (1975).

208. A. Lecloux and J. P. Pirard, *J. Colloid Interface Sci.*, **70,** 265 (1979).

209. D. H. Everett, *Proc. 10th Colston Symp.*, Butterworths, London, 1958, p. 95.

210. R. M. Barrer, N. McKenzie, and J. S. S. Reay, *J. Colloid Sci.*, **11,** 479 (1956).

210a. A. Seri-Levy and D. Avnir, *Langmuir*, **9,** 3067 (1993).

211. S. Brunauer, R. Sh. Mikhail, and E. E. Bodor, *J. Colloid Interface Sci.*, **24,** 451 (1967).

212. J. Hagymassy, Jr., I. Odler, M. Yudenfreund, J. Skalny, and S. Brunauer, *J. Colloid Interface Sci.*, **38,** 20 (1972).

213. A. V. Kiselev, *Usp. Khim.*, **14,** 367 (1945).

214. K. S. Rao, *J. Phys. Chem.*, **45,** 517 (1941).

215. P. H. Emmett, *Chem. Rev.*, **43,** 69 (1948).

216. D. H. Everett, in *The Solid-Gas Interface*, Vol. 2, E. A. Flood, ed., Marcel Dekker, New York, 1966; D. H. Everett, *Trans. Faraday Soc.*, **51,** 1551 (1955).

217. B. D. Adkins and B. H. Davis, *Langmuir*, **3,** 722 (1987).

218. M. Parlar and Y. C. Yortsos, *J. Colloid Interface Sci.*, **124,** 162 (1988).

219. M. Suzuki, ed., *Fundamentals of Adsorption*, Vol. 80, Elsevier, New York, 1993.

220. V. K. LaMer, *J. Colloid Interface Sci.*, **23,** 297 (1967) (posthumous paper).

221. G. L. Kington and P. S. Smith, *Trans. Faraday Soc.*, **60,** 705 (1964).

221a. IUPAC Manual of Symbols and Terminology, Appendix 2, Part I, Colloid and Surface Chemistry, *Pure Appl. Chem.*, **54,** 2201 (1982).

222. P. J. M. Carrott, R. A. Roberts, and K. S. W. Sing, *Carbon*, **25,** 59 (1987).

223. D. Dollimore and G. R. Heal, *Surf. Techn.*, **6,** 231 (1978).

224. M. Jaroniec, R. Madey, J. Choma, B. McEnaney, and T. J. Mays, *Carbon*, **27,** 77 (1989).

225. M. M. Dubinin and O. Kadlec, *Carbon*, **25,** 321 (1987).

226. G. Amarasekera, M. J. Scarlett, and D. E. Mainwaring, *J. Phys. Chem.*, **100,** 7580 (1996).

(Ref.

227. D. Atkinson, P. J. M. Carrott, Y. Grillet, J. Rouquerol, and K. S. W. Sing, *Fundamentals of Adsorption*, I. Lipais, ed., Engineering Foundation, New York, 1987.

228. H. F. Stoeckii, *J. Colloid Interface Sci.*, **59**, 184 (1977).

229. P. Brauer, H. Poosch, M. V. Szombathely, M. Heuchel, and M. Jaroniec, in *Fundamentals of Absorption*, Vol. 80, M. Suzuki, ed., Elsevier, New York, 1993 (Ref. 219), p. 67.

230. R. Sh. Mikhail, S. Brunauer, and E. E. Bodor, *J. Colloid Interface Sci.*, **26**, 45 (1968).

231. K. Kaneko, *Langmuir*, **3**, 357 (1987).

232. J. J. Fripiat and H. Van Damme, *Bull. Soc. Chim. Belg.*, **94**, 825 (1985).

233. D. Avnir and M. Jaroniec, *Langmuir*, **5**, 1431 (1989).

234. B. Sahouli, S. Blacher, and F. Brouers, *Langmuir*, **12**, 2872 (1996).

235. P. J. M. Carrott and K. S. W. Sing, *Characterization of Porous Solids*, K. K. Unger et al., eds., Elsevier, Amsterdam, 1988.

236. R. A. Roberts, K. S. W. Sing, and V. Tripathi, *Langmuir*, **3**, 331 (1987).

237. M. M. Dubinin, *J. Colloid Interface Sci.*, **46**, 351 (1974).

CHAPTER XVIII

Chemisorption and Catalysis

1. Introduction

In this concluding chapter we take up some of those aspects of the adsorption of gases on solids in which the adsorbent–adsorbate bond approaches an ordinary bond in strength and in which the chemical nature of the adsorbate may be significantly different in the adsorbed state. Such adsorption is generally called chemisorption, although as was pointed out in the introduction to Chapter XVII, the distinction between physical adsorption and chemisorption is sometimes blurred, and many of the principles of physical adsorption apply to both kinds of adsorption. One experimental distinction is that we now deal almost entirely with submonolayer adsorption, since in chemisorption systems the heat of adsorption in the first layer ordinarily is much greater than that in succeeding layers. In fact, most chemisorption systems involve temperatures above the critical temperature of the adsorbate, so that the usual treatments of multilayer adsorption do not apply.

At one time the twin subjects of chemisorption and catalysis were so closely intertwined as to be virtually indistinguishable. Chemisorption was the mode of adsorption and heterogeneous or contact catalysis the interesting consequence. The industrial importance of catalytic systems tended to bias the research toward those systems of special catalytic relevance. The massive development in recent years of high-vacuum technology and associated spectroscopic and diffraction techniques (see Chapter VIII) has brought the field of chemisorption to maturity as a distinct field of surface chemistry with research interests undirected toward catalysis (although often relatable to it). And atomic force, scanning tunneling, and other microscopies (see Table VIII-1) now often permit the imaging of a surface with its adsorbed species with atomic scale resolution.

The molecular emphasis of modern chemisorption studies has benefited the field of catalysis by giving depth and scope to the surface chemistry of catalytic processes. To paraphrase King [1], quantitative answers have become possible to the following questions:

1. Where are the atoms and molecules on the surface, and why?
2. What structural changes accompany surface processes?
3. What is the nature of surface chemical bonds, and what are their energies?
4. What factors control bond making and breaking?

Finally, there is a drawing together of catalysis involving metal surfaces and organometallic chemistry; metal cluster compounds now being studied are approaching indistinguishability from polyatomic metal patches on a supporting substrate.

The plan of this chapter is as follows. We discuss chemisorption as a distinct topic, first from the molecular and then from the phenomenological points of view. Heterogeneous catalysis is then taken up, but now first from the phenomenological (and technologically important) viewpoint and then in terms of current knowledge about surface structures at the molecular level. Section XVIII-9F takes note of the current interest in photodriven surface processes.

As on previous occasions, the reader is reminded that no very extensive coverage of the literature is possible in a textbook such as this one and that the emphasis is primarily on principles and their illustration. Several monographs are available for more detailed information (see General References). Useful reviews are on: future directions and ammonia synthesis [2], surface analysis [3], surface mechanisms [4], dynamics of surface reactions [5], single-crystal versus actual catalysts [6], oscillatory kinetics [7], fractals [8], surface electrochemistry [9], particle size effects [10], and supported metals [11, 12].

2. Chemisorption: The Molecular View

It is now a practice to use a variety of surface characterization techniques in the study of chemisorption and catalysis. The examples given here are illustrative; most references in this section as well as throughout the chapter will contain results from several techniques.

A. LEED Structures

The technique of low-energy electron diffraction, LEED (Section VIII-2D), has provided a considerable amount of information about the manner in which a chemisorbed layer rearranges itself. Somorjai [13] has summarized LEED results for a number of systems. Some examples are collected in Fig. XVIII-1. Figure XVIII-1a shows how N atoms are arranged on a Fe(100) surface [14] (relevant to ammonia synthesis); even H atoms may be located, as in Fig. XVIII-1b [15]. Figure XVIII-1c illustrates how the structure of the adsorbed layer, or *adlayer*, can vary with exposure [16].† There may be a series of structures, as with NO on Ru($10\bar{1}0$) [17] and HCl on Cu(110) [18]. Surface structures of

---

†In high-vacuum studies, the amount of adsorbate present is determined primarily by how many gas-phase molecules have hit the surface (as corrected by the sticking coefficient). A common measure of such "exposure" is the *langmuir*, L, defined as 1×10^{-6} torr/sec (see Problem XVIII-2). While an experimentally direct and useful unit, the Langmuir suffers in that its physical meaning depends on the temperature of the gas phase. An alternative measure of exposure is molecules per unit area (e.g., Ref. 16), alternatively expressed as degree of monolayer coverage, ML.

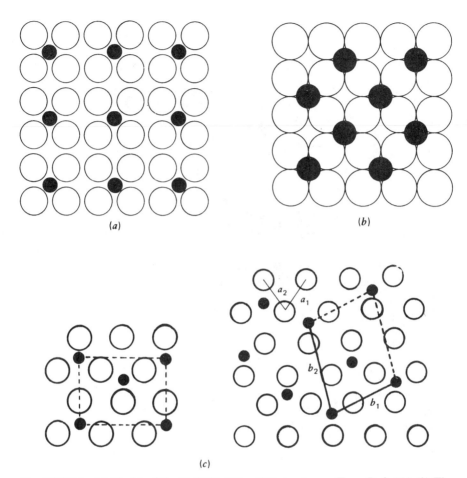

Fig. XVIII-1. (*a*) Model of the Fe(100) $C(2 \times 2)N$ structures. (From Ref. 14.) (*b*) The $C(2 \times 2)$ structure for H atoms on Pd(100) at $\theta = 0.5$. (From Ref. 15.) (*c*) Structures for Cl on Cr(110). Left: α phase, exposure 7.5×10^{18} molecules/m^2; right; β-phase, exposure 3.5×10^{18} molecules/m^2. (From Ref. 16.) In the above case the adatoms are shown as filled circles.

more complex molecules have also been deduced, as for C_2H_2 on Pd(111) [19] and benzene on Rh(111) [20], as well as for coadsorbed species, as shown in Fig. XVIII-1*d* for CO and ethylidyne (see Ref. 21) and in the case of H atoms and CO [22], both for the case of a Rh(111) surface.

Restructuring of a surface may occur as a phase change with a transition temperature as with the Si(001) surface [23]. It may occur on chemisorption, as in the case of oxygen atoms on a stepped Cu surface [24]. The reverse effect may occur! The surface layer for a Pt(100) face is not that of a terminal (100) plane but is reconstructed to hexagonal symmetry. On CO adsorption, the reconstruction is lifted, as shown in Fig. XVI-8.

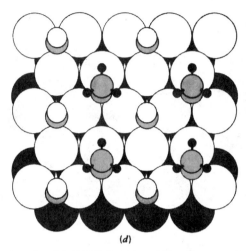

(d)

Fig. XVIII-1. (*Continued*)   (*d*) Rh(111) with a $C(4 \times 2)$ array of adsorbed CO and a layer of ethylidyne (C—CH$_3$). (Courtesy of G. A. Somorjai; see Ref. 21 for a similar example.)

B. Surface Microscopies

We confine ourselves here to scanning probe microscopies (see Section VIII-2B): scanning tunneling microscopy (STM) and atomic force microscopy (AFM), in which successive profiles of a surface (see Fig. VIII-1) are combined to provide a contour map of a surface. It is conventional to display a map in terms of dark to light areas, in order of increasing height above the surface; ordinary contour maps would be confusing to the eye.

Figure XVIII-2 shows how a surface reaction may be followed by STM, in this case the reaction on a Ni(110) surface: O(surface) + H$_2$S(g) = H$_2$O(g) + S(surface). Figure XVIII-2*a* shows the oxygen atom covered surface before any reaction, and Fig. XVIII-2*b*, the surface after exposure to 3 L of H$_2$S during which Ni islands and troughs have formed on which sulfur chemisorbs. The technique is powerful in the wealth of detail provided; on the other hand, there is so much detail that it is difficult to relate it to macroscopic observation (such as the kinetics of the reaction).

Atomic force microscopy yields magnifications and atomic resolution maps rather similar to those obtained by STM. There are two useful differences, however, AFM works with nonconducting surfaces, while STM requires conducting or semiconducting surfaces. Further, in AFM one can scan at constant force, and the force chosen can be varied. Scans at different force levels may show differences since projecting surface atoms may be pushed inward to varying degrees depending on the force level (e.g., Ref. 25). Both STM and AFM may be used at a solid–liquid interface, but more often the latter. An example shown in Fig. XVIII-3 is that of DNA at the mica–water interface.

Fig. XVIII-2. Successive STM images of (a) Ni(110) with a chemisorbed layer of oxygen atoms and (b) after exposure to 3 L of H_2S. The area shown in 85 × 91 Å. [From F. Besenbacher, P. T. Sprunger, L. Ruan, L. Olesen, I. Stensgaard, and E. Lægsgaard, *Top. Catal.*, **1**, 325 (1994).]

Rather sophisticated techniques have developed for both microscopies, too involved to discuss here. Details may be found in monographs such as Ref. 25, and references cited in Section VIII-2.

C. Spectroscopy of Chemisorbed Species

Vibrational Spectroscopy. Infrared absorption spectra may be obtained using convention IR or FTIR instrumentation; the catalyst may be present as a compressed disk, allowing transmission spectroscopy. If the surface area is high, there can be enough chemisorbed species for their spectra to be recorded. This approach is widely used to follow actual catalyzed reactions; see, for example, Refs. 26 (metal oxide catalysts) and 27 (zeolitic catalysts). Diffuse reflectance infrared reflection spectroscopy (DRIFTS) may be used on films [e.g., Ref. 28—SiO_2 films on Mo(110)]. Laser Raman spectroscopy (e.g., Refs. 29, 30) and infrared emission spectroscopy may give greater detail [31].

For species chemisorbed on well-defined surfaces, especially metals, high-vacuum spectroscopy gives better resolution and the method of choice is electron-energy-loss spectroscopy (EELS) or the high-resolution version, HREELS. As illustrated in Fig. XVIII-4, intermediate species can be identified. Figure XVIII-5 illustrates how HREELS can distinguish between various bonding geometries for adsorbed NO_2. Actual catalyzed reactions may be followed since it is possible to go from ambient to high vacuum conditions in less than a minute [32]. Additional, representative references are an overall review [33], CH_3 and CH radicals on Ni(III) [34], $ClCH_2I$ on Pt(III) [35], and H on Ni(III) [36].

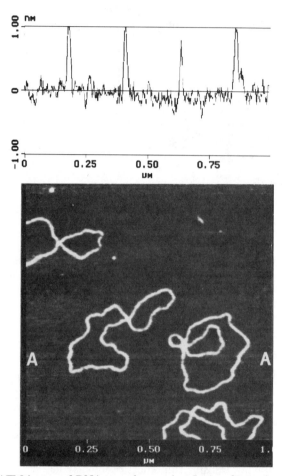

Fig. XVIII-3. AFM image of DNA strands on mica. Lower figure: image obtained in the contact mode under water. The contrast shown covers height variations in the range of 0–2 nm. Upper figure: observed profile along the line A–A of the lower figure. (From S. N. Magnov and M.-H. Whangbo, *Surface Analysis with STM and AFM*, VCH, New York, 1996.)

Electronic spectra of surfaces can give information about what species are present and their valence states. X-ray photoelectron spectroscopy (XPS) and its variant, ESCA, are commonly used. Figure VIII-11 shows the application to an Al surface and Fig. XVIII-6, to the more complicated case of Mo supported on TiO_2 [37]; Fig. XVIII-7 shows the detection of photochemically produced Br atoms on Pt(111) [38]. Other spectroscopies that bear on the chemical state of adsorbed species include (see Table VIII-1): photoelectron spectroscopy (PES) [39–41], angle resolved PES or ARPES [42], and Auger electron spectroscopy (AES) [43–47]. Spectroscopic detection of adsorbed *hydrogen* is difficult, and

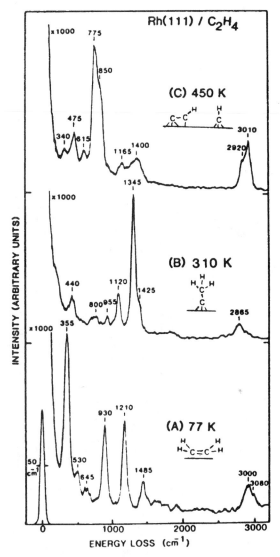

Fig. XVIII-4. HREELS vibrational spectra from ethylene chemisorbed on Rh(111). There is a sequential dehydrogenation on heating. (From Ref. 13, p. 64.)

an interesting alternative makes use of the $^1H(^{15}N, \alpha\gamma)^{12}C$ nuclear reaction [48].

The nature of reaction products and also the orientation of adsorbed species can be studied by atomic beam methods such as electron-stimulated desorption (ESD) [49,50], photon-stimulated desorption (PDS) [51], and ESD ion angular distribution ESDIAD [51–54]. (Note Fig. VIII-13). There are molecular beam scattering experiments such

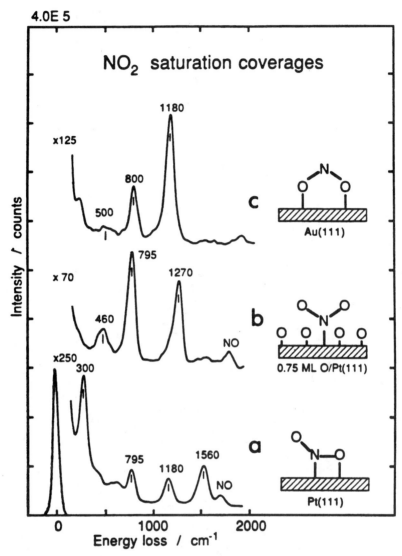

4.0E 5

NO₂ saturation coverages

Fig. XVIII-5. HREELS spectra for NO₂ on Pt(111) adsorbed in different bonding geometries. [From M. E. Bartram and B. E. Koel, *J. Vac. Sci. Tech.*, **A6**, 782 (1988).]

as secondary-ion mass spectroscopy SIMS [55–57], and various ion scattering methods (e.g., Ref. 57a). Nearest-neighbor distributions can be obtained by an angular-distribution version of AES called *ADAM* (angular-distribution Auger microscopy) [58] and, more commonly, by extended x-ray absorption fine structure (EXAFS) [59–63]. Surface states of iron-containing catalysts (important in Fischer–Tropsch syntheses—see Section XVIII-9B) may be probed by means of the Mössbauer effect [64–67]. Magnetic resonance can be used to differentiate reactants, products, and their various surface states:

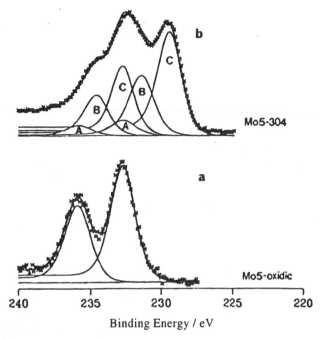

Fig. XVIII-6. Curve-fitted Mo XPS 3d spectra of a 5 wt% Mo/TiO$_2$ catalyst: (*a*) in the oxidic +6 valence state; (*b*) after reduction at 304°C. Doublets *A*, *B*, and *C* refer to Mo oxidation states +6, +5, and +4, respectively [37]. (Reprinted with permission from American Chemical Society; copyright 1974.)

^{1}H [68,69], ^{13}C [70–73], and ^{129}Xe [74,75]. Surface radicals have been studied by means of electron spin resonance (ESR) [76–70,80 (review)].

The preceding material, abounding in references (and multiple authors!), mainly makes evident the mass of information available through surface science techniques. All is not rosy, of course; see Ref. 37 for a critical analysis of difficulties and of uncertainties in interpretation. One advantage of surface spectroscopy, however, is that, unlike the various microscopies, one gets essentially an average or envelope of the entire surface region. At this point in the previous edition, an analogy was made to modern ships that can be taken to an exact destination and on time, versus the uncertainties for the ancient mariner. Like the modern ship, contemporary surface science is more and more giving us the ability to go accurately from reactant(s) to desired product and with detailed knowledge of what is happening on the molecular scale.

D. Work Function and Related Measurements

The work function across a phase boundary, discussed in Sections V-9B and VIII-2C, is strongly affected by the presence of adsorbed species. Conversely,

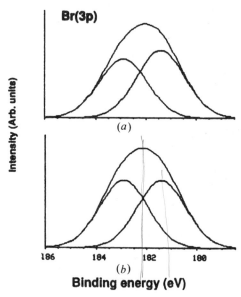

Fig. XVIII-7. Br(3p) XPS spectra of a 2.2-L dose of DBr on Pt(111) before (*a*) and (*b*) after UV irradiation. The 182.8-eV peak is due to DBr and the 181.5 peak, to atomic Br. On irradiation the latter peak increases relative to the former [38]. (Reprinted with permission from American Chemical Society, copyright 1992.)

work function changes can be diagnostic of changes in types of surface binding. Somorjai [81] summarizes a number of results of measurements of $\Delta\Phi$ due to chemisorption—values range from -1.5 V for CO on iron to 1.6 V for oxygen on nickel. The variation with θ can be instructive. In the case of oxygen adatoms on Pt(III), the linear increase in $\Delta\Phi$ with θ indicates that the surface dipole moment per adatom is independent of coverage, as shown in Fig. XVIII-8*a*. In the case of potassium, however, the leveling off with θ shown in Fig. XVIII-8*b* indicates that depolarization occurs.

Different types of chemisorption sites may be observed, each with a characteristic $\Delta\Phi$ value. Several adsorbed states appear to exist for CO chemisorbed on tungsten, as noted. These states of chemisorption probably have to do with different types of chemisorption bonding, maybe involving different types of surface sites. Much of the evidence has come initially from desorption studies, discussed immediately following.

E. *Programmed Desorption*

A powerful technique in studying both adsorption and desorption rates is that of programmed desorption. The general procedure (see Refs. 36, 84) is to expose a clean metal filament or a surface to a known, low pressure of gas that flows steadily over it. The pressure may be quite low, for example, 10^{-7} mm Hg or less, so that even nonactivated adsorption can take some minutes for

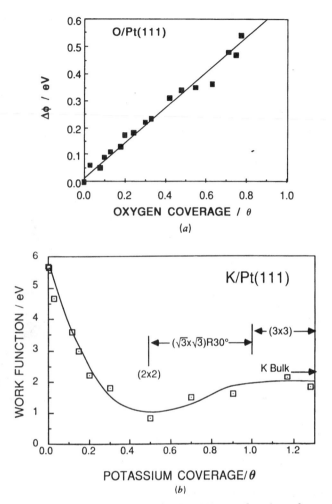

Fig. XVIII-8. (a) Work function change for Pt(111) as a function of oxygen adatom coverage. From Ref. 82. (b) Same, for potassium. The corresponding sequence of LEED structures is indicated. [Reprinted with permission from R. G. Windham, M. E. Bartram, and B. E. Koel, *J. Phys. Chem.*, **92**, 2862 (1988) (Ref. 83). Copyright 1988, American Chemical Society.]

complete monolayer coverage to be achieved. In *flash desorption*, the surface is heated suddenly, and the accumulated adsorbed material desorbs, the amount recovered thus determining how much and hence with what rate adsorption had occurred. The sudden heating may be done very quickly and on a small region of the surface, which might be a well-defined crystalline surface, by using a laser pulse, that is, by *pulsed laser-induced (thermal) desorption* (PLID) [85]. If the system is a reactive one, desorption will include reaction products whose amounts can be determined by mass spectrometry.

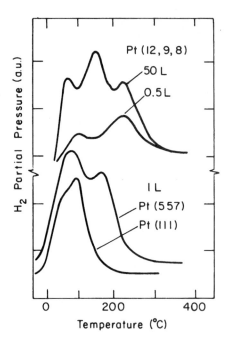

Fig. XVIII-9. TPD spectra for hydrogen chemisorbed on flat (111), stepped (557), and kinked (11, 9, 8) single-crystal surface of Pt. (Reprinted from Ref. 87, p. 427 by courtesy of Marcel Dekker Inc.)

If the heating is quickly to a high temperature, or a "flashing," all adsorbed gas is removed indiscriminately. If, however, the heating is gradual, then separate, successive desorptions may be observed. Thus, as illustrated in Fig. XVIII-9, hydrogen leaves flat, stepped, and kinked Pt surfaces in stages, indicating the presence of different adsorption sites. The presences of successive desorption stages is fairly common. No less than four are found for H_2 chemisorbed on Pd(110), as shown in Fig. XVIII-10; notice how successive maxima appear with increasing exposure. Xu and Koel [87a] report three different states for NO desorption from Pt(III), possibly due to different bonding geometries (as in Fig. XVIII-5). Yates [88] has reviewed the subject.

Temperature-programmed desorption (TPD) is amenable to simple kinetic analysis. The rate of desorption of a molecular species from a uniform surface is given by Eq. XVII-4, which may be put in the form

$$R = -\frac{d\theta}{dt} = A\theta e^{-E/RT} \qquad \text{(XVIII-1)}$$

where A is the frequency factor and E the desorption energy. The situation is simplified experimentally by imposing a uniform rate of heating on the surface, such that $T = T^0 + \beta t$. Differentiation of Eq. XVIII-1 yields

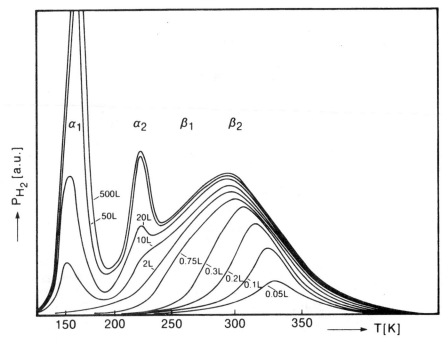

Fig. XVIII-10. TPD spectra for H_2 on Pd(110) for varying exposures (in langmuirs). (From Ref. 86.)

$$\frac{dR}{dt} = A \; \frac{d\theta}{dt} \; e^{-E/RT} + (A\theta)e^{-E/RT}\left(-\frac{E}{R}\right)\left(-\frac{\beta}{T^2}\right) \tag{XVIII-2}$$

On setting $dR/dt = 0$ (and using Eq. XVIII-1), we obtain an equation due to Redhead [89]:

$$\frac{E}{RT_m^2} = \frac{A}{\beta} \; e^{-E/RT_m} \tag{XVIII-3}$$

where T_m is the temperature of maximum desorption rate (see Refs. 89,90). Thus E may be calculated from T_m with the usual assumption that $A = 10^{13}$ sec^{-1}.

For example, the thermal desorption peak for Pt(111) in Fig. XVIII-9 is at about 100°C, and for this experiment, $\beta = 8$ K/sec. Using Eq. XVIII-3, one obtains by successive approximations that $E/R \cong 11,000$, or $E \cong 22$ kcal/mol.

The above illustration requires the value of A to be assumed. The assumption may be avoided by determing T_m for more than one heating rate. Thus E/R may be obtained from the slope of a plot of $\ln(\beta/T_m^2)$ versus $1/T_m$ [91,92]. Variations of this approach are found in Refs. 93–95. The rate law may be more complex than Eq. XVIII-1, and analysis of such situations is also possible; see Refs. 82, 96–98.

A reacting system may also be studied by programmed desorption, now called *temperature-programmed reaction* (TPR). Examples include the reaction of chemisorbed CO with H_2 to give methane on a Ni catalyst [99] and on Pt(111) [100] and more complex systems such as the reaction of $^{13}CO_2$ with carbon to give labeled $^{12}CO_2$ and ^{12}CO [101] and the production of H_2 through the surface decomposition of benzene [102] and phenyl thiolate [103]. Quantitative interpretation is now complicated since T_m may depend both on the activation energy for reaction and on the energy of desorption of the product.

3. Chemisorption Isotherms

In considering isotherm models for chemisorption, it is important to remember the types of systems that are involved. As pointed out, conditions are generally such that physical adsorption is not important, nor is multilayer adsorption, in determining the equilibrium state, although the former especially can play a role in the kinetics of chemisorption.

Because of the relatively strong adsorption bond supposed to be present in chemisorption, the fundamental adsorption model has been that of Langmuir (as opposed to that of a two-dimensional nonideal gas). The Langmuir model is therefore basic to the present discussion, but for economy in presentation, the reader is referred to Section XVII-3 as prerequisite material. However, the Langmuir equation (Eq. XVII-5) as such,

$$\theta = \frac{bP}{1 + bP} \tag{XVIII-4}$$

is not often obeyed in chemisorption systems. Ordinarily, complications appear, and the following material is largely concerned with the necessary specializations of the Langmuir model that are needed. These involve a review of ways of treating surface heterogeneity and lateral interactions and the new isotherm forms that arise if adsorption requires the presence of two adjacent sites or if dissociation occurs on adsorption.

A. Variable Heat of Adsorption

It is not surprising, in view of the material of the preceding section, that the heat of chemisorption often varies from the degree of surface coverage. It is convenient to consider two types of explanation (actual systems involving some combination of the two). First, the surface may be heterogeneous, so that a site energy distribution is involved (Section XVII-14). As an example, the variation of the calorimetric differential heat of adsorption of H_2 on ZnO is shown in Fig.

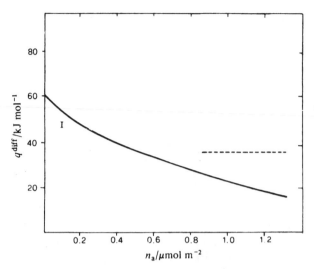

Fig. XVIII-11. Calorimetric differential heat of adsorption of H_2 on ZnO. Dashed line: differential heat of desorption. (From Ref. 104.)

XVIII-11 (the paradox of desorption heat exceeding adsorption heat is explainable in terms of a partial irreversibility of the adsorption–desorption process).

If the differential distribution function is exponential in Q (Section XVII-14A), the resulting $\Theta(P, T)$ is that known as the Freundlich isotherm

$$\Theta(P, T) = AP^c \qquad \text{(XVIII-5)}$$

where c is generally less than unity and is therefore often written as $1/n$. (In discussing situations involving surface heterogeneity, we use Θ to denote the average surface coverage and θ that of a given patch of uniform Q.) The linear form of Eq. XVIII-5 is

$$\log v = \log(v_m A) - n \log P \qquad \text{(XVIII-6)}$$

and quite a few experimental isotherms obey this form over some region of values.

The Freundlich equation is defective as a model because of the physically unrealistic $f(Q)$; consequences of this are that Henry's law is not approached at low P, nor is a limiting adsorption reached at high P. These difficulties can be patched by supposing that

$$\Theta = \frac{AP^c}{1 + AP^c} \qquad \text{(XVIII-7)}$$

for which $f(Q)$ is nearly Gaussian in shape and, moreover, now can be normalized.

A second supposition is that

$$Q = Q_0(1 - \alpha\Theta_Q) \qquad\qquad\qquad (\text{XVIII-8})$$

This leads to an equation that, in the middle range of Θ values, approximates to

$$\Theta_Q = \frac{RT}{Q_0\alpha} \ln P + \text{const} \qquad\qquad\qquad (\text{XVIII-9})$$

a form known as the Temkin isotherm [105] (see Section XVII-14A). Again, this is an equation that may be useful for fitting the middle region of an adsorption isotherm. However, it is often quite difficult to distinguish between these various equations. Thus the middle range of the adsorption of nitrogen on iron powder could be fitted to the Langmuir, the Freundlich, or the Temkin equation [106].

It would seem better to transform chemisorption isotherms into corresponding site energy distributions in the manner reviewed in Section XVII-14 than to make choices of analytical convenience regarding the $f(Q)$ function. The second procedure tends to give equations whose fit to data is empirical and deductions from which can be spurious.

Surface heterogeneity may merely be a reflection of different types of chemisorption and chemisorption sites, as in the examples of Figs. XVIII-9 and XVIII-10. The presence of various crystal planes, as in powders, leads to heterogeneous adsorption behavior; the effect may vary with particle size, as in the case of O_2 on Pd [107]. Heterogeneity may be deliberate; many catalysts consist of combinations of active surfaces, such as bimetallic alloys. In this last case, the surface properties may be intermediate between those of the pure metals (but one component may be in surface excess as with any solution) or they may be distinctly different. In this last case, one speaks of various "effects": *ensemble*, *dilution*, *ligand*, and *kinetic* (see Ref. 108 for details).

The second general cause of a variable heat of adsorption is that of adsorbate–adsorbate interaction. In physical adsorption, the effect usually appears as a lateral *attraction*, ascribable to van der Waals forces acting between adsorbate molecules. A simple treatment led to Eq. XVII-53.

Such attractive forces are relatively weak in comparison to chemisorption energies, and it appears that in chemisorption, *repulsion* effects may be more important. These can be of two kinds. First, there may be a short-range repulsion affecting nearest-neighbor molecules only, as if the spacing between sites is uncomfortably small for the adsorbate species. A repulsion between the electron clouds of adjacent adsorbed molecules would then give rise to a short-range repulsion, usually represented by an exponential term of the type employed in Eq. VII-17. In the treatment of lateral interaction given in Section XVII-3D, the increment to the differential heat of adsorption was $z\omega\theta$, where z is the number of nearest neighbors and ω is the interaction energy, negative in the case of repulsion. As discussed by Fowler and Guggenheim [109], a more elaborate approach allows for the fact that, at equilibrium, neighboring sites will be occupied more often than the statistical expectation if the situation is energetically favored (ω positive), and conversely, if there is lateral repulsion (ω negative), nearest-neighbor sites will be less frequently occupied than otherwise expected.

A second type of repulsion could be long range. If adsorption bond formation polarizes the adsorbate, or strongly orients an existing dipole, the adsorbate film will consist of similarly aligned dipoles that will have a mutual electrostatic repulsion. The presence of such dipoles can be inferred from the change in surface potential difference ΔV on adsorption, as mentioned in Section IV-3B. The resulting interaction should show both a short-range Coulomb repulsion due to adjacent dipoles whose positive and negative charges would repel each other, and which could be considered as part of ω, and a long-range repulsion due to the dipole field, which is inverse cube in distance.

The short-range repulsion is very difficult to estimate quantitatively, and comparisons of predicted versus observed shapes of Q versus θ plots (e.g., Ref. 106) encounter the difficulty that a treatment in terms of ω alone is relatively meaningless; some knowledge is needed of surface heterogeneity. Note, for example, the formal resemblance of the dependence of Q on Θ in Eq. XVIII-8, supposing surface heterogeneity, and Eq. XVII-53, supposing lateral interaction. If there is a single adsorption site, it may be possible to determine a variation of Q with θ from the shape of programmed desorption curves (see Ref. 110 for an example).

B. Effect of Site and Adsorbate Coordination Number

Since in chemisorption systems it is reasonable to suppose that the strong adsorbent-adsorbate interaction is associated with specific adsorption sites, a situation that may arise is that the adsorbate molecule occupies or blocks the occupancy of a second adjacent site. This means that each molecule effectively requires two adjacent sites. An analysis [106] suggests that in terms of the kinetic derivation of the Langmuir equation, the rate of adsorption should now be

$$\text{Rate of adsorption} = k_2 P \, \frac{z}{z - \theta} \, (1 - \theta)^2 \qquad \text{(XVIII-10)}$$

while the rate of desorption is still

$$\text{Rate of desorption} = k_1 \theta \qquad \text{(XVIII-11)}$$

where z is again the number of nearest neighbors to a site. On equating the two rates, one obtains a quadratic form of the Langmuir equation,

$$bP = \frac{z - \theta}{z} \frac{\theta}{(1 - \theta)^2} \qquad \text{(XVIII-12)}$$

If the adsorbed molecule occupies two sites because it dissociates, the desorption rate takes on the form

$$\text{Rate of desorption} = k_1 \frac{(z - 1)^2}{z(z - \theta)} \theta^2 \qquad \text{(XVIII-13)}$$

so that the isotherm becomes

$$(b'P)^{1/2} = \frac{\theta}{1 - \theta} \qquad \text{(XVIII-14)}$$

where

$$b' = \frac{k_2}{k_1} \left(\frac{z}{z - 1} \right)^2$$

It should be cautioned that the correctness of the factors involving z has not really been verified experimentally and that the algebraic forms involved are modelistic. Additional analyses have been made by Ruckenstein and Dadyburjor [111].

Halsey and Yeates [112] add an interaction term, writing Eq. XVIII-14 in the form

$$\left(\frac{P}{P_0} \right)^{1/2} = \frac{\theta}{1 - \theta} \exp\left(\frac{2w}{kT} \frac{\theta - 1}{2} \right) \qquad \text{(XVIII-15)}$$

where w is an interaction energy. Equation XVIII-15 was fit by data for the adsorption of H_2 on Ni(110) and Pt(111), for example, with respective w/kT values of -1.5 and 2.3 (note the observation of both positive and negative w values).

The preceding treatments are based on the concept of localized rather than mobile adsorption. The distinction may be difficult experimentally; note Ref. 109 and the discussion in connection with Fig. XVII-25. There are also conceptual subtleties; see Section XVIII-5.

C. Adsorption Thermodynamics

The thermodynamic treatment that was developed for physical adsorption applies, of course, to chemisorption, and the reader is therefore referred to Section XVII-12. As in physical adsorption the chief use that is made of adsorption thermodynamics is in the calculation of heats of adsorption from temperature dependence data, that is, the obtaining of q_{st} values. As in physical adsorption, these should be the same as the calorimetric differential heats of adsorption (except for the small difference RT), probably even for heterogeneous surfaces. There is, however, much more danger in chemisorption work that the data do not represent an equilibrium adsorption.

Entropies of adsorption are obtainable in the same manner as discussed in Chapter XVII.

4. Kinetics of Chemisorption

A. Activation Energies

It was noted in Section XVII-1 that chemisorption may become slow at low temperatures so that even though it is favored thermodynamically, the only process actually observed may be that of physical adsorption. Such slowness implies an activation energy for chemisorption, and the nature of this effect has been much discussed.

The classic explanation for the presence of an activation energy in the case where dissociation occurs on chemisorption is that of Lennard–Jones [113] and is illustrated in Fig. XVIII-12 for the case of O_2 interacting with an Ag(110) surface. The curve labeled O_2 represents the variation of potential energy as the molecule approaches the surface; there is a shallow minimum corresponding to the energy of physical adsorption and located at the sum of the van der Waals radii for the surface atom of Ag and the O_2 molecule. The curve labeled O + O, on the other hand, shows the potential energy variation for two atoms of oxygen. At the right, it is separated from the first curve by the O_2 dissociation energy of some 120 kcal/mol. As the atoms approach the surface, chemical bond formation develops, leading to the deep minimum located at the sum of the covalent radii for Ag and O. The two curves cross, which means that O_2 can first become physically adsorbed and then undergo a concerted dissociation and chemisorption process, leading to chemisorbed O atoms (see Ref. 113a for a more general diagram). In this type of sequence, the molecularly adsorbed species is known as a *precursor state* (see Refs. 115 and 116).

Clearly, the presence or absence of an activation energy for chemisorption will depend on the relative position and shapes of the two potential energy curves, and relatively minor displacements could result in there being no activation energy at all. Also, Fig. XVIII-12 simplifies the situation in that the O—O interatomic distance is not indicated, although it also affects the energy; that is, the barrier to chemisorption may involve an energy of stretching the O—O bond to match the distance between sites. Thus for the case or the adsorption of hydrogen on various carbon surfaces, the picture can be taken to be that of a hydrogen molecule approaching a pair of surface carbon atoms, with simultaneous H—H bond stretching and C—H bond formation as the final state of

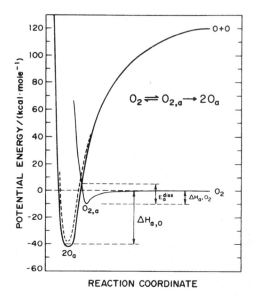

Fig. XVIII-12. Potential energy curves for O_2 and for O interacting with an Ag(110) surface. The dashed curve shows the effect expected if chemisorbed Cl is present, for $\theta < 0.25$. (From Ref. 114.)

chemisorbed hydrogen atoms is attained:

$$\begin{array}{cc} \text{H}-\text{H} & \text{H}--\text{H} \\ \vdots \quad \vdots & | \quad | \\ -\text{C}-\text{C}- & \text{C}-\text{C} \end{array}$$

An early calculation by Sherman and Eyring [117] led to a theoretical variation of the activation energy with the C—C distance, which showed a minimum (of 7 kcal/mol) at a spacing of 3.5 Å. The qualitative explanation for the minimum was that if the C—C distance is too large, then the H—H bond must be stretched considerably before much gain due to incipient C—H bond formation can occur. If the C—C distance is too short, H—H repulsion again raises the activation energy. Calculated activation energies based on suitably chosen C—C distances can be made to agree with experiment [106]. A complication is that surface relaxation can significantly alter the calculated activation energies [118].

A more elaborate theoretical approach develops the concept of surface molecular orbitals and proceeds to evaluate various overlap integrals [119]. Calculations for hydrogen on Pt(111) planes were consistent with flash desorption and LEED data. In general, the greatly increased availability of LEED structures for chemisorbed films has allowed correspondingly detailed theoretical interpretations, as, for example, of the commonly observed $(C2 \times 2)$ structure [120] (note also Ref. 121).

The idea of a physically adsorbed precursor state turns out to be difficult to evaluate experimentally. The process of forming such a state has been called *trapping*, and the probability of an impinging molecule losing enough kinetic energy to trap in a molecularly adsorbed state is called the *condensation coefficient, c*. One would expect c to *decrease* with *increasing* kinetic energy of the impinging molecule, and this has been observed [122]; and by using molecular beam techniques, the angle of incident of the beam may be varied, and now c should depend on $E_{kin} \cos^2 \theta$ where θ is the angle of incidence and E_{kin} is the overall kinetic energy (see Problem XVIII-32). Experimentally, the θ dependence may be $\cos \theta^n$, where n is much less than 2 (see Ref. 122), and the explanation is not clear. The complexities of the model are developed by Xu and Koel [115] for the case of bimetallic catalysts; some half dozen probabilities or coefficients are involved.

The precursor state or trapping model may in fact not be generally correct. Bottari and Greene [123] and, more recently, Ceyer [113a] have considered that there is conversion of kinetic to internal energy when a molecule impacts a surface, leading to chemical reaction. The latter author finds that experimentally, the probability of dissociation of CH_4 (and of CD_4) on a Ni(111) surface *increases* with increasing normal translational energy of the impinging molecule.

B. Rates of Adsorption

Mention was made in Section XVIII-2E of programmed desorption; this technique gives specific information about both the adsorption and the desorption of specific molecular states, at least when applied to single-crystal surfaces. The kinetic theory involved is essentially that used in Section XVII-3A. It will be recalled that the adsorption rate was there taken to be simply the rate at which molecules from the gas phase would strike a site area σ^0 times the fraction of unoccupied sites. If the adsorption is activated, the fraction of molecules hitting and sticking that can proceed to a chemisorbed state is given by $\exp(-E_a^*/RT)$. The adsorption rate constant of Eq. XVII-13 becomes

$$k_2 = \frac{Nc\sigma^0 \exp(-E_a^*/RT)}{(2\pi MRT)^{1/2}} = A \exp \frac{-E_a^*}{RT} \qquad \text{(XVIII-16)}$$

The rate of adsorption is then

$$R_a = \frac{d\theta}{dt} = k_2 f(\theta)P \qquad \text{(XVIII-17)}$$

where $f(\theta)$ is the fraction of available surface taken to be $1 - \theta$ in the simple Langmuir derivation, but capable of taking on other forms as, for example, if the adsorbing molecule must find two adjacent unoccupied sites (see Ref. 124). If a molecule occupies more than one site and is essentially irreversibly adsorbed, then arriving molecules cannot be expected to arrive in such a way as to dove-tail into complete surface coverage—a "*jamming*" limit is reached. In the case of a four-site molecule, probability calculations give this limit as $1 - \exp(-\frac{4}{3})$ or approximately $\frac{3}{4}$ [125]. A related effect occurs if one reduces the size of catalytically active crystallites, known as the *ensemble* effect [126]. The practical *sticking coefficient*, *s*, the fraction of molecules hitting a surface which go on to become chemisorbed may then be written

$$s = c f(\theta) e^{-E_a^*/RT} \qquad \text{(XVIII-18)}$$

where *c* is the condensation coefficient mentioned earlier in this section (for more detailed treatments, see Refs. 127 and 13, p. 336). This treatment assumes the precursor model and it can, alternatively, be put in the framework of absolute rate theory in which the equilibrium constant for forming the activated or transition state is invoked. This transition state would have the configuration of the system at the potential maximum in Fig. XVIII-12, and in the formal development it turns out that the sticking coefficient *s* is replaced by an expression involving the partition function of the transition state [121,128] (see also Ref. 116).

A variation of E_a^* with θ is not uncommon, and if the empirical relation $E_0^* + \alpha\theta$ is used, Eq. XVII-17 becomes

$$\frac{d\theta}{dt} = A f(\theta) P \exp\left(-\frac{E_0^* + \alpha\theta}{RT}\right) \qquad \text{(XVIII-19)}$$

so that at a given temperature, the rate should vary according to $f(\theta)e^{-\alpha\theta}$. Equation XVIII-19 is a form of what is known as the *Elovich* equation [120].

Where E_a^* is appreciable, adsorption rates may be followed by ordinary means. In a rather old but still informative study, Scholten and co-workers [130] were able to follow the adsorption of N_2 on an iron catalyst gravimetrically, and reported the rate law

$$\frac{d\theta}{dt} = 21.9 P_{N_2} e^{(132.4\theta/R)} e^{-(5250 + 77,500\theta)/RT} \qquad \text{(XVIII-20)}$$

for the range $\theta = 0.07$–0.22. Here, pressure is in centimeters of mercury, time is in minutes, and energy quantities are in calories. Equation XVIII-20 can be put in the form of Eq. XVIII-19; $f(\theta)$ is presumably $(1 - \theta)^2$ since chemisorption on two sites is presumed, but over the small range of θ values involved, this in turn can be approximated by $e^{-2.22\theta}$, which amounts to a negligible amendment to the other exponential terms. In this same range of θ values, the frequency factor (*A* in Eq. XVIII-16) increased by 10^5-fold, indicating that some type of progressive change in the sticking coefficient or, alternatively, in the partition function for the transition state, was occurring.

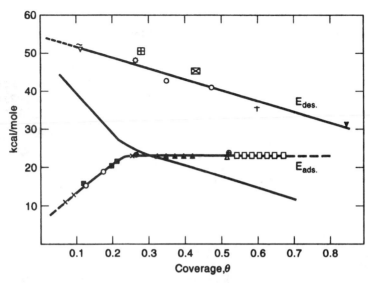

Fig. XVIII-13. Activation energies of adsorption and desorption and heat of chemisorption for nitrogen on a single promoted, intensively reduced iron catalyst; Q is calculated from $Q = E_{des} - E_{ads}$. (From Ref. 130.)

Activation energies for desorption were also obtained, and the overall picture is that shown in Fig. XVIII-13. Note the changes in slopes of the plots of Q and of E_{ads} around $\theta = 0.22$. Scholten et al. interpreted the region $\theta > 0.22$ as being one of mobile adsorption. It is possible, however, that nitrogen was now ceasing to undergo dissociation or was now adsorbing on crystal planes that allowed more vibrational and rotational degrees of freedom.

The general picture presented by the preceding example is fairly representative; that is, both the energy of activation and the frequency factor tend to increase as Q decreases or θ increases. The physical explanation of this correlation or *compensation effect* would seem to be that active sites have a relatively negative entropy of adsorption, perhaps because the adsorbate is very highly localized, but the high energy of adsorption assists in the bond deformations needed for chemisorption to occur. Thus in Fig. XVIII-12, a potential energy curve for (O + O) having a deeper minimum and hence large Q value would intersect the O_2 curve lower, to give a lower activation energy for chemisorption.

C. Rates of Desorption

It might be thought that since chemisorption equilibrium was discussed in Section XVIII-3 and chemisorption rates in Section XVIII-4B, the matter of desorption rates is determined by the principle of microscopic reversibility (or, detailed balancing) and, indeed, this principle is used (see Ref. 127 for example). Rates of adsorption and of desorption are often determined far from equilibrium, however, and moreover, may be represented by semi-empirical rate laws valid over different limited ranges of conditions; the forward and reverse rate laws so determined will not necessarily combine to give a correct chemisorption isotherm and do have to be discussed separately. Also, in contemporary literature, desorption rates are often studied by means of temperature programmed desorption (Section XVIII-2E) rather than isothermally, and hence constitute a distinct topic of its own. What follows here are some general points plus various specific examples.

First, desorption will always be activated since the minimum E_d^* is that equal to the energy of adsorption Q. Thus

$$E_d^* = E_a^* + Q \qquad \text{(XVIII-21)}$$

This means that desorption activation energies can be much larger than those for adsorption and very dependent on θ since the variation of Q with θ now contributes directly. The rate of desorption may be written, following the kinetic treatment of the Langmuir model,

$$R_d = -\frac{d\theta}{dt} = \frac{1}{\tau_0}\, e^{-Q/RT} e^{-E_a^*/RT} f'(\theta) \qquad \text{(XVIII-22)}$$

which is the same as Eq. XVIII-1 with $E = Q + E_a^*$, $1/\tau_0$ replacing A, and the general function $f'(\theta)$ replacing θ. The function could be θ^2, for example, if two surface atoms must associate to desorb. This was the case for the associative desorption of O_2 from Pt, for which $-dS_0/dt = 2.4 \times 10^{-2}\, S_0^2 \exp[(50,900 - 10.000\theta_0)/RT]$, where S_0 is the oxygen atom coverage in atoms/cm^2 (131) and energies are in calories. Hydrogen desorption from the ordered alloy Cu$_3$Pt(111) showed $E_{des} = 12,000-2400\,\theta$, again in calories [110]. On the stepped Pt(112) surface, however, two states were found, one desorbing similarly to that for Pt(111) surface, however, two states were found, one desorbing similarly to that for Pt(111) surfaces, but the other showing an *attractive* rather than a repulsive contribution to the activation energy for desorption [132].

In the case of nitrogen on iron, the experimental desorption activation energies are also shown in Fig. XVIII-13; the desorption rate was given by the empirical expression

$$-\frac{d\theta}{dt} = 4.8 \times 10^{14} \theta^2 e^{-10.64\theta} e^{(-55,000 + 29,200\theta)/RT} \qquad \text{(XVIII-23)}$$

(same units as before). Note the presence of the term in θ^2; that is $e^{-10.64\theta}$ could represent an empirical compensation to θ^2 for the statistics of finding two adjacent sites. The general picture, however, is that of an adsorbed state consisting of nitrogen atoms, which associate to desorb as N_2.

One might expect the frequency factor A for desorption to be around 10^{13} sec^{-1} (note Eq. XVII-2). Much smaller values are sometimes found, as in the case of the desorption of Cs from Ni surfaces [133], for which the adsorption lifetime obeyed the equation $\tau = 1.7 \times 10^{-3} \exp(3300/RT)$ sec (R in calories per mole per degree Kelvin). A suggested explanation was that surface diffusion must occur to desorption sites for desorption to occur. Conversely, A factors in the range of 10^{16} sec^{-1} have been observed and can be accounted for in terms of strong surface orientational forces [134].

As LEED studies have shown, the structure of a chemisorbed phase can change with θ. In terms of transition state theory, we can write $A = (1/\tau_0) e^{\Delta S/R}$, and a common observation is that while E may change with a phase change, ΔS will tend to change also, and similarly. The result, again known as a *compensation effect*, is that the product $Ae^{-E/RT}$ remains relatively constant (see Ref. 135).

We have seen various kinds of explanations of why E_{des} may vary with θ. The subject may, in a sense, be bypassed and an energy distribution function obtained much as in Section XVII-14A. In doing this, Cerefolini and Re [149] used a rate law in which the amount desorbed is linear in the logarithm of time (the Elovich equation).

5. Surface Mobility

The matter of surface mobility has come up at several points in the preceding material. The subject has been a source of confusion—see Ref. 112. Actually, two kinds of concepts seem to have been invoked. The first is that invoked in the discussion of physical adsorption, which has to do with whether the adsorbate can move on the surface so freely that its state is essentially that of a two-dimensional nonideal gas. For an adsorbate to be mobile in this sense, surface barriers must be small compared to kT. This type of mobile adsorbed layer seems unlikely to be involved in chemisorption.

In general, it seems more reasonable to suppose that in chemisorption specific sites are involved and that therefore definite potential barriers to lateral motion should be present. The adsorption should therefore obey the statistical thermodynamics of a localized state. On the other hand, the kinetics of adsorption and of catalytic processes will depend greatly on the frequency and nature of such surface jumps as do occur. A film can be fairly mobile in this kinetic sense and yet not be expected to show any significant deviation from the configurational entropy of a localized state.

Mobility of this second kind is illustrated in Fig. XVIII-14, which shows NO molecules diffusing around on terraces with intervals of being "trapped" at steps. Surface diffusion can be seen in field emission microscopy (FEM) and can be measured by observing the growth rate of patches or fluctuations in emission from a small area [136,138] (see Section VIII-2C), field ion microscopy [138], Auger and work function measurements, and laser-induced desorption

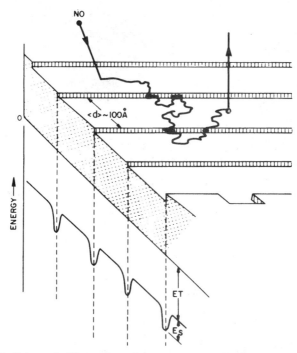

Fig. XVIII-14. Schematic illustration of the movement of NO molecules on a Pt(111) surface. Molecules diffuse around on terraces, get trapped at steps, escape, and repeat the process many times before eventually desorbing. [Reprinted with permission from M. Cardillo, Langmuir, **1**, 4 (1985) (Ref. 140). Copyright 1985, American Chemical Society.]

(LID); see Ref. 139. In this last method, a small area, about 0.03 cm radius, is depleted by a laser beam, and the number of adatoms, $N(t)$, that have diffused back is found as a function of time. From Fick's second law of diffusion:

$$N(t) = \frac{2}{a} \left(\frac{\mathcal{D}t}{\pi} \right)^2 \qquad \text{(XVIII-24)}$$

where a is the radius of the depleted spot and $\mathcal{D}$ is the surface diffusion coefficient. $\mathcal{D}$ may also be obtained from the rate of growth of LEED spots [143]. Table XVIII-1 summarizes some results, where $\mathcal{D} = \mathcal{D}_0 \exp(E_d/RT)$. Note the considerable range in $\mathcal{D}_0$ and E_d values. Simple theory gives

$$\mathcal{D}_0 = \frac{1}{4}A\lambda^2 \qquad \text{(XVIII-25)}$$

where the frequency factor A is on the order of 10^{13} sec^{-1} and λ, the diffusional

TABLE XVIII-1
Surface Diffusion Coefficients

System	Method and Conditions	$\mathcal{D}_0$ (cm^2/sec)	E_d (kcal/mol)	Reference
O/W(110)	FEM			
	$\theta < 0.2$	1×10^{-7}	14	136
	$\theta = 0.56$	1×10^{-4}	22	136
CO/W(110)	FEM			
	α phase	Small	Small	137
	β phase	1×10^{-5}	23	137
Xe/W(110)	FEM	7×10^{-8}	1.1	141
H/Ni(100)	LID	2.5×10^{-3}	3.5	142
CO/Ni(100)	LID, $\theta = 0.4$	0.05	4.9	139
O/Pd(100)	LEED	—	12.5	143
Ni/Ni(110)	Tracer	300	38	144

jump distance, on the order of angstroms. Interestingly however, diffusion of H atoms has been suggested to occur at low temperatures by means of a tunneling mechanism [145]. Some general rules are that E_d is about one-fourth of the desorption energy and that surface diffusion becomes important at about half the temperature at which evaporation does [84]. Theoretical calculations have been made for the diffusion of O on Ru(001) [146].

Notice in Table XVIII-1 a value for the *self-diffusion* of Ni on Ni(111) measured using radioactive Ni. More gross processes can occur. Supported Ni crystallites (on alumina) may show spreading and wetting phenomena due to complex interactions with the substrate [146].

As usual, things become more complicated when studied in detail. Note that for O/W(110) $\mathcal{D}_0$ varies with θ; the situation is shown more fully in Fig. XVIII-15. The authors speculate that variations in $\mathcal{D}_0$ and E have to do with a $p(2 \times 1)$ structure at low oxygen coverage, with O atoms occupying alternate rows of W atoms, the empty rows becoming occupied above $\theta = 0.5$. The consequence is that O—O interactions shift from being mostly attractive to being in part repulsive.

A third definition of surface mobility is essentially a rheological one; it represents the extension to films of the criteria we use for bulk phases and, of course, it is the basis for distinguishing states of films on liquid substrates. Thus as discussed in Chapter IV, solid films should be ordered and should show elastic and yield point behavior; liquid films should be coherent and show viscous flow; gaseous films should be in rapid equilibrium with all parts of the surface.

Neither the thermodynamic nor the rheological description of surface mobility has been very useful in the case of chemisorbed films. From the experimental point of view, the first is complicated by the many factors that can affect adsorption entropies and the latter by the lack of any methodology.

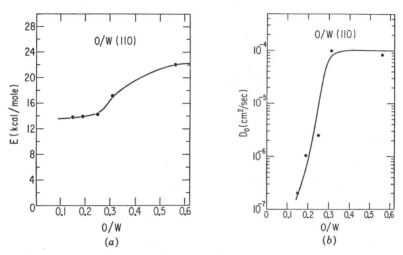

Fig. XVIII-15. Oxygen atom diffusion on a W(100) surface: (a) variation of the activation energy for diffusion with θ; and (b) variation of $\mathcal{D}_0$. (From Ref. 136. Reprinted with kind permission of Elsevier Science-NL, Sara Burgerhartstraat 25, 1055 KV Amsterdam, The Netherlands.)

6. The Chemisorption Bond

A. Some General Aspects

Various aspects of the experimental approach to the chemisorption bond are illustrated in the preceding sections. Modern spectroscopic and surface diffraction techniques provide a wealth of information about the chemisorbed state. Analysis of LEED intensity data permits the estimation of adsorbate–adsorbent bond lengths [147], usually 5–10% longer than in molecules having a similar bond. Bond lengths may also be obtained from XPD and SEXAFS data (see Table VIII-1) [148]. A bond energy can be obtained from temperature-programmed desorption data if coupled with knowledge of the activation energy for adsorption (Eq. XVIII-21); see Ref. 149 for the case of a heterogeneous surface. The traditional approach to obtaining bond energies is, of course, through isosteric heats of adsorption, although complications are that equilibrium may be difficult to reach and/or the surface may be heterogeneous. Some literature data compiled by Shustorovich, Baetzold, and Muetterties [150] are shown in Table XVIII-2. For hydrogen atom–metal bonds Q averages about 62 kcal/mol, corresponding to about 20 kcal/mol for desorption as H_2. Bond energies for CO and NO run somewhat higher. Values can vary depending on the surface preparation and, of course, on the crystal plane involved if the surface is a well-defined one. Older compilations may be found in Refs. 81 and 84, and more recent ones, in Somorjai [13].

There are various qualitative and traditional ways of estimating or predicting

TABLE XVIII-2
Some Heats of Chemisorption[a]

System	Q (kcal/mol)	System	Q (kcal/mol)
H/W(110)	68	CO/W(110)	27
H/Fe(110)	64	CO/Ni(111)	27
H/Ni(100)	63	CO/Pt(111)	32
H/Ni(111)	63	NO/Ni(111)	25
H/Pt(111)	57	NO/Pt(111)	27

[a]On atomically flat metal surfaces. See Ref. 150 for the individual literature citations.

bond energies. Fair estimates can be obtained by invoking model compounds. Thus, per CO, the heat of formation of $Ni(CO)_4(g)$ from $Ni(g)$ and $CO(g)$ is 35 kcal/mol, while the chemisorption Q is 42 kcal/mol [81,84]. The heat of formation of $WO_3(g)$ from the gaseous elements in 193 kcal/mol, as compared to a Q of 194 kcal/mol for the chemisorption of oxygen on tungsten. The Q for chemisorption of H_2 on carbon to give hydrogen atoms bound to carbon can be estimated as $Q = 2E_{C—H} - E_{H—H}$, where $E_{C—H}$ is one-fourth the bond energy for methane and $E_{H—H}$ is 104 kcal/mol.

The electronegativity system may be used. For example, in the chemisorption (with dissociation) of hydrogen on tungsten,

$$2W + H_2 = 2W—H \qquad (XVIII-26)$$

with

$$Q = 2E_{W—H} - E_{H—H} \qquad (XVIII-27)$$

assuming that no W—W bonds need be broken. The H—H bond energy is known, and the value of $E_{W—H}$ can be obtained from the relationship

$$E_{W—H} = \frac{1}{2}(E_{W—W} - E_{H—H}) + 23(X_W - X_H)^2 \qquad (XVIII-28)$$

where the X's are the respective electronegativities. The energy of adsorption is then

$$Q = E_{W—W} + 46(X_W - X_H)^2 \qquad (XVIII-29)$$

The W—W bond energy should be about one-sixth of the sublimation energy (note Section III-1B), and there are various schemes for estimating electronegativities, of which Mulliken's [151,152] is perhaps the most fundamental.

Chemisorption bonding to metal and metal oxide surfaces has been treated extensively by quantum-mechanical methods. Somorjai and Bent [153] give a general discussion of the surface chemical bond, and some specific theoretical treatments are found in Refs. 154–157; see also a review by Hoffman [158]. One approach uses the variation method (see physical chemistry textbooks):

$$E_{\text{bond}} \cong \frac{\int \phi \mathbf{H}^* \phi \, d\tau}{\int \phi^* \phi \, d\tau} \tag{XVIII-30}$$

where ϕ is an approximate wave function, $\mathbf{H}$ the Hamiltonian for the system, and $d\tau$ the element of coordinate space. Equation XVIII-30 gives what is known as the expectation value of E_{bond} and can be abbreviated $E = \langle \phi^* | \mathbf{H} | \phi \rangle$. As Messmer notes [159], one can look for an operator O such that $O\phi = \psi$, where ψ is the wavefunction that makes Eq. XVIII-30 exact. The operator may act either on ϕ or on $\mathbf{H}$, providing two different avenues of approach.

Figure XVIII-16 sketches what can happen in the case of a metal surface [160]. In Fig. XVIII-16a two atoms or molecular entities each with a frontier orbital having two electrons potentially can interact to give bonding and anti-bonding levels, as shown in the middle; however, since both levels are filled, there is no net bonding. The same atom or molecular entity can bond to a metal surface if the antibonding level is above the Fermi level of the metal, so that the antibonding electrons dump into the sea of metal electrons. Alternatively, one unfilled orbital each from the atoms or molecular entities can hypothetically form bonding and antibonding orbitals, both now empty, as shown in Fig. XVIII-16c. If, now, the bonding level lies below the Fermi level, electron transfer can occur, resulting in bonding as shown in Fig. XVIII-16d.

An increasingly popular approach because of its computational efficiency is known as *density functional theory* (DFT).† The method was originally developed for an interacting gas in a weak external field and calculates the effect of the field using the one-particle density, thus separating the effect from that of particle–particle interactions [161]. The expression for the total energy further separates out an electron exchange–correlation functional. The application to chemical bonding with its high electron density and rapidly varying local fields would seem to involve irresponsible approximations, yet actually works quite well (see Refs. 148, 162 for examples). One example is the calculation of the interaction of carbon monoxide with functional groups in zeolites [163], and of H_2 on various metal surfaces. See Refs. 163a-d and 164 for recent examples of the results of DFT calculations. Generally speaking, the method does better for bond energies and vibrational frequencies than for structure determinations. DFT, somewhat related to the Xα method, currently seems to be more in the

---

†A *function* is a mathematical expression involving variables that are in principle measurable; a *functional* is an expression involving functions as variables. For example, in the variation method (Eq. XVIII-30, the functions ϕ are variables.

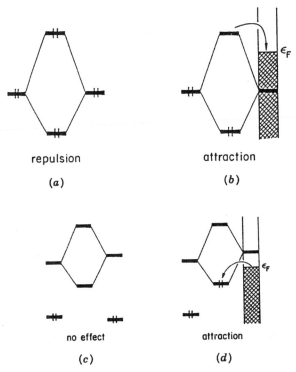

Fig. XVIII-16. A four-electron two-orbital interaction that (*a*) has no net bonding in the free molecule but can be bonding to a metal surface if (*b*) the Fermi level is below the antibonding level. In the lower part of the figure, a zero-electron two-orbital situation (*c*) has no bonding but there can be bonding to a metal surface as in (*d*) if the Fermi level is above the bonding level. (From Ref. 160.)

domain of physicists than of chemists and is too involved to be discussed in detail here.

B. Metals

We consider first some experimental observations. In general, the initial heats of adsorption on metals tend to follow a common pattern, similar for such common adsorbates as hydrogen, nitrogen, ammonia, carbon monoxide, and ethylene. The usual order of decreasing Q values is Ta > W > Cr > Fe > Ni > Rh > Cu > Au; a traditional illustration may be found in Refs. 81, 84, and 165. It appears, first, that transition metals are the most active ones in chemisorption and, second, that the activity correlates with the percent of d character in the metallic bond. What appears to be involved is the ability of a metal to use d orbitals in forming an adsorption bond. An old but still illustrative example is shown in Fig. XVIII-17, for the case of ethylene hydrogenation.

The bonding of CO to metals has been studied extensively using various

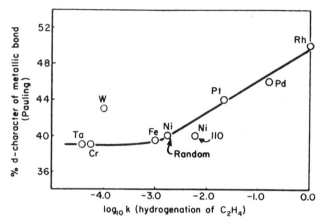

Fig. XVIII-17. Correlation of catalytic activity toward ethylene dehydrogenation and percent d character of the metallic bond in the metal catalyst. (From Ref. 166.)

semiempirical wave-mechanical methods such as INDO (incomplete neglect of differential overlap) and extended Hückel (see Refs. 167–169). Detailed calculations of this type can explain why CO may either be side-on- or upright-bonded, for example [170,171]. Similar calculations have been made for chemisorbed NH_3 [172,173], and a correlation diagram is shown in Fig. XVIII-18. Five different calculational methods for N_2 on Ni were compared, however, with the conclusion that more sophisticated approaches are in general needed than those illustrated in Fig. XVIII-18 [174]. Other molecular orbital treatments are those for NO on Ni(111) [175] and the H atom on ZnO [176]. The H atom interaction with small Ni, Pd, and Pt clusters has been examined by the self-consistent field $X\alpha$ method, and the results helped to explain the varia-

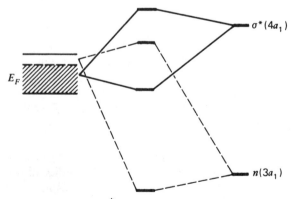

Fig. XVIII-18. Interaction of the σ^* and n molecular orbitals with the Pt d band; E_F is the Fermi level. (From Ref. 172.)

tions in catalytic activity and in hydrogen solubility [177]. The extended Hückel method was used to study both the strongly bound subsurface and the surface bound oxygen on (111) clusters of the same metals [178]. Finally, a number of theoretical calculations have been made on the binding of CH_3, CH_2, and CH fragments [179] and of acetylene [180,181] to metal surfaces.

C. Semiconductors

Some aspects of adsorption on oxides and other semiconductors can be treated in terms of the electrical properties of the solid, and these are reviewed briefly here. More details can be found in Refs. 84 and 182.

In many crystals there is sufficient overlap of atomic orbitals of adjacent atoms so that each group of a given quantum state can be treated as a crystal orbital or band. Such crystals will be electrically conducting if they have a partly filled band; but if the bands are all either full or empty, the conductivity will be small. Metal oxides constitute an example of this type of crystal; if exactly stoichiometric, all bands are either full or empty, and there is little electrical conductivity. If, however, some excess metal is present in an oxide, it will furnish electrons to an empty band formed of the $3s$ or $3p$ orbitals of the oxygen ions, thus giving electrical conductivity. An example is ZnO, which ordinarily has excess zinc in it.

If adsorption of oxygen on such an oxide involves the process

$$O_2 + 4e^- = 2O^{-2}(ads) \qquad (XVIII\text{-}31)$$

adsorption will tend to be limited to the extent that excess zinc is present, that is, it will be small and moreover will reduce the conductivity by removing electrons from the conduction band; both predictions are confirmed experimentally. This type of adsorption has been called *depletive*. The situation is illustrated qualitatively in Fig. XVIII-19 for the case where a surface electron acceptor state or adsorbate is present. Since the system remains electrically neutral, positive donor ions accumulate near the surface to complete an electrical double layer.

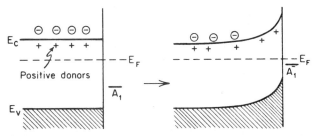

Fig. XVIII-19. Band bending with a negative charge on the surface states: E_v, E_f, and E_c are the energies of the valance band, the Fermi level, and the conduction level, respectively. (From Ref. 186.)

On the other hand, an oxide such as NiO is oxygen-rich, in the sense that occasional Ni^{2+} ions are missing, electroneutrality being preserved by some of the nickel being in the +3 valence state. These Ni^{3+} ions take electrons from the otherwise filled conduction bands, thus again providing the condition needed for electrical conductivity. Oxygen adsorption according to Eq. XVIII-31 can draw on the electrons in the slightly depleted band (or, alternatively, can produce unlimited additional Ni^{3+}) and so should be able to proceed to monolayer formation. Furthermore, since adsorption will make for more vacancies in a nearly filled band, electrical conductivity should rise. Again, the predictions are borne out experimentally [183].

In general, then, anion-forming adsorbates should find p-type semiconductors (such as NiO) more active than insulating materials and these, in turn, more active than n-type semiconductors (such as ZnO). It is not necessary that the semiconductor type be determined by an excess or deficiency of a native ion; impurities, often deliberately added, can play the same role. Thus if Li^+ ions are present in NiO, in lattice positions, additional Ni^{3+} ions must also be present to maintain electroneutrality; these now compete for electrons with oxygen and reduce the activity toward oxygen adsorption.

A quantitative treatment for the depletive adsorption of iogenic species on semiconductors is that known as the boundary layer theory [84,184], in which it is assumed that, as a result of adsorption, a charged layer is formed. Double-layer theory is applied, and it turns out that the change in surface potential due to adsorption of such a species is proportional to the square of the amount adsorbed. The important point is that very little adsorption, e.g., a θ of about 0.003, can produce a volt or more potential change. See Ref. 185 for a review.

Irradiation of a semiconductor with light of quantum energy greater than the band gap can lead to electron–hole separation. This can affect adsorption and lead to photocatalyzed or *photoassisted* reactions [187]. See Section XVIII-9F for some specifics.

D. Acid–Base Systems

Still another type of adsorption system is that in which either a proton transfer occurs between the adsorbent site and the adsorbate or a Lewis acid–base type of reaction occurs. An important group of solids having acid sites is that of the various silica–aluminas, widely used as cracking catalysts. The sites center on surface aluminum ions but could be either proton donor (Brønsted acid) or Lewis acid in type. The type of site can be distinguished by infrared spectroscopy, since an adsorbed base, such as ammonia or pyridine, should be either in the ammonium or pyridinium ion form or in coordinated form. The type of data obtainable is illustrated in Fig. XVIII-20, which shows a portion of the infrared spectrum of pyridine adsorbed on a Mo(IV)–Al_2O_3 catalyst. In the presence of some surface water both Lewis and Brønsted types of adsorbed pyridine are seen, as marked in the figure. Thus the features at 1450 and 1620 cm^{-1} are attributed to pyridine bound to Lewis acid sites, while those at 1540

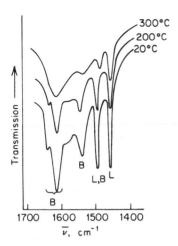

Fig. XVIII-20. Spectra of pyridine adsorbed on a water-containing molybdenum oxide (IV)-Al$_2$O$_3$ catalyst: L and B indicate features attributed to pyridine adsorbed on Lewis and Brønsted acid sites, respectively. (Reprinted with permission from Ref. 191. Copyright 1976 American Chemical Society.)

and above 1600 cm^{-1} are attributed to pyH$^+$. The proportion of Brønsted sites increased with increasing surface water. Some further examples and discussion may be found in Refs. 188 and 189, especially on the matter of how to selectively block either Lewis or Brønsted sites. Also, it is interesting that a given surface (e.g., alumina) may have Lewis acid-Lewis base *pairs* [190].

The chemisorption of molecules not ordinarily regarded as acids or bases may still be viewed in terms of the Lewis acid–base concept. Thus, in the formation of a metal carbonyl complex, CO acts as a Lewis base and the metal as a Lewis acid. In the case of chemisorption on metals, the actual situation can be assessed by determining the change in work function on adsorption. In the case of CO on Ni(111), this change is positive, indicating that net electron transfer *from* the metal *to* the CO has occurred, so that the metal is acting here as a Lewis *base*. For adsorbed C$_2$H$_2$ and NH$_3$, however, the change in work function is negative and the metal is now acting as a Lewis *acid* [192]. A useful extension of the Lewis acid–base idea was made by Pearson [193] in terms of "hard" and "soft" acids and bases. The rule of similarity is that hard acids prefer hard (less polarizable) bases, and soft acids prefer soft (more polarizable) bases. This rule has been useful in the acid–base treatment of chemisorption; see Ref. 194.

A new dimension to acid–base systems has been developed with the use of zeolites. As illustrated in Fig. XVIII-21, the alumino–silicate faujasite has an open structure of interconnected cavities. By exchanging H$^+$ for alkali metal (or NH$_4^+$ and then driving off ammonia), acid zeolites can be obtained whose acidity is comparable to that of sulfuric acid and having excellent catalytic properties (see Section XVIII-9D). Using spectral shifts, zeolites can be put on a relative acidity scale [195]. An important added feature is that the size of the channels and cavities, which can be controlled, gives selectivity in that only

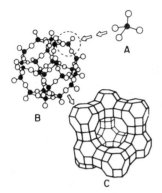

Fig. XVIII-21. The framework structure of faujasite. (*a*) Tetrahedral arrangement of silicon (or aluminum) atoms sharing oxygen atoms. (*b*) Sodalite unit consisting of 24 SiO_4^- and AlO_4^- tetrahedral. (*c*) Zeolite superstructure consisting of tetrahedral arrangement of sodalite units connected by oxygen bridges forming hexagonal prisms. (From Ref. 196.)

certain reactants or products can get in or out. See Section XVIII-9D for additional discussion.

7. Mechanisms of Heterogeneous Catalysis

The sequence of events in a surface-catalyzed reaction comprises (1) diffusion of reactants to the surface (usually considered to be fast); (2) adsorption of the reactants on the surface (slow if activated); (3) surface diffusion of reactants to active sites (if the adsorption is mobile); (4) reaction of the adsorbed species (often rate-determining); (5) desorption of the reaction products (often slow); and (6) diffusion of the products away from the surface. Processes 1 and 6 may be rate-determining where one is dealing with a porous catalyst [197]. The situation is illustrated in Fig. XVIII-22 (see also Ref. 198; notice in the figure the variety of processes that may be present).

A. Adsorption or Desorption as the Rate-Determining Step

Process 2, the adsorption of the reactant(s), is often quite rapid for nonporous adsorbents, but not necessarily so; it appears to be the rate-limiting step for the water–gas reaction, $CO + H_2O = CO_2 + H_2$, on Cu(111) [200]. On the other hand, process 4, the desorption of products, must always be activated at least by Q, the heat of adsorption, and is much more apt to be slow. In fact, because of this expectation, certain seemingly paradoxical situations have arisen. For example, the catalyzed exchange between hydrogen and deuterium on metal surfaces may be quite rapid at temperatures well below room temperature and under circumstances such that the rate of desorption of the product HD appeared to be so slow that the observed reaction should not have been able to occur! To be more specific, the originally proposed mechanism, due to Bonhoeffer and Farkas [201], was that of Eq. XVIII-32. That is,

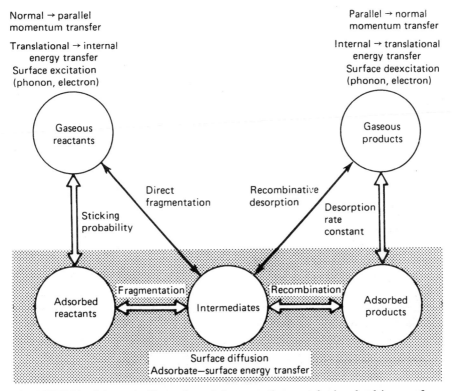

Fig. XVIII-22. Schematic illustration of the steps that may be involved in a surface-mediated reaction: initial adsorption, subsequent thermalization, diffusion and surface reaction, and desorption. (From Ref. 199; copyright 1984 by the AAAS.)

$$\frac{1}{2}H_2 + \frac{1}{2}D_2 + MM \rightarrow \overset{\overset{\displaystyle H \quad D}{\displaystyle | \quad |}}{M \ M} \rightarrow MM + HD \qquad \text{(XVIII-32)}$$

hydrogen and deuterium each chemisorbed as atoms, with exchange taking place through the random recombination of H and D followed by desorption of HD. It was this last step that could be impossibly slow.

The above situation led to the proposal by Rideal [202] of what has become an important alternative mechanism for surface reactions, illustrated by Eq. XVIII-33. Here, reaction takes place between chemisorbed atoms and a *colliding or physical adsorbed molecule* (see Ref. 203).

Similar equations were written by Eley [204] for the exchange of $^{28}N_2$ with $^{30}N_2$ catalyzed by Fe or W, and mechanisms such as Eq. XVIII-33 have come to be known as *Eley–Rideal* mechanisms. Mechanisms such as that of Eq. XVIII-32 are now most commonly called *Langmuir–Hinshelwood mechanisms* (see

$$
\left.
\begin{array}{l}
\hspace{3.2cm}\overset{\overset{\displaystyle H}{\displaystyle \cdot\cdot}}{} \\[-0.2em]
\overset{\displaystyle D}{\underset{\displaystyle |}{}}\quad \overset{\displaystyle H\,D}{\underset{\displaystyle \cdot\ \cdot}{}}\quad \overset{\displaystyle H}{\underset{\displaystyle |}{}} \\[-0.2em]
H_2 + MM \longrightarrow MM \longrightarrow MM + HD, \\[1.2em]
\hspace{3.2cm}\overset{\displaystyle H}{\underset{\displaystyle \cdot\ \ \cdot}{}} \\[-0.2em]
\overset{\displaystyle D}{\underset{\displaystyle |}{}}\ \ \overset{\displaystyle H}{\underset{\displaystyle \cdot}{}}\ \overset{\displaystyle D}{\underset{\displaystyle \cdot}{}}\ \ \overset{\displaystyle H}{\underset{\displaystyle |}{}} \\[-0.2em]
H_2 + M \longrightarrow M \longrightarrow M + HD.
\end{array}
\right\}
\qquad (XVIII\text{-}33)
$$

Ref. 205). The two mechanisms may sometimes be distinguished on the basis of the expected rate law (see Section XVIII-8); one or the other may be ruled out if unreasonable adsorption entropies are implied (see Ref. 206). Molecular beam studies, which can determine the residence time of an adsorbed species, have permitted an experimental decision as to which type of mechanism applies (Langmuir–Hinshelwood in the case of $CO + O_2$ on Pt(111)—note Problem XVIII-26) [207,208].

Some detailed calculations have been made by Tully [209] on the trajectories for Rideal-type processes. Thus the collision of an oxygen atom with a carbon atom bound to Pt results in a CO that departs with essentially all of the reaction energy as vibrational energy (see Ref. 210 for a later discussion).

B. Reaction within the Adsorbed Film as the Rate-Determining Step

The Langmuir–Hinshelwood picture is essentially that of Fig. XVIII-14. If the process is unimolecular, the species meanders around on the surface until it receives the activation energy to go over to product(s), which then desorb. If the process is bimolecular, two species diffuse around until a reactive encounter occurs. The reaction will be diffusion controlled if it occurs on every encounter (see Ref. 211); the theory of surface diffusional encounters has been treated (see Ref. 212); the subject may also be approached by means of Monte Carlo/molecular dynamics techniques [213]. In the case of activated bimolecular reactions, however, there will in general be many encounters before the reactive one, and the rate law for the surface reaction is generally written by analogy to the mass action law for solutions. That is, for a bimolecular process, the rate is taken to be proportional to the product of the two surface concentrations. It is interesting, however, that essentially the same rate law is obtained if the adsorption is strictly localized and species react only if they happen to adsorb on adjacent sites (note Ref. 214). (The *apparent* rate law, that is, the rate law in terms of gas pressures, depends on the form of the adsorption isotherm, as discussed in the next section.)

Various specific rate laws and mechanisms are described in the next section; the present discussion is in more general terms. For example, why is it that a contact catalyst is able to serve as such, that is, why is it able to provide a reaction path that is faster than the homogeneous one. A very simple explanation in the case of bimolecular reactions is simply that the concentration of the reacting species may be much higher in the surface film than it is in the gas phase. A catalyst may thus be effective purely because of the concentration factor, and it is not necessary that the surface reaction itself be any different in character than the homogeneous one (note Problem XVIII-10).

In many cases, however, well-designed catalysts provide intrinsically different reaction paths, and the specific nature of the catalyst surface can be quite important. This is clearly the case with unimolecular reactions for which the surface concentration effect is not applicable.

One specific factor that is important in catalytic activity is the precise nature of the atomic spacings of the catalyst sites. This point was mentioned in Section XVIII-4A, and there are many specific examples of activity depending on what crystal planes are present; face-centered-cubic metals tend to be more active than body-centered ones, and for the former, (111) planes are more active than (110) or (100) ones. Somorjai has discussed a number of such cases, as in Ref. 215. On the theoretical side, transition state structures have been calculated for metal atom insertion into methane CH bonds, leading to predicted orders of activity for Fe and Ni, (100) planes now being the most active [216]. The importance of the morphology of a catalyst surface has been recognized for a long time, of course, and some early ideas are those of Balandin [217] and Taylor [218]. A contemporary example is the effect of faceting of the Pt/W(111) surface on the product ratios in the catalyzed n-butane hydrogenolysis [219].

A catalyst may play an active role in a different sense. There are interesting temporal oscillations in the rate of the Pt-catalyzed oxidation of CO. Ertl and co-workers have related the effect to back-and-forth transitions between Pt surface structures [220] (note Fig. XVI-8). See also Ref. 221 and citations therein. More recently Ertl and co-workers have produced spiral as well as plane waves of surface reconstruction in this system [222] as well as reconstruction waves on the Pt tip of a field emission microscope as the reaction of H_2 with O_2 to form water occurred [223]. Theoretical simulations of these types of effects have been reviewed [224].

There are other complexities. Molecules may chemisorb in various configurations using different combinations of surface sites, giving rise to more than one type of reaction and a mix of rate laws; adsorption isotherms are affected—and also the apparent rate laws [225]. Contemporary catalysts may involve supported metal clusters (note Section XVIII-6B) of size ranging from a few to hundreds of angstroms, and both their reactivity and selectivity can depend markedly on the cluster size (see Ref. 226). Possibly related is that there is evidence that active sites in a polycrystalline catalyst may be reconstructed at grain boundaries [227,228]. Also, heating in a gaseous atmosphere can change the faceting of small metal particles. Finally, fractal geometry treatments have been mentioned in several places in this text, and it should be no surprise that fractal scaling laws have been deduced for heterogeneous and dispersed metal catalysts [229–231].

8. Influence of the Adsorption Isotherm on the Kinetics of Heterogeneous Catalysis

The course of a surface reaction can in principle be followed directly with the use of various surface spectroscopic techniques plus equipment allowing the rapid transfer of the surface from reaction to high-vacuum conditions; see Campbell [232]. More often, however, the experimental observables are the changes with time of the concentrations of reactants and products in the *gas* phase. The rate law in terms of surface concentrations might be called the "true" rate law and the one analogous to that for a homogeneous system. What is observed, however, is an apparent rate law giving the dependence of the rate on the various gas pressures. The true and the apparent rate laws can be related if one assumes that adsorption equilibrium is rapid compared to the surface reaction.

The problem of treating surface encounters where the surface is heterogeneous or where complex lateral interaction effects are present is virtually insurmountable, and as a consequence some very drastic simplifying assumptions have to be made. In extreme form, these amount to assuming that the film is sufficiently mobile that adsorbed molecules undergo many encounters before desorbing, so that mass action rate expressions can be written; and the simple Langmuir equation is obeyed, that is, the *Langmuir–Hinshelwood* model. Some of its elementary applications are outlined in the following. See Ref. 233 for a more detailed discussion.

It will be recalled (Eq. XVII-11) that in the case of competitive adsorption the Langmuir equation takes the form

$$S_i = S \frac{b_i P_i}{1 + \sum b_j P_j} \qquad \text{(XVIII-34)}$$

where

$$b_j = b_{0j} e^{Q_j/RT} \qquad \text{(XVIII-35)}$$

If a surface reaction is bimolecular in species A and B, the assumption is that the rate is proportional to $S_A \times S_B$. We now proceed to apply this interpretation to a few special cases.

A. Unimolecular Surface Reactions

We suppose the type reaction to be

$$A \rightarrow C + D$$

and that the surface reaction proceeds according to the rate law

$$\frac{dn_A}{dt} = -kS_A \qquad \text{(XVIII-36)}$$

or

$$\frac{dn_A}{dt} = -kS\,\frac{b_A P_A}{1 + b_A P_A + b_C P_C + b_D P_D} \qquad \text{(XVIII-37)}$$

If the products C and D are weakly adsorbed, Eq. XVIII-37 reduces to

$$\frac{dn_A}{dt} = -kS\,\frac{b_A P_A}{1 + b_A P_A} = -\frac{k' P_A}{1 + b_A P_A} \qquad \text{(XVIII-38)}$$

which means that the apparent rate law should show a behavior similar to that of the Langmuir equation, that is, at low P_A the rate should be proportional to P_A but should reach a limiting rate kS at high P_A.

If one or more of the products is strongly adsorbed, Eq. XVIII-37 takes on another limiting form, of the type

$$\frac{dn_A}{dt} = -kS\,\frac{b_A P_A}{1 + b_C P_C} = -\frac{k' P_A}{1 + b_C P_C} \qquad \text{(XVIII-39)}$$

(where product C is more strongly adsorbed than A or D; in the limit of very strong adsorption of C, the right-hand side of Eq. XVIII-38 becomes $-k' P_A / b_C P_C$).

The above equations can apply when the rate-determining step is first order even though the complete reaction mechanism is complicated. Thus for the reaction $NO + CO = \frac{1}{2}N_2 + CO_2$ on Rh(100), the proposed mechanism was [234]

$$CO(g) \overset{K_1}{=} CO(ads) \qquad \text{(XVIII-40)}$$

$$NO(g) \overset{K_2}{=} NO(ads) \qquad \text{(XVIII-41)}$$

$$NO(ads) \overset{k_3}{\to} N(ads) + O(ads) \qquad \text{(XVIII-42)}$$

$$2N(ads) \overset{k_4}{\to} N_2(g) \qquad \text{(XVIII-43)}$$

$$CO(ads) + O(ads) \overset{k_5}{\to} CO_2(g) \qquad \text{(XVIII-44)}$$

If reaction XVIII-42 is the slow step, the Langmuir–Hinshelwood rate law is

$$R = \frac{d(CO_2)}{dt} = \frac{k_3 K_2 P_{NO}}{1 + K_1 P_{CO} + K_2 P_{NO}} \qquad \text{(XVIII-45)}$$

Just as the surface and apparent kinetics are related through the adsorption isotherm, the surface or true activation energy and the apparent activation energy are related through the heat of adsorption. The apparent rate constant k' in these equations contains two temperature-dependent quantities, the true rate constant k and the parameter b_A. Thus

$$k' = kb_A S_A = kb_{0A} S_A e^{Q_A} \qquad \text{(XVIII-46)}$$

If the slight temperature dependencies of S_A and of b_{0A} are neglected, then

$$\frac{d \ln k'}{dt} = \frac{E_{app}}{RT^2} = \frac{E_{true} - Q_A}{RT^2} \qquad \text{(XVIII-47)}$$

or

$$E_{app} = E_{true} - Q_A$$

The apparent activation energy is then less than the actual one for the surface reaction per se by the heat of adsorption. Most of the algebraic forms cited are complicated by having a composite denominator, itself temperature dependent, which must be allowed for in obtaining k' from the experimental data. However, Eq. XVIII-47 would apply directly to the low-pressure limiting form of Eq. XVIII-38. Another limiting form of interest results if one product dominates the adsorption so that the rate law becomes

$$\frac{dn_A}{dt} = -\frac{k' P_A}{P_C} \qquad \text{(XVIII-48)}$$

If follows that

$$E_{app} = E_{true} - Q_A + Q_C \qquad \text{(XVIII-49)}$$

It should not be inferred from the foregoing that the heat of adsorption effect is the only one modifying the activation energy of a catalyzed reaction from that for the homogeneous one. The true or surface activation energy may itself be quite different from that for the homogeneous reaction. As an example, the true activation energy for the tungsten-catalyzed decomposition of ammonia is only 39 kcal/mol, as compared to the value of about 90 kcal/mol for the gas phase reaction.

B. Bimolecular Surface Reactions

Continuing the formal development of the influence of the adsorption isotherm on the apparent reaction kinetics, we next consider the case of a reaction that is bimolecular on the surface,

$$A + B \rightarrow C + D$$

and whose surface reaction rate law is

$$\frac{dn_A}{dt} = -kS_AS_B \qquad \text{(XVIII-50)}$$

The general expression for the apparent rate law is now

$$\frac{dn_A}{dt} = -kS \frac{b_Ab_BP_AP_B}{(1 + b_AP_A + b_BP_B + \sum b_{prod}P_{prod})^2} \qquad \text{(XVIII-51)}$$

Only two of the many possible special cases need be considered. Thus if the products and reactants are weakly adsorbed,

$$\frac{dn_A}{dt} = -k'P_AP_B \qquad \text{(XVIII-52)}$$

If A is weakly adsorbed as well as the products but B is strongly adsorbed, one finds

$$\frac{dn_A}{dt} = -\frac{k'P_A}{P_B} \qquad \text{(XVIII-53)}$$

so that retardation by a *reactant* is possible. The hydrogenation of pyridine on metal oxide catalysts shows retardation both by pyridine and by the reaction products, for example [235]. Other examples of complex rate laws are those for the dehydrogenation of alcohols on *liquid* metals [236], the oxidative dehydrogenation dimerization of propylene over Bi_2O_3 [237] and benzene hydrogenation over a supported iron catalyst [238]. One mechanism for the Fischer–Tropsch reaction

$$nCO + (2n + 1)H_2 = nH_2O + C_nH_{2n+2} \qquad \text{(XVIII-54)}$$

is (see Ref. 239)

$$CO(g) \overset{K_1}{=} C(ads) + O(ads) \qquad\qquad (XVIII\text{-}55)$$

$$O(ads) + H_2(g) \overset{K_2}{=} H_2O \qquad\qquad (XVIII\text{-}56)$$

$$C(ads) + H_2 \overset{k}{\rightarrow} CH_2(ads) \qquad\qquad (XVIII\text{-}57)$$

followed by fast steps. The corresponding rate expression proposed is

$$R = \frac{kK_1K_2P_{CO}P_{H_2}^2}{P_{H_2O} + K_1K_2P_{CO}P_{H_2}} \qquad\qquad (XVIII\text{-}58)$$

(see Problem XVIII-24).

Rate laws have also been observed that correspond to there being two kinds of surface, one adsorbing reactant A and the other reactant B and with the rate proportional to $S_A \times S_B$. For traditional discussions of Langmuir–Hinshelwood rate laws, see Refs. 240–242. Many catalytic systems involve a series of intermediates, and the simplifying assumption of steady-state equilibrium is usually made. See Boudart and co-workers [243–245] for a contemporary discussion of such complexities.

9. Mechanisms of a Few Catalyzed Reactions

A great deal of tax money is spent in support of fundamental research, and this is often defended as having an intrinsic virtue. To take the present topic as an example, however, the study of just how molecules adsorb and react on a surface is fascinating and challenging, yet the tax-paying public should not be asked merely to support the esoteric pleasures of a privileged few. The public should expect the occasional major practical advance whose benefits more than pay for the overall cost of all research. The benefits in the present case come from the discovery and development of catalytic processes of major importance to an industrial society.

It is appropriate that this chapter conclude with a short discussion of a few selected, widely used catalyzed reactions. The reactions chosen—ammonia synthesis, Fischer–Tropsch reactions, ethylene dehydrogenation, the catalytic cracking of hydrocarbons, the oxidation of CO, and photoassisted heterogeneous reactions represent the writers' choice of a balanced group of systems of major impact. It is an interesting tribute to the endurance of research problems that the first four systems were chosen in the first, 1960, edition of this book!† Many variations and many new catalysts have appeared since then, of course, and a host of other types of reactions. These range from the catalysis by ice crystals of ozone-depleting reactions to a wide variety of photoas-

---

†While many of the questions are the same, the answers have changed!

sisted ones that are useful in water purification or that hopefully may lead to efficient solar energy conversion. The ultimate catalysis would be that of deuterium fusion—a highly exoergic reaction of enormous homogeneous activation energy, the bypassing of which is, *in principle*, possible with a suitable heterogeneous catalyst. Such a process would be called *cold fusion*. Success has been proclaimed [246], explained [247], and denied [248a,248b], with the cold fusion proponents currently scientifically marginalized (but note Ref. 249).

To proceed with the topic of this section, Refs. 250 and 251 provide oversights of the application of contemporary surface science and bonding theory to catalytic situations. The development of bimetallic catalysts is discussed in Ref. 252. Finally, Weisz [253] discusses "windows on reality"; the acceptable range of rates for a given type of catalyzed reaction is relatively narrow. The reaction becomes impractical if it is too slow, and if it is too fast, mass and heat transport problems become limiting.

A. Ammonia Synthesis

The first industrial synthesis of NH_3 from H_2 and N_2 started up in 1913 (!) and was known as the *Haber–Bosch process*. Essentially the same catalyst is used today, with improvements. The catalyst is prepared by fusing Fe_3O_4 with a few percent of added K_2O and Al_2O_3 and then heating in a N_2–H_2 mixture, whereby the iron oxide is reduced to mainly metallic iron. The Al_2O_3 acts as a "structural" promoter in ensuring that a high surface area, porous mass is obtained, with the iron present as small crystallites (the manner in which these crystallites form and sinter is important—note Ref. 254). The K_2O acts as an "electronic" promoter, covering most of the internal surface [255] and changing its electronegativity. Poisons include CO_2 (probably due to adsorption on the K_2O), CO (probably due to adsorption on iron sites), and H_2 and O_2. Some useful general discussions are those by Ertl [256], Sinfelt [257], and Weinberg et al. [258]. Important older work is that of Emmett (see Ref. 259 and also Ref. 260). Boudart [261] gives a personalized discussion emphasizing Temkin's contributions.

Surface science techniques, including the ability to transfer a system rapidly from reaction to high-vacuum conditions, have established that Fe(111) surfaces are by far the most reactive [262,263]. The contemporary wisdom is that the slow step is the dissociative chemisorption of N_2 on a Fe site having C_7 coordination (seven nearest neighbors) (see Refs. 13, 264). The general mechanism is of the Langmuir–Hinshelwood type. Hydrogen also adsorbs dissociatively, and the surface reactions N(ads) + H(ads) = NH(ads), NH(ads) + H(ads) = NH_2(ads), and NH_2(ads) + H(ads) = NH_3(ads) then occur in sequence followed by desorption of product NH_3. The energetic scheme is shown in Fig. XVIII-23 [256].

The observed rate law depends on the type of catalyst used; with promoted iron catalysts a rather complex dependence on nitrogen, hydrogen, and ammonia pressures is observed, and it has been difficult to obtain any definitive form from experimental data (although note Eq. XVIII-20). A useful alternative approach

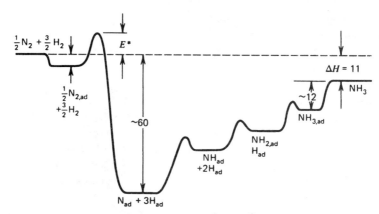

Fig. XVIII-23. Potential energy diagram for $\frac{1}{2}N_2 + \frac{3}{2}H_2 = NH_3$. (From Ref. 256.)

has been to study the kinetics of the *decomposition* of ammonia, and using (111), (110), and (100) surfaces both of tungsten [265] and rhodium [266]. Again, there was crystal face specificity, and the rate laws depended on the catalyst and on conditions. For tungsten, it was

$$-\frac{d(NH_3)}{dt} = a + bP_{NH_3}^{2/3} \qquad (XVIII\text{-}59)$$

the values of a and b being different for (111), (100), and (110) faces of their catalyst, single-crystal tungsten. They concluded that most of the surface was covered by the species W—N, the a term in Eq. XVIII-59 being due to the slow step: $2W—N \rightarrow W_2N + \frac{1}{2}N_2$. The second term in the rate law could not be fit by the Temkin–Pyzhev [267] mechanism, and instead one involving equilibrium between surface species $W_2N_3H_2$ and WNH and gaseous ammonia was proposed.

B. Fischer–Tropsch Reactions

The Fischer–Tropsch reaction is essentially that of Eq. XVIII-54 and is of great importance partly by itself and also as part of a coupled set of processes whereby steam or oxygen plus coal or coke is transformed into methane, olefins, alcohols, and gasolines. The first step is to produce a mixture of CO and H_2 (called "water–gas" or "synthesis gas") by the high-temperature treatment of coal or coke with steam. The water–gas shift reaction $CO + H_2O = CO_2 + H_2$ is then used to adjust the CO/H_2 ratio for the feed to the Fischer–Tropsch or "synthesis" reactor. This last process was disclosed in 1913 and was extensively developed around 1925 by Fischer and Tropsch [268].

The "classic" catalyst consists of $Co–ThO_2–MgO$ mixtures supported on Kieselguhr (see Ref. 269); group VIII metals, especially Ni, generally are active,

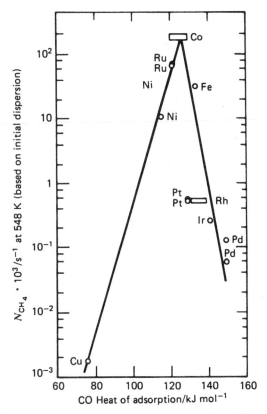

Fig. XVIII-24. Turnover frequencies for methanation using silica-supported metals. (From Ref. 270.)

as illustrated in Fig. XVIII-24) [270] for silica-supported metals. Activity is dependent on the nature of the support, and a different order is obtained, for example, if the metals are on alumina [271, 272]. This type of effect has been called a *strong metal-support interaction* [272] (see also 273, 274). The reaction system is a complex one, and product ratios are also sensitive to the metal-support used; variants include carbon-supported iron (see Refs. 275, 276), zeolite [277,278], and pillared clays [279]. Bimetallic particles may be used [280], and promoters such as K are common (see Ref. 271a). Reference 280 gives a summary of the various types of modifications. See Somorjai [13, pp. 483ff] for an extensive discussion of Fischer–Tropsch reactions.

The mechanism of the Fischer–Tropsch reactions has been the object of much study (note Eqs. XVIII-55–XVIII-57) and the subject of much controversy. Fischer and Tropsch proposed one whose essential feature was that of a metal carbide—patents have been issued on this basis. It is currently believed that a particular form of active adsorbed carbon atoms is involved, which is then methanated through a series of steps such as

$$H_2(g) = 2H(ads) \qquad\qquad (XVIII\text{-}60)$$

$$C(ads) + H(ads) \rightarrow CH(ads) \qquad\qquad (XVIII\text{-}61)$$

$$CH(ads) + H(ads) \rightarrow CH_2(ads), \text{ etc.} \qquad\qquad (XVIII\text{-}62)$$

This active carbon is considered to be produced through the reaction

$$2CO = C + CO_2 \qquad\qquad (XVIII\text{-}63)$$

known as the *Boudouard* reaction [13,281]. An alternative proposal is that the active C(ads) is produced by the surface dissociation of CO(ads) [282], followed by the methanation steps. At any rate, appreciable coverages of the CH_x(ads) intermediates may be present (see Refs. 272, 281). Chain growth then occurs by such processes as [270]

$$CH_2(ads) + CH_2(ads) \rightarrow C_2H_5(ads) \qquad\qquad (XVIII\text{-}64)$$

$$C_2H_5(ads) + H(ads) \rightarrow C_2H_6(ads) \qquad\qquad (XVIII\text{-}65)$$

and oxygen-containing products by reactions such as

$$CH_3(ads) + CO(ads) \rightarrow CH_3CO(ads) \overset{H(ads)}{\longrightarrow} CH_3COH \qquad (XVIII\text{-}66)$$

$$H(ads) + CO(ads) \rightarrow HCO(ads) \overset{H(ads)}{\longrightarrow} H_2CO(ads) \rightarrow \rightarrow CH_3OH$$
$$(XVIII\text{-}67)$$

Sequences such as the above allow the formulation of rate laws but do not reveal molecular details such as the nature of the transition states involved. Molecular orbital analyses can help, as in Ref. 270; it is expected, for example, that increased strength of the metal—CO bond means decreased C$=$O bond strength, which should facilitate process XVIII-55. The complexity of the situation is indicated in Fig. XVIII-24, however, which shows catalytic activity to go through a *maximum* with increasing heat of chemisorption of CO. Temperature-programmed reaction studies show the presence of more than one kind of site [99,100,283], and ESDIAD data show both the location and the orientation of adsorbed CO (on Pt) to vary with coverage [284].

A topic of current interest is that of *methane activation* to give ethane or selected oxidation products such as methanol or formaldehyde. Oxide catalysts are used, and there may be mechanistic connections with the Fischer–Tropsch system (see Ref. 285).

C. Hydrogenation of Ethylene

The catalytic hydrogenation of ethylene occurs on various metal catalysts, such as nickel, including active or skeletal forms produced by dissolving out silicon or aluminum from $NiSi_2$ or $NiAl$, $NiAl_2$ alloys, reduced copper, platinum, rhodium, iron, chromium, and other metals. It is perhaps one of the most studied catalytic reactions, yet even today there is not a consensus of opinion as to the detailed mechanism (see Ref. 286). A widely accepted outline of the sequence of steps is essentially that originally proposed by Horiuti and Polanyi in 1934 [287] (see Ref. 288):

$$H_2(g) \rightarrow 2H(ads) \tag{XVIII-68}$$

$$C_2H_4(g) \rightarrow CH_2-CH_2(ads) \tag{XVIII-69}$$

$$CH_2-CH_2(ads) + H(ads) \rightarrow CH_3-CH_2(ads) \overset{H(ads)}{\rightarrow} C_2H_6 \tag{XVIII-70}$$

where CH_2-CH_2 (ads) may be a precursor species occupying two adjacent adsorption sites. The experimental rate is proportional to $P_{H_2}^x P_{C_2H_4}^y$, where the exponents are between 0 and 1 depending on the metal.

Studies to determine the nature of intermediate species have been made on a variety of transition metals, and especially on Pt, with emphasis on the Pt(111) surface. Techniques such as TPD (temperature-programmed desorption), SIMS, NEXAFS (see Table VIII-1) and RAIRS (reflection absorption infrared spectroscopy) have been used, as well as all kinds of isotopic labeling (see Refs. 286 and 289). On Pt(111) the surface is covered with C_2H_3, *ethylidyne*, tightly bound to a three-fold hollow site, see Fig. XVIII-25, and Ref. 290. A current mechanism is that of the figure, in which ethylidyne acts as a kind of surface catalyst, allowing surface H atoms to add to a second, perhaps physically adsorbed layer of ethylene; this is, in effect, a kind of Eley–Rideal mechanism.

As an interesting footnote, Ceyer [291] has found that ethylene is hydrogenated by hydrogen *absorbed* in the bulk region just below a Ni(111) surface. In this case, the surface ethylenes are assumed to by lying flat, with the dissolved H atoms approaching the double bond from *underneath*.

D. Catalytic Cracking of Hydrocarbons and Related Reactions

A number of related reactions of hydrocarbons are catalyzed by acidic oxide types of materials. These include the *cracking* of high molecular weight

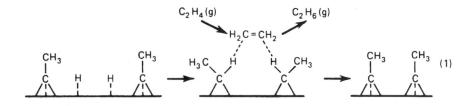

Fig. XVIII-25. The ethylidyne mechanism for ethylene hydrogenation. (See Ref. 292.)

hydrocarbons to lower molecular weight ones to produce gasolines from oils; *reforming*, which involves isomerizations and molecular weight redistributions through hydrogenation–dehydrogenation steps; and *alkylation*, which may amount to the reverse of cracking. The Houdry process, an early cracking procedure, was announced in 1933 [293] and made use of activated bentonite clay as the catalyst. Current catalysts also include synthetic aluminosilicates prepared by precipitation of alumina in the presence of silica gel, followed by filtration, washing, drying, and calcining. The activity is associated with the presence of acid sites [294–298], although in the discussion relating to Fig. XVIII-20 it was pointed out that there can be some question as to whether for a given system the sites are Brønsted or Lewis in type.

As mentioned in Section XVIII-6D, zeolites, including those in the acid form (and which may be more acidic than pure sulfuric acid!), have become important industrial catalysts for cracking and reforming. The special feature here is the added ability to select reactants and products on the basis of size and shape (note Fig. XVIII-21). A fairly recent development is the use of smectite clays; these are minerals having layer lattice structures in which two-dimensional oxyanions are separated by layers of hydrated cations. A variety of stable cations, which may themselves have acidic properties, may be intercalated by ion exchange to give "pillared" clays which may now have structural cavities similar to those in zeolites, as illustrated in Fig. XVIII-26 (see Refs. 299, 300). It is such pillared clays that have useful properties as cracking catalysts. Interesting inclusion ions are various Cr(III) hydrolytic oligamers; these allow variation in the cavity size and also introduce new potential catalytic sites [301].

The general features of the cracking mechanism involve carbonium ion formation by a reaction of the type

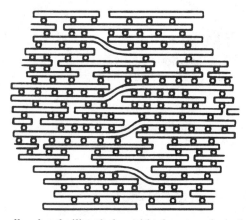

Fig. XVIII-26. A well-ordered pillared clay with almost exclusively zeolitelike microporosity. (From Ref. 299.)

$$RH + H^+(\text{acid site}) \rightarrow R^+(\text{ads}) + H_2 \qquad (XVIII\text{-}71)$$

where the acid site might, for example, be in the Brønsted form. The resulting carbonium ions then undergo various rearrangement and cleavage reactions such as

$$CH_3-CH_2-\overset{+}{C}H_2 \rightarrow \underset{\underset{H^+}{|}}{CH_3-CH=CH_2} \rightarrow CH_3-\overset{+}{C}H-CH_3$$

$$\overset{+}{R}CH-CH_2-CH_3 \rightarrow R-CH=CH_2 + \overset{+}{C}H_3 \qquad (XVIII\text{-}72)$$

$$R-CH=CH_2 + \overset{+}{C}H_3 \rightarrow R-CH(CH_3)-\overset{+}{C}H_2$$

and terminating by the reverse of reaction XVIII-71. See Ref. 302 for some specific examples of such sequences.

Because of the great industrial importance of these processes, a great deal of performance information is now available, and rather precise control of the degree of unsaturation, isomerization, and aromatic character of the product can be achieved, far more so than in the early, noncatalytic thermal cracking procedures. However, as can be imagined, the detailed kinetics are quite complicated, and much remains to be done in the way of fundamental research, including ways to reduce "coke," that is, higher molecular weight products that accumulate in the catalyst (e.g., Ref. 303).

E. Oxidation of CO

We consider next perhaps the best understood catalyzed reaction: the oxidation of CO over group VIII metal catalysts. The reaction is an important environmental one since it involves the conversion of CO to CO_2 in automobile catalytic converters. The mechanism is straightforward:

$$CO(g) = CO(\text{ads}) \qquad (XVIII\text{-}73)$$

$$O_2(g) \rightarrow 2O(\text{ads}) \qquad (XVIII\text{-}74)$$

$$CO(\text{ads}) + O(\text{ads}) \rightarrow CO_2(g) \qquad (XVIII\text{-}75)$$

It is Langmuir–Hinshelwood in type, and the usually observed rate law is

$$\frac{dCO}{dt} = k\left[\frac{P_{O_2}}{P_{CO}}\right] e^{-Q_{d,CO}/RT} \qquad (XVIII\text{-}76)$$

where $Q_{d,CO}$ is the CO desorption energy, although for large P_{O_2}/P_{CO} ratios the rate becomes first-order in P_{CO}. The system is reviewed in Ref. 304. What happens is that on metals such as Rh, Pt, Pd, Ru, and Ir, CO adsorbs strongly end-on, forming a metal–carbon bond, and the surface is nearly covered with CO; oxygen adsorbs dissociatively on a bare spot and an oxygen atom reacts quickly with an adjacent CO to give CO_2, which immediately desorbs. Experimental activation energies are around 25 kcal/mol and do not vary much with metal, indicating that the reaction is not very structure-sensitive. Some data are shown in Fig. XVIII-27. It is a tribute to Irving Langmuir that he anticipated most of the above conclusions in his early work. (See Ref. 305 and comments in Ref. 306.)

As with any system, there are complications in the details. The CO sticking probability is high and constant until a θ of about 0.5, but then drops rapidly [306a]. Practical catalysts often consist of nanometer size particles supported on an oxide such as alumina or silica. Different crystal facets behave differently and RAIRS spectroscopy reveals that CO may adsorb with various kinds of bonding and on various kinds of sites (three-fold hollow, bridging, linear) [307]. See Ref. 309 for a discussion of some debates on the matter. In the case of Pd crystallites on α-Al_2O_3, it is proposed that CO impinging on the support

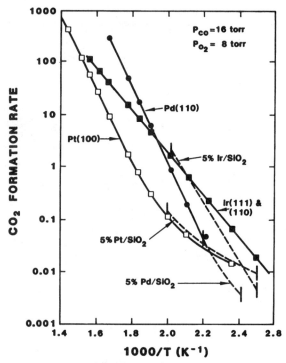

Fig. XVIII-27. Specific rates of CO oxidation on single crystal and supported catalysts as a function of temperature. (From Ref. 308. Reprinted with permission from American Chemical Society, copyright 1988.)

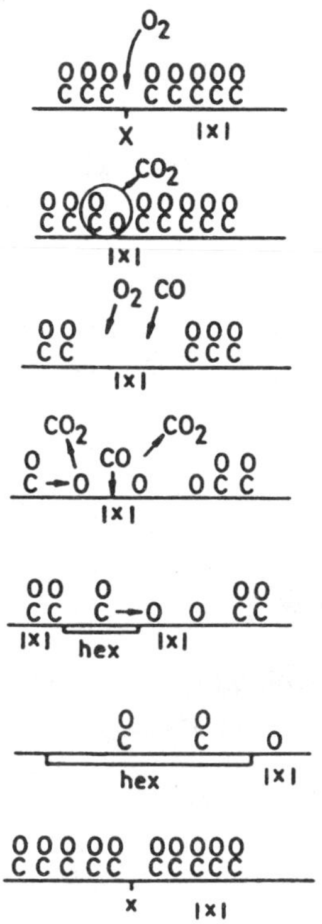

O_2
O O O | O O O O O
C C C | C C C C
  x |x|

CO_2
O O (O) O O O O O
C C C O C C C C C
|x|

O_2 CO
O O / / O O O
C C   C C C
|x|

CO_2 CO_2
O \ CO /   O O
C→O | O   O C C
|x|

O O O   O O
C C C→O O C C
|x| hex |x|

    O   O
    C   C   O
  hex |x|

O O O O O O O O O O
C C C C C C C C C
    x |x|

Fig. XVIII-28. Proposed sequence of events in CO oxidation on Pt(100). On admission of CO, the surface becomes covered in patches, inducing the 1×1 surface structure, on bare spots of which oxygen adsorbs strongly with dissociation. Surface O atoms react with CO, removing it as CO_2, and as enough bare surface forms, it reverts to the reconstructed hexagonal structure. (From Ref. 306. Reprinted with permission from American Chemical Society, copyright 1987.)

close to a particle, can diffuse to the metal surface to then undergo oxidation, thus adding to the effective surface area [310].

Perhaps the most fascinating detail is the surface reconstruction that occurs with CO adsorption (see Refs. 311 and 312 for more general discussions of chemisorption-induced reconstructions of metal surfaces). As shown in Fig. XVI-8, for example, the Pt(100) bare surface reconstructs itself to a hexagonal pattern, but on CO adsorption this reconstruction is lifted [306]; CO adsorption on Pd(110) reconstructs the surface to a missing-row pattern [309]. These reconstructions are reversible and as a result, oscillatory behavior can be observed. Returning to the Pt(100) case, as CO is adsorbed patches of the simple 1×1 structure (the structure of an undistorted (100) face) form. Oxygen adsorbs on any bare 1×1 spots, reacts with adjacent CO to remove it as CO_2, and at a certain point, the surface reverts to the hexagonal structure. The presumed sequence of events is shown in Fig. XVIII-28.

With certain critical P_{CO}/P_{O_2} ratios, structural oscillations can be observed [306]. Patterns of stationary and/or traveling waves can actually be seen by means of photoemission electron microscopy (see Ref. 313, and note Section XVIII-7B. Such behavior can be modeled mathematically (e.g., Refs. 214, 314).

F. Photochemical and Photoassisted Processes at Surfaces

Some mention should be made of a topic of current interest, that of photoassisted and photochemical processes at surfaces. By photoassisted is meant a process that can occur in the dark but occurs with less energy consumption if the surface is irradiated. An example is the electrolytic decomposition of water using TiO_2 as one electrode; the requisite voltage is much reduced on illumination (see Ref. 315). The photodecomposition can be induced without applied voltage if short enough wavelength light is used and works even better if platinized Pt has been deposited on the TiO_2 [316]. (See also Ref. 317.)

Heterogeneous photochemical reactions fall in the general category of photochemistry—often specific adsorbate excited states are involved (see, e.g., Ref. 318.) Photodissociation processes may lead to reactive radical or other species; electronic excited states may be produced that have their own chemistry so that there is specificity of reaction. The term *photocatalysis* has been used but can be stigmatized as an oxymoron; light cannot be a *catalyst*—it is not recovered unchanged.

A few illustrative examples are the following. Photohydrogenation of acetylene and ethylene occurs on irradiation of TiO_2 exposed to the gases, but only if TiOH surface groups are present as a source of hydrogen [319]. The photoinduced conversion of CO_2 to CH_4 in the presence of Ru and Os colloids has been reported [320]. Platinized TiO_2 powder shows, in the presence of water, photochemical oxidation of hydrocarbons [321,322]. Some of the postulated reactions are:

$$(TiO_2) + h\nu \longrightarrow e^- + hole^+$$
$$H_2O + hole^+ \longrightarrow \cdot OH + H^+$$
$$H^+ + e^- \longrightarrow H\cdot$$
$$O_2 + e^- \longrightarrow \cdot O_2^- \xrightarrow{H\cdot} \cdot HO_2^-$$
$$HO_2^- + hole^+ \longrightarrow \cdot HO_2$$
$$2 \cdot HO_2^- \longrightarrow O_2 + H_2O_2 \xrightarrow{\cdot O_2^-} \cdot OH + OH^- + O_2$$
$$RH + \cdot OH \text{ (or } \cdot HO_2) \longrightarrow ROH + \cdot H$$
$$RH + hole^+ \longrightarrow RH^+ \xrightarrow{hole^+} RH^{2+} \qquad \text{(XVIII-77)}$$

Again with platinized TiO_2, ultraviolet irradiation can lead to oxidation of aqueous CN^- [323] and to the water–gas shift reaction, $CO + H_2O = H_2 + CO_2$ [324]. Some mechanistic aspects of the photooxidation of water (to O_2) at the TiO_2–aqueous interface are discussed by Bocarsly et al. [325].

A large variety of organic oxidations, reductions, and rearrangements show photochemistry at interfaces, usually of a semiconductor. The subject has been reviewed [326,327]; some specific examples are the photo-Kolbe reaction (decarboxylation of acetic acid) using Pt supported on anatase [328], the photocatalyzed oxidation of alcohols by heteropolytungstates [329], and amine-*N*-alkylation photocatalyzed by platinized semiconductor particles [330]. The photooxidation of 2-propanol to acetone occurs on MnO_2 as support, which supplies the oxygen but can be regenerated [331]. Surface fluorination of diamond occurs on irradiation of adsorbed C_4F_9I and CF_3I [332]. Irradiation of DBr on Pt(III) surface is thought to lead to DBr^-, which can then react with coadsorbed ethylene to product ethane [333]. The photochemistry of adsorbed coordination compounds has been studied, examples being the photoaquation of $Rh(NH_3)_5{}^{2+}$ in zeolite Y [334] and the photoinduced attachment of $Fe(CO)_5$ to silica when adsorbed on porous vicor glass [335]. An interesting use of the adsorbed state to confine nascent radical pairs has been described [336]. Finally, a rapidly developing field is that of laser-induced surface chemistry—desorption, migration, chemical reaction. The photodesorption of SO_2 from Ag(111) and of ammonia from Cu(III) surfaces [337,338] are thought to be due to excitation of the substrate rather than of the adsorbate. (See Refs. 339 and 340 for reviews.) As a more recent example, *trans*-stilbene adsorbed on $Al_2O_3(0001)$ undergoes photodimerization, perhaps through eximer formation, but no photoisomerization [341]. Finally, zeolites are of great current importance in catalysis, and a recent study revealed that TiO_2 anchored in *Y*-zeolite exhibits photocatalytic reactivity for the decomposition of NO into N_2, O_2, and N_2O [342].

While hopes are high, heterogeneous photochemical systems seem not yet to have found major practical application. The photovoltaic cell or "solar cell" is the only system with important (although specialized) commercial use (see Ref. 343).

10. Problems

1. Give four specific experimental tests, measurements, or criteria that would be considered good evidence for characterizing adsorption in a given system as either physical adsorption or chemisorption.

2. The langmuir as a unit is defined in Section XVIII-2A. Considering the specific case of O_2 at 200°C, how many langmuirs (L) exposure would be required to give 10% coverage on, say, W, if every molecule hitting the surface was chemisorbed? Take the molecular area of O_2 to be 14.1 Å^2. Investigators may alternatively report exposure in molecules/m^2 or may give coverage as a θ value. Discuss the merits and demerits of each usage.

3. Suppose that the rate law of Eq. XVIII-1 showed θ^n rather than just θ. Derive the equation corresponding to Eq. XVIII-3.

4. If the desorption rate is second-order, as is often the case for hydrogen on a metal surface, so that θ^2 appears in Eq. XVIII-1, an equation analogous to Eq. XVIII-3 can be derived by the Redhead procedure. Derive this equation. In a particular case, H_2 on $Cu_3Pt(III)$ surface, A was taken to be 1×10^{-5} cm^2/atom, the maximum desorption rate was at 225 K, θ at the maximum was 0.5. Monolayer coverage was 4.2×10^4 atoms/cm^2, and $\beta = 5.5$ K/sec. Calculate the desorption enthalpy (from Ref. 110).

5. The temperature-programmed desorption of NO_2 shows a maximum rate at 330 K using a heating rate of 7.53°C/sec. Take A to be 10^{13} sec^{-1}, and assume that Eq. XVIII-1 applies. (Note Ref. 344.)

6. Repeat the calculation of Problem 5 for the case of $T_m = 680$ K and a heating rate of 40 K/sec.

7. Redhead [89] gives the approximate equation $E/RT_m = \ln(AT_m/\beta) - 3.64$. Check the usefulness of this equation by comparing with the answers to Problems 5 and 6.

8. The assumption of an A value (as in Problems 5 and 6) may be avoided if T_m is found as a function of β. Equation XVIII-3 may be put in the form $2 \ln T_m - \ln \beta = E/RT_m + \ln(E/RA)$. In the case of desorption of CO and CO_2 from ZnO, T_m values were 559 and 588 K for β values of 0.1 and 0.9 K sec^{-1}, respectively [92]. Calculate E and A.

9. A useful observation is that E is very nearly proportional to T_m over the range usually employed. Use Eq. XVIII-3 to demonstrate this. Suggestion: Find T_m for various x, $x = E/RT_m$, and thence E for various T_m. One can conclude from this an approximate universal value of x. Explain and give this value.

10. It was pointed out that a bimolecular reaction can be accelerated by a catalyst just from a concentration effect. As an illustrative calculation, assume that A and B react in the gas phase with 1:1 stoichiometry and according to a bimolecular rate law, with the second-order rate constant k equal to 10^{-5} l mol^{-1} sec^{-1} at 0°C. Now, assuming that an equimolar mixture of the gases is condensed to a liquid film on a catalyst surface and the rate constant in the condensed liquid solution is taken to be the same as for the gas phase reaction, calculate the ratio of half times for reaction in the gas phase and on the catalyst surface at 0°C. Assume further that the density of the liquid phase is 1000 times that of the gas phase.

11. The self-diffusion of Ni on a Ni(111) surface is given by the equation $\mathcal{D}$ (cm^2/sec) $= 300 \exp(-E^*/RT)$. Calculate $\mathcal{D}$ at 850 K assuming that E^* is 35 kcal/mol and thence how far, on the average, a given surface atom should diffuse in 1 hr.

12. What is the physical meaning of P_0 in Eq. XVIII-15 (it is *not* P^0)?

13. Calculate the entropy of adsorption ΔS_2 for several values of θ for the case of nitrogen on an iron catalyst. Use the data of Scholten and co-workers given in Section XVIII-4B.

14. The quantity Θ_Q in Eq. XVIII-8 corresponds to $F(Q)$ of Eq. XVII-129. Derive $\Theta(P, T)$ using these relationships and show how the result can be reduced to Eq. XVIII-9.

15. A rate of chemisorption obeys the rate law

$$\frac{d\theta}{dt} = Ae^{-b\theta}$$

Show whether a plot of θ versus log t should be linear (a form of the Elovich equation).

16. Estimate the heat of dissociative adsorption of I_2 on copper.

17. The Pt-catalyzed decomposition of NO (into N_2 and O_2) is found to obey the experimental rate law

$$\frac{dP_{NO}}{dt} = - \frac{kP_{NO}}{P_{O_2}}$$

Assuming that adsorbed gases obey the Langmuir equation, derive this rate law starting with some reasonable assumed mechanism for the surface reaction. Assuming further that the heat of adsorption of NO is 20 kcal/mol and that of O_2 is 25 kcal/mol, show what the actual activation energy for the surface reaction should be, given that the apparent activation energy (i.e., from the temperature dependence of k above) is 15 kcal/mol.

18. Some early observations on the catalytic oxidation of SO_2 to SO_3 on platinized asbestos catalysts led to the following observations: (1) the rate was proportional to the SO_2 pressure and was inversely proportional to the SO_3 pressure; (2) the apparent activation energy was 30 kcal/mol; (3) the heats of adsorption for SO_2, SO_3, and O_2 were 20, 25, and 30 kcal/mol, respectively. By using appropriate Langmuir equations, show that a possible explanation of the rate data is that there are two kinds of surfaces present, S_1 and S_2, and that the rate-determining step is

$$SO_2(\text{ads on } S_1) + O_2(\text{ads on } S_2) \rightarrow \text{intermediates}$$

On this basis, what would you expect the dependence of rate on pressure to be (*a*) at low oxygen pressure and (*b*) during the initial stages of reaction when negligible SO_3 is present? Finally, calculate the true activation energy, assuming the preceding ratedetermining step.

19. Calculate A (as in Eq. XVIII-19 for N_2 adsorption, using Eq. XVIII-20.

20. Calculate the magnification factor in Fig. XVIII-2.

21. Calculate the magnification factor in Fig. XVIII-3. Assuming the scanning tip to be 2 Å wide and the tip force to be 3 nN, calculate the pressure exerted on the DNA strand.

22. Write the expression for the equilibrium constant K for the adsorption of N_2 on an iron catalyst, using Eqs. XVIII-20 and XVIII-23, and calculate its value at $\theta = 0.1$ and θ (equilibrium) at $P = 0.1$ atm. Comment on the results.

23. Using Fig. XVIII-15, calculate $\mathcal{D}$ for O atom diffusion on W(100) as a function of θ. Comment on your results.

24. Show that Eq. XVIII-58 follows from the mechanism of Eqs. XVIII-55 to XVIII-57.

25. Use Table XVIII-2 to estimate the Q for the process H [on W(110)] $\rightarrow$ $H_2(g)$.

26. Derive the probable rate law for the reaction $CO + \frac{1}{2}O_2 = CO_2$ as catalyzed by a metal surface assuming (*a*) an Eley–Rideal mechanism and (*b*) a Langmuir–Hinshelwood mechanism.

27. Derive the steady-state rate law corresponding to the reaction sequence of Eqs. XVIII-40–XVIII-44, that is, without making the assumption that any one step is much slower than the others. See Ref. 234.

28. According to Schwab, the kinetics of the decomposition of ammonia on Pt are

(a) At low N_2 pressure: $\dfrac{dP_{NH_3}}{dt} = -\dfrac{kP_{NH_3}}{P_{H_2}}$

(b) At low H_2 pressure: $\dfrac{dP_{NH_3}}{dt} = -\dfrac{k'P_{NH_3}}{P_{N_2}}$

where P denotes partial pressure. Write a simple Langmuir–Hinshelwood mechanism that gives these limiting rate laws.

29. Write a possible reaction sequence for the photochemical oxidation of aqueous CN^- ion on TiO_2.

30. Hydrogen atoms chemisorbed on a metal surface may be bonded to just one metal atom or may be bonded to two atoms in a symmetrical bridge. In each case, there are three normal modes. Sketch what these are, and indicate any degeneracies (assume the metal atoms to be infinitely heavy).

31. Suggest a rate law for the Boudouard reaction; note Ref. 281.

32. Explain on what basis the condensation coefficient, c, might be expected to depend on $\cos^2 \theta$, where θ is the angle from the normal of an incident beam of particles (note Ref. 122).

33. Derive Eq. XVIII-76 and also the more complete form that would show under what conditions the reaction could become first-order in P_{CO}.

34. Calculate the energy release in kilocalories per mole (kcal/mol) of He for the cold fusion reaction

$$^2D_1 + {}^2D_1 = {}^4He_2$$

Calculate also the activation energy for the reaction, again in kcal/mol, assuming that the Coulomb repulsion maximizes at 3×10^{-13} cm separation of the nuclear centers. Assuming a successful cold-fusion device, how many fusions per second would generate one horsepower (1 hp) if the conversion of heat into work were 10% efficient?

35. STM and AFM profiles distort the shape of a particle because the side of the tip rides up on the particle. This effect can be corrected for. Consider, say, a spherical gold particle on a smooth surface. The sphere may be truncated, that is, the center may be a distance q above the surface, where $q < r$, the radius of the sphere. Assume the tip to be a cone of cone angle α. The observed profile in the vertical plane containing the center of the sphere will be a rounded hump of base width $2d$ and height h. Calculate q and r for the case where $\alpha = 32°$ and d and h are 275 nm and 300 nm, respectively. Note Chapter XVI, Ref. 133a. Can you show how to obtain the relevent equation?

General References

D. Avnir, ed., *The Fractal Approach to Heterogeneous Chemistry*, Wiley, New York, 1989.

A. T. Bell and M. L. Hair, eds., *Vibrational Spectroscopy for Adsorbed Species*, Washington, DC, American Chemical Society, 1980.

M. Boudart and G. Djega-Mariadassou, *Kinetics of Heterogeneous Catalytic Reactions*, Princeton University Press, Princeton, NJ, 1984.

C. R. Brundle and H. Morawitz, *Vibrations at Surfaces*, Elsevier, Amsterdam, 1983.

F. Cardon, W. P. Gomes, and W. Dekeyser, eds., *Photovoltaics and Photoelectrochemical Solar Energy Conversion*, Plenum, New York, 1981.

G. Ertl and J. Kuppers, *Low Energy Electrons and Surface Chemistry*, Verlag Chemie, Berlin, 1985.

M. Grunze and H. J. Kreuzer, eds., *Kinetics of Interface Reactions*, Springer-Verlag, Berlin, 1987.

N. B. Hannay, ed., *Treatise on Solid State Chemistry*, Vol. 6A, *Surfaces I* and Vol. 6B, *Surfaces II*, Plenum, New York, 1976.

D. O. Hayward and B. M. W. Trapnell, *Chemisorption*, Butterworths, London, 1964.

R. Hoffmann, *Solids and Surfaces. A Chemist's View of Bonding in Extended Structures*, VCH Publishers, New York, 1988.

H. Ibach and D. L. Mills, *Electron Energy Loss Spectroscopy and Surface Vibrations*, Academic, New York, 1982.

B. Imelik, ed., *Catalysis by Zeolites*, Elsevier, Amsterdam, 1980.

D. A. King and D. P. Woodruff, eds., *The Chemical Physics of Solid Surfaces and Heterogeneous Catalysis*, Elsevier, Amsterdam, 1982.

Ira N. Levine, *Quantum Chemistry*, 4th ed., Prentice-Hall, Englewood Cliffs, NJ, 1991. (Source for density functional theory.)

R. J. Madix, ed., *Surface Reactions.* Springer-Verlag, New York, 1994.

R. Masel, *Principles of Adsorption and Reaction on Solid Surfaces*, Wiley-Interscience, New York, 1996.

J. A. Rabo, ed., *Zeolite Chemistry and Catalysis*, ACS Monograph 171, American Chemical Society, Washington, DC, 1976.

T. N. Rhodin and G. Ertl, *The Nature of the Surface Chemical Bond*, North-Holland, Amsterdam, 1979.

B. W. Rossiter and R. C. Baetzold, eds., *Physical Methods of Chemistry*, Vol. IXA, *Investigations of Surfaces and Interfaces*, Wiley, New York, 1993.

J. H. Sinfelt, *Bimetallic Catalysts: Discoveries, Concepts, and Applications*, Wiley, New York, 1983.

G. A. Somorjai, *Chemistry in Two-Dimensions*, Cornell University Press, Ithaca, NY, 1981.

G. A. Somorjai, *Introduction to Surface Chemistry and Catalysis*, Wiley-Interscience, New York, 1994.

H. H. Storch, H. Golumbic, and R. R. Anderson, *The Fischer–Tropsch and Related Systems*, Wiley, New York, 1951.

B. K. Teo and D. C. Joy, *EXAFS Spectroscopy, Techniques and Applications*, Plenum, New York, 1981.

C. L. Thomas, *Catalytic Processes and Proven Catalysts*, Academic, New York, 1970.

S. N. Wagonov and M.-H. Whangbo, *Surface Analysis with STM and AFM*, VCH Publishers, New York, 1996.

A. Zangwill, *Physics at Surfaces*, Cambridge University Press, New York, 1988.

Textual References

1. D. A. King, *Top. Catal.*, **1**, 315 (1994).

2. G. A. Somorjai and J. M. Thomas, eds., *Top. Catal.*, **1** (1994).

3. D. M. Hercules, M. Houalla, A. Proctor, and J. N. Fiedor, *Anal. Chim. Acta*, **283**, 42 (1993).

4. M. K. Weldon and C. M. Friend, *Chem. Rev.*, **96**, 1391 (1996).

5. C. T. Ceyer, D. J. Gladstone, M. McGonigal, and M. T. Schulberg, in *Investigations of Surfaces and Interfaces, Part A*, B. W. Rossiter and R. C. Baetzold, eds.; *Physical Methods of Chemistry Series*, 2nd ed., Vol. IXA, Wiley, New York, 1993.

6. D. W. Goodman, *Surf. Sci.*, **299–300**, 837 (1994).

7. R. Imbihl and G. Ertl, *Chem. Rev.*, **95**, 697 (1995).

8. A. Harrison, *Fractals in Chemistry*, Oxford Chemistry Primers, Oxford Science Publishers, Oxford, 1995.

9. A. T. Hubbard, *Heterogen. Chem. Rev.*, **1**, 3 (1994).

10. J. R. Anderson, *Sci. Prog. Oxf.*, **69**, 461 (1985).

11. M. Boudart, *Supported Metals as Heterogeneous Catalysts, the Science of Precious Metals Applications*, International Precious Metals Institute, Allentown, PA, 1989.

12. C. T. Campbell, *J. Vac. Sci. Tech.*, **6**, 1108 (1988).

13. G. A. Somorjai, *Introduction to Surface Chemistry and Catalysis*, Wiley-Interscience, New York, 1994.

14. R. Imbihl, R. J. Behm, and G. Ertl, *Surf. Sci.*, **123**, 129 (1982).

15. R. J. Behm, K. Christmann, and G. Ertl, *Surf. Sci.*, **99**, 320 (1980).

16. J. S. Foord and R. M. Lambert, *Surf. Sci.*, **185**, L483 (1987).

17. T. W. Orent and R. S. Hansen, *Surf. Sci.*, **67**, 325 (1977).

18. J. L. Stickney, C. B. Ehlers, and B. W. Gregory, *Langmuir*, **4**, 1368 (1988).

19. J. A. Gates and L. L. Kesmodel, *J. Chem. Phys.*, **76**, 4281 (1982).

20. B. E. Kole and G. A. Somorjai, in *Catalysis: Science and Technology*, Vol. 38, J. R. Anderson and M. Boudart, eds., Springer-Verlag, New York, 1985.

21. B. E. Bent, C. M. Mate, J. E. Crowell, B. E. Koel, and G. A. Somorjai, *J. Phys. Chem.*, **91**, 1493 (1987).

22. E. D. Williams, P. A. Thiel, W. H. Weinberg, and J. T. Yates, Jr., *J. Chem. Phys.*, **72**, 3496 (1980).

23. T. Tabata, T. Aruga, and Y. Murata, *Surf. Sci.*, **179**, L63 (1987).

24. R. C. Baetzold, *Surf. Sci.*, **95**, 286 (1980).

25. S. N. Magonov and M.-H. Whangbo, *Surface Analysis with STM and AFM*, VCH Publishers, New York, 1996, p. 168.

26. G. A. M. Hussein, N. Sheppard, M. I. Zaki, and R. B. Fahim, *J. Chem. Soc., Faraday Trans. 1*, **85**, 1723 (1989).

27. J. Valyon and W. K. Hall, *J. Phys. Chem.*, **97**, 1204 (1993).

28. D. W. Goodman, *Chem. Rev.*, **95**, 523 (1995).

29. R. M. Quincy, M. Houalla, and D. M. Hercules, *Fresenius J. Anal. Chem.*, **346**, 676 (1993).

30. W. L. Parker, A. R. Siedle, and R. M. Hexter, *Langmuir*, **4**, 999 (1988).

31. S. Chiang, R. G. Tobin, P. L. Richards, and P. A. Thiel, *Phys. Rev. Lett.*, **52**, 648 (1984).

32. C. T. Campbell and B. E. Koel, *Surf. Sci.*, **186**, 393 (1987).

33. M. A. Chesters and N. Sheppard, in *Spectroscopy of Surfaces*, R. J. H. Clark and R. E. Hester, eds., Wiley, New York, 1988, Chapter 7.

34. Q. Y. Yang, K. J. Maynard, A. D. Johnson, and S. T. Ceyer, *J. Chem. Phys.*, **102**, 7734 (1995).

35. X.-L. Zhou, Z.-M. Liu, J. Kiss, D. W. Sloan, and J. M. White, *J. Am. Chem. Soc.*, **117**, 3565 (1995).

36. K. J. Maytrrd, A. D. Hohnson, S. P. Daley, and S. T. Ceyer, *Faraday Disc. Chem. Soc.*, **91**, 327 (1991).

37. D. M. Hercules, A. Proctor, and M. Houalla, *Acc. Chem. Res.*, **27**, 387 (1994).

38. J. Kiss, D. J. Alberas, and J. M. White, *J. Am. Chem. Soc.*, **114**, 10486 (1992).

39. W. T. Tysoe, G. L. Nyberg, and R. M. Lambert, *J. Chem. Soc., Chem. Commun.*, 623 (1983).

40. J. G. Tobin, S. W. Robey, L. E. Klebanoff, and D. A. Shirley, *Phys. Rev. B*, **35**, 9056 (1987).

41. Fu-Ming Pan and P. C. Stair, *Surf. Sci.*, **177**, 1 (1986).

42. J. C. Hansen and J. G. Tobin, *J. Vac. Sci. Techn.*, **7**, 2083, 2475 (1989).

43. S. Akhter and J. M. White, *Surf. Sci.*, **180**, 19 (1987).

44. M. T. Paffett, C. T. Campbell, and T. N. Taylor, *J. Chem. Phys.*, **85**, 6176 (1986).

45. K. W. Nebesny and N. R. Armstrong, *Langmuir*, **1**, 469 (1985).

46. M. Komiyama, H. Tsukamoto, and Y. Ogino, *J. Solid State Chem.*, **64**, 134 (1986).

47. T. C. Frank and J. L. Falconer, *Appl. Surf. Sci.*, **14**, 359 (1982).

48. Y. Iwata, F. Fujimoto, E. Vilalta, A. Ootuka, K. Komaki, K. Kobayashi, H. Yamashita, and Y. Murata, *Jpn. J. Appl. Phys.*, **26**, L1026 (1987).

49. R. H. Stulen and P. A. Thiel, *Surf. Sci.*, **157**, 99 (1985).

50. F. P. Netzer, D. L. Doering, and T. E. Madey, *Surf. Sci.*, **143**, L363 (1984).

51. T. E. Madey and R. Stockbauer, in *Solid State Physics: Surfaces, Methods of Experimental Physics*, Vol. 22, R. L. Park and M. G. Lagally, eds., Academic, New York, 1985.

52. R. H. Stulen, *Prog. Surf. Sci.*, **32**, 1 (1989).

53. R. D. Ramsler and J. T. Yates, *Surf. Sci. Rep.*, **12**, 243 (1991).

54. T. E. Hadey, H.-S. Tao, U. Diebold, S. M. Shivaprasad, A. L. Johnson, A. Poradzisz, N. D. Shinn, J. A. Yarmoff, V. Chakarian, and D. Shuh, *Desorption Induced by Electron Transitions DIET V, Springer Series in Surface Sciences*, Vol. 31, A. R. Burns, E. B. Stechel, and D. R. Jennison, eds., Springer-Verlag, Berlin, 1993.

55. M. P. Kaminsky, N. Winograd, and G. L. Geoffroy, *J. Am. Chem. Soc.*, **108**, 1315 (1986).

56. C. T. Rettner, J. Kimman, F. Fabre, D. J. Auerbach, J. A. Barker, and J. C. Tully, *J. Vac. Sci. Tech.*, **A5**, 508 (1987).

57. C. M. Greenlief, J. M. Robinson, and V. L. Shannon, *Investigations of Surfaces*

and Interfaces, Part A. Physical Methods of Chemistry, Vol. 9A, 2nd ed., B. M. Rossiter and R. C. Baetzold, eds., Wiley, New York, 1993.

57a. Y. D. Li, L. Q. Jiang, and B. E. Koel, *Phys. Rev. B*, **49**, 2813 (1994).

58. D. G. Frank, O. M. R. Chyan, T. Golden, and A. T. Hubbard, *J. Phys. Chem.*, **98**, 1985 (1994).

59. M. Sano, T. Maruo, H. Yamatera, M. Suzuki, and Y. Saito, *J. Am. Chem. Soc.*, **109**, 52 (1987).

60. T. Mizushima, K. Tohji, Y. Udagawa, M. Harada, M. Ishikawa, and A. Ueno, *J. Catal.*, **112**, 282 (1988).

61. F. Zaera, D. A. Fischer, R. G. Carr, and J. L. Gland, *J. Chem. Phys.*, **89**, 5335 (1988).

62. B. S. Clausen, L. Gråbæk, G. Steffensen, P. L. Hansen, and H. Topsøe, *Catal. Lett.*, **20**, 23 (1993).

63. M. J. Fay, A. Proctor, D. P. Hoffmann, M. Houalla, and D. M. Hercules, *Mikrochim. Acta*, **109**, 281 (1992).

64. J. W. Niemantsverdriet, A. M. van der Kraan, and W. N. Delgass, *J. Catal.*, **89**, 138 (1984).

65. L. M. Tau and C. O. Bennett, *J. Catal.*, **89**, 285 (1984).

66. R. R. Gatte and J. Phillips. *J. Phys. Chem.*, **91**, 5961 (1987).

67. V. Schünemann, H. Treviño, W. M. H. Sachtler, K. Fogash, and J. A. Dumesic, *J. Phys. Chem.*, **99**, 1317 (1995).

68. C. F. Tirendi, G. A. Mills, C. Dybowski, and G. Neue, *J. Phys. Chem.*, **96**, 5045 (1992).

69. T. W. Root and T. M. Duncan, *Chem. Phys. Lett.*, **137**, 57 (1987).

70. M. Engelsberg, C. S. Yannoni, M. A. Jacintha, C. Dybowski, and R. E. de Souza, *J. Phys. Chem.*, **98**, 2397 (1994).

71. T. W. Root and T. M. Duncan, *J. Catal.*, **101**, 527 (1986).

72. J. A. Robbins, *J. Phys. Chem.*, **90**, 3381 (1986).

73. M. T. Aronson, R. J. Gorte, W. E. Farneth, and D. White, *J. Am. Chem. Soc.*, **111**, 840 (1989).

74. R. Shoemaker and T. Apple, *J. Phys. Chem.*, **91**, 4024 (1987).

75. P. J. Barrie and J. Klinowski, *Prog. Nucl. Magn. Reson.*, **24**, 91 (1992).

76. J. H. Lunsford, *Langmuir*, **5**, 12 (1989).

77. S. Contarini, J. Michalik, M. Narayana, and L. Kevan, *J. Phys. Chem.*, **90**, 4587 (1986).

78. E. Giamello, Z. Sojka, M. Che, and A. Zecchina, *J. Phys. Chem.*, **90**, 6084 (1986).

79. M. Che, L. Bonneviot, C. Louis, and M. Kermarec, *Mat. Chem. Phys.*, **13**, 201 (1985).

80. J. H. Lunsford, in *Catalysis: Science and Technology*, Vol. 8, J. R. Anderson and M. Boudart, eds., Springer-Verlag, Berlin, 1987, Chapter 5.

81. G. A. Somorjai, *Principles of Surface Chemistry*, Prentice-Hall, Englewood Cliffs, NJ, 1972; G. A. Somorjai and L. L. Kesmodel, *MTP International Review of Science*, Butterworths, London, 1975.

82. D. H. Parker, M. E. Bartram, and B. E. Koel, *Surf. Sci.*, **217**, 489 (1989).

83. R. G. Windham, M. E. Bartram, and B. E. Koel, *J. Phys. Chem.*, **92**, 2862 (1988).

84. D. O. Hayward and B. M. W. Trapnell, *Chemisorption*, Butterworths, London, 1964.

85. D. Burgess, Jr., P. C. Stair, and E. Weitz, *J. Vac. Sci. Tech.*, **A4**, 1362 (1986); D. R. Burgess, Jr., I. Hussla, P. C. Stair, R. Viswanathan, and E. Weitz, *Rev. Sci. Instrum.*, **55**, 1771 (1984).

86. R. J. Behm, V. Penka, M. G. Cattania, K. Christmann, and G. Ertl, *J. Chem. Phys.*, **78**, 7486 (1983).

87. F. Zaera and G. A. Somorjai, in *Hydrogen Effects in Catalysis: Fundamentals and Practical Applications*, Z. Paal and P. G. Menon, eds., Marcel Dekker, New York, 1988.

87a. C. Xu and B. E. Koel, *Surf. Sci.*, **310**, 198 (1994).

88. J. T. Yates, Jr., *Methods of Experimental Physics*, Vol. 22, Academic Press, New York, 1985, p. 425.

89. P. A. Redhead, *Vacuum*, **12**, 203 (1962).

90. J. B. Miller, H. R. Siddiqui, S. M. Gates, J. N. Russell, Jr., J. T. Yates, Jr., J. C. Tully, and M. J. Cardillo, *J. Chem. Phys.*, **87**, 6725 (1987).

91. K. B. Kester and J. L. Falconer, *J. Catal.*, **89**, 380 (1984).

92. D. L. Roberts and G. L. Griffin, *J. Catal.*, **95**, 617 (1985).

93. J. L. Falconer and R. J. Madix, *Surf. Sci.*, **48**, 393 (1975).

94. P. Malet and G. Munuera, in *Adsorption at the Gas–Solid Interfaces*, J. Rouquerol and K. S. W. Sing, eds., Elsevier, Amsterdam, 1982.

95. K. D. Rendulic and B. A. Sexton, *J. Catal.*, **78**, 126 (1982).

96. J. M. Criado, P. Malet, G. Munuera, and V. Rives-Arnau, *Thermochimica Acta*, **38**, 37 (1980).

97. D. D. Eley and P. B. Moore, *Surf. Sci.*, **111**, 325 (1981).

98. R. J. Madix, G. Ertl, and K. Christmann, *Chem. Phys. Lett.*, **62**, 38 (1979).

99. P. G. Glugla, K. M. Bailey, and J. L. Falconer, *J. Catal.*, **115**, 24 (1989).

100. B. A. Sexton, *Surf. Sci.*, **102**, 271 (1981).

101. J. M. Saber, J. L. Falconer, and L. F. Brown, *J. Catal.*, **90**, 65 (1984).

102. A. K. Myers, G. R. Schoofs, and J. B. Benziger, *J. Phys. Chem.*, **91**, 2230 (1987).

103. J. T. Roberts and C. M. Friend, *J. Chem. Phys.*, **88**, 7172 (1988).

104. B. Fubini, E. Giambello, G. Della Gatta, and G. Venturello, *J. Chem. Soc., Faraday Trans. I*, **78**, 153 (1982).

105. See B. M. W. Trapnell, *Chemisorption*, Academic, New York, 1955, p. 124.

106. E. Aisexton, *Surf. Sci.*, **88**, 299 (1979).

107. P. Chou and M. A. Vannice, *J. Catal.*, **105**, 342 (1987).

108. U. Schneider, H. Busse, R. Linke, G. R. Castro, and K. Wandelt, *J. Vac. Soc. Tech. A*, **12**, 2069 (1994).

109. R. Fowler and E. A. Guggenheim, *Statistical Thermodynamics*, Cambridge University Press, Cambridge, England, 1952, p. 437.

110. R. Linke, U. Schneider, H. Busse, C. Becker, U. Schröder, G. R. Castro, and K. Wandelt, *Surf. Sci.*, **307–309**, 407 (1994).

111. E. Ruckenstein and D. B. Dadyburjor, *Chem. Eng. Commun.*, **14,** 59 (1982).

112. G. D. Halsey and A. T. Yeates, *J. Phys. Chem.*, **83,** 3236 (1979).

113. J. E. Lennard-Jones, *Trans. Faraday Soc.*, **28,** 333 (1932).

113a. S. T. Ceyer, *Science*, **249,** 133 (1990).

114. C. T. Campbell and B. E. Koel, *J. Catal.*, **92,** 272 (1985).

115. C. Xu and B. E. Koel, *J. Chem. Phys.*, **100,** 664 (1994).

116. D. J. Doren and J. C. Tully, *Langmuir*, **4,** 256 (1988).

117. A. Sherman and H. Eyring, *J. Am. Chem. Soc.*, **54,** 2661 (1932).

118. R. Dovesi, C. Pisani, F. Ricca, and C. Roetti, *Chem. Phys. Lett.*, **44,** 104 (1976).

119. W. H. Weinberg and R. P. Merrill, *Surf. Sci.*, **33,** 493 (1972).

120. T. L. Einstein and J. R. Schrieffer, *Phys. Rev. B*, **7,** 3629 (1973); idem. *J. Vac. Sci. Tech.*, **9,** 956 (1972).

121. J. C. Buchholz and G. A. Somorjai, *Acc. Chem. Res.*, **9,** 333 (1976).

122. C. B. Mullins and W. H. Weinberg, in *Surface Reactions*, R. J. Madix, ed., Springer-Verlag, New York, 1994, pp. 249–255.

123. F. J. Bottari and E. F. Greene, *J. Phys. Chem.*, **88,** 4238 (1984).

124. K. J. Vette, T. W. Orent, D. K. Hoffman, and R. S. Hansen, *J. Chem. Phys.*, **60,** 4854 (1974).

125. P. Schaaf, J. Talbot, H. M. Rabeony, and H. Reiss, *J. Phys. Chem.*, **92,** 4826 (1988).

126. C. T. Campbell, M. T. Paffett, and A. F. Voter, *J. Vac. Sci. Tech.*, **A4,** 1342 (1986).

127. H. A. Michelsen, C. T. Rettner, and D. J. Auerbach, in *Surface Reactions*, R. J. Madix, ed., Springer-Verlag, New York, 1994, pp. 208ff.

128. S. Glasstone, K. J. Laidler, and H. Eyring, *The Theory of Rate Processes*, McGraw-Hill, New York, 1941.

129. S. Y. Elovich and G. M. Zhabrova, *Zh. Fiz. Khim.*, **13,** 1761 (1939).

130. J. J. F. Scholten, P. Zweitering, J. A. Konvalinka, and J. H. de Boer, *Trans. Faraday Soc.*, **55,** 2116 (1959).

131. C. T. Campbell, G. Ertl, H. Kuipers and J. Segner, *Surf. Sci.*, **107,** 220 (1981).

132. A. Winkler, X. Guo, H. R. Sizziqui, P. L. Hagans, and J. T. Yates, Jr., *Surf. Sci.*, **201,** 419 (1988).

133. M. B. Liu and P. G. Wahlbeck, *J. Phys. Chem.*, **80,** 1484 (1976).

134. C. W. Muhlhausen, L. R. Williams, and J. C. Tully, *J. Chem. Phys.*, **83,** 2594 (1985).

135. P. J. Estrup, E. F. Greene, M. J. Cardillo, and J. C. Tully, *J. Phys. Chem.*, **90,** 4099 (1986).

136. J. R. Chen and R. Gomer, *Surf. Sci.*, **79,** 413 (1979).

137. J. R. Chen and R. Gomer, *Surf. Sci.*, **81,** 589 (1979).

138. V. T. Binh, Ed., *Surface Mobilities on Solid Materials*, Plenum, New York, 1983.

139. B. Roop, S. A. Costello, D. R. Mullins, and J. M. White, *J. Chem. Phys.*, **86,** 3003 (1987).

140. M. J. Cardillo, *Langmuir*, **1,** 4 (1985).

141. J. R. Chen and R. Gomer, *Surf. Sci.*, **94,** 456 (1980).

142. D. R. Mullins, B. Roop, S. A. Costello, and J. M. White, *Surf. Sci.*, **186,** 67 (1987).

143. S. L. Chang and P. A. Thiel, *Phys. Rev. Lett.*, **59,** 296 (1987).

144. J. R. Wolfe and H. W. Weart, in *The Structure and Chemistry of Solid Surfaces*, G. A. Somorjai, ed., Wiley, New York, 1969.

145. R. DiFoggio and R. Gomer, *Phys. Rev. Lett.*, **44,** 1258 (1980).

146. A. B. Anderson and M. K. Awad, *Surf. Sci.*, **183,** 289 (1987).

147. J. C. Buchholz and G. A. Somorjai, *Acc. Chem. Res.*, **9,** 333 (1976).

148. A. Zangwill, *Physics at Surfaces*, Cambridge University Press, New York, 1988, pp. 244ff.

149. G. F. Cerefolini and N. Re, *J. Coll. Interface Sci.*, **174,** 428 (1995).

150. E. Shustorovich, R. C. Baetzold, and E. L. Muetterties, *J. Phys. Chem.*, **87,** 1100 (1983).

151. See M. C. Day, Jr., and J. Selbin, *Theoretical Inorganic Chemistry*, Reinhold, New York, 1962, p. 112.

152. R. S. Mulliken, *J. Chem. Phys.*, **2,** 782 (1934); **3,** 573 (1935).

153. G. A. Somorjai and B. E. Bent, *Prog. Colloid Polym. Sci.*, **70,** 38 (1985).

154. A. B. Anderson, Z. Y. Al-Saigh, and W. K. Hall, *J. Phys. Chem.*, **92,** 803 (1988).

155. R. C. Baetzold, *Langmuir*, **3,** 189 (1987).

156. J. Silvestre and R. Hoffmann, *Langmuir*, **1,** 621 (1985).

157. P. A. Thiel and T. E. Madey, *Surf. Sci. Rep.*, **7,** 211 (1987).

158. R. Hoffmann, *Solid and Surfaces: A Chemist's View of Bonding in Extended Structures*, VCH Publishers, New York, 1988.

159. R. P. Messmer, *Surf. Sci.*, **158,** 40 (1985).

160. R. Hoffman, *J. Phys.: Condens, Matter*, **5,** A1 (1993).

161. P. Hohenberg and W. Kohn, *Phys. Rev.*, **136,** B864 (1964).

162. U. von Barth, in *The Electronic Structure of Complex Systems*, P. Phariseau and W. M. Temmerman, eds., Plenum, New York, 1984.

163. K. J. Farnsworth and P. J. O'Malley, *J. Phys. Chem.*, **100,** 1814 (1996).

163a. T. Ziegler, *Can. J. Chem.*, **73,** 743 (1995).

163b. B. L. Trout, A. K. Chakraborty, and A. T. Bell, *J. Phys. Chem.*, **100,** 4173 (1996).

163c. S. Suhai, *J. Phys. Chem.*, **100,** 3950 (1996).

163d. Y. Yokomichi, T. Yanabe, H. Ohtsuka, and T. Kakumoto, *J. Phys. Chem.*, 100, 14424 (1996).

164. B. Hammer and J. K. Nørskov, *Surf. Sci.*, 343, 211 (1995). (*Note an erratum:* α for Table 1 should be 0.42, not 0.38.)

165. G. Somorjai, *Chemistry in Two Dimensions: Surfaces*, Cornell University Press, Ithaca, NY, 1981.

166. P. W. Selwood, *J. Am. Chem. Soc.*, **78,** 3893 (1956).

167. S. Sung and R. Hoffmann, *J. Am. Chem. Soc.*, **107,** 578 (1985).

168. J. A. Rodriguez and C. T. Campbell, *J. Phys. Chem.*, **91,** 2161 (1987).

169. T. N. Rhodin and D. L. Adams, in *Treatise on Solid State Chemistry*, Vol. 6A, *Surfaces I*, N. B. Hannay, ed., Plenum, New York, 1976.

170. S. P. Mehandru and A. B. Anderson, *Surf. Sci.*, **169,** L281 (1986).

171. S. P. Mehandru, A. B. Anderson, and P. N. Ross, *J. Catal.*, **100,** 210 (1986).

172. R. C. Baetzold, G. Apai, and E. Shustorovich, *Appl. Surf. Sci.*, **19,** 135 (1984).

173. R. C. Baetzold, *Phys. Rev. B*, **29,** 4211 (1984).

174. R. P. Messmer, *J. Vac. Sci. Tech.*, **A2,** 899 (1984).

175. S. Sung, R. Hoffmann, and P. A. Thiel, *J. Phys. Chem.*, **90,** 1380 (1986).

176. A. B. Anderson and J. A. Nichols, *J. Am. Chem. Soc.*, **108,** 4742 (1986).

177. R. P. Messmer, D. R. Salahub, K. H. Johnson, and C. Y. Yang, *Chem. Phys. Lett.*, **51,** 84 (1977).

178. T. Halachev and E. Ruckenstein, *J. Molec. Catal.*, **16,** 149 (1982).

179. C. Zheng, Y. Apeloig, and R. Hoffmann, *J. Am. Chem. Soc.*, **110,** 749 (1988).

180. S. P. Mohandru and A. B. Anderson, *J. Am. Chem. Soc.*, **107,** 844 (1985).

181. J. Silvestre and R. Hoffmann, *J. Vac. Sci. Technol.*, **A4,** 1336 (1986).

182. J. C. Tracy and P. W. Palmberg, *J. Chem. Phys.*, **51,** 4852 (1969).

183. W. E. Garner, F. S. Stone, and P. F. Tiley, *Proc. Roy. Soc.* (London), **A211,** 472 (1962).

184. F. S. Stone, *Adv. Catal.*, **13,** 1 (1962).

185. H. Yoneyama and G. B. Hoflund, *Prog. Surf. Sci.*, **21,** 5 (1986).

186. S. R. Morrison, in *Treatise on Solid State Chemistry*, Vol. 6B, *Surfaces II*, N. B. Hannay, ed., Plenum, New York, 1976.

187. M. A. Fox, C. Chen, K. Park, and J. N. Younathan, ACS Symposium Series, No. 278, *Organic Phototransformations in Nonhomogeneous Media*, M. A. Fox, ed., American Chemical Society, Washington, DC, 1985; *Homogeneous and Heterogeneous Catalysis*, E. Pelizzetti and N. Serpone, eds., Kluwer Academic, Hingham, MA, 1986.

188. K. H. Babb and M. G. White, *J. Catal.*, **98,** 343 (1986).

189. D. J. Rosenthal, M. G. White, and G. D. Parks, *AIChE J.*, **33,** 336 (1987).

190. R. L. Burwell, Jr., *J. Catal.*, **86,** 301 (1984).

191. T. Fransen, O. van der Meer, and P. Mars, *J. Phys. Chem.*, **80,** 2103 (1976).

192. P. C. Stair, *J. Am. Chem. Soc.*, **104,** 4044 (1982).

193. R. G. Pearson, *Chem. Brit.*, **3,** 103 (1967).

194. J. E. Deffeyes, A. H. Smith, and P. C. Stair, *Surf. Sci.*, **163,** 79 (1985).

195. B. Umansky, J. Engelhardt, and W. K. Hall, *J. Catal.*, **127,** 128 (1991).

196. A. W. Sleight, *Science*, **208,** 895 (1980).

197. P. Politzer and S. D. Kasten, *J. Phys. Chem.*, **80,** 385 (1976).

198. T. E. Madey, J. T. Yates, Jr., D. R. Sandstrom, and R. J. H. Voorhoeve, in *Treatise on Solid State Chemistry*, Vol. 6B, *Surfaces II*, N. B. Hannay, ed., Plenum, New York, 1976.

199. J. C. Tully and M. J. Cardillo, *Science*, **223,** 445 (1984).

200. C. T. Campbell, B. E. Koel, and K. A. Daube, *J. Vac. Sci. Tech.*, **A5,** 810 (1987).

201. K. F. Bonhoeffer and A. Farkas, *Z. Phys. Chem.*, **B12,** 231 (1931).

202. E. K. Rideal, *Proc. Cambridge Phil. Soc.*, **35,** 130 (1938).

203. D. D. Eley, *Phil. Trans. Roy. Soc.* (London), **A318,** 117 (1986).

204. D. D. Eley and S. H. Russell, *Proc. Roy. Soc.* (London), **A341,** 31 (1974).

205. H. Wise and B. J. Wood, *Adv. Atom. Mol. Phys.*, **3,** 29 (1967).

206. M. A. Vannice, S. H. Hyun, B. Kalpakci, and W. C. Liauh, *J. Catal.*, **56,** 362 (1979).

207. G. Ertl, *Ber. Bunsenges. Phys. Chem.*, **86,** 425 (1982).

208. T. Engel and G. Ertl, *J. Chem. Phys.*, **69,** 1267 (1978).

209. J. C. Tully, *Acc. Chem. Res.*, **14,** 188 (1981).

210. T. F. George, K. Lee, W. C. Murphy, M. Hutchinson, and H. Lee, in *Theory of Chemical Reaction Kinetics*, Vol. IV, M Baer, ed., CRC Press, Boca Raton, FL, 1985.

211. A. W. Adamson, *A Textbook of Physical Chemistry*, 3rd ed., Academic, New York, p. 624f.

212. D. L. Freeman and J. D. Doll, *J. Chem. Phys.*, **79,** 2343 (1983); ibid., **78,** 6002 (1983).

213. J. D. Doll and D. L. Freeman, *Surf. Sci.*, **134,** 769 (1983).

214. P. Meakin and D. J. Scalapino, *J. Chem. Phys.*, **87,** 731 (1987).

215. G. A. Somorjai, *Catal. Rev.*, **7,** 87 (1972).

216. A. B. Anderson and J. J. Maloney, *J. Phys. Chem.*, **92,** 809 (1988).

217. A. A. Balandin, *Z. Phys. Chem.*, **B3,** 167 (1929).

218. H. S. Taylor, *Proc. Roy. Soc.*, **A108,** 105 (1925).

219. R. A. Campbell, J. Guan, and T. E. Madey, *Cat. Lett.*, **27,** 273 (1994).

220. M. Eiswirth and G. Ertl, *Surf. Sci.*, **177,** 90 (1986).

221. N. A. Collins, S. Sundaresan and Y. J. Chabal, *Surf. Sci.*, **180,** 136 (1987).

222. M. Bär, S. Nettesheim, H. H. Rotermund, M. Eiswirth, and G. Ertl, *Phys. Rev. Lett.*, **74,** 1246 (1995).

223. V. Gorodetskii, J. Lauterbach, H.-H. Rotermund, J. H. Block, and G. Ertl, *Nature*, **370,** 276 (1994).

224. V. P. Zhdanov and P. R. Norton, *Langmuir*, **12,** 101 (1996).

225. D. B. Dadyburjor and E. Ruckenstein, *J. Phys. Chem.*, **85,** 3396 (1981); E. Ruckenstein and D. B. Dadyburjor, *Chem. Eng. Commun.*, **14,** 59 (1982).

226. J. F. Hamilton and R. C. Baetzold, *Science*, **205,** 1213 (1979); R. C. Baetzold and J. F. Hamilton, *Prog. Solid State Chem.*, **15,** 1 (1983).

227. D. D. Eley, A. H. Klepping, and P. B. Moore, *J. Chem. Soc., Faraday Trans. I*, **81,** 2981 (1985).

228. R. R. Rye, *Acc. Chem. Res.*, **8,** 347 (1975).

229. D. Farin and D. Avnir, *J. Am. Chem. Soc.*, **110,** 2039 (1988).

230. P. Meakin, *Chem. Phys. Lett.*, **123,** 428 (1986).

231. H. Van Damme, P. Levitz, and L. Gatineau, in *Chemical Reactions in Organic and Inorganic Constrained Systems*, R. Setton, ed., Kluwer Academic, Hingham, MA, 1986.

232. C. T. Campbell, *J. Catal.*, **94,** 436 (1985).

233. R. J. Madix, *The Chemical Physics of Solid Surfaces and Heterogeneous Catalysis*, Vol. 4, Elsevier, Amsterdam, 1981.

234. R. E. Hensershot and R. S. Hansen, *J. Catal.*, **98,** 150 (1986).

235. J. Sonnemans, J. M. Janus, and P. Mars, *J. Phys. Chem.*, 2107 (1976).

236. Y. Ogino, *Catal. Rev.-Sci. Eng.*, **23**, 505 (1981).

237. M. G. White and J. W. Hightower, *J. Catal.*, **82**, 185 (1983).

238. K. J. Yoon and M. A. Vannice, *J. Catal.*, **82**, 457 (1983).

239. D. B. Dadyburjor, *J. Catal.*, **82**, 489 (1983).

240. P. H. Emmett, *Catalysis*, Reinhold, New York, 1954.

241. G. Schwab, H. S. Taylor, and R. Spence, *Catalysis*, Van Nostrand, New York, 1937.

242. P. G. Ashmore, *Catalysis and Inhibition of Chemical Reactions*, Butterworths, London, 1963.

243. M. Boudart and K. Tamaru, *Catal. Lett.*, **9**, 15 (1991).

244. M. Boudart, in *Perspective in Catalysis*, J. A. Thomas and K. I. Zamaraev, eds., Blackwell Scientific Publications, London, 1992.

245. M. Boudart and G. Djega-Mariadassou, *Catal. Lett.*, **29**, 7 (1994).

246. M. Fleischman and S. Pons, *J. Electroanal. Chem.*, **261**, 301 (1989).

247. C. Walling and J. Simmons, *J. Phys. Chem.*, **93**, 4693 (1989).

248a. R. Dagani, *Chem. Eng. News*, June 14, 1993, p. 38.

248b. R. Dagani, *Chem. Eng. News*, April 29, 1996, p. 69.

249. E. F. Mallove, *Chem. Eng. News*, June 17, 1996, p. 4.

250. N. D. Spencer and G. A. Somorjai, *Rep. Prog. Phys.*, **46**, 1 (1983).

251. A. B. Anderson, in *Theoretical Aspects of Heterogeneous Catalysis*, J. B. Moffat, ed., Van Nostrand, New York, 1989.

252. J. H. Sinfelt, *Acc. Chem. Res.*, **20**, 134 (1987).

253. P. B. Weisz, *Chemtech*, July, 1982, p. 424.

254. I. Sushumna and E. Ruckenstein, *J. Catal.*, **94**, 239 (1985).

255. G. Ertl and D. Prigge, *J. Catal.*, **79**, 359 (1983).

256. G. Ertl, *J. Vac. Sci. Tech.*, **A1**, 1247 (1983); *Catal. Rev.-Sci. Eng.*, **21**, 201 (1980).

257. J. H. Sinfelt, *J. Phys. Chem.*, **90**, 4711 (1986).

258. W. Tsai, J. J. Vajo, and W. H. Weinberg, *J. Phys. Chem.*, **92**, 1245 (1988).

259. P. H. Emmett, *J. Chem. Educ.*, **7**, 2571 (1930); S. Brunauer and P. H. Emmett, *J. Am. Chem. Soc.*, **62**, 1732 (1940).

260. W. G. Frankenburg, in *Catalysis*, Vol. 3, P. H. Emmett, ed., Reinhold, New York, 1955, p. 171.

261. M. Boudart, *Top. Catal.*, **1**, 405 (1994).

262. F. Zaera, A. J. Gellman, and G. A. Somorjai, *Acc. Chem. Res.*, **19**, 24 (1986).

263. N. D. Spencer, R. C. Schoonmaker, and G. A. Somorjai, *J. Catal.*, **74**, 129 (1982).

264. G. A. Somorjai and N. Materer, *Top. Catal.*, **1**, 215 (1994).

265. J. McAllister and R. S. Hansen, *J. Chem. Phys.*, **59**, 414 (1973).

266. A. Vavere and R. S. Hansen, *J. Catal.*, **69**, 158 (1981).

267. M. Temkin and V. Pyzhev, *Acta Physicochim.* (USSR), **12**, 327 (1940).

268. R. B. Anderson, *Catalysis*, Vol. 4, Reinhold, New York, 1956, pp. 1, 20; see also H. H. Storch, *Adv. Catal.*, **1**, 115 (1948).

269. B. A. Sexton, A. E. Hughes, and T. W. Turney, *J. Catal.*, **97**, 390 (1986).

270. M. A. Vannice, in *Catalysis—Science and Technology*, J. R. Anderson and M. Boudart, eds., Springer-Verlag, New York, 1982.

271. R. C. Baetzold and J. R. Monnier, *J. Phys. Chem.*, **90,** 2944 (1986).

272. D. M. Stockwell and C. O. Bennett, *J. Catal.*, **110,** 354 (1988); D. M. Stockwell, J. S. Chung, and C. O. Bennett, ibid., **112,** 135 (1988).

273. L. M. Tau and C. O. Bennett, *J. Catal.*, **96,** 408 (1985).

274. S. Y. Wang, S. H. Moon, and M. A. Vannice, *J. Catal.*, **71,** 167 (1981).

275. H. J. Jung, M. A. Vannice, L. N. Mulay, R. M. Stanfeld, and W. N. Delgass, *J. Catal.*, **76,** 208 (1982).

276. H. J. Jung, P. L. Walker, Jr., and M. A. Vannice, *J. Catal.*, **75,** 416 (1982).

277. K. C. McMahon, S. L. Suib, B. G. Johnson, and C. H. Bartholomew, Jr., *J. Catal.*, **106,** 47 (1987).

278. S. L. Suib, K. C. McMahon, L. M. Tau, and C. O. Bennett, *J. Catal.*, **89,** 20 (1984).

279. E. G. Rightor and T. J. Pinnavaia, *Ultramicroscopy*, **22,** 159 (1987).

280. J. Venter, M. Kaminsky, G. L. Geoffroy, and M. A. Vannice, *J. Catal.*, **103,** 450 (1987).

281. M. T. Tavares, I. Alstrup, C. A. Bernardo, and J. R. Rostru-Nielsen, *J. Catal.*, **147,** 525 (1994).

282. I. Alstrup, *J. Catal.*, **151,** 216 (1995).

283. P. G. Glugla, K. M. Bailey, and J. L. Falconer, *J. Phys. Chem.*, **92,** 4474 (1988).

284. M. A. Henderson, A. Szabo, and J. T. Yates, Jr., *Chem. Phys. Lett.*, **168,** 51 (1990).

285. M. Baerns, J. R. H. Ross, and K. Van der Wiele, eds., *Methane Activation, Catalysis Today*, **4,** Feb., 1989.

286. F. Zaera, *Langmuir*, **12,** 88 (1996).

287. J. Horiute and M. Polanyi, *Trans. Faraday Soc.*, **30,** 1164 (1934).

288. R. L. Burwell, Jr., *Langmuir*, **2,** 2 (1986).

289. N. Sheppard, *Ann. Rev. Phys. Chem.*, **39,** 589 (1988).

290. F. Zaera and G. A. Somorjai, *J. Am. Chem. Soc.*, **106,** 2288 (1984); *J. Phys. Chem.*, **89,** 3211 (1985).

291. S. P. Daley, A. L. Utz, T. R. Trautman, and S. T. Ceyer, *J. Am. Chem. Soc.*, **116,** 6001 (1994).

292. T. P. Beebe and J. T. Yates, Jr., *J. Am. Chem. Soc.*, **108,** 663 (1986).

293. R. V. Shankland, *Adv. Catal.*, **6,** 271 (1954).

294. L. B. Roland, M. W. Tamele, and J. N. Wilson, in *Catalysis*, Vol. 7, P. H. Emmett, ed., Reinhold, New York, 1960, p. 1.

295. C. L. Thomas, *Catalytic Processes and Proven Catalysts*, Academic, New York, 1970.

296. G. C. Lau and W. F. Maier, *Langmuir*, **3,** 164 (1987).

297. J. R. Sohn, S. J. DeCanio, P. O. Fritz, and J. H. Lunford, *J. Phys. Chem.*, **90,** 4847 (1986).

298. R. J. Pellet, C. S. Blackwell, and J. A. Rabo, *J. Catal.*, **114,** 71 (1988).

299. M. L. Occelli, S. D. Landau, and T. J. Pinnavaia, *J. Catal.*, **90,** 260 (1984).

300. T. J. Pinnavaia, *Science*, **220,** April 22, 1983, p. 365.

301. A. Drijaca, J. R. Anderson, L. Spiccia, and T. W. Turney, *Inorg. Chem.*, **31,** 4894 (1992).

302. J. Engelhardt and W. K. Hall, *J. Catal.*, **151,** 1 (1995).

303. J. R. Anderson, Q.-N. Dong, Y.-F. Chang, and R. J. Western, *J. Catal.*, **127,** 113 (1991).

304. A. G. Sault and D. W. Goodman, in *Advances in Chemical Physics*, K. P. Kawley, ed., Wiley, New York, 1989.

305. I. Langmuir, *Trans. Faraday Soc.*, **17,** 672 (1922).

306. G. Ertl, *Langmuir*, **3,** 4 (1987).

306a. J. Liu, M. Xi, T. Nordmeyer, and F. Zaera, *J. Phys. Chem.*, **99,** 6167 (1995).

307. D. W. Goodman, *Surf. Rev. Lett.*, **1,** 449 (1994).

308. P. J. Berlowitz, C. H. F. Peden, and D. W. Goodman, *J. Phys. Chem.*, **92,** 5213 (1988).

309. S. Titmuss, A. Wander, and D. A. King, *Chem. Rev.*, **96,** 1291 (1996).

310. F. Rumpf, H. Poppa, and M. Boudart, *Langmuir*, **4,** 722 (1988).

311. G. A. Somorjai, *Chem. Rev.*, **96,** 1223 (1996).

312. M. Kiskinova, *Chem. Rev.*, **96,** 1431 (1996).

313. M. D. Graham, J. G. Kevrekidis, K. Asakura, J. Lauterbach, K. Krischer, H. H. Rotemund, and G. Ertl, *Science*, **264,** 80 (1994). See also K. Asakura, J. Lauterbach, H. H. Rotemund, and G. Ertl, *J. Chem. Phys.*, **102,** 8175 (1995).

314. K. Krischer, M. Eiswirth, and G. Ertl, *J. Chem. Phys.*, **96,** 9161 (1992).

315. M. S. Wrighton, A. B. Ellis, P. T. Wolczanski, D. I. Morse, H. B. Abrahamson, and D. S. Ginley, *J. Am. Chem. Soc.*, **98,** 2774 (1976).

316. S. Sato and J. M. White, *Chem. Phys. Lett.*, **72,** 83 (1980); *J. Catal.*, **69,** 128 (1981).

317. T. Yamase and T. Ikawa, *Inorg. Chim. Acta*, **45,** L55 (1980).

318. A. W. E. Chan, R. Hoffmann, and W. Ho, *Langmuir*, **8,** 1111 (1992).

319. A. H. Boonstra and C. A. H. A Mutsaers, *J. Phys. Chem.*, **79,** 2025 (1975).

320. I. Willner, R. Maidan, D. Mandler, H. Dörr, G. Dürr, and K. Zengerle, *J. Am. Chem. Soc.*, **109,** 6080 (1987).

321. I. Izumi, W. W. Dunn, K. O. Wilbourn, F. F. Fan, and A. J. Bard, *J. Phys. Chem.*, **84,** 3207 (1980).

322. M. D. Ward, J. F. Brazdil, S. P. Mehandru, and A. B. Anderson, *J. Phys. Chem.*, **91,** 6515 (1987).

323. K. Kogo, H. Yoneyama, and H. Tamura, *J. Phys. Chem.*, **84,** 1705 (1980).

324. S. Sato and J. M. White, *J. Am. Chem. Soc.*, **102,** 7206 (1980).

325. A. P. Norton, S. L. Bernasek, and A. B. Bocarsly, *J. Phys. Chem.*, **92,** 6009 (1988).

326. M. A. Fox, *Acc. Chem. Res.*, **16,** 314 (1983).

327. M. Anpo and Y. Kubokawa, *Rev. Chem. Intermediates*, **8,** 105 (1987).

328. S. Sato, *J. Phys. Chem.*, **87,** 3531 (1983).

329. M. A. Fox, R. Cardona, and E. Gaillard, *J. Am. Chem. Soc.*, **109,** 6347 (1987).

330. B. Ohtani, H. Osaki, S. Nishimoto, and T. Kigiya, *J. Am. Chem. Soc.*, **108,** 308 (1986).

331. H. Cao and S. L. Suib, *J. Am. Chem. Soc.*, **116,** 5334 (1994).

332. V. S. Smentkowski and J. R. Yates, Jr., *Surf. Sci.*, **370,** 209 (1997).

333. J. Kiss, D. J. Alberas, and J. M. White, *J. Am. Chem. Soc.*, **114,** 10486 (1992).

334. M. J. Camara and J. H. Lunsford, *Inorg. Chem.*, **22,** 2498 (1983).

335. M. S. Darsillo, H. D. Garney, and M. S. Paquette, *J. Am. Chem. Soc.*, **109,** 3275 (1987).

336. G. A. Epling and E. Florio, *J. Am. Chem. Soc.*, **103,** 1237 (1981).

337. Z.-J. Sun, S. Gravelle, R. S. Mackay, K.-Y. Zhu, and J. M. White, *J. Chem. Phys.*, **99,** 10021 (1993).

338. T. Hertel, M. Wolf, and G. Ertl, *J. Chem. Phys.*, **102,** 3414 (1995).

339. T. F. George, J. Lint, A. C. Beri, and W. C. Murphy, *Prog. Surf. Sci.*, **16,** 139 (1984).

340. J. Lin, W. C. Murphy, and T. F. George, *I & EC Prod. Res. Dev.*, **23,** 334 (1984).

341. R. M. Slayton, N. R. Franklin, and N. J. Tro, *J. Phys. Chem.*, **100,** 15551 (1996).

342. H. Yamashita, Y. Ichihasi, and M. Anpo, *J. Phys. Chem.*, **100,** 16041 (1996).

343. E. A. Perez-Albuerne and Y. Tyan, *Science*, **208,** 902 (1980).

344. M. E. Bartram, R. G. Windham, and B. E. Koel, *Surf. Sci.*, **184,** 47 (1987).

Index

757